Algebraic Topology

Springer
New York
Berlin
Heidelberg
Barcelona
Budapest
Hong Kong
London
Milan
Paris
Singapore
Tokyo

Edwin H. Spanier

Algebraic Topology

Springer

Edwin H. Spanier
University of California
Department of Mathematics
Berkeley, CA 94720

Mathematics Subject Classifications (1991): 55-01

Library of Congress Cataloging in Publication Data

Spanier, Edwin Henry, 1921–
Algebraic topology.

Includes index.
1. Algebraic topology. I. Title.
QA612.S6 514'.2 81-18415
ISBN 0-387-90646-0 AACR2

This book was originally published by McGraw-Hill, 1966.

Printed in the United States of America.
First Corrected Springer Edition.
9 8 7 6 5 4 3

ISBN 0-387-94426-5 Springer-Verlag New York Berlin Heidelberg (softcover) SPIN 10689898
ISBN 0-387-90646-0 Springer-Verlag New York Berlin Heidelberg (hardcover)
ISBN 3-540-90646-0 Springer-Verlag Berlin Heidelberg New York (hardcover)

Algebraic Topology

PREFACE TO THE SECOND SPRINGER PRINTING

IN THE MORE THAN TWENTY YEARS SINCE THE FIRST APPEARANCE OF *Algebraic Topology* the book has met with favorable response both in its use as a text and as a reference. It was the first comprehensive treatment of the fundamentals of the subject. Its continuing acceptance attests to the fact that its content and organization are still as timely as when it first appeared. Accordingly it has not been revised.

Many of the proofs and concepts first presented in the book have become standard and are routinely incorporated in newer books on the subject. Despite this, *Algebraic Topology* remains the best complete source for the material which every young algebraic topologist should know. Springer-Verlag is to be commended for its willingness to keep the book in print for future topologists.

For the current printing all of the misprints known to me have been corrected and the bibliography has been updated.

Berkeley, California
December 1989

Edwin H. Spanier

PREFACE

THIS BOOK IS AN EXPOSITION OF THE FUNDAMENTAL IDEAS OF ALGEBRAIC topology. It is intended to be used both as a text and as a reference. Particular emphasis has been placed on naturality, and the book might well have been titled *Functorial Topology*. The reader is not assumed to have prior knowledge of algebraic topology, but he is assumed to know something of general topology and algebra and to be mathematically sophisticated. Specific prerequisite material is briefly summarized in the Introduction.

Since *Algebraic Topology* is a text, the exposition in the earlier chapters is a good deal slower than in the later chapters. The reader is expected to develop facility for the subject as he progresses, and accordingly, the further he is in the book, the more he is called upon to fill in details of proofs. Because it is also intended as a reference, some attempt has been made to include basic concepts whether they are used in the book or not. As a result, there is more material than is usually given in courses on the subject.

The material is organized into three main parts, each part being made up of three chapters. Each chapter is broken into several sections which treat individual topics with some degree of thoroughness and are the basic organizational units of the text. In the first three chapters the underlying theme is the fundamental group. This is defined in Chapter One, applied in Chapter Two in the study of covering spaces, and described by means of generators and relations in Chapter Three, where polyhedra are introduced. The concept of functor and its applicability to topology are stressed here to motivate interest in the other functors of algebraic topology.

Chapters Four, Five, and Six are devoted to homology theory. Chapter Four contains the first definitions of homology, Chapter Five contains further algebraic concepts such as cohomology, cup products, and cohomology operations, and Chapter Six contains a study of topological manifolds. With each new concept introduced applications are presented to illustrate its utility.

The last three chapters study homotopy theory. Basic facts about homotopy groups are considered in Chapter Seven, applications to obstruction theory are presented in Chapter Eight, and some computations of homotopy groups of spheres are given in Chapter Nine. Main emphasis is on the application to geometry of the algebraic tools introduced earlier.

There is probably more material than can be covered in a year course. The core of a first course in algebraic topology is Chapter Four. This contains elementary facts about homology theory and some of its most important applications. A satisfactory one-semester first course for graduate students can be based on the first four chapters, either omitting or treating briefly Secs. 5 and 6 of Chapter One, Secs. 7 and 8 of Chapter Two, Sec. 8 of Chapter Three, and Sec. 8 of Chapter Four. A second one-semester course can be based on Chapters Five, Six, Seven, and Eight or on Chapters Five, Seven, Eight, and Nine. For students with knowledge of homology theory and related algebraic concepts a course in homotopy theory based on the last three chapters is quite feasible.

Each chapter is followed by a collection of exercises. These are grouped into sets, each set being devoted to a single topic or a few related topics. With few exceptions, none of the exercises is referred to in the body of the text or in the sequel. There are various types of exercises. Some are examples of the general theory developed in the preceding chapter, some treat special cases of general topics discussed later, and some are devoted to topics not discussed in the text at all. There are routine exercises as well as more difficult ones, the latter frequently with hints of how to attack them. Occasionally a topic related to material in the text is developed in a set of exercises devoted to it.

Examples in the text are usually presented with little or no indication of why they have the stated properties. This is true both of examples illustrating new concepts and of counterexamples. The verification that an example has the desired properties is left to the reader as an exercise.

The symbol ■ is used to denote the end of a proof. It is also used at the end of a statement whose proof has been given before the statement or which follows easily from previous results. Bibliographical references are by footnotes in the text. Items in each section and in each exercise set are numbered consecutively in a single list. References to items in a different section are by triples indicating, respectively, the chapter, the section or exercise set, and the number of the item in the section. Thus 3.2.2 is item 2 in Sec. 2 of Chapter Three (and 3.2 of the Introduction is item 2 in Sec. 3 of the Introduction).

The idea of writing this book originated with the existence of lecture notes based on two courses I gave at the University of Chicago in 1955. It is a pleasure to acknowledge here my indebtedness to the authors of those notes, Guido Weiss for notes of the first course, and Edward Halpern for notes of the second course. In the years since then, the subject has changed substantially and my plans for the book changed along with it, so that the present volume differs in many ways from the original notes.

The final manuscript and galley proofs were read by Per Holm. He made a number of useful suggestions which led to improvements in the text. For his comments and for his friendly encouragement at dark moments, I am sincerely grateful to him. The final manuscript was typed by Mrs. Ann Harrington and Mrs. Ollie Cullers, to both of whom I express my thanks for their patience and cooperation.

I thank the Air Force Office of Scientific Research for a grant enabling me to devote all my time during the academic year 1962–63 to work on this book. I also thank the National Science Foundation for supporting, over a period of years, my research activities some of which are discussed here.

Edwin H. Spanier

LIST OF SYMBOLS

CONTENTS

2 COVERING SPACES AND FIBRATIONS 60

3 POLYHEDRA 106

4 HOMOLOGY 154

ALGEBRAIC TOPOLOGY

INTRODUCTION

THE READER OF THIS BOOK IS ASSUMED TO HAVE A GRASP OF THE ELEMENTARY concepts of set theory, general topology, and algebra. Following are brief summaries of some concepts and results in these areas which are used in this book. Those listed explicitly are done so either because they may not be exactly standard or because they are of particular importance in the subsequent text.

1 SET THEORY[1]

The terms "set," "family," and "collection" are synonyms, and the term "class" is reserved for an aggregate which is not assumed to be a set (for example, the class of all sets). If X is a set and $P(x)$ is a statement which is either true or false for each element $x \in X$, then

[1] As a general reference see P. R. Halmos, *Naïve Set Theory*, D. Van Nostrand Company, Inc., Princeton, N.J., 1960.

$$\{x \in X \mid P(x)\}$$

denotes the subset of X for which $P(x)$ is true.

If $J = \{j\}$ is a set and $\{A_j\}$ is a family of sets indexed by J, their *union* is denoted by $\cup A_j$ (or by $\cup_{j \in J} A_j$), their *intersection* is denoted by $\cap A_j$ (or by $\cap_{j \in J} A_j$), their *cartesian product* is denoted by $\times A_j$ (or by $\times_{j \in J} A_j$), and their *set sum* (sometimes called their *disjoint union*) is denoted by $\vee A_j$ (or by $\vee_{j \in J} A_j$) and is defined by $\vee A_j = \cup (j \times A_j)$. In case $J = \{1,2, \ldots ,n\}$, we also use the notation $A_1 \cup A_2 \cup \cdots \cup A_n$, $A_1 \cap A_2 \cap \cdots \cap A_n$, $A_1 \times A_2 \times \cdots \times A_n$, and $A_1 \vee A_2 \vee \cdots \vee A_n$, respectively, for the union, intersection, cartesian product, and set sum.

A *function* (or *map*) f from A to B is denoted by $f: A \to B$. The set of all functions from A to B is denoted by B^A. If $A' \subset A$, there is an *inclusion map* $i: A' \to A$, and we use the notation $i: A' \subset A$ to indicate that A' is a subset of A and i is the inclusion map. The inclusion map from a set A to itself is called the *identity map of* A and is denoted by 1_A. If $J' \subset J$, there is an inclusion map

$$i_{J'}: \bigvee_{j \in J'} A_j \subset \bigvee_{j \in J} A_j$$

An *equivalence relation* in a set A is a relation $\sim$ between elements of A which is *reflexive* (that is, $a \sim a$ for all $a \in A$), *symmetric* (that is, $a \sim a'$ implies $a' \sim a$ for $a, a' \in A$), and *transitive* (that is, $a \sim a'$ and $a' \sim a''$ imply $a \sim a''$ for $a, a', a'' \in A$). The *equivalence class* of $a \in A$ with respect to $\sim$ is the subset $\{a' \in A \mid a \sim a'\}$. The set of all equivalence classes of elements of A with respect to $\sim$ is denoted by $A/\sim$ and is called a *quotient set of* A. There is a *projection map* $A \to A/\sim$ which sends $a \in A$ to its equivalence class. If J' is a nonempty subset of J, there is also a *projection map*

$$p_{J'}: \bigtimes_{j \in J} A_j \to \bigtimes_{j \in J'} A_j$$

(which is a projection map in the sense above).

Given functions $f: A \to B$ and $g: B \to C$, their *composite* $g \circ f$ (also denoted by gf) is the function from A to C defined by $(g \circ f)(a) = g(f(a))$ for $a \in A$. If $A' \subset A$ and $f: A \to B$, the *restriction of f to A'* is the function $f \mid A': A' \to B$ defined by $(f \mid A')(a') = f(a')$ for $a' \in A'$ (thus $f \mid A' = f \circ i$, where $i: A' \subset A$), and the function f is called an *extension of $f \mid A'$ to A*.

An *injection* (or *injective function*) is a function $f: A \to B$ such that $f(a_1) = f(a_2)$ implies $a_1 = a_2$ for $a_1, a_2 \in A$. A *surjection* (or *surjective function*) is a function $f: A \to B$ such that $b \in B$ implies that there is $a \in A$ with $f(a) = b$. A *bijection* (also called a *bijective function* or a *one-to-one correspondence*) is a function which is both injective and surjective.

A *partial order* in a set A is a relation $\leq$ between elements of A which is reflexive and transitive (note that it is not assumed that $a \leq a'$ and $a' \leq a$ imply $a = a'$). A *total order* (or *simple order*) in A is a partial order in A such that for $a, a' \in A$ either $a \leq a'$ or $a' \leq a$ and which is *antisymmetric* (that is, $a \leq a'$ and $a' \leq a$ imply $a = a'$). A *partially ordered set* is a set with a partial order, and a *totally ordered set* is a set with a total order.

1 ZORN'S LEMMA *A partially ordered set in which every simply ordered subset has an upper bound contains maximal elements.*

A *directed set* Λ is a set with a partial-order relation $\leq$ such that for $\alpha, \beta \in \Lambda$ there is $\gamma \in \Lambda$ with $\alpha \leq \gamma$ and $\beta \leq \gamma$. A *direct system of sets* $\{A^\alpha, f_\alpha{}^\beta\}$ consists of a collection of sets $\{A^\alpha\}$ indexed by a directed set $\Lambda = \{\alpha\}$ and a collection of functions $f_\alpha{}^\beta: A^\alpha \to A^\beta$ for every pair $\alpha \leq \beta$ such that

(*a*) $f_\alpha{}^\alpha = 1_{A^\alpha}: A^\alpha \subset A^\alpha$ for all $\alpha \in \Lambda$
(*b*) $f_\alpha{}^\gamma = f_\beta{}^\gamma \circ f_\alpha{}^\beta: A^\alpha \to A^\gamma$ for $\alpha \leq \beta \leq \gamma$ in Λ

The *direct limit* of the direct system, denoted by $\lim_\rightarrow \{A^\alpha\}$, is the set of equivalence classes of $\bigvee A^\alpha$ with respect to the equivalence relation $a^\alpha \sim a^\beta$ if there is γ with $\alpha \leq \gamma$ and $\beta \leq \gamma$ such that $f_\alpha{}^\gamma a^\alpha = f_\beta{}^\gamma a^\beta$. For each α there is a map $i_\alpha: A^\alpha \to \lim_\rightarrow \{A^\alpha\}$, and if $\alpha \leq \beta$, then $i_\alpha = i_\beta \circ f_\alpha{}^\beta$.

2 *Given a direct system of sets $\{A^\alpha, f_\alpha{}^\beta\}$ and given a set B and for every $\alpha \in \Lambda$ a function $g_\alpha: A^\alpha \to B$ such that $g_\alpha = g_\beta \circ f_\alpha{}^\beta$ if $\alpha \leq \beta$, there is a unique map $g: \lim_\rightarrow \{A^\alpha\} \to B$ such that $g \circ i_\alpha = g_\alpha$ for all $\alpha \in \Lambda$.*

3 *With the same notation as in theorem 2, the map g is a bijection if and only if both the following hold:*

(*a*) $B = \bigcup g_\alpha(A^\alpha)$
(*b*) $g_\alpha(a^\alpha) = g_\beta(a^\beta)$ *if and only if there is γ with $\alpha \leq \gamma$ and $\beta \leq \gamma$ such that $f_\alpha{}^\gamma(a^\alpha) = f_\beta{}^\gamma(a^\beta)$*

Let $\{A_j\}$ be a collection of sets indexed by $J = \{j\}$. Let Λ be the collection of finite subsets of J and define $\alpha \leq \beta$ for $\alpha, \beta \in \Lambda$ if $\alpha \subset \beta$. Then Λ is a directed set and there is a direct system $\{A^\alpha\}$ defined by $A^\alpha = \bigvee_{j \in \alpha} A_j$, and if $\alpha \leq \beta$, then $f_\alpha{}^\beta: A^\alpha \to A^\beta$ is the injection map. Let $g_\alpha: A^\alpha \to \bigvee_{j \in J} A_j$ be the injection map.

4 *With the above notation, there is a bijection $g: \lim_\rightarrow \{A^\alpha\} \to \bigvee_{j \in J} A_j$ such that $g \circ i_\alpha = g_\alpha$ (that is, any set sum is the direct limit of its finite partial set sums).*

An *inverse system of sets* $\{A_\alpha, f_\alpha{}^\beta\}$ consists of a collection of sets $\{A_\alpha\}$ indexed by a directed set $\Lambda = \{\alpha\}$ and a collection of functions $f_\alpha{}^\beta: A_\beta \to A_\alpha$ for $\alpha \leq \beta$ such that

(*a*) $f_\alpha{}^\alpha = 1_{A_\alpha}: A_\alpha \subset A_\alpha$ for $\alpha \in \Lambda$
(*b*) $f_\alpha{}^\gamma = f_\alpha{}^\beta \circ f_\beta{}^\gamma: A_\gamma \to A_\alpha$ for $\alpha \leq \beta \leq \gamma$ in Λ

The *inverse limit* of the inverse system, denoted by $\lim_\leftarrow \{A_\alpha\}$, is the subset of $\times A_\alpha$ consisting of all points (a_α) such that if $\alpha \leq \beta$, then $a_\alpha = f_\alpha{}^\beta a_\beta$. For each α there is a map $p_\alpha: \lim_\leftarrow \{A_\alpha\} \to A_\alpha$, and if $\alpha \leq \beta$, then $p_\alpha = f_\alpha{}^\beta \circ p_\beta$.

5 *Given an inverse system of sets $\{A_\alpha, f_\alpha{}^\beta\}$ and given a set B and for every $\alpha \in \Lambda$ a function $g_\alpha: B \to A_\alpha$ such that $g_\alpha = f_\alpha{}^\beta \circ g_\beta$ if $\alpha \leq \beta$, there is a unique function $g: B \to \lim_\leftarrow \{A_\alpha\}$ such that $g_\alpha = p_\alpha \circ g$ for all $\alpha \in \Lambda$.*

6 *With the same notation as in theorem 5, the map g is a bijection if and only if both the following hold:*

(*a*) $g_\alpha(b) = g_\alpha(b')$ *for all* $\alpha \in \Lambda$ *implies* $b = b'$
(*b*) *Given* $(a_\alpha) \in \times A_\alpha$ *such that* $a_\alpha = f_\alpha{}^\beta a_\beta$ *if* $\alpha \le \beta$, *there is* $b \in B$ *such that* $g_\alpha(b) = a_\alpha$ *for all* $\alpha \in \Lambda$

Let $\{A^j\}$ be a collection of sets indexed by $J = \{j\}$. Let Λ be the collection of finite nonempty subsets of J, and define $\alpha \le \beta$ for $\alpha, \beta \in \Lambda$ if $\alpha \subset \beta$. Then Λ is a directed set and there is an inverse system $\{A_\alpha\}$ defined by $A_\alpha = \times_{j\in\alpha} A^j$, and if $\alpha \le \beta$, $f_\alpha{}^\beta\colon A_\beta \to A_\alpha$ is the projection map. For each $\alpha \in \Lambda$ let $g_\alpha\colon \times_{j\in J} A^j \to A_\alpha$ be the projection map.

7 *With the above notation, there is a bijection* $g\colon \times_{j\in J} A^j \to \lim_{\leftarrow} \{A_\alpha\}$ *such that* $g_\alpha = p_\alpha \circ g$ *(that is, any cartesian product is the inverse limit of its finite partial cartesian products).*

2 GENERAL TOPOLOGY[1]

A topological space, also called a space, is not assumed to satisfy any separation axioms unless explicitly stated. Paracompact, normal, and regular spaces will always be assumed to be Hausdorff spaces. A continuous map from one topological space to another will also be called simply a map.

Given a set X and an indexed collection of topological spaces $\{X_j\}_{j\in J}$ and functions $f_j\colon X \to X_j$, the *topology induced on* X by the functions $\{f_j\}$ is the smallest or coarsest topology such that each f_j is continuous.

1 *The topology induced on X by functions* $\{f_j\colon X \to X_j\}$ *is characterized by the property that if Y is a topological space, a function* $g\colon Y \to X$ *is continuous if and only if* $f_j \circ g\colon Y \to X_j$ *is continuous for each* $j \in J$.

A *subspace* of a topological space X is a subset A of X topologized by the topology induced by the inclusion map $A \subset X$. A *discrete subset* of a topological space X is a subset such that every subset of it is closed in X. The *topological product* of an indexed collection of topological spaces $\{X_j\}_{j\in J}$ is the cartesian product $\times X_j$, given the topology induced by the projection maps $p_j\colon \times X_j \to X_j$ for $j \in J$. If $\{X_\alpha\}_{\alpha\in\Lambda}$ is an inverse system of topological spaces (that is, X_α is a topological space for $\alpha \in \Lambda$ and $f_\alpha{}^\beta\colon X_\beta \to X_\alpha$ is continuous for $\alpha \le \beta$) their inverse limit $\lim_{\leftarrow} \{X_\alpha\}$ is given the topology induced by the functions $p_\alpha\colon \lim_{\leftarrow} \{X_\alpha\} \to X_\alpha$ for $\alpha \in \Lambda$.

Given a set X and an indexed collection of topological spaces $\{X_j\}_{j\in J}$ and functions $g_j\colon X_j \to X$, the *topology coinduced on* X by the functions $\{g_j\}$ is the largest or finest topology such that each g_j is continuous.

[1] As general references see J. L. Kelley, *General Topology*, D. Van Nostrand Company, Inc., Princeton, N.J., 1955, and S. T. Hu, *Elements of General Topology*, Holden-Day, Inc., San Francisco, 1964.

2 *The topology coinduced on X by functions $\{g_j: X_j \to X\}$ is characterized by the property that if Y is any topological space, a function $f: X \to Y$ is continuous if and only if $f \circ g_j: X_j \to Y$ is continuous for each $j \in J$.*

A *quotient space* of a topological space X is a quotient set X' of X topologized by the topology coinduced by the projection map $X \to X'$. If $A \subset X$, then X/A will denote the quotient space of X obtained by identifying all of A to a single point. The *topological sum* of an indexed collection of topological spaces $\{X_j\}_{j \in J}$ is the set sum $\vee X_j$, given the topology coinduced by the injection maps $i_j: X_j \to \vee X_j$ for $j \in J$. If $\{X^\alpha\}_{\alpha \in \Lambda}$ is a direct system of topological spaces (that is, X^α is a topological space for $\alpha \in \Lambda$ and $f_\alpha{}^\beta: X^\alpha \to X^\beta$ is continuous for $\alpha \leq \beta$) their direct limit $\lim_\rightarrow \{X^\alpha\}$ is given the topology coinduced by the functions $i_\alpha: X^\alpha \to \lim_\rightarrow \{X^\alpha\}$ for $\alpha \in \Lambda$.

Let $\mathcal{A} = \{A\}$ be a collection of subsets of a topological space X. X is said to have a *topology coherent with* $\mathcal{A}$ if the topology on X is coinduced from the subspaces $\{A\}$ by the inclusion maps $A \subset X$. (In the literature this topology is often called the *weak topology* with respect to $\mathcal{A}$.)

3 *A necessary and sufficient condition that X have a topology coherent with $\mathcal{A}$ is that a subset B of X be closed (or open) in X if and only if $B \cap A$ is closed (or open) in the subspace A for every $A \in \mathcal{A}$.*

4 *If $\mathcal{A}$ is an arbitrary open covering or a locally finite closed covering of X, then X has a topology coherent with $\mathcal{A}$.*

5 *Let X be a set and let $\{A_j\}$ be an indexed collection of topological spaces each contained in X and such that for each j and j', $A_j \cap A_{j'}$ is a closed (or open) subset of A_j and of $A_{j'}$ and the topology induced on $A_j \cap A_{j'}$ from A_j equals the topology induced on $A_j \cap A_{j'}$ from $A_{j'}$. Then the topology coinduced on X by the collection of inclusion maps $\{A_j \subset X\}$ is characterized by the properties that A_j is a closed (or open) subspace of X for each j and X has a topology coherent with the collection $\{A_j\}$.*

The topology on X in theorem 5 will be called *the topology coherent with* $\{A_j\}$. A *compactly generated space* is a Hausdorff space having a topology coherent with the collection of its compact subsets (this is the same as what is sometimes referred to as a Hausdorff *k-space*).

6 *A Hausdorff space which is either locally compact or satisfies the first axiom of countability is compactly generated.*

7 *If X is compactly generated and Y is a locally compact Hausdorff space, $X \times Y$ is compactly generated.*

If X and Y are topological spaces and $A \subset X$ and $B \subset Y$, then $\langle A;B \rangle$ denotes the set of continuous functions $f: X \to Y$ such that $f(A) \subset B$. Y^X denotes the space of continuous functions from X to Y, given the *compact-open topology* (which is the topology generated by the subbase $\{\langle K;U \rangle\}$, where K is a compact subset of X and U is an open subset of Y). If $A \subset X$

and $B \subset Y$, we use $(Y,B)^{(X,A)}$ to denote the subspace of Y^X of continuous functions $f: X \to Y$ such that $f(A) \subset B$. Let $E: Y^X \times X \to Y$ be the *evaluation map* defined by $E(f,x) = f(x)$. Given a function $g: Z \to Y^X$, the composite

$$Z \times X \xrightarrow{g \times 1} Y^X \times X \xrightarrow{E} Y$$

is a function from $Z \times X$ to Y.

8 THEOREM OF EXPONENTIAL CORRESPONDENCE *If X is a locally compact Hausdorff space and Y and Z are topological spaces, a map $g: Z \to Y^X$ is continuous if and only if $E \circ (g \times 1): Z \times X \to Y$ is continuous.*

9 EXPONENTIAL LAW *If X is a locally compact Hausdorff space, Z is a Hausdorff space, and Y is a topological space, the function $\psi: (Y^X)^Z \to Y^{Z \times X}$ defined by $\psi(g) = E \circ (g \times 1)$ is a homeomorphism.*

10 *If X is a compact Hausdorff space and Y is metrized by a metric d, then Y^X is metrized by the metric d' defined by*

$$d'(f,g) = \sup \{d(f(x),g(x)) \mid x \in X\}$$

3 GROUP THEORY[1]

A homomorphism is called a *monomorphism, epimorphism, isomorphism,* respectively, if it is injective, surjective, bijective. If $\{G_j\}_{j \in J}$ is an indexed collection of groups, their *direct product* is the group structure on the cartesian product $\times G_j$ defined by $(g_j)(g_j') = (g_j g_j')$. If $\{G_\alpha\}$ is an inverse system of groups (that is, G_α is a group for each α and $f_\alpha{}^\beta: G_\beta \to G_\alpha$ is a homomorphism for $\alpha \leq \beta$), their inverse limit $\lim_{\leftarrow} \{G_\alpha\}$ (which is a set) is a subgroup of $\times G_\alpha$.

Let A be a subset of a group G. G is said to be *freely generated* by A and A is said to be a *free generating set* or *free basis* for G if, given any function $f: A \to H$, where H is a group, there exists a unique homomorphism $\varphi: G \to H$ which is an extension of f. A group is said to be *free* if it is freely generated by some subset. For any set A a *free group generated by* A is a group $F(A)$ containing A as a free generating set. Such groups $F(A)$ exist, and any two are canonically isomorphic.

1 *Any group is isomorphic to a quotient group of a free group.*

A *presentation* of a group G consists of a set A of *generators,* a set $B \subset F(A)$ of *relations,* and a function $f: A \to G$ such that the extension of f to a homomorphism $\varphi: F(A) \to G$ is an epimorphism whose kernel is the nor-

[1] As a general reference for elementary group theory see G. Birkhoff and S. MacLane, *A Survey of Modern Algebra,* The Macmillan Company, New York, 1953. For a discussion of free groups see R. H. Crowell and R. H. Fox, *Introduction to Knot Theory,* Ginn and Company, Boston, 1963.

mal subgroup of $F(A)$ generated by B. If A and B are both finite sets, the presentation is said to be *finite* and G is said to be *finitely presented.*

4 MODULES[1]

We are mainly interested in R modules where R is a principal ideal domain. However, we shall begin with some properties of R modules where R is a commutative ring with a unit which acts as the identity on every module. If $\varphi: A \to B$ is a homomorphism of R modules, then we have R modules

$$\ker \varphi = \{a \in A \mid \varphi(a) = 0\} \subset A$$
$$\operatorname{im} \varphi = \{b \in B \mid b = \varphi(a) \text{ for some } a \in A\} \subset B$$
$$\operatorname{coker} \varphi = B/\operatorname{im} \varphi$$

1 NOETHER ISOMORPHISM THEOREM *Let A and B be submodules of a module C and let $A + B$ be the submodule of C generated by $A \cup B$. The inclusion map $A \subset A + B$ sends $A \cap B$ into B and induces an isomorphism of $A/(A \cap B)$ with $(A + B)/B$.*

If $\{A_j\}_{j\in J}$ is an indexed collection of R modules, their direct product $\times A_j$ is an R module and their *direct sum* $\bigoplus A_j$ is an R module ($\bigoplus A_j$ is the submodule of $\times A_j$ consisting of those elements having only a finite number of nonzero coordinates). The inverse limit $\lim_{\leftarrow} \{A_\alpha\}$ of an inverse system of R modules (and homomorphisms $f_\alpha{}^\beta: A_\beta \to A_\alpha$ for $\alpha \le \beta$) is an R module, and the direct limit of a direct system of R modules (and homomorphisms) is an R module.

2 *Any R module is isomorphic to the direct limit of its finitely generated submodules directed by inclusion.*

If A and B are R modules, their *tensor product* $A \otimes B$ (also written $A \otimes_R B$) is an R module. For $a \in A$ and $b \in B$, there is a corresponding element $a \otimes b \in A \otimes B$. $A \otimes B$ is generated by the elements $\{a \otimes b \mid a \in A, b \in B\}$ with the relations (for $a, a' \in A$, $b, b' \in B$, and $r, r' \in R$)

$$(ra + r'a') \otimes b = r(a \otimes b) + r'(a' \otimes b)$$
$$a \otimes (rb + r'b') = r(a \otimes b) + r'(a \otimes b')$$

In case A or B is also an R' module, then so is $A \otimes_R B$.

3 *For any R module A the homomorphisms $a \to a \otimes 1$ and $a \to 1 \otimes a$ define isomorphisms of A with $A \otimes R$ and $R \otimes A$.*

[1] As general references see H. Cartan and S. Eilenberg, *Homological Algebra,* Princeton University Press, Princeton, N.J., 1956 and S. MacLane, *Homology,* Springer-Verlag OHG, Berlin, 1963.

4 *For R modules A and B there is an isomorphism of* $A \otimes B$ *with* $B \otimes A$ *taking* $a \otimes b$ *to* $b \otimes a$.

5 *If A and B are R modules and B and C are R′ modules, there is an isomorphism of* $(A \otimes_R B) \otimes_{R'} C$ *with* $A \otimes_R (B \otimes_{R'} C)$ *(both being regarded as R and R′ modules) taking* $(a \otimes b) \otimes c$ *to* $a \otimes (b \otimes c)$.

If *A* and *B* are *R* modules, their *module of homomorphisms* Hom (A,B) [also written $\mathrm{Hom}_R(A,B)$] is an *R* module whose elements are *R* homomorphisms $A \to B$. In case *A* or *B* is also an *R′* module, then so is $\mathrm{Hom}_R(A,B)$.

6 *If A and B are R modules and B and C are R′ modules, there is an isomorphism of* $\mathrm{Hom}_{R'}(A \otimes_R B, C)$ *with* $\mathrm{Hom}_R(A, \mathrm{Hom}_{R'}(B,C))$ *(both being regarded as R and R′ modules) taking an R′ homomorphism* $\varphi: A \otimes_R B \to C$ *to the R homomorphism* $\varphi': A \to \mathrm{Hom}_{R'}(B,C)$ *such that* $\varphi'(a)(b) = \varphi(a \otimes b)$.

A subset S of an *R* module A is said to be a *basis for* A (and A is said to be *freely generated by* S) if any function $f: S \to B$, where *B* is an *R* module, admits a unique extension to a homomorphism $\varphi: A \to B$. If a module has a basis, it is said to be a *free module.* For any set S the *free module generated by* S, denoted by $F_R(S)$, is the module of all finitely nonzero functions from S to *R* (with pointwise addition and scalar multiplication) and with $s \in S$ identified with its characteristic function. $F_R(S)$ contains S as a basis, and any module containing S as a basis is canonically isomorphic to $F_R(S)$.

7 *Any R module is isomorphic to a quotient of a free R module.*

8 *If A′ is a submodule of A, with A/A′ free, then A is isomorphic to the direct sum* $A' \oplus (A/A')$.

We now assume that *R* is a principal ideal domain (that is, it is an integral domain in which every ideal is principal). If A is an *R* module, its *torsion submodule* Tor A is defined by

$$\mathrm{Tor}\, A = \{a \in A \mid ra = 0 \text{ for some nonzero } r \in R\}$$

A is said to be *torsion free* or *without torsion* if Tor A = 0.

9 *Over a principal ideal domain, a submodule of a free module is free.*

10 *Over a principal ideal domain, a finitely generated module is free if and only if it is torsion free.*

11 *Over a principal ideal domain, A/Tor A is torsion free.*

If A is a finitely generated module over a principal ideal domain *R*, its *rank* $\rho(A)$ is defined to be the number of elements in a basis of the quotient module A/Tor A.

12 *If A′ is a submodule of a finitely generated module A (over a principal ideal domain), then*

$$\rho(A) = \rho(A') + \rho(A/A')$$

Let $\varphi: A \to A$ be an endomorphism of a finitely generated module (over a principal ideal domain R). The *trace* of φ, $\operatorname{Tr} \varphi$, is the element of R which is the trace of the endomorphism φ' induced by φ on the free module $A/\operatorname{Tor} A$ [that is, if $A/\operatorname{Tor} A$ has a basis $a_1, \ldots, a_n$, then $\varphi'(a_i) = \sum r_{ij} a_j$ and $\operatorname{Tr} \varphi = \sum r_{ii}$].

13 *Let φ be an endomorphism of a finitely generated module A and let A' be a submodule of A such that $\varphi(A') \subset A'$. Then $\varphi \mid A'$ is an endomorphism of A' and there is induced an endomorphism φ'' of A/A'. Their traces satisfy the relation*

$$\operatorname{Tr} \varphi = \operatorname{Tr} (\varphi \mid A') + \operatorname{Tr} \varphi''$$

A module with a single generator is said to be *cyclic*. Over a principal ideal domain R such a module A is characterized, up to isomorphism, by the element $r_A \in R$ which generates the ideal of elements annihilating every element of A (r_A is unique up to multiplication by invertible elements of R).

14 STRUCTURE THEOREM FOR FINITELY GENERATED MODULES *Over a principal ideal domain every finitely generated module is the direct sum of a free module and cyclic modules $A_1, \ldots, A_q$ whose corresponding elements $r_1, \ldots, r_q \in R$ have the property that r_i divides r_{i+1} for $1 \le i \le q - 1$. The elements $r_1, \ldots, r_q$ are unique up to multiplication by invertible elements of R and, together with the rank of the module, characterize the module up to isomorphism.*

5 EUCLIDEAN SPACES

We use the following fixed notations:

$\varnothing$ = empty set
$\mathbf{Z}$ = ring of integers
$\mathbf{Z}_m$ = ring of integers modulo m
$\mathbf{R}$ = field of real numbers
$\mathbf{C}$ = field of complex numbers
$\mathbf{Q}$ = division ring of quaternions
$\mathbf{R}^n$ = *euclidean n-space*, with $\|x\| = \sqrt{\sum x_i^2}$ and $\langle x,y \rangle = \sum x_i y_i$
0 = *origin of* $\mathbf{R}^n$
I = *closed unit interval*
$\dot{I}$ = $\{0,1\} \subset I$
I^n = *n-cube* = $\{x \in \mathbf{R}^n \mid 0 \le x_i \le 1 \text{ for } 1 \le i \le n\}$
$\dot{I}^n$ = $\{x \in I^n \mid \text{for some } i, x_i = 0 \text{ or } x_i = 1\}$
E^n = *n-ball* = $\{x \in \mathbf{R}^n \mid \|x\| \le 1\}$
S^{n-1} = *$(n-1)$-sphere* = $\{x \in \mathbf{R}^n \mid \|x\| = 1\}$
P^n = *projective n-space* = quotient space of S^n with x and $-x$ identified for all $x \in S^n$

If x and y are points of a real vector space, the *closed line segment* joining them, denoted by $[x,y]$, is the set of points of the form $tx + (1 - t)y$ for $0 \le t \le 1$ (thus $I = [0,1]$). If $x \ne y$, the *line* determined by them is the set $\{tx + (1 - t)y \mid t \in \mathbf{R}\}$. A subset C of a real vector space is said to be an *affine variety* if whenever $x, y \in C$, with $x \ne y$, then the line determined by x and y is also in C. A subset C is said to be *convex* if $x, y \in C$ imply $[x,y] \subset C$. A *convex body*[1] in $\mathbf{R}^n$ is a convex subset of $\mathbf{R}^n$ containing a nonempty open subset of $\mathbf{R}^n$ (thus I^n and E^n are convex bodies in $\mathbf{R}^n$).

1 *If C is a convex body in $\mathbf{R}^n$ and C' is a convex body in $\mathbf{R}^m$, then $C \times C'$ is a convex body in $\mathbf{R}^n \times \mathbf{R}^m = \mathbf{R}^{n+m}$.*

2 *Any two compact convex bodies in $\mathbf{R}^n$ are homeomorphic.*

A subset S of a real vector space is said to be *affinely independent* if, given a finite number of distinct elements $x_0, x_1, \ldots, x_m \in S$ and $t_0, t_1, \ldots, t_m \in \mathbf{R}$ such that $\Sigma\, t_i = 0$ and $\Sigma\, t_i x_i = 0$, then $t_i = 0$ for $0 \le i \le m$ (this is equivalent to the condition that

$$x_1 - x_0, x_2 - x_0, \ldots, x_m - x_0$$

be linearly independent).

3 *There exist affinely independent subsets of $\mathbf{R}^n$ containing $n + 1$ points, but no subset of $\mathbf{R}^n$ containing more than $n + 1$ points is affinely independent.*

4 *Given points $x_0, x_1, \ldots, x_m \in \mathbf{R}^n$, the convex set generated by them is the set of all points of the form $\Sigma\, t_i x_i$, with $0 \le t_i \le 1$ and $\Sigma\, t_i = 1$. The set $\{x_0, x_1, \ldots, x_m\}$ is affinely independent if and only if every point x in the convex set generated by this set has a unique representation in the form $x = \Sigma\, t_i x_i$, with $0 \le t_i \le 1$ for $0 \le i \le m$ and $\Sigma\, t_i = 1$.*

SOME BOOKS ON ALGEBRAIC TOPOLOGY

Alexandroff, P. and H. Hopf: *Topologie*, Springer-Verlag, 1935.
Bott, R. and L.W. Tu: *Differential Forms in Algebraic Topology*, Springer-Verlag, 1982
Bourgin, D.G.: *Modern Algebraic Topology*, Macmillan, 1963.
Bredon, G.E.: *Sheaf Theory*, McGraw-Hill, 1967.
Cairns, S.S.: *Introductory Topology*, Ronald Press, 1962.
Dold, A.: *Lectures on Algebraic Topology*, Springer-Verlag, 1980.
Eilenberg, S. and N.E. Steenrod: *Foundations of Algebraic Topology*, Princeton University Press, 1952.
Godement, R.: *Topologie algébrique et théorie des faisceaux*, Hermann and Cie, 1958.

[1] For general properties of convex sets see F. A. Valentine, *Convex Sets*, McGraw-Hill Book Company, New York, 1964.

Gray, B.: *Homotopy Theory, An Introduction to Algebraic Topology*, Academic Press, 1975.

Greenberg, M.J. and J.R. Harper: *Algebraic Topology, A First Course* Benjamin/Cummings, 1981.

Hilton, P.J. and S. Wylie: *Homology Theory*, Cambridge University Press, 1960.

Hocking, J.G. and G.S. Young: *Topology*, Addison-Wesley, 1961.

Hu, S.T.: *Homotopy Theory*, Academic Press, 1959.

Lefschetz, S.: *Algebraic Topology*, American Math Society, 1942.

Lefschetz, S.: *Introduction to Topology*, Princeton Univ. Press, 1949.

Massey, W.S.: *Algebraic Topology, An Introduction*, Harcourt, Brace and World, 1967.

Massey, W.S.: *Homology and Cohomology Theory, An Approach Based on Alexander-Spanier Cochains*, Dekker, 1978.

Massey, W.S.: *Singular Homology Theory*, Springer-Verlag, 1980.

Maunder, C.R.F.: *Algebraic Topology*, Cambridge Univ. Press, 1980.

Munkres, J.R.: *Elements of Algebraic Topology*, Addison-Wesley, 1984.

Pontryagin, L.S.: *Foundations of Combinational Topology*, Graylock Press, 1952.

Schubert H.: *Topologie*, Teubner Verlagsgesellschaft, 1964.

Seifert, H. and W. Threlfall: *Lehrbuch der Topologie*, Teubner Verlagsgesellschaft, 1934.

Steenrod, N.E.: *The Topology of Fiber Bundles*, Princeton University Press, 1951.

Switzer, R.M.: *Algebraic Topology—Homotopy and Homology*, Springer-Verlag, 1975.

Vick, J.W.: *Homology Theory, An Introduction to Algebraic Topology*, Academic Press, 1973.

Wallace, A.H.: *An Introduction to Algebraic Topology*, Pergamon Press, 1957.

Wallace, A.H.: *Algebraic Topology, Homology and Cohomology*, Benjamin, 1970.

Wilder, R.L.: *Topology of Manifolds*, American Math Society, 1949.

CHAPTER ONE

HOMOTOPY AND THE FUNDAMENTAL GROUP

TOPOLOGY IS THE STUDY OF TOPOLOGICAL SPACES AND CONTINUOUS FUNCTIONS between them. A standard problem is the classification of such spaces and functions up to homeomorphism. A weaker equivalence relation, based on continuous deformation, leads to another classification problem. This latter classification problem is of fundamental importance in algebraic topology, since it is the one where the tools available seem to be most successful.

As a working definition for our purposes, algebraic topology may be regarded as the study of topological spaces and continuous functions by means of algebraic objects such as groups, rings, homomorphisms. The link from topology to algebra is by means of mappings, called functors. For this reason, Secs. 1.1 and 1.2 are devoted to the basic concepts of category and functor.

In Secs. 1.3 and 1.4 the concept of continuous deformation, known technically as homotopy, is introduced. We then define the homotopy category and certain functors on this category, all of which are important for the subject. Sections 1.5 and 1.6 are devoted to a study of conditions under which these functors on the homotopy category take values in the category of groups. As examples, the homotopy group functors are briefly mentioned.

The first functor considered in detail is the fundamental group functor, introduced and discussed in Secs. 1.7 and 1.8. This is an intuitively appealing example of the kind of functor considered in algebraic topology. Some applications of this functor are presented in the exercises at the end of the chapter. In Chapter Two this functor is used in a systematic study and classification of covering spaces.

1 CATEGORIES

An algebraic representation of topology is a mapping from topology to algebra. Such a representation converts a topological problem into an algebraic one to the end that, with sufficiently many representations, the topological problem will be solvable if (and only if) all the corresponding algebraic problems are solvable.

The definition of a representation, formally called a functor, is given in the next section. This section is devoted to the concept of category, because functors are functions, with certain naturality properties, from one or several categories to another.

A category may be thought of intuitively as consisting of sets, possibly with additional structure, and functions, possibly preserving additional structure. More precisely, a *category* $\mathcal{C}$ consists of

(*a*) A class of *objects*
(*b*) For every ordered pair of objects X and Y, a set $\hom(X,Y)$ of *morphisms* with *domain* X and *range* Y; if $f \in \hom(X,Y)$, we write $f: X \to Y$ or $X \xrightarrow{f} Y$
(*c*) For every ordered triple of objects X, Y, and Z, a function associating to a pair of morphisms $f: X \to Y$ and $g: Y \to Z$ their *composite*

$$gf = g \circ f: X \to Z$$

These satisfy the following two axioms:

Associativity. If $f: X \to Y$, $g: Y \to Z$, and $h: Z \to W$, then

$$h(gf) = (hg)f: X \to W$$

Identity. For every object Y there is a morphism $1_Y: Y \to Y$ such that if $f: X \to Y$, then $1_Y f = f$, and if $h: Y \to Z$, then $h 1_Y = h$.

If the class of objects is a set, the category is said to be *small.* For most of our purposes we could restrict our attention to small categories, but it would be inconvenient to have to specify a set of objects before obtaining a category. For example, we should like to consider categories whose objects are sets or groups, and we prefer to consider the class of all sets or groups, rather than some suitable set of sets or groups in each instance.

From the two axioms it follows that 1_Y is unique (see lemma 1 below), and it is called the *identity morphism* of Y. Given morphisms $f: X \to Y$ and $g: Y \to X$ such that $gf = 1_X$, g is called a *left inverse* of f and f is called a *right inverse* of g. A *two-sided inverse* (or simply an *inverse*) of f is a morphism which is both a left inverse of f and a right inverse of f. A morphism $f: X \to Y$ is called an *equivalence*, denoted by $f: X \approx Y$, if there is a morphism $g: Y \to X$ which is a two-sided inverse of f. If $g': Y \to X$ is a left inverse of f and $g'': Y \to X$ is a right inverse of f, then

$$g' = g' 1_Y = g'(fg'') = (g'f)g'' = 1_X g'' = g''$$

showing that $g' = g''$. Therefore we have the following lemma.

1 LEMMA *If $f: X \to Y$ has a left inverse and a right inverse, they are equal, and f is an equivalence.* ∎

In particular, it follows that an equivalence $f: X \approx Y$ has a unique inverse, denoted by $f^{-1}: Y \to X$, and f^{-1} is an equivalence. If there is an equivalence $f: X \approx Y$, X and Y are said to be *equivalent*, denoted by $X \approx Y$. Because the composite of equivalences is easily seen to be an equivalence, the relation $X \approx Y$ is an equivalence relation in any set of objects of $\mathcal{C}$.

We list some examples of categories.

2 The category of sets and functions [that is, the class of objects is the class of all sets, and for sets X and Y, hom (X,Y) equals the set of functions from X to Y]

3 The category of topological spaces and continuous maps

4 The category of groups and homomorphisms

5 The category of R modules and homomorphisms

6 The category of normed rings (over **R**) and continuous homomorphisms

7 The category of sets and injections (or surjections or bijections)

8 The category of *pointed sets* (a pointed set is a nonempty set with a distinguished element) and functions preserving distinguished elements

9 The category of *pointed topological spaces* (a pointed topological space is a nonempty topological space with a *base point*) and continuous maps preserving base points

10 The category of finite sets and functions

11 Given a partial order $\leq$ in X, there is a category whose objects are the elements of X and such that hom (x,x') is either the singleton consisting of the ordered pair (x,x') or empty, according to whether $x \leq x'$ or $x \nleq x'$

12 The category of groups and conjugacy classes of homomorphisms (that is, a morphism $G \to G'$ is an equivalence class of homomorphisms from G to G', two homomorphisms being equivalent if they differ by an inner automorphism of G')

A *subcategory* $\mathcal{C}' \subset \mathcal{C}$ is a category such that

(*a*) The objects of $\mathcal{C}'$ are also objects of $\mathcal{C}$
(*b*) For objects X' and Y' of $\mathcal{C}'$, $\hom_{\mathcal{C}'}(X',Y') \subset \hom_{\mathcal{C}}(X',Y')$
(*c*) If $f'\colon X' \to Y'$ and $g'\colon Y' \to Z'$ are morphisms of $\mathcal{C}'$, their composite in $\mathcal{C}'$ equals their composite in $\mathcal{C}$

$\mathcal{C}'$ is called a *full subcategory* of $\mathcal{C}$ if $\mathcal{C}'$ is a subcategory of $\mathcal{C}$ and for objects X' and Y' in $\mathcal{C}'$, $\hom_{\mathcal{C}'}(X',Y') = \hom_{\mathcal{C}}(X',Y')$. The category in example 7 above is a subcategory of the one in example 2, and the category in example 10 is a full subcategory of the one in example 2. The categories in examples 3, 4, 5, 6, and 8 are not subcategories of the category of sets, because each object of one of these categories consists of a set, together with an additional structure on it (hence, different objects in these categories may have the same underlying sets). In examples 11 and 12, the morphisms in the respective categories are not functions, and so neither of these categories is a subcategory of the category of sets.

A diagram of morphisms such as the square

$$\begin{array}{ccc} X & \xrightarrow{f} & Y \\ g\downarrow & & \downarrow h \\ X' & \xrightarrow{f'} & Y' \end{array}$$

is said to be *commutative* if any two composites of morphisms in the diagram beginning at the same place and ending at the same place are equal. This square is commutative if and only if $hf = f'g$.

Following are descriptions of some categories which are associated to a given category. Given a category $\mathcal{C}$, there is an associated category called the *category of morphisms of* $\mathcal{C}$. Its objects are morphisms $X \xrightarrow{f} Y$, and its morphisms with domain $X \xrightarrow{f} Y$ and range $X' \xrightarrow{f'} Y'$ are pairs of morphisms $g\colon X \to X'$ and $h\colon Y \to Y'$ such that the square

$$\begin{array}{ccc} X & \xrightarrow{f} & Y \\ g\downarrow & & \downarrow h \\ X' & \xrightarrow{f'} & Y' \end{array}$$

is commutative. In a similar way, diagrams of morphisms in $\mathcal{C}$ more general than $X \xrightarrow{f} Y$ are the objects of a suitable category associated to $\mathcal{C}$.

Let $\mathcal{C}$ be a category whose objects are sets with additional structures (such as distinguished elements or topologies) and whose morphisms are functions preserving the additional structures. For example, $\mathcal{C}$ might be any of the categories in examples 2 through 10. There is a category associated to $\mathcal{C}$, called the *category of pairs of* $\mathcal{C}$, whose objects are injective morphisms $i\colon A \to X$ (because each morphism in such a category is a function, it is meaningful to consider those which are injective) and whose morphisms are commutative squares

$$\begin{array}{ccc} A & \xrightarrow{i} & X \\ g\downarrow & & \downarrow h \\ B & \xrightarrow{j} & Y \end{array}$$

Thus the category of pairs of $\mathcal{C}$ is a full subcategory of the category of morphisms of $\mathcal{C}$. The notation (X,A) will denote the pair consisting of X and $i: A \subset X$, and the notation $f: (X,A) \to (Y,B)$ will mean that $f: X \to Y$ is a morphism of $\mathcal{C}$ such that $f(i(A)) \subset j(B)$. The category of pairs of $\mathcal{C}$, therefore, has as objects the pairs (X,A) and has as morphisms the morphisms $f: (X,A) \to (Y,B)$.

If $\mathcal{C}_1$ and $\mathcal{C}_2$ are categories, their *product* $\mathcal{C}_1 \times \mathcal{C}_2$ is the category whose objects are ordered pairs (Y_1, Y_2) of objects Y_1 in $\mathcal{C}_1$ and Y_2 in $\mathcal{C}_2$ and whose morphisms $(X_1,X_2) \to (Y_1, Y_2)$ are ordered pairs of morphisms (f_1,f_2), where $f_1: X_1 \to Y_1$ in $\mathcal{C}_1$ and $f_2: X_2 \to Y_2$ in $\mathcal{C}_2$. Similarly, there is a product of an arbitrary indexed family of categories.

Given a category $\mathcal{C}$, there is an *opposite* category $\mathcal{C}^*$ whose objects Y^* are in one-to-one correspondence with the objects Y of $\mathcal{C}$ and whose morphisms $f^*: Y^* \to X^*$ are in one-to-one correspondence with the morphisms $f: X \to Y$ [with $f^* g^*$ defined to equal $(gf)^*$ for $X \xrightarrow{f} Y \xrightarrow{g} Z$ in $\mathcal{C}$]. We identify $(\mathcal{C}^*)^*$ with $\mathcal{C}$, so that $(X^*)^* = X$ and $(f^*)^* = f$.

We next show how to interpret sums and products, as well as direct and inverse limits in arbitrary categories. An object X in a category $\mathcal{C}$ is said to be an *initial object* if for each object Y in $\mathcal{C}$ the set hom (X,Y) contains exactly one element. Dually, an object Z of $\mathcal{C}$ is said to be a *terminal object* if for each Y of $\mathcal{C}$ the set hom (Y,Z) contains exactly one element. Note that any two initial objects of $\mathcal{C}$ are equivalent and any two terminal objects of $\mathcal{C}$ are equivalent. In examples 2 and 3 the empty set is an initial object and any one-point set is a terminal object. In example 4 the trivial group is both an initial and a terminal object. In example 7 the category of sets and bijections has neither an initial object nor a terminal object.

Let $\{Y_j\}_{j \in J}$ be an indexed collection of objects of a category $\mathcal{C}$. Let $\mathcal{S}\{Y_j\}$ be the category whose objects are indexed collections of morphisms $\{f_j\}_{j \in J}$ of $\mathcal{C}$ having the same range and whose morphisms with domain $\{f_j: Y_j \to Z\}$ and range $\{f_j': Y_j \to Z'\}$ are morphisms $g: Z \to Z'$ of $\mathcal{C}$ such that $gf_j = f_j'$ for every $j \in J$. An initial object of $\mathcal{S}\{Y_j\}$ is called a *sum* of the collection $\{Y_j\}$. A given collection may or may not have a sum in $\mathcal{C}$. The set sum is a sum in the category of sets, the topological sum is a sum in the category of topological spaces, the free product is a sum in the category of groups, and the direct sum is a sum in the category of R modules. In the category of finite sets, in general only finite collections have a sum. Similarly, in the category of finitely generated R modules, in general only finite collections have a sum.

Dually, given an indexed collection of objects $\{Y_j\}_{j \in J}$ in $\mathcal{C}$, let $\mathcal{P}\{Y_j\}$ be the category whose objects are indexed collections of morphisms $\{g_j\}_{j \in J}$ of $\mathcal{C}$ having the same domain and whose morphisms with domain $\{g_j: X \to Y_j\}$ and range $\{g_j': X' \to Y_j\}$ are morphisms $f: X \to X'$ of $\mathcal{C}$ such that $g_j'f = g_j$ for every $j \in J$. A terminal object of $\mathcal{P}\{Y_j\}$ is called a *product* of the collection $\{Y_j\}$. The cartesian product of sets is a product in the category of sets, the topological product is a product in the category of topological spaces, and the direct product is a product in the category of groups, or R modules. In the category of finite sets (or finitely generated R modules), in general only finite collections have a product.

A *direct system* $\{Y^\alpha, f_\alpha{}^\beta\}$ in a category $\mathcal{C}$ consists of a collection of objects $\{Y^\alpha\}$ indexed by a directed set $\Lambda = \{\alpha\}$ and a collection of morphisms $\{f_\alpha{}^\beta\colon Y^\alpha \to Y^\beta\}$ in $\mathcal{C}$ for $\alpha \le \beta$ in Λ such that

(*a*) $f_\alpha{}^\alpha = 1_{Y^\alpha}$ for $\alpha \in \Lambda$
(*b*) $f_\alpha{}^\gamma = f_\beta{}^\gamma f_\alpha{}^\beta\colon Y^\alpha \to Y^\gamma$ for $\alpha \le \beta \le \gamma$ in Λ

There is then a category dir $\{Y^\alpha, f_\alpha{}^\beta\}$ whose objects are indexed collections of morphisms $\{g_\alpha\colon Y^\alpha \to Z\}_{\alpha\in\Lambda}$ such that $g_\alpha = g_\beta f_\alpha{}^\beta$ if $\alpha \le \beta$ in Λ and whose morphisms with domain $\{g_\alpha\colon Y^\alpha \to Z\}$ and range $\{g'_\alpha\colon Y^\alpha \to Z'\}$ are morphisms $h\colon Z \to Z'$ such that $hg_\alpha = g'_\alpha$ for $\alpha \in \Lambda$. An initial object of dir $\{Y^\alpha, f_\alpha{}^\beta\}$ is called a *direct limit* of the direct system $\{Y^\alpha, f_\alpha{}^\beta\}$. The direct limits of sets, topological spaces, groups, and *R* modules are examples of direct limits in their respective categories.

Dually, an *inverse system* $\{Y_\alpha, f_\alpha{}^\beta\}$ in $\mathcal{C}$ consists of a collection of objects $\{Y_\alpha\}$ indexed by a directed set $\Lambda = \{\alpha\}$ and a collection of morphisms $\{f_\alpha{}^\beta\colon Y_\beta \to Y_\alpha\}$ in $\mathcal{C}$ for $\alpha \le \beta$ in Λ such that

(*a*) $f_\alpha{}^\alpha = 1_{Y_\alpha}$ for $\alpha \in \Lambda$
(*b*) $f_\alpha{}^\gamma = f_\alpha{}^\beta f_\beta{}^\gamma\colon Y_\gamma \to Y_\alpha$ for $\alpha \le \beta \le \gamma$ in Λ

There is then a category inv $\{Y_\alpha, f_\alpha{}^\beta\}$ whose objects are indexed collections of morphisms $\{g_\alpha\colon X \to Y_\alpha\}_{\alpha\in\Lambda}$ such that $g_\alpha = f_\alpha{}^\beta g_\beta$ if $\alpha \le \beta$ in Λ and whose morphisms with domain $\{g_\alpha\colon X \to Y_\alpha\}$ and range $\{g'_\alpha\colon X' \to Y_\alpha\}$ are morphisms $h\colon X \to X'$ of $\mathcal{C}$ such that $g'_\alpha h = g_\alpha$ for $\alpha \in \Lambda$. A terminal object of inv $\{Y_\alpha, f_\alpha{}^\beta\}$ is called an *inverse limit* of the inverse system $\{Y_\alpha, f_\alpha{}^\beta\}$. The inverse limits of sets, topological spaces, groups, and *R* modules are examples of inverse limits in their respective categories.

By similar considerations it is possible to define a direct or inverse limit for an arbitrary indexed collection of objects in a category $\mathcal{C}$ and an indexed collection of morphisms in $\mathcal{C}$ between these objects. We omit the details.

2 FUNCTORS

Our main interest in categories is in the maps from one category to another. Those maps which have the natural properties of preserving identities and composites are called functors. This section is devoted to the definition of functors of one or more variables, some examples and applications, and the definition of natural transformations between functors.

Let $\mathcal{C}$ and $\mathcal{D}$ be categories. A *covariant functor* (or *contravariant functor*) *T* from $\mathcal{C}$ to $\mathcal{D}$ consists of an object function which assigns to every object *X* of $\mathcal{C}$ an object $T(X)$ of $\mathcal{D}$ and a morphism function which assigns to every morphism $f\colon X \to Y$ of $\mathcal{C}$ a morphism $T(f)\colon T(X) \to T(Y)$ [or $T(f)\colon T(Y) \to T(X)$] of $\mathcal{D}$ such that

(*a*) $T(1_X) = 1_{T(X)}$
(*b*) $T(gf) = T(g)T(f)$ [or $T(gf) = T(f)T(g)$]

We list some examples of functors.

1 There is a covariant functor from the category of topological spaces and continuous maps to the category of sets and functions which assigns to every topological space its underlying set. This functor is called a *forgetful functor* because it "forgets" some of the structure of a topological space.

2 There is a covariant functor from the category of sets and functions to the category of R modules and homomorphisms which assigns to every set the free R module generated by it.

3 Given a fixed R module M_0, there is a covariant functor (or contravariant functor) from the category of R modules and homomorphisms to itself which assigns to an R module M the R module $\mathrm{Hom}_R(M_0,M)$ [or $\mathrm{Hom}_R(M,M_0)$].

4 For any category $\mathcal{C}$ and object Y of $\mathcal{C}$ there is a covariant functor π_Y (or contravariant functor π^Y) from $\mathcal{C}$ to the category of sets and functions which assigns to an object Z (or X) of $\mathcal{C}$ the set $\pi_Y(Z) = \hom(Y,Z)$ [or $\pi^Y(X) = \hom(X,Y)$] and to a morphism $h: Z \to Z'$ [or $f: X \to X'$] the function

$$h_\#: \hom(Y,Z) \to \hom(Y,Z') \qquad [\text{or } f^\#: \hom(X',Y) \to \hom(X,Y)]$$

defined by $h_\#(g) = h \circ g$ for $g: Y \to Z$ [or $f^\#(g') = g' \circ f$ for $g': X' \to Y$]

5 There is a contravariant functor C from the category of compact Hausdorff spaces and continuous maps to the category of normed rings over $\mathbf{R}$ and continuous homomorphisms which assigns to X its normed ring of continuous real-valued functions.

6 There is a covariant functor H_0 from the category of topological spaces and continuous maps to the category of abelian groups and homomorphisms such that $H_0(X)$ is the free abelian group generated by the set of components of X, and if $f: X \to Y$, then $H_0(f): H_0(X) \to H_0(Y)$ is the homomorphism such that if C is a component of X and C' is the component of Y containing $f(C)$, then $H_0(f)C = C'$.

7 A direct system (or inverse system) in a category $\mathcal{C}$ is a covariant functor (or contravariant functor) from the category of a directed set (defined as in example 1.1.11) to $\mathcal{C}$.

8 For any category $\mathcal{C}$ there is a contravariant functor to its opposite category $\mathcal{C}^*$ which assigns to an object X of $\mathcal{C}$ the object X^* of $\mathcal{C}^*$ and to a morphism $f: X \to Y$ of $\mathcal{C}$ the morphism $f^*: Y^* \to X^*$.

Note that any contravariant functor on $\mathcal{C}$ corresponds to a covariant functor on $\mathcal{C}^*$, and vice versa. Therefore any functor can be regarded as covariant on a suitable category. Despite this, we shall find it convenient to consider contravariant as well as covariant functors on $\mathcal{C}$, rather than consider only covariant functors on two categories.

Any functor from the category of topological spaces and continuous maps to an algebraic category (such as the category of abelian groups and homomorphisms) is a representation of the topological category by an algebraic one. Algebraic topology is the study of such functors; we show that simple remarks about functors can be used to obtain necessary conditions for the solvability of topological problems.

9 THEOREM *Let T be a functor from a category $\mathcal{C}$ to a category $\mathcal{D}$. Then T maps equivalences in $\mathcal{C}$ to equivalences in $\mathcal{D}$.*

PROOF Assume that T is a covariant functor (the argument is similar if T is contravariant). Let $f\colon X \to Y$ be an equivalence in $\mathcal{C}$. Then $f^{-1}f = 1_X$. Therefore

$$1_{T(X)} = T(1_X) = T(f^{-1})T(f)$$

Similarly, $T(f)T(f^{-1}) = 1_{T(Y)}$. Therefore $T(f^{-1})$ is a two-sided inverse of $T(f)$, and $T(f)$ is an equivalence in $\mathcal{D}$. ■

In particular, if T is an algebraic functor on the category of topological spaces and continuous maps, a necessary condition that X be homeomorphic to Y is that $T(X)$ be equivalent to $T(Y)$. Thus the functor H_0 of example 6 shows that the real line $\mathbf{R}$ and the real plane $\mathbf{R}^2$ are not homeomorphic [if they were homeomorphic, then $\mathbf{R} - 0$ would be homeomorphic to $\mathbf{R}^2 - p$ for some $p \in \mathbf{R}^2$, but $H_0(\mathbf{R} - 0)$ is a free abelian group on two generators, while $H_0(\mathbf{R}^2 - p)$ is a free abelian group on one generator]. This is a trivial example. However, the homology functors H_q defined in Chapter 4 generalize H_0 and can be used in much the same way to prove that $\mathbf{R}^n$ and $\mathbf{R}^m$ are not homeomorphic if $n \neq m$.

In applications of algebraic functors to topological problems the algebra will frequently play an essential role. For example, let $T_0(X)$ be the functor obtained by composing the functor H_0 with the forgetful functor, which assigns to every abelian group its underlying set. The functor T_0 contains less information than the functor H_0 and does not give as strong a necessary condition for homeomorphism [for example, $T_0(\mathbf{R} - 0)$ and $T_0(\mathbf{R}^2 - p)$ are both countably infinite sets and are equivalent in the category of sets and functions]. For this reason it is important to provide functors with as much algebraic structure as possible. Later we shall consider functors which depend on a chosen topological space. These functors take values in the category of sets and functions, but some of them, depending on properties of the particular spaces which define them, are functors to the category of groups and homomorphisms. The added algebraic structure in such cases will prove useful.

To show how functors can be applied to another problem, let A be a subspace of a topological space X and let $f\colon A \to Y$ be continuous. The *extension problem* is to determine whether f has a continuous extension to X—that is, whether the dotted arrow in the triangle

$$\begin{array}{ccc} A & \subset & X \\ & {\scriptstyle f}\searrow \quad \swarrow & \\ & Y & \end{array}$$

corresponds to a continuous map making the diagram commutative.

10 THEOREM *Let T be a covariant functor (or contravariant functor) from the category of topological spaces and continuous maps to a category $\mathcal{C}$. A necessary condition that a map $f: A \to Y$ be extendable to X (where $i: A \subset X$) is that there exist a morphism $\varphi: T(X) \to T(Y)$ [or $\varphi: T(Y) \to T(X)$] such that $\varphi \circ T(i) = T(f)$ [or $T(f) = T(i) \circ \varphi$].*

PROOF Assume that $f': X \to Y$ is an extension of f. Then $f'i = f$. Therefore $T(f') \circ T(i) = T(f)$ [or $T(f) = T(i) \circ T(f')$], and $T(f')$ can be taken as the morphism φ. ■

The above result can be applied to prove that the identity map of $\dot{I}$ cannot be extended to a continuous map $I \to \dot{I}$. We use the functor H_0 and obtain the necessary condition that there must exist a homomorphism $\varphi: H_0(I) \to H_0(\dot{I})$ such that $\varphi \circ H_0(i) = H_0(1_{\dot{I}})$ (where $i: \dot{I} \subset I$). Because $H_0(\dot{I})$ is a free abelian group on two generators and $H_0(I)$ is a free abelian group on one generator, there is no such homomorphism φ. Again, this is a trivial example, but it illustrates the method, and the general homology functors H_q defined later can be used in the same way to show that there is no continuous map $E^{n+1} \to S^n$ that is the identity map on S^n.

Thus we see that a functor yields necessary conditions for the solvability of topological problems. There are situations in which these necessary conditions are also sufficient. For example, the functor C of example 5 gives a necessary and sufficient condition for homeomorphism—that is, two compact Hausdorff spaces X and Y are homeomorphic if and only if $C(X)$ and $C(Y)$ are isomorphic.[1] This is not a particularly useful result, however, because it seems to be no easier to determine whether or not two normed rings are isomorphic than it is to determine whether or not two compact Hausdorff spaces are homeomorphic. We seek functors to categories that are somewhat simpler than the category of topological spaces, so that the algebraic problems that arise in these categories can be effectively solved. One big problem of algebraic topology is to find, and compute, sufficiently many such functors that the solvability of a particular topological problem is equivalent to the solvability of the corresponding (and simpler) algebraic problems.

We shall also have occasion to compare functors with each other. This is done by means of a suitable definition of a map between functors. Let T_1 and T_2 be functors of the same variance (either both covariant or both contravariant) from a category $\mathcal{C}$ to a category $\mathcal{D}$. A *natural transformation* φ from T_1 to T_2 is a function from the objects of $\mathcal{C}$ to morphisms of $\mathcal{D}$ such that for every morphism $f: X \to Y$ of $\mathcal{C}$ the appropriate one of the following diagrams is commutative:

$$\begin{array}{ccc} T_1(X) & \xrightarrow{T_1(f)} & T_1(Y) \\ {\scriptstyle \varphi(X)}\downarrow & & \downarrow{\scriptstyle \varphi(Y)} \\ T_2(X) & \xrightarrow{T_2(f)} & T_2(Y) \end{array} \qquad \qquad \begin{array}{ccc} T_1(X) & \xleftarrow{T_1(f)} & T_1(Y) \\ {\scriptstyle \varphi(X)}\downarrow & & \downarrow{\scriptstyle \varphi(Y)} \\ T_2(X) & \xleftarrow{T_2(f)} & T_2(Y) \end{array}$$

T_1, T_2 *covariant* $\qquad$ T_1, T_2 *contravariant*

[1] See Theorem *D* on page 330 of G. F. Simmons, *Introduction to Topology and Modern Analysis,* McGraw-Hill Book Company, New York, 1963.

If φ is a natural transformation from T_1 to T_2 such that $\varphi(X)$ is an equivalence in $\mathcal{D}$ for each object X in $\mathcal{C}$, then φ is called a *natural equivalence.*

As an example of a natural transformation, let Y_1 and Y_2 be objects of a category $\mathcal{C}$ and let $g: Y_1 \to Y_2$ be a morphism in $\mathcal{C}$. There is a natural transformation $g^\#$ from the covariant functor π_{Y_2} to the covariant functor π_{Y_1} and a natural transformation $g_\#$ from the contravariant functor π^{Y_1} to the contravariant functor π^{Y_2}. If g is an equivalence in $\mathcal{C}$, both these natural transformations are natural equivalences.

It is also of interest to consider functors of several variables. Thus, if $\mathcal{C}_1$, $\mathcal{C}_2$, and $\mathcal{D}$ are categories, a covariant functor from $\mathcal{C}_1 \times \mathcal{C}_2$ to $\mathcal{D}$ is called a *functor of two arguments covariant* in each. A covariant functor from $\mathcal{C}_1 \times \mathcal{C}_2^*$ to $\mathcal{D}$, regarded as a function from ordered pairs (X_1,X_2), where X_1 is an object of $\mathcal{C}_1$ and X_2 is an object of $\mathcal{C}_2$, is called a *functor of two arguments covariant* in the first and *contravariant* in the second. In a similar fashion, functors of more arguments with mixed variance are defined.

If $\mathcal{C}$ is any category, there is a functor of two arguments in $\mathcal{C}$ to the category of sets and functions which is contravariant in the first argument and covariant in the second. This functor assigns to an ordered pair of objects X and Y of $\mathcal{C}$ the set hom (X,Y) and to an ordered pair of morphisms $f: X' \to X$ and $g: Y \to Y'$ in $\mathcal{C}$ the function $f^\# g_\# = g_\# f^\#$: hom $(X,Y) \to$ hom (X',Y').

3 HOMOTOPY

The problem of classifying topological spaces and continuous maps up to topological equivalence does not seem to be amenable to attack directly by computable algebraic functors, as described in Sec. 1.2. Many of the computable functors, because they are computable, are invariant under continuous deformation. Therefore they cannot distinguish between spaces (or maps) that can be continuously deformed from one to the other; the most that can be hoped for from such functors is that they characterize the space (or map) up to continuous deformation.

The intuitive concept of a continuous deformation will be made precise in this section in the concept of homotopy. This leads to the homotopy category which is fundamental for algebraic topology. Its objects are topological spaces and its morphisms are equivalence classes of continuous maps (two maps being equivalent if one can be continuously deformed into the other). For technical reasons we consider not just the homotopy category of topological spaces, but rather the larger homotopy category of pairs.

A *topological pair* (X,A) consists of a topological space X and a subspace $A \subset X$. If A is empty, denoted by $\varnothing$, we shall not distinguish between the pair $(X, \varnothing)$ and the space X. A *subpair* $(X',A') \subset (X,A)$ consists of a pair with $X' \subset X$ and $A' \subset A$. A *map* $f: (X,A) \to (Y,B)$ between pairs is a continuous function f from X to Y such that $f(A) \subset B$, and as in Sec. 1.1, there is a category of topological pairs and maps between them which contains as full subcategories the category of topological spaces and continuous maps, as well as the category of pointed topological spaces and continuous maps.

Given a pair (X,A), we let $(X,A) \times I$ denote the pair $(X \times I, A \times I)$. Let $X' \subset X$ and suppose that $f_0, f_1: (X,A) \to (Y,B)$ agree on X' (that is, $f_0 | X' = f_1 | X'$). Then f_0 is *homotopic to* f_1 *relative to* X', denoted by $f_0 \simeq f_1$ rel X', if there exists a map

$$F: (X,A) \times I \to (Y,B)$$

such that $F(x,0) = f_0(x)$ and $F(x,1) = f_1(x)$ for $x \in X$ and $F(x,t) = f_0(x)$ for $x \in X'$ and $t \in I$. Such a map F is called a *homotopy relative to* X' from f_0 to f_1 and is denoted by $F: f_0 \simeq f_1$ rel X'. If $X' = \emptyset$, we omit the phrase "relative to $\emptyset$." Clearly, $f_0 \simeq f_1$ rel X' implies $f_0 \simeq f_1$ rel X'' for any $X'' \subset X'$. A map from X to Y is said to be *null homotopic*, or *inessential*, if it is homotopic to some constant map.

For $t \in I$ define $h_t: (X,A) \to (X,A) \times I$ by $h_t(x) = (x,t)$. If $F: f_0 \simeq f_1$ rel X', then $Fh_0 = f_0$, $Fh_1 = f_1$, and $Fh_t | X' = f_0 | X'$ for all $t \in I$. Therefore the collection $\{Fh_t\}_{t \in I}$ is a continuous one-parameter family of maps from (X,A) to (Y,B), agreeing on X', which connects $f_0 = Fh_0$ to $f_1 = Fh_1$[1]. Hence $f_0 \simeq f_1$ rel X' corresponds to the intuitive idea of continuously deforming f_0 into f_1 by maps all of which agree on X'. Note that if $f_0 \simeq f_1$ rel X' there will usually be many maps F which are homotopies relative to X' from f_0 to f_1 (see example 3 below).

1 EXAMPLE Let $X = Y = \mathbf{R}^n$ and define $f_0(x) = x$ and $f_1(x) = 0$ for $x \in \mathbf{R}^n$ (that is, $f_0 = 1_{\mathbf{R}^n}$ and f_1 is the constant map of $\mathbf{R}^n$ to its origin). If $F: \mathbf{R}^n \times I \to \mathbf{R}^n$ is defined by

$$F(x,t) = (1-t)x$$

then $F: f_0 \simeq f_1$ rel 0.

2 EXAMPLE Let $X = Y = I$ and define $f_0(t) = t$ and $f_1(t) = 0$ for $t \in I$. If $F: I \times I \to I$ is defined by

$$F(t,t') = (1-t')t$$

then $F: f_0 \simeq f_1$ rel 0.

3 EXAMPLE Let $X = Y = E^2 = \{z \in \mathbf{C} | z = re^{i\theta},\ 0 \le r \le 1\}$ and let $A = B = S^1 = \{z \in \mathbf{C} | z = e^{i\theta}\}$. Define $f_0: (E^2,S^1) \to (E^2,S^1)$ to be the identity map and $f_1: (E^2,S^1) \to (E^2,S^1)$ to be the reflection in the origin [that is, $f_1(re^{i\theta}) = re^{i(\theta+\pi)}$]. Define a homotopy $F: f_0 \simeq f_1$ rel 0 by $F(re^{i\theta},t) = re^{i(\theta+t\pi)}$. Another homotopy $F': f_0 \simeq f_1$ rel 0 is defined by $F'(re^{i\theta},t) = re^{i(\theta-t\pi)}$.

[1] A one-parameter family $f_t: (X,A) \to (Y,B)$ for $t \in I$ is *continuous* if $f_t(x)$ is jointly continuous in t and x, in which case the function $(x,t) \to f_t(x)$ is a homotopy from f_0 to f_1. The corresponding function $t \to f_t$ from I to $(Y,B)^{(X,A)}$ is always continuous [where $(Y,B)^{(X,A)} = \{g: (X,A) \to (Y,B)\}$ topologized by the compact-open topology]. Conversely, in case X is a locally compact Hausdorff space, it follows from theorem 2.8 in the Introduction that for any continuous map $\varphi: I \to (Y,B)^{(X,A)}$ the one-parameter family $\varphi(t)$ is continuous and defines a homotopy from $\varphi(0)$ to $\varphi(1)$.

4 EXAMPLE Let X be an arbitrary space and let Y be a convex subset of $\mathbf{R}^n$. Let $f_0, f_1: X \to Y$ be maps which agree on some subspace $X' \subset X$. Then $f_0 \simeq f_1$ rel X', because the map $F: X \times I \to Y$ defined by

$$F(x,t) = tf_1(x) + (1 - t)f_0(x)$$

is a homotopy relative to X' from f_0 to f_1.

Example 4 is a generalization of examples 1 and 2. In example 3 the space E^2 is convex, but the homotopy between f_0 and f_1 cannot be taken to be a particular case of the homotopy in example 4, because it must keep S^1 mapped into itself at all stages, and S^1 is not convex.

To define the homotopy category we need the following easy results.

5 THEOREM *Homotopy relative to X' is an equivalence relation in the set of maps from (X,A) to (Y,B).*

PROOF *Reflexivity.* For $f: (X,A) \to (Y,B)$ define $F: f \simeq f$ rel X by $F(x,t) = f(x)$.

Symmetry. Given $F: f_0 \simeq f_1$ rel X', define $F': f_1 \simeq f_0$ rel X' by $F'(x,t) = F(x, 1 - t)$.

Transitivity. Given $F: f_0 \simeq f_1$ rel X' and $G: f_1 \simeq f_2$ rel X', define $H: f_0 \simeq f_2$ rel X' by

$$H(x,t) = \begin{cases} F(x,2t) & 0 \le t \le \frac{1}{2} \\ G(x, 2t - 1) & \frac{1}{2} \le t \le 1 \end{cases}$$

Note that H is continuous because its restriction to each of the closed sets $X \times [0,\frac{1}{2}]$ and $X \times [\frac{1}{2},1]$ is continuous. ∎

It follows that the set of maps from (X,A) to (Y,B) is partitioned into disjoint equivalence classes by the relation of homotopy relative to X'. These equivalence classes are called *homotopy classes relative to X'*. We use $[X,A; Y,B]_{X'}$ to denote this set of homotopy classes. Given $f: (X,A) \to (Y,B)$, we use $[f]_{X'}$ to denote the element of $[X,A; Y,B]_{X'}$ determined by f. Homotopy classes relative to the empty set will be denoted by omitting the subscript X'.

6 THEOREM *Composites of homotopic maps are homotopic.*

PROOF Let $f_0, f_1: (X,A) \to (Y,B)$ be homotopic relative to X' and let $g_0, g_1: (Y,B) \to (Z,C)$ be homotopic relative to Y', where $f_1(X') \subset Y'$. To show that $g_0f_0, g_1f_1: (X,A) \to (Z,C)$ are homotopic relative to X', let $F: f_0 \simeq f_1$ rel X' and $G: g_0 \simeq g_1$ rel Y'. Then the composite

$$(X,A) \times I \xrightarrow{F} (Y,B) \xrightarrow{g_0} (Z,C)$$

is a homotopy relative to X' from g_0f_0 to g_0f_1, and the composite

$$(X,A) \times I \xrightarrow{f_1 \times 1_I} (Y,B) \times I \xrightarrow{G} (Z,C)$$

is a homotopy relative to $f_1^{-1}(Y')$ from g_0f_1 to g_1f_1. Since $X' \subset f_1^{-1}(Y')$, we have shown that $g_0f_0 \simeq g_0f_1$ rel X' and $g_0f_1 \simeq g_1f_1$ rel X'. The result now follows from theorem 5. ∎

The last result shows that there is a *homotopy category of pairs* whose objects are topological pairs and whose morphisms are homotopy classes (relative to $\emptyset$). This category contains as full subcategories the *homotopy category of topological spaces* (also shortened to *homotopy category*) and the *homotopy category of pointed topological spaces*. There is a covariant functor from the category of pairs and maps to the homotopy category of pairs whose object function is the identity map and whose mapping function sends a map f to its homotopy class $[f]$. As pointed out at the beginning of the section, most of the algebraic functors we consider will be defined from the appropriate homotopy category. A diagram of topological pairs and maps is said to be *homotopy commutative* if it can be made a commutative diagram in the homotopy category (that is, when each map is replaced by its homotopy class).

As in example 1.2.4, for any pair (P,Q) there is a covariant functor $\pi_{(P,Q)}$ (or a contravariant functor $\pi^{(P,Q)}$) from the homotopy category of pairs to the category of sets and functions defined by $\pi_{(P,Q)}(X,A) = [P,Q;\ X,A]$ (or $\pi^{(P,Q)}(X,A) = [X,A;\ P,Q]$), and if $f: (X,A) \to (Y,B)$, then $\pi_{(P,Q)}([f]) = f_\#$ (or $\pi^{(P,Q)}([f]) = f^\#$), where $f_\#[g] = [fg]$ for $g: (P,Q) \to (X,A)$ (or $f^\#[h] = [hf]$ for $h: (Y,B) \to (P,Q)$). If $\alpha: (P,Q) \to (P',Q')$, there is a natural transformation $\alpha^\#$ from $\pi_{(P',Q')}$ to $\pi_{(P,Q)}$ and a natural transformation $\alpha_\#$ from $\pi^{(P,Q)}$ to $\pi^{(P',Q')}$.

A map $f: (X,A) \to (Y,B)$ is called a *homotopy equivalence* if $[f]$ is an equivalence in the homotopy category of pairs. A map $g: (Y,B) \to (X,A)$ is called a *homotopy inverse* of f if $[g] = [f]^{-1}$ in the homotopy category. Pairs (X,A) and (Y,B) are said to have the *same homotopy type* if they are equivalent in the homotopy category.

The simplest nonempty space is a one-point space. We characterize the homotopy type of such a space as follows. A topological space X is said to be *contractible* if the identity map of X is homotopic to some constant map of X to itself. A homotopy from 1_X to the constant map of X to $x_0 \in X$ is called a *contraction* of X to x_0. Examples 1 and 2 show that $\mathbf{R}^n$ and I are contractible, and example 4 shows that any convex subset of $\mathbf{R}^n$ is contractible. The following lemma may be regarded as a generalization of the result of example 4.

7 LEMMA *Any two maps of an arbitrary space to a contractible space are homotopic.*

PROOF Let Y be a contractible space and suppose $1_Y \simeq c$, where c is a constant map of Y to itself. Let $f_0, f_1: X \to Y$ be arbitrary. By theorem 6, $f_0 = 1_Y f_0 \simeq c f_0$, and similarly, $f_1 \simeq c f_1$. Since $c f_0 = c f_1$, it follows from theorem 5 that $f_0 \simeq f_1$. ∎

8 COROLLARY *If Y is contractible, any two constant maps of Y to itself are homotopic, and the identity map is homotopic to any constant map of Y to itself.* ∎

It is interesting to observe that lemma 7 cannot be strengthened to the case of relative homotopy. That is, if f_0 and f_1 are maps of X into a contractible space Y which agree on $X' \subset X$, it need not be true that $f_0 \simeq f_1$ rel X' (although example 4 shows this to be true for convex subsets of $\mathbf{R}^n$). The following example illustrates this and will be referred to again later.

9 EXAMPLE The *comb space* Y illustrated in the diagram

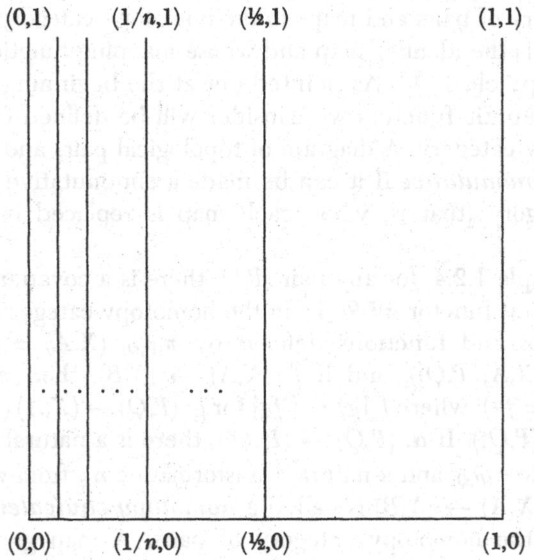

Comb space

is defined by

$$Y = \{(x,y) \in \mathbf{R}^2 \mid 0 \leq y \leq 1, x = 0, 1/n \quad \text{or} \quad y = 0, 0 \leq x \leq 1\}$$

Let $F: Y \times I \to Y$ be defined by $F((x,y), t) = (x, (1-t)y)$. Then F is a homotopy from 1_Y to the projection of Y to the x axis. Since the latter map is homotopic to a constant map, Y is contractible. Let $c: Y \to Y$ be the constant map of Y to the point (0,1). By corollary 8, $1_Y \simeq c$, but even though these two maps agree on (0,1), there is no homotopy relative to (0,1) between them.

The following theorem shows that contractible spaces are homotopically as simple as possible.

10 THEOREM *A space is contractible if and only if it has the same homotopy type as a one-point space.*

PROOF Assume that X is contractible and let $F: X \times I \to X$ be a contraction of X to a point $x_0 \in X$. Let P be the one-point space consisting of x_0 and let $f: X \to P$ and $j: P \subset X$. Then $fj = 1_P$ and $F: 1_X \simeq jf$. Therefore $[j] = [f]^{-1}$, and f is a homotopy equivalence from X to P.

Conversely, if X has the same homotopy type as a one-point space P, let $f: X \to P$ be a homotopy equivalence with homotopy inverse $g: P \to X$. Then $1_X \simeq gf$. Because gf is a constant map, X is contractible. ∎

11 COROLLARY *Two contractible spaces have the same homotopy type, and any continuous map between contractible spaces is a homotopy equivalence.*

PROOF The first part follows from theorem 10 and the transitivity of the relation of having the same homotopy type. The second part follows from the first part and lemma 7 (and from the obvious fact that any map homotopic to a homotopy equivalence is itself a homotopy equivalence). ∎

The next result establishes an important relation between homotopy and the extendability of maps.

12 THEOREM *Let p_0 be any point of S^n and let $f: S^n \to Y$. The following are equivalent:*

(a) *f is null homotopic*

(b) *f can be continuously extended over E^{n+1}*

(c) *f is null homotopic relative to p_0*

PROOF (a) ⟹ (b). Let $F: f \simeq c$, where c is the constant map of S^n to $y_0 \in Y$. Define an extension f' of f over E^{n+1} by

$$f'(x) = \begin{cases} y_0 & 0 \le \|x\| \le \tfrac{1}{2} \\ F(x/\|x\|, 2 - 2\|x\|) & \tfrac{1}{2} \le \|x\| \le 1 \end{cases}$$

Since $F(x,1) = y_0$ for all $x \in S^n$, the map f' is well-defined. f' is continuous because its restriction to each of the closed sets $\{x \in E^{n+1} \mid 0 \le \|x\| \le \tfrac{1}{2}\}$ and $\{x \in E^{n+1} \mid \tfrac{1}{2} \le \|x\| \le 1\}$ is continuous. Since $F(x,0) = f(x)$ for $x \in S^n$, $f' | S^n = f$ and f' is a continuous extension of f to E^{n+1}.

(b) ⟹ (c). If f has the continuous extension $f': E^{n+1} \to Y$, define $F: S^n \times I \to Y$ by

$$F(x,t) = f'((1 - t)x + tp_0)$$

Then $F(x,0) = f'(x) = f(x)$ and $F(x,1) = f'(p_0)$ for $x \in S^n$. Since $F(p_0,t) = f'(p_0)$ for $t \in I$, F is a homotopy relative to p_0 from f to the constant map to $f'(p_0)$.

(c) ⟹ (a). This is obvious. ∎

Combining theorem 12 with lemma 7, we obtain the following result.

13 COROLLARY *Any continuous map from S^n to a contractible space has a continuous extension over E^{n+1}.* ∎

4 RETRACTION AND DEFORMATION

This section is concerned mainly with inclusion maps. We consider whether such a map has a left inverse, a right inverse, or a two-sided inverse in either the category of topological spaces and continuous maps or the homotopy category.[1]

[1] Many of the results in this section can be found in R. H. Fox, On homotopy type and deformation retracts, *Annals of Mathematics*, vol. 44, pp. 40–50, 1943 (see also H. Samelson, Remark on a paper by R. H. Fox, *Annals of Mathematics*, vol. 45, pp. 448–449, 1944).

A subspace A of X is called a *retract* of X if the inclusion map $i: A \subset X$ has a left inverse in the category of topological spaces and continuous maps. Hence A is a retract of X if and only if there is a continuous map $r: X \to A$ such that $ri = 1_A$ [that is, $r(x) = x$ for $x \in A$]. Such a map r is called a *retraction* of X to A.

A subspace A of X is called a *weak retract* of X if the inclusion map $i: A \subset X$ has a left homotopy inverse (that is, a left inverse in the homotopy category). Thus A is a weak retract of X if and only if there is a continuous map $r: X \to A$ such that $ri \simeq 1_A$. Such a map r is called a *weak retraction* of X to A.

Any one-point subspace is a retract of any larger space containing it. A discrete space with more than one point is never a weak retract of a connected space containing it. If A is a retract of X, it is a weak retract of X. The converse is not true, as is shown by the following example.

1 EXAMPLE Let X be the closed unit square I^2 in $\mathbf{R}^2$ and let $A \subset X$ be the comb space of example 1.3.9. Then A and X are both contractible, and by corollary 1.3.11, the inclusion map $A \subset X$ is a homotopy equivalence. Therefore A is a weak retract of X. However, it can be shown that A is not a retract of X.

Despite the fact that, in general, a weak retract need not be a retract, these concepts do coincide when A is a suitable subspace of X. This occurs frequently enough to warrant special consideration and will prove of use later. Let (X,A) be a pair and Y be a space. (X,A) is said to have the *homotopy extension property with respect to* Y if, given maps $g: X \to Y$ and $G: A \times I \to Y$ such that $g(x) = G(x,0)$ for $x \in A$, there is a map $F: X \times I \to Y$ such that $F(x,0) = g(x)$ for $x \in X$ and $F|A \times I = G$. If g is regarded as a map of $X \times 0$ to Y, the existence of F is equivalent to the existence of a map represented by the dotted arrow which makes the following diagram commutative:

$$\begin{array}{ccccc} A \times 0 & & \subset & & A \times I \\ & & & \swarrow^{G} & \\ \cap & & Y & & \cap \\ & \nearrow^{g} & & \nwarrow & \\ X \times 0 & & \subset & & X \times I \end{array}$$

If (X,A) has the homotopy extension property with respect to Y and f_0, $f_1: A \to Y$ are homotopic, then if f_0 has an extension to X, so does f_1; for if $g: X \to Y$ is an extension of f_0 and $G: A \times I \to Y$ is a homotopy from f_0 to f_1, the homotopy extension property implies the existence of a map $F: X \times I \to Y$ which is an extension of G, therefore $F(x,1)$ is an extension of f_1. It follows that whether or not a map $A \to Y$ can be extended over X is a property of the homotopy class of that map. Therefore the homotopy extension property implies that the extension problem for maps $A \to Y$ is a problem in the homotopy category.

Of particular importance is the case when (X,A) has the homotopy extension property with respect to any space. More generally, a map $f: X' \to X$ is called a *cofibration* if, given maps $g: X \to Y$ and $G: X' \times I \to Y$ (where Y is arbitrary) such that $g(f(x')) = G(x',0)$ for $x' \in X'$, there is a map $F: X \times I \to Y$ such that $F(x,0) = g(x)$ for $x \in X$ and $F(f(x'), t) = G(x',t)$ for $x' \in X'$ and $t \in I$. If g is regarded as a map of $X \times 0$ to Y, the existence of F is equivalent to the existence of a map represented by the dotted arrow which makes the following diagram commutative:

$$
\begin{array}{ccc}
X' \times 0 & \subset & X' \times I \\
& G \\
f \times 1_0 \downarrow & \quad Y \quad & \downarrow f \times 1_I \\
& g \nearrow \quad \nwarrow \\
X \times 0 & \subset & X \times I
\end{array}
$$

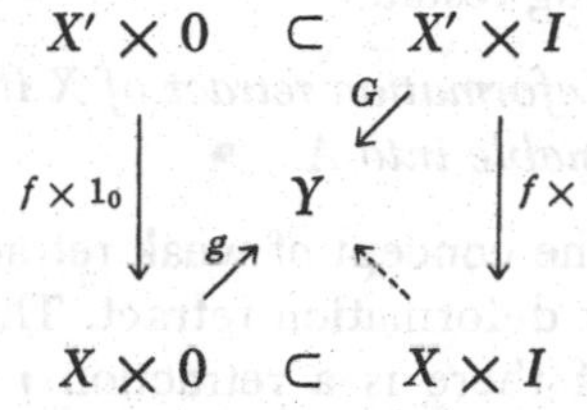

Thus an inclusion map $i: A \subset X$ is a cofibration if and only if (X,A) has the homotopy extension property with respect to any space.

2 **THEOREM** *If (X,A) has the homotopy extension property with respect to A, then A is a weak retract of X if and only if A is a retract of X.*

PROOF We show that any weak retraction $r: X \to A$ is, in fact, homotopic to a retraction. Let $i: A \subset X$; then $ri \simeq 1_A$. Let $G: A \times I \to A$ be a homotopy from ri to 1_A; then $G(x,0) = r(x)$ for $x \in A$. Because (X,A) has the homotopy extension property with respect to A, there is a map $F: X \times I \to A$ which extends G such that $F(x,0) = r(x)$ for $x \in X$. If $r': X \to A$ is defined by $r'(x) = F(x,1)$, then r' is a retraction of X to A, and F is a homotopy from r to r'. ∎

We can just as well consider inclusion maps with right homotopy inverses as those with left homotopy inverses. This leads to the following definitions. Given $X' \subset X$, a *deformation* D of X' in X is a homotopy

$$D: X' \times I \to X$$

such that $D(x',0) = x'$ for $x' \in X'$. If, moreover, $D(X' \times 1)$ is contained in a subspace A of X, D is said to be a *deformation of X' into A* and X' is said to be *deformable in X into A*. A space X is said to be *deformable* into a subspace A if it is deformable in itself into A. Thus a space X is contractible if and only if it is deformable into one of its points.

3 **LEMMA** *A space X is deformable into a subspace A if and only if the inclusion map $i: A \subset X$ has a right homotopy inverse.*

PROOF If i has a right homotopy inverse $f: X \to A$, then $if \simeq 1_X$. Let $F: X \times I \to X$ be a homotopy from 1_X to if; then $F(x,0) = x$, so F is a deformation of X, and $F(X \times 1) = if(X) \subset A$, so X is deformable into A.

Conversely, if X is deformable into A, let $D: X \times I \to X$ be a deformation such that $D(X \times 1) \subset A$. Let $f: X \to A$ be defined by the equation

$$if(x) = D(x,1) \qquad x \in X$$

Then $D: 1_X \simeq if$, showing that f is a right homotopy inverse of i. ∎

Note that an inclusion map $i: A \subset X$ never has a right inverse in the category of topological spaces and continuous maps except in the trivial case $A = X$.

We now consider inclusion maps which are homotopy equivalences. A subspace $A \subset X$ is called a *weak deformation retract* of X if the inclusion map $i: A \subset X$ is a homotopy equivalence. From lemma 1.1.1 and lemma 3 above we obtain the following result.

4 LEMMA *A is a weak deformation retract of X if and only if A is a weak retract of X and X is deformable into A.* ∎

As was the case with the concept of weak retract, there are more useful concepts than that of weak deformation retract. The subspace A is a *strong deformation retract* of X if there is a retraction r of X to A such that if $i: A \subset X$, then $1_X \simeq ir$ rel A. If $F: 1_X \simeq ir$ rel A, F is called a *strong deformation retraction* of X to A.

There is an intermediate concept useful in comparing the weak and strong forms already defined. A subspace A is called a *deformation retract* of X if there is a retraction r of X to A such that if $i: A \subset X$, then $1_X \simeq ir$. If $F: 1_X \simeq ir$, F is called a *deformation retraction* of X to A. A homotopy $F: X \times I \to X$ is a deformation retraction if and only if $F(x,0) = x$ for $x \in X$, $F(X \times 1) \subset A$, and $F(x,1) = x$ for $x \in A$. It is a strong deformation retraction if and only if it also satisfies the condition $F(x,t) = x$ for $x \in A$ and $t \in I$.

5 EXAMPLE It follows from example 1.3.4 that any one-point subset of a convex subset of $\mathbf{R}^n$ is a strong deformation retract of the convex set.

6 EXAMPLE S^n is a strong deformation retract of $\mathbf{R}^{n+1} - 0$. In fact the map $F: (\mathbf{R}^{n+1} - 0) \times I \to \mathbf{R}^{n+1} - 0$ defined by

$$F(x,t) = (1-t)x + \frac{tx}{\|x\|} \qquad x \in \mathbf{R}^{n+1} - 0,\ t \in I$$

is a strong deformation retraction of $\mathbf{R}^{n+1} - 0$ to S^n.

It is clear that a strong deformation retract is a deformation retract, and a deformation retract is a weak deformation retract. The following examples show that neither of these implications is reversible.

7 EXAMPLE As in example 1 above, let X be the closed unit square and A be the comb space. As pointed out in example 1, the inclusion map $A \subset X$ is a homotopy equivalence, but A is not a retract of X. Therefore A is a weak deformation retract of X which is not a deformation retract of X.

8 EXAMPLE Let X be the comb space and A be the one-point subspace of X consisting of the point $(0,1)$. Because X is contractible, there is a homotopy F from 1_X to the constant map of X to A. Such a map F is a deformation retraction of X to A. However, as was remarked in example 1.3.9, there is no homotopy relative to A from 1_X to the constant map to A; therefore A is a deformation retract of X which is not a strong deformation retract of X.

In the presence of suitable homotopy extension properties the three concepts of deformation retract coincide, and we shall now prove this.

9 LEMMA *If X is deformable into a retract A, then A is a deformation retract of X.*

PROOF Let $r: X \to A$ be a retraction and let $i: A \subset X$. Then r is a left homotopy inverse of i. Because X is deformable into A, it follows from lemma 3 that i has a right homotopy inverse. By lemma 1.1.1, r is also a right homotopy inverse of i. Since $1_X \simeq ir$, A is a deformation retract of X. ■

Combining lemma 9 with theorem 2 yields the following corollary.

10 COROLLARY *If (X,A) has the homotopy extension property with respect to A, then A is a weak deformation retract of X if and only if A is a deformation retract of X.* ■

11 THEOREM *If $(X \times I, (X \times 0) \cup (A \times I) \cup (X \times 1))$ has the homotopy extension property with respect to X and A is closed in X, then A is a deformation retract of X if and only if A is a strong deformation retract of X.*

PROOF If A is a deformation retract of X, let $F: X \times I \to X$ be a homotopy from 1_X to ir, where $r: X \to A$ is a retraction and $i: A \subset X$. A homotopy

$$G: [(X \times 0) \cup (A \times I) \cup (X \times 1)] \times I \to X$$

is defined by the equations

$$\begin{aligned} G((x,0), t') &= x && x \in X, t' \in I \\ G((x,t), t') &= F(x, (1-t')t) && x \in A; t, t' \in I \\ G((x,1), t') &= F(r(x), 1-t') && x \in X, t' \in I \end{aligned}$$

G is well-defined, because for $x \in A$

$$G((x,0), t') = x = F(x,0)$$

by the first two equations and

$$G((x,1), t') = F(x, 1-t') = F(r(x), 1-t')$$

by the last two equations. G is continuous because its restriction to each of the closed sets $(X \times 0) \times I$, $(A \times I) \times I$, and $(X \times 1) \times I$ is continuous. For $(x,t) \in (X \times 0) \cup (A \times I) \cup (X \times 1)$, $G((x,t), 0) = F(x,t)$ [because $F(x,0) = x$, and since r is a retraction, $F(r(x), 1) = ir(r(x)) = r(x) = F(x,1)$]. Therefore G restricted to $[(X \times 0) \cup (A \times I) \cup (X \times 1)] \times 0$ can be extended to $(X \times I) \times 0$. From the homotopy extension property in the hypothesis, G restricted to $[(X \times 0) \cup (A \times I) \cup (X \times 1)] \times 1$ can be extended to $(X \times I) \times 1$. Let $G': (X \times I) \times 1 \to X$ be such an extension, and define $H: X \times I \to X$

by $H(x,t) = G'((x,t), 1)$. Then we have the equations

$$\begin{aligned} H(x,0) &= G'((x,0), 1) = G((x,0), 1) &&= x && x \in X \\ H(x,1) &= G((x,1), 1) = F(r(x),0) &&= r(x) && x \in X \\ H(x,t) &= G((x,t), 1) = F(x,0) &&= x && x \in A, t \in I \end{aligned}$$

Therefore H is a homotopy relative to A from 1_X to ir, so A is a strong deformation retract of X. ∎

The next result asserts that any map is equivalent in the homotopy category to an inclusion map that is a cofibration. Let $f: X \to Y$ and let Z_f denote the quotient space obtained from the topological sum of $X \times I$ and Y by identifying $(x,1) \in X \times I$ with $f(x) \in Y$. Z_f is called the *mapping cylinder* of f and is depicted in the diagram

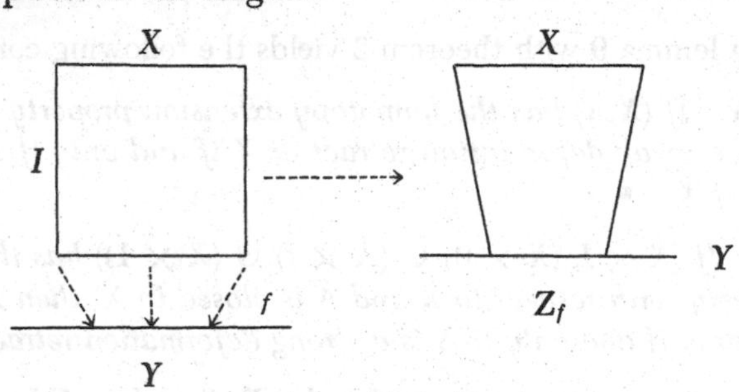

Mapping cylinder

We use $[x,t]$ to denote the point of Z_f corresponding to $(x,t) \in X \times I$ under the identification map and $[y]$ to denote the point of Z_f corresponding to $y \in Y$ (thus $[x,1] = [f(x)]$ for $x \in X$). There is an imbedding $i: X \to Z_f$ with $i(x) = [x,0]$ and an imbedding $j: Y \to Z_f$ with $j(y) = [y]$. X and Y are regarded as subspaces of Z_f by means of these imbeddings. A retraction $r: Z_f \to Y$ is defined by $r[x,t] = [f(x)]$ for $x \in X$ and $t \in I$ and $r[y] = [y]$ for $y \in Y$.

12 THEOREM *Given a map $f: X \to Y$, there is a commutative diagram*

$$\begin{array}{ccc} X & \xrightarrow{i} & Z_f \\ & {\scriptstyle f}\searrow \quad \swarrow{\scriptstyle r} & \\ & Y & \end{array}$$

such that (a) $1_{Z_f} \simeq jr$ rel Y (b) i is a cofibration

PROOF By definition, $ri = f$, and the triangle is commutative.

(*a*) A homotopy $F: Z_f \times I \to Z_f$ is defined by

$$\begin{aligned} F([x,t], t') &= [x, (1 - t')t + t'] && x \in X;\ t, t' \in I \\ F([y],t') &= [y] && y \in Y, t' \in I \end{aligned}$$

Then $F: 1_{Z_f} \simeq jr$ rel Y.

(F is continuous because $Z_f \times I$ has the topology coinduced by the maps $X \times I \times I \to Z_f \times I$ sending (x, t, t') to $([x, t], t')$ and $Y \times I \to Z_f \times I$ sending (y, t') to $([y], t')$.)

(*b*) Let $g\colon Z_f \to W$ and $G\colon X \times I \to W$ be such that $g([x,0]) = G(x,0)$ for $x \in X$. If $H\colon Z_f \times I \to W$ is defined by the equations

$$H([y],t') = g[y] \qquad y \in Y,\ t' \in I$$

$$H([x,t],\, t') = \begin{cases} g[x,\,(2t - t')/(2 - t')] & 0 \le t' \le 2t \le 2,\ x \in X \\ G(x,\,(t' - 2t)/(1 - t)) & 0 \le 2t \le t' \le 1,\ x \in X \end{cases}$$

then $H([x,t],\, 0) = g[x,t]$ and $H([y],0) = g[y]$, and $H|X \times I = G$. ∎

It follows that the map $i\colon X \subset Z_f$ is a cofibration equivalent in the homotopy category to the map $f\colon X \to Y$. The mapping cylinder can be used to prove the following amusing result.

13 THEOREM *Two spaces X and Y have the same homotopy type if and only if they can be imbedded as weak deformation retracts of the same space Z.*

PROOF If X and Y can be imbedded as weak deformation retracts of the same space Z, then X and Y each have the same homotopy type as Z. Therefore X and Y have the same homotopy type.

Conversely, if $f\colon X \to Y$ is a homotopy equivalence, it follows from theorem 12 that if Z_f is the mapping cylinder of f, then the composite $X \xrightarrow{i} Z_f \xrightarrow{r} Y$ is a homotopy equivalence. Because r is a homotopy equivalence, this implies that i is a homotopy equivalence. By theorem 12*a*, $j\colon Y \to Z_f$ is a homotopy equivalence. Therefore X and Y are imbedded as weak deformation retracts in Z_f. ∎

All the foregoing concepts can also be considered for pairs. For example, a pair $(X',A') \subset (X,A)$ is a *strong deformation retract* if there is a map $F\colon (X,A) \times I \to (X,A)$ such that $F(x,0) = x$ for $x \subset X$, $F(X \times 1) \subset X'$, $F(A \times 1) \subset A'$, and $F(x',t) = x'$ for $x' \in X'$ and $t \in I$. The *mapping cylinder* of a map $f\colon (X,A) \to (Y,B)$, where A is closed in X, is the pair (Z_{f_1},Z_{f_2}), where Z_{f_1} is the mapping cylinder of the map $f_1\colon X \to Y$ defined by f and Z_{f_2} is the mapping cylinder of the map $f_2\colon A \to B$ defined by f. A map $f\colon (X',A') \to (X,A)$ is a *cofibration* if, given maps $g\colon (X,A) \to (Y,B)$ and $G\colon (X',A') \times I \to (Y,B)$ [where (Y,B) is arbitrary] such that $G(x',0) = gf(x')$ for $x' \in X'$, there exists a map $F\colon (X,A) \times I \to (Y,B)$ such that $F(x,0) = g(x)$ for $x \in X$ and $G(x',t) = F(f(x'),\, t)$ for $x' \in X'$ and $t \in I$. All the results remain valid when suitably formulated for pairs.

5 H SPACES

In some cases it is possible to introduce a natural group structure in the set of homotopy classes of maps from one space (or pair) to another. In this section we consider spaces P such that $[X;P]$ admits a group structure for all X. It is not surprising that there is a close relation between natural group structures on $[X;P]$ for all X and "grouplike" structures on P.

We shall work in the homotopy category of pointed topological spaces, although much of what we do is also valid in the homotopy category of topological spaces. If X and Y are pointed topological spaces, $[X;Y]$ will denote the set of base-point-preserving homotopy classes of continuous maps $X \to Y$ (with all homotopies understood to be relative to the base point). Thus $[X;Y]$ is the set of morphisms from X to Y in the homotopy category of pointed topological spaces.

One method of obtaining a group structure on $[X;P]$ is to start with a group structure on P. Thus, let P be a topological group with identity element as base point. There is a law of composition in the set of all base-point-preserving continuous maps from X to P defined by pointwise multiplication of functions. That is, if $g_1, g_2: X \to P$, then $g_1g_2: X \to P$ is defined by $g_1g_2(x) = g_1(x)g_2(x)$, where the right-hand side is the group product in P. With this law of composition, the set of base-point-preserving continuous maps from X to P is a group (which is abelian if P is abelian). The law of composition carries over to give an operation on homotopy classes such that $[g_1][g_2] = [g_1g_2]$, and we have the following theorem.

1 THEOREM *If P is a topological group, π^P is a contravariant functor from the homotopy category of pointed topological spaces to the category of groups and homomorphisms.* ■

We give two examples.

2 S^1 is an abelian topological group (the multiplicative group of complex numbers of norm 1). Therefore $[X;S^1]$ is an abelian group, and if $f: X \to Y$, then $f^\#: [Y;S^1] \to [X;S^1]$ is a homomorphism.

3 S^3 is a topological group (the multiplicative group of quaternions of norm 1). Therefore $[X;S^3]$ is a group, and if $f: X \to Y$, then $f^\#: [Y;S^3] \to [X;S^3]$ is a homomorphism.

This group structure on $[X;P]$ was deduced from a group structure on the set of base-point-preserving continuous maps from X to P. There are situations in which $[X;P]$ admits a natural group structure, but the set of base-point-preserving continuous maps from X to P has no group structure. For example, if P is a pointed space having the same homotopy type as some topological group P', then π^P is naturally equivalent to $\pi^{P'}$. Therefore π^P can be regarded as a functor to the category of groups. The following definitions will be used to describe the additional structure needed on a pointed space P in order that π^P take values in the category of groups and homomorphisms.

If $f: X \to Y$ and $g: X \to Z$, we define

$$(f,g): X \to Y \times Z$$

to be the map $(f,g)(x) = (f(x),g(x))$ for $x \in X$.

An *H space* consists of a pointed topological space P together with a continuous multiplication

$$\mu: P \times P \to P$$

for which the (unique) constant map $c\colon P \to P$ is a *homotopy identity*, that is, each composite

$$P \xrightarrow{(c,1)} P \times P \xrightarrow{\mu} P \qquad \text{and} \qquad P \xrightarrow{(1,c)} P \times P \xrightarrow{\mu} P$$

is homotopic to 1_P. The multiplication μ is said to be *homotopy associative* if the square

$$\begin{array}{ccc} P \times P \times P & \xrightarrow{\mu \times 1} & P \times P \\ {\scriptstyle 1 \times \mu}\downarrow & & \downarrow{\scriptstyle \mu} \\ P \times P & \xrightarrow{\mu} & P \end{array}$$

is homotopy commutative, that is, $\mu \circ (\mu \times 1) \simeq \mu \circ (1 \times \mu)$. A continuous function $\varphi\colon P \to P$ is called a *homotopy inverse* for P and μ if each of the composites

$$P \xrightarrow{(1,\varphi)} P \times P \xrightarrow{\mu} P \qquad \text{and} \qquad P \xrightarrow{(\varphi,1)} P \times P \xrightarrow{\mu} P$$

is homotopic to $c\colon P \to P$.

A homotopy-associative *H* space with a homotopy inverse satisfies the group axioms up to homotopy. Such a pointed space is called an *H group*. Clearly, any topological group is an *H* group.

A multiplication μ in an *H* space is said to be *homotopy abelian* if the triangle

$$\begin{array}{ccc} P \times P & \xrightarrow{T} & P \times P \\ & {\scriptstyle \mu}\searrow \quad \swarrow{\scriptstyle \mu} & \\ & P & \end{array}$$

where $T(p_1,p_2) = (p_2,p_1)$, is homotopy commutative. An *H* group with homotopy-abelian multiplication is called an *abelian H* group.

If P and P' are *H* spaces with multiplications μ and μ', respectively, a continuous map $\alpha\colon P \to P'$ is called a *homomorphism* if the square

$$\begin{array}{ccc} P \times P & \xrightarrow{\mu} & P \\ {\scriptstyle \alpha \times \alpha}\downarrow & & \downarrow{\scriptstyle \alpha} \\ P' \times P' & \xrightarrow{\mu'} & P' \end{array}$$

is homotopy commutative.

4 THEOREM *A pointed space having the same homotopy type as an H space (or an H group) is itself an H space (or H group) in such a way that the homotopy equivalence is a homomorphism.*

PROOF Let $f\colon P \to P'$ and $g\colon P' \to P$ be homotopy inverses and let P be an H space with multiplication $\mu\colon P \times P \to P$. If $\mu'\colon P' \times P' \to P'$ is defined to be the composite

$$P' \times P' \xrightarrow{g \times g} P \times P \xrightarrow{\mu} P \xrightarrow{f} P'$$

then μ' is a continuous multiplication in P' and the composite $P' \xrightarrow{(1,c')} P' \times P' \xrightarrow{\mu'} P'$ equals the composite $P' \xrightarrow{g} P \xrightarrow{(1,c)} P \times P \xrightarrow{\mu} P \xrightarrow{f} P'$, which is homotopic to the composite $P' \xrightarrow{g} P \xrightarrow{f} P'$. Because $fg \simeq 1_{P'}$, the map $\mu' \circ (1,c')$ is homotopic to $1_{P'}$. Similarly, the map $\mu' \circ (c',1)$ is homotopic to $1_{P'}$. Therefore P' is an H space. Because the square

$$\begin{array}{ccc} P' \times P' & \xrightarrow{\mu'} & P' \\ {\scriptstyle g \times g}\downarrow & & \downarrow{\scriptstyle g} \\ P \times P & \xrightarrow{\mu} & P \end{array}$$

is homotopy commutative, g is a homomorphism (and so is f). If μ is homotopy associative or homotopy abelian, so is μ', and if $\varphi\colon P \to P$ is a homotopy inverse for P, then $f\varphi g\colon P' \to P'$ is a homotopy inverse for P'. ■

Given an H space P, for any pointed space X there is a law of composition in $[X;P]$ defined by $[g_1][g_2] = [\mu \circ (g_1,g_2)]$. If P is an H group, $[X;P]$ becomes a group with this law of composition, and if $f\colon X \to Y$, then $f^{\#}\colon [Y,P] \to [X;P]$ is a homomorphism. Therefore we have the following theorem.

5 THEOREM *If P is an H group, π^P is a contravariant functor from the homotopy category of pointed topological spaces with values in the category of groups and homomorphisms. If P is an abelian H group, this functor takes values in the category of abelian groups.* ■

It is interesting that the following converse of theorem 5 is also valid.

6 THEOREM *If P is a pointed space such that π^P takes values in the category of groups, then P is an H group (abelian if π^P takes values in the category of abelian groups). Furthermore, for any pointed space X, the group structure on $\pi^P(X)$ is the same as that given by theorem 5.*

PROOF Let $p_1\colon P \times P \to P$ and $p_2\colon P \times P \to P$ be the projections, and let $\mu\colon P \times P \to P$ be a map such that $[\mu] = [p_1] * [p_2]$, where $*$ is the law of composition in the group $[P \times P;\ P]$. For any maps $f, g\colon X \to P$, $(f,g)^{\#}\colon [P \times P;\ P] \to [X;P]$ is a homomorphism and

$$\begin{aligned}[\mu \circ (f,g)] &= (f,g)^{\#}[\mu] = (f,g)^{\#}([p_1] * [p_2]) \\ &= (f,g)^{\#}[p_1] * (f,g)^{\#}[p_2] = [f] * [g]\end{aligned}$$

This shows that the multiplication in $[X;P]$ is induced by the multiplication map μ.

Let X be a one-point space. The unique map $X \to P$ represents the identity element of the group $[X;P]$. Because the unique map $P \to X$ induces a homomorphism $[X;P] \to [P;P]$, it follows that the composite $P \to X \to P$, which is the constant map $c\colon P \to P$, represents the identity element of $[P;P]$. It follows that $\mu \circ (1_P,c) \simeq 1_P$ and $\mu \circ (c,1_P) \simeq 1_P$. Therefore P is an H space.

To prove that μ is homotopy associative, let $q_1, q_2, q_3\colon P \times P \times P \to P$ be the projections. Then

$$\begin{aligned}[\mu \circ (1 \times \mu)] &= (1 \times \mu)^\#[\mu] = (1 \times \mu)^\#[p_1] * (1 \times \mu)^\#[p_2] \\ &= [q_1] * [\mu(q_2,q_3)] = [q_1] * ([q_2] * [q_3])\end{aligned}$$

Similarly,

$$[\mu \circ (\mu \times 1)] = ([q_1] * [q_2]) * [q_3]$$

Because $[P \times P \times P;\ P]$ has an associative multiplication, $\mu \circ (1 \times \mu) \simeq \mu \circ (\mu \times 1)$.

To show that P has a homotopy inverse, let $\varphi\colon P \to P$ be such that $[1_P] * [\varphi] = [c]$; then $\mu(1_P,\varphi) \simeq c$. Also, $[\varphi] * [1_P] = [c]$, and so $\mu(\varphi,1_P) \simeq c$. Therefore φ is a homotopy inverse for P.

This proves that P is an H group and that the multiplication in π^P is induced from that on P. If $[P \times P;\ P]$ is an abelian group, a similar argument shows that P is an abelian H group. ■

The following complement to theorems 5 and 6 is easily established by similar methods.

7 THEOREM *Let $\alpha\colon P \to P'$ be a map between H groups. Then $\alpha_\#$ is a natural transformation from π^P to $\pi^{P'}$ in the category of groups if and only if α is a homomorphism.* ■

We describe a particularly useful example of an H group. Let Y be a pointed topological space with base point y_0. The *loop space* of Y (*based at* y_0), denoted by ΩY [or by $\Omega(Y,y_0)$], is defined to be the space of continuous functions $\omega\colon (I,\dot I) \to (Y,y_0)$ topologized by the compact-open topology. ΩY is regarded as a pointed space with base point ω_0 equal to the constant map of I to y_0. There is a map

$$\mu\colon \Omega Y \times \Omega Y \to \Omega Y$$

defined by

$$\mu(\omega,\omega')(t) = \begin{cases} \omega(2t) & 0 \le t \le \tfrac{1}{2} \\ \omega'(2t-1) & \tfrac{1}{2} \le t \le 1 \end{cases}$$

To prove that μ is continuous, let $E\colon \Omega Y \times I \to Y$ be the evaluation map. By theorem 2.8 in the Introduction, it suffices to show that the composite

$$\Omega Y \times \Omega Y \times I \xrightarrow{\mu \times 1} \Omega Y \times I \xrightarrow{E} Y$$

is continuous. The formula which defines μ shows that this composite is continuous on each of the closed sets $\Omega Y \times \Omega Y \times [0,\tfrac{1}{2}]$ and $\Omega Y \times \Omega Y \times [\tfrac{1}{2},1]$.

We construct a number of homotopies to show that ΩY is an H group.

Similar formulas will be used again in Sec. 1.7 to define homotopies of (nonclosed) paths in a topological space.

To prove that the map $\omega \to \mu(\omega,\omega_0)$ is homotopic to the identity map of ΩY, define $F: \Omega Y \times I \to \Omega Y$ by

$$F(\omega,t)(t') = \begin{cases} \omega\left(\dfrac{2t'}{t+1}\right) & 0 \le t' \le \dfrac{t+1}{2} \\ y_0 & \dfrac{t+1}{2} \le t' \le 1 \end{cases}$$

This formula shows that $E(F \times 1): (\Omega Y \times I) \times I \to Y$ is continuous; therefore F is continuous and is a homotopy from the map $\omega \to \mu(\omega,\omega_0)$ to $1_{\Omega Y}$. Similarly, the map $\omega \to \mu(\omega_0,\omega)$ is homotopic to $1_{\Omega Y}$. Therefore ΩY is an H space with multiplication μ.

To show that μ is homotopy associative, define

$$G: \Omega Y \times \Omega Y \times \Omega Y \times I \to \Omega Y$$

by the formula

$$E(G \times 1)(\omega,\omega',\omega'',t,t') = \begin{cases} \omega\left(\dfrac{4t'}{t+1}\right) & 0 \le t' \le \dfrac{t+1}{4} \\ \omega'(4t' - t - 1) & \dfrac{t+1}{4} \le t' \le \dfrac{t+2}{4} \\ \omega''\left(\dfrac{4t' - 2 - t}{2 - t}\right) & \dfrac{t+2}{4} \le t' \le 1 \end{cases}$$

Then $G: \mu \circ (\mu \times 1_{\Omega Y}) \simeq \mu \circ (1_{\Omega Y} \times \mu)$, showing that μ is homotopy associative.

We define a homotopy inverse $\varphi: \Omega Y \to \Omega Y$ by $\varphi(\omega)(t) = \omega(1 - t)$. Then we define $H: \Omega Y \times I \to \Omega Y$ by

$$E(H \times 1)(\omega,t,t') = \begin{cases} y_0 & 0 \le t' \le \dfrac{t}{2} \\ \omega(2t' - t) & \dfrac{t}{2} \le t' \le \dfrac{1}{2} \\ \omega(2 - 2t' - t) & \dfrac{1}{2} \le t' \le 1 - \dfrac{t}{2} \\ y_0 & 1 - \dfrac{t}{2} \le t' \le 1 \end{cases}$$

H is a homotopy from the map $\omega \to \mu(\omega,\varphi(\omega))$ to the constant map of ΩY to itself. Similarly, there is a homotopy from the map $\omega \to \mu(\varphi(\omega),\omega)$ to the constant map of ΩY. Therefore φ is a homotopy inverse for ΩY, and ΩY is an H group.

If $h: Y \to Y'$ preserves base points, there is a map

$$\Omega h: \Omega Y \to \Omega Y'$$

defined by $\Omega h(\omega)(t) = h(\omega(t))$. Clearly, Ωh is a homomorphism, and we summarize these remarks about loop spaces as follows.

8 THEOREM *The loop functor Ω is a covariant functor from the category of pointed topological spaces and continuous maps to the category of H groups and continuous homomorphisms.* ∎

The functor Ω also preserves homotopies. That is, if $h_0, h_1: Y \to Y'$ are homotopic by a homotopy h_t, then $\Omega h_0, \Omega h_1: \Omega Y \to \Omega Y'$ are homotopic by a homotopy Ωh_t, which is a continuous homomorphism for each $t \in I$.

6 SUSPENSION

This section deals primarily with results dual to those of Sec. 1.5. We consider pointed spaces Q such that π_Q is a covariant functor from the homotopy category of pointed spaces to the category of groups and homomorphisms, and this leads to the concept of H cogroup, dual to that of H group. An important example of an H cogroup is the suspension of a pointed space, a concept dual to that of the loop space. The homotopy groups of a space defined in the section are examples of groups of homotopy classes of maps from suspensions to the space.

If X and Y are pointed topological spaces, their sum in the category of pointed topological spaces will be denoted by $X \vee Y$. If X has base point x_0 and Y has base point y_0, $X \vee Y$ may be regarded as the subspace $X \times y_0 \cup x_0 \times Y$ of $X \times Y$. If $f: X \to Z$ and $g: Y \to Z$, we let $(f,g): X \vee Y \to Z$ be the map defined by the characteristic property of the sum [that is, $(f,g) \mid X = f$ and $(f,g) \mid Y = g$].

An *H cogroup* consists of a pointed topological space Q together with a continuous comultiplication

$$\nu: Q \to Q \vee Q$$

such that the following properties hold:

Existence of homotopy identity. If $c: Q \to Q$ is the (unique) constant map, each composite

$$Q \xrightarrow{\nu} Q \vee Q \xrightarrow{(c,1)} Q \qquad \text{and} \qquad Q \xrightarrow{\nu} Q \vee Q \xrightarrow{(1,c)} Q$$

is homotopic to 1_Q.

Homotopy associativity. The square

$$\begin{array}{ccc} Q & \xrightarrow{\nu} & Q \vee Q \\ {\scriptstyle \nu}\downarrow & & \downarrow{\scriptstyle 1 \vee \nu} \\ Q \vee Q & \xrightarrow{\nu \vee 1} & Q \vee Q \vee Q \end{array}$$

is homotopy commutative.

Existence of homotopy inverse. There exists a map $\psi: Q \to Q$ such that each composite

$$Q \xrightarrow{\nu} Q \vee Q \xrightarrow{(1,\psi)} Q \quad \text{and} \quad Q \xrightarrow{\nu} Q \vee Q \xrightarrow{(\psi,1)} Q$$

is homotopic to $c: Q \to Q$.

If X is any pointed space and Q is an H cogroup, there is a law of composition in $[Q;X]$ defined by $[f_1][f_2] = [(f_1,f_2) \circ \nu]$ which makes $[Q;X]$ a group.

An H cogroup is said to be *abelian* if the triangle

$$\begin{array}{ccc} & Q & \\ {}^{\nu}\swarrow & & \searrow^{\nu} \\ Q \vee Q & \xrightarrow{T'} & Q \vee Q \end{array}$$

where $T'(q_1,q_2) = (q_2,q_1)$ for $q_1, q_2 \in Q$, is homotopy commutative.

If Q and Q' are H cogroups with comultiplications ν and ν', respectively, a continuous map $\beta: Q \to Q'$ is called a *homomorphism* if the square

$$\begin{array}{ccc} Q & \xrightarrow{\nu} & Q \vee Q \\ {\scriptstyle\beta}\downarrow & & \downarrow{\scriptstyle\beta \vee \beta} \\ Q' & \xrightarrow{\nu'} & Q' \vee Q' \end{array}$$

is homotopy commutative.

The proofs of the following theorems are dual to the proofs of the corresponding statements about H groups (see theorems 1.5.4, 1.5.5, 1.5.6, and 1.5.7) and are omitted.

1 THEOREM *A pointed space having the same homotopy type as an H cogroup is itself an H cogroup in such a way that the homotopy equivalence is a homomorphism.* ■

2 THEOREM *If Q is an H cogroup, π_Q is a covariant functor from the homotopy category of pointed spaces with values in the category of groups and homomorphisms. If Q is an abelian H cogroup, this functor takes values in the category of abelian groups.* ■

3 THEOREM *If Q is a pointed space such that π_Q takes values in the category of groups, then Q is an H cogroup (abelian if π_Q takes values in the category of abelian groups). Furthermore, the group structure on $\pi_Q(X)$ is identical with that determined by the H cogroup structure of Q as in theorem 2.* ■

4 THEOREM *If $\beta: Q \to Q'$ is a map between H cogroups, then $\beta\#$ is a natural transformation from $\pi_{Q'}$ to π_Q in the category of groups if and only if β is a homomorphism.* ■

We describe an example of an H cogroup dual to the loop-space example of an H group. Let Z be a pointed topological space with base point z_0. The *suspension* of Z, denoted by SZ, is defined to be the quotient space of $Z \times I$ in which $(Z \times 0) \cup (z_0 \times I) \cup (Z \times 1)$ has been identified to a single point. This is sometimes called the *reduced suspension* in the literature, the term "suspension" being used for the suspension in the category of spaces (no base points). The latter is defined to be the quotient space of $Z \times I$ in which $Z \times 0$ is identified to one point and $Z \times 1$ is identified to another point.

If $(z,t) \in Z \times I$, we use $[z,t]$ to denote the corresponding point of SZ under the quotient map $Z \times I \to SZ$. Then $[z,0] = [z_0,t] = [z',1]$ for all z, $z' \in Z$ and $t \in I$. The point $[z_0,0] \in SZ$ is also denoted by z_0, and SZ is a pointed space with base point z_0. If $f\colon Z \to Z'$, then $Sf\colon SZ \to SZ'$ is defined by $Sf[z,t] = [f(z), t]$. Thus S is a covariant functor from the category of pointed spaces and continuous maps. To show that it is a covariant functor to the category of *H* cogroups and homomorphisms, we define a comultiplication

$$\nu\colon SZ \to SZ \vee SZ$$

by the formula

$$\nu([z,t]) = \begin{cases} ([z,2t], z_0) & 0 \leq t \leq \frac{1}{2} \\ (z_0, [z, 2t-1]) & \frac{1}{2} \leq t \leq 1 \end{cases}$$

and illustrate it in the diagram (where the dotted lines are collapsed to one point).

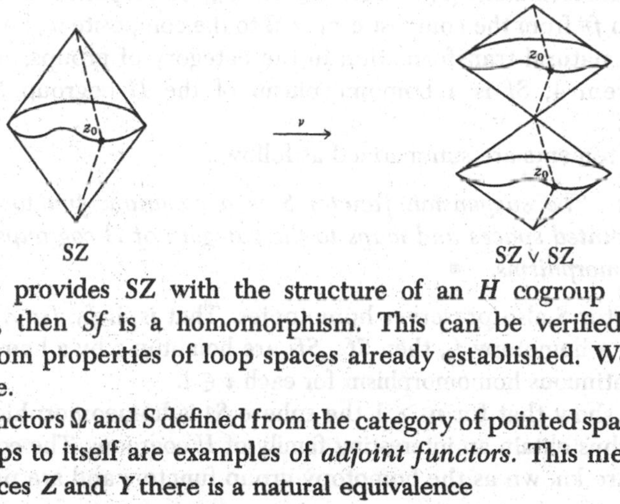

The map ν provides SZ with the structure of an *H* cogroup such that if $f\colon Z \to Z'$, then Sf is a homomorphism. This can be verified directly or deduced from properties of loop spaces already established. We follow the latter course.

The functors Ω and S defined from the category of pointed spaces and continuous maps to itself are examples of *adjoint functors*. This means that for pointed spaces Z and *Y* there is a natural equivalence

$$\hom(SZ,Y) \approx \hom(Z,\Omega Y)$$

where both sides are interpreted as the set of morphisms in the category of pointed spaces and continuous maps. This equivalence results from theorem 2.8 in the Introduction, and if $g\colon Z \to \Omega Y$, the corresponding $g'\colon SZ \to Y$ is defined by $g'[z,t] = g(z)(t)$ for $z \in Z$ and $t \in I$. It is obvious that if $h\colon Y \to Y'$, then $(\Omega h \circ g)' = h \circ g'$, and if $f\colon Z' \to Z$, then $(g \circ f)' = g' \circ Sf$. Therefore the equivalence $g \leftrightarrow g'$ comes from a natural equivalence from the functor $\hom(S\,\cdot\,,\,\cdot\,)$ to the functor $\hom(\,\cdot\,,\,\Omega\,\cdot\,)$.

This natural equivalence passes to morphisms in the homotopy category of pointed spaces. For pointed spaces a homotopy $G: Z \times I \to Y$ must map $z_0 \times I$ into y_0. Therefore it defines a map $F: Z \times I/z_0 \times I \to Y$. Because $S(Z \times I/z_0 \times I)$ can be identified with $SZ \times I/z_0 \times I$ by the homeomorphism

$$[(z,t), t'] \leftrightarrow ([z,t'], t) \qquad z \in Z;\ t, t' \in I$$

it follows that homotopies $F: Z \times I/z_0 \times I \to \Omega Y$ correspond bijectively to homotopies $F': SZ \times I/z_0 \times I \to Y$. Therefore the equivalence above gives rise to an equivalence

$$[SZ;Y] \approx [Z;\Omega Y]$$

such that if the maps $g: Z \to \Omega Y$ and $g': SZ \to Y$ are related by $g'[z,t] = g(z)(t)$, then $[g']$ corresponds to $[g]$. Hence there is a natural equivalence from the functor $[S\ \cdot\ ;\ \cdot]$ to the functor $[\cdot\ ;\ \Omega\ \cdot]$.

It follows from these remarks that for a fixed pointed space Z the functor π_{SZ} is naturally equivalent to the composite functor $\pi_Z \circ \Omega$. Here Ω is regarded as a covariant functor to the homotopy category of H groups and homomorphisms. Then the composite $\pi_Z \circ \Omega$ takes values in the category of groups and homomorphisms. By theorem 3, SZ is an H cogroup, and the map $\nu: SZ \to SZ \vee SZ$ defined above is the one which is the comultiplication in the H cogroup SZ (or is homotopic to it). In similar fashion, if $f: Z \to Z'$, the natural transformation $(Sf)^\#$ from $\pi_{SZ'}$ to π_{SZ} corresponds to the natural transformation $f^\#$ from the composite $\pi_{Z'} \circ \Omega$ to the composite $\pi_Z \circ \Omega$. Because the latter is a natural transformation in the category of groups, so is $(Sf)^\#$, and by theorem 4, Sf is a homomorphism of the H cogroup SZ to the H cogroup SZ'.

These statements are summarized as follows.

5 THEOREM *The suspension functor S is a covariant functor from the category of pointed spaces and maps to the category of H cogroups and continuous homomorphisms.* ■

The functor S also preserves homotopies. That is, if $f_0, f_1: Z \to Z'$ are homotopic by a homotopy f_t, then Sf_0, Sf_1 are homotopic by a homotopy Sf_t, which is a continuous homomorphism for each $t \in I$.

We now show that for $n \geq 1$ the sphere S^n is homeomorphic to a suspension, and thus obtain an interesting family of H cogroups. The corresponding functors are known as the homotopy group functors and are particularly important.

6 LEMMA *For $n \geq 0$, $S(S^n)$ is homeomorphic to S^{n+1}.*

PROOF Let $p_0 = (1,0, \ldots ,0)$ be the base point of S^n. We regard $\mathbf{R}^{n+1}$ as imbedded in $\mathbf{R}^{n+2}$ as the set of points in $\mathbf{R}^{n+2}$ whose $(n + 2)$nd coordinate is 0. Then S^n is imbedded as an equator in S^{n+1}.

$$S^n = \{z \in \mathbf{R}^{n+2} \mid \|z\| = 1 \quad \text{and} \quad z_{n+2} = 0\}$$

and E^{n+1} is also imbedded in E^{n+2}:

$$E^{n+1} = \{z \in \mathbf{R}^{n+2} \mid \|z\| \leq 1 \quad \text{and} \quad z_{n+2} = 0\}$$

Let H_+ and H_- be the two closed hemispheres of S^{n+1} defined by the equator S^n. Then

$$H_+ = \{z \in S^{n+1} \mid z_{n+2} \geq 0\} \quad \text{and} \quad H_- = \{z \in S^{n+1} \mid z_{n+2} \leq 0\}$$

and $S^{n+1} = H_+ \cup H_-$ and $S^n = H_+ \cap H_-$. Furthermore, the projection map $\mathbf{R}^{n+2} \to \mathbf{R}^{n+1}$ defines projection maps $p_+: H_+ \to E^{n+1}$ and $p_-: H_- \to E^{n+1}$, which are homeomorphisms. A map $f: S(S^n) \to S^{n+1}$ is defined by

$$f[z,t] = \begin{cases} p_-^{-1}(2tz + (1 - 2t)p_0) & 0 \leq t \leq \frac{1}{2} \\ p_+^{-1}((2 - 2t)z + (2t - 1)p_0) & \frac{1}{2} \leq t \leq 1 \end{cases}$$

and is verified to be a homeomorphism $f: S(S^n) \approx S^{n+1}$. ∎

For $n \geq 1$ the nth *homotopy group* functor π_n is the covariant functor on the homotopy category of pointed spaces defined by $\pi_n = \pi_{S^n}$. It follows from theorems 6 and 5 that these functors take values in the category of groups and homomorphisms.

In the last two sections of this chapter we give another definition of π_1 and study it in more detail. In Chapter 7 we return to the study of the higher homotopy groups π_n.

The following necessary and sufficient condition for a map $S^n \to X$ to represent the trivial element of $\pi_n(X)$ is an immediate consequence of theorem 1.3.12.

7 THEOREM *A map $\alpha: S^n \to X$ represents the trivial element of $\pi_n(X)$ for $n \geq 1$ if and only if α can be continuously extended over E^{n+1}.* ∎

Before leaving this section let us consider the interplay between two possible group structures on the set $[X;Y]$ for particular pointed spaces X and Y (for example, if X is an H cogroup and Y is an H group, this set can be given a group structure in two ways). It is a fact that under rather general circumstances two laws of composition on hom (X,Y) in a category are equal, and we establish this result.

8 THEOREM *Let X and Y be objects in a category and let $*$ and $*'$ be two laws of composition in* hom (X,Y) *such that*

(a) $*$ and $*'$ have a common two-sided identity element
(b) $*$ and $*'$ are mutually distributive

Then $*$ and $*'$ are equal, and each is commutative and associative.

PROOF Statement (a) means there is a map $f_0: X \to Y$ such that for any $f: X \to Y$

$$f * f_0 = f_0 * f = f = f *' f_0 = f_0 *' f$$

Statement (b) means that for $f_1, f_2, g_1, g_2: X \to Y$

$$(f_1 * f_2) *' (g_1 * g_2) = (f_1 *' g_1) * (f_2 *' g_2)$$

If $f, g: X \to Y$, then

$$f * g = (f *' f_0) * (f_0 *' g) = (f * f_0) *' (f_0 * g) = f *' g$$

and
$$g * f = (f_0 *' g) * (f *' f_0) = (f_0 * f) *' (g * f_0) = f *' g$$

Therefore $f * g = f *' g = g * f$. For associativity we have

$$(f * g) * h = (f * g) *' (f_0 * h) = (f *' f_0) * (g *' h) = f * (g * h) \quad \blacksquare$$

9 COROLLARY *If P is an H space and Q is any H cogroup, then [Q;P] is an abelian group and the group structure is defined by the multiplication map in P.*

PROOF This follows on observing that the two laws of composition defined in $[Q;P]$ by using the comultiplication in Q or the multiplication in P satisfy the hypotheses of theorem 8. ∎

Note that if P is just an H space (but not an H group), the law of composition in $[X;P]$ defined by the multiplication in P is in general not a group structure on $[X;P]$. However, if X is an H cogroup (for instance, a suspension), it follows from corollary 9 that this law of composition is a group structure on $[X;P]$, and in this case the resulting group structure on $[X;P]$ is the same no matter what multiplication map P is given (so long as it is an H space).

10 COROLLARY *If P is an H space, $\pi_n(P)$ is abelian for all $n \geq 1$ and the group structure in $\pi_n(P)$ is defined by the multiplication map in P.* ∎

For a double suspension $S(SZ)$ whose points are represented in the form $[[z,t],t']$, with $z \in Z$ and $t, t' \in I$, there are two laws of composition in the set of maps $S(SZ) \to X$. If $f, g: S(SZ) \to X$, we define

$$(f * g)[[z,t], t'] = \begin{cases} f[[z,2t], t'] & 0 \leq t \leq \frac{1}{2} \\ g[[z, 2t - 1], t'] & \frac{1}{2} \leq t \leq 1 \end{cases}$$

and

$$(f *' g)[[z,t], t'] = \begin{cases} f[[z,t], 2t'] & 0 \leq t' \leq \frac{1}{2} \\ g[[z,t], 2t' - 1] & \frac{1}{2} \leq t' \leq 1 \end{cases}$$

The corresponding operations in $[S(SZ);X]$ satisfy the hypotheses of theorem 8. Therefore they are equal, and $[S(SZ);X]$ is an abelian group. In particular, we have the following corollary.

11 COROLLARY *For $n \geq 2$, π_n is a functor to the category of abelian groups.* ∎

A similar argument can be applied to the loop space ΩP, where P is itself an H space. There is a multiplication map in ΩP, because it is a loop space, and another multiplication obtained from the original multiplication in P. The corresponding laws of composition in $[X;\Omega P]$ satisfy theorem 8. Therefore it follows that if P is an H space, $\pi^{\Omega P}$ is a contravariant functor to the category of abelian groups.

7 THE FUNDAMENTAL GROUPOID

This section concerns paths in a topological space. This leads to another description (in Sec. 1.8) of the first homotopy group π_1, introduced in Sec. 1.6. We shall have occasion to define a number of homotopies between paths in a topological space. These homotopies are generalizations (to nonclosed paths) of those used in Sec. 1.5 to prove that a loop space is an *H* group and are defined by the same formulas (except that the *t* and *t'* arguments are interchanged). It is clear that this repetition of formulas could have been eliminated by proving a suitably general result about path spaces instead of merely considering loop spaces in Sec. 1.5. However, each usage has its own value, and it is hoped that the repetition may be an aid to understanding the formulas.

A *groupoid* is a small category in which every morphism is an equivalence. We list without proof a number of facts about groupoids which are easy consequences of general properties of categories.

1 *The relation between objects A and B of a groupoid defined by the condition* hom $(A,B) \neq \emptyset$ *is an equivalence relation.* ■

The equivalence classes of this equivalence relation are called the *components* of the groupoid. The groupoid is said to be *connected* if it has just one component.

2 *For any object A of a groupoid, the law of composition which sends* $f, g: A \to A$ *to* $f \circ g: A \to A$ *is a group operation in* hom (A,A). ■

3 *There is a covariant functor from any groupoid to the category of groups and homomorphisms which assigns to an object A the group* hom (A,A) *and to a morphism* $f: A \to B$ *the homomorphism*

$$h_f\colon \hom(A,A) \to \hom(B,B)$$

defined by $h_f(g) = f \circ g \circ f^{-1}$ for $g: A \to A$. ■

Because any morphism $f: A \to B$ in a groupoid is an equivalence, $h_f\colon \hom(A,A) \to \hom(B,B)$ is an isomorphism. The following statement describes the collection of isomorphisms obtained by taking all possible morphisms $f: A \to B$.

4 *If A and B are in the same component of a groupoid, the collection of isomorphisms* $\{h_f \mid f: A \to B\}$ *is a conjugacy class of isomorphisms* $\hom(A,A) \to \hom(B,B)$. ■

5 *Let F be a covariant functor from one groupoid* $\mathcal{C}$ *to another* $\mathcal{C}'$. *Then F maps each component of* $\mathcal{C}$ *into some component of* $\mathcal{C}'$, *and there is a natural transformation* $F_*(A)$ *from the covariant functor* $\hom_{\mathcal{C}}(A,A)$ *on* $\mathcal{C}$ *to the covariant functor* $\hom_{\mathcal{C}'}(F(A), F(A))$ *on* $\mathcal{C}$ *defined by*

$$F_*(A)(f) = F(f)\colon F(A) \to F(A) \qquad f: A \to A \quad ■$$

With these general remarks about groupoids out of the way, we proceed to define the fundamental groupoid. A *path* ω in a topological space is defined to be a continuous map $\omega: I \to X$ [note that the path is the map, not just the image set $\omega(I)$]. The *origin* of the path is the point $\omega(0)$, and the *end* of the path is the point $\omega(1)$. We also say that ω is a *path from* $\omega(0)$ *to* $\omega(1)$. A *closed path*, or *loop, at* $x_0 \in X$ is a path ω such that $\omega(0) = x_0 = \omega(1)$. If ω and ω' are paths in X such that end ω = orig ω', there is a *product path* $\omega * \omega'$ in X defined by the formula

$$(\omega * \omega')(t) = \begin{cases} \omega(2t) & 0 \le t \le \frac{1}{2} \\ \omega'(2t-1) & \frac{1}{2} \le t \le 1 \end{cases}$$

Then orig $(\omega * \omega')$ = orig ω and end $(\omega * \omega')$ = end ω'.

We should like to form a category whose objects are the points of X, whose morphisms from x_1 to x_0 are the paths from x_0 to x_1, and with the composite defined to be the product path. With these definitions, neither axiom of a category is satisfied. That is, there need not be an identity morphism for each point, and it is generally not true that the associative law for product paths holds [that is, $\omega * (\omega' * \omega'')$ is usually different from $(\omega * \omega') * \omega''$]. A category can be obtained, however, if the morphisms are defined not to be the paths themselves, but instead, homotopy classes of paths.

Two paths ω and ω' in X are briefly said to be *homotopic*, denoted by $\omega \simeq \omega'$, if they are homotopic relative to $\dot{I}$. Thus a necessary condition that $\omega \simeq \omega'$ is that $\omega(0) = \omega'(0)$ and $\omega(1) = \omega'(1)$. For any $x_0, x_1 \in X$ the relation $\omega \simeq \omega'$ is an equivalence relation in the set of paths from x_0 to x_1. The resulting equivalence classes are called *path classes*, and if ω is a path in X, the path class containing it is denoted by $[\omega]$. Since two paths in the same path class have the same origin and the same end, we can speak of the origin and the end of a path class.

We shall construct a category whose objects are the points of X and whose morphisms from x_1 to x_0 are the path classes with x_0 as origin and x_1 as end. The following lemma shows that the path class of the product of two paths depends only on the path classes of the factors, and it will be used to define the composite in the category.

6 LEMMA *Let* $[\omega]$ *and* $[\omega']$ *be path classes in X with* end $[\omega]$ = orig $[\omega']$. *There is a well-defined path class* $[\omega] * [\omega'] = [\omega * \omega']$ *with* orig $([\omega] * [\omega'])$ = orig $[\omega]$ *and* end $([\omega] * [\omega'])$ = end $[\omega']$.

PROOF To prove that $\omega \simeq \omega_1$ and $\omega' \simeq \omega_1'$ imply $\omega * \omega' \simeq \omega_1 * \omega_1'$, let $F: I \times I \to X$ be a homotopy relative to $\dot{I}$ from ω to ω_1 and let $F': I \times I \to X$ be a homotopy relative to $\dot{I}$ from ω' to ω_1'. A homotopy $F * F': I \times I \to X$ is defined by the formula

$$(F * F')(t,t') = \begin{cases} F(2t,t') & 0 \le t \le \frac{1}{2} \\ F'(2t-1, t') & \frac{1}{2} \le t \le 1 \end{cases}$$

and illustrated in the diagram

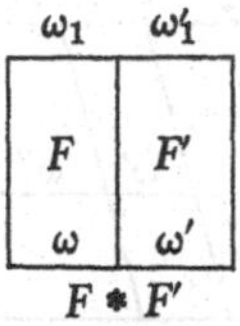

Then $F * F'\colon \omega * \omega' \simeq \omega_1 * \omega_1' \text{ rel } \dot{I}$. ∎

7 THEOREM *For each topological space X there is a category $\mathcal{P}(X)$ whose objects are the points of X, whose morphisms from x_1 to x_0 are the path classes with x_0 as origin and x_1 as end, and whose composite is the product of path classes.*

PROOF To prove the existence of identity morphisms, let $\varepsilon_x\colon I \to X$ be the constant map of I to x for any $x \in X$. We show that $[\varepsilon_x] = 1_x$. If ω is a path with $\omega(1) = x$, we must prove that $\omega * \varepsilon_x \simeq \omega$ (with a similar property for paths with origin at x). Such a homotopy $F\colon I \times I \to X$ is defined by the formula

$$F(t,t') = \begin{cases} \omega\left(\dfrac{2t}{t'+1}\right) & 0 \le t \le \dfrac{t'+1}{2} \\ x & \dfrac{t'+1}{2} \le t \le 1 \end{cases}$$

and pictured in the diagram

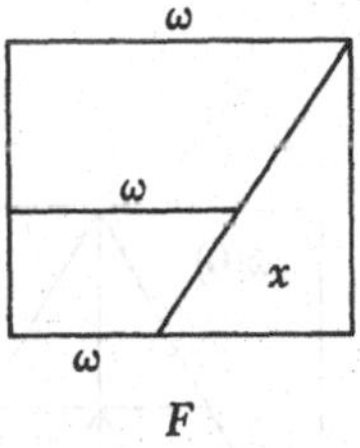

A similar homotopy shows that if $\omega(0) = x$, then $\varepsilon_x * \omega \simeq \omega$.

To prove the associativity of the composite of morphisms, let ω, ω', and ω'' be paths such that end ω = orig ω' and end ω' = orig ω''. We must prove that $(\omega * \omega') * \omega'' \simeq \omega * (\omega' * \omega'')$. Such a homotopy $G\colon I \times I \to X$ is defined by the formula

$$G(t,t') = \begin{cases} \omega\left(\dfrac{4t}{t'+1}\right) & 0 \le t \le \dfrac{t'+1}{4} \\ \omega'(4t - t' - 1) & \dfrac{t'+1}{4} \le t \le \dfrac{t'+2}{4} \\ \omega''\left(\dfrac{4t - 2 - t'}{2 - t'}\right) & \dfrac{t'+2}{4} \le t \le 1 \end{cases}$$

and pictured in the diagram

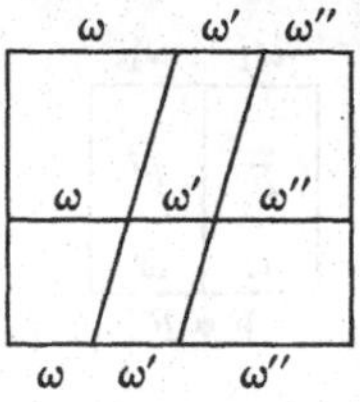

G ■

The category $\mathcal{P}(X)$ is called the *category of path classes* of X, or the *fundamental groupoid* of X, the latter because of the following theorem.

8 THEOREM *$\mathcal{P}(X)$ is a groupoid.*

PROOF Given a path ω in X, let $\omega^{-1}: I \to X$ be the path defined by $\omega^{-1}(t) = \omega(1-t)$. To prove that $[\omega^{-1}] = [\omega]^{-1}$ in $\mathcal{P}(X)$, we must show that $\omega * \omega^{-1} \simeq \varepsilon_{\omega(0)}$ [and also that $\omega^{-1} * \omega \simeq \varepsilon_{\omega(1)}$, which follows, however, from the first homotopy, because $\omega = (\omega^{-1})^{-1}$]. Such a homotopy $H: I \times I \to X$ is defined by the formula

$$H(t,t') = \begin{cases} \omega(0) & 0 \le t \le \dfrac{t'}{2} \\ \omega(2t - t') & \dfrac{t'}{2} \le t \le \dfrac{1}{2} \\ \omega(2 - 2t - t') & \dfrac{1}{2} \le t \le 1 - \dfrac{t'}{2} \\ \omega(0) & 1 - \dfrac{t'}{2} \le t \le 1 \end{cases}$$

and pictured in the diagram

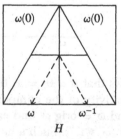

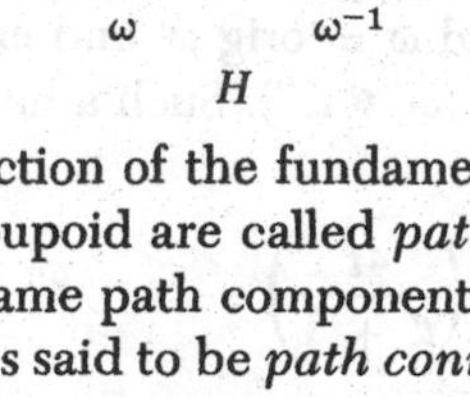

H ■

This completes the construction of the fundamental groupoid. The components of the fundamental groupoid are called *path components* of X. It is clear that x_0, and x_1 are in the same path component of X if and only if there is a path ω in X from x_0 to x_1. X is said to be *path connected* if its fundamental groupoid is connected. The following is an alternate characterization of the path components.

9 THEOREM *The path components of X are the maximal path-connected subspaces of X.*

PROOF Let A be a path component of X and let ω be a path in X such that $\omega(0) \in A$. We show that ω is a path in A. For each $t \in I$ define a path ω_t: $I \to X$ by $\omega_t(t') = \omega(tt')$ for $t' \in I$. Then ω_t is a path in X from $\omega(0)$ to $\omega(t)$. Therefore $\omega(t)$ is in the same path component of X as x_0, namely A. Since this is so for every $t \in I$, ω is a path in A.

A is path connected because if $x_0, x_1 \in A$ there is a path ω in X from x_0 to x_1. By the above result, ω is a path in A. Therefore any two points of A can be joined by a path in A, and A is path connected. Since any path in X that starts in A stays in A, A is a maximal path-connected subset of X. ∎

10 LEMMA *A path-connected space is connected.*

PROOF If ω is a path in X, then $\omega(I)$, being a continuous image of the connected space I, is connected. Therefore $\omega(0)$ and $\omega(1)$ lie in the same component of X. If X is path connected, any two points of X lie in the same component, and X is connected. ∎

The converse of lemma 10 is false, as is shown by the following example.

11 EXAMPLE Let X be the subspace of $\mathbf{R}^2$ defined by

$$X = \{(x,y) \in \mathbf{R}^2 \mid x > 0,\ y = \sin \frac{1}{x} \quad \text{or} \quad x = 0,\ -1 \le y \le 1\}$$

Then X is connected, but not path connected.

Given a map f: $X \to Y$, there is a covariant functor $f_\#$ from $\mathcal{P}(X)$ to $\mathcal{P}(Y)$ which sends an object x of $\mathcal{P}(X)$ to the object $f(x)$ of $\mathcal{P}(Y)$ and the morphism $[\omega]$ of $\mathcal{P}(X)$ to the morphism $f_\#[\omega] = [f \circ \omega]$ of $\mathcal{P}(Y)$. The functorial properties of $f_\#$ are easily verified. From the first part of statement 5, or by direct verification, it follows that f maps each path component of X into some path component of Y. Therefore there is a covariant functor π_0 from the category of topological spaces and maps to the category of sets and functions such that $\pi_0(X)$ equals the set of path components of X, and

$$\pi_0(f) = f_\#\colon \pi_0(X) \to \pi_0(Y)$$

maps the path component of x in X to the path component of $f(x)$ in Y. If F: $f_0 \simeq f_1$, then for any $x \in X$ there is a path ω_x in Y from $f_0(x)$ to $f_1(x)$ defined by $\omega_x(t) = F(x,t)$ for $t \in I$. Therefore $f_0(x)$ and $f_1(x)$ belong to the same path component of Y, and $f_{0\#} = f_{1\#}$. It follows that π_0 can be regarded as a covariant functor from the homotopy category to the category of sets and functions. This functor characterizes the functor π_X for a contractible space X as follows.

12 THEOREM *If X is a contractible space, then π_X and π_0 are naturally equivalent functors on the homotopy category.*

PROOF If X and X' have the same homotopy type, then π_X and $\pi_{X'}$ are naturally equivalent. It follows from corollary 1.3.11 that if P is a one-point space, π_X is naturally equivalent to π_P. It therefore suffices to prove that π_P

is naturally equivalent to π_0. $\pi_0(P)$ consists of the single path component P, and a natural transformation

$$\psi: \pi_P \to \pi_0$$

is defined by $\psi[f] = f_\#(P)$ for $[f] \in [P;X]$. Because X^P is in one-to-one correspondence with X in such a way that homotopies $P \times I \to X$ correspond to paths $I \to X$, it follows that ψ is a natural equivalence. ∎

The functor π_0 is closely related to the functor H_0 of example 1.2.6. In fact, for spaces X whose components and path components coincide, H_0 is the composite of π_0 with the covariant functor which assigns to every set the free abelian group generated by that set. In particular, π_0 could have been used to obtain the results of Sec. 1.2 that were obtained by using H_0.

8 THE FUNDAMENTAL GROUP

By choosing a fixed point $x_0 \in X$ and considering the path classes in X with x_0 as origin and end, a group called the fundamental group is obtained. We show now that this group is naturally equivalent to the first homotopy group π_1, defined in Sec. 1.6. The section closes with a calculation of the fundamental group of the circle.

Let X be a topological space and let $x_0 \in X$. The *fundamental group of X based at x_0*, denoted by $\pi(X,x_0)$, is defined to be the group of path classes with x_0 as origin and end. It follows from theorem 1.7.8 and statement 1.7.2 that this is a group, and if $f: (X,x_0) \to (Y,y_0)$, then $f_\#$ is a homomorphism from $\pi(X,x_0)$ to $\pi(Y,y_0)$. If, $f, f': (X,x_0) \to (Y,y_0)$ are homotopic, then

$$f_\# = f'_\#: \pi(X,x_0) \to \pi(Y,y_0).$$

Therefore, we have the following theorem.

1 THEOREM *There is a covariant functor from the homotopy category of pointed spaces to the category of groups which assigns to a pointed space its fundamental group and to a map f the homomorphism $f_\#$.* ∎

We show that the fundamental group functor π is naturally equivalent to π_1, defined in Sec. 1.6. Let $\lambda: I \to S(S^0)$ be defined by $\lambda(t) = [-1,t]$, where S^0 consists of the two points -1 and 1 and 1 is its basepoint. Then λ induces a bijection $\lambda^\#$ between the homotopy classes of maps $(S(S^0), 1) \to (X,x_0)$ and the path classes of closed paths in X at x_0 defined by

$$\lambda^\#[g] = [g\lambda] \qquad g: (S(S^0), 1) \to (X,x_0)$$

From the definition of the law of composition in $[S(S^0);X]$ and in $\pi(X,x_0)$, $\lambda^\#$ is seen to be a group isomorphism. Given a map $f: (X,x_0) \to (Y,y_0)$, $\lambda^\#$ commutes with $f_\#$. By lemma 1.6.6, $S(S^0)$ is homeomorphic to S^1.

2 THEOREM *The map $\lambda^\#$ is a natural equivalence of the first homotopy group functor π_1 with the fundamental group functor π.* ∎

It will sometimes be convenient to regard the elements of $\pi(X,x_0)$ as homotopy classes of maps $(S^1,p_0) \to (X,x_0)$, rather than as path classes.

Because any closed path at x_0 (and any homotopy between such paths) must lie in the path component A of X containing x_0, it follows that $\pi(A,x_0) \approx \pi(X,x_0)$. Hence the fundamental group can give information only about the path component of X containing x_0. From general groupoid considerations (see statements 1.7.3 and 1.7.4), if $[\omega]$ is a path class in X from x_0 to x_1, then $h_{[\omega]}$ is an isomorphism from $\pi(X,x_1)$ to $\pi(X,x_0)$.

3 THEOREM *The fundamental groups of a path-connected space based at different points are isomorphic by an isomorphism determined up to conjugacy.* ∎

Even though the fundamental groups based at different points of a path-connected space are isomorphic, we cannot identify them, because the isomorphism between them is not unique. If the fundamental group at some point (and hence all points) is abelian, the isomorphism is unique. In general, the fundamental group need not be abelian; however, the following consequence of theorem 2 and corollary 1.6.10 is a general result about the commutativity of fundamental groups.

4 THEOREM *The fundamental group of a path-connected H space is abelian, and if ω and ω' are closed paths at the base point, then*

$$[\omega] * [\omega'] = [\mu \circ (\omega,\omega')]$$

where μ is the multiplication map in the H space. ∎

A space X is said to be *n-connected* for $n \geq 0$ if every continuous map $f\colon S^k \to X$ for $k \leq n$ has a continuous extension over E^{k+1}. A 1-connected space is also said to be *simply connected.* Note that if $0 \leq m \leq n$, an n-connected space is m-connected. It follows from theorem 1.6.7 that a space X is n-connected if and only if it is path connected and $\pi_k(X,x)$ is trivial for every base point $x \in X$ and $1 \leq k \leq n$. From corollary 1.3.13 we have the following result.

5 LEMMA *A contractible space is n-connected for every $n \geq 0$.* ∎

Note that a space is 0-connected if and only if it is path connected, and a space is simply connected if and only if it is path connected, and $\pi(X,x_0) = 0$ for some (and hence all) points $x_0 \in X$.

From theorem 1 we know that two pointed spaces having the same homotopy type as pointed spaces have isomorphic fundamental groups. To prove a similar result for two path-connected spaces which have the same homotopy type as spaces (no base-point condition) we need some preliminary results.

6 LEMMA *Let $h: I \times I \to X$ and let α_0, α_1, β_0, and β_1 be the paths in X defined by restricting h to the edges of $I \times I$ [that is, $\alpha_i(t) = h(i,t)$ and $\beta_i(t) = h(t,i)$]. Then $(\alpha_0 * \beta_1) * (\alpha_1^{-1} * \beta_0^{-1})$ is a closed path in X at $h(0,0)$ which represents the trivial element of $\pi(X, h(0,0))$.*

PROOF Let α_0', α_1', β_0', and β_1' be the paths in $I \times I$ defined by $\alpha_i'(t) = (i,t)$ and $\beta_i'(t) = (t,i)$. Then $(\alpha_0' * \beta_1') * (\alpha_1'^{-1} * \beta_0'^{-1})$ is a closed path in $I \times I$ at $(0,0)$ and h maps this closed path into $(\alpha_0 * \beta_1) * (\alpha_1^{-1} * \beta_0^{-1})$. Since $I \times I$ is a convex subset of $\mathbf{R}^2$, it is contractible, and by lemma 5, it is simply connected. Therefore

$$(\alpha_0' * \beta_1') * (\alpha_1'^{-1} * \beta_0'^{-1}) \simeq \varepsilon_{(0,0)}$$

and

$$(\alpha_0 * \beta_1) * (\alpha^{-1} * \beta_0^{-1}) = h \circ ((\alpha_0' * \beta_1') * (\alpha_1'^{-1} * \beta_0'^{-1}))$$
$$\simeq h \circ \varepsilon_{(0,0)} = \varepsilon_{h(0,0)} \quad \blacksquare$$

7 THEOREM *Let $f: (X,x_0) \to (Y,y_0)$ and $g: (X,x_0) \to (Y,y_1)$ be homotopic as maps of X to Y. Then there is a path ω in Y from y_0 to y_1 such that*

$$f_\# = h_{[\omega]} \circ g_\#: \pi(X,x_0) \to \pi(Y,y_0)$$

PROOF Let $F: X \times I \to Y$ be a homotopy from f to g and let $\omega: I \to Y$ be defined by $\omega(t) = F(x_0,t)$. Then ω is a path in Y from y_0 to y_1. If ω' is any closed path in X at x_0, let $h: I \times I \to Y$ be defined by $h(t,t') = F(\omega'(t), t')$. Then $h(0,t') = F(x_0,t') = \omega(t')$, $h(t,1) = g\omega'(t)$, $h(1,t') = \omega(t')$, and $h(t,0) = f\omega'(t)$. By lemma 6 we have

$$(\omega * g\omega') * (\omega^{-1} * (f\omega')^{-1}) \simeq \varepsilon_{y_0}$$

This implies $[\omega] \circ g_\#[\omega'] \circ [\omega]^{-1} = f_\#[\omega']$, or $(h_{[\omega]} \circ g_\#)[\omega'] = f_\#[\omega']$. Since $[\omega']$ is an arbitrary element of $\pi(X,x_0)$, $h_{[\omega]} \circ g_\# = f_\#$. ■

8 THEOREM *Two path-connected spaces with the same homotopy type have isomorphic fundamental groups.*

PROOF Let $f: X \to Y$ be a homotopy equivalence with homotopy inverse $g: Y \to X$. Let $x_0 \in X$ and set $y_0 = f(x_0)$, $x_1 = g(y_0)$, and $y_1 = f(x_1)$. Let $f_0: (X,x_0) \to (Y,y_0)$ and $f_1: (X,x_1) \to (Y,y_1)$ be maps defined by f (that is, f_0 and f_1 are both equal to f but are regarded as maps of pairs), and let $g': (Y,y_0) \to (X,x_1)$ be defined by g. Then $g' \circ f_0: (X,x_0) \to (X,x_1)$ is homotopic, as a map of X to X, to $1_{(X,x_0)}: (X,x_0) \subset (X,x_0)$, and $f_1 \circ g': (Y,y_0) \to (Y,y_1)$ is homotopic, as a map of Y to Y, to $1_{(Y,y_0)}: (Y,y_0) \subset (Y,y_0)$. It follows from theorem 7 that there are paths ω in X from x_1 to x_0 and ω' in Y from y_1 to y_0 such that

$$h_{[\omega]} = (g' \circ f_0)_\# = g'_\# \circ f_{0\#} \quad \text{and} \quad h_{[\omega']} = (f_1 \circ g')_\# = f_{1\#} \circ g'_\#$$

Therefore we have a commutative diagram

$$\begin{array}{ccc} \pi(X,x_0) & \xrightarrow{h_{[\omega]}} & \pi(X,x_1) \\ {\scriptstyle f_{0\#}}\downarrow & \nearrow{\scriptstyle g'_\#} & \downarrow{\scriptstyle f_{1\#}} \\ \pi(Y,y_0) & \xrightarrow{h_{[\omega']}} & \pi(Y,y_1) \end{array}$$

$g'_\#$ is an epimorphism because $h_{[\omega]}$ is, and it is a monomorphism because $h_{[\omega']}$ is. Therefore $g'_\#$ is an isomorphism. ■

We close with an example of a space with a nontrivial fundamental group. For this purpose we compute $\pi(S^1,p_0)$ following a method used by Tucker[1], where $S^1 = \{e^{i\theta}\}$ and $p_0 = 1$.

The *exponential map ex*: $\mathbf{R} \to S^1$ is defined by $ex(t) = e^{2\pi i t}$. Then ex is continuous, $ex(t_1 + t_2) = ex(t_1)\, ex(t_2)$ (where the right-hand side is multiplication of complex numbers), and $ex(t_1) = ex(t_2)$ if and only if $t_1 - t_2$ is an integer. It follows that $ex|(-\frac{1}{2},\frac{1}{2})$ is a homeomorphism of the open interval $(-\frac{1}{2},\frac{1}{2})$ onto $S^1 - \{e^{\pi i}\}$. We let

$$lg\colon S^1 - \{e^{\pi i}\} \to (-\tfrac{1}{2},\tfrac{1}{2})$$

be the inverse of $ex|(-\frac{1}{2},\frac{1}{2})$.

A subset $X \subset \mathbf{R}^n$ will be called *starlike* from a point $x_0 \in X$ if, whenever $x \in X$, the closed line segment $[x_0,x]$ from x_0 to x lies in X.

9 LEMMA *Let X be compact and starlike from $x_0 \in X$. Given any continuous map $f\colon X \to S^1$ and any $t_0 \in \mathbf{R}$ such that $ex(t_0) = f(x_0)$, there exists a continuous map $f'\colon X \to \mathbf{R}$ such that $f'(x_0) = t_0$ and $ex(f'(x)) = f(x)$ for all $x \in X$.*

PROOF Clearly, we can translate X so that it is starlike from the origin; hence there is no loss of generality in assuming $x_0 = 0$. Since X is compact, f is uniformly continuous and there exists $\varepsilon > 0$ such that if $\|x - x'\| < \varepsilon$, then $\|f(x) - f(x')\| < 2$ [that is, $f(x)$ and $f(x')$ are not antipodes in S^1]. Since X is bounded, there exists a positive integer n such that $\|x\|/n < \varepsilon$ for all $x \in X$. Then for each $0 \le j < n$ and all $x \in X$

$$\left\|\frac{(j+1)x}{n} - \frac{jx}{n}\right\| = \frac{\|x\|}{n} < \varepsilon$$

and so

$$\left\|f\left(\frac{(j+1)x}{n}\right) - f\left(\frac{jx}{n}\right)\right\| < 2$$

It follows that the quotient $f((j+1)x/n)/f(jx/n)$ is a point of $S^1 - \{e^{\pi i}\}$. Let $g_j\colon X \to S^1 - \{e^{\pi i}\}$ for $0 \le j < n$ be the map defined by $g_j(x) =$

[1] See A. W. Tucker, Some topological properties of disk and sphere, *Proceedings of the Canadian Mathematical Congress*, 1945, pp. 285–309.

$f((j+1)x/n)/f(jx/n)$. Then, for all $x \in X$, we see that

$$f(x) = f(0)g_0(x)g_1(x) \cdots g_{n-1}(x)$$

We define $f'\colon X \to \mathbf{R}$ by

$$f'(x) = t_0 + lg(g_0(x)) + lg(g_1(x)) + \cdots + lg(g_{n-1}(x))$$

f' is the sum of $n+1$ continuous functions from X to $\mathbf{R}$, so it is continuous. Clearly, $f'(0) = t_0$ and $ex(f'(x)) = f(x)$. ∎

10 LEMMA *Let X be a connected space and let $f', g'\colon X \to \mathbf{R}$ be maps such that $ex \circ f' = ex \circ g'$ and $f'(x_0) = g'(x_0)$ for some $x_0 \in X$. Then $f' = g'$.*

PROOF Let $h = f' - g'\colon X \to \mathbf{R}$. Since $ex \circ f' = ex \circ g'$, $ex \circ h$ is the constant map of X to p_0. Therefore h is a continuous map of X to $\mathbf{R}$, taking only integral values. Because X is connected, h is constant, and since $h(x_0) = 0$, $h(x) = 0$ for all $x \in X$. ∎

Let $\alpha\colon I \to S^1$ be a closed path at p_0. Because I is starlike from 0 and $\alpha(0) = p_0 = ex(0)$, it follows from lemma 9 that there exists $\alpha'\colon I \to \mathbf{R}$ such that $\alpha'(0) = 0$ and $ex \circ \alpha' = \alpha$. By lemma 10, α' is uniquely characterized by these properties. Because $ex(\alpha'(1)) = p_0$, it follows that $\alpha'(1)$ is an integer. We define the *degree* of α by $\deg \alpha = \alpha'(1)$.

11 LEMMA *Let α and β be homotopic closed paths in S^1 at p_0. Then* $\deg \alpha = \deg \beta$.

PROOF Let $F\colon I \times I \to S^1$ be a homotopy relative to $\dot I$ from α to β. Because $I \times I$ is a starlike subset of $\mathbf{R}^2$ from $(0,0)$, it follows from lemma 9 that there is a map $F'\colon I \times I \to \mathbf{R}$ such that $F'(0,0) = 0$ and $ex \circ F' = F$. Since F is a homotopy relative to $\dot I$, $F(0,t') = F(1,t') = p_0$ for all $t \in I$. Therefore $F'(0,t')$ and $F'(1,t')$ take on only integral values for all $t' \in I$. It follows that $F'(0,t')$ must be constant and $F'(1,t')$ must be constant. Because $F'(0,0) = 0$, $F'(0,t') = 0$ for all $t' \in I$. Define $\alpha', \beta'\colon I \to \mathbf{R}$ by $\alpha'(t) = F'(t,0)$ and $\beta'(t) = F'(t,1)$. Then $\alpha'(0) = 0$ and $ex \circ \alpha' = \alpha$. Therefore $\deg \alpha = \alpha'(1) = F'(1,0)$. Similarly, $\beta'(0) = 0$ and $ex \circ \beta' = \beta$, so $\deg \beta = \beta'(1) = F'(1,1)$. Because $F'(1,t')$ is constant, $F'(1,0) = F'(1,1)$ and $\deg \alpha = \deg \beta$. ∎

It follows that there is a well-defined function deg from $\pi(S^1,p_0)$ to $\mathbf{Z}$ defined by

$$deg\, [\alpha] = \deg \alpha$$

where α is a closed path in S^1 at p_0.

12 THEOREM *The function deg is an isomorphism*

$$deg\colon \pi(S^1,p_0) \approx \mathbf{Z}$$

PROOF To prove that deg is a homomorphism, let α and β be two closed paths in S^1 at p_0 and let $\alpha\beta$ be the closed path which is their pointwise product in the group multiplication of S^1. We know from theorem 4 that $[\alpha] * [\beta] = [\alpha\beta]$. Let $\alpha', \beta'\colon I \to \mathbf{R}$ be such that $\alpha'(0) = 0$, $ex \circ \alpha' = \alpha$, $\beta'(0) = 0$, and $ex \circ \beta' = \beta$. Then $\alpha' + \beta'\colon I \to \mathbf{R}$ is such that $(\alpha' + \beta')(0) = 0$ and $ex \circ (\alpha' + \beta') = \alpha\beta$. Therefore

$$\begin{aligned} deg([\alpha] * [\beta]) &= deg[\alpha\beta] = (\alpha' + \beta')(1) \\ &= \deg \alpha + \deg \beta = deg[\alpha] + deg[\beta] \end{aligned}$$

showing that *deg* is a homomorphism.

The map *deg* is an epimorphism; for if n is an integer, there is a path α'_n in $\mathbf{R}$ defined by $\alpha'_n(t) = tn$. Let $\alpha_n = ex \circ \alpha'_n$. Then clearly, $deg[\alpha_n] = \alpha'_n(1) = n$.

The map *deg* is a monomorphism; for if $deg[\alpha] = 0$, there is a closed path α' in $\mathbf{R}$ at 0 such that $ex \circ \alpha' = \alpha$. Since $\mathbf{R}$ is simply connected (because it is contractible, and by lemma 5), $\alpha' \simeq \varepsilon_0$. Then $ex \circ \alpha' \simeq \varepsilon_{p_0}$. Therefore $\alpha \simeq \varepsilon_{p_0}$ and $[\alpha]$ is the identity element of $\pi(S^1,p_0)$. ∎

The method we have used to compute $\pi(S^1,p_0)$ will be generalized in Chapter 2 to give relations between the fundamental group of a space and the fundamental groups of its covering spaces.

It follows from theorem 2 that $\pi(S^1,p_0) \approx [S^1,p_0; S^1,p_0]$. Because S^1 is a topological group, the set $[S^1;S^1]$ (with no base-point condition) is also a group under pointwise multiplication of maps, and there is an obvious homomorphism

$$\gamma\colon [S^1,p_0; S^1,p_0] \to [S^1;S^1]$$

13 LEMMA *The homomorphism*

$$\gamma\colon [S^1,p_0; S^1,p_0] \to [S^1;S^1]$$

is an isomorphism.

PROOF To show that γ is an epimorphism, let $f\colon S^1 \to S^1$ and let $f(p_0) = e^{i\theta}$ for some $0 \le \theta < 2\pi$. Define a homotopy $F\colon S^1 \times I \to S^1$ by

$$F(z,t) = f(z)e^{-it\theta}$$

Then F is a homotopy from f to a map f' such that $f'(p_0) = p_0$. Therefore $\gamma[f']_{p_0} = [f'] = [f]$.

To show that γ is a monomorphism, assume that $f\colon (S^1,p_0) \to (S^1,p_0)$ is such that $\gamma[f]_{p_0} = [f]$ is trivial. Then $f\colon S^1 \to S^1$ is null homotopic. By theorem 1.3.12, f is null homotopic relative to p_0. Therefore $[f]_{p_0}$ is trivial. ∎

It follows from theorem 12 and lemma 13 that $[S^1;S^1] \approx \mathbf{Z}$. The isomorphism can be chosen so that for each integer n the map $z \to z^n$ from S^1 to itself represents a homotopy class corresponding to n.

EXERCISES

A CONTRACTIBLE SPACES

1 The *cone* over a topological space X with *vertex* v is defined to be the mapping cylinder of the constant map $X \to v$. Prove that X is contractible if and only if it is a retract of any cone over X.

2 Prove that S^n is a retract of E^{n+1} if and only if S^n is contractible.

3 If CX is a cone over X, prove that (CX,X) has the homotopy extension property with respect to any space.

4 Prove that a space Y is contractible if and only if, given a pair (X,A) having the homotopy extension property with respect to Y, any map $A \to Y$ can be extended over X.

5 Let Y be the comb space of example 1.3.9 and let y_0 be the point $(0,1) \in Y$. Let Y' be another copy of Y, with corresponding point y_0'. Let X be the space obtained by forming the disjoint union of Y and Y' and identifying y_0 with y_0'. Prove that X is n-connected for all n but not contractible. (*Hint:* Any deformation of X in itself must be a homotopy relative to y_0.)

B ADJUNCTION SPACES

1 Let A be a subspace of a space X and let $f: A \to Y$ be a continuous map. The *adjunction space* Z of X to Y by f is defined to be the quotient space of the disjoint union of X and Y by the identifications $x \in A$ equals $f(x) \in Y$ for all $x \in A$. Prove that if X and Y are normal spaces and A is closed in X, then Z is a normal space.

2 A space X is said to be *binormal* if $X \times I$ is a normal space. If X is a binormal space, Y is a normal space, and $f: X \to Y$ is continuous, prove that the mapping cylinder of f is a normal space.

3 Given a continuous map $f: A \to Y$, where A is a subspace of a space X, prove that f can be extended over X if and only if Y is a retract of the adjunction space of X to Y by f.

4 Let Z be the adjunction space of X to Y by a map $f: A \to Y$. Prove that (Z,Y) has the homotopy extension property with respect to a space W if (X,A) has the homotopy extension property with respect to W.

C ABSOLUTE RETRACTS AND ABSOLUTE NEIGHBORHOOD RETRACTS

A space Y is said to be an *absolute retract* (or *absolute neighborhood retract*) if, given a normal space X, closed subset $A \subset X$, and a continuous map $f: A \to Y$, then f can be extended over X (or f can be extended over some neighborhood of A in X).

1 Prove that a normal space Y is an absolute retract (or absolute neighborhood retract) if and only if, whenever Y is imbedded as a closed subset of a normal space Z, then Y is a retract of Z (or a retract of some neighborhood of Y in Z).

2 Prove that the product of arbitrarily many absolute retracts (or finitely many absolute neighborhood retracts) is itself an absolute retract (or absolute neighborhood retract).

3 Prove that $\mathbf{R}^n$ is an absolute retract for all n.

4 Prove that a retract of an absolute retract is an absolute retract and that a retract of some open subset of an absolute neighborhood retract is an absolute neighborhood retract.

5 Prove that E^n is an absolute retract and S^n is an absolute neighborhood retract for all n.

6 Prove that a binormal absolute neighborhood retract is *locally contractible* (that is, every neighborhood U of a point x contains a neighborhood V of x deformable to x in U).

7 Prove that a binormal absolute neighborhood retract is an absolute retract if and only if it is contractible.

D HOMOTOPY EXTENSION PROPERTY

1 Let A be a closed subset of a normal space X, let $f\colon X \to Y$ be continuous (where Y is arbitrary), and let $G\colon A \times I \to Y$ be a homotopy of $f \mid A$. If there exists a homotopy $G'\colon U \times I \to Y$ of $f \mid U$ which extends G, where U is an open neighborhood of A, show that there exists a homotopy $F\colon X \times I \to Y$ of f which extends G.

2 *Borsuk's homotopy extension theorem.* Let A be a closed subspace of a binormal space X. Then (X,A) has the homotopy extension property with respect to any absolute neighborhood retract Y.

3 Let A be a closed subset of a binormal space X and assume that the subspace $A \times I \cup X \times 0 \subset X \times I$ is an absolute neighborhood retract. Then (X,A) has the homotopy extension property with respect to any space Y.

4 Let A be a closed subset of X and B a subset of Y. Assume that (X,A) has the homotopy extension property with respect to B and that $(X \times I, X \times \dot{I} \cup A \times I)$ has the homotopy extension property with respect to Y. Prove that if $f\colon (X,A) \to (Y,B)$ is homotopic (as a map of pairs) to a map which sends all of X to B, then it is homotopic relative to A to such a map.

E COFIBRATIONS

1 Prove that any cofibration is an injective function.

2 Prove that a composite of cofibrations is a cofibration.

3 For a closed subspace A of X prove that the inclusion map $A \subset X$ is a cofibration if and only if $X \times 0 \cup A \times I$ is a retract of $X \times I$.

4 If A is a subspace of a Hausdorff space X, prove that if $A \subset X$ is a cofibration, then A is closed in X.

5 Assume that X is the union of closed subsets X_1 and X_2 and let A be a subset of X such that $X_1 \cap X_2 \subset A$. Prove that if $A \cap X_1 \subset X_1$ and $A \cap X_2 \subset X_2$ are cofibrations, so is $A \subset X$.

6 Let A be a closed subspace of a space X. Prove that the following are equivalent:[1]

(*a*) $A \subset X$ is a cofibration.

(*b*) There is a deformation $D\colon X \times I \to X$ rel A [that is, $D(x,0) = x$ and $D(a,t) = a$ for $x \in X$, $a \in A$, and $t \in I$] and a map $\varphi\colon X \to I$ such that $A = \varphi^{-1}(1)$ and $D(\varphi^{-1}(0,1] \times 1) \subset A$.

[1] If X is normal, the equivalence of (*a*) and (*c*) is proved in G. S. Young, A condition for the absolute homotopy extension property, *American Mathematical Monthly*, vol. 71, pp. 896–897, 1964.

(c) There is a neighborhood U of A deformable in X to A rel A [that is, there is a homotopy $H\colon U \times I \to X$ such that $H(x,0) = x$, $H(a,t) = a$, and $H(x,1) \in A$ for $x \in U$, $a \in A$, and $t \in I$] and a map $\varphi\colon X \to I$ such that $A = \varphi^{-1}(1)$ and $\varphi(x) = 0$ if $x \in X - U$.

7 If $A \subset X$ and $B \subset Y$ are cofibrations with A and B closed in X and Y, respectively, prove that $A \times B \subset X \times B \cup A \times Y$ and $X \times B \cup A \times Y \subset X \times Y$ are cofibrations.

F LOCAL SYSTEMS[1]

1 A *local system* on a space X is a covariant functor from the fundamental groupoid of X to some category. For any category $\mathcal{C}$ show that there is a category of local systems on X with values in $\mathcal{C}$. (Two local systems on X are said to be *equivalent* if they are equivalent objects in this category.)

2 Given a map $f\colon X \to Y$, show that f induces a covariant functor from the category of local systems on Y with values in $\mathcal{C}$ to the category of local systems on X with values in $\mathcal{C}$.

3 If A is an object of a category $\mathcal{C}$, let Aut A be the group of self-equivalences of A in $\mathcal{C}$. If $\varphi\colon A \approx B$ is an equivalence in $\mathcal{C}$, then show that $\bar\varphi$: Aut $A \to$ Aut B defined by $\bar\varphi(\alpha) = \varphi \circ \alpha \circ \varphi^{-1}$ is an isomorphism of groups.

4 If Γ is a local system on X and $x_0 \in X$, show that Γ induces a homomorphism

$$\bar\Gamma_{x_0}\colon \pi(X,x_0) \to \operatorname{Aut} \Gamma(x_0)$$

5 If X is path connected, prove that two local systems Γ and Γ' on X with values in $\mathcal{C}$ are equivalent if and only if there is an equivalence $\varphi\colon \Gamma(x_0) \approx \Gamma'(x_0)$, such that $\bar\varphi \circ \bar\Gamma_{x_0}$ is conjugate to $\bar\Gamma'_{x_0}$ in Aut $\Gamma'(x_0)$.

6 If X is path connected, given an object $A \in \mathcal{C}$ and a homomorphism $\alpha\colon \pi(X,x_0) \to$ Aut A, prove that there is a local system Γ on X with values in $\mathcal{C}$ such that $\Gamma(x_0) = A$ and $\bar\Gamma_{x_0} = \alpha$.

G THE FUNDAMENTAL GROUP

1 Prove that the fundamental group functor commutes with direct products.

2 If ω and ω' are paths in X from x_0 to x_1, prove that $\omega \simeq \omega'$ if and only if $\omega * \omega'^{-1} \simeq \varepsilon_{x_0}$.

3 Let a space X be the union of two open simply connected subsets U and V such that $U \cap V$ is nonempty and path connected. Prove that X is simply connected.

4 Prove that S^n is simply connected for $n \geq 2$.

5 If there exists a space with a nonabelian fundamental group, prove that the "figure eight" (that is, the union of two circles with a point in common) has a nonabelian fundamental group (cf. exercise 2.B.4).

6 Let $f\colon I \to \mathbf{R}^2$ be a continuously differentiable simple closed curve in the plane with a nowhere-vanishing tangent vector [that is, $f(0) = f(1)$, $f'(0) = f'(1)$, and $f'(t) \neq 0$ for

[1] See N. E. Steenrod, Homology with local coefficients, *Annals of Mathematics*, vol. 44, pp. 610–627, 1943.

$0 \leq t \leq 1$]. Let $\omega: I \to S^1$ be the closed path defined by $\omega(t) = f'(t)/\|f'(t)\|$. Prove that $[\omega]$ is a generator of $\pi(S^1)$.[1]

7 In $\mathbf{R}^2$, let X be the space consisting of the union of the circles C_n, where C_n has center $(1/n,0)$ and radius $1/n$ for all positive integers n. In $\mathbf{R}^3$ (with $\mathbf{R}^2$ imbedded as the plane $x_3 = 0$), let Y be the set of points on the closed line segments joining $(0,0,1)$ to X and let Y' be the reflection of Y through the origin of $\mathbf{R}^3$. Then Y and Y' are closed simply connected subsets of $\mathbf{R}^3$ such that $Y \cap Y'$ is a single point. Prove that $Y \cup Y'$ is not simply connected.[2]

H SOME APPLICATIONS OF THE FUNDAMENTAL GROUP

1 Prove that S^1 is not a retract of E^2.

2 Prove that S^1 and S^n for $n \geq 2$ are not of the same homotopy type.

3 Prove that $\mathbf{R}^2$ and $\mathbf{R}^n$ for $n > 2$ are not homeomorphic.

4 Let $p(z) = z^n + a_{n-1} z^{n-1} + \cdots + a_1 z + a_0$ be a polynomial of degree n, having complex coefficients and leading coefficient 1, and let $q(z) = z^n$. For $r > 0$ let $C_r = \{x \in \mathbf{R}^2 \mid \|x\| = r\}$. Prove that for r large enough, $p \mid C_r$ and $q \mid C_r$ are homotopic maps of C_r into $\mathbf{R}^2 - 0$.

5 *Fundamental theorem of algebra.* Prove that every complex polynomial has a root. (*Hint:* For any $r > 0$ the map $q \mid C_r\colon C_r \to \mathbf{R}^2 - 0$ is not null homotopic because it induces a nontrivial homomorphism of fundamental groups.)

[1] See H. Hopf, Uber die Drehung der Tangenten und Sehnen ebener Kurven, *Compositio Mathematica,* vol. 2, pp. 50–62, 1935. For generalizations see H. Whitney, On regular closed curves in the plane, *Compositio Mathematica,* vol. 4, pp. 276–284, 1937, and S. Smale, Regular curves on Riemannian manifolds, *Transactions of the American Mathematical Society,* vol. 87, pp. 495–512, 1958.

[2] See H. B. Griffiths, The Fundamental group of two spaces with a common point, *Quarterly Journal of Mathematics,* vol. 5, pp. 175–190, 1954.

CHAPTER TWO

COVERING SPACES AND FIBRATIONS

THE THEORY OF COVERING SPACES IS IMPORTANT NOT ONLY IN TOPOLOGY, BUT also in differential geometry, complex analysis, and Lie groups. The theory is presented here because the fundamental group functor provides a faithful representation of covering-space problems in terms of algebraic ones. This justifies our special interest in the fundamental group functor.

This chapter contains the theory of covering spaces, as well as an introduction to the related concepts of fiber bundle and fibration. These concepts will be considered again later in other contexts. Here we adopt the view that certain fibrations, namely, those having the property of unique path lifting, are generalized covering spaces. Because of this, we shall consider these fibrations in some detail.

Covering spaces are defined in Sec. 2.1, and fibrations are defined in Sec. 2.2, where it is proved that every covering space is a fibration. Section 2.3 deals with relations between the fundamental groups of the total space and base space of a fibration with unique path lifting, and Sec. 2.4 contains a solution of the lifting problem for such fibrations in terms of the fundamental group functor.

The lifting theorem is applied in Sec. 2.5 to classify the covering spaces of a connected locally path-connected space by means of subgroups of its fundamental group. This entails the construction of a covering space starting with the base space and a subgroup of its fundamental group. In Sec. 2.6 a converse problem is considered. The base space is constructed, starting with a covering space and a suitable group of transformations on it.

In Sec. 2.7 fiber bundles are introduced as natural generalizations of covering spaces. The main result of the section is that local fibrations are fibrations. This implies that a fiber bundle with paracompact base space is a fibration. Section 2.8 considers properties of general fibrations and the concept of fiber homotopy equivalence. These will be important in our later study of homotopy theory.

1 COVERING PROJECTIONS

A covering projection is a continuous map that is a uniform local homeomorphism. This and related concepts are introduced in this section, along with some examples and elementary properties.

Let $p: \tilde{X} \to X$ be a continuous map. An open subset $U \subset X$ is said to be *evenly covered* by p if $p^{-1}(U)$ is the disjoint union of open subsets of $\tilde{X}$ each of which is mapped homeomorphically onto U by p. If U is evenly covered by p, it is clear that any open subset of U is also evenly covered by p. A continuous map $p: \tilde{X} \to X$ is called a *covering projection* if each point $x \in X$ has an open neighborhood evenly covered by p. $\tilde{X}$ is called the *covering space* and X the *base space* of the covering projection.

The following are examples of covering projections.

1 Any homeomorphism is a covering projection.

2 If $\tilde{X}$ is the product of X with a discrete space, the projection $\tilde{X} \to X$ is a covering projection.

3 The map $ex: \mathbf{R} \to S^1$, defined by $ex(t) = e^{2\pi i t}$, (considered in Sec 1.8) is a covering projection.

4 For any positive integer n the map $p: S^1 \to S^1$, defined by $p(z) = z^n$, is a covering projection.

5 For any integer $n \geq 1$ the map $p: S^n \to P^n$, which identifies antipodal points, is a covering projection.

6 If G is a topological group, H is a discrete subgroup of G, and G/H is the space of left (or right) cosets, then the projection $G \to G/H$ is a covering projection.

A continuous map $f: Y \to X$ is called a *local homeomorphism* if each point $y \in Y$ has an open neighborhood mapped homeomorphically by f onto an open subset of X. If this is so, each point of Y has arbitrarily small neighborhoods with this property, and we have the following lemmas.

7 LEMMA *A local homeomorphism is an open map.* ∎

8 LEMMA *A covering projection is a local homeomorphism.*

PROOF Let $p: \tilde{X} \to X$ be a covering projection and let $\tilde{x} \in \tilde{X}$. Let U be an open neighborhood of $p(\tilde{x})$ evenly covered by p. Then $p^{-1}(U)$ is the disjoint union of open sets, each mapped homeomorphically onto U by p. Let $\tilde{U}$ be that one of these open sets which contains $\tilde{x}$. Then $\tilde{U}$ is an open neighborhood of $\tilde{x}$ such that $p \mid \tilde{U}$ is a homeomorphism of $\tilde{U}$ onto the open subset U of X. ∎

A local homeomorphism need not be a covering projection, as shown by the following example.

9 EXAMPLE Let $p: (0,3) \to S^1$ be the restriction of the map $ex: \mathbf{R} \to S^1$ of example 3 to the open interval $(0,3)$. Because p is the restriction of a local homeomorphism to an open subset, it is a local homeomorphism. It is also a surjection, but it is not a covering projection because the complex number $1 \in S^1$ has no neighborhood evenly covered by p.

The following is a consequence of lemmas 7 and 8 and the fact (immediate from the definition) that a covering projection is a surjection.

10 COROLLARY *A covering projection exhibits its base space as a quotient space of its covering space.* ∎

For locally connected spaces there is the following reduction of covering projections to the components of the base space.

11 THEOREM *If X is locally connected, a continuous map $p: \tilde{X} \to X$ is a covering projection if and only if for each component C of X the map*

$$p \mid p^{-1}C: p^{-1}C \to C$$

is a covering projection.

PROOF If p is a covering projection and C is a component of X, let $x \in C$ and let U be an open neighborhood of x evenly covered by p. Let V be the component of U containing x. Since X is locally connected, V is open in X, and hence open in C. Clearly, V is evenly covered by $p \mid p^{-1}C$. Therefore $p \mid p^{-1}C$ is a covering projection.

Conversely, assume that the map $p \mid p^{-1}C: p^{-1}C \to C$ is a covering projection for each component C of X. Let $x \in C$ and let U be an open neighborhood of x in C evenly covered by $p \mid p^{-1}C$. Since X is locally connected, C is open in X. Therefore U is also open in X and is clearly evenly covered by p. Hence p is a covering projection. ∎

In general, the representation of the inverse image of an evenly covered open set as a disjoint union of open sets, each mapped homeomorphically, is not unique (consider the case of an evenly covered discrete set); however, for connected evenly covered open sets there is the following characterization of these open subsets.

12 LEMMA *Let U be an open connected subset of X which is evenly covered by a continuous map $p: \tilde{X} \to X$. Then p maps each component of $p^{-1}(U)$ homeomorphically onto U.*

PROOF By assumption, $p^{-1}(U)$ is the disjoint union of open subsets, each mapped homeomorphically onto U by p. Since U is connected, each of these open subsets is also connected. Because they are open and disjoint, each is a component of $p^{-1}(U)$. ■

13 COROLLARY *Consider a commutative triangle*

$$\begin{array}{ccc} \tilde{X}_1 & \xrightarrow{p} & \tilde{X}_2 \\ & {\scriptstyle p_1}\searrow \quad \swarrow{\scriptstyle p_2} & \\ & X & \end{array}$$

where X is locally connected and p_1 and p_2 are covering projections. If p is a surjection, it is a covering projection.

PROOF If U is a connected open subset of X which is evenly covered by p_1 and p_2, it follows easily from lemma 12 that each component of $p_2{}^{-1}(U)$ is evenly covered by p. ■

14 THEOREM *If $p: \tilde{X} \to X$ is a covering projection onto a locally connected base space, then for any component $\tilde{C}$ of $\tilde{X}$ the map*

$$p \mid \tilde{C}: \tilde{C} \to p(\tilde{C})$$

is a covering projection onto some component of X.

PROOF *Let* $\tilde{C}$ be a component of $\tilde{X}$. We show that $p(\tilde{C})$ is a component of X. $p(\tilde{C})$ is connected; to show that it is an open and closed subset of X, let x be in the closure of $p(\tilde{C})$ and let U be an open connected neighborhood of x evenly covered by p. Because U meets $p(\tilde{C})$, $p^{-1}(U)$ meets $\tilde{C}$. Therefore some component $\tilde{U}$ of $p^{-1}(U)$ meets $\tilde{C}$. Since $\tilde{C}$ is a component of $\tilde{X}$, $\tilde{U} \subset \tilde{C}$. Then, by lemma 12, $p(\tilde{C}) \supset p(\tilde{U}) = U$. Therefore the closure of $p(\tilde{C})$ is contained in the interior of $p(\tilde{C})$, which implies that $p(\tilde{C})$ is open and closed. The same argument shows that if $x \in p(\tilde{C})$ and U is an open connected neighborhood of x in X evenly covered by p, then $U \subset p(\tilde{C})$ and $(p \mid \tilde{C})^{-1}(U)$ is the disjoint union of those components of $p^{-1}(U)$ that meet $\tilde{C}$. It follows from lemma 12 that U is evenly covered by $p \mid \tilde{C}$. Therefore $p \mid \tilde{C}: \tilde{C} \to p(\tilde{C})$ is a covering projection. ■

The following example shows that the converse of theorem 14 is false.

15 EXAMPLE Let $X = S^1 \times S^1 \times \cdots$ be a countable product of 1-spheres and for $n \geq 1$ let $\tilde{X}_n = \mathbf{R}^n \times S^1 \times S^1 \times \cdots$. Define $p_n: \tilde{X}_n \to X$ by

$$p_n(t_1, \ldots, t_n, z_1, z_2, \ldots) = (ex(t_1), \ldots, ex(t_n), z_1, z_2, \ldots)$$

Let $\tilde{X} = \vee \tilde{X}_n$ and define $p: \tilde{X} \to X$ so that $p \mid \tilde{X}_n = p_n$. The components of $\tilde{X}$ are the spaces $\tilde{X}_n$ and the map $p \mid \tilde{X}_n = p_n: \tilde{X}_n \to X$ is a covering projection. However, p is not a covering projection, because no open subset of X is evenly covered by p.

For later purposes we should like to have the analogues of theorems 11 and 14, in which "component" is replaced by "path component." For this we need the following definition: a topological space is said to be *locally path connected* if the path components of open subsets are open. The following are easy consequences of this definition.

16 *Any open subset of a locally path-connected space is itself locally path connected.* ∎

17 *A locally path-connected space is locally connected.* ∎

18 *In a locally path-connected space the components and path components coincide.* ∎

19 *A connected locally path-connected space is path connected.* ∎

From statements 17 and 18 we obtain the following extension of theorems 11 and 14.

20 THEOREM *If X is locally path connected, a continuous map $p: \tilde{X} \to X$ is a covering projection if and only if for each path component A of X*

$$p \mid p^{-1}A: p^{-1}A \to A$$

is a covering projection. In this case, if $\tilde{A}$ is any path component of $\tilde{X}$, then $p \mid \tilde{A}$ is a covering projection of $\tilde{A}$ onto some path component of X. ∎

2 THE HOMOTOPY LIFTING PROPERTY

The homotopy lifting property is dual to the homotopy extension property. It leads to the concept of fibration, which is dual to that of cofibration introduced in Sec. 1.4. In this section we define the concept of fibration and prove that a covering projection is a special kind of fibration. This special class of fibrations will be regarded as generalized covering projections, and our subsequent study of covering projections will be based on a study of the more general concept. At the end of the chapter we return to the general consideration of fibrations.

We begin with an important problem of algebraic topology, called the lifting problem, which is dual to the extension problem. Let $p: E \to B$ and $f: X \to B$ be maps. The *lifting problem* for f is to determine whether there is a continuous map $f'\colon X \to E$ such that $f = p \circ f'$—that is, whether the dotted arrow in the diagram

$$\begin{array}{ccc} & & E \\ & \nearrow & \downarrow p \\ X & \xrightarrow{f} & B \end{array}$$

corresponds to a continuous map making the diagram commutative. If there is such a map f', then f can be *lifted* to E, and we call f' a *lifting*, or *lift*, of f. In order that the lifting problem be a problem in the homotopy category, we need an analogue of the homotopy extension property, called the homotopy lifting property, defined as follows. A map $p\colon E \to B$ is said to have the *homotopy lifting property with respect to a space* X if, given maps $f'\colon X \to E$ and $F\colon X \times I \to B$ such that $F(x,0) = pf'(x)$ for $x \in X$, there is a map $F'\colon X \times I \to E$ such that $F'(x,0) = f'(x)$ for $x \in X$ and $p \circ F' = F$. If f' is regarded as a map of $X \times 0$ to E, the existence of F' is equivalent to the existence of a map represented by the dotted arrow that makes the following diagram commutative:

$$\begin{array}{ccc} X \times 0 & \xrightarrow{f'} & E \\ \downarrow \cap & \nearrow & \downarrow p \\ X \times I & \xrightarrow{F} & B \end{array}$$

If $p\colon E \to B$ has the homotopy lifting property with respect to X and $f_0, f_1\colon X \to B$ are homotopic, it is easy to see that f_0 can be lifted to E if and only if f_1 can be lifted to E. Hence, whether or not a map $X \to B$ can be lifted to E is a property of the homotopy class of the map. Thus the homotopy lifting property implies that the lifting problem for maps $X \to B$ is a problem in the homotopy category.

A map $p\colon E \to B$ is called a *fibration* (or *Hurewicz fiber space* in the literature) if p has the homotopy lifting property with respect to every space. E is called the *total space* and B the *base space* of the fibration. For $b \in B$, $p^{-1}(b)$ is called the *fiber over b*.

If $p\colon E \to B$ is a fibration, any path ω in B such that $\omega(0) \in p(E)$ can be lifted to a path in E. In fact, ω can be regarded as a homotopy $\omega\colon P \times I \to B$ where P is a one-point space, and a point $e_0 \in E$ such that $p(e_0) = \omega(0)$ corresponds to a map $f\colon P \to E$ such that $pf(P) = \omega(P,0)$. It follows from the homotopy lifting property of p that there exists a path $\tilde{\omega}$ in E such that $\tilde{\omega}(0) = e_0$ and $p \circ \tilde{\omega} = \omega$. Then $\tilde{\omega}$ is a lifting of ω.

1 EXAMPLE Let F be any space and let $p\colon B \times F \to B$ be the projection to the first factor. Then p is a fibration, and for any $b \in B$ the fiber over b is homeomorphic to F.

To prove that a covering projection is a fibration, we first establish the following *unique-lifting property* of covering projections for connected spaces.

2 THEOREM *Let $p: \tilde{X} \to X$ be a covering projection and let $f, g: Y \to \tilde{X}$ be liftings of the same map (that is, $p \circ f = p \circ g$). If Y is connected and f agrees with g for some point of Y, then $f = g$.*

PROOF Let $Y_1 = \{y \in Y \mid f(y) = g(y)\}$. We show that Y_1 is open in Y. If $y \in Y_1$, let U be an open neighborhood of $pf(y)$ evenly covered by p and let $\tilde{U}$ be an open subset of $\tilde{X}$ containing $f(y)$ such that p maps $\tilde{U}$ homeomorphically onto U. Then $f^{-1}(\tilde{U}) \cap g^{-1}(\tilde{U})$ is an open subset of Y containing y and contained in Y_1.

Let $Y_2 = \{y \in Y \mid f(y) \neq g(y)\}$. We show that Y_2 is also open in Y (if $\tilde{X}$ were assumed to be Hausdorff, this would follow from a general property of Hausdorff spaces). Let $y \in Y_2$ and let U be an open neighborhood of $pf(y)$ evenly covered by p. Since $f(y) \neq g(y)$, there are disjoint open subsets $\tilde{U}_1$ and $\tilde{U}_2$ of $\tilde{X}$ such that $f(y) \in \tilde{U}_1$ and $g(y) \in \tilde{U}_2$ and p maps each of the sets $\tilde{U}_1$ and $\tilde{U}_2$ homeomorphically onto U. Then $f^{-1}(\tilde{U}_1) \cap g^{-1}(\tilde{U}_2)$ is an open subset of Y containing y and contained in Y_2.

Since $Y = Y_1 \cup Y_2$ and Y_1 and Y_2 are disjoint open sets, it follows from the connectedness of Y that either $Y_1 = \varnothing$ or $Y_1 = Y$. By hypothesis, $Y_1 \neq \varnothing$, so $Y = Y_1$ and $f = g$. ∎

We are now ready to prove that a covering projection has the homotopy lifting property.

3 THEOREM *A covering projection is a fibration.*

PROOF Let $p: \tilde{X} \to X$ be a covering projection and let $f': Y \to \tilde{X}$ and $F: Y \times I \to X$ be maps such that $F(y,0) = pf'(y)$ for $y \in Y$. We show that for each $y \in Y$ there is an open neighborhood N_y of y in Y and a map $F'_y: N_y \times I \to \tilde{X}$ such that $F'_y(y',0) = f'(y')$ for $y' \in N_y$ and $pF'_y = F \mid N_y \times I$. Assume that we have such neighborhoods N_y and maps F'_y. If $y'' \in N_y \cap N_{y'}$, then $F'_y \mid y'' \times I$ and $F'_{y'} \mid y'' \times I$ are maps of the connected space $y'' \times I$ into $\tilde{X}$ such that for $t \in I$

$$p \circ (F'_y \mid y'' \times I)(y'',t) = F(y'',t) = p \circ (F'_{y'} \mid y'' \times I)(y'',t)$$

Because $(F'_y \mid y'' \times I)(y'',0) = f'(y'') = (F'_{y'} \mid y'' \times I)(y'',0)$, it follows from theorem 2 that $F'_y \mid y'' \times I = F'_{y'} \mid y'' \times I$. Since this is true for all $y'' \in N_y \cap N_{y'}$, it follows that $F'_y \mid (N_y \cap N_{y'}) \times I = F'_{y'} \mid (N_y \cap N_{y'}) \times I$. Hence there is a continuous map $F': Y \times I \to \tilde{X}$ such that $F' \mid N_y \times I = F'_y$, and F' is a lifting of F such that $F'(y,0) = f'(y)$ for $y \in Y$. Thus we have reduced the theorem to the construction of the open neighborhoods N_y and maps F'_y.

It follows from the fact that $p: \tilde{X} \to X$ is a covering projection (and the compactness of I) that for each $y \in Y$ there is an open neighborhood N_y of y and a sequence $0 = t_0 < t_1 < \cdots < t_m = 1$ of points of I such that for $i = 1, \ldots, m$, $F(N_y \times [t_{i-1},t_i])$ is contained in some open subset of X evenly covered by p. We show that there is a map $F'_y: N_y \times I \to \tilde{X}$ with the desired properties. It suffices to define maps

$$G_i\colon N_y \times [t_{i-1},t_i] \to \tilde X \qquad i = 1, \ldots, m$$

such that

$$\begin{aligned} p \circ G_i &= F \mid N_y \times [t_{i-1},t_i] \\ G_1(y',0) &= f'(y') \qquad & y' \in N_y \\ G_{i-1}(y',t_{i-1}) &= G_i(y',t_{i-1}) \qquad & y' \in N_y,\ i = 2, \ldots, m \end{aligned}$$

because, given such maps G_i, there is a map $F'_y\colon N_y \times I \to \tilde X$ such that $F'_y \mid N_y \times [t_{i-1},t_i] = G_i$ for $i = 1, \ldots, m$. Then F'_y has the desired properties.

The maps G_i are defined by induction on i. To define G_1, let U be an open subset of X evenly covered by p such that $F(N_y \times [t_0,t_1]) \subset U$. Let $\{\tilde U_j\}$ be a collection of disjoint open subsets of $\tilde X$ such that $p^{-1}(U) = \cup \tilde U_j$ and p maps $\tilde U_j$ homeomorphically onto U for each j. Let $V_j = f'^{-1}(\tilde U_j)$. Then $\{V_j\}$ is a collection of disjoint open sets covering N_y, and G_1 is defined to be the unique map such that for each j, G_1 maps $V_j \times [t_0,t_1]$ into $\tilde U_j$ to be a lifting of $F \mid V_j \times [t_0,t_1]$. This defines G_1.

Assume G_{i-1} defined for $1 < i \le m$. Let U' be an open subset of X evenly covered by p such that $F(N_y \times [t_{i-1},t_i]) \subset U'$. Let $\{\tilde U'_k\}$ be a collection of disjoint open subsets of $\tilde X$ such that $p^{-1}(U') = \cup \tilde U'_k$ and p maps $\tilde U'_k$ homeomorphically onto U' for each k. Let $V'_k = \{y' \in N_y \mid G_{i-1}(y',t_{i-1}) \in \tilde U'_k\}$. Then $\{V'_k\}$ is a collection of disjoint open sets covering N_y, and G_i is defined to be the unique map such that for each k, G_i maps $V'_k \times [t_{i-1},t_i]$ into $\tilde U'_k$ to be a lifting of $F \mid V'_k \times [t_{i-1},t_i]$. This defines G_i. ■

A map $p\colon E \to B$ is said to have *unique path lifting* if, given paths ω and ω' in E such that $p \circ \omega = p \circ \omega'$ and $\omega(0) = \omega'(0)$, then $\omega = \omega'$. It follows from theorem 2 that a covering projection has unique path lifting.

4 LEMMA *If a map has unique path lifting, it has the unique-lifting property for path-connected spaces.*

PROOF Assume that $p\colon E \to B$ has unique path lifting. Let Y be path connected and suppose that $f, g\colon Y \to E$ are maps such that $p \circ f = p \circ g$ and $f(y_0) = g(y_0)$ for some $y_0 \in Y$. We must show $f = g$. Let $y \in Y$ and let ω be a path in Y from y_0 to y. Then $f \circ \omega$ and $g \circ \omega$ are paths in E that are liftings of the same path in B and have the same origin. Because p has unique path lifting, $f \circ \omega = g \circ \omega$. Therefore

$$f(y) = (f \circ \omega)(1) = (g \circ \omega)(1) = g(y) \quad ■$$

The following theorem characterizes fibrations with unique path lifting.

5 THEOREM *A fibration has unique path lifting if and only if every fiber has no nonconstant paths.*

PROOF Assume that $p\colon E \to B$ is a fibration with unique path lifting. Let ω be a path in the fiber $p^{-1}(b)$ and let ω' be the constant path in $p^{-1}(b)$ such that $\omega'(0) = \omega(0)$. Then $p \circ \omega = p \circ \omega'$, which implies $\omega = \omega'$. Hence ω is a constant path.

Conversely, assume that $p: E \to B$ is a fibration such that every fiber has no nontrivial path and let ω and ω' be paths in E such that $p \circ \omega = p \circ \omega'$ and $\omega(0) = \omega'(0)$. For $t \in I$, let ω_t'' be the path in E defined by

$$\omega_t''(t') = \begin{cases} \omega((1 - 2t')t) & 0 \le t' \le 1/2 \\ \omega'((2t' - 1)t) & 1/2 \le t' \le 1 \end{cases}$$

Then ω_t'' is a path in E from $\omega(t)$ to $\omega'(t)$, and $p \circ \omega_t''$ is a closed path in B that is homotopic relative to $\dot{I}$ to the constant path at $p\omega(t)$. By the homotopy lifting property of p, there is a map $F': I \times I \to E$ such that $F'(t',0) = \omega_t''(t')$ and F' maps $0 \times I \cup I \times 1 \cup 1 \times I$ to the fiber $p^{-1}(p\omega(t))$. Because $p^{-1}(p\omega(t))$ has no nonconstant paths, F' maps $0 \times I$, $I \times 1$, and $1 \times I$ to a single point. It follows that $F'(0,0) = F'(1,0)$. Therefore $\omega_t''(0) = \omega_t''(1)$ and $\omega(t) = \omega'(t)$. ■

We have seen that a covering projection is a fibration with unique path lifting. It will be shown in Sec. 2.4 that if the base space satisfies some mild hypotheses, any fibration with unique path lifting is a covering projection. One reason for studying fibrations with unique path lifting as generalized covering projections is that the following two theorems are easily proved, but both are false for covering projections.

6 THEOREM *The composite of fibrations (with unique path lifting) is a fibration (with unique path lifting).* ■

7 THEOREM *The product of fibrations (with unique path lifting) is a fibration (with unique path lifting).* ■

An example shows that theorem 6 is false for covering projections.

8 EXAMPLE Let X and X_n, for $n \ge 1$, be a countable product of 1-spheres. Let $\tilde{X}_n = \mathbf{R}^n \times X$ and define $p_n: \tilde{X}_n \to X_n$ by

$$p_n(t_1, \ldots, t_n, z_1, z_2, \ldots) = (ex(t_1), \ldots, ex(t_n), z_1, z_2, \ldots)$$

Then p_n is a covering projection for $n \ge 1$. It follows from theorem 2.1.11 that $\vee p_n: \vee \tilde{X}_n \to \vee X_n$ is a covering projection. Since $\vee X_n$ is the product of X and the set of positive integers, there is a covering projection $\vee X_n \to X$ (see example 2.1.2). The composite

$$\vee \tilde{X}_n \to \vee X_n \to X$$

is not a covering projection (*cf.* example 2.1.15).

Similarly, theorem 7 is false for covering projections.

9 EXAMPLE For $n \ge 1$, let $p_n: \tilde{X}_n \to X_n$ be the covering projection $ex: \mathbf{R} \to S^1$. Then

$$\times p_n: \times \tilde{X}_n \to \times X_n$$

is not a covering projection.

It follows from theorem 6 that there is a category whose objects are topological spaces and whose morphisms are fibrations with unique path lifting. We shall now describe a category, depending on a given base space, which is of more use in studying covering projections or fibrations. For a given space X there is a category whose objects are maps $p: \tilde{X} \to X$, which are fibrations with unique path lifting, and whose morphisms are commutative triangles

$$\begin{array}{ccc} X_1 & \xrightarrow{f} & X_2 \\ & {\scriptstyle p_1}\searrow \quad \swarrow{\scriptstyle p_2} & \\ & X & \end{array}$$

If $p_j: \tilde{X}_j \to X$ is an indexed family of objects in this category, let $p: \vee \tilde{X}_j \to X$ be the map such that $p \mid \tilde{X}_j = p_j$. Then p is also an object in the category and is the sum of the collection $\{p_j\}$ in the category.

To show that this category also has products, given maps $p_j: \tilde{X}_j \to X$, let

$$\tilde{X} = \{(x_j) \in \times \tilde{X}_j \mid p_j(x_j) = p_{j'}(x_{j'}) \text{ for all } j, j'\}$$

and define $p: \tilde{X} \to X$ by $p((x_j)) = p_j(x_j)$. If each p_j is a fibration, so is p, and if each p_j has unique path lifting, so does p. Hence p is a product of $\{p_j\}$ in the category of fibrations with unique path lifting. This map p is called the *fibered product* of the maps $\{p_j\}$. We consider it in more detail in Sec. 2.8.

There is a similar category whose objects are covering projections with base space X and whose morphisms are commutative triangles. This category has finite sums and finite products, but neither arbitrary sums nor arbitrary products. In fact, for each n let

$$p_n: \mathbf{R}^n \times S^1 \times S^1 \times \cdots \to S^1 \times S^1 \times \cdots$$

be defined by $p_n(t_1, \ldots, t_n, z_1, z_2, \ldots) = (e^{2\pi i t_1}, \ldots, e^{2\pi i t_n}, z_1, z_2, \ldots)$, as in example 8. Then the collection $\{p_n\}$ has neither a sum nor a product in the category of covering projections with base space X.

3 RELATIONS WITH THE FUNDAMENTAL GROUP

In a fibration with unique path lifting the fundamental group of the total space is isomorphic to a subgroup of the fundamental group of the base space. The corresponding subgroup of the fundamental group will lead to a classification of fibrations with unique path lifting. In fact, we shall see in the next section that the fundamental group functor solves the lifting problem for fibrations with unique path lifting. The present section is devoted to consideration of the relation between the fundamental groups of the total space and the base space of a fibration with unique path lifting.

We begin with a localization property for fibrations which is an analogue of theorem 2.1.14.

1 LEMMA *Let $p: E \to B$ be a fibration. If A is any path component of E, then pA is a path component of B and $p \mid A: A \to pA$ is a fibration.*

PROOF Since pA is the continuous image of a path-connected space, it is path connected. It is a maximal path-connected subset of B, for if ω is a path in B that begins in pA, there is a lifting $\tilde{\omega}$ of ω that begins in A. Since A is a path component of E, $\tilde{\omega}$ is a path in A. Therefore $\omega = p \circ \tilde{\omega}$ is a path in pA. Hence pA is a maximal path-connected subset of B and, by theorem 1.7.9, a path component of B.

To show that $p \mid A: A \to pA$ has the homotopy lifting property, let $f': Y \to A$ and $F: Y \times I \to pA$ be maps such that $F(y,0) = pf'(y)$. Because p is a fibration, there is a map $F': Y \times I \to E$ such that $p \circ F' = F$ and $F'(y,0) = f'(y)$. For any $y \in Y$, F' must map $y \times I$ into the path component of E containing $F'(y,0)$. Therefore $F'(y \times I) \subset A$ for all y, and $F': Y \times I \to A$ is a lifting of F such that $F'(y,0) = f'(y)$. ∎

For locally path-connected spaces we have the following analogue of theorem 2.1.20, which reduces the study of fibrations to the study of fibrations with total space and base space path connected.

2 THEOREM *Let $p: E \to B$ be a map. If E is locally path connected, p is a fibration if and only if for each path component A of E, pA is a path component of B and $p \mid A: A \to pA$ is a fibration.*

PROOF If $p: E \to B$ is a fibration and A is a path component of E, it follows from lemma 1 that pA is a path component of B and $p \mid A: A \to pA$ is a fibration.

To prove the converse, let $f': Y \to E$ and $F: Y \times I \to B$ be such that $F(y,0) = f'(y)$. Let $\{A_j\}$ be the path components of E. Then $\{A_j\}$ are disjoint open subsets of E. Let $V_j = f'^{-1}(A_j)$. The collection $\{V_j\}$ is a disjoint open covering of Y. Therefore, to construct a map $F': Y \times I \to E$ such that $p \circ F' = F$ and $F'(y,0) = f'(y)$, it suffices to construct maps $F'_j: V_j \times I \to E$ for all j such that $p \circ F'_j = F \mid V_j \times I$ and $F'_j(y,0) = f'(y,0)$.

Because $F(y \times I)$ is contained in the path component of B containing $F(y,0) = pf'(y)$, it follows from the fact that pA_j is a path component of B that $F(V_j \times I) \subset pA_j$ for all j. Because $p \mid A_j: A_j \to pA_j$ is a fibration, there is a map $F'_j: V_j \times I \to A_j$ such that $pF'_j = F \mid V_j \times I$ and $F'_j(y,0) = f'(y)$ for $y \in V_j$. Therefore p has the homotopy lifting property. ∎

Since every path in a topological space lies in some path component of the space, it is clear that theorem 2 remains valid if the term "fibration" is replaced throughout by "fibration with unique path lifting."

The main result on fibrations with unique path lifting is embodied in the following statement.

3 LEMMA *Let $p: \tilde{X} \to X$ be a fibration with unique path lifting. If ω and ω' are paths in $\tilde{X}$ such that $\omega(0) = \omega'(0)$ and $p \circ \omega \simeq p \circ \omega'$, then $\omega \simeq \omega'$.*

PROOF Let $F: I \times I \to X$ be a homotopy relative to $\dot{I}$ from $p \circ \omega$ to $p \circ \omega'$ [that is, $F(t,0) = p\omega(t)$ and $F(t,1) = p\omega'(t)$, and $F(0,t) = p\omega(0)$ and $F(1,t) = p\omega(1)$]. By the homotopy lifting property of fibrations, there is a map $F': I \times I \to \tilde{X}$ such that $F'(t,0) = \omega(t)$ and $p \circ F' = F$. Then $F'(0 \times I)$ and $F'(1 \times I)$ are contained in $p^{-1}(p\omega(0))$ and $p^{-1}(p\omega(1))$, respectively. By theorem 2.2.5, $F'(0 \times I)$ and $F'(1 \times I)$ are single points. Hence F' is a homotopy relative to $\dot{I}$ from ω to some path ω'' such that $\omega''(0) = \omega(0)$ and $p \circ \omega'' = p \circ \omega'$. Since $\omega'(0) = \omega(0)$, it follows from the unique-path-lifting property of p that $\omega' = \omega''$ and $F': \omega \simeq \omega'$ rel $\dot{I}$. ∎

It follows from lemma 3 that if $p: \tilde{X} \to X$ is a fibration with unique path lifting, then for any two objects $\tilde{x}_0$ and $\tilde{x}_1$ in the fundamental groupoid of $\tilde{X}$, $p_\#$ maps hom $(\tilde{x}_0,\tilde{x}_1)$ injectively into hom $(p(\tilde{x}_0),p(\tilde{x}_1))$. In particular, if $\tilde{x}_0 = \tilde{x}_1$, we obtain the following theorem.

4 THEOREM *Let $p: \tilde{X} \to X$ be a fibration with unique path lifting. For any $\tilde{x}_0 \in \tilde{X}$ the homomorphism*

$$p_\#: \pi(\tilde{X},\tilde{x}_0) \to \pi(X,x_0)$$

is a monomorphism. ∎

This last result provides the basis for the reduction of problems concerning fibrations with unique path lifting to problems about the fundamental group. In order that the fundamental group be really representative of the space in question, we assume that the spaces involved are path connected. It follows from theorem 2 that this is no loss of generality for locally path-connected spaces.

5 LEMMA *Let $p: \tilde{X} \to X$ be a fibration with unique path lifting and assume that $\tilde{X}$ is a nonempty path-connected space. If $\tilde{x}_0$, $\tilde{x}_1 \in \tilde{X}$, there is a path ω in X from $p(\tilde{x}_0)$ to $p(\tilde{x}_1)$ such that*

$$p_\#\pi(\tilde{X},\tilde{x}_0) = h_{[\omega]}p_\#\pi(\tilde{X},\tilde{x}_1)$$

Conversely, given a path ω in X from $p(\tilde{x}_0)$ to x_1, there is a point $\tilde{x}_1 \in p^{-1}(x_1)$ such that

$$h_{[\omega]}p_\#\pi(\tilde{X},\tilde{x}_1) = p_\#\pi(\tilde{X},\tilde{x}_0)$$

PROOF For the first part, let $\tilde{\omega}$ be a path in $\tilde{X}$ from $\tilde{x}_0$ to $\tilde{x}_1$. Then $\pi(\tilde{X},\tilde{x}_0) = h_{[\tilde{\omega}]}\pi(\tilde{X},\tilde{x}_1)$. Therefore

$$p_\#\pi(\tilde{X},\tilde{x}_0) = h_{[p\circ\tilde{\omega}]}p_\#\pi(\tilde{X},\tilde{x}_1)$$

and so $p \circ \tilde{\omega}$ will do as the path from $p(\tilde{x}_0)$ to $p(\tilde{x}_1)$.

Conversely, given a path ω in X from $p(\tilde{x}_0)$ to x_1, let $\tilde{\omega}$ be a path in $\tilde{X}$ such that $\tilde{\omega}(0) = \tilde{x}_0$ and $p\tilde{\omega} = \omega$. If $\tilde{x}_1 = \tilde{\omega}(1)$, then

$$h_{[\omega]}p_\#\pi(\tilde{X},\tilde{x}_1) = p_\#(h_{[\tilde{\omega}]}\pi(\tilde{X},\tilde{x}_1)) = p_\#\pi(\tilde{X},\tilde{x}_0) \quad ∎$$

This easily implies the following result.

6 THEOREM *Let $p: \tilde{X} \to X$ be a fibration with unique path lifting and assume that $\tilde{X}$ is a nonempty path-connected space. For $x_0 \in p\tilde{X}$ the collection $\{p_\#\pi(\tilde{X},\tilde{x}_0) \mid \tilde{x}_0 \in p^{-1}(x_0)\}$ is a conjugacy class in $\pi(X,x_0)$. If ω is a path in $p\tilde{X}$ from x_0 to x_1, then $h_{[\omega]}$ maps the conjugacy class in $\pi(X,x_1)$ to the conjugacy class in $\pi(X,x_0)$.* ∎

Let $p: \tilde{X} \to X$ be a fibration and let ω be a path in X beginning at x_0. Define a map $F_\omega: p^{-1}(x_0) \times I \to X$ by $F_\omega(\tilde{x},t) = \omega(t)$ and let $i: p^{-1}(x_0) \subset \tilde{X}$. Then $pi(\tilde{x}) = F_\omega(\tilde{x},0)$ for $\tilde{x} \in p^{-1}(\tilde{x}_0)$. It follows from the homotopy lifting property of p that there exists a map $G_\omega: p^{-1}(x_0) \times I \to \tilde{X}$ such that $G_\omega(\tilde{x},0) = i(\tilde{x}) = \tilde{x}$ and $p \circ G_\omega = F_\omega$.

Suppose now that p has unique path lifting. We prove that the map $\tilde{x} \to G_\omega(\tilde{x},1)$ of $p^{-1}(x_0)$ to $p^{-1}(\omega(1))$ depends only on the path class of ω. If $\omega' \simeq \omega$ and $G'_{\omega'}: p^{-1}(x_0) \times I \to \tilde{X}$ is a map such that $G'_{\omega'}(\tilde{x},0) = \tilde{x}$ and $p \circ G'_{\omega'} = F_{\omega'}$, then for any $\tilde{x} \in p^{-1}(x_0)$, let $\tilde{\omega}$ and $\tilde{\omega}'$ be the paths in $\tilde{X}$ defined by $\tilde{\omega}(t) = G_\omega(\tilde{x},t)$ and $\tilde{\omega}'(t) = G'_{\omega'}(\tilde{x},t)$. Then $\tilde{\omega}$ and $\tilde{\omega}'$ begin at $\tilde{x}$ and

$$p \circ \tilde{\omega} = \omega \simeq \omega' = p \circ \tilde{\omega}'$$

It follows from lemma 3 that $\tilde{\omega} \simeq \tilde{\omega}'$. Then $G_\omega(\tilde{x},1) = G'_{\omega'}(\tilde{x},1)$ for every $\tilde{x} \in p^{-1}(x_0)$. Therefore there is a well-defined continuous map

$$f_{[\omega]}: p^{-1}(\omega(0)) \to p^{-1}(\omega(1))$$

defined by $f_{[\omega]}(\tilde{x}) = G_\omega(\tilde{x},1)$, where G_ω is as above. It is clear that if $\omega(1) = \omega'(0)$, then $f_{[\omega]*[\omega']} = f_{[\omega']} \circ f_{[\omega]}$.

7 THEOREM *Let $p: \tilde{X} \to X$ be a fibration with unique path lifting. There is a contravariant functor from the fundamental groupoid of X to the category of topological spaces and maps which assigns to $x \in X$ the fiber over x and to $[\omega]$ the function $f_{[\omega]}$.* ∎

The fact that $f_{[\omega]}$ is a homeomorphism for every $[\omega]$ leads to the following corollary.

8 COROLLARY *If $p: \tilde{X} \to X$ is a fibration with unique path lifting and X is path connected, then any two fibers are homeomorphic.* ∎

If X is path connected and $p: \tilde{X} \to X$ is a fibration with unique path lifting, the *number of sheets* of p (or the *multiplicity* of p) is defined to be the cardinal number of $p^{-1}(x)$ (which is independent of $x \in X$, by corollary 8). For a path-connected total space, the multiplicity is determined by the conjugacy class as follows.

9 THEOREM *Let $p: \tilde{X} \to X$ be a fibration with unique path lifting and assume $\tilde{X}$ and X to be nonempty path-connected spaces. If $\tilde{x}_0 \in \tilde{X}$, the multiplicity of p is the index of $p_\#\pi(\tilde{X},\tilde{x}_0)$ in $\pi(X,p(\tilde{x}_0))$.*

PROOF By theorem 7, $\pi(X,p(\tilde{x}_0))$ acts as a group of transformations on the right on $p^{-1}(p(\tilde{x}_0))$ by $\tilde{x} \circ [\omega] = f_{[\omega]}(\tilde{x})$ for $\tilde{x} \in p^{-1}(p(\tilde{x}_0))$. If $\tilde{x}_1, \tilde{x}_2 \in p^{-1}(p(\tilde{x}_0))$, let $\tilde{\omega}$ be a path in $\tilde{X}$ from $\tilde{x}_1$ to $\tilde{x}_2$. Then $[p \circ \tilde{\omega}] \in \pi(X,p(\tilde{x}_0))$ and $\tilde{x}_1 \circ [p\tilde{\omega}] = \tilde{x}_2$.

Therefore $\pi(X,p(\tilde{x}_0))$ acts transitively on $p^{-1}(p(\tilde{x}_0))$. The isotropy group of $\tilde{x}_0$ [that is, the subgroup of $\pi(X,p(\tilde{x}_0))$ leaving $\tilde{x}_0$ fixed] is clearly equal to $p_\# \pi(\tilde{X},\tilde{x}_0)$. From general considerations[1] there is a bijection between the set of right cosets of $p_\# \pi(\tilde{X},\tilde{x}_0)$ in $\pi(X,p(\tilde{x}_0))$ and $p^{-1}(p(\tilde{x}_0))$. ∎

10 EXAMPLE For $n \geq 2$ the covering $p: S^n \to P^n$ of example 2.1.5 has multiplicity 2. Because S^n is simply connected, $\pi(P^n) \approx \mathbf{Z}_2$ for $n \geq 2$.

A fibration $p: \tilde{X} \to X$ with unique path lifting is said to be *regular* if, given any closed path ω in X, either every lifting of ω is closed or none is closed.

11 THEOREM *Let $p: \tilde{X} \to X$ be a fibration with unique path lifting. p is regular if and only if $p_\# \pi(\tilde{X},\tilde{x}_0) = p_\# \pi(\tilde{X},\tilde{x}_1)$ whenever $p(\tilde{x}_0) = p(\tilde{x}_1)$.*

PROOF Assume that p is regular and let $\tilde{\omega}$ be a closed path in $\tilde{X}$ at $\tilde{x}_0$. Then $\tilde{\omega}$ is a closed lifting of $p\tilde{\omega}$. Therefore there is a closed lifting $\tilde{\omega}_1$ of $p\tilde{\omega}$ at $\tilde{x}_1$. It follows that $p_\#[\tilde{\omega}] = [p\tilde{\omega}] = p_\#[\tilde{\omega}_1]$. Therefore $p_\# \pi(\tilde{X},\tilde{x}_0) \subset p_\# \pi(\tilde{X},\tilde{x}_1)$. Since the roles of $\tilde{x}_0$ and $\tilde{x}_1$ can be interchanged, it follows that $p_\# \pi(\tilde{X},\tilde{x}_0) = p_\# \pi(\tilde{X},\tilde{x}_1)$.

Conversely, if $p_\# \pi(\tilde{X},\tilde{x}_0) = p_\# \pi(\tilde{X},\tilde{x}_1)$ whenever $p(\tilde{x}_0) = p(\tilde{x}_1)$, let ω be a closed path in X at $p(\tilde{x}_0)$ having a closed lifting $\tilde{\omega}$ at $\tilde{x}_0$. Then

$$[\omega] = p_\#[\tilde{\omega}] \in p_\# \pi(\tilde{X},\tilde{x}_0) = p_\# \pi(\tilde{X},\tilde{x}_1)$$

Therefore there is a closed path $\tilde{\omega}_1$ in $\tilde{X}$ at $\tilde{x}_1$ such that $p\tilde{\omega}_1 \simeq \omega$. If $\tilde{\omega}_1'$ is a lifting of ω such that $\tilde{\omega}_1'(0) = \tilde{x}_1$, then by the unique-path-lifting property of p, $\tilde{\omega}_1 = \tilde{\omega}_1'$. Therefore $\tilde{\omega}_1'$ is a closed lifting of ω at $\tilde{x}_1$ and p is regular. ∎

In case $\tilde{X}$ is a nonempty path-connected space, theorems 6 and 11 give the following result.

12 THEOREM *Let $p: \tilde{X} \to X$ be a fibration with unique path lifting and assume that $\tilde{X}$ is a nonempty path-connected space. Then p is regular if and only if for some $\tilde{x}_0 \in \tilde{X}_0$, $p_\# \pi(\tilde{X},\tilde{x}_0)$ is a normal subgroup of $\pi(X,p(\tilde{x}_0))$.* ∎

4 THE LIFTING PROBLEM

In this section we show that the fundamental group functor solves the lifting problem for fibrations with unique path lifting. As a consequence of this, the fundamental group functor provides a classification of covering projections, which is discussed in the next section.

Our first result is that any map of a contractible space to the base space of a fibration can be lifted.

1 LEMMA *Let $p: E \to B$ be a fibration. Any map of a contractible space to B whose image is contained in $p(E)$ can be lifted to E.*

[1] Whenever a group G acts transitively on the right on a set S there is induced a bijection between the set of right cosets of the isotropy group (of any $s \in S$) in G and the set S.

PROOF Let Y be contractible and let $f: Y \to B$ be a map such that $f(Y) \subset p(E)$. Because Y is contractible, f is homotopic to a constant map of Y to some point of $f(Y)$. $f(Y) \subset p(E)$, so this constant map can be lifted to E. The homotopy lifting property then implies that f can be lifted to E. ∎

Because we use the fundamental group functor, it will prove technically simpler to consider the lifting problem for spaces with base points.

2 **LEMMA** *Let $p: (\tilde{X},\tilde{x}_0) \to (X,x_0)$ be a fibration with unique path lifting. If y_0 is a strong deformation retract of Y, any map $(Y,y_0) \to (X,x_0)$ can be lifted to a map $(Y,y_0) \to (\tilde{X},\tilde{x}_0)$.*

PROOF Let $f: (Y,y_0) \to (X,x_0)$ be a map. f is homotopic relative to y_0 to the constant map $Y \to x_0$. The constant map can be lifted to the constant map $Y \to \tilde{x}_0$. By the homotopy lifting property, f can be lifted to a map $f': Y \to \tilde{X}$ such that f' is homotopic to the constant map $Y \to \tilde{x}_0$ by a homotopy which maps $y_0 \times I$ to $p^{-1}(x_0)$. Because $p^{-1}(x_0)$ has no nonconstant path by theorem 2.2.5, $f'(y_0) = \tilde{x}_0$. ∎

We shall apply lemma 2 to a contractible space in order to lift certain quotient spaces of the contractible space. The usual way to represent a space as the quotient space of a contractible space is to show it is a quotient space of its path space. Given $y_0 \in Y$, the *path space* $P(Y,y_0)$ is the space of continuous maps $\omega: (I,0) \to (Y,y_0)$ topologized by the compact-open topology. There is a function $\varphi: P(Y,y_0) \to Y$ defined by $\varphi(\omega) = \omega(1)$. If U is an open set in Y,

$$\varphi^{-1}(U) = \langle 1;U\rangle = \{\omega \in P(Y,y_0) \mid \omega(1) \in U\}$$

is an open set in $P(Y,y_0)$. Therefore φ is continuous.

3 **LEMMA** *The constant path at y_0 is a strong deformation retract of the path space $P(Y,y_0)$.*

PROOF A strong deformation retraction $F: P(Y,y_0) \times I \to P(Y,y_0)$ to the constant path at y_0 is defined by

$$F(\omega,t)(t') = \omega((1-t)t') \qquad \omega \in P(Y,y_0);\ t, t' \in I \quad ∎$$

We have shown that φ is a continuous map of the contractible path space $P(Y,y_0)$ to Y. If Y is path connected, φ is clearly surjective. If Y is also locally path connected, the following theorem shows that φ is a quotient projection.

4 **THEOREM** *A connected locally path-connected space Y is the quotient space of its path space $P(Y,y_0)$ by the map φ.*

PROOF We know that φ is continuous, and because a connected locally path-connected space is path connected, it is surjective. To complete the proof it suffices to show that φ is an open map. Let $\omega \in P(Y,y_0)$ and let $W = \bigcap_{1 \le i \le n}\langle K_i;U_i\rangle$ be a neighborhood of ω, where K_i is compact in I and U_i is open in Y. We enumerate the K's so that for some $0 \le k \le n$, $1 \in K_1 \cap \cdots \cap K_k$ and

$1 \notin K_{k+1} \cup \cdots \cup K_n$. Because $\omega(1) \in U_1 \cap \cdots \cap U_k$, there is a path-connected neighborhood V of $\omega(1)$ contained in $U_1 \cap \cdots \cap U_k$. Choose $0 < t' < 1$ such that $[t',1] \cap (K_{k+1} \cup \cdots \cup K_n) = \varnothing$ and $\omega([t',1]) \subset V$.

To prove that $\varphi(W) \supset V$, which completes the proof, let $y' \in V$ and let ω' be a path in V from $\omega(t')$ to y'. Define $\bar\omega\colon I \to Y$ by

$$\bar\omega(t) = \begin{cases} \omega(t) & 0 \le t \le t' \\ \omega'\left(\dfrac{t-t'}{1-t'}\right) & t' \le t \le 1 \end{cases}$$

For $i > k$, $\bar\omega(K_i) = \omega(K_i) \subset U_i$. For $i \le k$,

$$\bar\omega(K_i) = \bar\omega(K_i \cap [0,t']) \cup \bar\omega(K_i \cap [t',1]) \subset \omega(K_i) \cup \omega'(I) \subset U_i \cup V = U_i$$

Therefore $\bar\omega \in W$ and $\varphi(\bar\omega) = y'$. Hence $\varphi(W) \supset V$. ∎

We can put these results together to obtain the following result, called the *lifting theorem.*

5 THEOREM *Let $p\colon (\tilde X,\tilde x_0) \to (X,x_0)$ be a fibration with unique path lifting. Let Y be a connected locally path-connected space. A necessary and sufficient condition that a map $f\colon (Y,y_0) \to (X,x_0)$ have a lifting $(Y,y_0) \to (\tilde X,\tilde x_0)$ is that in $\pi(X,x_0)$*

$$f_\# \pi(Y,y_0) \subset p_\# \pi(\tilde X,\tilde x_0)$$

PROOF If $f'\colon (Y,y_0) \to (\tilde X,\tilde x_0)$ is a lifting of f, then $f = p \circ f'$ and

$$f_\# \pi(Y,y_0) = p_\# f'_\# \pi(Y,y_0) \subset p_\# \pi(\tilde X,\tilde x_0)$$

which shows that the condition is necessary.

We now prove that the condition is sufficient. It follows from lemmas 3 and 2 that if ω_0 is the constant path at y_0, the composite

$$(P(Y,y_0),\ \omega_0) \xrightarrow{\varphi} (Y,y_0) \xrightarrow{f} (X,x_0)$$

can be lifted to a map $\tilde f\colon (P(Y,y_0),\ \omega_0) \to (\tilde X,\tilde x_0)$. We show that if $f_\# \pi(Y,y_0) \subset p_\# \pi(\tilde X,\tilde x_0)$ and if $\omega, \omega' \in P(Y,y_0)$ are such that $\varphi(\omega) = \varphi(\omega')$, then $\tilde f(\omega) = \tilde f(\omega')$. Let $\bar\omega$ and $\bar\omega'$ be the paths in $P(Y,y_0)$ from ω_0 to ω and ω', respectively, defined by $\bar\omega(t)(t') = \omega(tt')$ and $\bar\omega'(t)(t') = \omega'(tt')$. Then $\tilde f \circ \bar\omega$ and $\tilde f \circ \bar\omega'$ are paths in $\tilde X$ from $\tilde x_0$ to $\tilde f(\omega)$ and $\tilde f(\omega')$, respectively, such that

$$p \circ \tilde f \circ \bar\omega = f \circ \varphi \circ \bar\omega = f \circ \omega \qquad \text{and} \qquad p \circ \tilde f \circ \bar\omega' = f \circ \omega'$$

Because $\omega * \omega'^{-1}$ is a closed path in Y at y_0 and $f_\# \pi(Y,y_0) \subset p_\# \pi(\tilde X,\tilde x_0)$, there is a closed path $\tilde\omega$ in $\tilde X$ at $\tilde x_0$ such that $(f \circ \omega) * (f \circ \omega')^{-1} \simeq p \circ \tilde\omega$. Then

$$p \circ (\tilde f \circ \bar\omega) = f \circ \omega \simeq (p \circ \tilde\omega) * (f \circ \omega') = p \circ (\tilde\omega * (\tilde f \circ \bar\omega'))$$

By lemma 2.3.3, $\tilde f \circ \bar\omega \simeq \tilde\omega * (\tilde f \circ \bar\omega')$. In particular, the endpoint of $\tilde f \circ \bar\omega$, which is $\tilde f(\omega)$, equals the endpoint of $\tilde f \circ \bar\omega'$, which is $\tilde f(\omega')$.

It follows that there is a function $f'\colon (Y,y_0) \to (\tilde X,\tilde x_0)$ such that $f' \circ \varphi = \tilde f$, and using theorem 4, we see that f' is continuous. Because

$$p \circ f' \circ \varphi = p \circ \tilde{f} = f \circ \varphi$$

and φ is surjective, $p \circ f' = f$. Therefore f' is a lifting of f. ∎

Let $p: E \to B$ be a fibration. A *section* of p is a map $s: B \to E$ such that $p \circ s = 1_B$ (thus a section is a right inverse of p). It follows easily from the homotopy lifting property that there is a section of p if and only if $[p]$ has a right inverse in the homotopy category. Because a section is a lifting of the identity map $B \subset B$, the following is an immediate consequence of theorem 5.

6 COROLLARY *Let $p: (\tilde{X},\tilde{x}_0) \to (X,x_0)$ be a fibration with unique path lifting. If X is a connected locally path-connected space, there is a section $(X,x_0) \to (\tilde{X},\tilde{x}_0)$ of p if and only if $p_\# \pi(\tilde{X},\tilde{x}_0) = \pi(X,x_0)$.* ∎

7 COROLLARY *Let $p: \tilde{X} \to X$ be a fibration with unique path lifting. If $\tilde{X}$ is a nonempty path-connected space and X is connected and locally path connected, then p is a homeomorphism if and only if for some $\tilde{x}_0 \in \tilde{X}$, $p_\# \pi(\tilde{X},\tilde{x}_0) = \pi(X,p(\tilde{x}_0))$.*

PROOF If p is a homeomorphism, $p_\# \pi(\tilde{X},\tilde{x}_0) = \pi(X,p(\tilde{x}_0))$. Conversely, if $p_\# \pi(\tilde{X},\tilde{x}_0) = \pi(X,p(\tilde{x}_0))$, then by theorem 2.3.9, p is a bijection. By corollary 6, it has a continuous right inverse. Therefore p is a homeomorphism. ∎

If $p: \tilde{X} \to X$ is a covering projection and $\tilde{X}$ is path connected, a necessary and sufficient condition that p be a homeomorphism is that $p_\# \pi(\tilde{X},\tilde{x}_0) = \pi(X,p(\tilde{x}_0))$ for some $\tilde{x}_0 \in \tilde{X}$. This condition on the fundamental groups implies that p is a bijection, and by lemmas 2.1.8 and 2.1.7, p is open; hence for covering projections corollary 7 is valid without the assumption that X be locally path connected. This is definitely false for fibrations with unique path lifting if X is not locally path connected, because p need not be open. The following example shows this.

8 EXAMPLE Let X be the subspace of $\mathbf{R}^2$ defined to be the union of the four sets

$$\begin{aligned} A_1 &= \{(x,y) \mid x = 0,\ -2 \le y \le 1\} \\ A_2 &= \{(x,y) \mid 0 \le x \le 1,\ y = -2\} \\ A_3 &= \{(x,y) \mid x = 1,\ -2 \le y \le 0\} \\ A_4 &= \{(x,y) \mid 0 < x \le 1,\ y = \sin 2\pi/x\} \end{aligned}$$

illustrated in the diagram

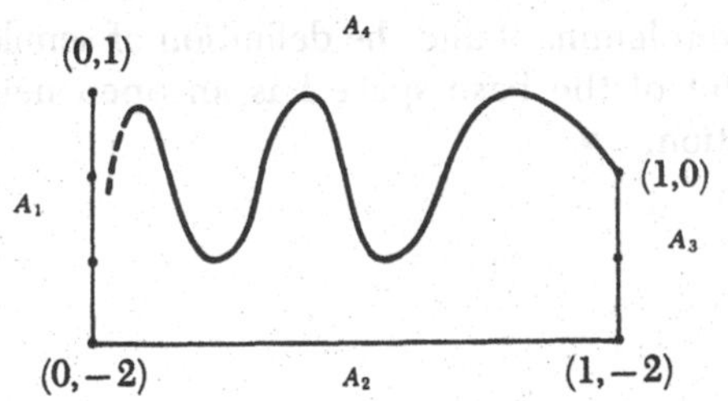

Let $\tilde{X}$ be the half-open interval $[0,4)$ and define $p: \tilde{X} \to X$ to map $[0,1]$ linearly onto A_1, $[1,2]$ linearly onto A_2, $[2,3]$ linearly onto A_3, and $[3,4)$ homeomorphically onto A_4 by the map $t \to (t-3, \sin(2\pi/(t-3)))$. Then $\tilde{X}$ and X are path connected and $p: \tilde{X} \to X$ is a fibration with unique path lifting. However, p is not a homeomorphism, although $\tilde{X}$ and X are both simply connected.

For locally path-connected spaces the lifting theorem provides the following criterion for determining whether an open path-connected subset of the base space is evenly covered by a fibration.

9 LEMMA *Let $p: \tilde{X} \to X$ be a fibration with unique path lifting. Assume that $\tilde{X}$ and X are locally path connected and let U be an open connected subset of X. Then U is evenly covered by p if and only if every lifting to $\tilde{X}$ of a closed path in U is a closed path.*

PROOF If U is evenly covered by p and $\tilde{\omega}$ is a path in $p^{-1}(U)$, then $\tilde{\omega}$ is a path in some component $\tilde{U}$ of $p^{-1}(U)$. By lemma 2.1.12, p maps $\tilde{U}$ homeomorphically onto U. Therefore, if $p \circ \tilde{\omega}$ is a closed path in U, $\tilde{\omega}$ is a closed path in $\tilde{U}$. Hence the condition is necessary.

It is also sufficient, because if $x_0 \in U$ and $\tilde{x}_0 \in p^{-1}(x_0)$, the hypothesis that every lifting of a closed path in U at x_0 is a closed path in $\tilde{X}$ implies that in $\pi(X,x_0)$

$$i_\# \pi(U,x_0) \subset p_\# \pi(\tilde{X},\tilde{x}_0) \qquad \text{where } i: (U,x_0) \subset (X,x_0)$$

By theorem 5, there is a lifting $i'_{\tilde{x}_0}: (U,x_0) \to (\tilde{X},\tilde{x}_0)$ of i. The collection $\{i'_{\tilde{x}_0}(U) \mid \tilde{x}_0 \in p^{-1}(x_0)\}$ consists of path-connected sets which, by lemma 2.2.4, are disjoint. We show that their union equals $p^{-1}(U)$. If $\tilde{x} \in p^{-1}(U)$, let ω be a path in U from $p(\tilde{x})$ to x_0 and let $\tilde{\omega}$ be a lifting of ω such that $\tilde{\omega}(0) = \tilde{x}$. Then $\tilde{\omega}(1) \in p^{-1}(x_0)$, and therefore $\tilde{\omega}$ is a path in $i'_{\tilde{\omega}(1)}(U)$. Hence $\tilde{x} \in i'_{\tilde{\omega}(1)}(U)$ and $\{i'_{\tilde{x}_0}(U) \mid \tilde{x}_0 \in p^{-1}(x_0)\}$ is a partition of $p^{-1}(U)$ into path-connected sets. Since $p^{-1}(U)$ is open and $\tilde{X}$ is locally path connected, $i'_{\tilde{x}_0}(U)$ is open in $\tilde{X}$ for each $\tilde{x}_0 \in p^{-1}(x_0)$. Clearly, p is a homeomorphism of $i'_{\tilde{x}_0}(U)$ onto U for each $\tilde{x}_0 \in p^{-1}(x_0)$, and U is evenly covered by p. ∎

A space X is said to be *semilocally* 1-*connected* if every point $x_0 \in X$ has a neighborhood N such that $\pi(N,x_0) \to \pi(X,x_0)$ is trivial.

10 THEOREM *Every fibration with unique path lifting whose base space is locally path connected and semilocally* 1-*connected and whose total space is locally path connected is a covering projection.*

PROOF It follows from lemma 9 and the definition of semilocally 1-connected space that each point of the base space has an open neighborhood evenly covered by the fibration. ∎

5 THE CLASSIFICATION OF COVERING PROJECTIONS

This section contains a classification of covering projections over a connected locally path-connected base space. It is based on the lifting theorem and reduces the problem of equivalence of covering projections to conjugacy of their corresponding subgroups of the fundamental group of the base space. A large part of the section is devoted to constructing a covering projection corresponding to a given subgroup of the fundamental group of the base space.

Let X be a connected space. The *category of connected covering spaces* of X has objects which are covering projections $p\colon \tilde{X} \to X$, where $\tilde{X}$ is connected, and morphisms which are commutative triangles

$$\begin{array}{ccc} \tilde{X}_1 & \xrightarrow{f} & \tilde{X}_2 \\ & {\scriptstyle p_1}\searrow \quad \swarrow{\scriptstyle p_2} & \\ & X & \end{array}$$

If X is locally path connected and $p\colon \tilde{X} \to X$ is an object of this category, then, by lemma 2.1.8, p is a local homeomorphism and $\tilde{X}$ is also locally path connected. We show that in this case every morphism in this category is a covering projection.

1 LEMMA *In the category of connected covering spaces of a connected locally path-connected space every morphism is itself a covering projection.*

PROOF Consider a commutative triangle

$$\begin{array}{ccc} \tilde{X}_1 & \xrightarrow{f} & \tilde{X}_2 \\ & {\scriptstyle p_1}\searrow \quad \swarrow{\scriptstyle p_2} & \\ & X & \end{array}$$

where p_1 and p_2 are covering projections and X is locally path connected. It follows from corollary 2.1.13 that f is a covering projection if it is surjective.

Because $\tilde{X}_2$ is connected and locally path connected, it is path connected. Let $\tilde{x}_1 \in \tilde{X}_1$ and $\tilde{x}_2 \in \tilde{X}_2$ be arbitrary and let $\tilde{\omega}_2$ be a path in $\tilde{X}_2$ from $f(\tilde{x}_1)$ to $\tilde{x}_2$. Because p_1 is a fibration, there is a path $\tilde{\omega}_1$ in $\tilde{X}_1$ beginning at $\tilde{x}_1$ such that $p_1 \circ \tilde{\omega}_1 = p_2 \circ \tilde{\omega}_2$. By the unique path lifting of p_2, $f \circ \tilde{\omega}_1 = \tilde{\omega}_2$. Therefore

$$f(\tilde{\omega}_1(1)) = \tilde{\omega}_2(1) = \tilde{x}_2$$

proving that f is surjective. ∎

The next result determines when there is a morphism from one object to another in the category of connected covering spaces of X.

2 THEOREM *Let $p_1\colon \tilde{X}_1 \to X$ and $p_2\colon \tilde{X}_2 \to X$ be objects in the category* *of connected covering spaces of a connected locally path-connected space X. The following are equivalent:*

(a) There is a covering projection $f: \tilde{X}_1 \to \tilde{X}_2$ *such that* $p_2 \circ f = p_1$.

(b) For all $\tilde{x}_1 \in \tilde{X}_1$ *and* $\tilde{x}_2 \in \tilde{X}_2$ *such that* $p_1(\tilde{x}_1) = p_2(\tilde{x}_2)$, $p_{1\#}\pi(\tilde{X}_1,\tilde{x}_1)$ *is conjugate in* $\pi(X,p_1(\tilde{x}_1))$ *to a subgroup of* $p_{2\#}\pi(\tilde{X}_2,\tilde{x}_2)$.

(c) There exist $\tilde{x}_1 \in \tilde{X}_1$ *and* $\tilde{x}_2 \in \tilde{X}_2$ *such that* $p_1(\tilde{x}_1) = p_2(\tilde{x}_2)$ *and* $p_{1\#}\pi(\tilde{X}_1,\tilde{x}_1)$ *is conjugate in* $\pi(X,p_1(\tilde{x}_1))$ *to a subgroup of* $p_{2\#}\pi(\tilde{X}_2,\tilde{x}_2)$.

PROOF (a) ⟹ (b) Given $f: \tilde{X}_1 \to \tilde{X}_2$ such that $p_2 \circ f = p_1$, if $\tilde{x}_1 \in \tilde{X}_1$ and $\tilde{x}_2 \in \tilde{X}_2$ are such that $p_1(\tilde{x}_1) = p_2(\tilde{x}_2)$, then

$$p_{1\#}\pi(\tilde{X}_1,\tilde{x}_1) = p_{2\#} \circ f_{\#}\pi(\tilde{X}_1,\tilde{x}_1) \subset p_{2\#}\pi(\tilde{X}_2,f(\tilde{x}_1))$$

Because $f(\tilde{x}_1)$ and $\tilde{x}_2$ lie in the same fiber of $p_2: \tilde{X}_2 \to X$, it follows from theorem 2.3.6 that $p_{2\#}\pi(\tilde{X}_2,f(\tilde{x}_1))$ and $p_{2\#}\pi(\tilde{X}_2,\tilde{x}_2)$ are conjugate in $\pi(X,p_1(\tilde{x}_1))$.

(b) ⟹ (c) The proof is trivial.

(c) ⟹ (a) Assume that $\tilde{x}_1 \in \tilde{X}_1$ and $\tilde{x}_2 \in \tilde{X}_2$ are such that $p_1(\tilde{x}_1) = p_2(\tilde{x}_2)$ and that $p_{1\#}\pi(\tilde{X}_1,\tilde{x}_1)$ is conjugate in $\pi(X,p_1(\tilde{x}_1))$ to a subgroup of $p_{2\#}\pi(\tilde{X}_2,\tilde{x}_2)$. By theorem 2.3.6, there is a point $\tilde{x}_2' \in \tilde{X}_2$ such that $p_2(\tilde{x}_2') = p_2(\tilde{x}_2)$ and such that $p_{1\#}\pi(\tilde{X}_1,\tilde{x}_1) \subset p_{2\#}\pi(\tilde{X}_2,\tilde{x}_2')$

Because $\tilde{X}_1$ is a connected locally path-connected space, the lifting theorem implies the existence of a map $f: (\tilde{X}_1,\tilde{x}_1) \to (\tilde{X}_2,\tilde{x}_2')$ such that $p_2 \circ f = p_1$. ■

3 COROLLARY *Two objects in the category of connected covering spaces of a connected locally path-connected space X are equivalent if and only if their fundamental groups (at some two points over the same point of X) map to conjugate subgroups of the fundamental group of X (at this point).* ■

We give two examples.

4 Because every nontrivial subgroup of $\pi(S^1) \approx \mathbf{Z}$ is infinite cyclic, by corollary 3 every connected covering space $\tilde{X} \to S^1$ is equivalent to $ex: \mathbf{R} \to S^1$ or to the map $S^1 \to S^1$ sending z to z^n for some positive integer n.

5 For $n \geq 2$, $\pi(P^n) \approx \mathbf{Z}_2$, and every connected covering space $\tilde{X} \to P^n$ is equivalent to the double covering $S^n \to P^n$ or to the trivial covering $P^n \subset P^n$.

A *universal covering space* of a connected space X is an object $p: \tilde{X} \to X$ of the category of connected covering spaces of X such that for any object $p': \tilde{X}' \to X$ of this category there is a morphism

$$\begin{array}{ccc} \tilde{X} & \xrightarrow{f} & \tilde{X}' \\ & {\scriptstyle p}\searrow \quad \swarrow{\scriptstyle p'} & \\ & X & \end{array}$$

in the category. It can be shown (see the paragraph following theorem 13 below) that a universal covering space is a regular covering space. The next result follows from this, theorem 2 and corollary 3.

6 COROLLARY *Two universal covering spaces of a connected locally path-connected space are equivalent.* ■

Another result also follows from theorem 2.

7 COROLLARY *A simply connected covering space of a connected locally path-connected space X is a universal covering space of X.* ∎

Having reduced the comparison of connected covering spaces of X to a comparison of their corresponding subgroups of the fundamental group of X, we shall determine which subgroups of the fundamental group correspond to covering spaces. This necessitates the construction of covering spaces. Let X be a space and let $\mathfrak{U}$ be an open covering of X. If $x_0 \in X$, let $\pi(\mathfrak{U},x_0)$ be the subgroup of $\pi(X,x_0)$ generated by homotopy classes of closed paths having a representative of the form $(\omega * \omega') * \omega^{-1}$, where ω' is a closed path lying in some element of $\mathfrak{U}$ and ω is a path from x_0 to $\omega'(0)$. The following statements are easily verified.

8 *If $\mathfrak{V}$ is an open covering of X that refines $\mathfrak{U}$, then $\pi(\mathfrak{V},x_0) \subset \pi(\mathfrak{U},x_0)$.* ∎

9 *$\pi(\mathfrak{U},x_0)$ is a normal subgroup of $\pi(X,x_0)$.* ∎

10 *If ω is a path in X, then $h_{[\omega]}\pi(\mathfrak{U},\omega(1)) = \pi(\mathfrak{U},\omega(0))$.* ∎

The connection of the groups $\pi(\mathfrak{U},x_0)$ with covering projections is explained by the following result.

11 LEMMA *Let $p: \tilde{X} \to X$ be a covering projection and let $\mathfrak{U}$ be a covering of X by open sets each evenly covered by p. For any $\tilde{x}_0 \in \tilde{X}$*

$$\pi(\mathfrak{U},p(\tilde{x}_0)) \subset p_\#\pi(\tilde{X},\tilde{x}_0)$$

PROOF If ω' is a closed path lying in some element of $\mathfrak{U}$, then, by lemma 2.4.9, any lifting of ω' is a closed path in $\tilde{X}$. Hence any path of the form $(\omega * \omega') * \omega^{-1}$, where ω' is a closed path lying in some element of $\mathfrak{U}$, can be lifted to a closed path [namely, to $(\tilde{\omega} * \tilde{\omega}') * \tilde{\omega}^{-1}$, where $\tilde{\omega}$ and $\tilde{\omega}'$ are suitable liftings of ω and ω', respectively]. Hence any element of $\pi(\mathfrak{U},p(\tilde{x}_0))$ has a representative which can be lifted to a closed path at $\tilde{x}_0$. ∎

The following theorem characterizes those fibrations with unique path lifting which are covering projections.

12 THEOREM *Let $p: \tilde{X} \to X$ be a fibration with unique path lifting, where X and $\tilde{X}$ are connected locally path-connected spaces. Then p is a covering projection if and only if there is an open covering $\mathfrak{U}$ of X and a point $\tilde{x}_0 \in \tilde{X}$ such that*

$$\pi(\mathfrak{U},p(\tilde{x}_0)) \subset p_\#\pi(\tilde{X},\tilde{x}_0)$$

PROOF If p is a covering projection, the desired result follows from lemma 11. Conversely, if there is such an open covering $\mathfrak{U}$ and point $\tilde{x}_0 \in \tilde{X}$, it follows from statements 9 and 10 that for any point $\tilde{x}_0' \in \tilde{X}$, $\pi(\mathfrak{U},p(\tilde{x}_0')) \subset p_\#\pi(\tilde{X},\tilde{x}_0')$. Using lemma 2.4.9, it follows that every element of $\mathfrak{U}$ is evenly covered by p. ∎

Lemma 11 gives a necessary condition for a subgroup of $\pi(X,x_0)$ to correspond to a covering space. The next result proves that this necessary condition is also sufficient.

13 THEOREM *Let X be a connected locally path-connected space and let $x_0 \in X$. Let H be a subgroup of $\pi(X,x_0)$ and assume that there is an open covering $\mathcal{U}$ of X such that $\pi(\mathcal{U},x_0) \subset H$. Then there is a covering projection $p\colon (\tilde{X},\tilde{x}_0) \to (X,x_0)$ such that $p_\#\pi(\tilde{X},\tilde{x}_0) = H$.*

PROOF Suppose such a covering projection exists, and suppose, moreover, that the space $\tilde{X}$ is path connected. The projection $\varphi\colon (P(X,x_0),\omega_0) \to (X,x_0)$ of the path space of (X,x_0) can then be lifted to a map $\varphi'\colon (P(X,x_0),\omega_0) \to (\tilde{X},\tilde{x}_0)$, which is surjective. If ω and ω' are elements of $P(X,x_0)$, then $\varphi'(\omega) = \varphi'(\omega')$ if and only if $\varphi(\omega) = \varphi(\omega')$ and $[\omega * \omega'^{-1}] \in p_\#\pi(\tilde{X},\tilde{x}_0) = H$. Therefore, for path connected $\tilde{X}$ there is a one-to-one correspondence between the points of $\tilde{X}$ and equivalence classes of $P(X,x_0)$ identifying ω with ω' if $\omega(1) = \omega'(1)$ and $[\omega * \omega'^{-1}] \in H$ (the group properties of H imply that this is an equivalence relation). Hence it is natural to try to construct $\tilde{X}$ by suitably topologizing these equivalence classes of $P(X,x_0)$. We could start with the compact-open topology on $P(X,x_0)$ and use the quotient topology on the set of equivalence classes, but it seems no simpler than merely topologizing the set of equivalence classes directly, as is done below.

We consider the set of all paths in X beginning at x_0. If ω and ω' are two such paths, set $\omega \sim \omega'$ if $\omega(1) = \omega'(1)$ and $[\omega * \omega'^{-1}] \in H$. This is an equivalence relation, and the equivalence class of ω will be denoted by $\langle\omega\rangle$. Let $\tilde{X}$ be the set of equivalence classes. There is a function $p\colon \tilde{X} \to X$ such that $p(\langle\omega\rangle) = \omega(1)$. If U is an open subset of X and ω is a path beginning at x_0 and ending in U, $\langle\omega,U\rangle$ will denote the subset of $\tilde{X}$ consisting of all the equivalence classes having a representative of the form $\omega * \omega'$, where ω' is a path in U beginning at $\omega(1)$.

We prove that the collection $\{\langle\omega,U\rangle\}$ is a base for a topology on $\tilde{X}$. If $\langle\omega'\rangle \in \langle\omega,U\rangle$, then $\omega' \sim \omega * \omega''$ for some path ω'' lying in U. If $\bar{\omega}$ is any path in U beginning at $\omega'(1)$, then

$$\omega' * \bar{\omega} \sim (\omega * \omega'') * \bar{\omega} \sim \omega * (\omega'' * \bar{\omega})$$

showing that $\langle\omega',U\rangle \subset \langle\omega,U\rangle$. Since $\omega \sim \omega' * \omega''^{-1}$, $\langle\omega\rangle \in \langle\omega',U\rangle$. The same argument shows that $\langle\omega,U\rangle \subset \langle\omega',U\rangle$, and so $\langle\omega,U\rangle = \langle\omega',U\rangle$. Therefore, if $\omega'' \in \langle\omega,U\rangle \cap \langle\omega',U'\rangle$, then $\langle\omega'', U \cap U'\rangle \subset \langle\omega,U\rangle \cap \langle\omega',U'\rangle$, and so the collection $\{\langle\omega,U\rangle\}$ is a base for a topology on $\tilde{X}$.

Let $\tilde{X}$ be topologized by the topology having $\{\langle\omega,U\rangle\}$ as a base. Then p is continuous; for if $p(\langle\omega\rangle) \in U$, then $p(\langle\omega,U\rangle) \subset U$. p is also open, because $p(\langle\omega,U\rangle)$ clearly equals the path component of U containing $\omega(1)$, and this is open because X is locally path connected.

Let $\mathcal{U}$ be an open covering of X such that $\pi(\mathcal{U},x_0) \subset H$ and let V be an open path-connected subset of X contained in some element of $\mathcal{U}$. We show that V is evenly covered by p, which will imply that p is a covering projection.

If $\langle\omega\rangle \in p^{-1}(V)$, then $\langle\omega,V\rangle \subset p^{-1}(V)$. The sets $\{\langle\omega,V\rangle \mid \langle\omega\rangle \in p^{-1}(V)\}$ are open and their union equals $p^{-1}(V)$. If $\langle\omega,V\rangle \cap \langle\omega',V\rangle \neq \emptyset$, let $\langle\omega''\rangle \in \langle\omega,V\rangle \cap \langle\omega',V\rangle$. Then $\langle\omega'',V\rangle = \langle\omega,V\rangle$ and $\langle\omega'',V\rangle = \langle\omega',V\rangle$. Hence the sets $\{\langle\omega,V\rangle \mid \langle\omega\rangle \in p^{-1}(V)\}$ are either identical or disjoint. To prove that V is evenly covered by p, it suffices to show that p maps each set $\langle\omega,V\rangle$ bijectively to V (because p has already been shown to be continuous and open). If $x \in V$, let ω' be a path in V from $\omega(1)$ to x. Then $\langle\omega * \omega'\rangle \in \langle\omega,V\rangle$ and $p(\langle\omega * \omega'\rangle) = x$, showing that p is surjective. Assume $p\langle\omega * \omega_1\rangle = p\langle\omega * \omega_2\rangle$. Then $\omega_1(1) = \omega_2(1)$, so $(\omega * \omega_1) * (\omega * \omega_2)^{-1}$ is a closed path in X at x_0. Also, $[(\omega * \omega_1) * (\omega * \omega_2)^{-1}] = [(\omega * (\omega_1 * \omega_2^{-1})) * \omega^{-1}]$
Since $\omega_1 * \omega_2^{-1}$ is a path in V and V is contained in some element of U, $[(\omega * (\omega_1 * \omega_2^{-1})) * \omega^{-1}] \in \pi(\mathfrak{U},x_0) \subset H$. Therefore $\omega * \omega_1 \sim \omega * \omega_2$ and $\langle\omega * \omega_1\rangle = \langle\omega * \omega_2\rangle$, showing that p is injective.

We have shown that $p\colon \tilde X \to X$ is a covering projection. Let $\tilde x_0 = \langle\omega_0\rangle$, where ω_0 is the constant path in X at x_0. It remains only to verify that $p_\#\pi(\tilde X,\tilde x_0) = H$. For this we need an explicit expression for the lift of a path in X that begins at x_0. Let ω be a path in X beginning at x_0, and for $t \in I$, define a path ω_t in X beginning at x_0 by $\omega_t(t') = \omega(tt')$. Let $\tilde\omega\colon I \to \tilde X$ be defined by $\tilde\omega(t) = \langle\omega_t\rangle$. We prove that $\tilde\omega$ is continuous. If $\tilde\omega(t_0) \in \langle\omega',U\rangle$, then $p\tilde\omega(t_0) = \omega(t_0) \in U$ and $\langle\omega',U\rangle = \langle\omega_{t_0},U\rangle$. Let N be any open interval in I containing t_0 such that $\omega(N) \subset U$. If $t \in N$, then $\omega_t \sim \omega_{t_0} * \omega_{t_0,t}$, where $\omega_{t_0,t}(t') = \omega(t_0 + t'(t - t_0))$. Therefore, for $t \in N$

$$\tilde\omega(t) = \langle\omega_t\rangle = \langle\omega_{t_0} * \omega_{t_0,t}\rangle \in \langle\omega_{t_0},U\rangle = \langle\omega',U\rangle$$

and so $\tilde\omega$ is continuous. Furthermore, $p\tilde\omega(t) = \omega_t(1) = \omega(t)$. Hence $\tilde\omega$ is a lift of ω beginning at $\tilde\omega(0) = \tilde x_0$ and ending at $\tilde\omega(1) = \langle\omega\rangle$.

If $[\omega] \in H$, then $\omega \sim \omega_0$ and $\langle\omega\rangle = \tilde x_0$. Therefore the lift $\tilde\omega$ of ω constructed above is a closed path in $\tilde X$ at $\tilde x_0$, proving that $H \subset p_\#\pi(\tilde X,\tilde x_0)$. On the other hand, if $\tilde\omega'$ is a closed path in $\tilde X$ at $\tilde x_0$ and $p\tilde\omega' = \omega$, let $\tilde\omega$ be the path in $\tilde X$ constructed above. Since $\tilde\omega$ is a lift of ω beginning at $\tilde x_0$, it follows from the unique path lifting of p that $\tilde\omega = \tilde\omega'$. Therefore $\tilde\omega(1) = \tilde\omega'(1) = \tilde x_0$. Since $\tilde\omega(1) = \langle\omega\rangle$, $\omega \sim \omega_0$, showing that $p_\#\pi(\tilde X,\tilde x_0) \subset H$. ∎

Note that if $p\colon \tilde X \to X$ is a universal covering space it is a regular covering.
In fact, if $\tilde x_0 \in \tilde X$ and $\mathscr{U}$ is a covering of X by open sets evenly covered by p than by 2.5.11 $\pi(\mathscr{U}, p(\tilde x_0)) \subset p_\#\pi(\tilde X, \tilde x_0) \subset \pi(X, p(\tilde x_0))$
By 2.5.13 there exists a connected covering $q\colon (\tilde Y, \tilde y) \to (X, p(\tilde x_0))$ such that $q_\#\pi(\tilde Y, \tilde y) = \pi(\mathscr{U}, p(\tilde x_0))$. Since $p\colon \tilde X \to X$ is universal there is a map $f\colon \tilde X \to \tilde Y$ such that $qf = p$. By 2.5.2, $p_\#\pi(\tilde X, \tilde x_0)$ is conjugate in $\pi(X, p(\tilde x_0))$ to a subgroup of $\pi(\mathscr{U}, p(\tilde x_0))$. By 2.5.9, $\pi(\mathscr{U}, p(\tilde x_0))$ is normal so we must have $p_\#\pi(\tilde X, \tilde x_0) \subset \pi(\mathscr{U}, p(\tilde x_0)$ and so $p_\#\pi(\tilde X, \tilde x_0) = \pi(\mathscr{U}, p(\tilde x_0))$ is normal.

A space X is semilocally 1-connected (defined in Sec. 2.4) if and only if there is an open covering $\mathfrak{U}$ of X such that $\pi(\mathfrak{U},x_0) = 0$. Hence we have the following result.

14 COROLLARY *A connected locally path-connected space X has a simply connected covering space if and only if X is semilocally 1-connected.* ∎

From corollaries 14 and 6 and theorem 2 we obtain the next result.

15 COROLLARY *Any universal covering space of a connected locally path-connected semilocally 1-connected space is simply connected.* ∎

Not every connected locally path-connected space has a universal covering space. We give two examples.

16 An infinite product of 1-spheres has no universal covering space.

17 Let X be the subspace of $\mathbf{R}^2$ equal to the union of the circumferences of circles C_n, with $n \geq 1$, where C_n has center at $(1/n, 0)$ and radius $1/n$. Then X is connected and locally path connected but has no universal covering space.

It is possible for a connected locally path-connected space to have a universal covering space that is not simply connected. We present an example.

18 EXAMPLE Let Y_1 be the cone with base X equal to the space of example 17 [Y_1 can be visualized as the set of line segments in $\mathbf{R}^3$ joining the points of X to the point (0,0,1)] and let y_1 be the point at which all the circles of X are tangent. Let (Y_2,y_2) be another copy of (Y_1,y_1). Let $Z = Y_1 \vee Y_2$. Then Z is connected and locally path connected but not simply connected (cf. exercise 1.G.7, a closed path oscillating back and forth from Y_1 to Y_2 around the decreasing circles C_n is not null homotopic). However, Y_1 and Y_2 are each closed contractible subsets of Z. By the lifting theorem, each of them can be lifted to any covering space of Z, so that y_1 is lifted arbitrarily and y_2 is lifted arbitrarily. Therefore any covering projection with base Z has a section. It follows that any connected covering space of Z is homeomorphic to Z.

In the category of fibrations with unique path lifting over a fixed path-connected base space (and with path-connected total spaces) there is always a universal object (that is, an object which has morphisms to any other object in the category). We sketch a proof of this fact. Let X be a path-connected space and let $\mathfrak{X}(X)$ be the collection of topological spaces whose underlying sets are cartesian products of X and the set of right cosets of some subgroup of the fundamental group of X. It follows from theorem 2.3.9 that any fibration whose base space is X and total space is path connected is equivalent to a fibration $\tilde{X} \to X$, where $\tilde{X} \in \mathfrak{X}(X)$. Since $\mathfrak{X}(X)$ is a set, those fibrations $\tilde{X} \to X$ with unique path lifting, where $\tilde{X}$ is a path-connected space in $\mathfrak{X}(X)$, constitute a set. We may form the fibered product of this set (as in Sec. 2.2). Any path component of this fibered product is then the desired universal fibration with unique path lifting.

If X is a connected locally path-connected space, it follows from theorem 13 that for any open covering $\mathfrak{U}$ of X there is a path-connected covering space of X whose fundamental group is isomorphic to $\pi(\mathfrak{U},x_0)$. This implies that if $\tilde{X}$ is a universal object in the category of path-connected fibrations over X with unique path lifting, then $\pi(\tilde{X},\tilde{x}_0)$ is isomorphic to a subgroup of $\bigcap_{\mathfrak{U}} \pi(\mathfrak{U},x_0)$. In particular, if $\bigcap_{\mathfrak{U}} \pi(\mathfrak{U},x_0) = 0$, then X has a simply connected fibration with unique path lifting that is a universal object in the category. Thus the spaces in examples 16 and 17 both have universal fibrations with unique path lifting that are simply connected. The space Z of example 18 is its own universal fibration with unique path lifting.

6 COVERING TRANSFORMATIONS

In this section we consider a problem inverse to the one of the last section, in which we constructed covering projections with given base space; we ask for covering projections with given covering space. On any regular covering space we prove that there is a group of covering transformations. The covering projection is then equivalent to the projection of the covering space onto the space of orbits of the group of covering transformations.

Let $p\colon \tilde{X} \to X$ be a fibration with unique path lifting. It is clear that there is a *group of self-equivalences* of this fibration (a self-equivalence is a homeomorphism $f\colon \tilde{X} \to \tilde{X}$ such that $p \circ f = p$). We denote this group by $G(\tilde{X} \mid X)$. In case $p\colon \tilde{X} \to X$ is a covering projection, $G(\tilde{X} \mid X)$ is also called the *group of covering transformations* of p. In general, there is a close analogy of $G(\tilde{X} \mid X)$ with the group of automorphisms of an extension field leaving a subfield pointwise fixed.

If $\tilde{X}$ is path connected, it follows from lemma 2.2.4 that two self-equivalences of $p\colon \tilde{X} \to X$ that agree at one point are identical. Hence we have the following lemma.

1 LEMMA *Let $p\colon \tilde{X} \to X$ be a fibration with unique path lifting. If $\tilde{X}$ is path connected and $\tilde{x}_0 \in \tilde{X}$, then the function $f \to f(\tilde{x}_0)$ is an injection of $G(\tilde{X} \mid X)$ into the fiber of p over $p(\tilde{x}_0)$.* ■

Theorem 2.3.9 established a bijection from the set of right cosets of $p_\#\pi(\tilde{X},\tilde{x}_0)$ in $\pi(X,p(\tilde{x}_0))$ to the fiber of p over $p(\tilde{x}_0)$. Combining the inverse of this bijection with the function of lemma 1 yields an injection ψ from $G(\tilde{X} \mid X)$ to the set of right cosets of $p_\#\pi(\tilde{X},\tilde{x}_0)$ in $\pi(X,p(\tilde{x}_0))$. ψ is defined explicitly as follows. For any $f \in G(\tilde{X} \mid X)$ let $\tilde{\omega}$ be a path in $\tilde{X}$ from $\tilde{x}_0$ to $f(\tilde{x}_0)$. Then $p \circ \tilde{\omega}$ is a closed path in X at $p(\tilde{x}_0)$, and the right coset $(p_\#\pi(\tilde{X},\tilde{x}_0))\,[p \circ \tilde{\omega}]$ is independent of the choice of $\tilde{\omega}$. The function ψ assigns to f this right coset.

Given $\tilde{x}_0 \in \tilde{X}$, let $N(p_\#\pi(\tilde{X},\tilde{x}_0))$ be the normalizer of $p_\#\pi(\tilde{X},\tilde{x}_0)$ in $\pi(X,p(\tilde{x}_0))$. Thus $N(p_\#\pi(\tilde{X},\tilde{x}_0))$ is the subgroup of $\pi(X,p(\tilde{x}_0))$ consisting of elements $[\omega] \in \pi(X,p(\tilde{x}_0))$ such that $p_\#\pi(\tilde{X},\tilde{x}_0)$ is invariant under conjugation by $[\omega]$. $N(p_\#\pi(\tilde{X},\tilde{x}_0))$ is the largest subgroup of $\pi(X,p(\tilde{x}_0))$ containing $p_\#\pi(\tilde{X},\tilde{x}_0)$ as a normal subgroup.

2 THEOREM *Let $p\colon \tilde{X} \to X$ be a fibration with unique path lifting. Let $\tilde{X}$ be path connected and let $\tilde{x}_0 \in \tilde{X}$. Then ψ is a monomorphism of $G(\tilde{X} \mid X)$ to the quotient group $N(p_\#\pi(\tilde{X},\tilde{x}_0))/p_\#\pi(\tilde{X},\tilde{x}_0)$. If $\tilde{X}$ is also locally path connected, ψ is an isomorphism.*

PROOF We already know that ψ is an injection. We show that ψ is a function from $G(\tilde{X} \mid X)$ to the set of right cosets of $p_\#\pi(\tilde{X},\tilde{x}_0)$ by elements of $N(p_\#\pi(\tilde{X},\tilde{x}_0))$.

If $\tilde\omega$ is a path in $\tilde X$ from $\tilde x_0$ to $f(\tilde x_0)$, there is a commutative square

$$\begin{array}{ccc} \pi(\tilde X,\tilde x_0) & \xleftarrow{h_{[\tilde\omega]}} & \pi(\tilde X,f(\tilde x_0)) \\ \downarrow p_\# & & \downarrow p_\# \\ \pi(X,p(\tilde x_0)) & \xleftarrow{h_{[p\circ\tilde\omega]}} & \pi(X,p(\tilde x_0)) \end{array}$$

Since $f\colon (\tilde X,\tilde x_0) \to (\tilde X,f(\tilde x_0))$ is a homeomorphism,

$$f_\#\pi(\tilde X,\tilde x_0) = \pi(\tilde X,f(\tilde x_0))$$

and since $p_\# f_\# = p_\#$,

$$\begin{aligned} h_{[p\circ\tilde\omega]}p_\#\pi(\tilde X,\tilde x_0) &= h_{[p\circ\tilde\omega]}p_\# f_\#\pi(\tilde X,\tilde x_0) = h_{[p\circ\tilde\omega]}p_\#\pi(\tilde X,f(\tilde x_0)) \\ &= p_\# h_{[\tilde\omega]}\pi(\tilde X,f(\tilde x_0)) = p_\#\pi(\tilde X,f(\tilde x_0)) \end{aligned}$$

Hence $[p\circ\tilde\omega] \in N(p_\#\pi(\tilde X,\tilde x_0))$. Because $\psi(f)$ is equal to the right coset $(p_\#\pi(\tilde X,\tilde x_0))\,[p\circ\tilde\omega]$, ψ is an injection of $G(\tilde X\mid X)$ into the set of right cosets of $p_\#\pi(\tilde X,\tilde x_0)$ by elements of $N(p_\#\pi(\tilde X,\tilde x_0))$.

We now verify that ψ is an homomorphism. If $f_1,f_2 \in G(\tilde X\mid X)$ let $\tilde\omega_1$ and $\tilde\omega_2$ be paths in $\tilde X$ from $\tilde x_0$ to $f_1(\tilde x_0)$ and $f_2(\tilde x_0)$, respectively. Then $f_1\circ\tilde\omega_2$ is a path from $f_1(\tilde x_0)$ to $f_1f_2(\tilde x_0)$, and $\tilde\omega_1 * (f_1\circ\tilde\omega_2)$ is a path from $\tilde x_0$ to $f_1f_2(\tilde x_0)$. Therefore $\psi(f_1f_2)$ is the right coset

$$(p_\#\pi(\tilde X,\tilde x_0))[(p\circ\tilde\omega_1) * (p\circ f_1\circ\tilde\omega_2)] = (p_\#\pi(\tilde X,\tilde x_0))[p\circ\tilde\omega_1] * [p\circ\tilde\omega_2]$$

and this equals $\psi(f_1)\psi(f_2)$.

Finally, we show that if $\tilde X$ is locally path connected, ψ is an epimorphism to the set of right cosets of $p_\#\pi(\tilde X,\tilde x_0)$ in $N(p_\#\pi(\tilde X,\tilde x_0))$. Assume that $[\omega] \in \pi(X,p(\tilde x_0))$ belongs to $N(p_\#\pi(\tilde X,\tilde x_0))$. Let $\tilde\omega$ be a lifting of ω ending at $\tilde x_0$ and let $\tilde x = \tilde\omega(0)$. Then

$$p_\#\pi(\tilde X,\tilde x_0) = h_{[\omega]}(p_\#\pi(\tilde X,\tilde x_0)) = p_\#(h_{[\tilde\omega]}\pi(\tilde X,\tilde x_0)) = p_\#\pi(\tilde X,\tilde x)$$

Because X is connected and locally path connected, the lifting theorem implies the existence of maps $f\colon (\tilde X,\tilde x_0) \to (\tilde X,\tilde x)$ and $g\colon (\tilde X,\tilde x) \to (\tilde X,\tilde x_0)$ such that $p\circ f = p$ and $p\circ g = p$. From the unique-lifting property (lemma 2.2.4), it follows that $f\circ g = 1_{\tilde X}$ and $g\circ f = 1_{\tilde X}$. Therefore $f \in G(\tilde X\mid X)$ and $\psi(f)$ equals the right coset $(p_\#\pi(\tilde X,\tilde x_0))[\omega]^{-1}$. ∎

Combining theorem 2 with theorem 2.3.12, we have the following corollary.

3 COROLLARY *Let $p\colon \tilde X \to X$ be a fibration with unique path lifting. If $\tilde X$ is connected and locally path connected and $\tilde x_0 \in \tilde X$, then p is regular if and only if $G(\tilde X\mid X)$ is transitive on each fiber of p, in which case*

$$\psi\colon G(\tilde X\mid X) \approx \pi(X,p(\tilde x_0))/p_\#\pi(\tilde X,\tilde x_0) \quad ∎$$

If $\tilde{X}$ is simply connected, any fibration $p: \tilde{X} \to X$ is regular, and we also have the next result.

4 COROLLARY *Let $p: \tilde{X} \to X$ be a fibration with unique path lifting, where $\tilde{X}$ is simply connected, locally path connected, and nonempty. Then the group of self-equivalences of p is isomorphic to the fundamental group of X.* ■

If $p: \tilde{X} \to X$ is a regular covering projection and $\tilde{X}$ is connected and locally path connected, then X is homeomorphic to the space of orbits of $G(\tilde{X} \mid X)$ (an *orbit* of a group of transformations G acting on a set S is an equivalence class of S with respect to the equivalence relation $s_1 \sim s_2$ if there is $g \in G$ such that $gs_1 = s_2$). We are interested in the converse problem —that is, in knowing what conditions on a group G of homeomorphisms of a topological space Y will ensure that the projection of Y onto the space of orbits Y/G is a regular covering projection whose group of covering transformations is equal to G.

A group G of homeomorphisms of a topological space Y is said to be *discontinuous* if the orbits of G in Y are discrete subsets of Y. G is *properly discontinuous* if for $y \in Y$ there is an open neighborhood U of y in Y such that if $g, g' \in G$ and gU meets $g'U$, then $g = g'$. G *acts without fixed points* if the only element of G having fixed points is the identity element. The following are clear.

5 *A properly discontinuous group of homeomorphisms is discontinuous and acts without fixed points.* ■

6 *A finite group of homeomorphisms acting without fixed points on a Hausdorff space is properly discontinuous.* ■

If G is the group of covering transformations of a covering projection, then a simple verification shows that G is properly discontinuous. We now show that any properly discontinuous group of homeomorphisms defines a covering projection.

7 THEOREM *Let G be a properly discontinuous group of homeomorphisms of a space Y. Then the projection of Y to the orbit space Y/G is a covering projection. If Y is connected, this covering projection is regular and G is its group of covering transformations.*

PROOF Let $p: Y \to Y/G$ be the projection. Then p is continuous. It is an open map, for if U is an open set in Y, then $p^{-1}(p(U)) = \cup \{gU \mid g \in G\}$ is open in Y, and therefore pU is open in Y/G. Let U be an open subset of Y such that whenever gU meets $g'U$, then $g = g'$. We show that $p(U)$ is evenly covered by p. The hypothesis on U ensures that $\{gU \mid g \in G\}$ is a disjoint collection of open sets whose union is $p^{-1}(p(U))$. It suffices to prove that $p \mid gU$ is a bijection from gU to $p(U)$. If $y \in U$, then $p(gy) = p(y)$, so $p(gU) = p(U)$. If $p(gy_1) = p(gy_2)$, with $y_1, y_2 \in U$, there is $g' \in G$ such that $gy_1 = g'gy_2$.

Therefore gU meets $g'gU$, and $g = g'g$. Hence $g' = 1_Y$ and $gy_1 = gy_2$. We have proved that p is a homeomorphism of gU onto $p(U)$. Since G is properly discontinuous, the sets $p(U)$ evenly covered by p constitute an open covering of Y/G.

Because $p(gy) = p(y)$, we see that G is contained in the group of covering transformations of p. Since G is transitive on the fibers of p, it follows from theorem 2.2.2 that if Y is connected, G equals the group of covering transformations. Since the group of covering transformations is transitive on each fiber, the covering projection is regular. ∎

8 COROLLARY *Let G be a properly discontinuous group of homeomorphisms of a simply connected space Y. Then the fundamental group of the orbit space Y/G is isomorphic to G.*

PROOF By theorem 7, G is the group of covering transformations of the regular covering projection $p\colon Y \to Y/G$. By theorem 2, ψ is a monomorphism of G into the fundamental group of Y/G. Because G is transitive on the fibers of p, ψ is an isomorphism. ∎

9 EXAMPLE Let $S^3 = \{(z_0,z_1) \in \mathbf{C}^2 \mid |z_0|^2 + |z_1|^2 = 1\}$ and let p and q be relatively prime integers. Define $h\colon S^3 \to S^3$ by

$$h(z_0,z_1) = (e^{2\pi i/p}z_0, e^{2\pi qi/p}z_1)$$

Then h is a homeomorphism of S^3 with period p (that is, $h^p = 1$), and $\mathbf{Z}_p$ acts on S^3 by

$$n(z_0,z_1) = h^n(z_0,z_1)$$

where n denotes the residue class of the integer n modulo p. In this way $\mathbf{Z}_p$ acts without fixed points on S^3. The orbit space of this action of $\mathbf{Z}_p$ on S^3 is called a *lens space* and is denoted by $L(p,q)$. By statement 6 and corollary 8, the fundamental group of $L(p,q)$ is isomorphic to $\mathbf{Z}_p$.

10 EXAMPLE Let $S^{2n+1} = \{(z_0,z_1, \ldots ,z_n) \in \mathbf{C}^{n+1} \mid \Sigma |z_i|^2 = 1\}$ and let $q_1, \ldots ,q_n$ be integers relatively prime to p. Define $h\colon S^{2n+1} \to S^{2n+1}$ by

$$h(z_0,z_1, \ldots ,z_n) = (e^{2\pi i/p}z_0, e^{2\pi i q_1/p}z_1, \ldots , e^{2\pi i q_n/p}z_n)$$

Then, as in example 9, h determines an action of $\mathbf{Z}_p$ on S^{2n+1} without fixed points; the orbit space is called a *generalized lens space* and is denoted by $L(p,q_1, \ldots ,q_n)$. Its fundamental group is isomorphic to $\mathbf{Z}_p$.

It is possible to use theorem 7 to show that the projection $Y \to Y/G$ is a regular fibration with unique path lifting even when it may not be a covering projection. Note that if G acts on Y without fixed points, so does any subgroup of G, and if G' is a normal subgroup of Y, then G/G' acts without fixed points on Y/G'.

11 THEOREM *Let G be a group of homeomorphisms acting without fixed points on a path-connected space Y and assume that there is a decreasing sequence of subgroups*

$$G = G_0 \supset G_1 \supset \cdots \supset G_n \supset G_{n+1} \supset \cdots$$

such that

(*a*) $\bigcap G_n = \{1_Y\}$
(*b*) G_{n+1} *is a normal subgroup of* G_n *for* $n \geq 0$
(*c*) G_n/G_{n+1} *is a properly discontinuous group of homeomorphisms on* Y/G_{n+1} *and the projection* $Y \to Y/G_n$ *is a closed map for* $n \geq 0$
(*d*) *Any orbit of* Y *under* G_n *for* $n \geq 0$ *is compact*

Then the projection $p: Y \to Y/G$ *is a regular fibration with unique path lifting whose group of self-equivalences is* G.

PROOF Since $Y/G_n = (Y/G_{n+1})/(G_n/G_{n+1})$, it follows from (*c*) and theorem 7 that the projection

$$p_{n+1}: Y/G_{n+1} \to Y/G_n$$

is a regular covering projection for $n \geq 0$. Let

$$\tilde{Y} = \{(y_n) \in \times (Y/G_n) \mid p_{n+1}(y_{n+1}) = y_n \text{ for } n \geq 0\}$$

and define $\tilde{p}: \tilde{Y} \to Y/G$ by $\tilde{p}((y_n)) = y_0$. It is easy to verify that $\tilde{p}$ is a fibration with unique path lifting (it is the fibered product of the maps $\{p_1 \circ \cdots \circ p_j\}$).

For $n \geq 0$ there is a continuous closed projection map $\varphi_n: Y \to Y/G_n$ such that $p_{n+1} \circ \varphi_{n+1} = \varphi_n$. Therefore there is a continuous closed map $\varphi: Y \to \tilde{Y}$ defined by $\varphi(y) = (\varphi_n(y))$ and such that $\tilde{p} \circ \varphi = p$. To prove that φ is a homeomorphism, it suffices to show that it is a bijection. If $\varphi(y) = \varphi(y')$, then for $n \geq 0$ there is $g_n \in G_n$ such that $y = g_n y'$. Then $g_n y' = g_m y'$ for all m and n, and because G acts without fixed points, $g_m = g_n$ for all m and n. Therefore $g_n \in G_m$ for all m, and by (*a*), $g_n = 1_Y$. It follows that $y = y'$, and hence that φ is injective.

If $(y_n) \in \tilde{Y}$, then $\varphi_n^{-1} y_n$ is an orbit of Y under G_n. By (*d*), $\varphi_n^{-1} y_n$ is compact. Since

$$\varphi_n^{-1} y_n = \varphi_{n+1}^{-1} p_{n+1}^{-1} y_n \supset \varphi_{n+1}^{-1} y_{n+1}$$

the collection $\{\varphi_n^{-1} y_n\}$ consists of compact sets having the finite-intersection property. Therefore $\bigcap \varphi_n^{-1} y_n \neq \emptyset$. If $y \in \bigcap \varphi_n^{-1} y_n$, then $\varphi(y) = (y_n)$, showing that φ is surjective.

We have shown that $\varphi: Y \to \tilde{Y}$ is a homeomorphism. Therefore $p: Y \to Y/G$ is a fibration with unique path lifting. Since each element of G is a self-equivalence of p, the group of self-equivalences of p is transitive on each fiber. By corollary 3, p is a regular fibration and G is the group of self-equivalences of p. ∎

7 FIBER BUNDLES

A covering space is locally the product of its base space and a discrete space. This is generalized by the concept of fiber bundle, defined in this section, because the total space of a fiber bundle is locally the product of its base space and its fiber. The main result is that the bundle projection of a fiber bundle is a fibration.[1]

A *fiber bundle* $\xi = (E,B,F,p)$ consists of a *total space* E, a *base space* B, a *fiber* F, and a *bundle projection* $p: E \to B$ such that there exists an open covering $\{U\}$ of B and, for each $U \in \{U\}$, a homeomorphism $\varphi_U: U \times F \to p^{-1}(U)$ such that the composite

$$U \times F \xrightarrow{\varphi_U} p^{-1}(U) \xrightarrow{p} U$$

is the projection to the first factor. Thus the bundle projection $p: E \to B$ and the projection $B \times F \to B$ are locally equivalent. The *fiber over* $b \in B$ is defined to equal $p^{-1}(b)$, and we note that F is homeomorphic to $p^{-1}(b)$ for every $b \in B$. Usually there is also given a structure group G for the bundle consisting of homeomorphisms of F, and we define this concept next.

Let G be a group of homeomorphisms of F. Given a space F' and a collection $\Phi = \{\varphi\}$ of homeomorphisms $\varphi: F \to F'$, define $\varphi g: F \to F'$ for $\varphi \in \Phi$ and $g \in G$ by $\varphi g(y) = \varphi(gy)$ for $y \in F$. The collection Φ is called a G *structure* on F' if

(*a*) Given $\varphi \in \Phi$ and $g \in G$, then $\varphi g \in \Phi$
(*b*) Given $\varphi_1, \varphi_2 \in \Phi$, there is $g \in G$ such that $\varphi_1 = \varphi_2 g$

Condition (*a*) implies that G acts on the right on Φ, and condition (*b*) implies that this action of G is transitive on Φ. A fiber bundle (E,B,F,p) is said to have *structure group* G if each fiber $p^{-1}(b)$ has a G structure $\Phi(b)$ such that there exists an open covering $\{U\}$ of B and, for each $U \in \{U\}$, a homeomorphism $\varphi_U: U \times F \to p^{-1}(U)$ such that for $b \in U$, the map $F \to p^{-1}(b)$ sending x to $\varphi_U(b,x)$ is in $\Phi(b)$. It is clear that a given fiber bundle can always be given the structure of a fiber bundle with structure group the group of all homeomorphisms of F. It is also clear that a given fiber bundle can sometimes be given the structure of a fiber bundle with two different structure groups of homeomorphisms of F.

An *n-plane bundle*, or *real vector bundle*, is a fiber bundle whose fiber is $\mathbf{R}^n$ and whose structure group is the general linear group $GL(\mathbf{R}^n)$, which consists of all linear automorphisms of $\mathbf{R}^n$. A *complex n-plane bundle*, or *complex vector bundle*, is a fiber bundle whose fiber is $\mathbf{C}^n$ and whose structure group is $GL(\mathbf{C}^n)$.

We give some examples.

1 For spaces B and F the *product bundle* is the fiber bundle $(B \times F, B, F, p)$, where $p: B \times F \to B$ is projection to the first factor (it has the trivial group as structure group).

2 Given that $p: \tilde{X} \to X$ is a covering projection and X is a connected and locally path connected space, if $x_0 \in X$, then $(\tilde{X},X,p^{-1}(x_0),p)$ is a fiber bundle (and if X is path connected, it can be given the structure of a fiber bundle with

[1] For the general theory of fiber bundles see N. E. Steenrod, *The Topology of Fibre Bundles*, Princeton University Press, Princeton, N.J., 1951.

structure group $\pi(X,x_0)$, where $\pi(X,x_0)$ acts on $p^{-1}(x_0)$ by $[\omega]\tilde{x} = \tilde{x}[\omega]^{-1}$, with the right-hand side as in the proof of theorem 2.3.9).

3 Given that M is a differentiable n-manifold and $T(M)$ is the set of all tangent vectors to M, there is a fiber bundle $(T(M),M,\mathbf{R}^n,p)$, where $p\colon T(M) \to M$ assigns to each tangent vector its origin. This is called the *tangent bundle* and is denoted by $\tau(M)$. Because it can be given the structure group $GL(\mathbf{R}^n)$, it is an n-plane bundle, and if M is a complex manifold of complex dimension m, then $\tau(M)$ is a complex m-plane bundle.

4 Given that H is a closed subgroup of a Lie group G and that G/H is the quotient space of left cosets and $p\colon G \to G/H$ the projection, then $(G,G/H,H,p)$ is a fiber bundle (having structure group H acting on itself by left translation).

5 Represent S^n as the union of closed hemispheres E_-^n and E_+^n with intersection S^{n-1} and let G be a group of homeomorphisms of a space F. Given a map $\varphi\colon S^{n-1} \to G$ such that the map $S^{n-1} \times F \to F$ sending (x,y) to $\varphi(x)y$ is continuous, let E_φ be the space obtained from $(E_-^n \times F) \vee (E_+^n \times F)$ by identifying $(x,y) \in E_-^n \times F$ with $(x,\varphi(x)y) \in E_+^n \times F$ for $x \in S^{n-1}$ and $y \in F$. These identifications are compatible with the projections $E_-^n \times F \to E_-^n$ and $E_+^n \times F \to E_+^n$. Therefore there is a map $p_\varphi\colon E_\varphi \to S^n$ such that each of the composites

$$E_-^n \times F \to E_\varphi \xrightarrow{p_\varphi} S^n \qquad \text{and} \qquad E_+^n \times F \to E_\varphi \xrightarrow{p_\varphi} S^n$$

is projection to the first factor. Then $(E_\varphi,S^n,F,p_\varphi)$ is a fiber bundle (having structure group G) which is said to be defined by the *characteristic map* φ.

6 Let $P_n(\mathbf{C})$ be the n-dimensional complex projective space coordinatized by homogeneous coordinates. If $z_0, z_1, \ldots, z_n \in \mathbf{C}$ are not all zero, let $[z_0,z_1, \ldots ,z_n] \in P_n(\mathbf{C})$ be that point of $P_n(\mathbf{C})$ having homogeneous coordinates $z_0, z_1, \ldots, z_n$. Regard S^{2n+1} as the set $\{(z_0,z_1, \ldots ,z_n) \in \mathbf{C}^{n+1} \mid \Sigma |z_i|^2 = 1\}$ and define $p\colon S^{2n+1} \to P_n(\mathbf{C})$ by $p(z_0,z_1, \ldots ,z_n) = [z_0,z_1, \ldots ,z_n]$. If $U_i \subset P_n(\mathbf{C})$ is the subset of points having a nonzero ith homogeneous coordinate, it is easy to see that $p^{-1}(U_i)$ is homeomorphic to $U_i \times S^1$. Therefore there is a fiber bundle $(S^{2n+1},P_n(\mathbf{C}),S^1,p)$ (having structure group S^1 acting on itself by left translation), and this is called the *Hopf bundle.*

7 If $\mathbf{Q}$ is the division ring of quaternions, there is a similar map $p\colon S^{4n+3} \to P_n(\mathbf{Q})$ and a *quaternionic Hopf bundle* $(S^{4n+3},P_n(\mathbf{Q}),S^3,p)$ (having structure group S^3 acting on itself by left translation).

The structure group will not be important for our purposes. Thus we define an *n-sphere bundle* to be a fiber bundle whose fiber is S^n [usually it is also required that it have as structure group the *orthogonal group* $O(n + 1)$ of all isometries in $GL(\mathbf{R}^{n+1})$]. If ξ is an n-sphere bundle, we shall denote its total space by $\dot{E}_\xi$. The mapping cylinder of the bundle projection $\dot{E}_\xi \to B$ is the total space E_ξ of a fiber bundle (E_ξ,B,E^{n+1},p_ξ), where $p_\xi\colon E_\xi \to B$ is the retraction of the mapping cylinder to B (and $p_\xi \mid \dot{E}_\xi\colon \dot{E}_\xi \to B$ is the original bundle projection).

If $\xi = (E,B,\mathbf{R}^{n+1},p)$ is an $(n+1)$-plane bundle having structure group $0(n+1)$, it is possible to introduce a norm in each fiber $p^{-1}(b)$. The subset $E' \subset E$ of all elements in E having unit norm is the total space of an n-sphere bundle $(E', B, S^n, p \mid E')$ called the *unit n-sphere bundle* of ξ. If the base space B of an $(n+1)$-plane bundle is a paracompact Hausdorff space, the bundle can always be given $0(n+1)$ as structure group. In particular, there is a *unit tangent bundle* of a paracompact differentiable manifold.

Two fiber bundles (E_1,B,F,p_1) and (E_2,B,F,p_2) with the same fiber and same base are said to be *equivalent* if there is a homeomorphism $h: E_1 \to E_2$ such that $p_2 \circ h = p_1$. If they both have structure group G, they are *equivalent over* G if there is a homeomorphism h as above, with the additional property that if $\varphi \in \Phi_1(b)$, then $h \circ \varphi \in \Phi_2(b)$ for $b \in B$. A fiber bundle is said to be *trivial* if it is equivalent to the product bundle of example 1 (or, equivalently, if it can be given the trivial group as structure group).

In view of example 2, fiber bundles are related to covering spaces in much the same way that fibrations are related to fibrations with unique path lifting. The rest of this section is devoted to a proof of the fact that in a fiber bundle (E,B,F,p) whose base space B is a paracompact Hausdorff space the map p is a fibration.

A map $p: E \to B$ is called a *local fibration* if there is an open covering $\{U\}$ of B such that $p \mid p^{-1}(U): p^{-1}(U) \to U$ is a fibration for every $U \in \{U\}$. It is clear that a fibration is a local fibration[1] and that any bundle projection is a local fibration.

Given a map $p: E \to B$, we define a subspace $\bar{B} \subset E \times B^I$ by

$$\bar{B} = \{(e,\omega) \in E \times B^I \mid \omega(0) = p(e)\}$$

There is a map $\bar{p}: E^I \to \bar{B}$ defined by $\bar{p}(\bar{\omega}) = (\bar{\omega}(0), p \circ \bar{\omega})$ for $\bar{\omega}: I \to E$. A *lifting function* for p is a map

$$\lambda: \bar{B} \to E^I$$

which is a right inverse of $\bar{p}$. Thus a lifting function assigns to each point $e \in E$ and path ω in B starting at $p(e)$ a path $\lambda(e,\omega)$ in E starting at e that is a lift of ω. The relation between lifting functions and fibrations is contained in the following theorem.

8 THEOREM *A map $p: E \to B$ is a fibration if and only if there exists a lifting function for p.*

PROOF The proof involves repeated use of theorem 2.8 in the Introduction. If p is a fibration, let $f': \bar{B} \to E$ and $F: \bar{B} \times I \to B$ be defined by $f'(e,\omega) = e$

[1] Our proof of the converse for paracompact Hausdorff spaces B can be found in W. Hurewicz, On the concept of fibre space, *Proceedings of the National Academy of Sciences, U.S.A.*, vol. 41, pp. 956–961 (1955). Another proof can be found in W. Huebsch, On the covering homotopy theorem, *Annals of Mathematics*, vol. 61, pp. 555–563 (1955). Generalizations and related questions are treated in A. Dold, Partitions of unity in the theory of fibrations, *Annals of Mathematics*, vol. 78, pp. 223–255 (1963).

and $F((e,\omega), t) = \omega(t)$. Then

$$F((e,\omega), 0) = \omega(0) = p(e) = (p \circ f')(e,\omega)$$

By the homotopy lifting property of p, there is a map F': $\bar{B} \times I \to E$ such that $F'((e,\omega), 0) = f'(e,\omega) = e$ and $p \circ F' = F$. F' defines a lifting function λ for p by $\lambda(e,\omega)(t) = F'((e,\omega), t)$.

Conversely, if λ is a lifting function for p, let f': $X \to E$ and F: $X \times I \to B$ be such that $F(x,0) = pf'(x)$. Let g: $X \to B^I$ be defined by $g(x)(t) = F(x,t)$. There is a map F': $X \times I \to E$ such that $F'(x,t) = \lambda(f'(x),g(x))(t)$. Because $F'(x,0) = f'(x)$ and $p \circ F' = F$, p has the homotopy lifting property. ∎

Let p: $E \to B$ and let W be a subset of B^I. Let $\tilde{W}$ be defined by

$$\tilde{W} = \{(e,\omega,s) \in E \times W \times I \mid \omega(s) = p(e)\}$$

An *extended lifting function* over W is a map

$$\Lambda: \tilde{W} \to E^I$$

such that $p(\Lambda(e,\omega,s)(t)) = \omega(t)$ and $\Lambda(e,\omega,s)(s) = e$. Thus an extended lifting function is a function which lifts paths to paths that pass through a given point of E at a given parameter value. It is reasonable to expect the following relation between the existence of lifting functions and extended lifting functions.

9 LEMMA *A map p: $E \to B$ has a lifting function if and only if there is an extended lifting function over B^I.*

PROOF If Λ is an extended lifting function over B^I, a lifting function λ for p is defined by $\lambda(e,\omega) = \Lambda(e,\omega,0)$.

To prove the converse, given a path ω in B, let ω_s and ω^s be the paths in B defined by

$$\omega_s(t) = \begin{cases} \omega(s-t) & 0 \le t \le s \\ \omega(0) & s \le t \le 1 \end{cases}$$

$$\omega^s(t) = \begin{cases} \omega(s+t) & 0 \le t \le 1-s \\ \omega(1) & 1-s \le t \le 1 \end{cases}$$

The maps $(\omega,s) \to \omega_s$ and $(\omega,s) \to \omega^s$ are continuous maps $B^I \times I \to B^I$. Given a lifting function λ: $\bar{B} \to E^I$ for p, we define an extended lifting function Λ over B^I by

$$\Lambda(e,\omega,s)(t) = \begin{cases} \lambda(e,\omega_s)(s-t) & 0 \le t \le s \\ \lambda(e,\omega^s)(t-s) & s \le t \le 1 \end{cases} \quad ∎$$

The main step in proving that a local fibration is a fibration is the fitting together of extended lifting functions over various open subsets of B^I. For this we need an additional concept. A covering $\{W\}$ of a space X is said to be *numerable* if it is locally finite and if for each W there is a function f_W: $X \to [0,1]$ such that $W = \{x \in X \mid f_W(x) \ne 0\}$.

10 LEMMA   *Let $p\colon E \to B$ be a map. If there is a numerable covering $\{W_j\}$ of B^I such that for each j there is an extended lifting function over W_j, then there is a lifting function for p.*

PROOF Let the indexing set be $J = \{j\}$ and for each j let $f_j\colon B^I \to I$ be a map such that $W_j = \{\omega \in B^I \mid f_j(\omega) \neq 0\}$. For any subset $\alpha \subset J$ let $W_\alpha = \bigcup_{j \in \alpha} W_j$ and define $f_\alpha\colon B^I \to \mathbf{R}$ by

$$f_\alpha(\omega) = \Sigma_{j \in \alpha} f_j(\omega)$$

(this is a finite sum and is continuous because $\{W_j\}$ is locally finite). Then $f_\alpha(\omega) \geq 0$ for $\omega \in B^I$ and

$$W_\alpha = \{\omega \in B^I \mid f_\alpha(\omega) \neq 0\}$$

We define $\bar{B}_\alpha = \{(e,\omega) \in \bar{B} \mid \omega \in W_\alpha\}$.

Consider the set of pairs (α, λ_α), where $\alpha \subset J$ and $\lambda_\alpha\colon \bar{B}_\alpha \to E^I$ is a lifting function over $\bar{B}_\alpha$ [that is, $\lambda_\alpha(e,\omega)(0) = e$ and $p\lambda_\alpha(e,\omega)(t) = \omega(t)$]. We define a partial order $\leq$ in this set by $(\alpha, \lambda_\alpha) \leq (\beta, \lambda_\beta)$ if $\alpha \subset \beta$ and $\lambda_\alpha(e,\omega) = \lambda_\beta(e,\omega)$ whenever $(e,\omega) \in \bar{B}_\alpha$ and $f_\alpha(\omega) = f_\beta(\omega)$ [so if $(e,\omega) \in \bar{B}_\alpha$ and $\lambda_\alpha(e,\omega) \neq \lambda_\beta(e,\omega)$, then $\omega \in W_j$ for some $j \in \beta - \alpha$].

To prove that every simply ordered subset $\{\alpha_i, \lambda_{\alpha_i}\}$ has an upper bound, let $\beta = \bigcup \alpha_i$. We shall define $\lambda_\beta\colon \bar{B}_\beta \to E^I$ so that $(\alpha_i, \lambda_{\alpha_i}) \leq (\beta, \lambda_\beta)$ for all i. Let U be any open subset of W_β meeting only finitely many W_j with $j \in \beta$, say $W_{j_1}, \ldots, W_{j_r}$ (W_β can be covered by such sets U). Choose i so that $j_1, \ldots, j_r$ all belong to α_i. Then if $\alpha_i \subset \alpha_k$, $f_{\alpha_i} \mid U = f_{\alpha_k} \mid U$. Because $(\alpha_i, \lambda_{\alpha_i}) \leq (\alpha_k, \lambda_{\alpha_k})$, it follows that $\lambda_{\alpha_i}(e,\omega) = \lambda_{\alpha_k}(e,\omega)$ for $(e,\omega) \in \bar{B}_{\alpha_i}$, with $\omega \in U$. Therefore there exists a map $\lambda_\beta\colon \bar{B}_\beta \to E^I$ such that $\lambda_\beta(e,\omega) = \lambda_{\alpha_i}(e,\omega)$ for α_i sufficiently large. We now show that $(\alpha_i, \lambda_{\alpha_i}) \leq (\beta, \lambda_\beta)$. If $(e,\omega) \in \bar{B}_{\alpha_i}$ and $\lambda_{\alpha_i}(e,\omega) \neq \lambda_\beta(e,\omega)$, there exists α_k such that $(\alpha_i, \lambda_{\alpha_i}) \leq (\alpha_k, \lambda_{\alpha_k})$ and $\lambda_{\alpha_i}(e,\omega) \neq \lambda_{\alpha_k}(e,\omega)$. This implies $\omega \in W_j$ for some $j \in \alpha_k - \alpha_i$. Therefore $\omega \in W_j$ for some $j \in \beta - \alpha_i$, hence $(\alpha_i, \lambda_{\alpha_i}) \leq (\beta, \lambda_\beta)$.

By Zorn's lemma, there is a maximal element (α, λ_α). To complete the proof we need only show that $\alpha = J$. If $\alpha \neq J$, let $j_0 \in J - \alpha$ and let $\beta = \alpha \cup \{j_0\}$. Define $g\colon W_\beta \to \mathbf{R}$ by $g(\omega) = f_\alpha(\omega)/f_\beta(\omega)$. Then $0 \leq g(\omega) \leq 1$, $g(\omega) \neq 0 \Leftrightarrow \omega \in W_\alpha$, and $g(\omega) \neq 1 \Leftrightarrow \omega \in W_{j_0}$. Define $\mu\colon \bar{B}_{j_0} \to E$ by

$$\mu(e,\omega) = \begin{cases} e & 0 \leq g(\omega) \leq \tfrac{1}{3} \\ \lambda_\alpha(e,\omega)(2g(\omega) - \tfrac{2}{3}) & \tfrac{1}{3} \leq g(\omega) \leq \tfrac{2}{3} \\ \lambda_\alpha(e,\omega)(g(\omega)) & \tfrac{2}{3} \leq g(\omega) \leq 1 \end{cases}$$

Then μ is continuous. Let Λ be an extended lifting function over W_{j_0} and define $\lambda_\beta\colon \bar{B}_\beta \to E^I$ by

$$\lambda_\beta(e,\omega)(t) = \begin{cases} \Lambda(e,\omega,0)(t) & & 0 \leq g(\omega) \leq \tfrac{1}{3} \\[4pt] \left. \begin{array}{ll} \lambda_\alpha(e,\omega)(t) & 0 \leq t \leq 2g(\omega) - \tfrac{2}{3} \\ \Lambda(\mu(e,\omega), \omega, 2g(\omega) - \tfrac{2}{3})(t) & 2g(\omega) - \tfrac{2}{3} \leq t \leq 1 \end{array} \right\} & \tfrac{1}{3} \leq g(\omega) \leq \tfrac{2}{3} \\[8pt] \left. \begin{array}{ll} \lambda_\alpha(e,\omega)(t) & 0 \leq t \leq g(\omega) \\ \Lambda(\mu(e,\omega), \omega, g(\omega))(t) & g(\omega) \leq t \leq 1 \end{array} \right\} & \tfrac{2}{3} \leq g(\omega) \leq 1 \end{cases}$$

Then λ_β is a well-defined lifting function over W_β. Moreover, for $(e,\omega) \in \bar{B}_\alpha$, if $\lambda_\alpha(e,\omega) \neq \lambda_\beta(e,\omega)$, then $g(\omega) \neq 1$ and $\omega \in W_{j_0}$. Since $j_0 \in \beta - \alpha$, this means that $(\alpha,\lambda_\alpha) < (\beta,\lambda_\beta)$, contradicting the maximality of (α,λ_α). ∎

In case p has unique path lifting, lemma 10 would hold for any open covering $\{W_j\}$ of B^I such that there is a lifting function over W_j for each j (because the uniqueness of lifted paths enables the extended liftings to be amalgamated to a lifting for p). This was used in the proof of the theorem that a covering projection is a fibration (theorem 2.2.3), which was valid without any assumption on the base space.

11 LEMMA *Given a map $p: E \to B$ and subsets $U_1, \ldots, U_k$ of B such that there is an extended lifting function over $U_1{}^I, U_2{}^I, \ldots, U_k{}^I$, let W be the subset of B^I defined by*

$$W = \left\{\omega \in B^I \mid \omega\left(\left[\frac{i-1}{k}, \frac{i}{k}\right]\right) \subset U_i \text{ for } i = 1, \ldots, k\right\}$$

Then there is an extended lifting function over W.

PROOF Let Λ_i be an extended lifting function over $U_i{}^I$ for $i = 1, \ldots, k$. Given a path $\omega \in W$, let ω_i be the path equal to ω on $[(i-1)/k, i/k]$ and constant on $[0, (i-1)/k]$ and on $[i/k, 1]$. Given $(e,\omega,s) \in \tilde{W}$ such that $(n-1)/k \leq s \leq n/k$, define $e_i \in E$ for $i = 0, \ldots, k$ inductively so that

$$e_{n-1} = \Lambda_n(e,\omega_n,s)\left(\frac{n-1}{k}\right)$$

$$e_n = \Lambda_n(e,\omega_n,s)\left(\frac{n}{k}\right)$$

and

$$e_{i-1} = \Lambda_i\left(e_i,\omega_i,\frac{i}{k}\right)\left(\frac{i-1}{k}\right) \qquad 0 < i < n-1$$

$$e_{i+1} = \Lambda_{i+1}\left(e_i,\omega_{i+1},\frac{i}{k}\right)\left(\frac{i+1}{k}\right) \qquad n < i+1 \leq k$$

An extended lifting function Λ over W is defined by

$$\Lambda(e,\omega,s)(t) = \begin{cases} \Lambda_i\left(e_i,\omega_i,\frac{i}{k}\right)(t) & \frac{i-1}{k} \leq t \leq \frac{i}{k} \leq \frac{n-1}{k} \\ \Lambda_n(e,\omega_n,s)(t) & \frac{n-1}{k} \leq t \leq \frac{n}{k} \\ \Lambda_{i+1}\left(e_i,\omega_{i+1},\frac{i}{k}\right)(t) & \frac{n}{k} \leq \frac{i}{k} \leq t \leq \frac{i+1}{k} \end{cases}$$

We are now ready for the main result on the passage from a local fibration to a fibration.

12 THEOREM *Given a map $p: E \to B$ and a numerable covering $\mathcal{U}$ of B such that for $U \in \mathcal{U}$, $p \mid p^{-1}(U): p^{-1}(U) \to U$ is a fibration, then p is a fibration.*

PROOF Let $\mathcal{U} = \{U_j\}$ and for $k \geq 1$, given a set of indices $j_1, \ldots, j_k$, let $W_{j_1 j_2 \ldots j_k}$ be the subset of B^I defined by

$$W_{j_1 j_2 \ldots j_k} = \left\{\omega \in B^I \mid \omega\left(\left[\frac{i-1}{k}, \frac{i}{k}\right]\right) \subset U_{j_i}, i = 1, \ldots, k\right\}$$

It is then clear that the collection $\{W_{j_1 j_2 \ldots j_k}\}$ (with k varying) is an open covering of B^I, and by lemma 11, each set $W_{j_1 j_2 \ldots j_k}$ has an extended lifting function. For k fixed the collection $\{W_{j_1 j_2 \ldots j_k}\}$ is locally finite. In fact, if $\omega \in B^I$, for each $i = 1, \ldots, k$ there is a neighborhood V_i of $\omega([(i-1)/k, i/k])$ meeting only finitely many U_j. Then $\bigcap_{1 \le i \le k} \langle [(i-1)/k, i/k], V_i \rangle$ is a neighborhood of ω meeting only finitely many $\{W_{j_1 j_2 \ldots j_k}\}$.

For each j let $f_j\colon B \to I$ be a continuous map such that $f_j(b) \neq 0$ if and only if $b \in U_j$. Define $\bar{f}_{j_1 \ldots j_k}\colon B^I \to I$ by

$$\bar{f}_{j_1 \ldots j_k}(\omega) = \inf \left\{ f_{j_i}\omega(t) \;\middle|\; \frac{i-1}{k} \le t \le \frac{i}{k}, i = 1, \ldots, k \right\}$$

Then $\bar{f}_{j_1 \ldots j_k}(\omega) \neq 0$ if and only if $\omega \in W_{j_1 \ldots j_k}$.

Unfortunately, the collection $\{W_{j_1 j_2 \ldots j_k}\}$ (all k) is not locally finite, otherwise the proof would be complete by lemma 10. This difficulty is circumvented by modifying the sets $W_{j_1 j_2 \ldots j_k}$. Since for fixed m the collection $\{W_{j_1 \ldots j_k}\}$ with $k < m$ is locally finite, the sum of the functions $\bar{f}_{j_1 \ldots j_k}$ with $k < m$ is a continuous real-valued function g_m on B^I. Define

$$f'_{j_1 \ldots j_m} = \inf(\sup(0, \bar{f}_{j_1 \ldots j_m} - m g_m), 1)$$

Then $f'_{j_1 \ldots j_m}\colon B^I \to I$ and we define $W'_{j_1 \ldots j_m} = \{\omega \in B^I \mid f'_{j_1 \ldots j_m}(\omega) \neq 0\}$. Clearly, $W'_{j_1 \ldots j_m} \subset W_{j_1 \ldots j_m}$; therefore there is an extended lifting function over $W'_{j_1 \ldots j_m}$. To complete the proof, it follows from lemma 10 that we need only verify that $\{W'_{j_1 \ldots j_k}\}$ (with k varying) is a locally finite covering of B^I.

For $\omega \in B^I$, let m be the smallest integer such that $\bar{f}_{j_1 \ldots j_m}(\omega) \neq 0$ for some $j_1, \ldots, j_m$. Then $g_m(\omega) = 0$ and $f'_{j_1 \ldots j_m}(\omega) = \bar{f}_{j_1 \ldots j_m}(\omega) \neq 0$. Therefore $\omega \in W'_{j_1 \ldots j_m}$, proving that $\{W'_{j_1 \ldots j_m}\}$ is a covering of B^I. To show that it is locally finite, assume N chosen so that $N > m$ and $\bar{f}_{j_1 \ldots j_m}(\omega) > 1/N$. Then $g_N(\omega) > 1/N$ and $N g_N(\omega) > 1$. Hence $N g_N(\omega') > 1$ for all ω' in some neighborhood V of ω. Therefore all functions $f'_{j_1 \ldots j_k}$ with $k \ge N$ vanish on V. But this means that the corresponding set $W'_{j_1 \ldots j_k}$ is disjoint from V. Since the collection $\{W'_{j_1 \ldots j_k}\}$ with $k < N$ is locally finite, the collection $\{W'_{j_1 \ldots j_k}\}$ (all k) is locally finite. ∎

The fact that any open covering of a paracompact Hausdorff space has a numerable refinement, leads to our next theorem.

13 THEOREM *If B is a paracompact Hausdorff space, a map $p\colon E \to B$ is a fibration if and only if it is a local fibration.* ∎

A bundle projection is a local fibration. Therefore, we have the following corollary.

14 COROLLARY *If (E,B,F,p) is a fiber bundle with base space B paracompact and Hausdorff, then p is a fibration.* ∎

8 FIBRATIONS

This section contains a general discussion of fibrations. We establish a relation between cofibrations and fibrations which allows the construction of fibrations from cofibrations by means of function spaces. We also prove that every map is equivalent, up to homotopy, to a map that is a fibration (this dualizes a similar result concerning cofibrations). The section contains definitions of the concepts of fiber homotopy type and induced fibration and a proof of the result that homotopic maps induce fiber-homotopy-equivalent fibrations.

We begin with an analogue of theorem 2.7.8 for cofibrations. Given a map $f: X' \to X$, let $\bar{X}$ be the quotient space of the sum $(X' \times I) \vee (X \times 0)$, obtained by identifying $(x',0) \in X' \times I$ with $(f(x'),0) \in X \times 0$ for all $x' \in X'$. We use $[x',t]$ and $[x,0]$ to denote the points of $\bar{X}$ corresponding to $(x',t) \in X' \times I$ and $(x,0) \in X \times 0$, respectively. Then $[x',0] = [f(x'),0]$. There is a map

$$\bar{i}: \bar{X} \to X \times I$$

defined by

$$\begin{aligned} \bar{i}[x',t] &= (f(x'),t) \qquad & x' \in X', t \in I \\ \bar{i}[x,0] &= (x,0) \qquad & x \in X \end{aligned}$$

A *retracting function* for f is a map

$$\rho: X \times I \to \bar{X}$$

which is a left inverse of $\bar{i}$. In case f is a closed inclusion map, so is $\bar{i}$, and a retracting function for f is a retraction of $X \times I$ to the subspace $X' \times I \cup X \times 0$.

1 THEOREM *A map $f: X' \to X$ is a cofibration if and only if there exists a retracting function for f.*

PROOF If f is a cofibration, let $g: X \to \bar{X}$ and $G: X' \times I \to \bar{X}$ be the maps defined by $g(x) = [x,0]$ and $G(x',t) = [x',t]$. Because

$$G(x',0) = [x',0] = [f(x'),0] = gf'(x)$$

it follows from the fact that f is a cofibration that there exists a map $\rho: X \times I \to \bar{X}$ such that $\rho(x,0) = g(x)$ and $\rho(f(x'),t) = G(x',t)$. Then ρ is a retracting function for f.

Conversely, given maps $g: X \to Y$ and $G: X' \times I \to Y$ such that $G(x',0) = gf(x')$ for $x' \in X'$, define

$$\bar{G}: \bar{X} \to Y$$

by $\bar{G}[x',t] = G(x',t)$ and $\bar{G}[x,0] = g(x)$. If $\rho: X \times I \to \bar{X}$ is a retracting function for f, the map $F = \bar{G} \circ \rho: X \times I \to Y$ has the properties $F(x,0) = g(x)$ and $F(f(x'),t) = G(x',t)$, showing that f is a cofibration. ∎

This leads to the following construction of fibrations from cofibrations.

2 THEOREM *Let $f: X' \to X$ be a cofibration, where X' and X are locally compact Hausdorff spaces, and let Y be any space. Then the map $p: Y^X \to Y^{X'}$ defined by $p(g) = g \circ f$ is a fibration.*

PROOF Let $\rho: X \times I \to \bar{X}$ be a retracting function for f (which exists by theorem 1). Then ρ defines a map

$$\rho': Y^{\bar{X}} \to Y^{X \times I}$$

such that $\rho'(g) = g \circ \rho$ for $g: \bar{X} \to Y$. Because X' and X are locally compact

Hausdorff spaces, so is $\bar{X}$, and by theorem 2.9 in the introduction, $Y^{X \times I} \approx (Y^X)^I$ and

$$Y^{\bar{X}} \approx \{(g,G) \in Y^X \times (Y^{X'})^I \mid g \circ f = G(0)\}$$

Therefore ρ' corresponds to a lifting function for p: $Y^X \to Y^{X'}$, and by theorem 2.7.8, p is a fibration. ■

3 COROLLARY *For any space Y let p: $Y^I \to Y \times Y$ be the map $p(\omega) = (\omega(0),\omega(1))$ for ω: $I \to Y$. Then p is a fibration.*

PROOF Because $\dot{I} \times I \cup I \times 0$ is a retract of $I \times I$, the inclusion map $\dot{I} \subset I$ is a cofibration [equivalently, the pair $(I,\dot{I})$ has the homotopy extension property with respect to any space]. The result follows from theorem 2 and the observation that $Y^{\dot{I}}$ is homeomorphic to $Y \times Y$ under the map $g \to (g(0),g(1))$ for g: $\dot{I} \to Y$. ■

Let f: $B' \to B$ and p: $E \to B$ be maps and let E' be the subset of $B' \times E$ defined by

$$E' = \{(b',e) \in B' \times E \mid f(b') = p(e)\}$$

E' is called the *fibered product* of B' and E (more precisely, the fibered product of f and p; *cf.* Sec. 2.2). Note that there are maps p': $E' \to B'$ and f'; $E' \to E$ defined by $p'(b',e) = b'$ and $f'(b',e) = e$. E' and the maps p' and f' are characterized as the product of f: $B' \to B$ and p: $E \to B$ in the category whose objects are continuous maps with range B and whose morphisms are commutative triangles

$$\begin{array}{ccc} X_1 & \xrightarrow{h} & X_2 \\ & {\scriptstyle g_1}\searrow \quad \swarrow{\scriptstyle g_2} & \\ & B & \end{array}$$

The following properties are easily verified.

4 *If p is injective (or surjective), so is p'.* ■

5 *If p: $B \times F \to B$ is the trivial fibration, then p': $E' \to B'$ is equivalent to the trivial fibration $B' \times F \to B'$.* ■

6 *If p is a fibration (with unique path lifting), so is p'.* ■

7 *If p is a fibration, f can be lifted to E if and only if p' has a section.* ■

Note that since the fibered product is symmetric in B and E (or rather, in f and p), there is a similar set of statements where p and p' are replaced by f and f'.

If p: $E \to B$ is a fibration (or covering projection) and f: $B' \to B$ is a map, then, by property 6 (or property 5), p': $E' \to B'$ is a fibration (or covering projection) and is called the *fibration induced* from p by f (or *covering projection induced* from p by f). If $\xi = (E,B,F,p)$ is a fiber bundle and f: $B' \to B$ is a map, it follows from property 5 that there is a fiber bundle (E',B',F,p'). This is called the *fiber bundle induced* from ξ by f and is denoted by $f^*\xi$. In the case of an inclusion map i: $B' \subset B$ we use $E \mid B'$ to denote the fibered product of B' and E, and if ξ is a fiber bundle with base space B, $\xi \mid B'$ will denote the fiber bundle with base space B' induced by i. Observe that $\xi \mid B'$ is equivalent to $(p^{-1}(B'), B', F, p \mid p^{-1}(B'))$.

8 COROLLARY *For any space Y and point $y_0 \in Y$, let p: $P(Y,y_0) \to Y$ be the map sending each path starting at y_0 to its endpoint. Then p is a fibration whose fiber over y_0 is the loop space ΩY.*

PROOF Let f: $Y \to Y \times Y$ be defined by $f(y) = (y_0,y)$ and let $\bar{p}$: $Y^I \to Y \times Y$ be the fibration of corollary 3. The fibration induced by f is equivalent to the map p: $P(Y,y_0) \to Y$, where $p(\omega) = \omega(1)$, and $p^{-1}(y_0)$ the fiber over y_0, is by definition, the loop space ΩY. ■

It follows from corollary 3 that the map p': $Y^I \to Y$ defined by $p'(\omega) = \omega(0)$ [or by $p'(\omega) = \omega(1)$] is a fibration, because it is the composite of fibrations $Y^I \to Y \times Y \to Y$. If p: $E \to B$ is any map and p': $B^I \to B$ is the fibration defined by $p'(\omega) = \omega(0)$, then the fibered product of E and B^I is just the space $\bar{B}$ used to define the concept of lifting function for p.

These remarks about fibered products and induced fibrations have analogues for cofibrations. Given maps f_1: $X \to X_1$ and f_2: $X \to X_2$, the *cofibered sum* of X_1 and X_2 is the quotient space X' of $X_1 \vee X_2$ obtained by identifying $f_1(x)$ with $f_2(x)$ for all $x \in X$. There are maps i_1: $X_1 \to X'$ and i_2: $X_2 \to X'$, and these characterize X' as the sum of f_1 and f_2 in the category whose objects are maps with domain X and whose morphisms are commutative triangles. If f_1: $X \to X_1$ is a cofibration, so is i_2: $X_2 \to X'$, and this is called the *cofibration induced* from f_1 by f_2.

The map h_0: $X' \to X' \times I$ defined by $h_0(x') = (x',0)$ is a cofibration for any space X', and if f: $X' \to X$ is any map, the cofibered sum of $X' \times I$ and X is just the space $\tilde{X}$ used to define the concept of retracting function for f.

Let p: $E \to B$ be a fibration. Maps f_0, f_1: $X \to E$ are said to be *fiber homotopic*, denoted by $f_0 \underset{p}{\simeq} f_1$, if there is a homotopy F: $f_0 \simeq f_1$ such that $pF(x,t) = pf_0(x)$ for $x \in X$ and $t \in I$ (in which case $p \circ f_0 = p \circ f_1$). This is an equivalence relation in the set of maps $X \to E$. The equivalence classes are denoted by $[X;E]_p$, and if f: $X \to E$, $[f]_p$ denotes its fiber homotopy class. The concept of fiber homotopy is dual to the concept of relative homotopy.

We use induced fibrations to prove that any map is, up to homotopy equivalence, a fibration. Let f: $X \to Y$ and let p': $Y^I \to Y$ be the fibration defined by $p'(\omega) = \omega(0)$. Let p: $P_f \to X$ be the fibration induced from p' by f. It is called the *mapping path fibration* of f and is dual to the mapping cylinder. There is a section s: $X \to P_f$ of p defined by $s(x) = (x,\omega_{f(x)})$, where $\omega_{f(x)}$ is the constant path in Y at $f(x)$. There is also a map p'': $P_f \to Y$ defined by $p''(x,\omega) = \omega(1)$. We then have the following dual of theorem 1.4.12.

9 THEOREM *Given a map f: $X \to Y$, there is a commutative diagram*

$$\begin{array}{ccc} X & \xrightarrow{s} & P_f \\ & f \searrow \quad \swarrow p'' & \\ & Y & \end{array}$$

such that

(*a*) $1_{P_f} \underset{p}{\simeq} s \circ p$

(*b*) p'' *is a fibration*

PROOF The triangle is commutative by the definition of the maps involved.

(*a*) Define $F: P_f \times I \to P_f$ by $F((x,\omega), t) = (x,\omega_{1-t})$, where $\omega_{1-t}(t') = \omega((1-t)t')$. Then F is a fiber homotopy from 1_{P_f} to $s \circ p$.

(*b*) Let $g: W \to P_f$ and $G: W \times I \to Y$ be such that $G(w,0) = p''g(w)$ for $w \in W$. Then there exist maps $g': W \to X$ and $g'': W \to Y^I$ such that $g''(w)(0) = fg'(w)$ and $g(w) = (g'(w),g''(w))$ for $w \in W$. We define a lifting $G': W \times I \to P_f$ of G beginning with g by $G'(w,t) = (g'(w), \bar{g}(w,t))$, where $\bar{g}(w,t) \in Y^I$ is defined by

$$\bar{g}(w,t)(t') = \begin{cases} g''(w)(2t'/(2-t)) & 0 \le 2t' \le 2 - t \le 2,\ w \in W \\ G(w,2t'+t-2) & 1 \le 2 - t \le 2t' \le 2,\ w \in W \end{cases}$$

Since p'' has the homotopy lifting property, it is a fibration. ■

It follows that the fibration $p'': P_f \to Y$ is equivalent (by means of $s: X \to P_f$ and $p: P_f \to X$) in the homotopy category of maps with range Y to the original map $f: X \to Y$. In replacing f by an equivalent fibration, we replaced X by a space P_f of the same homotopy type, whereas in Sec. 1.4, when f was replaced by an equivalent cofibration, the space Y was replaced by a space Z_f of the same homotopy type.

Two fibrations $p_1: E_1 \to B$ and $p_2: E_2 \to B$ are said to be *fiber homotopy equivalent* (or to have the *same fiber homotopy type*) if there exist maps $f: E_1 \to E_2$ and $g: E_2 \to E_1$ preserving fibers in the sense that $p_2 \circ f = p_1$ and $p_1 \circ g = f_2$ and such that $g \circ f \underset{p_1}{\simeq} 1_{E_1}$ and $f \circ g \underset{p_2}{\simeq} 1_{E_2}$. Each of the maps f and g is called a *fiber homotopy equivalence*. The rest of this section is concerned with fiber homotopy equivalence.

We begin with the following result concerning liftings of homotopic maps.

10 THEOREM *Let $p: E \to B$ be a fibration and let $F_0, F_1: X \times I \to E$ be maps. Given homotopies $H: p \circ F_0 \simeq p \circ F_1$ and $G: F_0 \mid X \times 0 \simeq F_1 \mid X \times 0$ such that $H(x,0,t) = pG(x,0,t)$, there is a lifting $H': X \times I \times I \to E$ of H which is a homotopy from F_0 to F_1 and is an extension of G.*

PROOF Let $A = (I \times 0) \cup (0 \times I) \cup (I \times 1) \subset I \times I$ and define $f: X \times A \to E$ by

$$\begin{aligned} f(x,t,0) &= F_0(x,t) \\ f(x,0,t) &= G(x,0,t) \\ f(x,t,1) &= F_1(x,t) \end{aligned}$$

Then $H \mid X \times A = p \circ f$. Because there is a homeomorphism of $I \times I$ with itself taking A onto $I \times 0$, there is a homeomorphism of $X \times I \times I$ with itself taking $X \times A$ onto $X \times I \times 0$. It follows from the homotopy lifting property of p that there is a lifting $H': X \times I \times I \to E$ of H such that $H' \mid X \times A = f$. ■

Taking H and G to be constant homotopies, we obtain the following corollary.

11 COROLLARY *Let $p: E \to B$ be a fibration and let $F_0, F_1: X \times I \to E$ be liftings of the same map such that $F_0 \mid X \times 0 = F_1 \mid X \times 0$. Then $F_0 \underset{p}{\simeq} F_1$* rel $X \times 0$. ■

Let $p: E \to B$ be a fibration and let $\omega: I \to B$ be a path in its base space. By the homotopy lifting property of p, there exists a map $F: p^{-1}(\omega(0)) \times I \to E$ such that $pF(x,t) = \omega(t)$ and $F(x,0) = x$ for $x \in p^{-1}(\omega(0))$ and $t \in I$. Let $f: p^{-1}(\omega(0)) \to p^{-1}(\omega(1))$ be the map $f(x) = F(x,1)$. It follows from theorem 10 that if $\omega \simeq \omega'$ are homotopic paths in B and if $F, F': p^{-1}(\omega(0)) \times I \to E$ are such that $pF(x,t) = \omega(t)$, $pF'(x,t) = \omega'(t)$, and $F(x,0) = x = F'(x,0)$ for $x \in p^{-1}(\omega(0))$ and $t \in I$, then the maps $f, f': p^{-1}(\omega(0)) \to p^{-1}(\omega(1))$ defined by $f(x) = F(x,1)$ and $f'(x) = F'(x,1)$ are homotopic. Hence there is a well-defined homotopy class $[f] \in [p^{-1}(\omega(0)); p^{-1}(\omega(1))]$ corresponding to a path class $[\omega]$ in B. We let $h[\omega] = [f]$.

The following is the form theorem 2.3.7 takes for an arbitrary fibration.

12 THEOREM *Let $p: E \to B$ be a fibration. There is a contravariant functor from the fundamental groupoid of B to the homotopy category which assigns to $b \in B$ the fiber over b and to a path class $[\omega]$ the homotopy class $h[\omega]$.*

PROOF If ω_b is the constant path at b, let $F: p^{-1}(b) \times I \to E$ be the map $F(x,t) = x$. The corresponding map $f: p^{-1}(b) \to p^{-1}(b)$ defined by $f(x) = F(x,1)$ is the identity map. Hence

$$h[\omega_b] = [1_{p^{-1}(b)}]$$

showing that h preserves identities.

Let ω and ω' be paths in B such that $\omega(1) = \omega'(0)$. Given a map $F: p^{-1}(\omega(0)) \times I \to E$ such that $F(x,0) = x$ and $pF(x,t) = \omega(t)$ for $x \in p^{-1}(\omega(0))$ and $t \in I$, and given $F': p^{-1}(\omega(1)) \times I \to E$ such that $F'(x',0) = x'$ and $pF'(x',t) = \omega'(t)$ for $x' \in p^{-1}(\omega'(0))$ and $t \in I$, let $f: p^{-1}(\omega(0)) \to p^{-1}(\omega'(0))$ be defined by $f(x) = F(x,1)$ and let $F'': p^{-1}(\omega(0)) \times I \to E$ be defined by

$$F''(x,t) = \begin{cases} F(x,2t) & 0 \le t \le \tfrac{1}{2}, x \in p^{-1}(\omega(0)) \\ F'(f(x), 2t-1) & \tfrac{1}{2} \le t \le 1, x \in p^{-1}(\omega(0)) \end{cases}$$

Then $pF''(x,t) = (\omega * \omega')(t)$ and $F''(x,0) = x$ for $x \in p^{-1}(\omega(0))$ and $t \in I$. Let $f': p^{-1}(\omega'(0)) \to p^{-1}(\omega'(1))$ be defined by $f'(x') = F'(x',1)$. Then $F''(x,1) = f'(f(x))$ for $x \in p^{-1}(\omega(0))$, which shows that

$$h[\omega * \omega'] = h[\omega'] * h[\omega]$$

Therefore h is a contravariant functor. ■

This yields the following analogue of corollary 2.3.8 for an arbitrary fibration.

13 COROLLARY *If $p: E \to B$ is a fibration with a path-connected base space, any two fibers have the same homotopy type.* ■

The following result asserts that homotopic maps induce fiber-homotopy-equivalent fibrations.

14 THEOREM *Let $p: E \to B$ be a fibration and let $f_0, f_1: X \to B$ be homotopic. The fibrations induced from p by f_0 and by f_1 are fiber homotopy equivalent.*

PROOF Let $p_0: E_0 \to X$ and $p_1: E_1 \to X$ be the fibrations induced from p by f_0 and f_1, respectively, and let $f'_0: E_0 \to E$ and $f'_1: E_1 \to E$ be the corresponding maps such that $p \circ f'_0 = f_0 \circ p_0$ and $p \circ f'_1 = f_1 \circ p_1$. Given a homotopy $F: X \times I \to B$ from f_0 to f_1, there are maps $F'_0: E_0 \times I \to E$ and $F'_1: E_1 \times I \to E$ such that $p \circ F'_0 = F \circ (p_0 \times 1_I)$ and $p \circ F'_1 = F \circ (p_1 \times 1_I)$ and $F'_0 \mid E_0 \times 0 = f'_0$ and $F'_1 \mid E_1 \times 1 = f'_1$. Let $g_0: E_0 \to E_1$ and $g_1: E_1 \to E_0$ be the fiber preserving maps defined by the property $F'_0(x,1) = f'_1 g_0(x)$ for $x \in E_0$ and $F'_1(y,0) = f'_0 g_1(y)$ for $y \in E_1$. Then

$$p \circ F'_0 \circ (g_1 \times 1_I) = F \circ (p_0 \times 1_I) \circ (g_1 \times 1_I) = F \circ (p_1 \times 1_I)$$

and
$$F'_0 \circ (g_1 \times 1_I) \mid E_1 \times 0 = F'_1 \mid E_1 \times 0$$

It follows from theorem 10 that $F'_1 \underset{p}{\simeq} F'_0 \circ (g_1 \times 1_I)$. In a similar fashion $F'_0 \underset{p}{\simeq} F'_1 \circ (g_0 \times 1_I)$. This implies that $g_0 g_1 \underset{p_1}{\simeq} 1_{E_1}$ and $g_1 g_0 \underset{p_0}{\simeq} 1_{E_0}$. ∎

Clearly, a constant map induces a trivial fibration, and we have the following result.

15 COROLLARY *If $p: E \to B$ is a fibration and B is contractible, then p is fiber homotopy equivalent to the trivial fibration $B \times p^{-1}(b_0) \to B$ for any $b_0 \in B$.* ∎

Let B be a space which is the join of some space Y with S^0. Then $B = C_-Y \cup C_+Y$, where C_-Y and C_+Y are cones over Y and $C_-Y \cap C_+Y = Y$. Let $y_0 \in Y$ and let $p: E \to B$ be a fibration with fiber $F_0 = p^{-1}(y_0)$. It follows from corollary 15 that there are fiber homotopy equivalences $f_-: C_-Y \times F_0 \to p^{-1}(C_-Y)$ and $g_+: p^{-1}(C_+Y) \to C_+Y \times F_0$. A *clutching function* $\mu: Y \times F_0 \to F_0$ for p is a function μ defined by the equation

$$g_+ f_-(y,z) = (y, \mu(y,z)) \qquad y \in Y, z \in F_0$$

where $f_-: C_-Y \times F_0 \to p^{-1}(C_-Y)$ and $g_+: p^{-1}(C_+Y) \to C_+Y \times F_0$ are fiber homotopy equivalences. If C_-Y and C_+Y are contractible to y_0 relative to y_0, it follows from theorem 10 that f_- and g_+ can be chosen so that $z \to f_-(y_0,z)$ is homotopic to the map $F_0 \subset p^{-1}(C_-Y)$ and $z \to g_+(z)$ is homotopic to the map $z \to (y_0,z)$ of F_0 to $C_+Y \times F_0$. In this case the clutching function μ corresponding to f_- and g_+ has the property that the map $z \to \mu(y_0,z)$ is homotopic to the identity map $F_0 \subset F_0$.

Let E_φ be the fiber bundle over S^n defined by a characteristic map $\varphi: S^{n-1} \to G$, as in example 2.7.5 (where G is a group of homeomorphisms of the fiber F). Then $E_-^n = C_-S^{n-1}$ and $E_+^n = C_+S^{n-1}$, and it is easy to verify that f_- and g_+ can be chosen so that the corresponding clutching function $\mu: S^{n-1} \times F \to F$ is the map $\mu(x,z) = \varphi(x)z$.

EXERCISES

A LOCAL CONNECTEDNESS

1 Prove that a space X is locally path connected if and only if for any neighborhood U of x in X there exists a neighborhood V of x such that every pair of points in V can be joined by a path in U.

2 If X is a space, let $\tilde{X}$ denote the set X retopologized by the topology generated by path components of open sets of X. Prove that $\tilde{X}$ is locally path connected and that the identity map of X is a continuous function $j: \tilde{X} \to X$ having the property that for any locally path-connected space Y a function $f: Y \to \tilde{X}$ is continuous if and only if $j \circ f: Y \to X$ is continuous.

3 For any space X let $\tilde{X}$ and $j: \tilde{X} \to X$ be as in exercise 2. Prove that $j_{\#}: \pi(\tilde{X},x_0) \approx \pi(X,x_0)$.

B COVERING SPACES

1 Let X be the union of two closed simply connected and locally path-connected subsets A and B such that $A \cap B$ consists of a single point. Prove that if $p: \tilde{X} \to X$ is a nonempty path-connected fibration with unique path lifting, then p is a homeomorphism.

2 Let $\tilde{X} = \{(x,y) \in \mathbf{R}^2 \mid x \text{ or } y \text{ an integer}\}$ and let

$$X = S^1 \vee S^1 = \{(z_1,z_2) \in S^1 \times S^1 \mid z_1 = 1 \text{ or } z_2 = 1\}$$

Prove that the map $p: \tilde{X} \to X$ such that $p(x,y) = (e^{2\pi ix},e^{2\pi iy})$ is a covering projection.

3 With $p: \tilde{X} \to X$ as in exercise 2 above, let $Y \subset \tilde{X}$ be defined by

$$Y = \{(x,y) \in \tilde{X} \mid 0 \le x \le 1, 0 \le y \le 1\}$$

Prove that Y is a retract of $\tilde{X}$ and that $(p \mid Y)_{\#}$ maps a generator of $\pi(Y)$ to the commutator of the two elements of $\pi(X)$ corresponding to the two circles of X.

4 Prove that $\pi(S^1 \vee S^1)$ is nonabelian.

C THE COVERING SPACE ex: $\mathbf{R} \to S^1$

1 For an arbitrary space X prove that a map $f: X \to S^1$ can be lifted to a map $\tilde{f}: X \to \mathbf{R}$ such that $f = ex \circ \tilde{f}$ if and only if f is null homotopic.

2 Let X be a connected locally path-connected space with base point $x_0 \in X$. Prove that the map

$$[X,x_0; S^1,1] \to \operatorname{Hom}(\pi(X,x_0), \pi(S^1,1))$$

which assigns to $[f]$ the homomorphism

$$f_{\#}: \pi(X,x_0) \to \pi(S^1,1)$$

is a monomorphism (the set of homotopy classes being a group by virtue of the group structure on S^1).

3 Prove that any two maps from a simply connected locally path-connected space to S^1 are homotopic.

4 Prove that any map of the real projective space P^n for $n \ge 2$ to S^1 is null homotopic.

5 Prove that there is no map $f: S^n \to S^1$ for $n \geq 2$ such that $f(-x) = -f(x)$.

6 *Borsuk-Ulam theorem.* Prove that if $f: S^2 \to \mathbf{R}^2$ is a map such that $f(-x) = -f(x)$, then there exists a point $x_0 \in S^2$ such that $f(x_0) = 0$.

D COVERING SPACES OF TOPOLOGICAL GROUPS

1 Let H be a subgroup of a topological group and let G/H be the homogeneous space of right cosets. Prove that the projection $G \to G/H$ is a covering projection if and only if H is discrete.

2 Prove that a connected locally path-connected covering space of a topological group can be given a group structure that makes it a topological group and makes the projection map a homomorphism.

A *local homomorphism* φ from one topological group G to another G' is a continuous map from some neighborhood U of e in G to G' such that if $g_1, g_2, g_1g_2 \in U$, then $\varphi(g_1g_2) = \varphi(g_1)\varphi(g_2)$. A *local isomorphism* from G to G' is a homeomorphism φ from some neighborhood U of e to some neighborhood U' of e' such that φ and φ^{-1} are both local homomorphisms (in which case G and G' are said to be *locally isomorphic*).

3 Prove that a continuous homomorphism $\varphi: G \to G'$ between connected topological groups is a covering projection if and only if there exists a neighborhood U of e in G such that $\varphi \mid U$ is a local isomorphism from G to G'.

4 Let φ be a local homomorphism from a connected topological group G to a topological group G' defined on a connected neighborhood U of e in G. Let $\tilde{G}$ be the subgroup of $G \times G'$ generated by the graph of φ (that is, generated by $\{(g,g') \in G \times G' \mid g' = \varphi(g), g \in U\}$). $\tilde{G}$ is topologized by taking as a base for neighborhoods of (e,e') the graph of $\varphi \mid N$ as N varies over neighborhoods of e in U. Prove that $\tilde{G}$ is a connected topological group, the projection $p_1: \tilde{G} \to G$ is a covering projection, and the projection $p_2: \tilde{G} \to G'$ is continuous.

5 Prove that two connected locally path-connected topological groups are locally isomorphic if and only if there is a topological group which is a covering space of each of them.

6 If G is a simply connected locally path-connected topological group and φ is a local homomorphism from G to a topological group G', prove that there is a continuous homomorphism $\varphi': G \to G'$ which agrees with φ on some neighborhood of e in G.

E FIBRATIONS

1 If $p: E \to B$ is a fibration, prove that $p(E)$ is a union of path components of B.

2 If a fibration has path-connected base and some fiber is path connected, prove that its total space is also path connected.

3 Let $p: E \to B$ be a fibration and let X be a locally compact Hausdorff space. Define $p': E^X \to B^X$ by $p'(g) = p \circ g$ for $g: X \to E$. Prove that p' is a fibration.

4 Let $p: E \to B$ be a fibration and let $b_0 \in p(E)$, $F = p^{-1}(b_0)$. Let X be a space regarded as a subset of some cone CX. Prove that the map

$$p_\#: [CX,X; E,F] \to [CX,X; B,b_0]$$

is a bijection.

5 Let $p: E \to B$ be a fibration and let $e_0 \in E$, $b_0 = p(e_0)$, and $F = p^{-1}(b_0)$. If B is simply connected, prove that $\pi(F,e_0) \to \pi(E,e_0)$ is an epimorphism.

6 Let $p: E \to B$ be a fibration and let $e_0 \in E$ and $b_0 = p(e_0)$. If $p^{-1}(b_0)$ is simply connected, prove that

$$p_\#: \pi(E,e_0) \approx \pi(B,b_0)$$

7 Let $p: E \to B$ be a fibration and $b_0 \in p(E)$. If E is simply connected, prove that there is a bijection between $\pi(B,b_0)$ and the set of path components of $p^{-1}(b_0)$.

CHAPTER THREE
POLYHEDRA

IN CHAPTER TWO THE FUNDAMENTAL GROUP FUNCTOR WAS USED TO CLASSIFY covering spaces. We now consider the problem of computing the fundamental group of a specific space. We shall show that the fundamental groups of many spaces (the class of polyhedra) can be described by means of generators and relations.

A polyhedron is a topological space which admits a triangulation by a simplicial complex. Thus we start with a study of the category of simplicial complexes. A simplicial complex consists of an abstract scheme of vertices and simplexes (each simplex being a finite set of vertices). Associated to such a simplicial complex is a topological space built by piecing together convex cells with identifications prescribed by the abstract scheme. Since the topological properties of these spaces are determined by the abstract scheme, the study of simplicial complexes and polyhedra is often called combinatorial topology.

A compact polyhedron admits a triangulation by a finite simplicial complex. Thus these spaces are effectively described in finite terms and serve as a useful class of spaces for questions involving computability of functors.

Sections 3.1 and 3.2 are devoted to definitions and elementary topological properties of polyhedra. Section 3.3 introduces the concept of subdivision of a simplicial complex, and it is shown that a compact polyhedron admits arbitrarily fine triangulations. This result is used in Sec. 3.4 to prove the simplicial-approximation theorem, which asserts that continuous maps from compact polyhedra to arbitrary polyhedra can be approximated by simplicial maps.

The technique of simplicial approximation is used in Sec. 3.5 to prove that the set of homotopy classes of continuous maps from a compact polyhedron to an arbitrary polyhedron can be described abstractly in terms of triangulations of the polyhedra. In Sec. 3.6 this result provides an abstract description of the fundamental group of a polyhedron as the edge-path group of a triangulation, which is used in Sec. 3.7 to obtain a system of generators and relations for the fundamental group of a polyhedron. It is also shown in Sec. 3.7 that the fundamental group functor provides a faithful representation of the homotopy category of connected one-dimensional polyhedra. Section 3.8 consists of applications of the results on the fundamental group, some examples of polyhedra, and a description of the fundamental group of an arbitrary surface.

1 SIMPLICIAL COMPLEXES

This section contains definitions of the category of simplicial complexes and of covariant functors from this category to the category of topological spaces.

A *simplicial complex* K consists of a set $\{v\}$ of *vertices* and a set $\{s\}$ of finite nonempty subsets of $\{v\}$ called *simplexes* such that

(*a*) Any set consisting of exactly one vertex is a simplex.
(*b*) Any nonempty subset of a simplex is a simplex.

A simplex s containing exactly $q + 1$ vertices is called a *q-simplex*. We also say that the *dimension* of s is q and write $\dim s = q$. If $s' \subset s$, then s' is called a *face* of s (a *proper face* if $s' \neq s$), and if s' is a p-simplex, it is called a *p-face* of s. If s is a q-simplex, then s is the only q-face of s, and a face s' of s is a proper face if and only if $\dim s' < q$. It is clear that any simplex has only a finite number of faces. Because any face of a face of s is itself a face of s, the simplexes of K are partially ordered by the face relation (written $s' \leq s$ if s' is a face of s).

It follows from condition (*a*) that the 0-simplexes of K correspond bijectively to the vertices of K. It follows from condition (*b*) that any simplex is determined by its 0-faces. Therefore K can be regarded as equal to the set of its simplexes, and we shall identify a vertex of K with the 0-simplex corresponding to it.

We list some examples.

1   The empty set of simplexes is a simplicial complex denoted by $\varnothing$.

2   For any set A the set of all finite nonempty subsets of A is a simplicial complex.

3   If s is a simplex of a simplicial complex K, the set of all faces of s is a simplicial complex denoted by $\bar{s}$.

4   If s is a simplex of a simplicial complex K, the set of all proper faces of s is a simplicial complex denoted by $\dot{s}$.

5   If K is a simplicial complex, its *q-dimensional skeleton* K^q is defined to be the simplicial complex consisting of all p-simplexes of K for $p \leq q$.

6   Given a set X and a collection $\mathcal{W} = \{\,W\,\}$ of subsets of X, the *nerve* of $\mathcal{W}$, denoted by $K(\mathcal{W})$, is the simplicial complex whose simplexes are finite nonempty subsets of $\mathcal{W}$ with nonempty intersection. Thus the vertices of $K(\mathcal{W})$ are the nonempty elements of $\mathcal{W}$.

7   If K_1 and K_2 are simplicial complexes, their *join* $K_1 * K_2$ is the simplicial complex defined by

$$K_1 * K_2 = K_1 \vee K_2 \cup \{s_1 \vee s_2 \mid s_1 \in K_1, s_2 \in K_2\}$$

Thus the set of vertices of $K_1 * K_2$ is the set sum of the set of vertices of K_1 and the set of vertices of K_2.

8   There is a simplicial complex whose set of vertices is $\mathbf{Z}$ and whose set of simplexes is

$$\{\{n\} \mid n \in \mathbf{Z}\} \cup \{\{n, n+1\} \mid n \in \mathbf{Z}\}$$

9   For $n \geq 1$ regard $\mathbf{Z}^n$ as partially ordered by the ordering of its coordinates (that is, given x, $x' \in \mathbf{Z}^n$, then $x \leq x'$ if for the ith coordinates $x_i \leq x_i$ in $\mathbf{Z}$). There is a simplicial complex whose set of vertices is $\mathbf{Z}^n$ and whose simplexes are finite nonempty totally ordered subsets $\{x^0, \ldots, x^q\}$ of $\mathbf{Z}^n$ (that is, $x^0 \leq x^1 \leq \cdots \leq x^q$) such that for all $1 \leq i \leq n$, $x_i{}^q - x_i{}^0 = 0$ or 1.

If K is a simplicial complex, its *dimension*, denoted by dim K, is defined to equal -1 if K is empty, to equal n if K contains an n-simplex but no $(n+1)$-simplex, and to equal ∞ if K contains n-simplexes for all $n \geq 0$. Thus dim $K = \sup \{\dim s \mid s \in K\}$. K is said to be *finite* if it contains only a finite number of simplexes. If K is finite, then dim $K < \infty$; however, if dim $K < \infty$, K need not be finite (example 8 is an infinite simplicial complex whose dimension is 1).

A *simplicial map* $\varphi\colon K_1 \to K_2$ is a function φ from the vertices of K_1 to the vertices of K_2 such that for any simplex $s \in K_1$ its image $\varphi(s)$ is a simplex of K_2. For any K there is an identity simplicial map $1_K\colon K \to K$ corresponding to the identity vertex map. Given simplicial maps $K_1 \xrightarrow{\varphi} K_2 \xrightarrow{\psi} K_3$, the composite simplicial map $\psi \circ \varphi: K_1 \to K_3$ corresponds to the composite vertex map. Therefore there is a category of simplicial complexes and simplicial maps.

A *subcomplex* L of a simplicial complex K, denoted by $L \subset K$, is a subset of K (that is, $s \in L \Rightarrow s \in K$) that is a simplicial complex. It is clear that a subset L of K is a subcomplex if and only if any simplex in K that is a face of a simplex of L is a simplex of L. If $L \subset K$, there is a simplicial inclusion map $i: L \subset K$.

A subcomplex $L \subset K$ is said to be *full* if each simplex of K having all its vertices in L itself belongs to L. There is a subcomplex N of K consisting of all simplexes of K with no vertex in L. Clearly, N is the largest subcomplex of K disjoint from L. If $s = \{v_0,v_1, \ldots ,v_q\}$ is any simplex of K, then either no vertex of s is in L (in which case $s \in N$), or every vertex belongs to L (in which case, if L is full, $s \in L$), or the vertices can be enumerated so that $v_i \in L$ if $i \le p$ and $v_i \notin L$ if $i > p$, where $0 \le p < q$. In the latter case, $s = s' \cup s''$, where $s' = \{v_0, \ldots ,v_p\}$ is in L, if L is full, and $s'' = \{v_{p+1}, \ldots ,v_q\}$ is in N. Therefore we have the following result.

10 LEMMA *If L is a full subcomplex of K and N is the largest subcomplex of K disjoint from L, any simplex of K is either in N, or in L, or of the form $s' \cup s''$ for some $s' \in L$ and $s'' \in N$.* ∎

There is a category of *simplicial pairs* (K,L) (that is, K is a simplicial complex and L is a subcomplex, possibly empty) and simplicial maps $\varphi: (K_1,L_1) \to (K_2,L_2)$ (that is, φ is a simplicial map $K_1 \to K_2$ such that $\varphi(L_1) \subset L_2$). The category of simplicial complexes is a full subcategory of the category of simplicial pairs. There is also a category of *pointed simplicial complexes* K (that is, K is a simplicial complex together with a distinguished *base vertex*) and simplicial maps preserving base vertices which is a full subcategory of the category of simplicial pairs. Following are some examples.

11 For any q the q-dimensional skeleton K^q is a subcomplex of K, and if $p \le q$, K^p is a subcomplex of K^q.

12 For any $s \in K$ there are subcomplexes $\dot{s} \subset \bar{s} \subset K$.

13 If $\{L_j\}_{j \in J}$ is a family of subcomplexes of K, then $\cap L_j$ and $\cup L_j$ are also subcomplexes of K.

14 Given that $A \subset X$, $\mathcal{W} = \{W\}$ is a collection of subsets of X, and $K_A(\mathcal{W})$ is the collection of finite nonempty subsets of $\mathcal{W}$ whose intersection meets A in a nonempty subset, then $K_A(\mathcal{W})$ is a subcomplex of the nerve $K(\mathcal{W})$.

We now define a covariant functor from the category of simplicial complexes and simplicial maps to the category of topological spaces and continuous maps. Given a nonempty simplicial complex K, let $|K|$ be the set of all functions α from the set of vertices of K to I such that

(*a*) For any α, $\{v \in K \mid \alpha(v) \neq 0\}$ is a simplex of K (in particular, $\alpha(v) \neq 0$

for only a finite set of vertices).

(b) For any α, $\Sigma_{v \in K} \alpha(v) = 1$.

If $K = \emptyset$, we define $|K| = \emptyset$.

The real number $\alpha(v)$ is called the vth *barycentric coordinate of* α. There is a metric d on $|K|$ defined by

$$d(\alpha,\beta) = \sqrt{\Sigma_{v \in K} [\alpha(v) - \beta(v)]^2}$$

and the topology on $|K|$ defined by this metric is called the *metric topology*. The set $|K|$ with the metric topology is denoted by $|K|_d$.

We shall define another topology on $|K|$. For $s \in K$ the *closed simplex* $|s|$ is defined by

$$|s| = \{\alpha \in |K| \mid \alpha(v) \neq 0 \Rightarrow v \in s\}$$

If s is a q-simplex, $|s|$ is in one-to-one correspondence with the set $\{x \in \mathbf{R}^{q+1} \mid 0 \leq x_i \leq 1, \Sigma x_i = 1\}$. Furthermore, the metric topology on $|K|_d$ induces on $|s|$ a topology that makes it a topological space $|s|_d$ homeomorphic to the above compact convex subset of $\mathbf{R}^{q+1}$. If s_1, $s_2 \in K$, then clearly $s_1 \cap s_2$ is either empty (in which case $|s_1| \cap |s_2| = \emptyset$) or a face of s_1 and of s_2 (in which case $|s_1 \cap s_2| = |s_1| \cap |s_2|$). Therefore, in either case $|s_1|_d \cap |s_2|_d$ is a closed set in $|s_1|_d$ and in $|s_2|_d$, and the topology induced on this intersection from $|s_1|_d$ equals the topology induced on it from $|s_2|_d$. It follows from theorem 2.5 in the Introduction that there is a topology on $|K|$ coherent with $\{|s|_d \mid s \in K\}$. This topology will be called the *coherent topology*. The *space* of K, also denoted by $|K|$, is the set $|K|$ with the coherent topology. (What we call here the coherent topology is known in the literature as the weak topology.) Note that $|\bar{s}| = |s|_d$; we shall also use $|s|$ to denote the space $|\bar{s}|$.

Because a subset $A \subset |K|$ is closed (or open) in the coherent topology if and only if $A \cap |s|$ is closed (or open) in $|s|$ for every $s \in K$, we have the following theorem and its corollary.

15 THEOREM *A function f: $|K| \to X$, where X is a topological space, is continuous in the coherent topology if and only if $f \mid |s|$: $|s| \to X$ is continuous for every $s \in K$.* ∎

16 COROLLARY *A function f: $|K| \to X$ is continuous in the coherent topology if and only if $f \mid |K^q|$: $|K^q| \to X$ is continuous for every $q \geq 0$.* ∎

It follows from theorem 15 that the identity map of the set $|K|$ is a continuous map $|K| \to |K|_d$. Note that $L \subset K \Rightarrow |L| \subset |K|$ and $|L|_d$ is a closed subset of $|K|_d$ (which implies that $|L|$ is a closed subset of $|K|$). Furthermore, if $\{L_j\}_{j \in J}$ is a collection of subcomplexes of K, then $\cup |L_j| = |\cup L_j|$ and $\cap |L_j| = |\cap L_j|$.

The coherent topology has the following property.

17 THEOREM *For any simplicial complex K, its space $|K|$ is a normal Hausdorff space.*

PROOF Because $|K|_d$ is a Hausdorff space and $i\colon |K| \to |K|_d$ is continuous, $|K|$ is a Hausdorff space. To prove that $|K|$ is normal it suffices to show that if A is a closed subset of $|K|$, any continuous map $f\colon A \to I$ can be continuously extended over $|K|$. By theorem 15, the existence of such an extension of f is equivalent to the existence of an indexed family of continuous maps $\{f_s\colon |s| \to I \mid s \in K\}$ such that

(a) If s' is a face of s, then $f_s \mid |s'| = f_{s'}$
(b) $f_s \mid (A \cap |s|) = f \mid (A \cap |s|)$

The existence of the family $\{f_s\}$ is proved by induction on dim s. If s is a 0-simplex, $|s|$ is a single point, and either $|s| \in A$, in which case we define $f_s = f \mid |s|$, or $|s| \notin A$, in which case we define f_s arbitrarily.

Let $q > 0$ and assume f_s defined for all simplexes s with dim $s < q$ to satisfy conditions (a) and (b). Given a q-simplex s, define $f_s'\colon |\dot{s}| \cup (A \cap |s|) \to I$ by the conditions

$$f_s' \mid |s'| = f_{s'} \qquad s' \text{ a face of } s$$
$$f_s' \mid (A \cap |s|) = f \mid (A \cap |s|)$$

Because $\{f_{s'}\}_{\dim s' \leq q}$ satisfies conditions (a) and (b), f_s' is a continuous map of the closed subset $|\dot{s}| \cup (A \cap |s|)$ of $|s|$ to I. By the Tietze extension theorem, there exists a continuous extension $f_s\colon |s| \to I$ of f_s'. ∎

The same technique can be used to prove that $|K|$ is perfectly normal (that is, every closed subset of $|K|$ is the set of zeros of some continuous real-valued function on $|K|$) and paracompact.

For $s \in K$ the *open simplex* $\langle s \rangle \subset |K|$ is defined by

$$\langle s \rangle = \{\alpha \in |K| \mid \alpha(v) \neq 0 \Leftrightarrow v \in s\}$$

Although a closed simplex is a closed set in $|K|$, an open simplex need not be open in $|K|$. However, the open simplex $\langle s \rangle$ is an open subset of $|s|$ because $\langle s \rangle = |s| - |\dot{s}|$. Every point $\alpha \in |K|$ belongs to a unique open simplex (namely, the open simplex $\langle s \rangle$, where $s = \{v \in K \mid \alpha(v) \neq 0\}$). Therefore the open simplexes constitute a partition of $|K|$.

If A is a nonempty subset of $|K|$ that is contained in some closed simplex $|s|$, there is a unique smallest simplex $s \in K$ such that $A \subset |s|$. This smallest simplex is called the *carrier* of A in K. If $A \subset \langle s \rangle$, then the carrier of A is necessarily s. In particular any point α of $|K|$ has as carrier the simplex s such that $\alpha \in \langle s \rangle$.

18 LEMMA Let $A \subset |K|$; *then A contains a discrete subset (in the coherent topology) that consists of exactly one point from each open simplex meeting A.*

PROOF For each $s \in K$ such that $A \cap \langle s \rangle \neq \emptyset$ let $\alpha_s \in A \cap \langle s \rangle$ and let $A' = \{\alpha_s\}$. Because any closed simplex can contain at most a finite subset of A', it follows that every subset of A' is closed in the coherent topology and A' is discrete. ∎

Because a compact subset of any topological space can contain no infinite discrete set, we have the following result.

19 COROLLARY *Every compact subset of $|K|$ is contained in the union of a finite number of open simplexes.* ■

A finite simplicial complex has a compact space. The converse follows from corollary 19.

20 COROLLARY *A simplicial complex K is finite if and only if $|K|$ is compact.* ■

We establish the following analogue of theorem 15 for homotopies.

21 THEOREM *A function $F: |K| \times I \to X$ is continuous if and only if $F \mid (|s| \times I): |s| \times I \to X$ is continuous for every $s \in K$.*

PROOF Because $|K|$ has the topology coherent with the collection of its closed simplexes, and each closed simplex is a closed compact subset of $|K|$, it follows that $|K|$ is compactly generated. By theorem 2.7 in the Introduction, $|K| \times I$ is also compactly generated. It follows from corollary 19 that every compact subset of $|K| \times I$ is contained in $|L| \times I$ for some finite subcomplex $L \subset K$. Therefore $|K| \times I$ has the topology coherent with the collection $\{|L| \times I \mid L \subset K,\ L \text{ finite}\}$. It is clear that this topology is identical with the topology coherent with $\{|s| \times I \mid s \in K\}$ (because if L is finite, $|L| \times I$ has the topology coherent with $\{|s| \times I \mid s \in L\}$). ■

If $\varphi: K_1 \to K_2$ is a simplicial map, then there is a continuous map $|\varphi|_d: |K_1|_d \to |K_2|_d$ defined by

$$|\varphi|_d(\alpha)(v') = \Sigma_{\varphi(v)=v'}\, \alpha(v) \qquad v' \in K_2$$

The same formula defines a continuous map $|\varphi|: |K_1| \to |K_2|$, and there is a commutative square

$$\begin{array}{ccc} |K_1| & \to & |K_1|_d \\ {\scriptstyle |\varphi|}\downarrow & & \downarrow{\scriptstyle |\varphi|_d} \\ |K_2| & \to & |K_2|_d \end{array}$$

An easy verification shows that $|\ |$ and $|\ |_d$ are covariant functors from the category of simplicial complexes to the category of topological spaces, and $|K| \to |K|_d$ is a natural transformation between them. These functors can also be regarded as defined on the category of simplicial pairs to the category of pairs of topological spaces.

A *triangulation* (K,f) of a topological space X consists of a simplicial complex K and a homeomorphism $f: |K| \to X$. If X has a triangulation, X is called a *polyhedron*. Similarly a *triangulation* $((K,L), f)$ *of a pair* (X,A) consists of a simplicial pair (K,L) and a homeomorphism $f: (|K|, |L|) \to (X,A)$. If

(X,A) has a triangulation, (X,A) is called a *polyhedral pair.* In general, a given polyhedron will have triangulations (K_1,f_1) and (K_2,f_2), for which K_1 and K_2 are not isomorphic simplicial complexes.

Following are some examples.

22 For any $n \geq 1$, $(E^{n+1},S^n)$is homeomorphic to $(|\bar{s}|,|\dot{s}|)$, where s is an $(n + 1)$-simplex. Therefore (E^{n+1},S^n) is a polyhedral pair.

23 Given that K is the simplicial complex of example 8 and $f\colon |K| \to \mathbf{R}$ is defined so that $f(|\{n\}|) = n$ and $f \mid |\{n, n + 1\}|$ is a homeomorphism of $|\{n, n + 1\}|$ onto the closed interval $[n, n + 1]$, then (K,f) is a triangulation of $\mathbf{R}$, and $\mathbf{R}$ is a polyhedron.

24 For $n \geq 1$, given that K is the simplicial complex of example 9 and $f\colon |K| \to \mathbf{R}^n$ is defined by the equation $(f(\alpha))_i = \Sigma_{x \in Z^n} \alpha(x)(x)_i$, then (K,f) is a triangulation of $\mathbf{R}^n$, and $\mathbf{R}^n$ is a polyhedron.

Given a vertex $v \in K$, its *star* is defined by

$$\operatorname{st} v = \{\alpha \in |K| \mid \alpha(v) \neq 0\}$$

Because $\alpha \to \alpha(v)$ is a continuous map from $|K|_d$ to I, st v is open in $|K|_d$, and hence also in $|K|$. It is immediate from the definition that

$$\begin{aligned} \alpha \in \operatorname{st} v &\Leftrightarrow \text{carrier } \alpha \text{ has } v \text{ as vertex} \\ &\Leftrightarrow \alpha \in \langle s \rangle \qquad \text{where } s \text{ has } v \text{ as vertex} \end{aligned}$$

Therefore st $v = \bigcup\{\langle s \rangle \mid v \text{ is vertex of } s\}$.

25 LEMMA *Let $L \subset K$ and let $v_0, v_1, \ldots, v_q$ be vertices of K. Then $v_0, v_1, \ldots, v_q$ are vertices of a simplex of L if and only if*

$$\bigcap_{0 \leq i \leq q} \operatorname{st} v_i \cap |L| \neq \varnothing$$

PROOF If there is a simplex $s \in L$ with vertices $v_0, \ldots, v_q$, then $\langle s \rangle \subset \operatorname{st} v_i$ for every i, and $\langle s \rangle \subset |L|$. Therefore $\bigcap \operatorname{st} v_i \cap |L| \neq \varnothing$. Conversely, if $\bigcap \operatorname{st} v_i \cap |L| \neq \varnothing$, let $\alpha \in \bigcap \operatorname{st} v_i \cap |L|$. Then $\alpha(v_i) \neq 0$ for $0 \leq i \leq q$, and carrier α is a simplex s of L whose vertices include $v_0, \ldots, v_q$. Then the set $\{v_0, \ldots, v_q\}$ is a face of s and must belong to L, because L is a complex. ∎

This yields the following relation between K and the open covering of $|K|$ of vertex stars.

26 THEOREM *Let $\mathcal{U} = \{\operatorname{st} v \mid v \in K\}$. The vertex map φ from K to $K(\mathcal{U})$ defined by $\varphi(v) = \operatorname{st} v$ is a simplicial isomorphism $\varphi\colon K \approx K(\mathcal{U})$, and for any $L \subset K$, $\varphi \mid L\colon L \approx K_{|L|}(\mathcal{U})$.* ∎

2 LINEARITY IN SIMPLICIAL COMPLEXES

The linear structure in the set of all functions from any set to $\mathbf{R}$ defines linearity in the space of a simplicial complex. This section is devoted to a study of such linearity. We show that a closed simplex $|s|$ is homeomorphic to the cone with base $|\dot{s}|$. This implies that a closed simplex can be parametrized by "polar coordinates," which are convenient for the construction of maps. We use them to prove that a polyhedral pair has the homotopy extension property with respect to any space.

We also consider linear imbeddings in euclidean space of the space of a simplicial complex; this entails a discussion of locally finite simplicial complexes. Such complexes are characterized by the property that their spaces are locally compact or the equivalent property that the coherent and metric topologies coincide on their spaces.

Let K be a simplicial complex and let $\alpha_1, \ldots, \alpha_p$ be points of a closed simplex $|s|$. Given real numbers $t_1, \ldots, t_p$ such that $0 \le t_i \le 1$ for $i = 1, \ldots, p$ and such that $\Sigma t_i = 1$, the function $\alpha = \Sigma t_i \alpha_i$ is again a point of $|s|$. Therefore each closed simplex has a linear structure such that convex combinations of its points are again points of the closed simplex. Conversely, if $\alpha = \Sigma t_i \alpha_i$ has a simplex s as carrier (so that $\alpha \in \langle s \rangle$), then each $\alpha_i \in |s|$. Therefore we have the following lemma.

1 LEMMA *A convex combination of points of $|K|$ is again a point of $|K|$ if and only if the points all lie in some closed simplex.* ∎

We shall find it convenient to identify the vertices of K with their characteristic functions. That is, if v is a vertex of K, we regard v as also being the function from vertices $v' \in K$ defined by

$$v(v') = \begin{cases} 0 & v \neq v' \\ 1 & v = v' \end{cases}$$

If $\alpha \in |K|$, then we can write $\alpha = \Sigma_{v \in K}\, \alpha(v)v$, the sum on the right being a convex combination of points of $|K|$.

Let X be a topological space which is a subset of some real vector space. We assume that X has a topology coherent with its intersections with finite-dimensional subspaces each such intersection being topologized as a subspace of the finite-dimensional topological linear space in which it lies. For example, X is euclidean space or X is the space of a simplicial complex. A continuous map $f\colon |K| \to X$ is said to be *linear on* K if it is linear in terms of barycentric coordinates. That is, f is linear if for every $\alpha \in |K|$, $\Sigma_{v \in K}\, \alpha(v)f(v)$ is a point of X and

$$f(\alpha) = \Sigma_{v \in K}\, \alpha(v)f(v)$$

It is then clear that a linear map is uniquely determined by the vertex map f_0 from vertices of K to X such that $f_0(v) = f(v)$. Conversely, a vertex map f_0 from vertices of K to X may be extended to a linear map $f\colon |K| \to X$ if and only if for every simplex $s \in K$ all convex combinations of elements in $f_0(s)$ lie in X.

If $\varphi: K_1 \to K_2$ is a simplicial map, then the definition of $|\varphi|$ shows that

$$|\varphi|(\alpha) = \Sigma\, \alpha(v)|\varphi|(v)$$

Therefore $|\varphi|$ is linear.

Let X be a topological space. The *cone* $X * w$ with *base* X and *vertex* w is defined to be the mapping cylinder of the constant map $X \to w$. The points of $X * w$ are parametrized by $[x,t]$ with $x \in X$ and $t \in I$, where $x \in X$ is identified with $[x,0]$ and $[x,1]$ is identified with w for all $x \in X$. Because w is a strong deformation retract of $X * w$, a cone is contractible.

2 LEMMA *For any simplex s of K the cone $|\dot{s}| * w$ is homeomorphic to $|s|$.*

PROOF Choose a point $w_0 \in \langle s \rangle$ and define a map $f\colon |\dot{s}| * w \to |s|$ by $f([\alpha,t]) = tw_0 + (1-t)\alpha$. Then f is continuous (because the linear operations in $|s|$ are continuous). To show that f is injective, assume $f([\alpha,t]) = f([\beta,t'])$ for $\alpha, \beta \in |\dot{s}|$ and $t, t' \in I$. Then

$$tw_0 + (1-t)\alpha = t'w_0 + (1-t')\beta$$

Let s have vertices $v_0, v_1, \ldots, v_q$ and suppose that $\alpha = \Sigma \alpha_i v_i$, $\beta = \Sigma \beta_i v_i$, and $w_0 = \Sigma\, \gamma_i v_i$. Because $\alpha, \beta \in |\dot{s}|$, there is j such that $\alpha_j = 0$ and there is k such that $\beta_k = 0$. Then

$$t\gamma_j = t'\gamma_j + (1-t')\beta_j \qquad \text{and} \qquad (t-t')\gamma_j = (1-t')\beta_j$$

Because $\gamma_j \neq 0$, $t \geq t'$. Similarly, $t\gamma_k + (1-t)\alpha_k = t'\gamma_k$ and so $t' \geq t$. Therefore $t = t'$. It follows then that $(1-t)\,\alpha = (1-t)\beta$, and if $t \neq 1$, $\alpha = \beta$. Therefore either $t = t'$ and $\alpha = \beta$ or $t = t' = 1$. In either case $[\alpha,t] = [\beta,t']$, and f is injective.

We now show that f is surjective. Clearly, $f([\alpha,0]) = \alpha$ and $f([\alpha,1]) = w_0$, and so f maps onto $|\dot{s}|$ and w_0. To show that every point of $\langle s \rangle - w_0$ is on a unique line segment from w_0 to some point of $|\dot{s}|$, let $\alpha \in \langle s \rangle$, with $\alpha \neq w_0$, and suppose that $\alpha = \Sigma \alpha_i v_i$. Consider the function $\varphi(t') = (1+t')\alpha - t'w_0$. $\varphi(0) = \alpha \in \langle s \rangle$, and as t' increases, the barycentric coordinates of $\varphi(t')$ change continuously. Because $\alpha \neq w_0$, there is some i such that $\alpha_i < \gamma_i$. Therefore

$$\varphi(t')(v_i) = \alpha_i - t'(\gamma_i - \alpha_i)$$

is a monotonically decreasing function of t'. By continuity, there exists a unique $t' > 0$ such that $\varphi(t')(v_i) = 0$. Hence there exists a $t'_0 > 0$ which is the smallest t' for which $\varphi(t'_0)(v_i) = 0$ for any $0 \leq i \leq q$. Then $\varphi(t'_0) \in |\dot{s}|$ and

$$\alpha = \frac{t'_0}{1+t'_0} w_0 + \frac{1}{1+t'_0} \varphi(t'_0)$$

shows that $\alpha = f([\varphi(t'_0), t'_0/(1+t'_0)])$, and f is surjective.

Because f is a continuous bijection from a compact space to a Hausdorff space, it is a homeomorphism. ■

The *barycenter* $b(s)$ of the simplex $s = \{v_0,v_1, \ldots ,v_q\}$ is defined to be the point

$$b(s) = \Sigma_{0\le i\le q} \frac{1}{q+1} v_i$$

Clearly, $b(s) \in \langle s\rangle$, and so the carrier of $b(s)$ is s. By lemma 2, $|s|$ is homeomorphic to $|\dot{s}| * w$ in such a way that w corresponds to $b(s)$. If $\alpha \in |\dot{s}|$ and $t \in I$, the point $tb(s) + (1-t)\alpha$ of $|s|$ will be parametrized by *polar coordinates* $[\alpha,t]$, where $[\alpha,t]$ denotes the point of $|\dot{s}| * w$ corresponding to the given point of $|s|$. Then $[\alpha,0] = \alpha$ and $[\alpha,1] = b(s)$ for all $\alpha \in |\dot{s}|$. We use polar coordinates for the following homotopy.

3 LEMMA *For any simplex s, $|s| \times 0 \cup |\dot{s}| \times I$ is a strong deformation retract of $|s| \times I$.*

PROOF If s is a 0-simplex, $|\dot{s}| = \varnothing$ and we know the point $|s| \times 0$ is a strong deformation retract of the closed interval $|s| \times I$. If dim $s > 0$, we define a deformation retraction

$$F: |s| \times I \times I \to |s| \times I$$

to $|s| \times 0 \cup |\dot{s}| \times I$ by the formula in polar coordinates

$$F([\alpha,t], t', t'') = \begin{cases} \left(\left[\alpha, (1-t'')t + \dfrac{t''(2t-t')}{2-t'}\right], (1-t'')t'\right) & t' \le 2t \\ \left([\alpha, (1-t'')t], (1-t'')t' + \dfrac{t''(t'-2t)}{1-t}\right) & 2t \le t' \end{cases}$$

and diagram it for the cases of a 1-simplex and a 2-simplex:

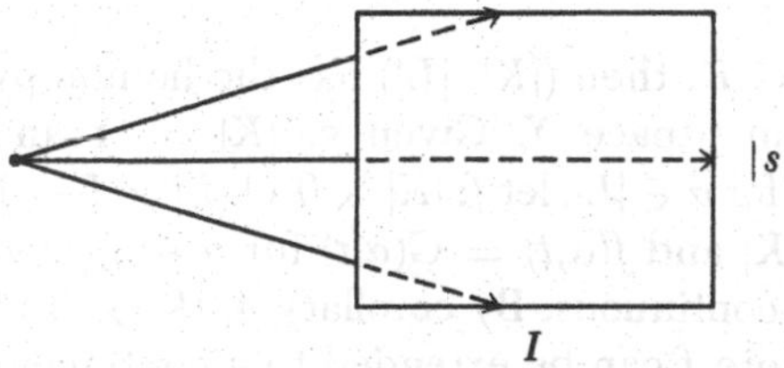

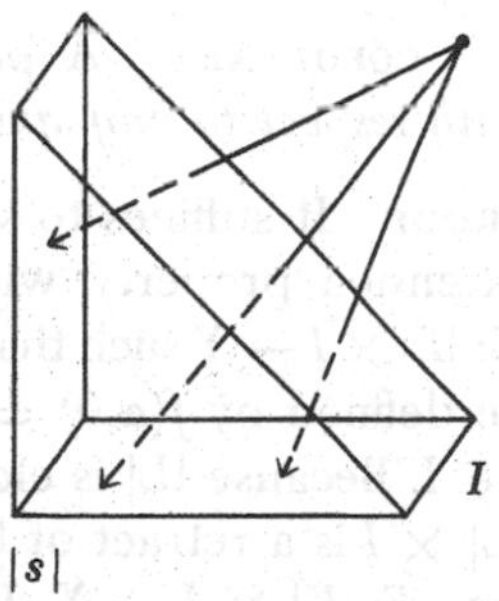

■

4 COROLLARY *For any subcomplex $L \subset K$ the subspace $|K| \times 0 \cup |L| \times I$ is a strong deformation retract of $|K| \times I$.*

PROOF Let $X^n = |K| \times 0 \cup |K^n \cup L| \times I$ for $n \ge -1$. We first show that for each $n \ge 0$ the space X^{n-1} is a strong deformation retract of X^n. For each n-simplex $s \in K - L$ let F_s: $|s| \times I \times I \to |s| \times I$ be a strong deformation retraction of $|s| \times I$ to $|s| \times 0 \cup |\dot{s}| \times I$ (which exists, by lemma 3). For $n \ge 0$ define a map

$$F_n: X^n \times I \to X^n$$

by the conditions

$$F_n \mid |s| \times I \times I = F_s \qquad \text{for an } n\text{-simplex } s \in K - L$$
$$F_n(x,t) = x \qquad x \in X^{n-1}, t \in I$$

Then F_n is well-defined and continuous (because for every simplex s the restriction $F_n \mid |s| \times I \times I$ is continuous), and F_n is a strong deformation retraction of X^n to X^{n-1}.

Let $f_n: X^n \to X^{n-1}$ be the retraction defined by $f_n(x) = F_n(x,1)$ for $x \in X^n$. Let $a_n = 1/n$ for $n \geq 1$, and define $G_n: X^n \times I \to X^n$ by induction on n so that

$$G_0(x,t) = \begin{cases} x & 0 \leq t \leq a_2 \\ F_0\left(x, \dfrac{t - a_2}{1 - a_2}\right) & a_2 \leq t \leq 1 \end{cases}$$

and for $n \geq 1$

$$G_n(x,t) = \begin{cases} x & 0 < t \leq a_{n+2} \\ F_n\left(x, \dfrac{t - a_{n+2}}{a_{n+1} - a_{n+2}}\right) & a_{n+2} \leq t \leq a_{n+1} \\ G_{n-1}(f_n(x),t) & a_{n+1} \leq t \leq 1 \end{cases}$$

By induction on n, it is easily verified that G_n is a strong deformation retraction of X^n to X^{-1} such that $G_n \mid X^{n-1} \times I = G_{n-1}$. Therefore there is a map

$$G: |K| \times I \times I \to |K| \times I$$

such that $G \mid X^n \times I = G_n$. Then G is a strong deformation retraction of $|K| \times I$ to $|K| \times 0 \cup |L| \times I$. ∎

5 COROLLARY *A polyhedral pair has the homotopy extension property with respect to any space.*

PROOF It suffices to show that if $L \subset K$, then $(|K|, |L|)$ has the homotopy extension property with respect to any space Y. Given $g: |K| \to Y$ and $G: |L| \times I \to Y$ such that $G(\alpha,0) = g(\alpha)$ for $\alpha \in |L|$, let $f: |K| \times 0 \cup |L| \times I \to Y$ be defined by $f(\alpha,0) = g(\alpha)$ for $\alpha \in |K|$ and $f(\alpha,t) = G(\alpha,t)$ for $\alpha \in |L|$ and $t \in I$. Because $|L|$ is closed in $|K|$, f is continuous. By corollary 4, $|K| \times 0 \cup |L| \times I$ is a retract of $|K| \times I$. Therefore f can be extended to a continuous map $F: |K| \times I \to Y$. Then $F(\alpha,0) = g(\alpha)$ for $\alpha \in |K|$ and $F \mid |L| \times I = G$. ∎

Let us now consider linear imbeddings of $|K|$ in euclidean space.

6 LEMMA *A linear map $f: |s| \to \mathbf{R}^n$ is an imbedding if and only if it maps the vertex set of s to an affinely independent set in $\mathbf{R}^n$.*

PROOF Let $f(v_i) = p_i$, where $s = \{v_i\}$. We show that the set $\{p_i\}$ is affinely dependent if and only if f is not injective. $\{p_i\}$ is affinely dependent if and only if there exist α_i not all zero such that $\Sigma\alpha_i p_i = 0$ and $\Sigma\alpha_i = 0$. Assume the points p_i enumerated so that $\alpha_i \geq 0$ for $i \leq j_0$ and $\alpha_i < 0$ for $i > j_0$.

Then $\sum_{i \le j_0} \alpha_i p_i = \sum_{i > j_0} (-\alpha_i) p_i$. If $a = \sum_{i \le j_0} \alpha_i = \sum_{i > j_0} -\alpha_i$, then $\sum_{i \le j_0} (\alpha_i/a) p_i = \sum_{i > j_0} (-\alpha_i/a) p_i$. It follows from the linearity of f that $f(\sum_{i \le j_0} (\alpha_i/a) v_i) = f(\sum_{i > j_0} (-\alpha_i/a) v_i)$, showing that f is not injective.

Conversely, if f is not injective, then $f(\sum \alpha_i v_i) = f(\sum \beta_i v_i)$, where $\alpha_{j_0} \neq \beta_{j_0}$ for some j_0. Then $\sum (\alpha_i - \beta_i) p_i = 0$ and $\sum (\alpha_i - \beta_i) = 0$. Because $\alpha_{j_0} - \beta_{j_0} \neq 0$, the set $\{p_i\}$ is affinely dependent. ∎

A simplicial complex K is said to be *locally finite* if every vertex v of K belongs to only finitely many simplexes of K.

7 LEMMA *If K is locally finite, every point of $|K|_d$ has a neighborhood of the form $|L|_d$, where L is a finite subcomplex of K.*

PROOF Let $\alpha \in |K|_d$. Then $\alpha \in \operatorname{st} v$ for some vertex v of K. Because v is a vertex of only finitely many simplexes $\{s_i\}$ of K, $\operatorname{st} v$ is contained in the compact set $\bigcup |s_i|$. Let $L = \{s \in K \mid s \text{ is a face of } s_i \text{ for some } i\}$. Then L is a finite subcomplex of K, and $\alpha \in \operatorname{st} v \subset |L|_d$. ∎

8 THEOREM *For a simplicial complex K, the following are equivalent:*

(a) K is locally finite.
(b) $|K|$ is locally compact.
(c) $|K| \to |K|_d$ is a homeomorphism.
(d) $|K|$ is metrizable.
(e) $|K|$ satisfies the first axiom of countability.

PROOF (a) ⟹ (b). By lemma 7, if α is a point of $|K|_d$, there is a finite subcomplex $L \subset K$ such that α is in the interior of $|L|_d$. Then α is in the interior of $|L|$ in $|K|$. Therefore $|L|$ is a compact neighborhood of α in $|K|$.

(b) ⟹ (c). To show that $|K| \to |K|_d$ is an open map, let U be an open subset of $|K|$ with compact closure $\bar{U}$ in $|K|$. It suffices to show that U is open in $|K|_d$. Because $\bar{U}$ is compact, there is a finite subcomplex $L \subset K$ such that $\bar{U} \subset |L|$ (by corollary 3.1.19). Let K_1 be the subcomplex of K defined by

$$K_1 = \{s \in K \mid |s| \cap U = \varnothing\}$$

If $s \in K - K_1$, then $|s| \cap U$ is a nonempty open subset of $|s|$. Therefore $\langle s \rangle \cap U \neq \varnothing$ and $\langle s \rangle \cap |L| \neq \varnothing$. The fact that the open simplexes of K form a partition of K implies that $s \in L$, and we have shown that $K = K_1 \cup L$. Now, $|K|_d - |K_1|_d$ is an open subset of $|K|_d$. Because L is finite, $|L| \to |L|_d$ is a homeomorphism. Therefore U is open in $|L|_d$, and so it is open in $|L|_d - |K_1|_d$. Because $|L|_d - |K_1|_d = |K|_d - |K_1|_d$, U is open in $|K|_d$.

(c) ⟹ (d). Because $|K|_d$ is metrizable, if $|K|$ and $|K|_d$ are homeomorphic, then $|K|$ is also metrizable.

(d) ⟹ (e). Every metrizable space satisfies the first axiom of countability.

(e) ⟹ (a). Assume that K is not locally finite and let v be a vertex of an infinite set of simplexes $\{s_i\}_{i=1,2,\ldots}$ of K. Assume that v has a countable base of neighborhoods $\{U_i\}_{i=1,2,\ldots}$ in $|K|$. Without loss of generality, we may assume $U_i \supset U_{i+1}$ for all $i \geq 1$. For each i, $\langle s_i \rangle \cap U_i \neq \emptyset$, because v, being a vertex of s_i, is in the closure of $\langle s_i \rangle$. Let $\alpha_i \in \langle s_i \rangle \cap U_i$. Then the sequence $\{\alpha_i\}$ has v as a limit point (because each U_i contains all α_j with $j \geq i$), but in the coherent topology the set $\{\alpha_i\}$ is discrete, because it meets every closed simplex $|s|$ in a finite set. ∎

A *realization* of a simplicial complex K in $\mathbf{R}^n$ is a linear imbedding of $|K|$ in $\mathbf{R}^n$. The following theorem characterizes those complexes K which have realizations in some euclidean space.

9 THEOREM *If K has a realization in $\mathbf{R}^n$, then K is countable and locally finite, and* $\dim K \leq n$. *Conversely, if K is countable and locally finite, and* $\dim K \leq n$, *then K has a realization as a closed subset in* $\mathbf{R}^{2n+1}$.

PROOF Let $f: |K| \to \mathbf{R}^n$ be a linear imbedding. If K is uncountable, it follows from lemma 3.1.18 that $|K|$ contains an uncountable discrete set A'. Then $f(A')$ is an uncountable discrete subset of $\mathbf{R}^n$, which is impossible because $\mathbf{R}^n$ is separable. Therefore K is countable. Clearly $|K|$ is metrizable and, by theorem 8, K is locally finite. It follows from lemma 6 and theorem 5.3 in the Introduction that $\dim K \leq n$.

To prove the converse statement, let $\{p_i\}$ be a sequence of points in $\mathbf{R}^{2n+1}$ such that

(*a*) Every set of $2n + 2$ of the points p_i is affinely independent.
(*b*) If C is any compact subset of $\mathbf{R}^{2n+1}$, there exists j such that C is disjoint from the convex subset of $\mathbf{R}^{2n+1}$ generated by the set $\{p_i \mid i \geq j\}$.

For example, let $H_1 \supset H_2 \supset \cdots$ be a decreasing sequence of closed half-spaces of $\mathbf{R}^{2n+1}$ such that $\bigcap H_i = \emptyset$, and assuming p_i defined for $i < q$, inductively choose p_q to be a point of H_q not lying on any of the finite number of affine varieties determined by $2n + 1$ or fewer points of the set $\{p_i \mid 1 \leq i \leq q - 1\}$.)

Assume that K is countable and locally finite and $\dim K \leq n$, and let $\{v_i\}_{i=1,2,\ldots}$ be an enumeration of the vertices of K. Define $f: |K| \to \mathbf{R}^{2n+1}$ to be the linear map such that $f(v_i) = p_i$. Because of condition (*a*), it follows that for any $s \in K$, $f \mid |s|$ is a linear imbedding of $|s|$ in $|K|$, and if s and $s' \in K$, then

$$f(|s| \cap |s'|) = f(|s|) \cap f(|s'|)$$

Therefore f is injective. Because of condition (*b*), if C is any compact subset of $\mathbf{R}^{2n+1}$, there is j such that $f^{-1}(C) \subset \bigcup \{\text{st } v_i \mid i \leq j\}$. Since K is locally finite, this implies that $f^{-1}(C) \subset |L|$ for some finite subcomplex $L \subset K$. Therefore $f^{-1}(C)$ is compact in $|K|$. If A is closed in $|K|$ and C is compact in $\mathbf{R}^{2n+1}$, then $f(A) \cap C = f(A \cap f^{-1}(C))$ is closed in C [because $A \cap f^{-1}(C)$ is a closed subset of the compact subset $f^{-1}(C)$ of $|K|$ and $f \mid f^{-1}(C)$ is a homeomorphism of

$f^{-1}(C)$ to $f(f^{-1}(C))$]. Therefore f is a closed map and is a linear imbedding of $|K|$ as a closed subset in $\mathbf{R}^{2n+1}$. ∎

3 SUBDIVISION

Our main interest in simplicial complexes is in the polyhedra they describe. To study a polyhedron it is important to consider its different triangulations and their interrelationships. This section is devoted to proving the existence of "small" triangulations of a polyhedron, which are used in the next section in proving that arbitrary continuous maps between polyhedra can be approximated by simplicial maps.

Let K be a simplicial complex. A *subdivision* of K is a simplicial complex K' such that

(*a*) The vertices of K' are points of $|K|$.
(*b*) If s' is a simplex of K', there is some simplex s of K such that $s' \subset |s|$ (that is, s' is a finite nonempty subset of $|s|$).
(*c*) The linear map $|K'| \to |K|$ mapping each vertex of K' to the corresponding point of $|K|$ is a homeomorphism.

Note that conditions (*a*) and (*b*) assert that every simplex s' of K' has a carrier $s \in K$. If K' is a subdivision of K, we identify $|K'|$ and $|K|$ by the linear homeomorphism of condition (*c*). The following fact is immediate from the definition.

1 *Any subdivision of a subdivision of K is itself a subdivision of K.* ∎

The next fact is also true (but somewhat more difficult to prove).

2 *If K' and K'' are subdivisions of K, there is a subdivision K''' of K that is a subdivision of K' and of K''.* ∎

Thus, statements 1 and 2 assert that the subdivisions of K form a directed set with respect to the partial ordering defined by the relation of subdivision.

3 LEMMA *Let K and K' be simplicial complexes satisfying conditions (a) and (b). If $s \in K$ is the carrier of $s' \in K'$, then $\langle s' \rangle \subset \langle s \rangle$.*

PROOF Let $v_0', \ldots, v_p'$ be the vertices of s' and let $v_0, \ldots, v_q$ be the vertices of the carrier s of s'. Because $s' \subset |s|$, for $0 \le i \le p$, $v_i' = \Sigma \alpha_{ij} v_j$. Because s is the smallest such simplex, for $0 \le j \le q$ there exists $0 \le i \le p$ such that $\alpha_{ij} \ne 0$. Let $\beta \in \langle s' \rangle$. Then

$$\beta = \sum_i \beta_i v_i' = \sum_j \left(\sum_i \beta_i \alpha_{ij}\right) v_j$$

and because $\beta_i > 0$ for all i, $\sum_i \beta_i \alpha_{ij} > 0$ for all j. Therefore $\beta \in \langle s \rangle$ and $\langle s' \rangle \subset \langle s \rangle$. ∎

4 THEOREM *Let K' and K be simplicial complexes satisfying conditions (a) and (b). Then K' is a subdivision of K if and only if for $s \in K$ the set $\{\langle s'\rangle \mid s' \in K', \langle s'\rangle \subset \langle s\rangle\}$ is a finite partition of $\langle s\rangle$.*

PROOF Assume that K' and K satisfy conditions (*a*) and (*b*) and the condition that $\{\langle s'\rangle \mid s' \in K', \langle s'\rangle \subset \langle s\rangle\}$ is a finite partition of $\langle s\rangle$ for $s \in K$. Because any simplex $s \in K$ has only a finite number of faces, it follows that

$$K'(s) = \{s' \in K' \mid \text{there exists a face } s_1 \text{ of } s \text{ such that } \langle s'\rangle \subset \langle s_1\rangle\}$$

is a finite subcomplex of K', and the linear map $h_s\colon |K'(s)| \to |s|$ that maps each vertex of $K'(s)$ to itself is a homeomorphism. Therefore there is a continuous map $g\colon |K| \to |K'|$ such that $g \mid |s| = h_s^{-1}$ for $s \in K$, which is an inverse of the linear map $h\colon |K'| \to |K|$. Therefore h is a homeomorphism, and K' and K satisfy condition (*c*).

Conversely, if K' is a subdivision of K, then $\{s' \mid s' \in K'\}$ is a partition of $|K'| = |K|$. For $s \in K$, consider the sets $\langle s'\rangle \cap \langle s\rangle$ for $s' \in K'$. By lemma 3, either $\langle s'\rangle \cap \langle s\rangle = \emptyset$ or $\langle s'\rangle \subset \langle s\rangle$. Therefore $\{\langle s'\rangle \mid s' \in K', \langle s'\rangle \subset \langle s\rangle\}$ is a partition of $\langle s\rangle$. Because $|s|$ is compact, it follows from corollary 3.1.19 that this set is a finite partition of $\langle s\rangle$. ∎

We use this result to show that any subdivision of K simultaneously subdivides every subcomplex of K.

5 COROLLARY *Let K' be a subdivision of K and let L be a subcomplex of K. There is a unique subcomplex L' of K' which is a subdivision of L.*

PROOF If L' is a subcomplex of K' that is a subdivision of L, then $L' = \{s' \in K' \mid \langle s'\rangle \subset |L|\}$, which proves the uniqueness of L'. To prove the existence of L', we prove that $\{s' \in K' \mid \langle s'\rangle \subset |L|\}$ has the desired properties. It is clear that this set is a subcomplex L' of K' and that L' and L satisfy conditions (*a*) and (*b*) above. We use theorem 4 to show that L' is a subdivision of L. If $s \in L$, by theorem 4 the set $\{\langle s'\rangle \mid s' \in K', \langle s'\rangle \subset \langle s\rangle\}$ is a finite partition of $\langle s\rangle$. By definition of L',

$$\{\langle s'\rangle \mid s' \in K', \langle s'\rangle \subset \langle s\rangle\} = \{\langle s'\rangle \mid s' \in L', \langle s'\rangle \subset \langle s\rangle\}$$

Therefore, by theorem 4, L' is a subdivision of L. ∎

The subdivision L' of L in corollary 5 is called the ***subdivision** of L **induced by** K'* and is denoted by $K' \mid L$.

From the definition of subdivision two facts are immediate.

6 *If $f\colon |K| \to X$ is linear on K and K' is a subdivision of K, then f is also linear on K'.* ∎

7 *If $((K,L), f)$ is a triangulation of (X,A) and K' is a subdivision of K, then $((K',K' \mid L), f)$ is also a triangulation of (X,A).* ∎

For any simplicial complex we construct a particular subdivision, called the barycentric subdivision. For this we need the following lemma, which shows how to extend a subdivision of $\dot{s}$ to a subdivision of $\bar{s}$ for any simplex s.

8 LEMMA *Let s be a simplex of some complex and let K' be a subdivision of $\dot{s}$. For any $w_0 \in \langle s \rangle$, $K' * w_0$ is a subdivision of $\bar{s}$.*

PROOF In the statement of lemma 8, w_0 is regarded as a simplicial complex having a single vertex and $K' * w_0$ is the join defined in example 3.1.7. It is clear that $K' * w_0$ satisfies requirements (*a*) and (*b*) for a subdivision of $\bar{s}$. It follows from lemma 3.2.2 that any point of $|s|$ either equals w_0, belongs to $|\dot{s}|$, or belongs to a unique open simplex of the form $\langle s' \cup \{w_0\} \rangle$, where $s' \in K'$. Therefore the open simplexes of $|K' * w_0|$ constitute a finite partition of $|s|$, and by theorem 4, $K' * w_0$ is a subdivision of $\bar{s}$. ∎

The subdivision of $\bar{s}$ obtained by applying lemma 8 is pictured below for a 2-simplex s.

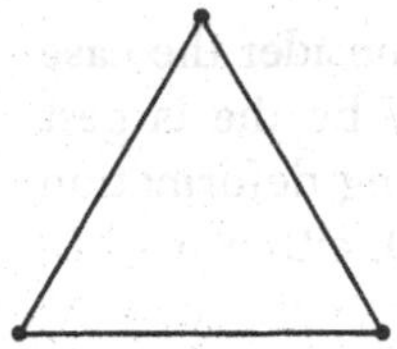

s = *triangle and its faces*

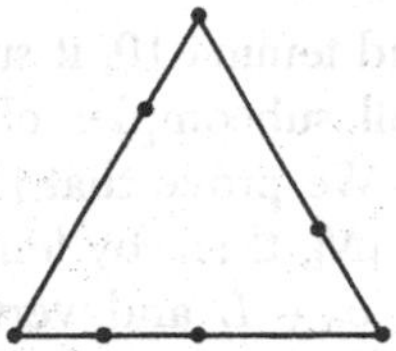

K' = *pictured subdivision of the boundary of the triangle*

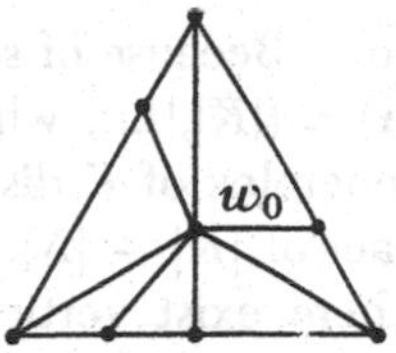

$K' * w_0$ = *pictured triangles and their faces*

We are now ready to prove the existence of the barycentric subdivision. Let K be a simplicial complex. We define sd K to be the simplicial complex whose vertices are the barycenters of the simplexes of K and whose simplexes are finite nonempty collections of barycenters of simplexes which are totally ordered by the face relation in K. Thus the simplexes of sd K are finite sets $\{b(s_0), \ldots, b(s_q)\}$ such that s_{i-1} is a face of s_i for $i = 1, \ldots, q$. We shall always assume the vertices of a simplex of sd K to be enumerated in this order.

It is clear that sd K is a simplicial complex and that if L is a subcomplex of K, then sd L is a subcomplex of sd K. Furthermore, if $b(s_q)$ is the last vertex of a simplex $s' \in$ sd K, then $s' \subset |s_q|$, and since s_q is the carrier of $b(s_q)$, s_q is the carrier of s'. Therefore sd K and K satisfy conditions (*a*) and (*b*).

9 THEOREM sd K *is a subdivision of* K.

PROOF We show that sd K and K satisfy the hypotheses of theorem 4. If $s \in K$, then, by lemma 3 and the remarks above,

$$\{s' \in \text{sd } K \mid \langle s' \rangle \subset \langle s \rangle\} = \{s' \in \text{sd } K \mid \text{last vertex of } s' = b(s)\}$$
$$= \{s' \in \text{sd } \bar{s} \mid \langle s' \rangle \subset \langle s \rangle\}$$

Therefore we need only show that sd $\bar{s}$ is a subdivision of $\bar{s}$ for any $s \in K$. We do this by induction on dim s. If dim $s = 0$, sd $\bar{s} = \bar{s}$ is a subdivision of $\bar{s}$. For $q > 0$, assume that sd $\bar{s}_1$ is a subdivision of $\bar{s}_1$ for every simplex s_1 with dim $s_1 < q$, and let s be a q-simplex. By the inductive assumption, sd $\dot{s}$ is a subdivision of $\dot{s}$. The definition of the barycentric subdivision shows that sd $\bar{s}$ = sd $\dot{s} * b(s)$. By lemma 8, this is a subdivision of $\bar{s}$. ∎

The subdivision sd K is called the *barycentric subdivision of* K. The *iterated barycentric subdivisions* $\mathrm{sd}^n K$ are defined for $n \geq 0$ inductively, so that

$$\mathrm{sd}^0 K = K$$
$$\mathrm{sd}^n K = \mathrm{sd}\,(\mathrm{sd}^{n-1} K) \qquad n \geq 1$$

10 LEMMA *If L is a subcomplex of K, sd L is a full subcomplex of* sd K.

PROOF Let $\{b(s_0), \ldots, b(s_q)\}$ be a simplex of sd K all of whose vertices belong to sd L. Then s_{i-1} is a face of s_i for $i = 1, \ldots, q$ and each $s_i \in L$. Therefore $\{b(s_0), \ldots, b(s_q)\} \in \mathrm{sd}\, L$. ■

11 COROLLARY *Let (X,A) be a polyhedral pair. Then A is a strong deformation retract of some neighborhood of A in X.*

PROOF Because of statement 7 and lemma 10, it suffices to consider the case $(X,A) = (|K|,|L|)$, where L is a full subcomplex of K. Let N be the largest subcomplex of K disjoint from L. We prove that $|L|$ is a strong deformation retract of $|K| - |N|$. If $\alpha \in |K| - |N|$, then, by lemma 3.1.10, either $\alpha \in |L|$ or there exist vertices $v_0, \ldots, v_p \in L$ and vertices $v_{p+1}, \ldots, v_q \in N$, with $0 \leq p$ and $p + 1 \leq q$, such that $\alpha \in \langle v_0, \ldots, v_q \rangle$. In the latter case, $\alpha = \sum_{0 \leq i \leq q} \alpha_i v_i$, with $\alpha_i > 0$, and we define $a = \sum_{0 \leq i \leq p} \alpha_i$. Then $0 < a < 1$ and we let $\alpha_i' = \alpha_i / a$ for $0 \leq i \leq p$ and $\alpha_i'' = \alpha_i/(1-a)$ for $p + 1 \leq i \leq q$. Then $\alpha = a\alpha' + (1-a)\alpha''$, where $\alpha' = \sum_{0 \leq i \leq p} \alpha_i' v_i$ is in $|L|$ and $\alpha'' = \sum_{p+1 \leq i \leq q} \alpha_i'' v_i$ is in $|N|$. A strong deformation retraction $F\colon (|K| - |N|) \times I \to |K| - |N|$ of $|K| - |N|$ to $|L|$ is defined by

$$F(\alpha,t) = \begin{cases} \alpha & \alpha \in |L|,\ t \in I \\ t\alpha' + (1-t)\alpha & \alpha \in |K| - (|N| \cup |L|),\ t \in I \end{cases}$$

F is continuous because $F \mid |L| \times I$ is continuous, and for any simplex of K of the form $s' \cup s''$, where $s' \in L$ and $s'' \in N$, $F \mid [|s' \cup s''| \cap (|K| - |N|)] \times I$ is continuous. ■

Let X be a polyhedron and let $\mathcal{U}$ be an open covering of X. A triangulation (K,f) of X is said to be *finer* than $\mathcal{U}$ if for every vertex $v \in K$ there is $U \in \mathcal{U}$ such that $f(\mathrm{st}\, v) \subset U$. A simplicial complex K is said to be *finer* than an open covering $\mathcal{U}$ of $|K|$ if the triangulation $(K, 1_{|K|})$ of $|K|$ is finer than $\mathcal{U}$ (that is, for each vertex $v \in K$ there is $U \in \mathcal{U}$ such that $\mathrm{st}\, v \subset U$). We show that if $\mathcal{U}$ is any open covering of a compact polyhedron, there are triangulations finer than $\mathcal{U}$.

A metric on $|K|$ is said to be *linear on* K if it is induced from the norm in $\mathbf{R}^n$ by a realization of K in $\mathbf{R}^n$. Any finite simplicial complex has linear metrics, and if K' is any subdivision of K, a metric that is linear on $|K|$ is also linear on $|K'|$.

12 LEMMA *Given a metric linear on an m-simplex s, then for any $s' \in \mathrm{sd}\, \bar{s}$*

$$\mathrm{diam}\, |s'| \leq \frac{m}{m+1}\, \mathrm{diam}\, |s|$$

PROOF Let $\{p_j \mid 0 \le j \le m\}$ be points of $\mathbf{R}^n$ and assume that y is a convex combination of $\{p_j\}$ (that is, $y = \Sigma t_j p_j$, where $\Sigma t_j = 1$ and $t_j \ge 0$) and let $x \in \mathbf{R}^n$. Then

$$\|x - y\| \le \|x - \Sigma t_j p_j\| = \|\Sigma t_j(x - p_j)\| \le \Sigma t_j \|x - p_j\|$$

Therefore $\|x - y\| \le \sup \|x - p_j\|$. If x is also a convex combination of $\{p_j\}$, then $\|x - y\| \le \sup \|p_i - p_j\|$.

Regard $|s|$ as imbedded linearly in $\mathbf{R}^n$, with vertices $p_0, p_1, \ldots, p_m$. Then, by the above result, diam $|s| \le \sup \|p_i - p_j\|$, and if s' is a simplex of sd $\bar{s}$, diam $|s'| \le \sup \{\|p' - p''\| \mid p', p'' \in s'\}$. Therefore we need only show that if $p' = (p_0 + \cdots + p_q)/(q + 1)$ and $p'' = (p_0 + \cdots + p_r)/(r + 1)$, where $q \le r$, then $\|p' - p''\| \le [m/(m + 1)] \sup \|p_i - p_j\|$. Again by the result above,

$$\|p' - p''\| \le \sup \{\|p_i - p''\| \mid 0 \le i \le q\}$$

and also, for $0 \le i \le q$,

$$\|p_i - p''\| = \|p_i - \frac{1}{r + 1} \sum_{0 \le j \le r} p_j\| \le \frac{1}{r + 1} \sum_{0 \le j \le r} \|p_i - p_j\|$$

$$\le \frac{r}{r + 1} \sup \|p_i - p_j\|$$

Therefore

$$\|p' - p''\| \le \frac{r}{r + 1} \sup \{\|p_i - p_j\| \mid 0 \le i \le q, 0 \le j \le r\}$$

$$\le \frac{r}{r + 1} \operatorname{diam} |s|$$

Because $r \le m$, $r/(r + 1) \le m/(m + 1)$ and diam $|s'| \le [m/(m + 1)]$ diam $|s|$. ∎

Given a metric on $|K|$, we define *mesh* of K by

$$\operatorname{mesh} K = \sup \{\operatorname{diam} |s| \mid s \in K\}$$

13 COROLLARY *If K is an m-dimensional complex and $|K|$ has a metric linear on K, then*

$$\operatorname{mesh} (\operatorname{sd} K) \le \frac{m}{m + 1} \operatorname{mesh} K \quad ∎$$

This gives us the important result toward which we have been heading.

14 THEOREM *Let $\mathfrak{U}$ be an open covering of a compact polyhedron X. Then X has triangulations finer than $\mathfrak{U}$.*

PROOF Let (K,f) be a triangulation of X. We shall show that there exists an integer N such that if $n \ge N$, then $(\operatorname{sd}^n K, f)$ is finer than $\mathfrak{U}$. Let $|K|$ be provided with a metric linear on K and let $\varepsilon > 0$ be a Lebesque number of the open covering $f^{-1}\mathfrak{U} = \{f^{-1}U \mid U \in \mathfrak{U}\}$ with respect to this metric [thus, if

$A \subset |K|$ and diam $A \leq \varepsilon$, then $f(A)$ is contained in some element of $\mathcal{U}$]. Such a number $\varepsilon > 0$ exists because $|K|$ is compact. Let $m = \dim K$ and choose N so that $[m/(m+1)]^N$ mesh $K \leq \varepsilon/2$ (such an N exists because $\lim_{n\to\infty} [m/(m+1)]^n = 0$). If $n \geq N$, then, by corollary 13, mesh sd^n $K \leq \varepsilon/2$. If v' is any vertex of sd^n K, diam (st v') ≤ 2 mesh sd^n $K \leq \varepsilon$. Therefore f(st v') is contained in some element of $\mathcal{U}$, and (sd^n K, f) is finer than $\mathcal{U}$ if $n \geq N$. ■

This last result is true even if X is not compact. More precisely, if (K,f) is a triangulation of a polyhedron X and $\mathcal{U}$ is an open covering of X, there exist subdivisions K' of K such that (K',f) is finer than $\mathcal{U}$.[1] However, when X is not compact K' cannot generally be chosen to be an iterated barycentric subdivision of K, and so the proof for this case is more complicated than the proof of theorem 14. We need only the form proven in theorem 14, however, and so omit further consideration of the more general case.

4 SIMPLICIAL APPROXIMATION

A continuous map between the spaces of simplicial complexes can be suitably approximated by simplicial maps. This section contains a definition and characterization of the approximations and a proof of their existence for maps of a compact polyhedron into any polyhedron. Finally, we apply the result obtained to deduce some connectivity properties of spheres.

Let K_1 and K_2 be simplicial complexes and let $f: |K_1| \to |K_2|$ be continuous. A simplicial map $\varphi: K_1 \to K_2$ is called a *simplicial approximation* to f if $f(\alpha) \in \langle s_2 \rangle$ implies $|\varphi|(\alpha) \in |s_2|$ (or, equivalently, $f(\alpha) \in |s_2|$ implies $|\varphi|(\alpha) \in |s_2|$) for $\alpha \in |K_1|$ and $s_2 \in K_2$. Note that if v is a vertex of K_1 such that $f(v)$ is a vertex of K_2, then $|\varphi|(v) = f(v)$. Therefore we obtain the following result.

1 LEMMA *Let $f: |K_1| \to |K_2|$ be a map and suppose that for some subcomplex $L_1 \subset K_1$, $f \mid |L_1|$ is induced by a simplicial map $L_1 \to K_2$. If $\varphi: K_1 \to K_2$ is a simplicial approximation to f, then $|\varphi| \mid |L_1| = f \mid |L_1|$.* ■

In particular, the only simplicial approximation to a map $|\varphi|: |K_1| \to |K_2|$ induced by a simplicial map $\varphi: K_1 \to K_2$ is φ itself. One sense in which a simplicial approximation is an approximation is the following.

2 LEMMA *Let $\varphi: K_1 \to K_2$ be a simplicial approximation to a map $f: |K_1| \to |K_2|$ and let $A \subset |K_1|$ be the subset of $|K_1|$ on which $|\varphi|$ and f agree. Then $|\varphi| \simeq f$ rel A.*

PROOF A homotopy relative to A from $|\varphi|$ to f is defined by the equation

$$F(\alpha,t) = tf(\alpha) + (1-t)(|\varphi|(\alpha)) \qquad \alpha \in |K_1|, t \in I$$

[1] See theorem 35 in J. H. C. Whitehead, Simplicial spaces, nuclei, and m-groups, *Proceedings of the London Mathematical Society*, vol. 45, pp. 243–327 (1939).

The right-hand side is well-defined, because if $f(\alpha) \in \langle s_2 \rangle$, then $|\varphi|(\alpha) \in |s_2|$, and so $F(\alpha,t) \in |s_2|$ for $t \in I$. The continuity of F is easily verified. Clearly, if $\alpha \in A$, then $F(\alpha,t) = f(\alpha)$ for all $t \in I$. Therefore $F\colon |\varphi| \simeq f$ rel A. ∎

The following theorem is a useful characterization of simplicial approximations.

3 THEOREM *A vertex map φ from K_1 to K_2 is a simplicial approximation to $f\colon |K_1| \to |K_2|$ if and only if for every vertex $v \in K_1$*

$$f(\text{st } v) \subset \text{st } \varphi(v)$$

PROOF Assume that φ is a simplicial approximation to f. Let $\alpha \in \text{st } v$ and suppose $f(\alpha) \in \langle s_2 \rangle$. Then $\alpha(v) \neq 0$ and $|\varphi|(\alpha) \in |s_2|$. Because φ is simplicial, $|\varphi|(\alpha)(\varphi(v)) \neq 0$. Therefore $\varphi(v)$ is a vertex of $|s_2|$, and $f(\alpha) \in \text{st } \varphi(v)$. Since this is so for every $\alpha \in \text{st } v$, $f(\text{st } v) \subset \text{st } \varphi(v)$.

Conversely, assume that φ is a vertex map such that $f(\text{st } v) \subset \text{st } \varphi(v)$ for every vertex $v \in K_1$. We show that φ is a simplicial map. If $\{v_i\}$ are vertices of a simplex of K_1, then $\bigcap \text{st } v_i \neq \varnothing$ (by lemma 3.1.25) and

$$\varnothing \neq f(\bigcap \text{st } v_i) \subset \bigcap f(\text{st } v_i) \subset \bigcap \text{st } \varphi(v_i)$$

By lemma 3.1.25, $\{\varphi(v_i)\}$ are vertices of some simplex of K_2. Therefore φ is a simplicial map $K_1 \to K_2$.

To show that φ is a simplicial approximation to f, assume $\alpha \in \langle s_1 \rangle$ and $f(\alpha) \in \langle s_2 \rangle$ and let v be any vertex of s_1. Then $\alpha \in \text{st } v$ and, by hypothesis, $f(\alpha) \in \text{st } \varphi(v)$. Therefore $\varphi(v)$ is a vertex of s_2. This is so for every vertex v of s_1. Because φ is simplicial, $|\varphi|(|s_1|) \subset |s_2|$. Hence $|\varphi|(\alpha) \in |s_2|$, and φ is a simplicial approximation to f. ∎

We are also interested in simplicial approximations $\varphi\colon (K_1,L_1) \to (K_2,L_2)$ to maps $f\colon (|K_1|,|L_1|) \to (|K_2|,|L_2|)$. The following corollary shows that any simplicial approximation $K_1 \to K_2$ to a map $f\colon (|K_1|,|L_1|) \to (|K_2|,|L_2|)$ is automatically a simplicial approximation when regarded as a map of pairs.

4 COROLLARY *Let $f\colon |K_1| \to |K_2|$ be a map such that $f(|L_1|) \subset |L_2|$ for $L_1 \subset K_1$ and $L_2 \subset K_2$ and let $\varphi\colon K_1 \to K_2$ be a simplicial approximation to f. Then $\varphi \mid L_1$ maps L_1 to L_2 and is a simplicial approximation to $f \mid |L_1|$.*

PROOF By theorem 3, it suffices to show that if v is a vertex of L_1, then $\varphi(v)$ is a vertex of L_2 such that

$$f(\text{st } v \cap |L_1|) \subset \text{st } \varphi(v) \cap |L_2|$$

Since φ is a simplicial approximation to f, $f(\text{st } v) \subset \text{st } \varphi(v)$, and if v is a vertex of L_1, then $f(v) \in \langle s_2 \rangle$ for some $s_2 \in L_2$ [because $f(|L_1|) \subset |L_2|$]. Therefore $\varphi(v)$ is a vertex of L_2 and

$$f(\text{st } v \cap |L_1|) \subset f(\text{st } v) \cap |L_2| \subset \text{st } \varphi(v) \cap |L_2| \quad ∎$$

It follows from corollary 4 that any simplicial approximation to a map

f: $(|K_1|,|L_1|) \to (|K_2|,|L_2|)$ is a simplicial map φ: $(K_1,L_1) \to (K_2,L_2)$. From lemma 2, it follows that $f \simeq |\varphi|$ as a map of pairs.

5 COROLLARY *The composite of simplicial approximations to maps is a simplicial approximation to the composite of the maps.*

PROOF Let φ: $K_1 \to K_2$ be a simplicial approximation to f: $|K_1| \to |K_2|$ and let ψ: $K_2 \to K_3$ be a simplicial approximation to g: $|K_2| \to |K_3|$. Then, by theorem 3, for a vertex $v \in K_1$

$$gf(\text{st } v) \subset g(\text{st } \varphi(v)) \subset \text{st } \psi\varphi(v)$$

and $\psi\varphi$: $K_1 \to K_3$ is thus a simplicial approximation to gf: $|K_1| \to |K_3|$. ■

Theorem 3 leads to the following necessary and sufficient condition for the existence of a simplicial approximation to a map.

6 THEOREM *A map f: $|K_1| \to |K_2|$ admits simplicial approximations $K_1 \to K_2$ if and only if K_1 is finer than the open covering $\{ f^{-1}(\text{st } v) \mid v$ is a vertex of $K_2\}$.*

PROOF By theorem 3, there exist simplicial approximations to f if and only if for each vertex $v_1 \in K_1$ there is a vertex $v_2 \in K_2$ such that st $v_1 \subset f^{-1}(\text{st } v_2)$. This is equivalent to the condition that K_1 is finer than $\{f^{-1}(\text{st } v)\}$ $v \in K_2$. ■

If K' is a subdivision of K, then for vertices $v' \in K'$ and $v \in K$

$$v' \in \text{st}_K v \Leftrightarrow \text{st}_{K'} v' \subset \text{st}_K v$$

Combining this fact with theorem 3 yields the following corollary.

7 COROLLARY *Let K' be a subdivision of K. A vertex map φ from K' to K is a simplicial approximation to the identity map $|K'| \subset |K|$ if and only if $v' \in$ st $\varphi(v')$ for every vertex $v' \in K'$.* ■

In particular, if K' is a subdivision of K, there exist simplicial approximations $K' \to K$ to the identity map $|K'| \subset |K|$. Combining theorems 6 and 3.3.14 and corollary 4, we obtain the following *simplicial-approximation theorem.*

8 THEOREM *Let (K_1,L_1) be a finite simplicial pair and let f: $(|K_1|,|L_1|) \to (|K_2|,|L_2|)$ be a map. There exists an integer N such that if $n \geq N$ there are simplicial approximations* $(\text{sd}^n K_1, \text{sd}^n L_1) \to (K_2,L_2)$ *to f.* ■

As remarked at the end of Sec. 3.3, theorem 3.3.14 is also valid for an arbitrary polyhedron X. Therefore, if K_1 is arbitrary and f: $|K_1| \to |K_2|$ is a map, there exists a subdivision K_1' of K_1 and a simplicial approximation $K_1' \to K_2$ to f: $|K_1'| \to |K_2|$. If K_1 is not finite, however, K_1' cannot generally be taken to be an iterated barycentric subdivision of K_1.

9 EXAMPLE If $\dot{s}$ is the complex consisting of all proper faces of a 2-simplex s, then $|\dot{s}|$ is homeomorphic to S^1, and therefore $[|\dot{s}|;|\dot{s}|]$ is an infinite set.

Because $\dot{s}$ is a finite complex, there are only a finite number of simplicial maps $\text{sd}^n\ \dot{s} \to \dot{s}$ for any n. Therefore for any n there exist maps $|\dot{s}| \to |\dot{s}|$ having no simplicial approximation $\text{sd}^n\ \dot{s} \to \dot{s}$.

10 EXAMPLE Let $\dot{s}$ be as in example 9 and let its vertices be v_0, v_1, v_2. Define $f\colon |\dot{s}| \to |\dot{s}|$ to be the map linear on sd $\dot{s}$ such that

$$f(v_0) = b\{v_0,v_1\} \qquad f(v_1) = b\{v_1,v_2\} \qquad f(v_2) = b\{v_2,v_0\}$$
$$f(b\{v_0,v_1\}) = v_1 \qquad f(b\{v_1,v_2\}) = v_2 \qquad f(b\{v_2,v_0\}) = v_0$$

Then $f \simeq |1_{\dot{s}}|$, but there is no simplicial approximation $\dot{s} \to \dot{s}$ to f. There are exactly eight simplicial approximations φ: sd $\dot{s} \to \dot{s}$ to f [φ is unique on $b\{v_0,v_1\}$, $b\{v_1,v_2\}$, and $b\{v_2,v_0\}$, and $\varphi(v_0) = v_0$ or v_1, $\varphi(v_1) = v_1$ or v_2, and $\varphi(v_2) = v_2$ or v_0].

As an application of the technique of simplicial approximation, we deduce the following useful result.

11 THEOREM *S^n is $(n-1)$-connected for $n \geq 1$.*

PROOF By theorem 1.6.7, it suffices to prove that if $m < n$, any map $S^m \to S^n$ is null homotopic. Let s_1 be an $(m+1)$-simplex and s_2 an $(n+1)$-simplex. Then S^m and S^n are homeomorphic, respectively, to $|\dot{s}_1|$ and $|\dot{s}_2|$. By theorem 8 and lemma 2, it suffices to show that if $\varphi\colon \text{sd}^i\ \dot{s}_1 \to \dot{s}_2$ is any simplicial map, then $|\varphi|$ is null homotopic. Because $\dim(\text{sd}^i\ \dot{s}_1) = m < n$, φ maps $\text{sd}^i\ \dot{s}_1$ into the m-dimensional skeleton of $\dot{s}_2$. Therefore there is some $\alpha \in |\dot{s}_2|$ such that

$$|\varphi|(|\text{sd}^i\ \dot{s}_1|) \subset |\dot{s}_2| - \alpha$$

Because $|\dot{s}_2| - \alpha$ is homeomorphic to S^n minus a point, which is homeomorphic to $\mathbf{R}^n$, it is contractible. Therefore $|\varphi|$ is null homotopic. ∎

In particular, we have the following result.

12 COROLLARY *For $n > 1$, S^n is simply connected.* ∎

Because S^n is locally path connected, corollary 12 and the lifting theorem imply that any continuous map $f\colon S^n \to S^1$ can be factored through the covering map $ex\colon \mathbf{R} \to S^1$. Since $\mathbf{R}$ is contractible, this implies the following corollary.

13 COROLLARY *For $n > 1$ any continuous map $S^n \to S^1$ is null homotopic.* ∎

5 CONTIGUITY CLASSES

In the last section it was shown that any continuous map between the spaces of simplicial complexes has simplicial approximations defined on sufficiently fine subdivisions of the domain complex. In general, simplicial approximations to a given continuous map are not unique, and in this section we investigate this nonuniqueness.

We shall define an analogue of homotopy, called contiguity, in the category of simplicial pairs and simplicial maps. Different simplicial approximations to the same continuous map will be shown to the contiguous. The main result of the section is the existence of a bijection between the set of homotopy classes of continuous maps (from the space of a finite simplicial complex to the space of an arbitrary complex) and the direct limit of a certain sequence of contiguity classes of simplicial maps.

Let (K_1,L_1) and (K_2,L_2) be simplicial pairs. Two simplicial maps φ, φ': $(K_1,L_1) \to (K_2,L_2)$ are *contiguous* if, given a simplex $s \in K_1$ (or $s \in L_1$), $\varphi(s) \cup \varphi'(s)$ is a simplex of K_2 (or of L_2). Obviously, this is a reflexive and symmetric relation in the set of simplicial maps $(K_1,L_1) \to (K_2,L_2)$, but in general it is not transitive. There is, however, an equivalence relation, denoted by $\varphi \sim \varphi'$, in this set of simplicial maps that is defined by $\varphi \sim \varphi'$ if and only if there exists a finite sequence $\varphi_0, \varphi_1, \ldots, \varphi_n$ such that $\varphi_0 = \varphi$ and $\varphi_n = \varphi'$ and such that φ_{i-1} and φ_i are contiguous for $i = 1, 2, \ldots, n$. The corresponding equivalence classes are called *contiguity classes*, and the set of contiguity classes of simplicial maps from (K_1,L_1) to (K_2,L_2) is denoted by $[K_1,L_1; K_2,L_2]$. If φ: $(K_1,L_1) \to (K_2,L_2)$ is a simplicial map, its contiguity class is denoted by $[\varphi]$.

We shall see that contiguity classes are algebraic analogues of homotopy classes. We begin by showing that contiguity classes can be composed.

1 LEMMA *Composites of contiguous simplicial maps are contiguous.*

PROOF Assume that φ, φ': $(K_1,L_1) \to (K_2,L_2)$ are contiguous and ψ, ψ': $(K_2,L_2) \to (K_3,L_3)$ are contiguous. If s is a simplex of K_1 (or L_1), $\varphi(s) \cup \varphi'(s)$ is a simplex of K_2 (or L_2). Therefore

$$\psi(\varphi(s) \cup \varphi'(s)) \cup \psi'(\varphi(s) \cup \varphi'(s))$$

is a simplex of K_3 (or L_3). This implies that the subset $\psi\varphi(s) \cup \psi'\varphi'(s)$ is a simplex of K_3 (or L_3) and that $\psi\varphi$, $\psi'\varphi'$: $(K_1,L_1) \to (K_3,L_3)$ are contiguous. ∎

It follows easily from lemma 1 that if $\varphi \sim \varphi'$ and $\psi \sim \psi'$, then $\psi\varphi \sim \psi\varphi' \sim \psi'\varphi'$. Therefore there is a well-defined composite of contiguity classes

$$[\psi] \circ [\varphi] = [\psi\varphi]$$

for $(K_1,L_1) \xrightarrow{\varphi} (K_2,L_2) \xrightarrow{\psi} (K_3,L_3)$. Thus there is a *contiguity category* whose objects are simplicial pairs and whose morphisms are contiguity classes of simplicial pairs. There are full subcategories of the contiguity category determined by the pairs $(K, \varnothing)$ or by the pointed simplicial complexes.

2 LEMMA *Contiguous simplicial maps which agree on a subcomplex define continuous maps which are homotopic relative to the space of the subcomplex.*

PROOF Assume that $\varphi, \varphi'\colon (K_1,L_1) \to (K_2,L_2)$ are contiguous and agree on $L \subset K_1$. Define a homotopy $F\colon (|K_1| \times I, |L_1| \times I) \to (|K_2|,|L_2|)$ rel $|L|$ from $|\varphi|$ to $|\varphi'|$ by the equation

$$F(\alpha,t) = (1 - t)(|\varphi|(\alpha)) + t(|\varphi'|(\alpha)) \qquad \alpha \in |K_1|, t \in I \quad \blacksquare$$

Since homotopy is an equivalence relation, if $\varphi \sim \varphi'$, then $|\varphi| \simeq |\varphi'|$. Therefore we have the following result.

3 COROLLARY *There is a covariant functor from the contiguity category of simplicial pairs to the homotopy category of topological pairs which assigns to (K,L) the pair $(|K|,|L|)$ and to $[\varphi]$ the homotopy class $[|\varphi|]$.* $\blacksquare$

The next result considers different simplicial approximations to the same continuous map.

4 LEMMA *Two simplicial approximations $(K_1,L_1) \to (K_2,L_2)$ to the same continuous map are contiguous.*

PROOF Let $\varphi, \varphi'\colon (K_1,L_1) \to (K_2,L_2)$ be simplicial approximations to $f\colon (|K_1|,|L_1|) \to (|K_2|,|L_2|)$ and let $\{v_i\}$ be a simplex of K_1. Then $\bigcap \text{st } v_i \neq \emptyset$, and by theorem 3.4.3,

$$\emptyset \neq f(\bigcap \text{st } v_i) \subset \bigcap f(\text{st } v_i) \subset \bigcap (\text{st } \varphi(v_i) \cap \text{st } \varphi'(v_i))$$

Therefore $\{\varphi(v_i)\} \cup \{\varphi'(v_i)\}$ is a simplex of K_2. If $\{v_i\}$ is a simplex of L_1, a similar argument shows that $\{\varphi(v_i)\} \cup \{\varphi'(v_i)\}$ is a simplex of L_2. Therefore φ and φ' are contiguous. $\blacksquare$

Since it was necessary to subdivide in order to obtain simplicial approximations to arbitrary continuous maps, we should also expect to subdivide to make contiguity classes correspond to homotopy classes. An example will illustrate the relation between homotopy and contiguity.

5 EXAMPLE Let s be a 2-simplex with vertices v_0, v_1, v_2 and let $K_1 = K_2 = \dot{s}$. Any vertex map from K_1 to K_2 is a simplicial map. Therefore there are exactly 27 simplicial maps $K_1 \to K_2$. Of these 27, there are 21 which map K_1 into a proper subcomplex of K_2, and these constitute one contiguity class. Of the remaining 6, each is the only element of its contiguity class, the 3 even permutations of the vertices defining homotopic continuous maps corresponding to one generator of the group

$$[|K_1|;|K_2|] \approx [S^1;S^1] \approx \mathbf{Z}$$

and the 3 odd permutations corresponding to the other generator of this group. Therefore $[K_1;K_2]$ consists of 7 elements, and the image

$$[K_1;K_2] \to [|K_1|;|K_2|]$$

consists of 3 elements.

This example shows that simplicial maps which define homotopic continuous maps need not be in the same contiguity class. The next result shows that a finite simplicial complex can be subdivided so that homotopic simplicial maps from it to some other complex can be simplicially approximated on the subdivision by maps in the same contiguity class; it is the analogue for homotopy of the simplicial-approximation theorem.

6 THEOREM *Let $f, f'\colon (|K_1|,|L_1|) \to (|K_2|,|L_2|)$ be homotopic, where K_1 is finite. Then there exists N such that f and f' have simplicial approximations*

$$\varphi, \varphi'\colon (\mathrm{sd}^N K_1, \mathrm{sd}^N L_1) \to (K_2,L_2)$$

respectively, in the same contiguity class.

PROOF Let $F\colon (|K_1| \times I, |L_1| \times I) \to (|K_2|,|L_2|)$ be a homotopy from f to f'. Because $|K_1|$ is compact, there exists a sequence $0 = t_0 < t_1 < \cdots < t_n = 1$ of points of I such that for $\alpha \in |K_1|$ and $i = 1, 2, \ldots, n$ there is a vertex $v \in K_2$ such that $F(\alpha,t_{i-1})$ and $F(\alpha,t_i)$ both belong to st v. Let $f_i\colon (|K_1|,|L_1|) \to (|K_2|,|L_2|)$ be defined by $f_i(\alpha) = F(\alpha,t_i)$. Then $f = f_0$ and $f' = f_n$, and for $i = 1,2, \ldots, n$ the set

$$\mathcal{U}_i = \{f_i^{-1}(\mathrm{st}\ v) \cap f_{i-1}^{-1}(\mathrm{st}\ v) \mid v \in K_2\}$$

is an open covering of $|K_1|$. Let N be chosen large enough so that $\mathrm{sd}^N K_1$ is finer than $\mathcal{U}_1, \mathcal{U}_2, \ldots, \mathcal{U}_n$ (which is possible, by theorem 3.3.14). For $i = 1, 2, \ldots, n$ let φ_i be a vertex map from $\mathrm{sd}^N K_1$ to K_2 such that

$$f_i(\mathrm{st}\ v) \cup f_{i-1}(\mathrm{st}\ v) \subset \mathrm{st}\ \varphi_i(v)$$

for each vertex $v \in K_1$ (such a vertex map φ_i exists because $\mathrm{sd}^N K_1$ is finer than $\mathcal{U}_i$). By theorem 3.4.3,

$$\varphi_i\colon (\mathrm{sd}^N K_1, \mathrm{sd}^N L_1) \to (K_2,L_2)$$

is a simplicial approximation to f_i and to f_{i-1}. Because φ_i and φ_{i+1} are simplicial approximations to f_i, it follows from lemma 4 that φ_i and φ_{i+1} are contiguous for $i = 1, 2, \ldots, n-1$. Therefore $\varphi_1 \sim \varphi_n$, and also φ_1 is a simplicial approximation to $f_0 = f$ and φ_n is a simplicial approximation to $f_n = f'$. ■

Unlike the simplicial-approximation theorem, this last result is definitely false if K_1 is not a finite simplicial complex. That is, given homotopic maps $f, f'\colon |K_1| \to |K_2|$, there need not be a subdivision K_1' of K_1 such that f and f' have simplicial approximations $K_1' \to K_2$ in the same contiguity class.

7 EXAMPLE Let $K_1 = K_2$ equal the simplicial complex of example 3.1.8, with space homeomorphic to $\mathbf{R}$. Let $\varphi\colon K_1 \to K_2$ be the identity simplicial map and $\varphi'\colon K_1 \to K_2$ be the constant simplicial map sending every vertex of K_1 to the vertex 0 of K_2. Since $\mathbf{R}$ is contractible, $|\varphi| \simeq |\varphi'|$. However, if K_1' is any subdivision of K_1, a simplicial approximation $\psi\colon K_1' \to K_2$ to $|\varphi|$ must be surjective to the vertices of K_2 and a simplicial approximation $\psi'\colon K_1' \to K_2$ to $|\varphi'|$ must be the constant map to 0. Since two contiguous maps $K_1' \to K_2$ either both map onto a finite set of vertices or neither does, ψ and ψ' are not in the same contiguity class.

We show that if K_1 is finite the set of homotopy classes of maps $[|K_1|,|L_1|; |K_2|,|L_2|]$ is the direct limit of the set of contiguity classes

$$[\mathrm{sd}^n K_1, \mathrm{sd}^n L_1; K_2,L_2]$$

Note that simplicial approximations $(\mathrm{sd}\, K_1, \mathrm{sd}\, L_1) \to (K_1,L_1)$ to the identity map $(|\mathrm{sd}\, K_1|, |\mathrm{sd}\, L_1|) \subset (|K_1|,|L_1|)$ exist, by corollary 3.4.7, and any two are contiguous, by lemma 4. Because the composites of contiguous simplicial maps are contiguous by lemma 1, there is a well-defined map

$$sd\colon [K_1,L_1; K_2,L_2] \to [\mathrm{sd}\, K_1, \mathrm{sd}\, L_1; K_2,L_2]$$

defined by

$$sd[\varphi] = [\varphi\lambda]$$

where $\lambda\colon (\mathrm{sd}\, K_1, \mathrm{sd}\, L_1) \to (K_1,L_1)$ is any simplicial approximation to the identity $(|\mathrm{sd}\, K_1|, |\mathrm{sd}\, L_1|) \subset (|K_1|,|L_1|)$ and $\varphi\colon (K_1,L_1) \to (K_2,L_2)$ is an arbitrary simplicial map. By iteration there is obtained a sequence

$$\cdots \to [\mathrm{sd}^n K_1, \mathrm{sd}^n L_1; K_2,L_2] \xrightarrow{sd} [\mathrm{sd}^{n+1} K_1, \mathrm{sd}^{n+1} L_1; K_2,L_2] \to \cdots$$

which begins with $[K_1,L_1; K_2,L_2]$ on the left and extends indefinitely on the right. The direct limit $\lim_\to \{[\mathrm{sd}^n K_1, \mathrm{sd}^n L_1; K_2,L_2]\}$ is a functor of two arguments contravariant in (K_1,L_1) and covariant in (K_2,L_2). For finite K_1 this functor is naturally equivalent to the functor $[|K_1|,|L_1|; |K_2|,|L_2|]$.

8 THEOREM *If K_1 is a finite simplicial complex, there is a natural equivalence*

$$\lim_\to \{[\mathrm{sd}^n K_1, \mathrm{sd}^n L_1; K_2,L_2]\} \approx [|K_1|,|L_1|; |K_2|,|L_2|]$$

PROOF A function from the direct limit to $[|K_1|,|L_1|; |K_2|,|L_2|]$ consists of a sequence of functions

$$f_n\colon [\mathrm{sd}^n K_1, \mathrm{sd}^n L_1; K_2,L_2] \to [|K_1|,|L_1|; |K_2|,|L_2|]$$

for $n \geq 0$ such that $f_n = f_{n+1} \circ sd$ for $n \geq 0$. Such a sequence f_n is defined by $f_n[\varphi] = [|\varphi|]$ for $\varphi\colon (\mathrm{sd}^n K_1, \mathrm{sd}^n L_1) \to (K_2,L_2)$, because if

$$\lambda_n\colon (\mathrm{sd}^{n+1} K_1, \mathrm{sd}^{n+1} L_1) \to (\mathrm{sd}^n K_1, \mathrm{sd}^n L_1)$$

is a simplicial approximation to the identity map

$$(|\mathrm{sd}^{n+1}K_1|, |\mathrm{sd}^{n+1}L_1|) \subset (|\mathrm{sd}^n K_1|, |\mathrm{sd}^n L_1|)$$

then, by lemma 3.4.2, $|\lambda_n| \simeq 1$, and

$$f_{n+1}\, sd[\varphi] = [|\varphi\lambda_n|] = [|\varphi|] = f_n[\varphi]$$

The sequence $\{f_n\}$ defines a natural transformation

$$f\colon \lim_\to \{[\mathrm{sd}^n K_1, \mathrm{sd}^n L_1; K_2,L_2]\} \to [|K_1|,|L_1|; |K_2|,|L_2|]$$

and we show that f is a bijection.

It follows easily from the simplicial-approximation theorem that $\{f_n\}$ satisfies (*a*) of theorem 1.3 of the Introduction; for if $g\colon (|K_1|,|L_1|) \to (|K_2|,|L_2|)$

is a map and φ: $(\text{sd}^n K_1, \text{sd}^n L_1) \to (K_2,L_2)$ is a simplicial approximation to g, then $|\varphi| \simeq g$, and

$$f_n[\varphi] = [|\varphi|] = [g]$$

To show that $\{f_n\}$ satisfies (*b*) of theorem 1.3 of the Introduction, assume

$$\varphi, \varphi': (\text{sd}^n K_1, \text{sd}^n L_1) \to (K_2,L_2)$$

are such that $|\varphi| \simeq |\varphi'|$. By theorem 6, there exists $m \geq n$ such that $|\varphi|$ and $|\varphi'|$ have simplicial approximations

$$\psi, \psi': (\text{sd}^m K_1, \text{sd}^m L_1) \to (K_2,L_2)$$

in the same contiguity class. Let

$$\lambda_{m,n}: (\text{sd}^m K_1, \text{sd}^m L_1) \to (\text{sd}^n K_1, \text{sd}^n L_1)$$

be the composite $\lambda_{m,n} = \lambda_n \lambda_{n+1} \cdots \lambda_{m-1}$. Then $\lambda_{m,n}$ is a simplicial approximation to the identity map, and because φ is a simplicial approximation to $|\varphi|$, $\varphi\lambda_{m,n}$ is also a simplicial approximation to $|\varphi|$, by corollary 3.4.5. By lemma 4, $\varphi\lambda_{m,n}$ is contiguous to ψ. Similarly, $\varphi'\lambda_{m,n}$ is contiguous to ψ'. Since ψ and ψ' are in the same contiguity class, so are $\varphi\lambda_{m,n}$ and $\varphi'\lambda_{m,n}$. This means that $sd^{m-n}[\varphi] = sd^{m-n}[\varphi']$ in $[\text{sd}^m K_1, \text{sd}^m L_1; K_2,L_2]$. ■

For finite K_1 the last result furnishes an algebraic description of the set $[|K_1|,|L_1|; |K_2|,|L_2|]$. As an application, note that if K_2 is a countable complex, there are only a countable number of simplicial maps $(\text{sd}^n K_1, \text{sd}^n L_1) \to (K_2,L_2)$ for $n \geq 0$. Therefore $[\text{sd}^n K_1, \text{sd}^n L_1; K_2,L_2]$ is countable for $n \geq 0$. Because the direct limit of a sequence of countable sets is countable, we obtain the following result.

9 COROLLARY *Let (X,A) and (Y,B) be polyhedral pairs with X compact and Y the space of a countable complex. Then $[X,A; Y,B]$ is a countable set.* ■

6 THE EDGE-PATH GROUPOID

It was shown in the last section that for finite K_1, $[|K_1|;|K_2|]$ is describable as a limit in which K_1 is subdivided but K_2 is not. In particular, for any simplicial complex K the set of path classes of $|K|$ from v_0 to v_1 is determined by the simplicial structure of K. This is made explicit in the present section, where we define a simplicial analogue of the fundamental groupoid of a space. In the next section the fundamental group of a polyhedron is presented in terms of generators and relations.

An *edge* of a simplicial complex K is an ordered pair of vertices (v,v') which belong to some simplex of K. The first vertex v is called the *origin* of the edge, and the second vertex v' is called the *end* of the edge. An *edge path* ζ of K is a finite nonempty sequence $e_1e_2 \cdots e_r$ of edges of K such that end

$e_i = \text{orig } e_{i+1}$ for $i = 1, \ldots, r - 1$. We define $\text{orig } \zeta = \text{orig } e_1$ and $\text{end } \zeta = \text{end } e_r$. A *closed edge path* at $v_0 \in K$ is an edge path ζ such that $\text{orig } \zeta = v_0 = \text{end } \zeta$. If ζ_1 and ζ_2 are edge paths of K such that $\text{end } \zeta_1 = \text{orig } \zeta_2$, we define the *product edge path* $\zeta_1\zeta_2$ to be the edge path consisting of the sequence of edges of ζ_1 followed by the sequence of edges of ζ_2. Then $\text{orig } \zeta_1\zeta_2 = \text{orig } \zeta_1$ and $\text{end } \zeta_1\zeta_2 = \text{end } \zeta_2$. It is clear that if $\text{end } \zeta_1 = \text{orig } \zeta_2$ and $\text{end } \zeta_2 = \text{orig } \zeta_3$, then $\zeta_1(\zeta_2\zeta_3) = (\zeta_1\zeta_2)\zeta_3$. The product of edge paths thus satisfies associativity; however, there are no left or right identity elements for the product. To obtain a category (as was done for paths in a topological space) it is necessary to define an equivalence relation in the set of edge paths of K.

Two edge paths ζ and ζ' in K are *simply equivalent* if there exist vertices v, v', and v'' of K belonging to some simplex of K such that the unordered pair $\{\zeta,\zeta'\}$ equals one of the following:

The unordered pair $\{(v,v''), (v,v')(v',v'')\}$
The unordered pair $\{\zeta_1(v,v''), \zeta_1(v,v')(v',v'')\}$ for some edge path ζ_1 in K with $\text{end } \zeta_1 = v$
The unordered pair $\{(v,v'')\zeta_2, (v,v')(v',v'')\zeta_2\}$ for some edge path ζ_2 in K with $\text{orig } \zeta_2 = v''$
The unordered pair $\{\zeta_1(v,v'')\zeta_2, \zeta_1(v,v')(v',v'')\zeta_2\}$ for some edge paths ζ_1 and ζ_2 in K with $\text{end } \zeta_1 = v$ and $\text{orig } \zeta_2 = v''$.

Two edge paths ζ and ζ' will be said to be *equivalent*, denoted by $\zeta \sim \zeta'$, if there is a finite sequence of edge paths $\zeta_0, \zeta_1, \ldots, \zeta_n$ such that $\zeta = \zeta_0$ and $\zeta' = \zeta_n$, and ζ_{i-1} and ζ_i are simply equivalent for $i = 1, \ldots, n$. This is an equivalence relation, and the following statements are easily verified.

1 *$\zeta \sim \zeta'$ implies that ζ and ζ' have the same origin and the same end.* ∎

2 *$\zeta_1 \sim \zeta_1'$, $\zeta_2 \sim \zeta_2'$ and* $\text{end } \zeta_1 = \text{orig } \zeta_2$ *imply $\zeta_1\zeta_2 \sim \zeta_1'\zeta_2'$.* ∎

3 *If* $\text{orig } \zeta = v_1$ *and* $\text{end } \zeta = v_2$, *then $(v_1,v_1)\zeta \sim \zeta \sim \zeta(v_2,v_2)$.* ∎

If ζ is an edge path, $[\zeta]$ denotes its equivalence class. It follows from statement 1 that $\text{orig } [\zeta]$ and $\text{end } [\zeta]$ are well-defined (by $\text{orig } [\zeta] = \text{orig } \zeta$ and $\text{end } [\zeta] = \text{end } \zeta$). From statement 2 there is a well-defined composite $[\zeta_1] \circ [\zeta_2] = [\zeta_1\zeta_2]$ defined if $\text{end } \zeta_1 = \text{orig } \zeta_2$. We then have the following simplicial analogue of theorem 1.7.7.

4 THEOREM *There is a category $\mathcal{E}(K)$ whose objects are the vertices of K and whose morphisms from v_1 to v_0 are the equivalence classes $[\zeta]$ with* $\text{orig } [\zeta] = v_0$ *and* $\text{end } [\zeta] = v_1$ *and whose composite is $[\zeta_1] \circ [\zeta_2]$.*

PROOF The existence of identity morphisms follows from statement 3, and the associativity of the composite is a consequence of the associativity of the product of edge paths. ∎

We now show that $\mathcal{E}(K)$ is a groupoid. If $e = (v,v')$ is an edge of K, we define $e^{-1} = (v',v)$, and if $\zeta = e_1 \cdots e_r$ is an edge path in K, we define $\zeta^{-1} = e_r^{-1} \cdots e_1^{-1}$. The following statements are then easily verified.

5 $(\zeta^{-1})^{-1} = \zeta$. ■

6 orig ζ^{-1} = end ζ *and* end ζ^{-1} = orig ζ. ■

7 $\zeta_1 \sim \zeta_2$ *implies* $\zeta_1^{-1} \sim \zeta_2^{-1}$. ■

8 *If* orig $\zeta = v_1$ *and* end $\zeta = v_2$, *then* $\zeta\zeta^{-1} \sim (v_1,v_1)$ *and* $\zeta^{-1}\zeta \sim (v_2,v_2)$. ■

It follows that in $\mathcal{E}(K)$, $[\zeta^{-1}] = [\zeta]^{-1}$, and so $\mathcal{E}(K)$ is a groupoid. This groupoid is called the *edge-path groupoid* of K. If v_0 is a vertex of K, the operation $[\zeta] \circ [\zeta']$ in the set of elements of $\mathcal{E}(K)$ with origin and end at v_0 gives a group denoted by $E(K,v_0)$ and is called the ***edge-path group of K with base vertex*** v_0.

To compare $\mathcal{E}(K)$ [and $E(K,v_0)$] with $\mathcal{P}(|K|)$ [and $\pi(|K|,v_0)$], for $r \geq 1$ let I_r denote the subdivision of I into r equal subintervals; that is, I_r is the simplicial complex

$$I_r = \left\{\left\{\frac{i}{r}\right\} \middle| 0 \leq i \leq r\right\} \cup \left\{\left\{\frac{i-1}{r}, \frac{i}{r}\right\} \middle| 1 \leq i \leq r\right\}$$

Given an edge path $\zeta = e_1 \cdots e_r$ in K with r edges, let $\varphi_\zeta: I_r \to K$ be the simplicial map defined by

$$\varphi_\zeta\left(\frac{i}{r}\right) = \begin{cases} \text{orig } e_{i+1} & 0 \leq i \leq r-1 \\ \text{end } e_i & 1 \leq i \leq r \end{cases}$$

Then $|\varphi_\zeta|: I \to |K|$ is a path in $|K|$, and it is easily seen that the following statements hold.

9 orig $|\varphi_\zeta|$ = orig ζ *and* end $|\varphi_\zeta|$ = end ζ. ■

10 $\zeta \sim \zeta'$ *implies* $|\varphi_\zeta| \simeq |\varphi_{\zeta'}|$ rel $\dot{I}$. ■

11 *If* end ζ_1 = orig ζ_2, *then* $|\varphi_{\zeta_1\zeta_2}| \simeq |\varphi_{\zeta_1}| * |\varphi_{\zeta_2}|$ rel $\dot{I}$. ■

It follows that there is a natural transformation ρ from $\mathcal{E}(K)$ to $\mathcal{P}(|K|)$ which assigns to $v \in K$ the point $v \in |K|$ and to a morphism $[\zeta]$ in $\mathcal{E}(K)$ the morphism $[|\varphi_\zeta|]$ in $\mathcal{P}(|K|)$. We shall prove that for vertices v_0, $v_1 \in K$, ρ is a bijection

$$\rho: \hom_{\mathcal{E}}(v_1,v_0) \approx \hom_{\mathcal{P}}(v_1,v_0)$$

This can be obtained from theorem 3.5.8, but there is also a direct proof (which seems no longer than a proof based on theorem 3.5.8).

12 LEMMA *For any* v_0, $v_1 \in K$ *the function*

$$\rho: \hom_{\mathcal{E}}(v_1,v_0) \to \hom_{\mathcal{P}}(v_1,v_0)$$

is surjective.

PROOF Given a path $\omega: I \to |K|$ from v_0 to v_1, because $I = |I_1|$, it follows from theorem 3.4.8 that there is a simplicial map

$$\varphi\colon \operatorname{sd}^n I_1 \to K$$

which is a simplicial approximation to ω. Since $\operatorname{sd}^n I_1 = I_{2^n}$, there is an edge path $\zeta = e_1 \cdots e_{2^n}$ defined by $e_i = (\varphi((i-1)/2^n), \varphi(i/2^n))$ such that $\varphi = \varphi_\zeta$ for this ζ. Because $\varphi(0) = \omega(0)$ and $\varphi(1) = \omega(1)$, it follows from lemma 3.4.2 that $|\varphi| \simeq \omega$ rel $\dot{I}$. Therefore $[\omega] = [|\varphi|] = [|\varphi_\zeta|] = \rho[\zeta]$. ■

We shall need some preliminary lemmas before showing that ρ is injective.

13 LEMMA *For any simplex s two edge paths in $\bar{s}$ with the same origin and the same end are equivalent.*

PROOF It suffices to prove that if ζ is any edge path in $\bar{s}$ from orig $\zeta = v$ and end $\zeta = v'$, then ζ is equivalent to the edge path consisting of the single edge (v,v'). This is proved by induction on the number of edges of ζ. ■

14 LEMMA *Let ζ and ζ' be edge paths in K, each with r edges, such that $\varphi_\zeta, \varphi_{\zeta'}\colon I_r \to K$ are contiguous. Then $\zeta \sim \zeta'$.*

PROOF Let $\zeta = e_1 \cdots e_r$, where $e_i = (v_{i-1},v_i)$, and let $\zeta' = e_1' \cdots e_r'$, where $e_i' = (v_{i-1}',v_i')$. For $0 \le i \le r$ let $\bar{e}_i = (v_i,v_i')$ (this is an edge of K because φ_ζ and $\varphi_{\zeta'}$ are contiguous). From the definition of equivalence

$$\zeta \sim e_1\bar{e}_1\bar{e}_1{}^{-1}e_2\bar{e}_2 \cdots \bar{e}_{r-1}^{-1}e_r$$

Because φ_ζ and $\varphi_{\zeta'}$ are contiguous, for each $1 \le i \le r$ there is some simplex s_i of K such that e_i, e_i', $\bar{e}_{i-1}$, and $\bar{e}_i$ all are edges of $\bar{s}_i$. By lemma 13, $e_1\bar{e}_1 \sim e_1'$ and $\bar{e}_{i-1}^{-1}e_i\bar{e}_i \sim e_i'$ for $2 \le i \le r-1$, and $\bar{e}_{r-1}^{-1}e_r \sim e_r'$. Therefore

$$e_1\bar{e}_1\bar{e}_1^{-1}\,e_2\bar{e}_2 \cdots \bar{e}_{r-1}^{-1}e_r \sim e_1'e_2' \cdots e_r' = \zeta' \quad ■$$

15 LEMMA *Let $\zeta = e_1 \cdots e_r$ be an edge path of K and let $\lambda\colon I_{2r} \to I_r$ be a simplicial approximation to the identity map $|I_{2r}| \subset |I_r|$. Then $\varphi_\zeta\lambda = \varphi_{\zeta'}\colon I_{2r} \to K$ for some $\zeta' = e_1' \cdots e_{2r}'$ and $\zeta \sim \zeta'$.*

PROOF Let $e_i = (v_{i-1},v_i)$ for $0 \le i \le r$. Then $e_{2i-1}'e_{2i}' = (v_{i-1},\bar{v}_i)(\bar{v}_i,v_i)$ for a vertex $\bar{v}_i$ which equals either v_{i-1} or v_i. By lemma 13, $e_{2i-1}'e_{2i}' \sim e_i$ and $\zeta' \sim \zeta$. ■

We are now ready for the main result on the edge-path groupoid.

16 THEOREM *For vertices $v_0, v_1 \in K$ the function*

$$\rho\colon \hom_{\mathcal{E}}(v_1,v_0) \to \hom_{\mathcal{P}}(v_1,v_0)$$

is a bijection.

PROOF In view of lemma 12, it only remains to prove that ρ is injective. Assume that ζ and ζ' are edge paths from v_0 to v_1 such that $|\varphi_\zeta| \simeq |\varphi_{\zeta'}|$ rel $\dot{I}$. By juxtaposing trivial edges (v_1,v_1) at the end of ζ or ζ' sufficiently often, we can replace ζ and ζ' by equivalent edge paths having an equal number of edges. Hence there is no loss of generality in assuming ζ and ζ' both to have r edges. Then $\varphi_\zeta, \varphi_{\zeta'}: I_r \to K$ are such that $|\varphi_\zeta| \simeq |\varphi_{\zeta'}|$ rel $\dot{I}$. By theorem 3.5.6, there exists m such that if λ: sd^m $I_r \to I_r$ is a simplicial approximation to the identity, then $\varphi_\zeta\lambda$ and $\varphi_{\zeta'}\lambda$ are in the same contiguity class. Now $\varphi_\zeta\lambda = \varphi_{\zeta_1}$ and $\varphi_{\zeta'}\lambda = \varphi_{\zeta_1'}$ for edge paths ζ_1 and ζ_1' in K. By lemma 15 (and induction on m), $\zeta \sim \zeta_1$ and $\zeta' \sim \zeta_1'$. From lemma 14 it follows that $\zeta_1 \sim \zeta_1'$. Therefore $\zeta \sim \zeta'$. ∎

If $\varphi: K_1 \to K_2$ is a simplicial map, there is a covariant functor $\varphi_\#: \mathcal{E}(K_1) \to \mathcal{E}(K_2)$ defined by

$$\varphi_\#[\zeta] = [\varphi(\zeta)]$$

where, if $\zeta = (v_0,v_1)(v_1,v_2) \cdots (v_{r-1},v_r)$, then $\varphi(\zeta) = (\varphi(v_0),\varphi(v_1)) \cdots (\varphi(v_{r-1}),\varphi(v_r))$. It is trivial to verify that commutativity holds in the square

$$\begin{array}{ccc} \mathcal{E}(K_1) & \xrightarrow{\varphi_\#} & \mathcal{E}(K_2) \\ {\scriptstyle\rho}\downarrow & & \downarrow{\scriptstyle\rho} \\ \mathcal{P}(|K_1|) & \xrightarrow{|\varphi|_\#} & \mathcal{P}(|K_2|) \end{array}$$

Therefore we have the following result.

17 COROLLARY *On the category of pointed simplicial complexes K with base vertex v_0, ρ is a natural equivalence of the covariant functor $E(K,v_0)$ with the covariant functor $\pi(|K|,v_0)$.* ∎

Note that $\mathcal{E}(K)$ is determined by the 2-skeleton of K; that is, the edge paths of K are determined by pairs of vertices of K which are vertices of a simplex, and the equivalences between edge paths are determined by triples of vertices which are vertices of a simplex. Hence $\mathcal{E}(K^2) \approx \mathcal{E}(K)$.

18 COROLLARY *For any pointed simplicial complex (K,v_0), the inclusion map $K^2 \subset K$ induces an isomorphism*

$$\pi(|K^2|,v_0) \approx \pi(|K|,v_0) \quad ∎$$

If s is a simplex of K, any two of its vertices belong to the same component of $\mathcal{E}(K)$. Therefore the components $\{E_j\}$ of $\mathcal{E}(K)$ define a partition of K into subcomplexes $\{K_j\}$, called the *components* of K, defined by $K_j = \{s \in K \mid s$ has some vertex in $E_j\}$. K is said to be *connected* if it contains exactly one component.

19 THEOREM *If $\{K_j\}$ are the components of K, then $\{|K_j|\}$ are the path components of $|K|$.*

PROOF If v is a vertex of K, then st v is path connected and so belongs to the same path component of $|K|$ as v. It follows from theorem 16 that two vertices of K are in the same component of $\mathcal{P}(|K|)$ if and only if they are in the same component of $\mathcal{E}(K)$. Therefore, if $\{E_j\}$ is the set of components of $\mathcal{E}(K)$, the path components of $|K|$ are the sets $\{P_j\}$ defined by $P_j = \cup\{\text{st } v \mid v \in E_j\}$. Clearly, $P_j = |K_j|$. ∎

7 GRAPHS

We show how a system of generators and relations for the edge-path group $E(K,v_0)$ can be determined. This provides a method for finding generators and relations of the fundamental group of a polyhedron. Since every edge path of K is an edge path of the one-dimensional skeleton of K, we start with a description of the edge path group of a one-dimensional complex.

A one-dimensional simplicial complex is called a *graph*. A *tree* is defined to be a simply connected graph.

1 LEMMA *A graph is a tree if and only if it is contractible.*

PROOF Since a contractible space is simply connected, a contractible graph is a tree. Conversely, let K be a tree and let α_0 be a point of $|K|$. We shall define a homotopy $F: |K| \times I \to |K|$ from the identity map 1 of $|K|$ to the constant map c of $|K|$ to α_0. Since $|K|$ is path connected, for each vertex v of K there is a path ω_v in $|K|$ from v to α_0. We now define F on $v \times I$ by $F(v,t) = \omega_v(t)$. For every 1-simplex s of K the map F is defined on the subset $(|s| \times 0) \cup (|s| \times 1) \cup (|\dot{s}| \times I)$ of $|s| \times I$. Since $|K|$ is simply connected and $(|s| \times I, (|s| \times 0) \cup (|s| \times 1) \cup (|\dot{s}| \times I))$ is homeomorphic to (E^2,S^1), it follows that F can be extended over $|s| \times I$. In this way we obtain a map $F: |K| \times I \to |K|$ whose restriction to each $|s| \times I$ is continuous. Then F is continuous and $F: 1 \simeq c$. ∎

2 LEMMA *Let K be a connected simplicial complex. Then K contains a maximal tree, and any maximal tree contains all the vertices of K.*

PROOF The collection of trees contained in K is partially ordered by inclusion. Let $\{K_j\}$ be a simply ordered set of trees in K and let $T = \cup K_j$. We show that T is a tree. Since K_j is one-dimensional, T is one-dimensional. Since $\{K_j\}$ is a simply ordered set of trees, it follows that any finite subcomplex of T is contained in some K_j. To show that T is simply connected, let $f: S^i \to |T|$, where $i = 0$ or 1. Then $f(S^i)$ is compact and is therefore contained in $|K_j|$ for some j. Since $|K_j|$ is simply connected, the map $f: S^i \to |K_j|$ can be extended to a map $f': E^{i+1} \to |K_j| \subset |T|$, and $|T|$ is simply connected.

As a result, every simply ordered set of trees in K has a tree as upper bound. Zorn's lemma can be applied to yield a maximal tree in K. If T is a maximal tree of K and does not contain all the vertices of K, it follows from the connectedness of K that there is a 1-simplex $\{v_1,v_2\} \in K$ with $v_1 \in T$ and $v_2 \notin T$. Let $T_1 = T \cup \{\{v_2\}, \{v_1,v_2\}\}$. Since v_1 is a strong deformation retract of $|\{v_1,v_2\}|$, $|T|$ is a strong deformation retract of $|T_1|$. Therefore $|T_1|$ is contractible, and so T_1 is a tree strictly larger than T, contradicting the maximality of T. ∎

It follows from lemma 2 that if K is a connected complex and T is a maximal tree in K, then $K - T$ consists of simplexes of dimension ≥ 1. Because $|T|$ is contractible, any edge path in K is determined by its part in $K - T$, as we shall see below. This motivates the following definition.

Let T be a maximal tree of the connected complex K. Let G be the group generated by the edges (v,v') of K with the relations

(*a*) If (v,v') is an edge of T, then $(v,v') = 1$.
(*b*) If v, v', and v'' are vertices of a simplex of K, then $(v,v')(v',v'') = (v,v'')$.

3 THEOREM *With the notation above,* $E(K,v_0) \approx G$.

PROOF Since T is connected and contains the vertices of K, for each vertex v of K there is an edge path ζ_v in T such that orig $\zeta_v = v_0$ and end $\zeta_v = v$. If (v,v') is an edge of K, the edge path $\zeta_v(v,v')\zeta_{v'}^{-1}$ is a closed edge path in K at v_0. Therefore there is a homomorphism α of the free group F generated by the edges of K into $E(K,v_0)$ defined by $\alpha(v,v') = [\zeta_v(v,v')\zeta_{v'}^{-1}]$.

We show that α can be factored through G. If (v,v') is an edge of T, then $\zeta_v(v,v')\zeta_{v'}^{-1}$ is a closed edge path in T. Because T is simply connected, $[\zeta_v(v,v')\zeta_{v'}^{-1}] = 1$ and α sends relations of type (*a*) into 1. If v, v' and v'' are vertices of a simplex of K, then

$$\begin{aligned}[\zeta_v(v,v')\zeta_{v'}^{-1}] \circ [\zeta_{v'}(v',v'')\zeta_{v''}^{-1}] &= [\zeta_v(v,v')(v',v'')\zeta_{v''}^{-1}] \\ &= [\zeta_v(v,v'')\zeta_{v''}^{-1}]\end{aligned}$$

Therefore $\alpha((v,v')(v',v'')) = \alpha(v,v'')$, and so there is a homomorphism $\alpha'\colon G \to E(K,v_0)$ such that $\alpha'(v,v') = \alpha(v,v') = [\zeta_v(v,v')\zeta_{v'}^{-1}]$.

To prove that α' is an isomorphism we construct an inverse $\beta'\colon E(K,v_0) \to G$ as follows. For each closed edge path $\zeta = e_1 \cdots e_r$ let $\beta(\zeta) = e_1 \cdots e_r$, where the right-hand side is interpreted as a product in G. If ζ and ζ' are simply equivalent, then because of the relations of type (*b*), $\beta(\zeta) = \beta(\zeta')$. Therefore there is a homomorphism $\beta'\colon E(K,v_0) \to G$ such that $\beta'[\zeta] = \beta(\zeta)$.

We show that α' and β' are inverses of each other. Given an edge path $\zeta = (v_0,v_1)(v_1,v_2) \cdots (v_r,v_0)$, then $\alpha'\beta'[\zeta] = [\zeta']$, where

$$\begin{aligned}\zeta' &= \zeta_{v_0}(v_0,v_1)\zeta_{v_1}^{-1}\zeta_{v_1}(v_1,v_2)\zeta_{v_2}^{-1} \cdots \zeta_{v_r}(v_r,v_0)\zeta_{v_0}^{-1} \\ &\sim \zeta_{v_0}(v_0,v_1)(v_1,v_2) \cdots (v_r,v_0)\zeta_{v_0}^{-1}\end{aligned}$$

Since ζ_{v_0} is a closed edge path in the simply connected complex T, $\zeta_{v_0} \sim 1$ and $\zeta' \sim \zeta$. Therefore $\alpha'\beta'$ is the identity on $E(K,v_0)$.

Consider $\beta'\alpha'(v,v') = \beta(\zeta_v)(v,v')\beta(\zeta_{v'}^{-1})$. Since ζ_v and $\zeta_{v'}^{-1}$ are paths in T, they are products of edges in T. Hence $\beta(\zeta_v)$ and $\beta(\zeta_{v'}^{-1})$ are both equal to 1 by relations of type (*a*). Therefore $\beta'\alpha'(v,v') = (v,v')$, and since $\{(v,v')\}$ generate G, $\beta'\alpha' = 1$ on G. ∎

In case K is finite, there are only a finite number of edges of K, and G is finitely generated. Similarly, there are only a finite number of relations of type (*a*) or (*b*). G is thereby represented as the quotient group of a finitely generated free group by a finitely generated subgroup. Hence we have the following corollary.

4 COROLLARY *If K is a finite connected simplicial complex, then $E(K,v_0)$ is finitely presented.* ∎

5 COROLLARY *If K is a connected graph, $E(K,v_0)$ is a free group, and if T is a maximal tree in K, a set of generators of $E(K,v_0)$ is in one-to-one correspondence with the 1-simplexes of $K - T$.*

PROOF Consider the group G. Because of relations of type (*a*), we need only consider edges of K not in T. Every such edge e corresponds to an order of the vertices of some 1-simplex of $K - T$. The relations of type (*b*) imply that the oppositely ordered edge equals e^{-1} in G. Therefore the group G is generated by edges one for each 1-simplex of $K - T$. There are no relations on these generators of G, for if v, v', and v'' are vertices of a simplex of K, then at least two of them are equal. If $v = v'$ or $v' = v''$, the corresponding relation of type (*b*) is the trivial relation $1(v',v'') = (v',v'')$ or $(v,v')1 = (v,v')$. If $v = v''$, the corresponding relation is $(v,v')(v',v) = 1$, which, in terms of our generators, becomes $ee^{-1} = 1$. ∎

6 EXAMPLE Let $J = \{j\}$ be a set and let X be the pointed space which is the sum (in the category of pointed spaces) of pointed 1-spheres $\{S_j^1\}_{j \in J}$. Each S_j^1 can be triangulated by $\dot{s}_j$, where s_j is a 2-simplex $s_j = \{v_j,v_j',v_j''\}$ and v_j corresponds to the base point of S_j^1. Then X can be triangulated by the complex K with vertices

$$\{v\} \cup \{v_j',v_j''\}_{j \in J}$$

and 1-simplexes

$$\{\{v,v_j'\}\}_{j \in J} \cup \{\{v,v_j''\}\}_{j \in J} \cup \{\{v_j',v_j''\}\}_{j \in J}$$

Let T be the maximal tree in K such that $K - T$ consists of the collection $\{\{v_j',v_j''\}\}_{j \in J}$. By corollary 5, $E(K,v)$ is a free group on generators in one-to-one correspondence with J. Therefore there is an isomorphism of $\pi(X,x_0)$, where x_0 corresponds to v, with the free group generated by J.

7 EXAMPLE Let X be the complement in $\mathbf{R}^2$ of a set of p disjoint closed disks or points. Then X has the same homotopy type as the sum of p pointed 1-spheres. Therefore the fundamental group of X is a free group with p generators.

For connected graphs the fundamental group functor is a faithful representation of the category of their underlying spaces and homotopy classes by means of groups and homomorphisms. This is summarized in the following theorem.

8 THEOREM *Let K_1 and K_2 be connected graphs and let v_0 be a vertex of K_1. Then*

(*a*) *Any continuous map $f\colon |K_1| \to |K_2|$ is homotopic to a continuous map $f'\colon |K_1| \to |K_2|$ such that $f'(v_0)$ is a vertex of K_2.*
(*b*) *If v_0' is any vertex of K_2 and $h\colon \pi(|K_1|,v_0) \to \pi(|K_2|,v_0')$ is an arbitrary homomorphism, there exists a continuous map $f\colon (|K_1|,v_0) \to$*

$(|K_2|,v_0')$ such that $h = f_\#$.

(c) Let v_0' and v_0'' be vertices of K_2 and assume that $f_1, f_2\colon |K_1| \to |K_2|$ are maps such that $f_1(v_0) = v_0'$ and $f_2(v_0) = v_0''$. Then $f_1 \simeq f_2$ if and only if there is a path ω in $|K_2|$ from v_0' to v_0'' such that the following triangle is commutative:

$$\begin{array}{ccc} & \pi(|K_1|,v_0) & \\ {\scriptstyle f_{1\#}}\swarrow & & \searrow{\scriptstyle f_{2\#}} \\ \pi(|K_2|,v_0') & \xleftarrow{h_{[\omega]}} & \pi(|K_2|,v_0'') \end{array}$$

PROOF Since K_2 is connected, it is path connected, and *(a)* follows from the fact that the pair $(|K_1|,v_0)$ has the homotopy extension property with respect to $|K_2|$ (by corollary 3.2.5).

To prove *(b)*, let T be a maximal tree in K_1. Let $\{s_j\}$ be the simplexes of $K_1 - T$ and for each j let $e_j = (v_j,v_j')$ be an edge whose vertices are the vertices of s_j in some order. For each vertex v in K_1 there is an edge path ζ_v in T from v_0 to v. For each j let

$$\omega_j = |\zeta_{v_j} e_j \zeta_{v_j'}^{-1}|$$

Then $[\omega_j] \in \pi(|K_1|,v_0)$, and by corollaries 5 and 3.6.17, the set $\{\omega_j\}$ is a system of free generators of $\pi(|K_1|,v_0)$. For each j let ω_j' be a closed path in $|K_2|$ at v_0' such that $h[\omega_j] = [\omega_j']$. We define a continuous map

$$f\colon (|K_1|,v_0) \to (|K_2|,v_0')$$

by $f(|T|) = v_0'$, and for each j we define $f \mid |s_j|$ by

$$f(tv_j' + (1-t)v_j) = \omega_j'(t)$$

where the points of $|s_j|$ are written in the form $tv_j' + (1-t)v_j$ for $t \in I$. f is continuous because its restriction to $|T|$ and to each $|s_j|$ is continuous. Clearly, $f_\#[\omega_j] = [\omega_j'] = h[\omega_j]$; therefore $f_\# = h$.

To prove *(c)*, note that if $f_1 \simeq f_2$, there is a path ω in $|K_2|$ from v_0' to v_0'' such that, by theorem 1.8.7, $f_{1\#} = h_{[\omega]} f_{2\#}$. Conversely, if $f_{1\#} = h_{[\omega]} f_{2\#}$, let T be a maximal tree in K_1 and for each vertex v of K_1 let ζ_v be an edge path in T from v_0 to v. We shall define $F\colon |K_1| \times I \to |K_2|$, a homotopy from f_1 to f_2, in several stages. First we set $F(x,0) = f_1(x)$ and $F(x,1) = f_2(x)$ for $x \in |K_1|$. Then F has been defined on $(|K_1| \times 0) \cup (|K_1| \times 1)$. If v is a vertex of K_1, we define $F(v,t) = ((f_1(|\zeta_v^{-1}|) * \omega) * f_2|\zeta_v|)(t)$ for $t \in I$. Then $F(v,0) = f_1(v)$ and $F(v,1) = f_2(v)$, and F is thus defined on $|K_1^0| \times I$ to be consistent with its previous definition on $(|K_1| \times 0) \cup (|K_1| \times 1)$. It only remains to extend F over $|s| \times I$ for each 1-simplex $s \in K_1$. Let v and v' be the vertices of s in some order. Then $|s| \times I$ is a square with the following product, arbitrarily associated, as boundary

$$(|(v,v')| \times 0) * (v' \times I) * (|(v',v)| \times 1) * (v \times I)^{-1}$$

F can be extended over $|s| \times I$ if and only if F maps this product into a null homotopic path of $|K_2|$. By the definition of F, the above path is sent into a path homotopic to the following product associated arbitrarily

$$\begin{aligned}
&f_1|(v,v')| * (f_1|\zeta_{v'}^{-1}| * \omega * f_2|\zeta_{v'}|) * f_2|(v',v)| * (f_2|\zeta_v^{-1}| * \omega^{-1} * f_1|\zeta_v|) \\
&\quad \simeq f_1|(v,v')| * f_1|\zeta_{v'}^{-1}| * (\omega * f_2(|\zeta_{v'}| * |(v',v)| * |\zeta_v^{-1}|) * \omega^{-1}) * f_1|\zeta_v| \\
&\quad \simeq f_1|(v,v')| * f_1|\zeta_{v'}^{-1}| * f_1(|\zeta_{v'}| * |(v',v)| * |\zeta_v^{-1}|) * f_1|\zeta_v| \\
&\quad \simeq \varepsilon_{f_1(v)}
\end{aligned}$$

Therefore F can be extended over $|s| \times I$, and the resulting map $F\colon |K_1| \times I \to |K_2|$ will be continuous, because for each closed simplex $|s|$ of K_1 its restriction to $|s| \times I$ is continuous. Then $F\colon f_1 \simeq f_2$. ■

It follows from theorem 8*b* that if $f\colon (|K_1|,v_0) \to (|K_2|,v_0')$ induces an isomorphism $f_\#\colon \pi(|K_1|,v_0) \approx \pi(|K_2|,v_0')$, then there is a continuous map $g\colon (|K_2|,v_0') \to (|K_1|,v_0)$ such that $g_\# = (f_\#)^{-1}$. By theorem 8*c*, it follows that $gf \simeq 1_{|K_1|}$ and $fg \simeq 1_{|K_2|}$. Hence we have the next result.

9 COROLLARY *Let K_1 and K_2 be connected graphs with v_0 a vertex of K_1 and v_0' a vertex of K_2. A continuous map $f\colon (|K_1|,v_0) \to (|K_2|,v_0')$ is a homotopy equivalence if and only if f induces an isomorphism $f_\#\colon \pi(|K_1|,v_0) \approx \pi(|K_2|,v_0')$.* ■

The step-by-step extension procedure used to construct the homotopy F to prove theorem 8*c* is a standard method for constructing continuous maps on the space of a complex. The map is constructed on one skeleton at a time and extended over the next skeleton.

8 EXAMPLES AND APPLICATIONS

This section contains assorted results concerning the fundamental group. We begin with some applications to the theory of free groups; in particular, we show that any subgroup of a free group is free. Next we consider the effect on the fundamental group of attaching 2-cells to a space. We use the result obtained to prove that any group is isomorphic to the fundamental group of some space. Finally, we describe how the fundamental group of a surface can be represented by means of generators and relations.

If K is a simplicial complex and $\alpha \in |K|$ has carrier s (that is, $\alpha \in \langle s \rangle$), then for any subdivision K' of $\dot{s}$ the simplicial complex $K' * \alpha$ is a subdivision of $\bar{s}$ (by lemma 3.3.8). It follows that a modified barycentric subdivision of K can be constructed whose vertices are α and the barycenters of simplexes of K other than s. Therefore there is a subdivision of K having α as a vertex, and we have the following result.

1 LEMMA *If $\alpha \in |K|$, there is a subdivision K' of K having α as a vertex.* ∎

2 THEOREM *A polyhedron is locally contractible.*

PROOF In view of lemma 1, it suffices to prove that if v is a vertex of a simplicial complex K, every neighborhood U of v in $|K|$ contains a neighborhood V of v deformable in U to v. Let U be a neighborhood of v and let $A = \operatorname{st} v$. Define $F: A \times I \to |K|$ by

$$F(\alpha,t) = tv + (1 - t)\alpha$$

Then F is a deformation of A in $|K|$ to the point v, and $F(v \times I) = v \in U$. Therefore there is some neighborhood V of v in A such that $F(V \times I) \subset U$. Because $A = \operatorname{st} v$ is open in $|K|$, V is a neighborhood of v in $|K|$. Since $F \mid V \times I$ is a deformation of V in U to v, $|K|$ is locally contractible. ∎

It follows from theorem 2 that the theory of covering projections applies to polyhedra, and corresponding to any subgroup of the fundamental group of a polyhedron there is a covering projection. We show that any covering projection of a polyhedron corresponds to a simplicial map.

3 THEOREM *Let $p: \tilde{X} \to X$ be a covering projection, where X is a polyhedron. Then $\tilde{X}$ is a polyhedron, of the same dimension as X, in such a way that p corresponds to a simplicial map.*

PROOF Assume that $p: \tilde{X} \to |K|$ is a covering projection. For any simplex $s \in K$ the closed simplex $|s|$ is simply connected. It follows from the lifting theorem that the inclusion map $|s| \subset |K|$ can be lifted to a map $|s| \to \tilde{X}$, and it follows from the unique lifting theorem that two such liftings are either identical or have disjoint images. Hence there are as many liftings of $|s|$ as sheets of $\tilde{X}$ over $|s|$.

Define a simplicial complex $\tilde{K}$ to have the collection $\{p^{-1}(v) \mid v \text{ is a vertex of } K\}$ as vertex set and to have simplexes $\{\tilde{s}\}$, where $\tilde{s} = \{\tilde{v}_0, \ldots, \tilde{v}_q\}$ is a simplex of $\tilde{K}$ if and only if there is a simplex $s = \{v_0, \ldots, v_q\}$ in K and a lifting $f_{\tilde{s}}: |s| \to \tilde{X}$ of $|s|$ such that $f_{\tilde{s}}(v_i) = \tilde{v}_i$ for $0 \le i \le q$ [in which case $s = p(\tilde{s})$ and $f_{\tilde{s}}$ are both unique]. Then $\tilde{K}$ is a simplicial complex and has the same dimension as K. If $\tilde{s}_1$ is a face of $\tilde{s}$, then $p(\tilde{s}_1)$ is a face of $p(\tilde{s})$ and $f_{\tilde{s}} \mid |p(\tilde{s}_1)| = f_{\tilde{s}_1}$. Therefore the collection $\{f_{\tilde{s}}\}_{\tilde{s} \in \tilde{K}}$ defines a continuous map $f: |\tilde{K}| \to \tilde{X}$ such that

$$f(\Sigma \alpha_i \tilde{v}_i) = f_{\tilde{s}}(\Sigma \alpha_i p(\tilde{v}_i)) \qquad \Sigma \alpha_i \tilde{v}_i \in |\tilde{s}|$$

Let $\varphi: \tilde{K} \to K$ be the simplicial map $\varphi(\tilde{v}) = p(\tilde{v})$. Then there is a commutative triangle

$$\begin{array}{ccc} |\tilde{K}| & \xrightarrow{f} & \tilde{X} \\ & {\scriptstyle |\varphi|} \searrow \quad \swarrow {\scriptstyle p} & \\ & |K| & \end{array}$$

To complete the proof it suffices to prove that $(\tilde{K},f)$ is a triangulation of $\tilde{X}$ (that is, that f is a homeomorphism). If v is a vertex of K, then st v, being contractible, is evenly covered by p. For $\tilde{v} \in p^{-1}(v)$ let $U_{\tilde{v}}$ be the component of $p^{-1}(\text{st } v)$ containing $\tilde{v}$. Then $p \mid U_{\tilde{v}}$ is a homeomorphism of $U_{\tilde{v}}$ onto st v. By the definition of $\tilde{K}$ and φ, $|\varphi| \mid \text{st } \tilde{v}$ is a homeomorphism of st $\tilde{v}$ onto st v for $\tilde{v} \in p^{-1}(v)$. From the commutativity of the above triangle, $f \mid \text{st } \tilde{v}$ is a homeomorphism of st $\tilde{v}$ onto $U_{\tilde{v}}$ for $\tilde{v} \in p^{-1}(v)$. Since $|\varphi|^{-1}(\text{st } v) = \cup\{\text{st } \tilde{v} \mid \tilde{v} \in p^{-1}(v)\}$, $f \mid |\varphi|^{-1}(\text{st } v)$ is a homeomorphism of $|\varphi|^{-1}(\text{st } v)$ onto $p^{-1}(\text{st } v)$. Since this is so for every vertex v of K, f is a homeomorphism of $|\tilde{K}|$ onto $\tilde{X}$. ■

The following corollary is an interesting application of these results.

4 COROLLARY *Any subgroup of a free group is free.*

PROOF Let F be a free group. It follows from example 3.7.6 that there is a polyhedron (in fact, a wedge of 1-spheres) X with base point x_0 such that $\pi(X,x_0) \approx F$. Let F' be any subgroup of F. Under the above isomorphism F' corresponds to some subgroup $H \subset \pi(X,x_0)$. Let $p\colon \tilde{X} \to X$ be a covering projection such that $\tilde{X}$ is path connected, $p(\tilde{x}_0) = x_0$, and $p_\# \pi(\tilde{X},\tilde{x}_0) = H$. By theorem 3, $\tilde{X}$ is homeomorphic to the space of a connected graph. By corollary 3.7.5, $\pi(\tilde{X},\tilde{x}_0)$ is a free group. ■

If K is a finite connected graph, it follows from corollary 3.7.5 that $E(K,v_0)$ is a free group on $1 - n_0 + n_1$ generators, where n_0 is the number of vertices of K and n_1 is the number of 1-simplexes of K. If $p\colon \tilde{X} \to |K|$ is a covering projection of multiplicity m, the number of q-simplexes in the corresponding triangulation $(\tilde{K},f)$ of $\tilde{X}$ (given by theorem 3) equals mn_q, where n_q is the number of q-simplexes of K. Therefore the method used to prove corollary 4 also yields the following result.

5 COROLLARY *Let F be a free group on n generators and let F' be a subgroup of F of index m. Then F' is a free group on $1 - m + mn$ generators.* ■

We now investigate the effect on the fundamental group of the process of attaching cells. Let A be a closed subset of a space X. X is said to be *obtained from A by adjoining n-cells* $\{e_j^n\}$, where $n \geq 0$, if

(*a*) For each j, e_j^n is a subset of X.
(*b*) If $\dot{e}_j^n = e_j^n \cap A$, then for $j \neq j'$, $e_j^n - \dot{e}_j^n$ is disjoint from $e_{j'}^n - \dot{e}_{j'}^n$.
(*c*) X has a topology coherent with $\{A, e_j^n\}$ and $X = A \cup \bigcup_j e_j^n$.
(*d*) For each j there is a map

$$f_j\colon (E^n,S^{n-1}) \to (e_j^n,\dot{e}_j^n)$$

such that $f_j(E^n) = e_j^n$, f_j maps $E^n - S^{n-1}$ homeomorphically onto $e_j^n - \dot{e}_j^n$, and e_j^n has the topology coinduced by f_j and the inclusion map $\dot{e}_j^n \subset e_j^n$.

Note that if $n = 0$, X is the topological sum of A and a discrete space. A map $f_j\colon (E^n,S^{n-1}) \to (e_j^n,\dot{e}_j^n)$ satisfying condition (*d*) above is called a *characteristic map* for e_j^n, and $f_j \mid S^{n-1}: S^{n-1} \to A$ is called an *attaching map* for e_j^n. X is characterized by A and the collection $\{f_j \mid S^{n-1}\}$ of attaching maps. Given A and an indexed collection of maps $\{g_j: S^{n-1} \to A\}$, there is a space X obtained from A by attaching n-cells $\{e_j^n\}$ by the maps g_j. X is defined as the quotient space of the topological sum $\bigvee E_j^n \vee A$, where $E_j^n = E^n$ for each j, by the identifications $z \in S_j^{n-1}$ equals $g_j(z) \in A$. Then the inclusion map $(E_j^n, S_j^{n-1}) \subset (\bigvee E_j^n \vee A, \bigvee S_j^{n-1} \vee A)$ followed by the projection to (X,A) is a characteristic map $f_j: (E_j^n, S_j^{n-1}) \to (X,A)$ for an n-cell $e_j^n = f_j(E_j^n)$.

Following are two examples.

6 If K is a simplicial complex, $|K^q|$ is obtained from $|K^{q-1}|$ by adjoining q-cells $\{|s| \mid s \text{ is a } q\text{-simplex of } K\}$.

7 For $i = 1, 2$, or 4 let F_i be $\mathbf{R}$, $\mathbf{C}$, or $\mathbf{Q}$, respectively, and for $q \geq 0$ let $P_q(F_i)$ be the real, complex, or quaternionic projective space of dimension q. $P_q(F_i)$ is imbedded in $P_{q+1}(F_i)$ by the map $[t_0, t_1, \ldots, t_q] \to [t_0, t_1, \ldots, t_q, 0]$ for $t_j \in F_i$. Then $P_{q+1}(F_i)$ is obtained from $P_q(F_i)$ by adjoining a single $(q+1)i$-cell. If $E^{(q+1)i}$ is identified with the space $\{(t_0, t_1, \ldots, t_q) \in F_i^{q+1} \mid \Sigma |t_j|^2 \leq 1\}$, then a characteristic map $f: (E^{(q+1)i}, S^{(q+1)i-1}) \to (P_{q+1}(F_i), P_q(F_i))$ for this single cell is defined by the equation

$$f(t_0, t_1, \ldots, t_q) = [t_0, t_1, \ldots, t_q, 1 - \Sigma |t_j|^2]$$

8 LEMMA *Let X be obtained from A by adjoining n-cells for $n \geq 2$. Then for any point $x_0 \in A$ the inclusion map $i: (A,x_0) \subset (X,x_0)$ induces an epimorphism*

$$i_\#: \pi(A,x_0) \to \pi(X,x_0)$$

PROOF Let X be obtained from A by adjoining the n-cells $\{e_j^n\}$, and for each j let $y_j \in e_j^n - \dot{e}_j^n$ and let B_j be a neighborhood of y_j in $e_j^n - \dot{e}_j^n$ homeomorphic to E^n. Let $\omega: (I,\dot{I}) \to (X,x_0)$ be a closed path at x_0. We show that ω is homotopic to a path in $U = X - \{y_j\}_j$. By the compactness of I, we can subdivide I by points $0 = t_0 < t_1 < \cdots < t_n = 1$ such that for $0 \leq i < n$ either $\omega[t_i,t_{i+1}] \subset U$ or $\omega[t_i,t_{i+1}] \subset B_j$ for some j. If $\omega[t_i,t_{i+1}] \cup \omega[t_{i+1},t_{i+2}] \subset B_j$, we can omit the point t_{i+1} from the subdivision of I to obtain another subdivision of I with the same property. Continuing in this way we can obtain a subdivision such that if $\omega[t_i,t_{i+1}] \subset B_j$, then neither $\omega[t_{i-1},t_i]$ nor $\omega[t_{i+1},t_{i+2}]$ is contained in B_j. It follows that $\omega(t_i) \neq y_j$ and $\omega(t_{i+1}) \neq y_j$. For each such i, because $B_j - y_j$ is path connected and B_j is simply connected, $\omega \mid [t_i,t_{i+1}]$ is homotopic rel $\{t_i,t_{i+1}\}$ to a path contained in $B_j - y_j$. Since altogether there are only a finite number of such subintervals of I, $\omega \simeq \omega'$, where $\omega'(I) \subset U$.

Because S^{n-1} is a strong deformation retract of E^n minus a point, it follows that $\dot{e}_j^n$ is a strong deformation retract of $e_j^n - y_j$. Therefore A is a strong deformation retract of U and $\omega' \simeq \omega''$, where $\omega''(I) \subset A$. Then $i_\#[\omega''] = [\omega]$. ∎

9 COROLLARY *For all $n \geq 0$, $P_n(\mathbf{C})$ and $P_n(\mathbf{Q})$ are simply connected.*

PROOF Because $P_0(\mathbf{C})$ and $P_0(\mathbf{Q})$ are each one-point spaces, the result follows by induction on q, using lemma 8 and the fact that $P_{q+1}(\mathbf{C})$ is obtained from $P_q(\mathbf{C})$ by adjoining a $2(q+1)$-cell and $P_{q+1}(\mathbf{Q})$ is obtained from $P_q(\mathbf{Q})$ by adjoining a $4(q+1)$-cell. ∎

We want to compute the kernel of $i_\#$ for the case $n = 2$. Given any map $g: S^1 \to A$, where A is path connected, and given a point $x_0 \in A$, a normal subgroup of $\pi(A,x_0)$ is determined as follows. If $g(p_0) = x_1$ and ω is a path in A from x_0 to x_1, then $h_{[\omega]}g_\#(\pi(S^1,p_0))$ is a cyclic subgroup of $\pi(A,x_0)$, and for a different choice of ω we obtain a conjugate subgroup in $\pi(A,x_0)$. Therefore the normal subgroup of $\pi(A,x_0)$ generated by $h_{[\omega]}g_\#(\pi(S^1,p_0))$ is independent of the choice of the path ω. Similar statements apply to a collection of maps $\{g_j: S^1 \to A\}$. There is a well-defined normal subgroup of $\pi(A,x_0)$ determined by these maps.

10 THEOREM *Let A be a connected polyhedron and let X be obtained from A by attaching 2-cells to A by maps $\{g_j: S^1 \to A\}$. If N is the normal subgroup of $\pi(A,x_0)$ determined by the maps $\{g_j\}$, then*

$$i_\#: \pi(A,x_0) \to \pi(X,x_0)$$

is an epimorphism with kernel N.

PROOF By lemma 8, $i_\#$ is a surjection. Let $p: \tilde{A} \to A$ be a covering projection such that $\tilde{A}$ is path connected, $p(\tilde{x}_0) = x_0$, and $p_\#(\pi(\tilde{A},\tilde{x}_0)) = N$. Because N is normal in $\pi(A,x_0)$, p is a regular covering projection. Because N is the subgroup determined by the maps $\{g_j\}$, each map g_j lifts to a map $\tilde{g}_j: S^1 \to \tilde{A}$. Let $\tilde{X}$ be the space obtained from $\tilde{A}$ by attaching 2-cells for all the lifted maps $\{\tilde{g}_j\}$ and extend p to a map $p': \tilde{X} \to X$ such that p' maps each 2-cell of $\tilde{X}$ homeomorphically onto its corresponding 2-cell of X. Then p' is easily seen to be a covering projection.

We know from the definition of N that $i_\#(N) = 1$. Assume that $[\omega] \in \pi(A,x_0)$ is in the kernel of $i_\#$. Let $\tilde{\omega}$ be any lifting of ω in $\tilde{A}$ such that $\tilde{\omega}(0) = \tilde{x}_0$. Then $\tilde{\omega}$ is a lifting of ω in $\tilde{X}$. Because ω is null homotopic in X, $\tilde{\omega}$ is a closed path in $\tilde{X}$. Therefore $\tilde{\omega}$ is a closed path in $\tilde{A}$, and so

$$[\omega] = p_\#[\tilde{\omega}] \in N \quad ∎$$

Note that for the proof of theorem 10 it was not necessary that A be a connected polyhedron. It would have been sufficient to assume A path connected, locally path connected, and semilocally 1-connected.

11 COROLLARY *For any group G there is a space X with $\pi(X,x_0) \approx G$.*

PROOF Represent G as the quotient group of a free group F and a normal subgroup N. There is a polyhedron A such that $\pi(A,x_0) \approx F$ (in fact, as in example 3.7.6, A can be taken to be a wedge of 1-spheres). For each $\lambda \in N$

let $g_\lambda\colon (S^1,p_0) \to (A,x_0)$ be a map such that $[g_\lambda]$ corresponds to λ under the isomorphism $\pi(A,x_0) \approx F$. Let X be the space obtained from A by attaching 2-cells by the maps $\{g_\lambda\}$. By theorem 10, there is an isomorphism $\pi(X,x_0) \approx G$. ∎

We now specialize to the case of a surface. These are the spaces of finite two-dimensional pseudomanifolds without boundary. An *n-dimensional pseudomanifold without boundary* (or *absolute n-circuit*) is a simplicial complex K such that

(*a*) Every simplex of K is a face of some n-simplex of K.
(*b*) Every $(n-1)$-simplex of K is the face of exactly two n-simplexes of K.
(*c*) If s and s' are n-simplexes of K, there is a finite sequence $s = s_1, s_2, \ldots, s_m = s'$ of n-simplexes of K such that s_i and s_{i+1} have an $(n-1)$-face in common for $1 \le i < m$.

We define a *surface* to be the space of a finite two-dimensional pseudomanifold without boundary in which the star of every vertex is homeomorphic to $\mathbf{R}^2$. It can be shown[1] that every surface has a normal form consisting of a polygon in the plane with identifications of its edges. These fall into classes, those with $h \ge 0$ handles and those with k crosscaps. The surface with 0 handles is the polygon with identifications of its edges pictured as

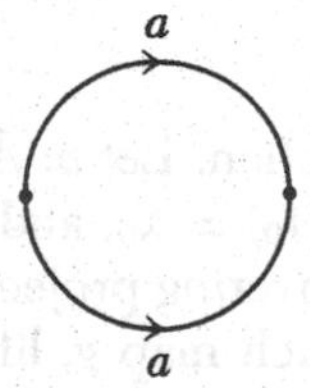

Surface with 0 handles

Topologically it is homeomorphic to the 2-sphere S^2. For $h > 0$ the surface with h handles is pictured as

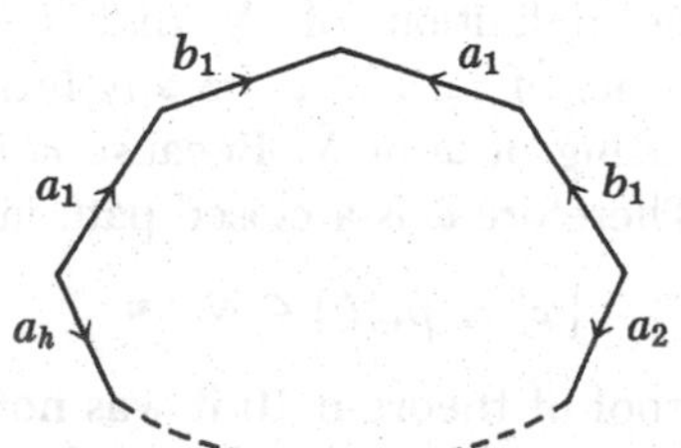

Surface with $h > 0$ handles

The surface with one handle is topologically the *torus*.

[1] See S. Lefschetz, *Introduction to Topology*, Princeton University Press, Princeton, N.J., 1949, and H. Seifert and W. Threlfall, *Lehrbuch der Topologie*, B. G. Teubner, Verlagsgesellschaft, Leipzig, 1934.

For $k \geq 1$, the surface with k crosscaps is pictured as

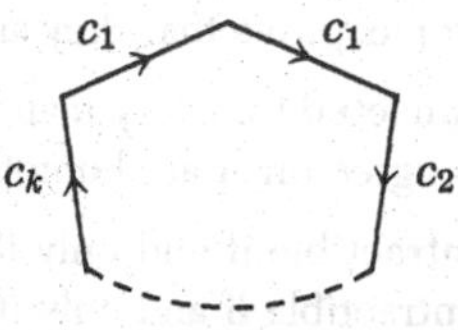

Surface with k crosscaps

The surface with one crosscap is topologically the real projective plane P^2, and the surface with two crosscaps is topologically the *Klein bottle.*

The normal form represents a surface with $h \geq 1$ handles as a wedge of $2h$ 1-spheres with a single 2-cell attached by a suitable map. If A is the wedge of $2h$ 1-spheres, then $\pi(A)$ is a free group on $2h$ generators, which generators we denote by a_i and b_i, where $1 \leq i \leq n$. If X is the surface with h handles, X is obtained from A by attaching a single 2-cell to A by a map $g: S^1 \to A$ such that $g_\#$ maps a generator of $\pi(S^1)$ to the element $a_1b_1a_1^{-1}b_1^{-1} \cdots a_hb_ha_h^{-1}b_h^{-1} \in \pi(A)$. Theorem 10 then provides a description of $\pi(X)$ in terms of generators and relations. Similar remarks apply to a surface with $k \geq 1$ crosscaps. The result is summarized below.

12 *The fundamental group of a surface is*

(*a*) *Trivial for the surface with no handles.*
(*b*) *A group with generators $a_1, b_1, \ldots, a_h, b_h$ and the single relation $a_1b_1a_1^{-1}b_1^{-1} \cdots a_hb_ha_h^{-1}b_h^{-1} = 1$ for a surface with $h \geq 1$ handles.*
(*c*) *A group with generators $c_1, c_2, \ldots, c_k$ and the single relation $c_1^2c_2^2 \cdots c_k^2 = 1$ for a surface with $k \geq 1$ crosscaps.* ∎

EXERCISES

A TOPOLOGICAL PROPERTIES OF POLYHEDRA

1 Prove that a compact polyhedron is an absolute neighborhood retract. (*Hint:* Assume $X = |K|$ and let K be a subcomplex of a simplex s. Use induction on the number of simplexes in $s - K$ and the fact that a retract of an open subset of an absolute neighborhood retract is an absolute neighborhood retract.)

2 Give an example of a space X and closed subset $A \subset X$ such that A and X are both polyhedra but (X,A) is not a polyhedral pair.

3 Prove that an open subset of a compact polyhedron is a polyhedron. [*Hint:* Since $|K| - U$ is a G_δ, there exists a sequence of open subsets V_i of $|K|$ such that $\bigcap V_i = |K| - U$. By induction on n, construct a sequence of subdivisions K_n and subcomplexes $L_n \subset K_n$ such that (*a*) K_n is finer than the covering $\{U,V_n\}$, (*b*) L_n is the largest subcomplex of K_n such that $|L_n| \subset U$, and (*c*) K_{n+1} is a subdivision of K_n containing L_n as subcomplex. Then $L = \bigcup L_n$ is a simplicial complex such that $|L| = |K| - U$.]

4 Let Y be an n-connected space and K be a simplicial complex. Prove that any continuous map $|K| \to Y$ is homotopic to a map which sends $|K^n|$ to a single point. If $f_0, f_1: (|K|,|K^n|) \to (Y,y_0)$ are homotopic, prove that they are homotopic relative to $|K^{n-1}|$.

5 Let Y be a space which is n-connected for every n and let (X,A) be a polyhedral pair. Prove that two maps $X \to Y$ which agree on A are homotopic relative to A.

6 Prove that a polyhedron is contractible if and only if it is n-connected for every n. If it has finite dimension m, it is contractible if and only if it is m-connected.

B EXAMPLES

1 Prove that P^n is a polyhedron for all n.

2 Let K be the simplicial complex consisting of vertices $v_1, v_2, \ldots, v_p$ and simplexes $\{v_1,v_2\}, \{v_2,v_3\}, \ldots, \{v_{p-1},v_p\}$, and $\{v_p,v_1\}$ and let I be the simplicial complex with 0 and 1 as vertices and $\{0,1\}$ as 1-simplex. Then $K * I$ is a simplicial complex with vertices $v_1, \ldots, v_p$, 0, and 1. If q is an integer relatively prime to p and v_i is defined for all integers i to be equal to v_j if $i \equiv j \bmod p$, then let X be the space obtained from $|K * I|$ by identifying the 2-simplex $\{v_i,v_{i+1},0\}$ linearly with the 2-simplex $\{v_{i+q},v_{i+q+1},1\}$ for all i. Prove that X is homeomorphic to the lens space $L(p,q)$ and that X is a polyhedron.

3 Prove that the generalized lens space $L(p, q_1, \ldots, q_n)$ is a polyhedron.

4 If X and Y are polyhedra and one of them is locally compact, prove that $X * Y$ and $X \times Y$ are also polyhedra.

C PSEUDOMANIFOLDS

A simplicial complex is said to be *homogeneously n-dimensional* if every simplex is a face of some n-simplex of the complex. An *n-dimensional pseudomanifold* is a simplicial complex K such that

(a) K is homogeneously n-dimensional.

(b) Every $(n-1)$-simplex of K is the face of at most two n-simplexes of K.

(c) If s and s' are n-simplexes of K, there is a finite sequence $s = s_1, s_2, \ldots, s_m = s'$ of n-simplexes of K such that s_i and s_{i+1} have an $(n-1)$-face in common for $1 \le i < m$.

The *boundary* of an n-dimensional pseudomanifold K, denoted by $\dot{K}$, is defined to be the subcomplex of K generated by the set of $(n-1)$-simplexes which are faces of exactly one n-simplex of K. (If $\dot{K}$ is empty, then K is an n-dimensional pseudomanifold without boundary, as defined in Sec. 3.8.)

1 Prove that an n-simplex is an n-dimensional pseudomanifold whose boundary, as a pseudomanifold, is $\dot{s}$.

2 If K is a pseudomanifold and L is a subdivision of K, prove that L is a pseudomanifold and $\dot{L} = L \mid \dot{K}$.

3 If K is a finite 1-dimensional pseudomanifold, prove that $\dot{K}$ is either empty or consists of exactly two vertices.

4 Give an example of an n-dimensional pseudomanifold K such that $\dot{K}$ is neither empty nor an $(n-1)$-dimensional pseudomanifold.

D SIMPLICIAL MAPS

In the first four exercises K will be a finite n-dimensional pseudomanifold, where $n > 0$, with nonempty boundary $\dot{K}$, K' will be a simplicial subdivision of K, and $\varphi: K' \to K$ will be a simplicial map such that $\varphi \mid \dot{K}'$ maps $\dot{K}'$ to $\dot{K}$ and is a simplicial approximation to the identity map $|\dot{K}'| \subset |\dot{K}|$. Furthermore, s^{n-1} will be a fixed $(n-1)$-simplex of $\dot{K}$ and s^n will be the unique n-simplex of K having s^{n-1} as a face.

1 For each n-simplex s' of K' let $\alpha(s')$ be the number of $(n-1)$-faces of s' mapped onto s^{n-1} by φ. Prove that $\alpha(s') = 1$ if and only if φ maps s' onto s^n and that $\alpha(s') = 0$ or 2 otherwise.

2 Prove that the number of n-simplexes of K' mapped onto s^n by φ has the same parity as the number of $(n-1)$-simplexes of $\dot{K}'$ mapped onto s^{n-1} by φ. [*Hint:* They both have the same parity as $\Sigma\, \alpha(s')$, the summation being over all n-simplexes s' of K'.]

3 *Sperner lemma.* Prove that the number of n-simplexes of K' mapped onto s^n by φ is odd. (*Hint:* Use induction on n.)

4 Prove that $|\dot{K}|$ is not a retract of $|K|$.

5 *Brouwer fixed-point theorem.* Prove that every continuous map of E^n to itself has a fixed point.

E SIMPLICIAL MAPPING CYLINDERS

Let $\varphi: K \to L$ be a simplicial map between simplicial complexes whose vertex sets are disjoint. We assume that the vertices of K are simply ordered. The *simplicial mapping cylinder* M of φ is the simplicial complex whose vertex set is the union of the vertex sets of K and L and whose simplexes are the simplexes of K and of L and all subsets of sets of the form $\{v_0, \ldots, v_k, \varphi(v_k), \ldots, \varphi(v_p)\}$, where $\{v_0, v_1, \ldots, v_p\}$ is a simplex of K and $v_0 < v_1 < \cdots < v_p$ in the simple ordering of the vertices of K.

1 Prove that the inclusion maps $i: K \subset M$ and $j: L \subset M$ are simplicial maps. If $\rho: M \to L$ is defined by $\rho(v) = \varphi(v)$ for v a vertex of K and $\rho(v') = v'$ for v' a vertex of L, then prove that ρ is a simplicial map such that $\varphi = \rho \circ i$ and $\rho \circ j = 1_L$.

2 If K is finite, prove that $j \circ \rho$ and 1_M are contiguous.

3 Prove that $|L|$ is a deformation retract of $|M|$.

F EDGE-PATH GROUPS

1 Prove that if K is a simplicial complex, there is a one-to-one correspondence between equivalence classes of local systems on $|K|$ with values in $\mathcal{C}$ and natural equivalence classes of covariant functors from the edge-path groupoid of K to $\mathcal{C}$.

2 *Van Kampen's theorem for simplicial complexes.*[1] Let K be a connected simplicial complex with connected subcomplexes L_1 and L_2 such that $L_1 \cap L_2$ is connected and $K = L_1 \cup L_2$. Let v_0 be a vertex of $L_1 \cap L_2$ and let $i_1: (L_1 \cap L_2, v_0) \subset (L_1, v_0)$ and $i_2: (L_1 \cap L_2, v_0) \subset (L_2, v_0)$. Prove that $E(K, v_0)$ is isomorphic to the quotient group of the free product of $E(L_1, v_0)$ with $E(L_2, v_0)$ by the normal subgroup generated by the set

$$\{(i_{1\#}[\zeta]) \circ (i_{2\#}[\zeta]^{-1}) \mid [\zeta] \in E(L_1 \cap L_2, v_0)\}$$

3 If G is a finitely presented group, prove that there is a finite connected two-dimensional simplicial complex K whose edge-path group is isomorphic to G.

[1] For the topological case see P. Olum, Non-abelian cohomology and Van Kampen's theorem, *Annals of Mathematics*, vol. 68, pp. 658–668, 1958.

4 Let X be a space with base point $x_0 \in X$. Prove that there exists a polyhedron Y, with base point $y_0 \in Y$, and a continuous map $f\colon (Y,y_0) \to (X,x_0)$ such that $f_\#\colon \pi(Y,y_0) \approx \pi(X,x_0)$.

G NERVES OF COVERINGS

If $\mathcal{U} = \{U\}$ in an open covering of a space X and $K(\mathcal{U})$ is its nerve, a *canonical map* $f\colon X \to |K(\mathcal{U})|$ is a continuous map such that $f^{-1}(\text{st } U) \subset U$ for every $U \in \mathcal{U}$.

1 If $\mathcal{U}$ is a locally finite open covering of X, prove that there is a one-to-one correspondence between canonical maps $X \to |K(\mathcal{U})|$ and partitions of unity subordinate to $\mathcal{U}$.

2 If $\mathcal{U}$ is a locally finite open covering of X, prove that any two canonical maps $X \to |K(\mathcal{U})|$ are homotopic.

If $\mathcal{U}$ and $\mathcal{V}$ are open coverings of X, with $\mathcal{V}$ a refinement of $\mathcal{U}$, a *canonical projection* from $\mathcal{V}$ to $\mathcal{U}$ is a function φ which assigns to each $V \in \mathcal{V}$ an element $\varphi(V) \in \mathcal{U}$ *such that* $V \subset \varphi(V)$.

3 Prove that a canonical projection from $\mathcal{V}$ to $\mathcal{U}$ defines a simplicial map $K(\mathcal{V}) \to K(\mathcal{U})$ and any two canonical projections from $\mathcal{V}$ to $\mathcal{U}$ define contiguous simplicial maps $K(\mathcal{V}) \to K(\mathcal{U})$.

4 If $\varphi\colon K(\mathcal{V}) \to K(\mathcal{U})$ is a canonical projection and $f\colon X \to |K(\mathcal{V})|$ is a canonical map, prove that the composite $|\varphi| \circ f\colon X \to |K(\mathcal{U})|$ is a canonical map.

5 Let X be a paracompact space and let $g\colon X \to |K|$ be a continuous map (where K is a simplicial complex). Prove that there exists a locally finite open covering $\mathcal{U}$ of X and a simplicial map $\varphi\colon K(\mathcal{U}) \to K$ such that for any canonical map $f\colon X \to |K(\mathcal{U})|$ the composite $|\varphi| \circ f$ is homotopic to g. [*Hint:* Choose $\mathcal{U}$ to be any locally finite open refinement of the open covering $\{g^{-1}(\text{st } v) \mid v \text{ a vertex of } K\}$, and for $U \in \mathcal{U}$ choose $\varphi(U)$ a vertex of K such that $U \subset g^{-1}(\text{st } \varphi(U))$.]

6 Let X be a compact Hausdorff space and let K be a simplicial complex. Prove that there is a bijection

$$\lim_{\to} \{[K(\mathcal{U});K]\} \approx [X;|K|]$$

where the direct limit is with respect to the family of finite open coverings of X directed by refinement with maps induced by canonical projections and the bijection is induced by canonical maps.

H DIMENSION THEORY

A topological space X is said to have *dimension* $\leq n$, abbreviated $\dim X \leq n$, if every open covering of X has an open refinement whose nerve is a simplicial complex of dimension $\leq n$. If $\dim X \leq n$ but $\dim X \nleq n - 1$, then X is said to have *dimension* n, denoted by $\dim X = n$. If $\dim X \nleq n$ for any n, we write $\dim X = \infty$.

1 If A is a closed subset of X, prove that $\dim A \leq \dim X$.

2 If K is a finite simplicial complex with $\dim K \leq n$, prove that $\dim |K| \leq n$.

3 If s is an n-simplex, prove that $\dim |s| = n$. (*Hint:* Let $\mathcal{U}$ be the open covering of $|s|$ of stars of the vertices of s and assume that there is a refinement $\mathcal{V}$ of $\mathcal{U}$ such that $\dim K(\mathcal{V}) \leq n - 1$. Let K' be a subdivision of s finer than $\mathcal{V}$. There are simplicial maps $K' \to K(\mathcal{V}) \to s$ whose composite λ is a simplicial approximation to the identity map $|K'| \subset |s|$.)

4 Let X be a paracompact space with $\dim X \leq n$. Prove that any map $X \to S^m$, with $m > n$, is null homotopic.

5 Let X be a compact metric space and let C be the space of maps $f: X \to \mathbf{R}^{2n+1}$ topologized by the metric

$$d(f,g) = \sup \{\|f(x) - g(x)\| \mid x \in X\}$$

Prove that C is a complete metric space, and if

$$C_m = \{f \in C \mid \operatorname{diam} f^{-1}(z) < \frac{1}{m} \text{ for all } z \in \mathbf{R}^{2n+1}\}$$

then show that C_m is an open subset of C for every positive integer m and $\bigcap C_m$ is the set of homeomorphisms of X into $\mathbf{R}^{2n+1}$.

6 If X is a compact metric space of dimension $\leq n$, prove that C_m is a dense subset of C for every positive integer m. [*Hint:* Let $\mathcal{U}$ be a finite open covering of X by sets of diameter $< 1/m$ such that $\dim K(\mathcal{U}) \leq n$ and let $h: |K(\mathcal{U})| \to \mathbf{R}^{2n+1}$ be a realization of $K(\mathcal{U})$. If $f: X \to |K(\mathcal{U})|$ is any canonical map, then $h \circ f \in C_m$. Given $g: X \to \mathbf{R}^{2n+1}$ and given $\varepsilon > 0$, show that it is possible to choose $\mathcal{U}$ and h as above, so that $d(h \circ f, g) < \varepsilon$.]

7 If X is a compact metric space of dimension $\leq n$, prove that X can be embedded in $\mathbf{R}^{2n+1}$ (in fact, the set of homeomorphisms of X into $\mathbf{R}^{2n+1}$ is dense in C).

CHAPTER FOUR

HOMOLOGY

THIS CHAPTER INTRODUCES THE CONCEPT OF HOMOLOGY THEORY, WHICH IS OF fundamental importance in algebraic topology. A homology theory involves a sequence of covariant functors H_n to the category of abelian groups, and we shall define homology theories on two categories–the singular homology theory on the category of topological pairs and the simplicial homology theory on the category of simplicial pairs. The former is topologically invariant by definition and is formally easier to work with, while the latter is easier to visualize geometrically and by definition is effectively computable for finite simplicial complexes. The two theories are related by the basic result that the singular homology of a polyhedron is isomorphic to the simplicial homology of any of its triangulating simplicial complexes.

The functor H_n measures the number of "n-dimensional holes" in the space (or simplicial complex), in the sense that the n-sphere S^n has exactly one n-dimensional hole and no m-dimensional holes if $m \neq n$. A 0-dimensional hole is a pair of points in different path components, and so H_0 measures path connectedness. The functors H_n measure higher dimensional connectedness, and some of the applications of homology are to prove higher dimensional analogues of results obtainable in low dimensions by using connectedness considerations.

Sections 4.1 and 4.2 are devoted to the definition of the category of chain complexes and to an appropriate concept of homotopy in this category. Homology theory is introduced as a sequence of covariant functors naturally defined from the category of chain complexes to the category of abelian groups.

Simplicial homology theory is defined by means of a covariant functor from the category of simplicial complexes to the category of chain complexes. We study it in detail in Sec. 4.3, where it is shown that two different definitions (one based on oriented simplexes, the other on ordered simplexes) are isomorphic. In similar fashion, singular homology theory is defined via a covariant functor from the category of topological spaces to the category of chain complexes. Its basic properties are considered in Sec. 4.4, where it is shown that "small" singular simplexes suffice to define singular homology.

Section 4.5 introduces the concept of exact sequence. All the homology functors H_n occur together in the exact sequences of homology, and it is for this reason that we consider all these functors simultaneously, rather than one at a time. Section 4.6 is devoted to the exact Mayer-Vietoris sequences connecting the homology of the union of two spaces (or simplicial complexes), the homology of the spaces, and the homology of their intersection. We use these to prove the isomorphism of the simplicial homology groups of a simplicial complex with the singular homology groups of its corresponding space.

Section 4.7 contains some applications of homology theory. We prove that euclidean spaces of different dimensions are not homeomorphic. We also prove the Brouwer fixed-point theorem and the more general Lefschetz fixed-point theorem. Finally, we prove Brouwer's generalization of the Jordan curve theorem (that an $(n - 1)$-sphere imbedded in S^n separates S^n into two components), and we establish the invariance of domain. Section 4.8 contains a discussion of the axiomatic characterization of homology given by Eilenberg and Steenrod, as well as some related concepts.

1 CHAIN COMPLEXES

This section introduces the category of chain complexes and chain maps and the homology functor on this category. We also define covariant functors from the category of simplicial complexes and from the category of topological spaces to the category of chain complexes. The composites of these and the homology functor define homology functors on the category of simplicial complexes and on the category of topological spaces.

A *differential group* C consists of an abelian group C and an endomorphism $\partial: C \to C$ such that $\partial\partial = 0$. The endomorphism ∂ is called the *differential*, or *boundary operator* of C. There is a category whose objects are differential groups and whose morphisms are homomorphisms commuting with the differentials.

For a differential group C there is a subgroup of *cycles* $Z(C) = \ker \partial$ and a subgroup of *boundaries* $B(C) = \operatorname{im} \partial$. Because $\partial\partial = 0$, $B(C) \subset Z(C)$. The *homology group* $H(C)$ is defined to be the quotient group

$$H(C) = Z(C)/B(C)$$

The elements of $H(C)$ are called *homology classes*. If z is a cycle, its homology class in $H(C)$ is denoted by $\{z\}$. Two cycles z_1 and z_2 are *homologous*, denoted by $z_1 \sim z_2$, if their difference is a boundary, that is, if $\{z_1\} = \{z_2\}$.

If $\tau: C \to C'$ is a homomorphism of differential groups commuting with the differentials, then τ maps cycles of C to cycles of C' and boundaries of C to boundaries of C'. Therefore τ induces a homomorphism

$$\tau_*: H(C) \to H(C')$$

such that $\tau_*\{z\} = \{\tau(z)\}$ for $z \in Z(C)$. Because $(\tau_1\tau_2)_* = \tau_{1*}\tau_{2*}$, there is a covariant functor from the category of differential groups to the category of groups which assigns to a differential group C its homology group $H(C)$ and to a homomorphism τ its induced homomorphism τ_*.

A *graded group* $C = \{C_q\}$ consists of a collection of abelian groups C_q indexed by the integers. Elements of C_q are said to have *degree* q. A *homomorphism* $\tau: C \to C'$ *of degree* d from one graded group to another consists of a collection $\tau = \{\tau_q: C_q \to C'_{q+d}\}$ of homomorphisms indexed by the integers. We shall omit the subscript in τ_q where there is no likelihood of confusion. It is obvious that the composite of homomorphisms of degrees d and d' is a homomorphism of degree $d + d'$, and that there thus is a category of graded groups and homomorphisms (with each homomorphism having some degree). It has a subcategory of graded groups and homomorphisms of fixed degree 0. Because the sum of two homomorphisms from C to C' of degree 0 is again a homomorphism from C to C' of degree 0, hom (C,C') is an abelian group [hom (C,C') being the set of morphisms in the category whose morphisms are homomorphisms of degree 0].

A *differential graded group* (sometimes abbreviated to DG group) is a graded group that has a differential compatible with the graded structure (that is, the differential is of degree r for some r). A *chain complex* is a differential graded group in which the differential is of degree -1. Thus a chain complex C consists of a sequence of abelian groups C_q and homomorphisms

$$\partial_q: C_q \to C_{q-1}$$

indexed by the integers such that the composite

$$C_{q+1} \xrightarrow{\partial_{q+1}} C_q \xrightarrow{\partial_q} C_{q-1}$$

is the trivial homomorphism. The elements of C_q are called *q-chains* of the complex. Most of the chain complexes we consider will have the additional property that $C_q = 0$ for $q < 0$. Such a complex is said to be *nonnegative*. A *free chain complex* is a chain complex in which C_q is a free abelian group for every q.

For a chain complex the group of cycles $Z(C)$ is a graded group consisting of the collection $\{Z_q(C) = \ker \partial_q\}$, and the group of boundaries $B(C)$ is a graded group consisting of $\{B_q(C) = \operatorname{im} \partial_{q+1}\}$. The homology group $H(C)$ is a graded group consisting of $\{H_q(C) = Z_q(C)/B_q(C)\}$.

A *chain map* $\tau: C \to C'$ (also called a *chain transformation*) between chain complexes is a homomorphism of degree 0 commuting with the differentials. Thus τ is a collection $\{\tau_q: C_q \to C'_q\}$ such that commutativity holds in each square

$$\begin{array}{ccc} C_q & \xrightarrow{\partial_q} & C_{q-1} \\ {\scriptstyle\tau_q}\downarrow & & \downarrow{\scriptstyle\tau_{q-1}} \\ C'_q & \xrightarrow{\partial'_q} & C'_{q-1} \end{array}$$

It is clear that there is a *category of chain complexes* whose objects are chain complexes and whose morphisms are chain maps. It is also clear that if C and C' are two objects in this category, hom (C,C') is an abelian group.

If $\tau: C \to C'$ is a chain map, its *induced homomorphism*

$$\tau_*: H(C) \to H(C')$$

is the homomorphism of degree 0 such that $(\tau_*)_q\{z\} = \{\tau_q(z)\}$ for $z \in Z_q(C)$. The following theorem is easily verified.

1 THEOREM *There is a covariant functor from the category of chain complexes to the category of graded groups and homomorphisms of degree 0 which assigns to a chain complex C its homology group $H(C)$ and to a chain map τ its induced homomorphism τ_*. For any two chain complexes the map $\tau \to \tau_*$ is a homomorphism from* hom (C,C') *to* hom $(H(C),H(C'))$. ∎

A *subcomplex* C' of a chain complex C, denoted by $C' \subset C$, is a chain complex such that $C'_q \subset C_q$ and $\partial'_q = \partial_q \mid C'_q$ for all q. There is then an inclusion map $i: C' \subset C$ consisting of the collection of inclusion maps $\{C'_q \subset C_q\}$. There is also a *quotient chain complex* $C/C' = \{C_q/C'_q\}$ with boundary operator induced from that of C by passing to the quotient. The collection of projections $\{C_q \to C_q/C'_q\}$ is the *projection chain map* $C \to C/C'$.

To describe a covariant functor from the category of simplicial complexes to the category of free chain complexes, let K be a simplicial complex. An *oriented q-simplex* of K is a q-simplex $s \in K$ together with an equivalence class of total orderings of the vertices of s, two orderings being equivalent if they differ by an even permutation of the vertices. If $v_0, v_1, \ldots, v_q$ are the vertices of s, then $[v_0,v_1, \ldots ,v_q]$ denotes the oriented q-simplex of K consisting of the simplex s together with the equivalence class of the ordering $v_0 < v_1 < \cdots < v_q$ of its vertices.

For $q < 0$ there are no oriented q-simplexes. For every vertex v of K there is a unique oriented 0-simplex $[v]$, and to every q-simplex, with $q \geq 1$, there correspond exactly two oriented q-simplexes. Let $C_q(K)$ be the abelian group generated by the oriented q-simplexes σ^q with the relations $\sigma_1{}^q + \sigma_2{}^q = 0$

if $\sigma_1{}^q$ and $\sigma_2{}^q$ are different oriented q-simplexes corresponding to the same q-simplex of K. Then $C_q(K) = 0$ for $q < 0$, and for $q \geq 0$ $C_q(K)$ is a free abelian group with rank equal to the number of q-simplexes of K. If K is empty, $C_q(K) = 0$ for all q.

We define homomorphisms $\partial_q\colon C_q(K) \to C_{q-1}(K)$ for $q \geq 1$ by defining them on the generators by

$$(a) \qquad \partial_q[v_0,v_1, \ldots ,v_q] = \sum_{0 \leq i \leq q} (-1)^i [v_0,v_1, \ldots ,\hat{v}_i, \ldots ,v_q]$$

where $[v_0,v_1, \ldots ,\hat{v}_i, \ldots ,v_q]$ denotes the oriented $(q-1)$-simplex obtained by omitting v_i. If $\sigma_1{}^q + \sigma_2{}^q = 0$ in $C_q(K)$, then it is easily verified that $\partial_q(\sigma_1{}^q) + \partial_q(\sigma_2{}^q) = 0$ in $C_{q-1}(K)$. Therefore ∂_q extends to a homomorphism from $C_q(K)$ to $C_{q-1}(K)$. For $q \leq 0$ we define ∂_q to be the trivial homomorphism from $C_q(K)$ to $C_{q-1}(K)$. It is not difficult to show that $\partial_q \partial_{q+1} = 0$ for all q. Therefore there is a free nonnegative chain complex $C(K) = \{C_q(K),\partial_q\}$, which is called the *oriented chain complex of* K. Its homology group, denoted by $H(K)$, is a graded group $\{H_q(K) = H_q(C(K))\}$, called the *oriented homology group of* K. $H_q(K)$ is called the *qth oriented homology group of* K.

If K is realized in some euclidean space, the oriented q-simplexes of K are q-simplexes of K together with orientations, in the sense of linear algebra, of the affine varieties spanned by them. The boundary of an oriented q-simplex is the sum of its oriented $(q-1)$-faces, with each face oriented by the orientation compatible with that of the q-simplex, as shown in the diagrams.

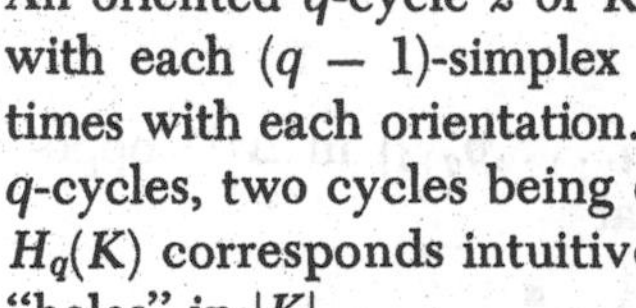

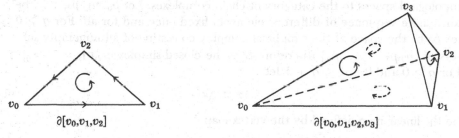

$\partial[v_0,v_1,v_2]$

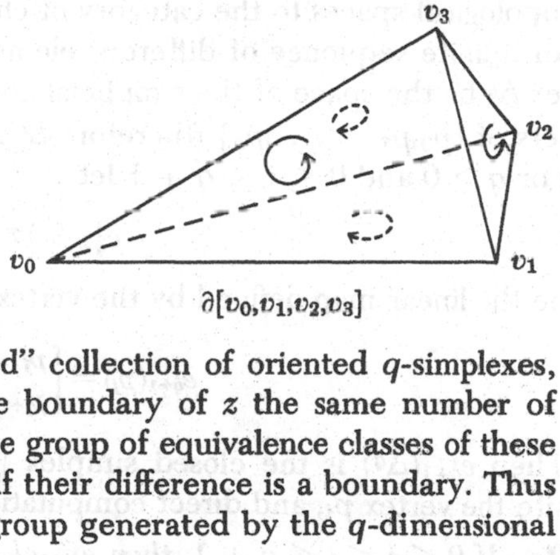

$\partial[v_0,v_1,v_2,v_3]$

An oriented q-cycle z of K is a "closed" collection of oriented q-simplexes, with each $(q-1)$-simplex lying in the boundary of z the same number of times with each orientation. $H_q(K)$ is the group of equivalence classes of these q-cycles, two cycles being equivalent if their difference is a boundary. Thus $H_q(K)$ corresponds intuitively to the group generated by the q-dimensional "holes" in $|K|$.

It is convenient to add more generators and more relations to the chain groups $C_q(K)$. If $v_0, v_1, \ldots, v_q$ are vertices (not necessarily distinct) of some simplex of K, we define $[v_0,v_1, \ldots ,v_q] \in C_q(K)$ to be 0 if the vertices are not distinct and to be the oriented q-simplex as defined above if they are distinct. Then equation (*a*) remains correct for these added generators (that is, if the vertices $v_0,v_1, \ldots ,v_q$ are not all distinct, the left-hand side of equation (*a*) is 0 and the right-hand side can also be verified to be 0).

If $\varphi: K_1 \to K_2$ is a simplicial map, there is an associated chain map $C(\varphi): C(K_1) \to C(K_2)$ defined by

(b) $$C(\varphi)([v_0,v_1, \ldots ,v_q]) = [\varphi(v_0), \varphi(v_1), \ldots ,\varphi(v_q)]$$

Note that if $v_0, v_1, \ldots , v_q$ are distinct vertices of some simplex of K_1, then $\varphi(v_0), \varphi(v_1), \ldots , \varphi(v_q)$ are vertices of some simplex of K_2 but are not necessarily distinct. Therefore the right-hand side of equation (b) above would not be defined unless we had defined $[v_0,v_1, \ldots ,v_q]$ as an element of C_q, whether or not the terms v_i are distinct. It is easy to verify that $C(\varphi)$ is a chain map.

2 THEOREM *There is a covariant functor C from the category of simplicial complexes to the category of chain complexes which assigns to K its oriented chain complex C(K).* ∎

The composite of the functor C and the homology functor is a covariant functor, called the *oriented homology functor*, from the category of simplicial complexes to the category of graded groups. To a simplicial complex K it assigns the graded group $H(K) = \{H_q(K) = H_q(C(K))\}$, and to a simplicial map $\varphi: K_1 \to K_2$ it assigns the homomorphism $\varphi_*: H(K_1) \to H(K_2)$ of degree 0 induced by $C(\varphi): C(K_1) \to C(K_2)$. If L is a subcomplex of K, and $i: L \subset K$, then $C(i): C(L) \to C(K)$ is a monomorphism by means of which we identify $C(L)$ with a subcomplex of $C(K)$.

We next describe the singular chain functor from the category of topological spaces to the category of chain complexes. Let $p_0, p_1, p_2, \ldots$ be an infinite sequence of different elements fixed once and for all. For $q \geq 0$ let Δ^q be the space of the simplicial complex consisting of all nonempty subsets of $\{p_0,p_1, \ldots ,p_q\}$ (therefore Δ^q is the closed simplex $|p_0,p_1, \ldots ,p_q|$). For $q \geq 0$ and $0 \leq i \leq q + 1$ let

$$e^i_{q+1}: \Delta^q \to \Delta^{q+1}$$

be the linear map defined by the vertex map

$$e^i_{q+1}(p_j) = \begin{cases} p_j & j < i \\ p_{j+1} & j \geq i \end{cases}$$

Then $e^i_{q+1}(\Delta^q)$ is the closed simplex $|p_0,p_1, \ldots ,\hat{p}_i, \ldots ,p_{q+1}|$ in Δ^{q+1} opposite the vertex p_i, and direct computation shows that

3 *If* $0 \leq j < i \leq q + 1$, *then* $e^i_{q+2}e^j_{q+1} = e^j_{q+2}e^{i-1}_{q+1}$. ∎

Let X be a topological space. For $q \geq 0$ a *singular q-simplex* σ *of* X is defined to be a continuous map

$$\sigma: \Delta^q \to X$$

For $q > 0$ and $0 \leq i \leq q$ the *ith face of* σ, denoted by $\sigma^{(i)}$, is defined to be the singular $(q - 1)$-simplex of X which is the composite

$$\sigma^{(i)} = \sigma \circ e_q^i: \Delta^{q-1} \to \Delta^q \to X$$

It follows from statement 3 that

4 *If $q > 1$ and $0 \le j < i \le q$, then $(\sigma^{(i)})^{(j)} = (\sigma^{(j)})^{(i-1)}$.* ∎

The *singular chain complex of X*, denoted by $\Delta(X)$, is defined to be the free nonnegative chain complex $\Delta(X) = \{\Delta_q(X), \partial_q\}$, where $\Delta_q(X)$ is the free abelian group generated by the singular q-simplexes of X for $q \ge 0$ [and $\Delta_q(X) = 0$ for $q < 0$], and for $q \ge 1$, ∂_q is defined by the equation

$$\partial_q(\sigma) = \sum_{0 \le i \le q} (-1)^i \sigma^{(i)}$$

This is a chain complex because $\partial_q \partial_{q+1} = 0$ is an immediate consequence of statement 4. If X is empty, $\Delta_q(X) = 0$ for all q.

If $f\colon X \to Y$ is continuous, there is a chain map

$$\Delta(f)\colon \Delta(X) \to \Delta(Y)$$

defined by $\Delta(f)(\sigma) = f \circ \sigma$ for a singular q-simplex $\sigma\colon \Delta^q \to X$. Then $\Delta(f)$ is a chain map, and we have the following result.

5 THEOREM *There is a covariant functor Δ from the category of topological spaces to the category of chain complexes which assigns to X its singular chain complex $\Delta(X)$.* ∎

The composite of the functor Δ and the homology functor is a covariant functor, called the *singular homology functor*, from the category of topological spaces to the category of graded groups. To a space X it assigns the graded group $H(X) = \{H_q(X) = H_q(\Delta(X))\}$ and to a map $f\colon X \to Y$ it assigns the homomorphism

$$f_*\colon H(X) \to H(Y)$$

of degree 0 induced by $\Delta(f)\colon \Delta(X) \to \Delta(Y)$. $H_q(X)$ is called the *qth singular homology group of X*. If A is a subspace of X and $i\colon A \subset X$, then the map $\Delta(i)\colon \Delta(A) \to \Delta(X)$ is a monomorphism by means of which we identify $\Delta(A)$ with a subcomplex of $\Delta(X)$.

The category of chain complexes has arbitrary sums and products of indexed collections. That is, if $\{C^j\}_{j \in J}$ is an indexed collection of chain complexes, there is a *sum chain complex* $\bigoplus C^j$ and a *product chain complex* $\times C^j$ in which $(\bigoplus C^j)_q = \bigoplus C^j_q$ and $(\times C^j)_q = \times C^j_q$ for every q. It follows that $Z_q(\bigoplus C^j) = \bigoplus Z_q(C^j)$ and $B_q(\bigoplus C^j) = \bigoplus B_q(C^j)$ and $Z_q(\times C^j) = \times Z_q(C^j)$ and $B_q(\times C^j) = \times B_q(C^j)$ for all q. Therefore $H(\bigoplus C^j) = \bigoplus H(C^j)$ and $H(\times C^j) = \times H(C^j)$.

6 THEOREM *On the category of chain complexes the homology functor commutes with sums and with products.* ∎

The category of chain complexes also has direct and inverse limits (whose qth chain groups are appropriate limits of the qth chain groups of the factors).

7 THEOREM *The homology functor commutes with direct limits.*

PROOF Let $\{C^\alpha,\tau_\alpha{}^\beta\}$ be a direct system of chain complexes and let $\{C,i_\alpha\}$ be the direct limit of this system (that is, $i_\alpha\colon C^\alpha \to C$, and if $\alpha \le \beta$, then $i_\alpha = i_\beta \circ \tau_\alpha{}^\beta\colon C^\alpha \to C^\beta \to C$). Then $\{H(C^\alpha),\tau_{\alpha*}^\beta\}$ is a direct system of graded groups, and we show that $\{H(C),i_{\alpha*}\}$ is the direct limit of this system.

We show that 1.3*a* of the Introduction is satisfied. Let $\{z\} \in H_q(C)$. Then $z = i_\alpha c^\alpha$ for some $c^\alpha \in (C^\alpha)_q$. Since

$$0 = \partial_q z = \partial_q i_\alpha c^\alpha = i_\alpha \partial_q{}^\alpha c^\alpha$$

there is β with $\alpha \le \beta$ such that $\tau_\alpha{}^\beta \partial_q{}^\alpha c^\alpha = 0$. Then $\tau_\alpha{}^\beta c^\alpha$ is a cycle of $(C^\beta)_q$ and $i_\beta \tau_\alpha{}^\beta c^\alpha = i_\alpha c^\alpha = z$. Therefore $i_{\beta*}\{\tau_\alpha{}^\beta c^\alpha\} = \{z\}$.

We show that 1.3*b* of the Introduction is also satisfied. Because we are dealing with the direct limit of groups, it suffices to show that if $\{z^\alpha\} \in H_q(C^\alpha)$ is in the kernel of $i_{\alpha*}$, then there is γ with $\alpha \le \gamma$ such that $\{z^\alpha\}$ is in the kernel of $\tau_{\alpha*}^\gamma$. If $i_{\alpha*}\{z^\alpha\} = 0$, then $i_\alpha z^\alpha = \partial_{q+1} c$ for some $c \in C_{q+1}$. Because $c = i_\beta c^\beta$ for some β, we have $i_\alpha z^\alpha = i_\beta \partial_{q+1}^\beta c^\beta$. Choose γ' so that $\alpha, \beta \le \gamma'$. Then $i_{\gamma'}(\tau_\alpha{}^{\gamma'} z^\alpha - \tau_\beta{}^{\gamma'} \partial_{q+1}^\beta c^\beta) = 0$. Therefore there is γ with $\gamma' \le \gamma$ such that $\tau_{\gamma'}{}^\gamma(\tau_\alpha{}^{\gamma'} z^\alpha - \tau_\beta{}^{\gamma'} \partial_{q+1}^\beta c^\beta) = 0$. Then $\tau_\alpha{}^\gamma z^\alpha = \partial_{q+1}^\gamma(\tau_\beta{}^\gamma c^\beta)$, so $\tau_{\alpha*}^\gamma\{z^\alpha\} = 0$. ■

It is false that the homology functor commutes with inverse limits. We present an example to show this.

8 EXAMPLE For any integer $n \ge 1$ let C_n be the chain complex with $(C_n)_q = 0$ if $q \ne 0$ or 1 and $(C_n)_1 \xrightarrow{(\partial_n)_1} (C_n)_0$ equal to $\mathbf{Z} \to \mathbf{Z}$, where the homomorphism is multiplication by 2. For each n let $\tau^n\colon C_{n+1} \to C_n$ be the chain map which is multiplication by 3 on each chain group, and for $n \le m$ define $\tau_n{}^m\colon C_m \to C_n$ to be the composite $\tau_n{}^m = \tau^n \tau^{n+1} \cdots \tau^{m-1}$. Then $\{C_n,\tau_n{}^m\}$ is an inverse system whose inverse limit is the trivial chain complex. Therefore $H_q(\lim_{\leftarrow}\{C_n,\tau_n{}^m\}) = 0$ for all q. However, $H_0(C_n) = \mathbf{Z}_2$ for all n and $\tau_{n*}^m\colon H_0(C_m) \approx H_0(C_n)$ for all $n \le m$. Therefore $\lim_{\leftarrow}\{H_0(C_n), \tau_{n*}^m\} \approx \mathbf{Z}_2$.

2 CHAIN HOMOTOPY

This section deals with homotopy in the category of chain complexes. For free chain complexes we prove that contractibility is equivalent to triviality of all the homology groups. This leads to discussion of a method for constructing chain maps and homotopies by a general procedure known as the method of acyclic models. The section closes with a definition of mapping cone of a chain map and its relation to the chain map.

Let $\tau, \tau'\colon C \to C'$ be chain maps. A *chain homotopy D* from τ to τ', denoted by $D\colon \tau \simeq \tau'$, is a homomorphism $D = \{D_q\}$ from C to C' of degree 1 such that for all q

$$\partial'_{q+1} D_q + D_{q-1} \partial_q = \tau_q - \tau'_q\colon C_q \to C'_q$$

If there is a chain homotopy from τ to τ', we say that τ is *chain homotopic* to τ' and write $\tau \simeq \tau'$. It is trivial that chain homotopy is an equivalence relation in the set of chain maps from C to C'. The corresponding set of equivalence classes is denoted by $[C;C']$, and if $\tau\colon C \to C'$ is a chain map, its equivalence class is denoted by $[\tau]$.

1 LEMMA *The composites of chain-homotopic chain maps are chain homotopic.*

PROOF Assume $D\colon \tau \simeq \tau'$, where $\tau, \tau'\colon C \to C'$, and $\bar D\colon \bar\tau \simeq \bar\tau'$, where $\bar\tau, \bar\tau'\colon C' \to C''$. Then

$$\bar\tau D + \bar D \tau'\colon C \to C' \to C''$$

is of degree 1 and is a chain homotopy from $\bar\tau\tau$ to $\bar\tau'\tau'$. ∎

It follows that there is a category whose objects are chain complexes and whose morphisms are chain homotopy classes. A chain map $\tau\colon C \to C'$ is called a *chain equivalence* if $[\tau]$ is an equivalence in the homotopy category of chain complexes. If there is a chain equivalence from C to C', we say that C and C' are *chain equivalent*.

2 THEOREM *If $\tau, \tau'\colon C \to C'$ are chain homotopic, then*

$$\tau_* = \tau'_*\colon H(C) \to H(C')$$

PROOF Assume $D\colon \tau \simeq \tau'$. For any $z \in Z_q(C)$

$$\partial'_{q+1} D_q(z) = \tau_q(z) - \tau'_q(z)$$

showing that $\tau_q(z) \sim \tau'_q(z)$ and $\tau_*\{z\} = \tau'_*\{z\}$. ∎

A *chain contraction* of a chain complex C is a homotopy from the identity chain map 1_C to the zero chain map 0_C of C to itself. If there is a chain contraction of C, C is said to be *chain contractible*. C is said to be *acyclic* if $H(C) = 0$ (that is, $H_q(C) = 0$ for all q).

3 COROLLARY *A contractible chain complex is acyclic.*

PROOF Assume that C is a chain complex such that $1_C \simeq 0_C$. By theorem 2, $(1_C)_* = (0_C)_*$. However, $(1_C)_* = 1_{H(C)}$ and $(0_C)_* = 0_{H(C)}$. Therefore $1_{H(C)} = 0_{H(C)}$, which can happen only if $H(C) = 0$. ∎

The converse of corollary 3 is false.

4 EXAMPLE Let C be the chain complex with $C_q = 0$ for $q \neq 0, 1, 2$ and with $C_2 \xrightarrow{\partial_2} C_1 \xrightarrow{\partial_1} C_0$ equal to $\mathbf{Z} \xrightarrow{\alpha} \mathbf{Z} \xrightarrow{\beta} \mathbf{Z}_2$, where $\alpha(n) = 2n$, $\beta(2m) = 0$, and $\beta(2m + 1) = 1$. Then C is acyclic but not contractible. In fact, if $D\colon 1_C \simeq 0_C$ were a chain contraction of C, then the homomorphism β would have a right inverse $D_0\colon \mathbf{Z}_2 \to \mathbf{Z}$, but any homomorphism $\mathbf{Z}_2 \to \mathbf{Z}$ is trivial.

If C is assumed to be a free chain complex, there is a converse of corollary 3.

5 THEOREM *A free chain complex is acyclic if and only if it is contractible.*

PROOF We show that if C is an acyclic free chain complex, it is contractible. For each q the map ∂_q is an epimorphism of C_q to $B_{q-1}(C) = Z_{q-1}(C)$. Because C_{q-1} is free, so is $Z_{q-1}(C)$, and there is a homomorphism

$$s_{q-1}: Z_{q-1}(C) \to C_q$$

which is a right inverse of ∂_q. Then $1_{C_q} - s_{q-1}\partial_q$ maps C_q to $Z_q(C)$, and we define $\{D_q\}$ by

$$D_q = s_q(1_{C_q} - s_{q-1}\partial_q): C_q \to C_{q+1}$$

Then

$$\partial_{q+1}D_q + D_{q-1}\partial_q = 1_{C_q} - s_{q-1}\partial_q + s_{q-1}(1_{C_{q-1}} - s_{q-2}\partial_{q-1})\partial_q = 1_{C_q}$$

which shows that $\{D_q\}$ is a chain contraction of C. ∎

The method of proof of theorem 5 is a standard one used to construct chain maps and homotopies from a free chain complex to an acyclic chain complex. We now extend it to obtain a general method of constructing chain maps and chain homotopies, called the *method of acyclic models.* Repeated application of this method will be made in subsequent discussions. We consider a special version of the method of acyclic models which suffices for our applications.[1]

A *category with models* consists of a category $\mathcal{C}$ and a set $\mathfrak{M}$ of objects of $\mathcal{C}$ called *models.* Given a covariant functor G from a category $\mathcal{C}$ with models $\mathfrak{M}$ to the category of abelian groups, a *basis* for G is an indexed collection $\{g_j \in G(M_j)\}_{j \in J}$, where $M_j \in \mathfrak{M}$ such that for any object X of $\mathcal{C}$ the indexed collection

$$\{G(f)(g_j)\}_{j \in J, f \in \hom(M_j, X)}$$

is a basis for $G(X)$. If G has a basis, it is called a *free functor* on $\mathcal{C}$ with models $\mathfrak{M}$. In this case, if $h \in \hom(X,Y)$, then $G(h)$ maps each basis element of $G(X)$ to some basis element of $G(Y)$. Hence G is the composite of the covariant functor which assigns to X the set $\{G(f)(g_j) \mid j \in J, f \in \hom(M_j,X)\}$ with the covariant functor of example 1.2.2, which assigns to every set the free abelian group generated by it.

Let G be a covariant functor from a category $\mathcal{C}$ with models $\mathfrak{M}$ to the category of chain complexes. G is said to be *free* if G_q is a free functor to the category of abelian groups.

6 EXAMPLE Let K be a simplicial complex and let $\mathcal{C}(K)$ be the category defined by the partially ordered set of subcomplexes of K (as in example 1.1.11). Let $\mathfrak{M}(K) = \{\bar{s} \mid s \in K\}$ be models for $\mathcal{C}(K)$. We show that the covariant functor C which assigns to each subcomplex of K its oriented chain complex is a free nonnegative functor on $\mathcal{C}(K)$ with models $\mathfrak{M}(K)$ to the category of

[1] A general treatment can be found in S. Eilenberg and S. Mac Lane, Acyclic models, *American Journal of Mathematics,* vol. 79, pp. 189–199 (1953).

chain complexes. For each model $\bar{s}$ of dimension q choose once and for all an oriented q-simplex $\sigma(s)$ which generates $C_q(\bar{s})$. Then the indexed collection $\{\sigma(s) \mid \dim s = q\}_{s \in K}$ is a basis for C_q. Hence C_q is free with models $\mathfrak{M}(K)$.

7 EXAMPLE Let $\mathcal{C}$ be the category of topological spaces with models $\mathfrak{M} = \{\Delta^q \mid q \geq 0\}$ and let Δ be the singular chain functor. Then Δ is free and nonnegative on $\mathcal{C}$ with models $\mathfrak{M}$. In fact, if $\xi_q\colon \Delta^q \subset \Delta^q$, then the singleton $\{\xi_q \in \Delta_q(\Delta^q)\}$ is a basis for Δ_q.

Let G be a covariant functor on a category $\mathcal{C}$ to the category of chain complexes. Then there are covariant functors $H_q(G)$, for all q, from $\mathcal{C}$ to the category of abelian groups that assign to an object X the group $H_q(G(X))$. If $\mathcal{C}$ is a category with models $\mathfrak{M}$, a functor G from $\mathcal{C}$ to the category of chain complexes is said to be *acyclic in positive dimensions* if $H_q(G(M)) = 0$ for $q > 0$ and $M \in \mathfrak{M}$. We now establish the main result dealing with the construction of chain maps and homotopies.

8 THEOREM *Let $\mathcal{C}$ be a category with models $\mathfrak{M}$ and let G and G' be covariant functors from $\mathcal{C}$ to the category of chain complexes such that G is free and nonnegative and G' is acyclic in positive dimensions. Then*

(a) Any natural transformation $H_0(G) \to H_0(G')$ is induced by a natural chain map $\tau\colon G \to G'$.
(b) Two natural chain maps $\tau, \tau'\colon G \to G'$ inducing the same natural transformation $H_0(G) \to H_0(G')$ are naturally chain homotopic.

PROOF For every object X of $\mathcal{C}$ we must define a chain map $\tau(X)\colon G(X) \to G'(X)$ [or a chain homotopy $D(X)\colon \tau(X) \simeq \tau'(X)$] such that if $h\colon X \to Y$ is a morphism in $\mathcal{C}$, then

$$\tau(Y)G(h) = G'(h)\tau(X) \qquad [\text{or } D(Y)G(h) = G'(h)D(X)]$$

For $q \geq 0$ let $\{g_j \in G_q(M_j)\}_{j \in J_q}$ be a basis for G_q, where $M_j \in \mathfrak{M}$ for each $j \in J_q$. Then $G_q(X)$ has the basis

$$\{G_q(f)(g_j)\}_{j \in J_q, f \in \hom(M_j, X)}$$

It follows that $\tau_q(X)$ [or $D_q(X)$] is determined by the collection $\{\tau_q(M_j)(g_j)\}_{j \in J_q}$ and the equation

$$\text{(a)} \qquad \tau_q(X)(\Sigma n_{ij} G_q(f_{ij})(g_j)) = \Sigma n_{ij} G'_q(f_{ij})\tau_q(M_j)(g_j)$$

or by the collection $\{D_q(M_j)g_j\}_{j \in J_q}$ and the equation

$$\text{(b)} \qquad D_q(X)(\Sigma n_{ij} G_q(f_{ij})(g_j)) = \Sigma n_{ij} G'_{q+1}(f_{ij}) D_q(M_j)(g_j)$$

We shall define $\tau_q(X)$ by induction on q so that

$$\text{(c)} \qquad \partial \tau_q(X) = \tau_{q-1}(X)\partial$$

and define $D_q(X)$ by induction on q so that

$$\text{(d)} \qquad \partial D_q(X) = \tau_q(X) - \tau'_q(X) - D_{q-1}(X)\partial$$

Having defined τ_i [or D_i] for $i < q$, where $q > 0$, it suffices to define $\tau_q(M_j)(g_j)$ for $j \in J_q$ so that

$$(e) \qquad \partial \tau_q(M_j)(g_j) = \tau_{q-1}(M_j)(\partial g_j)$$

and to define $D_q(M_j)(g_j)$ for $j \in J_q$ so that

$$(f) \qquad \partial D_q(M_j)(g_j) = \tau_q(M_j)(g_j) - \tau'_q(M_j)(g_j) - D_{q-1}(M_j)(\partial g_j)$$

since $\tau_q(X)$ [and $D_q(X)$] are then determined by equation (*a*) [or by (*b*)]. It will then be true that $\tau_q(X)$ [and $D_q(X)$] are natural and will satisfy equation (*c*) [and (*d*)].

Given a natural transformation $\varphi: H_0(G) \to H_0(G')$, the inductive definition of τ proceeds as follows. For $q = 0$ we define $\tau_0(M_j)(g_j)$ for $j \in J_0$ to be any element of $G'_0(M_j)$ such that $\{\tau_0(M_j)(g_j)\} = \varphi(M_j)\{g_j\}$. We use equation (*a*) to define $\tau_0(X)$ for all X. Then, for $g \in G_0(X)$, $\{\tau_0(X)(g)\} = \varphi(X)\{g\}$. In particular, for $j \in J_1$, $\tau_0(M_j)(\partial g_j)$ is a boundary in $G'_0(M_j)$. Hence we can define $\tau_1(M_j)(g_j) \in G'_1(M_j)$ so that $\partial \tau_1(M_j)(g_j) = \tau_0(M_j)(\partial g_j)$. We then use equation (*a*) to define $\tau_1(X)$ for all X. Assuming τ_i defined for $i < q$, where $q > 1$, so that equation (*c*) is satisfied, we observe that the right-hand side of equation (*e*) is a cycle of $G'_{q-1}(M_j)$. Because $q > 1$, $H_{q-1}(G'(M_j)) = 0$, and we define $\tau_q(M_j)(g_j)$ to satisfy equation (*e*). We next define $\tau_q(X)$ for all X to satisfy equation (*a*). This completes the definition of τ.

Given $\tau, \tau': G \to G'$ such that τ and τ' induce the same natural transformation $H_0(G) \to H_0(G')$, we define $D_0(M_j)(g_j)$ for $j \in J_0$ to be any element of $G'_1(M_j)$ whose boundary equals $\tau_0(M_j)(g_j) - \tau'_0(M_j)(g_j)$. Then $D_0(X)$ is defined for all X by equation (*b*). Assuming D_i defined for $i < q$, where $q > 0$, so that equation (*d*) is satisfied, we observe that the right-hand side of equation (*f*) is a cycle of $G'_q(M_j)$. Because $q > 0$, $H_q(G'(M_j)) = 0$, and this cycle is a boundary. We define $D_q(M_j)(g_j) \in G'_{q+1}(M_j)$ to satisfy equation (*f*), use equation (*b*) to define $D_q(X)$ for all X, and complete the definition of D. ∎

The last result provides another proof of theorem 5 for nonnegative complexes. In fact, if C is a free nonnegative chain complex, let $\mathcal{C}$ be the category consisting of one object X and one morphism 1_X and let C be regarded as a covariant functor on $\mathcal{C}$ with model $\{X\}$. Then C is a free nonnegative functor, and if C is an acyclic chain complex, the functor C is acyclic in positive dimensions. In this case, because 1_C and 0_C are chain transformations of C inducing the same homomorphism of $H_0(C) = 0$, it follows from theorem 8 that $1_C \simeq 0_C$, and C is contractible.

There is a useful algebraic object (related to the mapping cylinder of Sec. 1.4) which we now describe. Let $\tau: C \to C'$ be a chain map. The *mapping cone* of τ is the chain complex $\bar{C} = \{\bar{C}_q, \bar{\partial}_q\}$ defined by $\bar{C}_q = C_{q-1} \oplus C'_q$ and

$$\bar{\partial}_q(c,c') = (-\partial_{q-1}(c), \tau(c) + \partial'_q(c')) \qquad c \in C_{q-1},\ c' \in C'_q$$

The following result is trivial to verify.

9 LEMMA *$\bar{C}$ is a chain complex, and if C and C' are free chain complexes, so is $\bar{C}$.* ∎

The next theorem is the main reason for introducing mapping cones.

10 THEOREM *A chain map is a chain equivalence if and only if its mapping cone is chain contractible.*

PROOF Assume that $\tau: C \to C'$ is a chain equivalence. There exist $\tau': C' \to C$ and $D: C \to C$ and $D': C' \to C'$ such that $D: \tau'\tau \simeq 1_C$ and $D': \tau\tau' \simeq 1_{C'}$. Define $\bar{D}: \bar{C} \to \bar{C}$ by $\bar{D}(c,c') = (c_1,c_2)$, where

$$c_1 = D(c) + \tau' D' \tau(c) - \tau'\tau D(c) + \tau'(c')$$
$$c_2 = D'\tau D(c) - D'D'\tau(c) - D'(c')$$

A straightforward computation shows that $\bar{D}$ is a chain contraction of $\bar{C}$.

Conversely, assume that $\bar{D}$ is a chain contraction of $\bar{C}$. Define $\tau': C' \to C$ and $D: C \to C$ and $D': C' \to C'$ by the equations

$$(\tau'(c'), -D'(c')) = \bar{D}(0,c')$$
$$(D(c), \cdot) = \bar{D}(c,0)$$

Direct verification shows τ' to be a chain map and $D: \tau'\tau \simeq 1_C$ and $D': \tau\tau' \simeq 1_{C'}$, so τ is a chain equivalence. ∎

Combining this with theorem 5 and lemma 9 yields the following result.

11 COROLLARY *A chain map between free chain complexes is a chain equivalence if and only if its mapping cone is acyclic.* ∎

3 THE HOMOLOGY OF SIMPLICIAL COMPLEXES

This section begins with a discussion of augmented chain complexes and their reduced homology groups. Next we define the ordered chain complex of a simplicial complex and prove that it is chain equivalent to the oriented chain complex. We use this result to show that simplicial maps in the same contiguity class induce chain-homotopic chain maps. We also compute $H_0(K)$ in terms of the components of K. At the end of the section the relative homology groups and the Euler characteristic of a simplicial pair are defined.

In the category of nonempty simplicial complexes any simplicial complex P consisting of a single vertex is a terminal object. If K is a nonempty simplicial complex, the simplicial map $K \to P$ has a right inverse. Therefore the induced homology map $H(K) \to H(P)$ has a right inverse. Because $H_q(P) = 0$ if $q \neq 0$ and $H_0(P) \approx \mathbf{Z}$, it follows that there is an epimorphism $H_0(K) \to \mathbf{Z}$. Since $H_0(K) = C_0(K)/\partial_1 C_1(K)$, there is an epimorphism $\varepsilon: C_0(K) \to \mathbf{Z}$ such that $\varepsilon\partial_1 = 0$. Similarly, in the category of nonempty topological spaces X any one-point space is a terminal object. The same kind of considerations yield an epimorphism $\varepsilon: \Delta_0(X) \to \mathbf{Z}$ such that $\varepsilon\partial_1 = 0$. This motivates the following definition of augmentation.

An *augmentation* (*over* $\mathbf{Z}$) of a chain complex C is an epimorphism $\varepsilon: C_0 \to \mathbf{Z}$ such that $\varepsilon\partial_1: C_1 \to C_0 \to \mathbf{Z}$ is trivial. An *augmented chain complex* is a nonnegative chain complex C with augmentation. An augmentation ε of a chain complex can be regarded as an epimorphic chain map of C to the chain complex (also denoted by $\mathbf{Z}$) whose only nontrivial chain group is $\mathbf{Z}$ in degree 0. For this chain complex $\mathbf{Z}$, it is clear that $H_q(\mathbf{Z}) = 0$ for $q \neq 0$ and that $H_0(\mathbf{Z}) = \mathbf{Z}$. Therefore ε induces an epimorphism $\varepsilon_*: H_0(C) \to \mathbf{Z}$. Hence an augmented chain complex has a nontrivial homology group in degree 0.

The oriented chain complex $C(K)$ of a nonempty simplicial complex K is augmented by the homomorphism $\varepsilon: C_0(K) \to \mathbf{Z}$ defined by $\varepsilon([v]) = 1$ for every vertex v of K. The singular chain complex $\Delta(X)$ of a nonempty space X is augmented by the homomorphism $\varepsilon: \Delta_0(X) \to \mathbf{Z}$ defined by $\varepsilon(\sigma) = 1$ for every singular 0-simplex of X.

A chain map $\tau: C \to C'$ between augmented chain complexes *preserves augmentation* if $\varepsilon' \circ \tau = \varepsilon: C_0 \to \mathbf{Z}$. Note that τ preserves augmentation if and only if τ_* does—that is, if and only if $\varepsilon'_* \circ \tau_* = \varepsilon_*: H_0(C) \to \mathbf{Z}$. There is a category of augmented chain complexes and chain maps preserving augmentation. A chain homotopy in this category is any chain homotopy between chain maps in the category.

We want to divide out the functorial nontrivial part of $H_0(C)$ of an augmented chain complex C. The *reduced chain complex* $\tilde{C}$ *of* an augmented chain complex C is defined to be the chain complex defined by $\tilde{C}_q = C_q$ if $q \neq 0$, $\tilde{C}_0 = \ker \varepsilon$, and $\tilde{\partial}_q = \partial_q$ [note that $\partial_1(\tilde{C}_1) \subset \tilde{C}_0$ because $\varepsilon\partial_1 = 0$]. Thus $\tilde{C}$ is the kernel of the chain map $\varepsilon: C \to \mathbf{Z}$. If $\tau: C \to C'$ is a chain map preserving augmentation, τ induces a chain map $\tilde{C} \to \tilde{C}'$ between their reduced chain complexes. The homology group $H(\tilde{C})$ is called the *reduced homology group* of C and is denoted by $\tilde{H}(C)$. For a nonempty simplicial complex K we define $\tilde{H}(K) = \tilde{H}(C(K))$, and for a nonempty topological space X we define $\tilde{H}(X) = \tilde{H}(\Delta(X))$. Because the chain complex of an empty simplicial complex or an empty topological space has no augmentation, the reduced groups are not defined in this case. For that reason some of the arguments, which otherwise involve the reduced groups, require a special remark in the case of empty complexes or spaces.

Clearly, there is an inclusion chain map $\tilde{C} \subset C$.

1 LEMMA *If C is an augmented chain complex, then*

$$H_q(C) \approx \begin{cases} \tilde{H}_q(C) & q \neq 0 \\ \tilde{H}_0(C) \oplus \mathbf{Z} & q = 0 \end{cases}$$

PROOF Because $\mathbf{Z}$ is a free group, $C_0 \approx \tilde{C}_0 \oplus \mathbf{Z}$. Then $Z_q(C) = Z_q(\tilde{C})$ if $q \neq 0$, $Z_0(C) \approx Z_0(\tilde{C}) \oplus \mathbf{Z}$, and $B_q(C) = B_q(\tilde{C})$ for all q. ∎

It is clear that if $\tau: C \to C'$ is an augmentation-preserving chain map, the isomorphism of lemma 1 commutes with τ_*. It is also obvious that if C is a free augmented chain complex, $\tilde{C}$ is a free chain complex.

It follows from lemma 1 that if C is an augmented chain complex, $H_0(C) \neq 0$. Hence an augmented chain complex is never acyclic. The most that can be hoped for is that $\tilde{C}$ will be acyclic.

2 LEMMA *If C is an augmented chain complex, $\tilde{C}$ is chain contractible if and only if the augmentation ε is a chain equivalence of C with the chain complex* $\mathbf{Z}$.

PROOF Let $\bar{C}$ be the mapping cone of the chain map $\varepsilon: C \to \mathbf{Z}$. Then $\bar{C}_0 = \mathbf{Z}$ and $\bar{C}_q = C_{q-1}$ if $q > 0$, and $\bar{\partial}_1 = \varepsilon$ and $\bar{\partial}_q = -\partial_{q-1}$ for $q > 1$. By theorem 4.2.10, ε is a chain equivalence if and only if $\bar{C}$ is chain contractible.

We show that $\bar{C}$ is chain contractible if and only if $\tilde{C}$ is chain contractible. If $\bar{D}: \bar{C} \to \bar{C}$ is a chain contraction of $\bar{C}$, define $\tilde{D}: \tilde{C} \to \tilde{C}$ by $\tilde{D}_{q-1} = -\bar{D}_q \mid \tilde{C}_{q-1}$. Then $\tilde{D}$ is a chain contraction of $\tilde{C}$. Conversely, if $\tilde{D}$ is a chain contraction of $\tilde{C}$, define $\bar{D}: \bar{C} \to \bar{C}$ so that $\bar{D}_0: \mathbf{Z} \to C_0$ is a right inverse of $\varepsilon: C_0 \to \mathbf{Z}$, $\bar{D}_1: C_0 \to C_1$ is 0 on $\bar{D}_0(\mathbf{Z})$ and equal to $-\tilde{D}_0$ on $\tilde{C}_0$, and for $q > 1$, $\bar{D}_q: C_{q-1} \to C_q$ is equal to $-\tilde{D}_{q-1}$. Then $\bar{D}$ is a chain contraction of $\bar{C}$. ■

Let $\mathcal{C}$ be a category with models $\mathfrak{M}$. A functor G' from $\mathcal{C}$ to the category of augmented chain complexes (and chain maps preserving augmentation) is said to be *acyclic* if $\tilde{G}'(M)$ is acyclic for $M \in \mathfrak{M}$. For augmented chain complexes there is the following form of the acyclic-model theorem.

3 THEOREM *Let $\mathcal{C}$ be a category with models $\mathfrak{M}$ and let G and G′ be covariant functors from $\mathcal{C}$ to the category of augmented chain complexes such that G is free and G′ is acyclic. There exist natural chain maps preserving augmentation from G to G′, and any two are naturally chain homotopic.*

PROOF Let $\{g_j \in G_0(M_j)\}_{j \in J_0}$ be a basis for G_0. By lemma 1, $\varepsilon': H_0(G'(M_j)) \approx \mathbf{Z}$, and there is a unique $z_j \in H_0(G'(M_j))$ such that $\varepsilon'(z_j) = \varepsilon(g_j)$. A natural transformation $H_0(G) \to H_0(G')$ is defined by sending $\{\Sigma n_{ij} G_0(f_{ij})(g_j)\} \in H_0(G(X))$ to $\Sigma n_{ij} G_0'(f_{ij}) z_j \in H_0(G'(X))$ for $j \in J_0$ and $f_{ij} \in \hom(M_j, X)$ (where X is any object of $\mathcal{C}$), and this is the unique natural transformation $H_0(G) \to H_0(G')$ commuting with augmentation. The theorem now follows from theorem 4.2.8. ■

With the hypotheses of theorem 3 there is a unique natural transformation from $H(G)$ to $H(G')$ commuting with augmentation. It is the homomorphism induced by any natural chain map preserving augmentation from G to G′.

4 COROLLARY *Let G and G′ be free and acyclic covariant functors from a category $\mathcal{C}$ with models $\mathfrak{M}$ to the category of augmented chain complexes. Then G and G′ are naturally chain equivalent; in fact, any natural chain map preserving augmentation from G to G′ is a natural chain equivalence.*

PROOF Let $\tau: G \to G'$ be a natural chain map preserving augmentation (which exists, by theorem 3). Also by theorem 3, there is a natural chain map $\tau': G' \to G$ preserving augmentation and there are natural chain homotopies $D: \tau' \circ \tau \simeq 1_G$ and $D': \tau \circ \tau' \simeq 1_{G'}$. ■

We are ultimately interested in comparing the chain complex $C(K)$ of a simplicial complex K with the singular chain complex $\Delta(|K|)$ of the space of K.

For this purpose we introduce a chain complex $\Delta(K)$ intermediate between them. Let K be a simplicial complex. An *ordered q-simplex* of K is a sequence $v_0, v_1, \ldots, v_q$ of $q+1$ vertices of K which belong to some simplex of K. We use $(v_0,v_1, \ldots ,v_q)$ to denote the ordered q-simplex consisting of the sequence $v_0, v_1, \ldots, v_q$ of vertices. For $q < 0$ there are no ordered q-simplexes. An ordered 0-simplex (v) is the same as the oriented 0-simplex $[v]$. An ordered 1-simplex (v,v') is the same as an edge of K.

We define a free nonnegative chain complex, called the *ordered chain complex of K*, by $\Delta(K) = \{\Delta_q(K),\partial_q\}$, where $\Delta_q(K)$ is the free abelian group generated by the ordered q-simplexes of K [and $\Delta_q(K) = 0$ if $q < 0$] and ∂_q is defined by the equation

$$\partial_q(v_0,v_1, \ldots ,v_q) = \sum_{0 \le i \le q} (-1)^i(v_0, \ldots ,\hat{v}_i, \ldots ,v_q)$$

Then $\Delta(K)$ is a chain complex, and if K is nonempty, $\Delta(K)$ is augmented by the augmentation $\varepsilon(v) = 1$ for any vertex v of K. If $\varphi\colon K_1 \to K_2$ is a simplicial map, there is an augmentation-preserving chain map

$$\Delta(\varphi)\colon \Delta(K_1) \to \Delta(K_2)$$

such that $\Delta(\varphi)(v_0,v_1, \ldots ,v_q) = (\varphi(v_0), \varphi(v_1), \ldots ,\varphi(v_q))$. Therefore we have the following theorem.

5 THEOREM *There is a covariant functor Δ from the category of nonempty simplicial complexes to the category of free augmented chain complexes which assigns to K the ordered chain complex $\Delta(K)$.* ∎

If L is a subcomplex of K and $i\colon L \subset K$, then $\Delta(i)\colon \Delta(L) \to \Delta(K)$ is a monomorphism by means of which we identify $\Delta(L)$ with a subcomplex of $\Delta(K)$. If $\mathcal{C}(K)$ is the category defined by the partially ordered set of subcomplexes of K and $\mathfrak{M}(K) = \{\bar{s} \mid s \in K\}$, then Δ is a free functor on $\mathcal{C}(K)$ with models $\mathfrak{M}(K)$.

For any simplicial complex K there is a surjective chain map (preserving augmentation if K is nonempty)

$$\mu\colon \Delta(K) \to C(K)$$

such that $\mu(v_0,v_1, \ldots ,v_q) = [v_0,v_1 \ldots ,v_q]$. Then μ is a natural transformation from Δ to C on the category of simplicial complexes. We shall show that it is a chain equivalence for every simplicial complex. The following theorem will be used to show that Δ and C are acyclic functors on $\mathcal{C}(K)$ with models $\mathfrak{M}(K)$.

6 THEOREM *Let K be a simplicial complex and let w be the simplicial complex consisting of a single vertex. Then $\tilde{\Delta}(K * w)$ and $\tilde{C}(K * w)$ are chain contractible.*

PROOF Since the proofs are analogous, we give the details only in the ordered complex. According to lemma 2, it suffices to prove that $\varepsilon\colon \Delta(K * w) \to \mathbf{Z}$ is a chain equivalence. Define a homomorphism $\tau: \mathbf{Z} \to \Delta_0(K * w)$ by $\tau(1) = (w)$ and regard it as a chain map $\tau: \mathbf{Z} \to \Delta(K * w)$. Then $\varepsilon \circ \tau = 1_{\mathbf{Z}}$. To show that $1_{\Delta(K*w)} \simeq \tau \circ \varepsilon$, define a chain homotopy $D: 1_{\Delta(K*w)} \simeq \tau \circ \varepsilon$ by the equation

$$D(v_0,v_1, \ldots ,v_q) = (w,v_0,v_1, \ldots ,v_q) \quad \blacksquare$$

Because a q-simplex is the join of a $(q-1)$-face with the opposite vertex, we have the next result.

7 COROLLARY *For any simplex $s \in K$, $\tilde{\Delta}(\bar{s})$ and $\tilde{C}(\bar{s})$ are acyclic.* ■

8 THEOREM *For any simplicial complex K the natural chain map $\mu: \Delta(K) \to C(K)$ is a chain equivalence.*

PROOF If K is empty, $\Delta(K) = C(K)$ and μ is the identity, so the result is true in this case. If K is nonempty, it follows from corollary 7 that Δ and C are free acyclic functors on $\mathcal{C}(K)$ with models $\mathfrak{M}(K) = \{\bar{s} \mid s \in K\}$. By corollary 4, μ is a natural chain equivalence of Δ with C on $\mathcal{C}(K)$. In particular, $\mu: \Delta(K) \to C(K)$ is a chain equivalence. ■

The next result is that the functors Δ and C convert contiguity of simplicial maps into chain homotopy of chain maps. This result could also be proved by the method of acyclic models.

9 THEOREM *Let $\varphi, \varphi': K_1 \to K_2$ be in the same contiguity class. Then $\Delta(\varphi), \Delta(\varphi'): \Delta(K_1) \to \Delta(K_2)$ are chain homotopic, and in similar fashion $C(\varphi), C(\varphi'): C(K_1) \to C(K_2)$ are chain homotopic.*

PROOF Because chain homotopy is an equivalence relation, it suffices to prove the theorem for the case that φ and φ' are contiguous. An explicit chain homotopy $D: \Delta(\varphi) \simeq \Delta(\varphi')$ is defined by the formula

$$D(v_0,v_1, \ldots ,v_q) = \sum_{0 \le i \le q} (-1)^i (\varphi'(v_0), \ldots ,\varphi'(v_i), \varphi(v_i), \ldots ,\varphi(v_q))$$

That $C(\varphi)$ and $C(\varphi')$ are chain homotopic follows from the fact that $\Delta(\varphi)$ and $\Delta(\varphi')$ are chain homotopic and from theorem 8. ■

10 THEOREM *The homology groups of a complex are the direct sums of the homology groups of its components.*

PROOF If $\{K_j\}$ are the components of K, then $\bigoplus C(K_j) = C(K)$. The result follows from theorem 4.1.6. ■

If $\{K_\alpha\}$ is the collection of finite subcomplexes of K directed by inclusion, then $C(K) \approx \varinjlim \{C(K_\alpha)\}$. From theorem 4.1.7 we have the next result.

11 THEOREM *The homology groups of a simplicial complex are isomorphic to the direct limit of the homology groups of its finite subcomplexes.* ■

We are now ready to compute $H_0(K)$.

12 LEMMA *If K is a nonempty connected simplicial complex, then $\tilde{H}_0(K) = 0$.*

PROOF Let v_0 be a fixed vertex of K. For any vertex v of K there is an edge path $e_1e_2 \cdots e_r$ of K with origin at v_0 and end at v. Then $e_1 + e_2 + \cdots + e_r$ is a 1-chain $c_v \in \Delta_1(K)$ such that $\partial c_v = v - v_0$. Since $\varepsilon(\Sigma n_v v) = \Sigma n_v$, we see that if $\Sigma n_v v$ is any 0-chain of $\tilde{\Delta}_0(K)$, then $\Sigma n_v = 0$ and

$$\partial(\Sigma n_v c_v) = \Sigma n_v v - \Sigma n_v v_0 = \Sigma n_v v$$

Therefore $\tilde{H}_0(\Delta(K)) = 0$, and by theorem 8, $\tilde{H}_0(K) = 0$. ∎

13 COROLLARY *For any simplicial complex K, $H_0(K)$ is a free group whose rank equals the number of nonempty components of K.*

PROOF If K is empty, $H_0(K) = 0$, and the result is valid in this case. If K is nonempty and connected, it follows from lemmas 12 and 1 that $H_0(K) \approx \mathbf{Z}$. The general result then follows from theorem 10. ∎

If L is a subcomplex of K, there is a *relative oriented homology group* $H(K,L) = \{H_q(K,L) = H_q(C(K)/C(L))\}$ *of* K *modulo* L. If L is empty, $H(K,\varnothing) = H(K)$ is called the *absolute oriented homology group of* K. Similarly, there is a *relative ordered homology group* $H(\Delta(K)/\Delta(L))$ *of* K *modulo* L that generalizes the *absolute ordered homology group* $H(\Delta(K),\Delta(\varnothing))$. The relative homology groups $H(K,L)$ and $H(\Delta(K),\Delta(L))$ are covariant functors from the category of simplicial pairs to the category of graded groups.

If $H_q(K,L)$ is finitely generated (which will necessarily be true if $K - L$ contains only finitely many simplexes), it follows from the structure theorem (theorem 4.14 in the Introduction) that $H_q(K,L)$ is the direct sum of a free group and a finite number of finite cyclic groups $\mathbf{Z}_{n_1} \oplus \mathbf{Z}_{n_2} \oplus \cdots \oplus \mathbf{Z}_{n_k}$, where n_i divides n_{i+1} for $i = 1, \ldots, k-1$. The rank $\rho(H_q(K,L))$ is called the *qth Betti number of* (K,L), and the numbers $n_1, n_2, \ldots, n_k$ are called the *qth torsion coefficients of* (K,L). The qth Betti number and the qth torsion coefficients characterize $H_q(K,L)$ up to isomorphism.

A graded group C is said to be *finitely generated* if C_q is finitely generated for all q and $C_q = 0$ except for a finite set of integers q. It is obvious that if C is a finitely generated chain complex, $H(C)$ is a finitely generated graded group. Given a finitely generated graded group C, its *Euler characteristic* (also called the *Euler-Poincaré characteristic*), denoted by $\chi(C)$, is defined by

$$\chi(C) = \Sigma(-1)^q \rho(C_q)$$

14 THEOREM *Let C be a finitely generated chain complex. Then*

$$\chi(C) = \chi(H(C))$$

PROOF By definition, $Z_q(C) \subset C_q$ and the quotient group $C_q/Z_q(C) \approx B_{q-1}(C)$. By theorem 4.12 in the Introduction,

$$\rho(C_q) = \rho(Z_q(C)) + \rho(B_{q-1}(C))$$

Similarly, $H_q(C) = Z_q(C)/B_q(C)$, and again by theorem 4.12 of the Introduction,

$$\rho(Z_q(C)) = \rho(H_q(C)) + \rho(B_q(C))$$

Eliminating $\rho(Z_q(C))$, we have

$$\rho(C_q) = \rho(H_q(C)) + \rho(B_q(C)) + \rho(B_{q-1}(C))$$

Multiplying this equation by $(-1)^q$ and summing the resulting equations over q yields the result. ∎

If $H(K,L)$ is finitely generated, its Euler characteristic, called the *Euler characteristic of* (K,L), is denoted by $\chi(K,L)$.

15 COROLLARY *If $K - L$ is finite and if α_q equals the number of q-simplexes of $K - L$, then*

$$\chi(K,L) = \Sigma(-1)^q \alpha_q$$

PROOF If $K - L$ is finite, $C_q(K)/C_q(L)$ is a free group of rank α_q. The result follows from theorem 14. ∎

4 SINGULAR HOMOLOGY

In this section we define a natural transformation from the ordered chain complex to the singular chain complex of its space. This will be shown in Sec. 4.6 to be a chain equivalence for every simplicial complex K. We also give a proof, based on acyclic models, that homotopic continuous maps induce chain-homotopic chain maps on the singular chain complexes. There is then a computation of $H_0(X)$ in terms of the path components of X. The final result is that the subcomplex of the singular chain complex generated by "small" singular simplexes is chain equivalent to the whole singular chain complex.[1]

Let K be a simplicial complex. Given an ordered q-simplex $(v_0,v_1, \ldots ,v_q)$ of K, there is a singular q-simplex in $|K|$ which is the linear map $\Delta^q \to |K|$ sending p_i to v_i for $0 \le i \le q$. This imbeds $\Delta(K)$ in $\Delta(|K|)$, and we define an augmentation-preserving chain map

$$\nu: \Delta(K) \to \Delta(|K|)$$

to send $(v_0,v_1, \ldots ,v_q)$ to the linear singular simplex defined above. Then ν is a natural chain map from the covariant functor $\Delta(\cdot)$ to the covariant functor $\Delta(|\cdot|)$ on the category of simplicial complexes. It will be shown in Sec. 4.6 that ν is a natural chain equivalence. We prove now that it is a chain equivalence for the complex $\bar{s}$ of an arbitrary simplex s.

1 LEMMA *Let X be a star-shaped subset of some euclidean space. Then the reduced singular complex of X is chain contractible.*

[1] Our treatment is similar to that in S. Eilenberg, Singular homology theory, *Annals of Mathematics*, vol. 45, pp. 407–447 (1944).

PROOF Without loss of generality, X may be assumed to be star-shaped from the origin. We define a homomorphism $\tau: \mathbf{Z} \to \Delta_0(X)$ with $\tau(1)$ equal to the singular simplex $\Delta^0 \to X$ which is the constant map to 0. Then $\varepsilon \circ \tau = 1_{\mathbf{Z}}$. We define a chain homotopy $D: \Delta(X) \to \Delta(X)$ from $1_{\Delta(X)}$ to $\tau \circ \varepsilon$. If $\sigma: \Delta^q \to X$ is a singular q-simplex in X, let $D(\sigma): \Delta^{q+1} \to X$ be the singular $(q+1)$-simplex in X defined by the equation

$$D(\sigma)(tp_0 + (1-t)\alpha) = (1-t)\sigma(\alpha)$$

for $\alpha \in |p_1, \ldots, p_{q+1}|$ and $t \in I$. If $q > 0$, then $(D(\sigma))^{(0)} = \sigma$, and for $0 \le i \le q$, $(D(\sigma))^{(i+1)} = D(\sigma^{(i)})$. If $q = 0$, then $(D(\sigma))^{(0)} = \sigma$ and $(D(\sigma))^{(1)} = \tau(1)$. Therefore

$$\partial D + D\partial = 1_{\Delta(X)} - \tau \circ \varepsilon$$

and $D: 1_{\Delta(X)} \simeq \tau \circ \varepsilon$. By lemma 4.3.2, $\tilde{\Delta}(X)$ is chain contractible. ∎

2 COROLLARY *For any simplex s the chain map ν induces an isomorphism of the ordered homology group of $\bar{s}$ with the singular homology group of $|\bar{s}|$.*

PROOF Because ν preserves augmentation, ν induces a homomorphism $\tilde{\nu}_*$ from $\tilde{H}(\Delta(\bar{s}))$ to $\tilde{H}(|\bar{s}|)$, and under the isomorphism of lemma 4.3.1, $\nu_* = \tilde{\nu}_* \oplus 1_{\mathbf{Z}}$. By corollary 4.3.7, $\tilde{H}(\Delta(\bar{s})) = 0$. By lemma 1 and corollary 4.2.3, $\tilde{H}(|\bar{s}|) = 0$. Therefore ν_* is an isomorphism. ∎

We use lemma 1 to prove that if $f_0, f_1: X \to Y$ are homotopic, then $\Delta(f_0), \Delta(f_1): \Delta(X) \to \Delta(Y)$ are chain homotopic. We prove this first for the maps $h_0, h_1: X \to X \times I$, where $h_0(x) = (x,0)$ and $h_1(x) = (x,1)$.

3 THEOREM *The maps $h_0, h_1: X \to X \times I$ induce naturally chain-homotopic chain maps*

$$\Delta(h_0) \simeq \Delta(h_1): \Delta(X) \to \Delta(X \times I)$$

PROOF Let $\Delta'(X) = \Delta(X \times I)$. Then Δ and Δ' are covariant functors from the category of topological spaces to the category of augmented chain complexes and $\Delta(h_0)$ and $\Delta(h_1)$ are natural chain maps preserving augmentation from Δ to Δ'. Since Δ is free with models $\{\Delta^q\}$ and

$$\tilde{\Delta}'(\Delta^q) = \tilde{\Delta}(\Delta^q \times I)$$

is acyclic, by lemma 1, it follows from theorem 4.3.3 that $\Delta(h_0)$ and $\Delta(h_1)$ are naturally chain homotopic. ∎

This special case implies the general result.

4 COROLLARY *If $f_0, f_1: X \to Y$ are homotopic, then*

$$\Delta(f_0) \simeq \Delta(f_1): \Delta(X) \to \Delta(Y)$$

PROOF Let $F: X \times I \to Y$ be a homotopy from f_0 to f_1. Then $f_0 = Fh_0$ and $f_1 = Fh_1$. Therefore, using theorem 3,

$$\Delta(f_0) = \Delta(F)\Delta(h_0) \simeq \Delta(F)\Delta(h_1) = \Delta(f_1) \quad ∎$$

Since Δ^q is path connected for every q, any singular simplex $\sigma: \Delta^q \to X$ maps Δ^q to some path component of X. Hence, if $\{X_j\}$ is the set of path components of X, then $\Delta(X) = \bigoplus \Delta(X_j)$. By theorem 4.1.6, we have the following theorem.

5 THEOREM *The singular homology group of a space is the direct sum of the singular homology groups of its path components.* ■

Because Δ^q is compact, every singular simplex $\sigma: \Delta^q \to X$ maps Δ^q into some compact subset of X. Hence, if $\{X_\alpha\}$ is the collection of compact subsets of X directed by inclusion, then $\Delta(X) = \lim_\to \Delta(X_\alpha)$. By theorem 4.1.7, we have our next result.

6 THEOREM *The singular homology group of a space is isomorphic to the direct limit of the singular homology groups of its compact subsets.* ■

We now compute the 0-dimensional homology group of a space.

7 LEMMA *If X is a nonempty path-connected topological space, then $\tilde{H}_0(X) = 0$.*

PROOF Let x_0 be a fixed point of X. For any point $x \in X$ there is a path ω_x from x_0 to x. Because Δ^1 is homeomorphic to I, ω_x corresponds to a singular 1-simplex $\sigma_x: \Delta^1 \to X$ such that $\sigma_x{}^{(0)} = x$ and $\sigma_x{}^{(1)} = x_0$. A singular 0-simplex in X is identified with a point of X. Therefore a 0-chain (that is, a 0-cycle) of X is a sum $\Sigma n_x x$, where $n_x = 0$ except for a finite set of x's. Since $\varepsilon(\Sigma n_x x) = \Sigma n_x$, we see that if $\varepsilon(\Sigma n_x x) = 0$ [that is, if $\Sigma n_x x \in \tilde{\Delta}_0(X)$], then

$$\partial(\Sigma n_x \sigma_x) = \Sigma n_x x - (\Sigma n_x) x_0 = \Sigma n_x x$$

Therefore $\tilde{H}_0(X) = 0$. ■

8 COROLLARY *For any topological space X, $H_0(X)$ is a free group whose rank equals the number of nonempty components of X.*

PROOF If X is empty, $H_0(X) = 0$, and the result is valid in this case. If X is nonempty and path connected, it follows from lemmas 7 and 4.3.1 that $H_0(X) \approx \mathbf{Z}$. The general result now follows from theorem 5. ■

If A is a subspace of X, there is a *relative singular homology group* $H(X,A) = \{H_q(X,A) = H_q(\Delta(X)/\Delta(A))\}$ *of X modulo A*. $H(X,\varnothing) = H(X)$ is called the *absolute singular homology group of X*. The relative homology group is a covariant functor from the category of topological pairs to the category of graded groups. We show that this functor can be regarded as defined on the homotopy category of pairs.

9 THEOREM *If $f_0, f_1: (X,A) \to (Y,B)$ are homotopic, then*

$$f_{0*} = f_{1*}: H(X,A) \to H(Y,B)$$

PROOF Let $F: (X \times I, A \times I) \to (Y,B)$ be a homotopy from f_0 to f_1. Then $f_0 = Fh_0$ and $f_1 = Fh_1$, where $h_0, h_1: (X,A) \to (X \times I, A \times I)$ are defined by

$\bar{h}_0(x) = (x,0)$ and $\bar{h}_1(x) = (x,1)$. By theorem 3, there is a natural chain homotopy $D: \Delta(h_0) \simeq \Delta(h_1)$, where $h_0, h_1: X \to X \times I$ are maps defined by $\bar{h}_0$ and $\bar{h}_1$. Because D is natural, $D(\Delta(A)) \subset \Delta(A \times I)$. For $i = 0$ or 1 there is a commutative diagram

$$\begin{array}{ccccc} \Delta(A) & \subset & \Delta(X) & \to & \Delta(X)/\Delta(A) \\ {\scriptstyle \Delta(h_i)}\downarrow & & {\scriptstyle \Delta(h_i)}\downarrow & & \downarrow{\scriptstyle \Delta(\bar{h}_i)} \\ \Delta(A \times I) & \subset & \Delta(X \times I) & \to & \Delta(X \times I)/\Delta(A \times I) \end{array}$$

and a chain homotopy $\bar{D}: \Delta(\bar{h}_0) \simeq \Delta(\bar{h}_1)$ is obtained by passing to the quotient with D. By theorem 4.2.2,

$$\bar{h}_{0*} = \bar{h}_{1*}: H(X,A) \to H(X \times I, A \times I)$$

Then

$$f_{0*} = F_* \bar{h}_{0*} = F_* \bar{h}_{1*} = f_{1*} \quad ■$$

If $H_q(X,A)$ is finitely generated, its rank is called the *qth Betti number of* (X,A) and the orders of its finite cyclic summands given by the structure theorem are called the *qth torsion coefficients of* (X,A). If $H(X,A)$ is finitely generated, its Euler characteristic is called the *Euler characteristic of* (X,A), denoted by $\chi(X,A)$.

The remainder of this section is directed toward a proof that the subcomplex of the singular chain complex generated by small singular simplexes is chain equivalent to the singular chain complex. We begin by defining a subdivision chain map in singular theory. A singular simplex $\sigma: \Delta^q \to \Delta^n$ is said to be *linear* if $\sigma(\Sigma t_i p_i) = \Sigma t_i \sigma(p_i)$ for $t_i \in I$ with $\Sigma t_i = 1$. If σ is linear, so is $\sigma^{(i)}$ for $0 \le i \le q$. Therefore the set of linear simplexes in Δ^n generates a subcomplex $\Delta'(\Delta^n) \subset \Delta(\Delta^n)$.

A linear simplex σ in Δ^n is completely determined by the points $\sigma(p_i)$. If $x_0, x_1, \ldots, x_q \in \Delta^n$, we write $(x_0,x_1, \ldots ,x_q)$ to denote the linear simplex $\sigma: \Delta^q \to \Delta^n$ such that $\sigma(p_i) = x_i$. With this notation, it is clear that

$$\partial(x_0, \ldots, x_q) = \Sigma(-1)^i(x_0, \ldots ,\hat{x}_i, \ldots ,x_q)$$

Furthermore, the identity map $\xi_n: \Delta^n \subset \Delta^n$ is the linear simplex $\xi_n = (p_0,p_1, \ldots ,p_n)$.

Let b_n be the barycenter of Δ^n (that is, $b_n = \Sigma(1/(n+1))p_i$. For $q \ge 0$ a homomorphism

$$\beta_n: \Delta'_q(\Delta^n) \to \Delta'_{q+1}(\Delta^n)$$

is defined by the formula

$$\beta_n(x_0, \ldots ,x_q) = (b_n,x_0, \ldots ,x_q)$$

Let $\tau: \mathbf{Z} \to \Delta'_0(\Delta^n)$ be defined by $\tau(1) = (b_n)$. Direct computation shows that

10 $$\beta_n: 1_{\Delta'(\Delta^n)} \simeq \tau \circ \varepsilon \quad ■$$

For every topological space X we define an augmentation-preserving chain map

$$sd\colon \Delta(X) \to \Delta(X)$$

and a chain homotopy

$$D\colon \Delta(X) \to \Delta(X)$$

from sd to $1_{\Delta(X)}$, both of which are functorial in X. That is, if $f\colon X \to Y$, there are commutative squares

$$\begin{array}{ccc} \Delta(X) & \xrightarrow{sd} & \Delta(X) \\ {\scriptstyle \Delta(f)}\downarrow & & \downarrow{\scriptstyle \Delta(f)} \\ \Delta(Y) & \xrightarrow{sd} & \Delta(Y) \end{array} \qquad \begin{array}{ccc} \Delta(X) & \xrightarrow{D} & \Delta(X) \\ {\scriptstyle \Delta(f)}\downarrow & & \downarrow{\scriptstyle \Delta(f)} \\ \Delta(Y) & \xrightarrow{D} & \Delta(Y) \end{array}$$

Both sd and D are defined on q-chains by induction on q. If c is a 0-chain, we define $sd(c) = c$ and $D(c) = 0$. Assume sd and D defined on q-chains for $0 \le q < n$, where $n \ge 1$. We define sd and D on the universal singular n-simplex $\xi_n\colon \Delta^n \subset \Delta^n$ by the formulas

$$sd(\xi_n) = \beta_n(sd\, \partial(\xi_n))$$
$$D(\xi_n) = \beta_n(sd\,(\xi_n) - \xi_n - D\partial(\xi_n))$$

For any singular n-simplex $\sigma\colon \Delta^n \to X$ we define

$$sd(\sigma) = \Delta(\sigma)(sd(\xi_n))$$
$$D(\sigma) = \Delta(\sigma)(D(\xi_n))$$

Then sd and D have all the requisite properties.

If X is a metric space and $c = \Sigma n_\sigma \sigma$ is a singular q-chain of X, we define

$$\text{mesh } c = \sup \{\text{diam } \sigma(\Delta^q) \mid n_\sigma \neq 0\}$$

11 LEMMA *Let Δ^n have a linear metric and let c be a linear q-chain of Δ^n. Then*

$$\text{mesh } (sd\, c) \le \frac{q}{q+1} \text{ mesh } c$$

PROOF The proof is based on induction on q, using the inductive definition of sd. It suffices to show that if $\sigma = (x_0,x_1, \ldots ,x_q)$ is a linear q-simplex of Δ^n, then mesh $(sd\,\sigma) \le (q/(q+1))$ mesh σ. If $b = \Sigma\, (1/(q+1))x_i$, a computation similar to that of lemma 3.3.12 shows that the distance from b to any convex combination of the points $x_0, x_1, \ldots, x_q$ is less than or at most equal to $(q/(q+1))$ mesh $(x_0, \ldots ,x_q)$. Therefore

$$\text{mesh } (sd\,\sigma) \le \sup \left(\frac{q}{q+1} \text{ mesh } \sigma, \text{ mesh } (sd\,\partial\sigma)\right)$$

By induction

$$\text{mesh } (sd\, \partial\sigma) \le \frac{q-1}{q} \text{ mesh } \partial\sigma$$
$$\le \frac{q}{q+1} \text{ mesh } \sigma$$

which yields the result. ∎

We next define augmentation-preserving chain maps

$$sd^m: \Delta(X) \to \Delta(X)$$

for $m \ge 0$ by induction

$$sd^0 = 1_{\Delta(X)} \qquad \text{and} \qquad sd^m = sd(sd^{m-1}) \qquad m \ge 1$$

Then, from lemma 11, we obtain the following result.

12 COROLLARY *Let Δ^n have a linear metric and let $c \in \Delta'_q(\Delta^n)$. Then*

$$\text{mesh } (sd^m c) \le [q/(q+1)]^m \text{ mesh } c \quad ∎$$

Let $\mathcal{U} = \{A\}$ be a collection of subsets of a topological space X and let $\Delta(\mathcal{U})$ be the subcomplex of $\Delta(X)$ generated by singular q-simplexes $\sigma: \Delta^q \to X$ such that $\sigma(\Delta^q) \subset A$ for some $A \in \mathcal{U}$ [if $\sigma(\Delta^q) \subset A$, then $\sigma^{(i)}(\Delta^{q-1}) \subset A$, and so $\Delta(\mathcal{U})$ is a subcomplex of $\Delta(X)$]. Because sd and D are natural, $sd(\Delta(\mathcal{U})) \subset \Delta(\mathcal{U})$ and $D(\Delta(\mathcal{U})) \subset \Delta(\mathcal{U})$.

13 LEMMA *Let $\mathcal{U} = \{A\}$ be such that $X = \bigcup \{\text{int } A \mid A \in \mathcal{U}\}$. For any singular q-simplex σ of X there is $m \ge 0$ such that $sd^m \sigma \in \Delta(\mathcal{U})$.*

PROOF Because $X = \bigcup\{\text{int } A \mid A \in \mathcal{U}\}$, $\Delta^q = \bigcup\{\sigma^{-1}(\text{int } A) \mid A \in \mathcal{U}\}$. Let Δ^q be metrized by a linear metric and let $\lambda > 0$ be a Lebesgue number for the open covering $\{\sigma^{-1}(\text{int } A) \mid A \in \mathcal{U}\}$ of Δ^q relative to this metric. Choose $m \ge 0$ so that $[q/(q+1)]^m \text{ diam } \Delta^q \le \lambda$. By corollary 12, mesh $(\text{sd}^m \xi_q) \le \lambda$. Therefore every singular simplex of $sd^m \xi_q$ maps into $\sigma^{-1}(\text{int } A)$ for some $A \in \mathcal{U}$. Then $sd^m \sigma = \Delta(\sigma)\, sd^m \xi_q$ is a chain in $\Delta(\mathcal{U})$. ∎

We are now ready to prove the chain equivalence mentioned earlier.

14 THEOREM *Let $\mathcal{U} = \{A\}$ be such that $X = \bigcup\{\text{int } A \mid A \in \mathcal{U}\}$. Then the inclusion map $\Delta(\mathcal{U}) \subset \Delta(X)$ is a chain equivalence.*

PROOF For each singular simplex σ in X let $m(\sigma)$ be the smallest nonnegative integer such that $sd^{m(\sigma)}\sigma \in \Delta(\mathcal{U})$. Such an integer $m(\sigma)$ exists by lemma 13, and it is clear that $m(\sigma) = 0$ if and only if $\sigma \in \Delta(\mathcal{U})$. Furthermore, $m(\sigma^{(i)}) \le m(\sigma)$ for $0 \le i \le \deg \sigma$.

Define $\bar{D}: \Delta(X) \to \Delta(X)$ by $\bar{D}(\sigma) = \sum_{0 \le j \le m(\sigma)-1} D\, sd^j(\sigma)$. Then $\bar{D}(\sigma) = 0$ if and only if $\sigma \in \Delta(\mathcal{U})$. Also

$$\begin{aligned}\partial\bar{D}(\sigma) &= \textstyle\sum sd^{j+1}(\sigma) - \sum sd^j(\sigma) - \sum D\, sd^j(\partial\sigma)\\ &= sd^{m(\sigma)}(\sigma) - \sigma - \textstyle\sum_{0 \le j \le m(\sigma)-1} \sum_i (-1)^i D\, sd^j(\sigma^{(i)})\\ \bar{D}\partial(\sigma) &= \textstyle\sum_i (-1)^i \sum_{0 \le j \le m(\sigma^{(i)})-1} D\, sd^j(\sigma^{(i)})\end{aligned}$$

Therefore

$$\sigma + \partial \bar{D}(\sigma) + \bar{D}\partial(\sigma) = \Sigma_i (-1)^i \Sigma_{m(\sigma^{(i)}) \le j \le m(\sigma)-1} D\, sd^j(\sigma^{(i)}) + sd^{m(\sigma)}(\sigma)$$

is in $\Delta(\mathfrak{U})$. Define $\tau: \Delta(X) \to \Delta(\mathfrak{U})$ by $\tau(\sigma) = \sigma + \partial\bar{D}(\sigma) + \bar{D}\partial(\sigma)$. Then τ is a chain map preserving augmentation. Clearly, if $i: \Delta(\mathfrak{U}) \subset \Delta(X)$, then $\tau \circ i = 1_{\Delta(\mathfrak{U})}$ and $\bar{D}: i \circ \tau \simeq 1_{\Delta(X)}$. Therefore $[\tau] = [i]^{-1}$, and i is a chain equivalence. ∎

5 EXACTNESS

In this section we consider the relations among the homology groups of C', C, and C/C', where C' is a subcomplex of C. A concise way of summarizing these relations is by means of the concept of exact sequence. The basic result is the existence of an exact sequence connecting the homology of C', C, and C/C'.

A three-term sequence of abelian groups and homomorphisms

$$G' \xrightarrow{\alpha} G \xrightarrow{\beta} G''$$

is said to be *exact at* G if $\ker \beta = \operatorname{im} \alpha$. A sequence of abelian groups and homomorphisms indexed by integers (which may or may not terminate at either or both ends)

$$\cdots \to G_{n+1} \xrightarrow{\alpha_{n+1}} G_n \xrightarrow{\alpha_n} G_{n-1} \to \cdots$$

is said to be an *exact sequence* if every three-term subsequence of consecutive groups is exact at its middle group. Note that an exact sequence terminating at one end with a trivial group can be extended indefinitely on that end to an exact sequence by adjoining trivial groups and homomorphisms.

A *short exact sequence of abelian groups*, written

$$0 \to G' \xrightarrow{\alpha} G \xrightarrow{\beta} G'' \to 0$$

is a five-term exact sequence whose end groups are trivial. In such a short exact sequence α is a monomorphism and β is an epimorphism whose kernel is $\alpha(G')$. Therefore α is an isomorphism of G' with the subgroup $\alpha(G') \subset G$, and β induces an isomorphism from the quotient group $G/\alpha(G')$ to G''. The group G is called an *extension of G' by G''*.

Given an exact sequence

$$\cdots \to G_{n+1} \xrightarrow{\alpha_{n+1}} G_n \xrightarrow{\alpha_n} G_{n-1} \to \cdots$$

let $G'_n = \ker \alpha_n = \operatorname{im} \alpha_{n+1}$. Then the given sequence gives rise to short exact sequences

$$0 \to G'_n \to G_n \to G'_{n-1} \to 0$$

for every G_n not on one or the other end of the original sequence, and the composite $G_n \to G'_{n-1} \to G_{n-1}$ equals α_n.

A *homomorphism* γ from one sequence $\{G_n \xrightarrow{\alpha_n} G_{n-1}\}$ to another $\{H_n \xrightarrow{\beta_n} H_{n-1}\}$ with the same set of indices (that is, of the same length) is a sequence $\{\gamma_n: G_n \to H_n\}$ of homomorphisms such that the following diagram is commutative:

$$\begin{array}{ccccccccc} \cdots \to & G_{n+1} & \xrightarrow{\alpha_{n+1}} & G_n & \xrightarrow{\alpha_n} & G_{n-1} & \to \cdots \\ & \gamma_{n+1}\downarrow & & \gamma_n\downarrow & & \downarrow\gamma_{n-1} & \\ \cdots \to & H_{n+1} & \xrightarrow{\beta_{n+1}} & H_n & \xrightarrow{\beta_n} & H_{n-1} & \to \cdots \end{array}$$

There is a category of exact sequences with the same set of indices. In particular, there is a category of short exact sequences, and also a category of exact sequences (indexed by all the integers).

Note that a sequence of abelian groups and homomorphisms

$$\cdots \to C_{n+1} \xrightarrow{\partial_{n+1}} C_n \xrightarrow{\partial_n} C_{n-1} \to \cdots$$

is a chain complex if and only if im $\partial_{n+1} \subset$ ker ∂_n for all n. This is half of the condition of exactness at C_n. For a chain complex C, the group $H_n(C) =$ ker ∂_n/im ∂_{n+1} is a measure of the nonexactness of the sequence at C_n. Thus a chain complex is an exact sequence if and only if its graded homology group is trivial. In any case, the fact that the homology group measures the nonexactness of the chain complex suggests that there should be some relation between homology and exactness, and this is indeed so.

A *short exact sequence of chain complexes,* written

$$0 \to C' \xrightarrow{\alpha} C \xrightarrow{\beta} C'' \to 0$$

is a five-term sequence of chain complexes and chain maps such that for all q there is a short exact sequence of abelian groups

$$0 \to C'_q \xrightarrow{\alpha_q} C_q \xrightarrow{\beta_q} C''_q \to 0$$

A homomorphism from one short exact sequence of chain complexes to another consists of a commutative diagram of chain maps

$$\begin{array}{ccccccc} 0 \to & C' & \xrightarrow{\alpha} & C & \xrightarrow{\beta} & C'' & \to 0 \\ & \tau'\downarrow & & \tau\downarrow & & \downarrow\tau'' & \\ 0 \to & \bar{C}' & \xrightarrow{\bar{\alpha}} & \bar{C} & \xrightarrow{\bar{\beta}} & \bar{C}'' & \to 0 \end{array}$$

There is a category of short exact sequences of chain complexes and homomorphisms.

1 EXAMPLE Let C' be a subcomplex of a chain complex C and let $i: C' \subset C$ and $j: C \to C/C'$ be the inclusion and projection chain maps, respectively. There is a short exact sequence of chain complexes

$$0 \to C' \xrightarrow{i} C \xrightarrow{j} C/C' \to 0$$

Given a subcomplex $\bar{C}' \subset \bar{C}$ and a chain map $\tau: C \to \bar{C}$ such that $\tau(C') \subset \bar{C}'$, there is a homomorphism

$$\begin{array}{ccccccccc} 0 & \to & C' & \xrightarrow{i} & C & \xrightarrow{j} & C/C' & \to & 0 \\ & & \tau'\downarrow & & \tau\downarrow & & \downarrow\tau'' & & \\ 0 & \to & \bar{C}' & \xrightarrow{\bar{i}} & \bar{C} & \xrightarrow{\bar{j}} & \bar{C}/\bar{C}' & \to & 0 \end{array}$$

where $\tau' = \tau \mid C'$ and τ'' is induced from τ by passing to the quotient.

2 EXAMPLE If C is an augmented chain complex, there is a short exact sequence of chain complexes

$$0 \to \tilde{C} \to C \xrightarrow{\varepsilon} \mathbf{Z} \to 0$$

There is a covariant functor C from the category of simplicial pairs to the category of short exact sequences of chain complexes which assigns to (K,L) the short exact sequence

$$0 \to C(L) \to C(K) \to C(K)/C(L) \to 0$$

Similarly, there is a covariant functor Δ from the category of topological pairs to the category of short exact sequences of chain complexes which assigns to (X,A) the short exact sequence

$$0 \to \Delta(A) \to \Delta(X) \to \Delta(X)/\Delta(A) \to 0$$

There is also a covariant functor Δ from the category of simplicial pairs to the category of short exact sequences of chain complexes which assigns to (K,L) the short exact sequence

$$0 \to \Delta(L) \to \Delta(K) \to \Delta(K)/\Delta(L) \to 0$$

Then μ is a natural transformation from Δ to C and ν is a natural transformation from Δ to $\Delta(|\cdot|)$ (both natural transformations in the category of short exact sequences of chain complexes).

We define covariant functors H', H, and H'' from the category of short exact sequences of chain complexes

$$0 \to C' \xrightarrow{\alpha} C \xrightarrow{\beta} C'' \to 0$$

to the category of graded groups such that H', H, and H'' map the above sequence into $H(C')$, $H(C)$, and $H(C'')$, respectively.

3 LEMMA *On the category of short exact sequences of chain complexes*

$$0 \to C' \xrightarrow{\alpha} C \xrightarrow{\beta} C'' \to 0$$

there is a natural transformation $\partial_*: H'' \to H'$ *such that if* $\{z''\} \in H(C'')$, *then* $\partial_* \{z''\} = \{\alpha^{-1}\partial\beta^{-1}z''\} \in H(C')$.

PROOF There is a commutative diagram

$$\begin{array}{ccccccccc} 0 & \to & C'_{q+1} & \xrightarrow{\alpha} & C_{q+1} & \xrightarrow{\beta} & C''_{q+1} & \to & 0 \\ & & \partial'\downarrow & & \partial\downarrow & & \downarrow\partial'' & & \\ 0 & \to & C'_q & \xrightarrow{\alpha} & C_q & \xrightarrow{\beta} & C''_q & \to & 0 \\ & & \partial'\downarrow & & \partial\downarrow & & \downarrow\partial'' & & \\ 0 & \to & C'_{q-1} & \xrightarrow{\alpha} & C_{q-1} & \xrightarrow{\beta} & C''_{q-1} & \to & 0 \end{array}$$

in which each row is a short exact sequence of groups. If z'' is a q-cycle of C'', let $c \in C_q$ be such that $\beta(c) = z''$. Then

$$\beta(\partial c) = \partial''\beta(c) = \partial'' z'' = 0$$

Therefore there is a unique $c' \in C'_{q-1}$ such that $\alpha(c') = \partial c$. Then

$$\alpha(\partial' c') = \partial\alpha(c') = \partial\partial c = 0$$

Because α is a monomorphism, $\partial' c' = 0$. Hence c' is a $(q-1)$-cycle of C'.

We show that the homology class of c' in C' depends only on the homology class of z'' in C'', which will prove that there is a well-defined homomorphism $\partial_*\{z''\} = \{c'\}$. Let $c_1 \in C_q$ be such that $\beta(c_1) \sim z''$. Then there is $d'' \in C''_{q+1}$ such that $\beta(c_1) = \beta(c) + \partial'' d''$. Choose $d \in C_{q+1}$ such that $\beta(d) = d''$. Then

$$\beta(c_1) = \beta(c) + \partial''\beta(d) = \beta(c + \partial d)$$

Therefore there is a $d' \in C'_q$ such that $c_1 = c + \partial d + \alpha(d')$, and

$$\partial c_1 = \partial c + \partial\alpha(d') = \alpha(c') + \alpha(\partial' d') = \alpha(c' + \partial' d')$$

Hence $\alpha^{-1}(\partial c_1) = c' + \partial' d' \sim c'$ and $\{\alpha^{-1}(\partial c_1)\} = \{\alpha^{-1}(\partial c)\}$, showing that ∂_* is well-defined.

To prove that ∂_* is a natural transformation, assume given a commutative diagram of chain maps

$$\begin{array}{ccccccccc} 0 & \to & C' & \xrightarrow{\alpha} & C & \xrightarrow{\beta} & C'' & \to & 0 \\ & & \tau'\downarrow & & \tau\downarrow & & \downarrow\tau'' & & \\ 0 & \to & \bar{C}' & \xrightarrow{\bar\alpha} & \bar{C} & \xrightarrow{\bar\beta} & \bar{C}'' & \to & 0 \end{array}$$

where the horizontal rows are short exact sequences. Then

$$\begin{aligned} \tau'_*\partial_*\{z''\} &= \tau'_*\{\alpha^{-1}\partial\beta^{-1}z''\} = \{\tau'\alpha^{-1}\partial\beta^{-1}z''\} \\ &= \{\bar\alpha^{-1}\tau\partial\beta^{-1}z''\} = \{\bar\alpha^{-1}\bar\partial\bar\beta^{-1}\tau''z''\} = \bar\partial_*\tau''_*\{z''\} \end{aligned}$$ ∎

The natural transformation ∂_* is called the *connecting homomorphism* for homology because of its importance in the following *exactness theorem.*

4 THEOREM *There is a covariant functor from the category of short exact sequences of chain complexes to the category of exact sequences of groups which assigns to a short exact sequence*

$$0 \to C' \xrightarrow{\alpha} C \xrightarrow{\beta} C'' \to 0$$

the sequence

$$\cdots \xrightarrow{\partial_*} H_q(C') \xrightarrow{\alpha_*} H_q(C) \xrightarrow{\beta_*} H_q(C'') \xrightarrow{\partial_*} H_{q-1}(C') \xrightarrow{\alpha_*} \cdots$$

PROOF The sequence of homology groups is functorial on short exact sequences because ∂_* is a natural transformation. It only remains to verify that it is an exact sequence. This entails a proof of exactness at $H_q(C')$, $H_q(C)$, and $H_q(C'')$, each exactness requiring two inclusion relations. Therefore the proof of exactness has six parts. We shall prove exactness at $H_q(C'')$ and leave the other parts of the proof to the reader.

(*a*) im $\beta_* \subset$ ker ∂_*. Let $\{z\} \in H_q(C)$. Then

$$\partial_* \beta_* \{z\} = \partial_* \{\beta(z)\} = \{\alpha^{-1}\partial\beta^{-1}\beta(z)\} = \{\alpha^{-1}\partial z\} = \{\alpha^{-1}(0)\} = 0$$

(*b*) ker $\partial_* \subset$ im β_*. Let $\{z''\} \in$ ker ∂_*. Then there is $c \in C_q$ such that $\beta(c) = z''$ and $\alpha^{-1}\partial(c) = \partial'(d')$ for some $d' \in C'_q$. The difference $c - \alpha(d') \in C_q$ is such that

$$\partial(c - \alpha(d')) = \partial c - \alpha(\partial' d') = 0$$

Hence $\{c - \alpha(d')\} \in H_q(C)$ and

$$\beta_* \{c - \alpha(d')\} = \{\beta(c) - \beta\alpha(d')\} = \{z''\} \quad \blacksquare$$

Combining theorem 4 with example 2, we again obtain lemma 4.3.1. As an example of the utility of exactness, note that the following corollary is immediate from theorem 4.

5 COROLLARY *Given a short exact sequence of chain complexes*

$$0 \to C' \xrightarrow{\alpha} C \xrightarrow{\beta} C'' \to 0$$

(*a*) *C' is acyclic if and only if $\beta_* : H(C) \approx H(C'')$.*
(*b*) *C is acyclic if and only if $\partial_* : H(C'') \approx H(C')$.*
(*c*) *C'' is acyclic if and only if $\alpha_* : H(C') \approx H(C)$.* ■

In (*b*) above it should be noted that ∂_* has degree -1. It follows from corollary 5 that if two of the chain complexes C', C, and C'' are acyclic, so is the third.

6 COROLLARY *Given an exact sequence of abelian groups*

$$\cdots \to G_{n+1} \xrightarrow{\alpha_{n+1}} G_n \xrightarrow{\alpha_n} G_{n-1} \to \cdots$$

and a subsequence

$$\cdots \to G'_{n+1} \xrightarrow{\alpha'_{n+1}} G'_n \xrightarrow{\alpha'_n} G'_{n-1} \to \cdots$$

(that is, $G'_n \subset G_n$ and $\alpha'_n = \alpha_n \mid G'_n$), the subsequence is exact if and only if the quotient sequence

$$\cdots \to G_n/G'_n \to G_{n-1}/G'_{n-1} \to \cdots$$

is exact.

PROOF Let C be the chain complex consisting of the original exact sequence and let C' be the subcomplex consisting of the subsequence. Then the quotient chain complex C/C' is the quotient sequence. Because C is an exact sequence, C is acyclic, and $\partial_*: H_q(C/C') \approx H_{q-1}(C')$. Therefore C' is exact [that is, $H(C') = 0$] if and only if C/C' is exact [that is, $H(C/C') = 0$]. ∎

7 THEOREM *The direct limit of exact sequences is exact.*

PROOF Each exact sequence is an acyclic chain complex. The direct limit is also a chain complex, and it is acyclic, by theorem 4.1.7. Therefore the limit sequence is exact. ∎

This result is false if direct limit is replaced by inverse limit, because the homology functor fails to commute with inverse limits.

Let K be a simplicial complex and let $L_1 \subset L_2 \subset K$. By the Noether isomorphism theorem, there is a short exact sequence of chain complexes

$$0 \to C(L_2)/C(L_1) \xrightarrow{i} C(K)/C(L_1) \xrightarrow{j} C(K)/C(L_2) \to 0$$

By theorem 4, there is an exact sequence

$$\cdots \xrightarrow{\partial_*} H_q(L_2,L_1) \xrightarrow{i_*} H_q(K,L_1) \xrightarrow{j_*} H_q(K,L_2) \xrightarrow{\partial_*} H_{q-1}(L_2,L_1) \xrightarrow{i_*} \cdots$$

where i_* is induced by i: $(L_2,L_1) \subset (K,L_1)$, j_* is induced by j: $(K,L_1) \subset (K,L_2)$, and ∂_* is the connecting homomorphism. This sequence is called the *homology sequence of the triple* (K,L_2,L_1). It is functorial on triples. If $L_1 = \varnothing$, the resulting exact sequence

$$\cdots \xrightarrow{\partial_*} H_q(L_2) \xrightarrow{i_*} H_q(K) \xrightarrow{j_*} H_q(K,L_2) \xrightarrow{\partial_*} H_{q-1}(L_2) \xrightarrow{i_*} \cdots$$

is called the *homology sequence of the pair* (K,L_2). It is functorial on pairs.

Because there is an inclusion map of the triple $(K,L_2,\varnothing)$ into the triple (K,L_2,L_1), the next result follows.

8 LEMMA *The connecting homomorphism* $\partial_*: H_q(K,L_2) \to H_{q-1}(L_2,L_1)$ *of the triple* (K,L_2,L_1) *is the composite*

$$H_q(K,L_2) \xrightarrow{\partial_*} H_{q-1}(L_2) \xrightarrow{k_*} H_{q-1}(L_2,L_1)$$

of the connecting homomorphism of the pair (K,L_2) *followed by the homomorphism induced by* k: $(L_2,\varnothing) \subset (L_2,L_1)$. ∎

If L is a nonempty subcomplex of a simplicial complex, $\tilde{C}(L) \subset \tilde{C}(K)$, and by the Noether isomorphism theorem, $\tilde{C}(K)/\tilde{C}(L) \approx C(K)/C(L)$. Therefore there is a short exact sequence of chain complexes

$$0 \to \tilde{C}(L) \xrightarrow{i} \tilde{C}(K) \xrightarrow{j} C(K)/C(L) \to 0$$

The corresponding exact sequence

$$\cdots \xrightarrow{\partial_*} \tilde{H}_q(L) \xrightarrow{i_*} \tilde{H}_q(K) \xrightarrow{j_*} H_q(K,L) \xrightarrow{\partial_*} \tilde{H}_{q-1}(L) \xrightarrow{i_*} \cdots$$

is called the *reduced homology sequence of the pair* (K,L). It is not defined if $L = \varnothing$, because $C(L)$ has no augmentation in this case.

In the same way, there is a *singular homology sequence of a triple* (X,A,B) and *of a pair* (X,A). If A is nonempty, there is also a *reduced homology sequence of* (X,A). All these sequences are exact, and the analogue of lemma 8 is valid relating the connecting homomorphism of a triple to the connecting homomorphism of a pair.

9 LEMMA *Let s be an n-simplex. Then*

$$H_q(\bar{s},\dot{s}) \approx \begin{cases} 0 & q \neq n \\ \mathbf{Z} & q = n \end{cases}$$

PROOF $C_q(\dot{s}) = C_q(\bar{s})$ if $q \neq n$. Therefore $[C(\bar{s})/C(\dot{s})]_q = 0$ if $q \neq n$, and $[C(\bar{s})/C(\dot{s})]_n \approx \mathbf{Z}$. ■

Because $\tilde{H}(\bar{s}) = 0$, by corollary 4.3.7, it follows from the exactness of the reduced homology sequence of $(\bar{s},\dot{s})$ that $\partial_*\colon H_q(\bar{s},\dot{s}) \approx \tilde{H}_{q-1}(\dot{s})$ for all q. Therefore we have the next result.

10 COROLLARY *If s is an n-simplex, then*

$$\tilde{H}_q(\dot{s}) \approx \begin{cases} 0 & q \neq n-1 \\ \mathbf{Z} & q = n-1 \end{cases} \quad ■$$

We conclude by proving the following *five lemma* (so named because of the five-term exact sequences involved in its formulation).

11 LEMMA *Given a commutative diagram of abelian groups and homomorphisms*

$$\begin{array}{ccccccccc} G_5 & \xrightarrow{\alpha_5} & G_4 & \xrightarrow{\alpha_4} & G_3 & \xrightarrow{\alpha_3} & G_2 & \xrightarrow{\alpha_2} & G_1 \\ \downarrow{\scriptstyle\gamma_5} & & \downarrow{\scriptstyle\gamma_4} & & \downarrow{\scriptstyle\gamma_3} & & \downarrow{\scriptstyle\gamma_2} & & \downarrow{\scriptstyle\gamma_1} \\ H_5 & \xrightarrow{\beta_5} & H_4 & \xrightarrow{\beta_4} & H_3 & \xrightarrow{\beta_3} & H_2 & \xrightarrow{\beta_2} & H_1 \end{array}$$

in which each row is exact and γ_1, γ_2, γ_4, *and* γ_5 *are isomorphisms, then* γ_3 *is an isomorphism.*

PROOF The proof is straightforward. To show that γ_3 is a monomorphism, assume $\gamma_3(g_3) = 0$. Then $\gamma_2\alpha_3(g_3) = \beta_3\gamma_3(g_3) = 0$. Therefore $\alpha_3(g_3) = 0$. Hence there is $g_4 \in G_4$ such that $\alpha_4(g_4) = g_3$. Then $\beta_4\gamma_4(g_4) = 0$, and there is $h_5 \in H_5$ such that $\beta_5(h_5) = \gamma_4(g_4)$. There is $g_5 \in G_5$ with $\gamma_5(g_5) = h_5$. Then $\gamma_4(\alpha_5(g_5)) = \gamma_4(g_4)$, and so $g_4 = \alpha_5(g_5)$. Then $g_3 = \alpha_4\alpha_5(g_5) = 0$.

To show that γ_3 is an epimorphism let $h_3 \in H_3$. There is $g_2 \in G_2$ such that $\gamma_2(g_2) = \beta_3(h_3)$. Then $\gamma_1\alpha_2(g_2) = \beta_2\beta_3(h_3) = 0$. Therefore $\alpha_2(g_2) = 0$, and there is $g_3 \in G_3$ such that $\alpha_3(g_3) = g_2$. Then $\beta_3(h_3 - \gamma_3(g_3)) = 0$, and there is $h_4 \in H_4$ such that $\beta_4(h_4) = h_3 - \gamma_3(g_3)$. Let $g_4 \in G_4$ be such that $\gamma_4(g_4) = h_4$. Then $g_3 + \alpha_4(g_4) \in G_3$ and $\gamma_3(g_3 + \alpha_4(g_4)) = \gamma_3(g_3) + \beta_4(h_4) = h_3$. ■

Note that to prove γ_3 a monomorphism we merely needed γ_2 and γ_4 to be monomorphisms and γ_5 to be an epimorphism, and to prove γ_3 an epimorphism we merely needed γ_2 and γ_4 to be epimorphisms and γ_1 to be a monomorphism. This type of proof is called *diagram chasing* and will be omitted in the future.

We shall have several occasions to use the five lemma. We mention the following as a typical example. For any simplicial pair (K,L) the natural transformation μ from the ordered homology theory induces a homomorphism of the corresponding exact sequences

$$\begin{array}{ccccccccc} \cdots \to & H_q(\Delta(L)) & \to & H_q(\Delta(K)) & \to & H_q(\Delta(K)/\Delta(L)) & \to \cdots \\ & \downarrow \mu_* & & \downarrow \mu_* & & \downarrow \mu_* & \\ \cdots \to & H_q(L) & \to & H_q(K) & \to & H_q(K,L) & \to \cdots \end{array}$$

By theorem 4.3.8, μ_* is an isomorphism on the absolute groups. It follows from the five lemma that it is also an isomorphism on the relative groups.

12 COROLLARY *For any simplicial pair (K,L) the natural transformation μ induces an isomorphism from the ordered homology sequence of (K,L) to the oriented homology sequence of (K,L).* ■

6 MAYER-VIETORIS SEQUENCES

There is an exact sequence which relates the homology of the union of two sets to the homology of each of the sets and to the homology of their intersection. This sequence provides an inductive procedure for computing the homology of spaces which are built from pieces whose homology is known. We shall define this exact sequence as well as its analogue involving relative homology groups, and use them to prove that the natural transformation ν from $\Delta(K)$ to $\Delta(|K|)$ is a chain equivalence for any simplicial complex K.

Let K_1 and K_2 be subcomplexes of a simplicial complex K. Then $K_1 \cap K_2$ and $K_1 \cup K_2$ are subcomplexes of K, and $C(K_1)$, $C(K_2) \subset C(K)$. Clearly $C(K_1 \cap K_2) = C(K_1) \cap C(K_2)$ and $C(K_1) + C(K_2) = C(K_1 \cup K_2)$. Let $i_1\colon K_1 \cap K_2 \subset K_1$, $i_2\colon K_1 \cap K_2 \subset K_2$, $j_1\colon K_1 \subset K_1 \cup K_2$, and $j_2\colon K_2 \subset K_1 \cup K_2$. Then we have a short exact sequence of chain complexes

$$0 \to C(K_1 \cap K_2) \xrightarrow{i} C(K_1) \oplus C(K_2) \xrightarrow{j} C(K_1 \cup K_2) \to 0$$

where $i(c) = (C(i_1)c, -C(i_2)c)$ and $j(c_1,c_2) = C(j_1)c_1 + C(j_2)c_2$. The corresponding exact sequence of homology groups

$$\cdots \xrightarrow{\partial_*} H_q(K_1 \cap K_2) \xrightarrow{i_*} H_q(K_1) \oplus H_q(K_2) \xrightarrow{j_*} H_q(K_1 \cup K_2) \xrightarrow{\partial_*} H_{q-1}(K_1 \cap K_2) \xrightarrow{i_*} \cdots$$

is called the *Mayer-Vietoris sequence* of the subcomplexes K_1 and K_2. The homomorphisms i_* and j_* in the Mayer-Vietoris sequence are described by means of homomorphisms induced by inclusion maps by

$$i_* z = (i_{1*} z, -i_{2*} z) \qquad \text{and} \qquad j_*(z_1,z_2) = j_{1*} z_1 + j_{2*} z_2$$

for $z \in H(K_1 \cap K_2)$, $z_1 \in H(K_1)$, and $z_2 \in H(K_2)$.

If $K_1 \cap K_2 \neq \varnothing$, there is a commutative diagram of abelian groups and homomorphisms

$$\begin{array}{ccccccccc} 0 & \to & C_0(K_1 \cap K_2) & \xrightarrow{i} & C_0(K_1) \oplus C_0(K_2) & \xrightarrow{j} & C_0(K_1 \cup K_2) & \to & 0 \\ & & \downarrow{\scriptstyle \varepsilon} & & \downarrow{\scriptstyle \varepsilon \oplus \varepsilon} & & \downarrow{\scriptstyle \varepsilon} & & \\ 0 & \to & \mathbf{Z} & \xrightarrow{\alpha} & \mathbf{Z} \oplus \mathbf{Z} & \xrightarrow{\beta} & \mathbf{Z} & \to & 0 \end{array}$$

where $\alpha(n) = (n,-n)$ and $\beta(n,m) = n + m$. Since the rows are exact and the vertical homomorphisms are epimorphisms, it follows from corollary 4.5.6 that there is an exact sequence of the kernels

$$0 \to \tilde{C}_0(K_1 \cap K_2) \xrightarrow{i} \tilde{C}_0(K_1) \oplus \tilde{C}_0(K_2) \xrightarrow{j} \tilde{C}_0(K_1 \cup K_2) \to 0$$

and so there is a short exact sequence of chain complexes

$$0 \to \tilde{C}(K_1 \cap K_2) \xrightarrow{i} \tilde{C}(K_1) \oplus \tilde{C}(K_2) \xrightarrow{j} \tilde{C}(K_1 \cup K_2) \to 0$$

The corresponding exact sequence of reduced homology groups

$$\cdots \xrightarrow{\partial_*} \tilde{H}_q(K_1 \cap K_2) \xrightarrow{i_*} \tilde{H}_q(K_1) \oplus \tilde{H}_q(K_2) \xrightarrow{j_*} \tilde{H}_q(K_1 \cup K_2) \xrightarrow{\partial_*} \cdots$$

is called the *reduced Mayer-Vietoris sequence of* K_1 *and* K_2.

If (K_1,L_1) and (K_2,L_2) are simplicial pairs in K, there is also a short exact sequence

$$0 \to C(L_1 \cap L_2) \to C(L_1) \oplus C(L_2) \to C(L_1 \cup L_2) \to 0$$

which is a subsequence of the short exact sequence

$$0 \to C(K_1 \cap K_2) \to C(K_1) \oplus C(K_2) \to C(K_1 \cup K_2) \to 0$$

It follows from corollary 4.5.6 that the quotient sequence is a short exact sequence of chain complexes

$$0 \to C(K_1 \cap K_2)/C(L_1 \cap L_2) \to C(K_1)/C(L_1) \oplus C(K_2)/C(L_2) \to C(K_1 \cup K_2)/C(L_1 \cup L_2) \to 0$$

The corresponding exact sequence of homology groups

$$\cdots \xrightarrow{\partial_*} H_q(K_1 \cap K_2, L_1 \cap L_2) \xrightarrow{i_*} H_q(K_1,L_1) \oplus H_q(K_2,L_2) \xrightarrow{j_*} H_q(K_1 \cup K_2, L_1 \cup L_2) \xrightarrow{\partial_*} \cdots$$

is called the *relative Mayer-Vietoris sequence of* (K_1,L_1) *and* (K_2,L_2).

The relative Mayer-Vietoris sequence specializes to the exact sequence of a triple or a pair. In fact, given a triple (K,L_1,L_2), the relative Mayer-Vietoris sequence of (K,L_2) and (L_1,L_1) is easily seen to be the homology sequence of the triple (K,L_1,L_2) as defined in Sec. 4.5. In case $L_2 = \varnothing$, the relative Mayer-Vietoris sequence of $(K,\varnothing)$ and (L_1,L_1) is the homology sequence of the pair (K,L_1).

An inclusion map $(K_1,L_1) \subset (K_2,L_2)$ is called an *excision map* if $K_1 - L_1 = K_2 - L_2$. The exactness of the Mayer-Vietoris sequence is closely related (in fact, equivalent) to the following *excision property.*

1 THEOREM *Any excision map between simplicial pairs induces an isomorphism on homology.*

PROOF If $(K_1,L_1) \subset (K_2,L_2)$ is an excision map, then $K_2 = K_1 \cup L_2$ and $L_1 = K_1 \cap L_2$. By the Noether isomorphism theorem,

$$C(K_1)/C(L_1) \approx [C(K_1) + C(L_2)]/C(L_2) = C(K_2)/C(L_2) \quad \blacksquare$$

For the ordered chain complex it is still true that if K_1 and K_2 are subcomplexes of some simplicial complex, then $\Delta(K_1 \cup K_2) = \Delta(K_1) + \Delta(K_2)$. Therefore all the above results remain valid if the oriented homology is replaced throughout by the ordered homology.

An inclusion map $(X_1,A_1) \subset (X_2,A_2)$ between topological pairs is called an *excision map* if $X_1 - A_1 = X_2 - A_2$. It is *not* true that every excision map induces an isomorphism of the singular homology groups. Neither is it true that there is an exact Mayer-Vietoris sequence of any two subsets X_1 and X_2 of a topological space.

2 EXAMPLE Let $f\colon \mathbf{R} \to \mathbf{R}$ be defined by

$$f(x) = \begin{cases} \sin \dfrac{1}{x} & x > 0 \\ 0 & x \leq 0 \end{cases}$$

and let $X_1 = \{(x,y) \in \mathbf{R}^2 \mid y \geq f(x) \text{ or } x = 0, |y| \leq 1\}$ and $X_2 = \{(x,y) \in \mathbf{R}^2 \mid y \leq f(x) \text{ or } x = 0, |y| \leq 1\}$. Then X_1 and X_2 are closed path-connected subsets of $\mathbf{R}^2$ such that $X_1 \cup X_2 = \mathbf{R}^2$ and $X_1 \cap X_2$ consists of two path components. Therefore there is no homomorphism $\tilde H_1(X_1 \cup X_2) \to \tilde H_0(X_1 \cap X_2)$ which will make the sequence

$$\tilde H_1(X_1 \cup X_2) \to \tilde H_0(X_1 \cap X_2) \to \tilde H_0(X_1) \oplus \tilde H_0(X_2)$$

exact at $\tilde H_0(X_1 \cap X_2)$ [the ends are both trivial, but $\tilde H_0(X_1 \cap X_2) \neq 0$].

We can, however, develop a Mayer-Vietoris sequence in singular homology for certain subsets X_1 and X_2 of a topological space. Let X_1 and X_2 be subsets of some space. $\{X_1,X_2\}$ is said to be an *excisive couple of subsets* if the inclusion chain map $\Delta(X_1) + \Delta(X_2) \subset \Delta(X_1 \cup X_2)$ induces an isomorphism of homology. Our next result follows from theorem 4.4.14.

3 THEOREM *If* $X_1 \cup X_2 = \operatorname{int}_{X_1 \cup X_2} X_1 \cup \operatorname{int}_{X_1 \cup X_2} X_2$, *then* $\{X_1,X_2\}$ *is an excisive couple.* $\blacksquare$

In particular, if $A \subset X$, then $\{X,A\}$ is always an excisive couple. The relation between an excisive couple $\{X_1,X_2\}$ and excision maps is expressed as follows.

4 THEOREM $\{X_1,X_2\}$ *is an excisive couple if and only if the excision map* $(X_1,X_1 \cap X_2) \subset (X_1 \cup X_2,X_2)$ *induces an isomorphism of singular homology.*

PROOF We have a commutative diagram of chain maps induced by inclusions

$$\begin{array}{ccc} \Delta(X_1)/\Delta(X_1 \cap X_2) & \xrightarrow{\Delta(j)} & \Delta(X_1 \cup X_2)/\Delta(X_2) \\ {\scriptstyle i}\searrow & & \nearrow{\scriptstyle i'} \\ & [\Delta(X_1) + \Delta(X_2)]/\Delta(X_2) & \end{array}$$

where j is the excision map $j: (X_1, X_1 \cap X_2) \subset (X_1 \cup X_2, X_2)$. By the Noether isomorphism theorem, i is an isomorphism; therefore $j_* = i'_* i_*$ is an isomorphism if and only if i'_* is an isomorphism. Using the exactness of the homology sequence of a pair and the five lemma, i'_* is an isomorphism if and only if the inclusion map $\Delta(X_1) + \Delta(X_2) \subset \Delta(X_1 \cup X_2)$ induces an isomorphism of homology, which is by definition equivalent to the condition that $\{X_1,X_2\}$ be an excisive couple. ■

This yields the following *excision property for singular theory.*

5 COROLLARY *Let $U \subset A \subset X$ be such that $\bar U \subset \operatorname{int} A$. Then the excision map $(X - U, A - U) \subset (X,A)$ induces an isomorphism of singular homology.*

PROOF The hypothesis $\bar U \subset \operatorname{int} A$ implies $\operatorname{int}(X - U) \supset X - \bar U \supset X - \operatorname{int} A$. By theorem 3, $\{A, X - U\}$ is an excisive couple, and the result follows from this and from theorem 4. ■

For any subsets X_1 and X_2 of a space, $\Delta(X_1 \cap X_2) = \Delta(X_1) \cap \Delta(X_2)$, and there is a short exact sequence of singular chain complexes

$$0 \to \Delta(X_1 \cap X_2) \xrightarrow{i} \Delta(X_1) \oplus \Delta(X_2) \xrightarrow{j} \Delta(X_1) + \Delta(X_2) \to 0$$

This yields an exact sequence

$$\cdots \xrightarrow{\partial_*} H_q(X_1 \cap X_2) \xrightarrow{i_*} H_q(X_1) \oplus H_q(X_2) \xrightarrow{j_*} H_q(\Delta(X_1) + \Delta(X_2)) \xrightarrow{\partial_*} H_{q-1}(X_1 \cap X_2) \to \cdots$$

If $\{X_1,X_2\}$ is an excisive couple, the group $H_q(\Delta(X_1) + \Delta(X_2))$ can be replaced by the group $H_q(X_1 \cup X_2)$, and the resulting exact sequence is

$$\cdots \xrightarrow{\partial_*} H_q(X_1 \cap X_2) \xrightarrow{i_*} H_q(X_1) \oplus H_q(X_2) \xrightarrow{j_*} H_q(X_1 \cup X_2) \xrightarrow{\partial_*} H_{q-1}(X_1 \cap X_2) \to \cdots$$

where $i_*(z) = (i_{1*} z, -i_{2*} z)$ and $j_*(z_1,z_2) = j_{1*} z_1 + j_{2*} z_2$ for $z \in H(X_1 \cap X_2)$, $z_1 \in H(X_1)$, and $z_2 \in H(X_2)$. This is the *Mayer-Vietoris sequence of singular theory of an excisive couple* $\{X_1,X_2\}$. Similarly, if $X_1 \cap X_2 \neq \varnothing$, there is a reduced Mayer-Vietoris sequence of $\{X_1,X_2\}$.

If (X_1,A_1) and (X_2,A_2) are pairs in a space X, we say that $\{(X_1,A_1), (X_2,A_2)\}$ is an *excisive couple of pairs* if $\{X_1,X_2\}$ and $\{A_1,A_2\}$ are both excisive couples of subsets. In this case it follows from the five lemma that the map induced by inclusion

$$[\Delta(X_1) + \Delta(X_2)]/[\Delta(A_1) + \Delta(A_2)] \to [\Delta(X_1 \cup X_2)]/[\Delta(A_1 \cup A_2)]$$

induces an isomorphism of homology. Hence, if $\{(X_1,A_1), (X_2,A_2)\}$ is an excisive couple of pairs, there is an exact sequence

$$\cdots \xrightarrow{\partial_*} H_q(X_1 \cap X_2, A_1 \cap A_2) \xrightarrow{i_*} H_q(X_1,A_1) \oplus H_q(X_2,A_2) \xrightarrow{j_*} H_q(X_1 \cup X_2, A_1 \cup A_2) \xrightarrow{\partial_*} \cdots$$

called the *relative Mayer-Vietoris sequence of* $\{(X_1,A_1), (X_2,A_2)\}$.

The relative Mayer-Vietoris sequence specializes to the exact sequence of a triple (or a pair). In fact, given a triple (X,A,B), $\{(X,B), (A,A)\}$ is always an excisive couple of pairs, and the relative Mayer-Vietoris sequence of $\{(X,B), (A,A)\}$ is the homology sequence of the triple (X,A,B).

We use the Mayer-Vietoris sequence to compute the singular homology of a sphere.

6 THEOREM *For* $n \geq 0$

$$\tilde{H}_q(S^n) \approx \begin{cases} 0 & q \neq n \\ \mathbf{Z} & q = n \end{cases}$$

PROOF Let p and p' be distinct points of S^n. Because $S^n - p$ and $S^n - p'$ are contractible (each being homeomorphic to $\mathbf{R}^n$), $\tilde{H}(S^n - p) = 0 = \tilde{H}(S^n - p')$. Since $S^n - p$ and $S^n - p'$ are open subsets of S^n, it follows from theorem 3 that $\{S^n - p, S^n - p'\}$ is an excisive couple. From the exactness of the corresponding Mayer-Vietoris sequence, it follows that

$$\partial_*: \tilde{H}_q(S^n) \approx \tilde{H}_{q-1}(S^n - (p \cup p'))$$

Because $S^n - (p \cup p')$ has the same homotopy type as S^{n-1}, there is an isomorphism $\tilde{H}_{q-1}(S^n - (p \cup p')) \approx \tilde{H}_{q-1}(S^{n-1})$, and the result follows by induction and the trivial verification that for $n = 0$ the theorem is valid. ■

We now show that a couple consisting of polyhedral subsets of a polyhedron is excisive.

7 LEMMA *Let* K_1 *and* K_2 *be subcomplexes of a simplicial complex* K. *Then* $\{|K_1|,|K_2|\}$ *is an excisive couple.*

PROOF Let V be a neighborhood of $|K_1 \cap K_2|$ in $|K_1|$ having $|K_1 \cap K_2|$ as a strong deformation retract (such a V exists, by corollary 3.3.11). There is a commutative diagram

$$\begin{array}{ccccccc}
\cdots \to & H_q(|K_1 \cap K_2|) & \to & H_q(|K_1|) & \to & H_q(|K_1|, |K_1 \cap K_2|) & \to \cdots \\
& \downarrow i_* & & \downarrow 1 & & \downarrow j_* & \\
\cdots \to & H_q(V) & \to & H_q(|K_1|) & \to & H_q(|K_1|,V) & \to \cdots
\end{array}$$

Because $i: |K_1 \cap K_2| \subset V$ is a homotopy equivalence, $i_*: H(|K_1 \cap K_2|) \approx H(V)$. By the five lemma, $j_*: H(|K_1|, |K_1 \cap K_2|) \approx H(|K_1|,V)$.

Also, $V \cup |K_2|$ is a neighborhood of $|K_2|$ in $|K_1 \cup K_2|$ having $|K_2|$ as a strong deformation retract. Therefore a similar proof shows that

$$j'_*: H(|K_1 \cup K_2|, |K_2|) \approx H(|K_1 \cup K_2|, V \cup |K_2|)$$

By theorem 4, $\{|K_1|, |K_2|\}$ is an excisive couple if and only if the excision map $(|K_1|, |K_1 \cap K_2|) \subset (|K_1 \cup K_2|, |K_2|)$ induces an isomorphism of homology. In view of the isomorphisms j_* and j'_*, this will be so if and only if the excision map $(|K_1|,V) \subset (|K_1 \cup K_2|, V \cup |K_2|)$ induces an isomorphism of homology. Again by theorem 4, this is equivalent to the condition that $\{|K_1|, V \cup |K_2|\}$ be an excisive couple. This is so by theorem 3, since $|K_2| \subset \operatorname{int}(V \cup |K_2|)$ and $|K_1| - |K_2| \subset \operatorname{int}|K_1|$. ∎

8 THEOREM *For any simplicial pair (K,L) the natural transformation ν induces an isomorphism of the ordered homology sequence of (K,L) onto the singular homology sequence of $(|K|,|L|)$.*

PROOF It suffices to prove that for any simplicial complex K, $\nu_*: H(\Delta(K)) \approx H(|K|)$, because the theorem will follow from this and the five lemma. We prove this first for finite simplicial complexes by induction on the number of simplexes. If K contains one simplex, then $K = \bar{s}$, where s is a 0-simplex, and the result follows from corollary 4.4.2.

Assume the result inductively for simplicial complexes with fewer than m simplexes, where $m > 1$, and let K contain exactly m simplexes. Let s be a simplex of K of maximum dimension and let L be the subcomplex of K consisting of all simplexes other than s. Then $K = L \cup \bar{s}$ and $\dot{s} = L \cap \bar{s}$. Because L has exactly $m - 1$ simplexes, ν_* is an isomorphism $H(\Delta(L)) \approx H(|L|)$ and an isomorphism $H(\Delta(\dot{s})) \approx H(|\dot{s}|)$. By corollary 4.4.2, $\nu_*: H(\Delta(\bar{s})) \approx H(|\bar{s}|)$. By the exactness of the ordered Mayer-Vietoris sequence of L and $\bar{s}$ and the Mayer-Vietoris sequence of singular theory for $|L|$ and $|\bar{s}|$ (which exists, by lemma 7), it follows from the five lemma that $\nu_*: H(\Delta(K)) \approx H(|K|)$.

For infinite simplicial complexes K let $\{K_\alpha\}$ be the family of finite subcomplexes of K directed by inclusion. It follows from theorem 4.3.11 that $H(\Delta(K)) \approx \lim_{\rightarrow} H(\Delta(K_\alpha))$ and from theorem 4.4.6 that $H(|K|) \approx \lim_{\rightarrow} H(|K_\alpha|)$. The theorem now holds for K because ν_* is natural. ∎

We show next that for free chain complexes a chain map is a chain equivalence if and only if it induces an isomorphism in homology. First we establish an exact sequence containing the homomorphism induced by a chain map.

9 LEMMA *Let $\tau: C \to C'$ be a chain map and let $\bar{C}$ be the mapping cone of τ. There is an exact sequence*

$$\cdots \to H_{q+1}(\bar{C}) \to H_q(C) \xrightarrow{\tau_*} H_q(C') \to H_q(\bar{C}) \to \cdots$$

PROOF Let $\alpha: C' \to \bar{C}$ be the chain map defined by $\alpha(c) = (0,c)$. Then α imbeds C' as a subcomplex of $\bar{C}$ and the quotient complex $\bar{C}/C'$ is such that $(\bar{C}/C')_q \approx C_{q-1}$; the boundary operator of $\bar{C}/C'$ corresponds to the negative of the boundary operator of C under this isomorphism. The desired exact sequence is then obtained from the exact homology sequence of the short exact sequence of chain complexes

$$0 \to C' \xrightarrow{\alpha} \bar{C} \to \bar{C}/C' \to 0$$

by replacing $H_q(\bar{C}/C')$ by $H_{q-1}(C)$ and verifying that the connecting homomorphism $\partial_*: H_{q+1}(\bar{C}/C') \to H_q(C')$ corresponds to $\tau_*: H_q(C) \to H_q(C')$. ∎

10 THEOREM *If C and C′ are free chain complexes, a chain map $\tau: C \to C'$ is a chain equivalence if and only if $\tau_*: H(C) \approx H(C')$.*

PROOF By corollary 4.2.11, τ is a chain equivalence if and only if $\bar{C}$ is acyclic. By lemma 9 and corollary 4.5.5, $\bar{C}$ is acyclic if and only if $\tau_*: H(C) \approx H(C')$. ∎

Because $\Delta(K)/\Delta(L)$ and $\Delta(|K|)/\Delta(|L|)$ are free chain complexes, we have the following result.

11 COROLLARY *For any simplicial pair (K,L), v is a chain equivalence of $\Delta(K)/\Delta(L)$ with $\Delta(|K|)/\Delta(|L|)$.* ∎

If $\varphi: K_1 \to K_2$ is a simplicial map, there is a commutative diagram

$$\begin{array}{ccccc} H(K_1) & \overset{\mu_*}{\underset{\approx}{\longleftarrow}} & H(\Delta(K_1)) & \overset{\nu_*}{\underset{\approx}{\longrightarrow}} & H(|K_1|) \\ \varphi_* \downarrow & & \downarrow \Delta(\varphi)_* & & \downarrow |\varphi|_* \\ H(K_2) & \overset{\mu_*}{\underset{\approx}{\longleftarrow}} & H(\Delta(K_2)) & \overset{\nu_*}{\underset{\approx}{\longrightarrow}} & H(|K_2|) \end{array}$$

In particular, if K' is a subdivision of K and $\varphi: K' \to K$ is a simplicial approximation to the identity $|K'| \subset |K|$, then

$$|\varphi| \simeq 1_{|K|} \qquad \text{and} \qquad |\varphi|_* = 1_{H(|K|)}$$

From the commutativity of the above diagram we obtain our next result.

12 THEOREM *Let K' be a subdivision of K and let $\varphi: K' \to K$ be a simplicial approximation to the identity map $|K'| \subset |K|$. Then*

$$\varphi_*: H(K') \approx H(K) \quad ∎$$

By theorem 10, $C(\varphi): C(K') \to C(K)$ is a chain equivalence. It will be useful to construct a chain map $C(K) \to C(K')$ which is a chain homotopy inverse of $C(\varphi)$. If K' is a subdivision of K, an augmentation-preserving chain map $\tau: C(K) \to C(K')$ is called a *subdivision chain map* if $\tau: C(L) \subset C(K' \mid L)$ for every subcomplex $L \subset K$ [that is, if τ is a natural chain map from C to $C(K' \mid \cdot)$ on $\mathcal{C}(K)$].

13 THEOREM *If K' is a subdivision of K, there exist subdivision chain maps $\tau: C(K) \to C(K')$. If $\varphi: K' \to K$ is a simplicial approximation to the identity $|K'| \subset |K|$, then $\tau_* = \varphi_*^{-1}: H(K) \approx H(K')$.*

PROOF If s is any simplex of K, then $\tilde{C}(K' \mid \bar{s})$ is acyclic [because $\tilde{H}(K' \mid \bar{s}) \approx \tilde{H}(|\bar{s}|) = 0$]. Hence, on the category $\mathcal{C}(K)$ of subcomplexes of K with models $\mathfrak{M}(K) = \{\bar{s} \mid s \in K\}$, the functor C is free and $C(K' \mid \cdot)$ is acyclic. It follows from theorem 4.3.3 that there exist natural chain maps τ from C to $C(K' \mid \cdot)$ preserving augmentation.

If τ is any subdivision chain map and $\varphi: K' \to K$ is a simplicial approximation to the identity map $|K'| \subset K$, the composite

$$C(\varphi)\tau\colon C(K) \to C(K)$$

is a natural chain map over $\mathcal{C}(K)$ from C to C preserving augmentation. Since C is free and acyclic with models $\mathfrak{M}(K)$, it follows from theorem 4.3.3 that $C(\varphi)\tau \simeq 1_{C(K)}$. Therefore $\varphi_* \tau_* = 1_{H(K)}$. Since, by theorem 12, φ_* is an isomorphism, $\tau_* = \varphi_*^{-1}$. ∎

7 SOME APPLICATIONS OF HOMOLOGY

In this section we use homology for some of the applications mentioned earlier. We shall show that euclidean spaces of different dimensions are not homeomorphic, and also that S^n is not a retract of E^{n+1} (which is easily seen to be equivalent to the Brouwer fixed-point theorem). This leads to the general consideration of fixed points of maps, and we prove the Lefschetz fixed-point theorem. Finally, we shall consider separation properties of the sphere. Proofs are given of Brouwer's generalization of the Jordan curve theorem and of the invariance of domain.

1 THEOREM *If $n \neq m$, S^n and S^m are not of the same homotopy type.*

PROOF By theorem 4.6.6, $\tilde{H}_n(S^n) \neq 0$ and $\tilde{H}_n(S^m) = 0$. ∎

2 COROLLARY *If $n \neq m$, $\mathbf{R}^n$ and $\mathbf{R}^m$ are not homeomorphic.*

PROOF If $\mathbf{R}^n$ and $\mathbf{R}^m$ were homeomorphic, their one-point compactifications S^n and S^m would also be homeomorphic, in contradiction to theorem 1. ∎

In corollary 2 both $\mathbf{R}^n$ and $\mathbf{R}^m$ are contractible. Therefore they have the same homotopy type and cannot be distinguished by their homology groups. To distinguish them it was necessary to consider associated spaces having nonisomorphic homology. We chose to consider their one-point compactifications, but another proof could have been based on the fact that $\mathbf{R}^n$ minus a point has the same homotopy type as S^{n-1}.

These two results are applications of homology to the problem of classifying spaces up to topological equivalence. Our next application is to an extension problem.

3 LEMMA *Let (X,A) be a pair such that A is a retract of X. Then*

$$H(X) \approx H(A) \oplus H(X,A)$$

PROOF Given $i\colon A \subset X$ and $j\colon (X,\varnothing) \subset (X,A)$ and a retraction $r\colon X \to A$, then $ri = 1_A$. Therefore $r_* i_* = 1_{H(A)}$ and i_* is a monomorphism of $H(A)$ onto a direct summand of $H(X)$. The other summand is the kernel of r_*. From the exactness of the homology sequence of (X,A)

$$\cdots \to H_q(X,A) \xrightarrow{\partial_*} H_{q-1}(A) \xrightarrow{i_*} H_{q-1}(X) \xrightarrow{j_*} H_{q-1}(X,A) \xrightarrow{\partial_*} \cdots$$

because ker $i_* = 0$, ∂_* is the trivial map. Therefore j_* is an epimorphism. Since ker $j_* = \text{im } i_*$, j_* induces an isomorphism of ker r_* onto $H(X,A)$. ∎

Note that lemma 3 is still valid if A is a weak retract of X.

4 COROLLARY *For $n \geq 0$, S^n is not a retract of E^{n+1}.*

PROOF By theorem 4.6.6, $\tilde{H}_n(S^n) \neq 0$, but because E^{n+1} is contractible, $\tilde{H}_n(E^{n+1}) = 0$. Therefore $H(S^n)$ is not isomorphic to a direct summand of $H(E^{n+1})$. ∎

This implies the following *Brouwer fixed-point theorem.*

5 THEOREM *For $n \geq 0$ every continuous map from E^n to itself has a fixed point.*

PROOF For $n = 0$ there is nothing to prove. For $n > 0$ let $f\colon E^n \to E^n$ be continuous. If f has no fixed point, define a map $g\colon E^n \to S^{n-1}$ by $g(x)$ equal to the unique point of S^{n-1} on the ray from $f(x)$ to x, as shown in the figure.

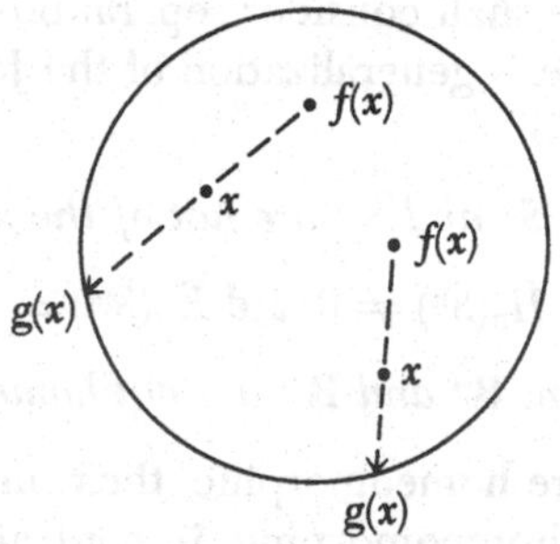

Then g is a retraction from E^n to S^{n-1}, in contradiction to corollary 4. ∎

We have, in fact, proved that corollary 4 implies theorem 5. The converse is also true, for if $r\colon E^{n+1} \to S^n$ were a retraction, the map $f\colon E^{n+1} \to E^{n+1}$ defined by $f(x) = -r(x)$ would have no fixed points.

There is an interesting generalization of theorem 5 which contains a criterion for showing that a certain map from X to itself has a fixed point even if not every map of X to itself has fixed points. This generalization also illustrates another type of application of homology in that it is based on an algebraic count of the number of fixed points, the algebraic count being formulated in homological terms. This type of application of homology occurs frequently. Generally it involves a set of singularities of X of a certain type (for example, the set of fixed points of a map $X \to X$, the set of discontinuties of a function $X \to Y$, the set of self-intersections of a local homeomorphism $X \to \mathbf{R}^n$, etc.) and measures the singular set by means of a homology class associated to it.

Let C be a finitely generated graded group and let $h\colon C \to C$ be an endomorphism of C of degree 0. The *Lefschetz number* $\lambda(h)$ is defined by the formula

$$\lambda(h) = \Sigma(-1)^q Tr(h_q)$$

where $h_q\colon C_q \to C_q$ is the endomorphism defined by h in degree q. The following *Hopf trace formula* equates the Lefschetz numbers of a chain map and its induced homology homomorphism.

6 THEOREM *Let C be a finitely generated chain complex and let $\tau\colon C \to C$ be a chain map. Then*

$$\lambda(\tau) = \lambda(\tau_*)$$

PROOF The proof is similar to the proof of the corresponding statement about the Euler characteristic (theorem 4.3.14), the Euler characteristic being the Lefschetz number of the identity map, with theorem 4.13 of the Introduction used in place of theorem 4.12. Details are left to the reader. ∎

Let $f\colon X \to X$ be a map, where X has finitely generated homology. The *Lefschetz number of f*, denoted by $\lambda(f)$, is defined to be the Lefschetz number of the homomorphism $f_*\colon H(X) \to H(X)$ induced by f. It counts the algebraic number of fixed homology classes of f_*. The following *Lefschetz fixed-point theorem* shows that $\lambda(f) \neq 0$ is a sufficient condition for f to have a fixed point.

7 THEOREM *Let X be a compact polyhedron and let $f\colon X \to X$ be a map. If $\lambda(f) \neq 0$, then f has a fixed point.*

PROOF We assume that f has no fixed point and prove $\lambda(f) = 0$. Without loss of generality, we may assume $X = |L|$ for some finite simplicial complex L. Because $|L|$ is a compact metric space, if f has no fixed point, there is $a > 0$ such that $d(\alpha,f(\alpha)) \geq a$ for all $\alpha \in |L|$. Let K be a subdivision of L with mesh $K < a/3$ and let K' be a subdivision of K for which there exists a simplicial map $\varphi\colon K' \to K$ which is a simplicial approximation to $f\colon |K| \to |K|$. Since $|\varphi|(\alpha)$ and $f(\alpha)$ belong to some simplex of K, $d(|\varphi|(\alpha),f(\alpha)) < a/3$ for $\alpha \in |K|$. If s is any simplex of K, $|s|$ is disjoint from $|\varphi|(|s|)$, for if $\alpha \in |s|$ is equal to $|\varphi|(\beta)$ for $\beta \in |s|$, then

$$d(\beta,f(\beta)) \leq d(\beta,\alpha) + d(|\varphi|(\beta),f(\beta)) < 2a/3$$

in contradiction to the choice of a.

Let $\tau\colon C(K) \to C(K')$ be a subdivision chain map (which exists, by theorem 4.6.13). Then $C(\varphi)\tau\colon C(K) \to C(K)$ is a chain map. If σ is an oriented q-simplex on a q-simplex s of K, then $C(\varphi)\tau(\sigma)$ is a q-chain on the largest subcomplex of K disjoint from s. Therefore $C(\varphi)\tau(\sigma)$ is a q-chain having coefficient 0 on σ. Since this is so for every σ, all the coefficients summed in forming $Tr(C(\varphi)\tau)_q$ are zero and $Tr((C(\varphi)\tau)_q) = 0$ for all q, which implies $\lambda(C(\varphi)\tau) = 0$. By theorem 6, $\lambda((C(\varphi)\tau)_*) = 0$. Let $\varphi'\colon K' \to K$ be a simplicial approximation to the identity map $|K'| \subset |K|$. There is a commutative diagram

$$\begin{array}{ccccc}
H(K) & \xleftarrow{\varphi'_*} & H(K') & \xrightarrow{\varphi_*} & H(K) \\
\uparrow\approx & & \uparrow\approx & & \uparrow\approx \\
H(\Delta(K)) & \xleftarrow{\Delta(\varphi')_*} & H(\Delta(K')) & \xrightarrow{\Delta(\varphi)_*} & H(\Delta(K)) \\
\downarrow\approx & & \downarrow\approx & & \downarrow\approx \\
H(|K|) & \xleftarrow{|\varphi'|_* = 1} & H(|K|) & \xrightarrow{|\varphi|_* = f_*} & H(|K|)
\end{array}$$

from which it follows that

$$\lambda(f_*) = \lambda(|\varphi|_*(|\varphi'|_*)^{-1}) = \lambda(\varphi_*(\varphi'_*)^{-1})$$

By theorem 4.6.13, $(\varphi'_*)^{-1} = \tau_*$ and $\lambda(\varphi_*(\varphi'_*)^{-1}) = \lambda(\varphi_*\tau_*) = \lambda([C(\varphi)\tau]_*)$. Therefore $\lambda(f) = 0$. ∎

This yields the following generalization of the Brouwer fixed-point theorem.

8 COROLLARY *Every continuous map from a compact contractible polyhedron to itself has a fixed point.*

PROOF If X is contractible, $\tilde{H}(X) = 0$, and for any $f\colon X \to X$, $\lambda(f) = 1$ [because f_* is the identity map on $H_0(X) \approx \mathbf{Z}$]. ∎

This result is false for noncompact polyhedra. In fact, $\mathbf{R}$ is a contractible polyhedron and any translation different from $1_{\mathbf{R}}$ fails to have a fixed point.

Given a continuous map $f\colon S^n \to S^n$, the *degree of* f is the unique integer $\deg f$ such that

$$f_*(z) = (\deg f)z \qquad z \in \tilde{H}_n(S^n)$$

The following fact is obvious.

9 *For any map $f\colon S^n \to S^n$, $\lambda(f) = 1 + (-1)^n \deg f$.* ∎

Since the antipodal map $S^n \to S^n$ has no fixed points, the next result follows from theorem 7 and statement 9.

10 COROLLARY *The antipodal map of S^n has degree $(-1)^{n+1}$.* ∎

11 COROLLARY *If n is even, there is no continuous map $f\colon S^n \to S^n$ such that x and $f(x)$ are orthogonal for all $x \in S^n$.*

PROOF Assume that such a map exists. Then a homotopy $F\colon f \simeq 1_{S^n}$ is defined by

$$F(x,t) = \frac{(1-t)f(x) + tx}{\|(1-t)f(x) + tx\|}$$

This is well-defined, because the condition that x and $f(x)$ be orthogonal implies $\|(1-t)f(x) + tx\|^2 = (1-t)^2 + t^2 \neq 0$ for $0 \le t \le 1$. Since $f \simeq 1_{S^n}$, $\lambda(f) = \lambda(1_{S^n}) = 1 + (-1)^n \neq 0$. Hence, by theorem 7, f must have a fixed point, in contradiction to the orthogonality of x and $f(x)$ for all x. ∎

This last result is equivalent to the statement that an even-dimensional sphere S^n has no continuous tangent vector field which is nonzero everywhere on S^n. For odd n such vector fields do exist because the map $f\colon S^{2m-1} \to S^{2m-1}$ defined by

$$f(x_1, \ldots, x_{2m}) = (-x_2, x_1, \ldots, -x_{2m}, x_{2m-1})$$

is continuous and has the property that x and $f(x)$ are orthogonal for all x.

Instead of considering vector fields, we consider one-parameter groups of homeomorphisms. A *flow* on X is a continuous map

$$\psi\colon \mathbf{R} \times X \to X$$

such that

$$\begin{aligned} &(a)\ \psi(t_1 + t_2, x) = \psi(t_1, \psi(t_2,x)) && t_1, t_2 \in \mathbf{R};\ x \in X \\ &(b)\ \psi(0,x) = x && x \in X \end{aligned}$$

For $t \in \mathbf{R}$ let $\psi_t\colon X \to X$ be defined by $\psi_t(x) = \psi(t,x)$. Then (*a*) and (*b*) imply $\psi_{-t} = (\psi_t)^{-1}$, and so ψ_t is a homeomorphism of X for all $t \in \mathbf{R}$. A *fixed point of the flow* is a point $x_0 \in X$ such that $\psi(t,x_0) = x_0$ for all $t \in \mathbf{R}$.

12 THEOREM *If X is a compact polyhedron with $\chi(X) \neq 0$, then any flow on X has a fixed point.*

PROOF Each ψ_t is homotopic to 1_X [by the homotopy $F\colon X \times I \to X$ defined by $F(x,t') = \psi((1 - t')t, x)$]. Therefore

$$\lambda(\psi_t) = \lambda(1_X) = \chi(X) \neq 0$$

Hence, by theorem 7, each ψ_t has fixed points. For $n \geq 1$ let A_n be the closed subset of X consisting of the fixed points of $\psi_{1/2^n}$. Then $A_{n+1} \subset A_n$, and $\{A_n\}$ is a decreasing sequence of nonempty closed subsets of the compact space X. Let $F = \bigcap A_n$. Then F is nonempty, and any point of F is fixed under ψ_t for all t of the form $1/2^n$ for $n \geq 1$. This implies that each point of F is fixed under ψ_t for all dyadic rationals $t = m/2^n$. Since the dyadic rationals are dense in $\mathbf{R}$, each point of F is fixed under ψ_t for all t. ∎

We now turn our attention to separation properties of the sphere.

13 LEMMA *If $A \subset S^n$ is homeomorphic to I^k for $0 \leq k \leq n$, then $\tilde{H}(S^n - A) = 0$.*

PROOF We prove this by induction on k. If $k = 0$, then A is a point and $S^n - A$ is homeomorphic to $\mathbf{R}^n$. Therefore $\tilde{H}(S^n - A) = 0$.

Assume the result for $k < m$, where $m \geq 1$, and let A be homeomorphic to I^m. Regard A as being homeomorphic to $B \times I$, where B is homeomorphic to I^{m-1}, by a homeomorphism $h\colon B \times I \to A$. Let $A' = h(B \times [0,\frac{1}{2}])$ and $A'' = h(B \times [\frac{1}{2},1])$. Then $A = A' \cup A''$ and $A' \cap A''$ is homeomorphic to $B \times \frac{1}{2}$. By the inductive assumption, $\tilde{H}(S^n - (A' \cap A'')) = 0$. Because $S^n - A'$ and $S^n - A''$ are open sets, they are excisive and from the exactness of the corresponding reduced Mayer-Vietoris sequence

$$i_*\colon \tilde{H}_q(S^n - A) \approx \tilde{H}_q(S^n - A') \oplus \tilde{H}_q(S^n - A'')$$

If $z \in \tilde{H}_q(S^n - A)$ is nonzero, then either $i'_* z \neq 0$ in $\tilde{H}_q(S^n - A')$ or $i''_* z \neq 0$ in $\tilde{H}_q(S^n - A'')$, where $i'\colon S^n - A \subset S^n - A'$ and $i''\colon S^n - A \subset S^n - A''$. Assume $i'_* z \neq 0$. We repeat the argument for A' and thus obtain a sequence of sets

$$A \supset A_1 \supset A_2 \supset \cdots$$

such that

(a) The inclusion $S^n - A \subset S^n - A_j$ maps z into a nonzero element of $\tilde{H}_q(S^n - A_j)$.
(b) $\cap A_i$ is homeomorphic to I^{m-1}

Because every compact subset of $S^n - \cap A_i$ is contained in $S^n - A_j$ for some j, it follows from theorem 4.4.6 that $\tilde{H}_q(S^n - \cap A_i) \approx \lim_{\rightarrow} \{\tilde{H}_q(S^n - A_j)\}$. This is a contradiction because, by condition (a), the element z determines a nonzero element of $\lim_{\rightarrow} \{\tilde{H}_q(S^n - A_j)\}$, but by condition (b) and the inductive assumption, $\tilde{H}_q(S^n - \cap A_i) = 0$. ∎

14 COROLLARY *Let B be a subset of S^n which is homeomorphic to S^k for $0 \le k \le n - 1$. Then*

$$\tilde{H}_q(S^n - B) \approx \begin{cases} 0 & q \neq n - k - 1 \\ \mathbf{Z} & q = n - k - 1 \end{cases}$$

PROOF We use induction on k. If $k = 0$, then B consists of two points and $S^n - B$ has the same homotopy type as S^{n-1}. Therefore

$$\tilde{H}_q(S^n - B) \approx \begin{cases} 0 & q \neq n - 1 \\ \mathbf{Z} & q = n - 1 \end{cases}$$

If $k \geq 1$, set $B = A_1 \cup A_2$, where A_1 and A_2 are closed hemispheres of S^k and assume the result valid for $k - 1$. Then A_1 and A_2 are homeomorphic to I^k and $A_1 \cap A_2$ is homeomorphic to S^{k-1}. Because $S^n - A_1$ and $S^n - A_2$ are open, $\{S^n - A_1, S^n - A_2\}$ is an excisive couple, and there is an exact reduced Mayer-Vietoris sequence

$$\to \tilde{H}_{q+1}(S^n - A_1) \oplus \tilde{H}_{q+1}(S^n - A_2) \to \tilde{H}_{q+1}(S^n - (A_1 \cap A_2)) \to$$
$$\tilde{H}_q(S^n - B) \to \tilde{H}_q(S^n - A_1) \oplus \tilde{H}_q(S^n - A_2) \to$$

By lemma 13, the groups at the ends vanish. The result then follows from the inductive assumption. ∎

For the special case of an $(n - 1)$-sphere imbedded in S^n, we obtain the following *Jordan-Brouwer separation theorem.*

15 THEOREM *An $(n - 1)$-sphere imbedded in S^n separates S^n into two components of which it is their common boundary.*

PROOF If $B \subset S^n$ is homeomorphic to S^{n-1}, then $\tilde{H}_0(S^n - B) \approx \mathbf{Z}$, by corollary 14. Therefore $S^n - B$ consists of two path components. Since $S^n - B$ is an open subset of S^n, it is locally path connected and its path components U and V, say, are its components.

Clearly, B contains the boundary of U and of V. To prove $B \subset \bar{U} \cap \bar{V}$, let $x \in B$ and let N be a neighborhood of x in S^n. Let $A \subset B \cap N$ be a subset such that $B - A$, is homeomorphic to I^{n-1}. Then $\tilde{H}(S^n - (B - A)) = 0$, by lemma 13, so $S^n - (B - A)$ is path connected. If $p \in U$ and $q \in V$, there is a path ω in $S^n - (B - A)$ from p to q. Because p and q are in different path components of $S^n - B$, ω meets A. Therefore A contains a point of $\bar{U}$ and a point of $\bar{V}$. Hence N meets $\bar{U}$ and $\bar{V}$, and $x \in \bar{U} \cap \bar{V}$. ∎

A related result is the following *Brouwer theorem on the invariance of domain.*

16 THEOREM *If U and V are homeomorphic subsets of S^n and U is open in S^n, then V is open in S^n.*

PROOF Let $h: U \to V$ be a homeomorphism and let $h(x) = y$. Let A be a neighborhood of x in U that is homeomorphic to I^n and with boundary B homeomorphic to S^{n-1}. Let $A' = h(A) \subset V$ and let $B' = h(B)$. By lemma 13, $S^n - A'$ is connected, and by theorem 15, $S^n - B'$ has two components. Because

$$S^n - B' = (S^n - A') \cup (A' - B')$$

and $S^n - A'$ and $A' - B'$ are connected, they are the components of $S^n - B'$. Therefore $A' - B'$ is an open subset of S^n. Since $y \in A' - B' \subset V$ and y was arbitrary, V is open in S^n. ∎

8 AXIOMATIC CHARACTERIZATION OF HOMOLOGY

A simple set of axioms characterizing homology on the class of compact polyhedral pairs has been given by Eilenberg and Steenrod[1]. This section describes the axiom system and related concepts. For compact polyhedral pairs, the axioms are categorical (that is, two theories satisfying them are isomorphic). Thus the axioms are basic theorems from which other properties of homology theories can be deduced. In many cases, proofs based on the axioms are simpler and more elegant than proofs which refer back to the original definition of the homology theory.

To formulate the axioms it is usual to start with a suitable category of topological pairs and maps (called "admissible categories" by Eilenberg and Steenrod). We shall not define these categories. The category of all topological pairs is such a category, and so are its full subcategories defined by the polyhedral pairs and defined by the compact polyhedral pairs. For our purposes we shall always regard a homology theory as defined on the category of all topological pairs, and we identify a space X with the pair $(X, \varnothing)$.

A *homology theory* H and ∂ consists of

(*a*) A covariant functor H from the category of topological pairs and maps to the category of graded abelian groups and homomorphisms of degree 0 [that is, $H(X,A) = \{H_q(X,A)\}$]

[1] See S. Eilenberg and N. E. Steenrod, "Foundations of Algebraic Topology," Princeton University Press, Princeton, N.J., 1952.

(*b*) A natural transformation ∂ of degree -1 from the functor H on (X,A) to the functor H on $(A, \varnothing)$ [that is, $\partial(X,A) = \{\partial_q(X,A): H_q(X,A) \to H_{q-1}(A)\}$].

These satisfy the following axioms.

1 HOMOTOPY AXIOM *If $f_0, f_1: (X,A) \to (Y,B)$ are homotopic, then*

$$H(f_0) = H(f_1): H(X,A) \to H(Y,B)$$

2 EXACTNESS AXIOM *For any pair (X,A) with inclusion maps $i: A \subset X$ and $j: X \subset (X,A)$ there is an exact sequence*

$$\cdots \xrightarrow{\partial_{q+1}(X,A)} H_q(A) \xrightarrow{H_q(i)} H_q(X) \xrightarrow{H_q(j)} H_q(X,A) \xrightarrow{\partial_q(X,A)} H_{q-1}(A) \xrightarrow{H_{q-1}(i)} \cdots$$

3 EXCISION AXIOM *For any pair (X,A), if U is an open subset of X such that $\bar{U} \subset \operatorname{int} A$, then the excision map $j: (X - U, A - U) \subset (X,A)$ induces an isomorphism*

$$H(j): H(X - U, A - U) \approx H(X,A)$$

4 DIMENSION AXIOM *On the full subcategory of one-point spaces, there is a natural equivalence of H with the constant functor; that is, if P is a one-point space, then*

$$H_q(P) \approx \begin{cases} 0 & q \neq 0 \\ \mathbf{Z} & q = 0 \end{cases}$$

Obviously, the homotopy axiom is equivalent to the condition that the homology theory can be factored through the homotopy category of topological pairs.

Singular homology theory is an example of a homology theory. In fact, the homotopy axiom is a consequence of theorem 4.4.9, the exactness axiom is a consequence of theorem 4.5.4, the excision axiom is a consequence of corollary 4.6.5, and the dimension axiom is a consequence of lemmas 4.4.1 and 4.3.1. Therefore, there exist homology theories.

Corresponding to any homology theory there are reduced groups defined as follows. If X is a nonempty space, let $c: X \to P$ be the unique map from X to some one-point space P. The reduced group $\tilde{H}(X)$ is defined to be the kernel of the homomorphism

$$H(c): H(X) \to H(P)$$

Because c has a right inverse, so does $H(c)$. Therefore

$$H(X) \approx \tilde{H}(X) \oplus H(P)$$

and the reduced groups have properties similar to those of the reduced singular groups.

Given a triple $B \subset A \subset X$, let $k: A \subset (A,B)$ and define $\partial(X,A,B)$: $H(X,A) \to H(A,B)$ to be the composite

$$\partial(X,A,B) = H(k)\partial(X,A): H(X,A) \to H(A) \to H(A,B)$$

5 THEOREM *For any triple (X,A,B), with inclusion maps $i: (A,B) \subset (X,B)$ and $j: (X,B) \subset (X,A)$, there is an exact sequence*

$$\cdots \to H_q(A,B) \xrightarrow{H_q(i)} H_q(X,B) \xrightarrow{H_q(j)} H_q(X,A) \xrightarrow{\partial_q(X,A,B)} H_{q-1}(A,B) \to \cdots$$

PROOF The proof involves diagram chasing based on the exactness axiom 2. We prove exactness at $H_q(A,B)$ and leave the other parts of the proof to the reader.

(*a*) im $\partial_{q+1}(X,A,B) \subset \ker H_q(i)$. $H_q(i)\partial_{q+1}(X,A,B)$ is the composite

$$H_{q+1}(X,A) \xrightarrow{\partial_{q+1}(X,A)} H_q(A) \xrightarrow{H_q(k)} H_q(A,B) \xrightarrow{H_q(i)} H_q(X,B)$$

which also equals the composite

$$H_{q+1}(X,A) \xrightarrow{\partial_{q+1}(X,A)} H_q(A) \xrightarrow{H_q(i')} H_q(X) \xrightarrow{H_q(i'')} H_q(X,B)$$

where $i': A \subset X$ and $i'': X \subset (X,B)$. By axiom 2, $H_q(i')\partial_{q+1}(X,A) = 0$. Therefore $H_q(i)\partial_{q+1}(X,A,B) = 0$.

(*b*) $\ker H_q(i) \subset \operatorname{im} \partial_{q+1}(X,A,B)$. Let $z \in H_q(A,B)$ be such that $H_q(i)z = 0$. Then $\partial_q(X,B)H_q(i)z = 0$, and because $\partial_q(A,B) = \partial_q(X,B)H_q(i)$, $\partial_q(A,B)z = 0$. By axiom 2, there is $z' \in H_q(A)$ such that $H_q(k)z' = z$. Because the composite

$$H_q(A) \xrightarrow{H_q(i')} H_q(X) \xrightarrow{H_q(i'')} H_q(X,B)$$

equals the composite $H_q(i)H_q(k)$, it follows that

$$H_q(i'')H_q(i')z' = H_q(i)H_q(k)z' = H_q(i)z = 0$$

By axiom 2, there is $z'' \in H_q(B)$ such that if $j': B \subset X$, then $H_q(i')z' = H_q(j')z''$. Given $j'': B \subset A$, then $H_q(j') = H_q(i')H_q(j'')$. Therefore $H_q(i')(z' - H_q(j'')z'') = 0$. Again by axiom 2, there is $\bar{z} \in H_{q+1}(X,A)$ such that $\partial_{q+1}(X,A)\bar{z} = z' - H_q(j'')z''$. Then, because $H_q(k)H_q(j'') = 0$,

$$\partial_{q+1}(X,A,B)\bar{z} = H_q(k)\partial_{q+1}(X,A)\bar{z} = H_q(k)z' - H_q(k)H_q(j'')z'' = z$$

which shows that z is in im $\partial_{q+1}(X,A,B)$. ∎

The exact sequence of theorem 5 is called the *homology sequence of the triple* (X,A,B). If $B = \varnothing$, it reduces to the homology sequence of the pair (X,A).

Let H and ∂ and H' and ∂' be homology theories. A *homomorphism* from H and ∂ to H' and ∂' is a natural transformation h from H to H' commuting with ∂ and ∂'. That is, for every (X,A) there is a commutative diagram

$$\begin{array}{ccc} H(X,A) & \xrightarrow{\partial(X,A)} & H(A) \\ {\scriptstyle h(X,A)}\downarrow & & \downarrow{\scriptstyle h(A)} \\ H'(X,A) & \xrightarrow{\partial'(X,A)} & H'(A) \end{array}$$

in which the vertical maps are homomorphisms of degree 0. In view of the dimension axiom, a homomorphism h induces a homomorphism $h_0: \mathbf{Z} \to \mathbf{Z}$

that characterizes h on one-point spaces. The main result proved by Eilenberg and Steenrod is that corresponding to any homomorphism h_0: $\mathbf{Z} \to \mathbf{Z}$ there exists a unique homomorphism h from H and ∂ to H' and ∂', on the category of compact polyhedral pairs, which induces h_0. We shall not prove this, but shall content ourselves with proving that a homomorphism h which is an isomorphism for one-point spaces is an isomorphism for any compact polyhedral pair. This will illustrate how the axioms can be used and will suffice for our later applications.

The following is an easy consequence of the exactness axiom and the five lemma (or of theorem 5 and axiom 2).

6 LEMMA *Let $A' \subset A \subset X$. Then $H(A') \approx H(A)$ if and only if $H(X,A') \approx H(X,A)$ (both maps induced by inclusion).* ∎

We now prove a stronger excision property. A map f: $(X,A) \to (Y,B)$ is called a *relative homeomorphism* if f maps $X - A$ homeomorphically onto $Y - B$. Following are some examples.

7 An excision map $(X - U, A - U) \subset (X,A)$, where $U \subset A$, is a relative homeomorphism.

8 If X is obtained from A by adjoining an n-cell e and f: $(E^n,S^{n-1}) \to (e,\dot{e})$ is a characteristic map for e, then f is a relative homeomorphism.

9 THEOREM *Let X be a compact Hausdorff space and let A be a closed subset of X which is a strong deformation retract of one of its closed neighborhoods in X. Let f: $(X,A) \to (Y,B)$ be a relative homeomorphism, where Y is a Hausdorff space and B is closed in Y. Then, for any homology theory $H(f)$: $H(X,A) \approx H(Y,B)$.*

PROOF Let N be a closed neighborhood of A in X such that A is a strong deformation retract of N and let U be an open subset of X such that $A \subset U \subset \bar{U} \subset N$ (U exists because X is a normal space). Let F: $N \times I \to N$ be a strong deformation retraction of N to A.

Define $N' = f(N) \cup B$, $U' = f(U) \cup B$, and F': $N' \times I \to N'$ by

$$F'(y,t) = y \qquad y \in B, t \in I$$
$$F'(y,t) = fF(f^{-1}(y),t) \qquad y \in f(N), t \in I$$

Then F' is well-defined and continuous on each of the closed sets $B \times I$ and $f(N) \times I$. Therefore F' is continuous and is easily verified to be a strong deformation retraction of N' to B. Because $X - \bar{U}$ is open in $X - A$, $Y - (f(\bar{U}) \cup B)$ is open in $Y - B$, and because B is closed, it is open in Y. Therefore $f(\bar{U}) \cup B$ is closed in Y, and $\bar{U}' \subset f(\bar{U}) \cup B \subset N'$. Because $X - U$ is a closed, and hence compact, subset of X, $f(X - U) = Y - U'$ is a compact subset of Y. Because Y is a Hausdorff space, $Y - U'$ is closed in Y, and U' is open in Y. We have $B \subset U' \subset \bar{U}' \subset N'$ and a commutative diagram

$$\begin{array}{ccccc} H(X,A) & \xrightarrow{\approx} & H(X,N) & \xleftarrow{\approx} & H(X-U, N-U) \\ {\scriptstyle H(f)}\downarrow & & \downarrow & & \downarrow\approx \\ H(Y,B) & \xrightarrow{\approx} & H(Y,N') & \xleftarrow{\approx} & H(Y-U', N'-U') \end{array}$$

where the vertical maps are induced by f and the horizontal maps are induced by inclusion maps. Because A and B are deformation retracts of N and N', respectively, $H(A) \approx H(N)$ and $H(B) \approx H(N')$. It follows from lemma 6 that the left-hand horizontal maps are isomorphisms. The right-hand horizontal maps are isomorphisms by the excision axiom. The right-hand vertical map is an isomorphism because it is induced by a homeomorphism. From the commutativity of the diagram, it follows that $H(f)$ is an isomorphism. ∎

10 THEOREM *Let h be a homomorphism from H and ∂ to H' and ∂' which is an isomorphism for one-point spaces. Then, for any compact polyhedral pair (X,A), $h(X,A)$: $H(X,A) \approx H'(X,A)$.*

PROOF By the five lemma, it suffices to prove $h(X)$: $H(X) \approx H'(X)$ for any compact polyhedron X. Hence, let K be a finite simplicial complex. We need only prove that $h(|K|)$: $H(|K|) \approx H'(|K|)$. We prove this by induction on the number of simplexes of K. If K has just one simplex, $|K|$ is a one-point space, and $h(|K|)$ is an isomorphism by hypothesis.

Assume that K has m simplexes, where $m > 0$, and that h is an isomorphism for the space of any simplicial complex with fewer than m simplexes. Assume $\dim K = n$ and let s be an n-simplex of K. Let L be the subcomplex consisting of all simplexes of K different from s. By the five lemma and the exactness axiom, $h(|K|)$ is an isomorphism if and only if $h(|K|,|L|)$ is an isomorphism. If j: $(|s|,|\dot{s}|) \subset (|K|,|L|)$, it follows from theorem 9 that $H(j)$ and $H'(j)$ are isomorphisms. Hence we need only prove that $h(|s|,|\dot{s}|)$ is an isomorphism.

If $n = 0$, $(|s|,|\dot{s}|)$ is a one-point space, and $h(|s|,|\dot{s}|)$ is an isomorphism by hypothesis. If $n > 0$, because $|s|$ has the same homotopy type as a one-point space, $h(|s|)$ is an isomorphism. By the five lemma and the exactness axiom, $h(|s|,|\dot{s}|)$ is an isomorphism if and only if $h(|\dot{s}|)$ is an isomorphism. Because $\dot{s}$ is a proper subcomplex of K, $h(|\dot{s}|)$ is an isomorphism by the inductive hypothesis. ∎

To extend this result to arbitrary polyhedral pairs (not merely compact ones), we add an additional axiom. A pair (X,A) with X compact and A closed in X is called a *compact pair.*

11 AXIOM OF COMPACT SUPPORTS *Given any pair (X,A) and given $z \in H_q(X,A)$, there is a compact pair $(X',A') \subset (X,A)$ such that z is in the image of $H(X',A') \to H(X,A)$.*

A homology theory H and ∂ satisfying axiom 11 is called a *homology theory with compact supports* (Eilenberg and Steenrod use the term "homology theory with compact carriers"). It is clear that singular homology theory is a homology theory with compact supports. We shall see that any homology theory with compact supports satisfies the analogue of theorem 4.4.6. The following lemma is the main point in proving this.

12 LEMMA *Let H be a homology theory with compact supports and let (X',A') be a compact pair in (X,A). Given $z \in H_q(X',A')$ in the kernel of $H_q(X',A') \to H_q(X,A)$, there is a compact pair (X'',A''), with $(X',A') \subset (X'',A'') \subset (X,A)$, such that z is in the kernel of $H(X',A') \to H(X'',A'')$.*

PROOF In the proof all unlabeled maps are induced by inclusion. z is in the kernel of the composite

$$H_q(X',A') \xrightarrow{H_q(i)} H_q(X,A') \to H_q(X,A)$$

By theorem 5, $H_q(i)z$ is in the image of $H_q(A,A') \to H_q(X,A')$. By axiom 11, there is a compact space A'' such that $A' \subset A'' \subset A$ and such that $H_q(i)z$ is in the image of the composite $H_q(A'',A') \to H_q(A,A') \to H_q(X,A')$. By theorem 5, the composite $H_q(A'',A') \to H_q(X,A') \to H_q(X,A'')$ is trivial. Therefore z is in the kernel of $H_q(X',A') \to H_q(X,A'')$ for some compact A'' containing A'.

Because z is in the kernel of the composite

$$H_q(X',A') \xrightarrow{H_q(j)} H_q(X' \cup A'', A'') \to H_q(X,A'')$$

it follows from theorem 5, that $H_q(j)z$ is in the image of

$$\partial_{q+1}: H_{q+1}(X, X' \cup A'') \to H_q(X' \cup A'', A'')$$

By axiom 11, there is a compact X'' containing $X' \cup A''$ such that $H_q(j)z$ is in the image of the composite

$$H_{q+1}(X'', X' \cup A'') \to H_{q+1}(X, X' \cup A'') \xrightarrow{\partial_{q+1}} H_q(X' \cup A'', A'')$$

This composite is also equal to the map $\partial_{q+1}: H_{q+1}(X'', X' \cup A'') \to H_q(X' \cup A'', A'')$. By theorem 5, the composite

$$H_{q+1}(X'', X' \cup A'') \xrightarrow{\partial_{q+1}} H_q(X' \cup A'', A'') \to H_q(X'',A'')$$

is trivial. Therefore, z is in the kernel of $H_q(X',A') \to H_q(X'',A'')$. ∎

For any pair (X,A) the family of compact pairs (X',A') contained in (X,A) is directed by inclusion. For any homology theory H and ∂ the groups $\{H(X',A') \mid (X',A') \text{ compact} \subset (X,A)\}$ constitute a direct system, and the maps $H(X',A') \to H(X,A)$ define a homomorphism $i: \varinjlim \{H(X',A')\} \to H(X,A)$.

13 THEOREM *A homology theory H and ∂ has compact supports if and only if for any pair (X,A), $i: \varinjlim \{H(X',A')\} \approx H(X,A)$, where (X',A') varies over the family of compact pairs contained in (X,A).*

PROOF It is clear that axiom 11 is equivalent to the condition that i be an epimorphism. Hence, if i is an isomorphism, H and ∂ has compact supports. Conversely, if H has compact supports, i is an epimorphism, and lemma 12 implies that i is also a monomorphism. ∎

14 THEOREM *Let h be a homomorphism from H and ∂ to H′ and ∂' that is an isomorphism for one-point spaces. If H and ∂ and H′ and ∂' have compact supports, h is an isomorphism for any polyhedral pair.*

PROOF This follows from theorems 10 and 13 and from the fact that for any polyhedral pair (X,A) the compact polyhedral pairs (X',A') contained in it are cofinal in the family of all compact pairs in (X,A). ■

EXERCISES

A CHAIN HOMOTOPY CLASSES

1 For chain complexes C and C' show that $[C;C']$ is an abelian group (with group operation $[\tau_1] + [\tau_2] = [\tau_1 + \tau_2]$) and that there is a homomorphism

$$\varphi: [C;C'] \to \text{Hom}\,(H(C),H(C'))$$

such that $\varphi[\tau] = \tau_*$.

2 If C is a free chain complex, prove that the homomorphism φ is an epimorphism.

3 If C is a free chain complex and $H(C)$ is also free, prove that φ is an isomorphism.

B EULER CHARACTERISTICS

1 Let (X,A) be a pair and assume that two of the three graded groups $H(A)$, $H(X)$, and $H(X,A)$ are finitely generated. Prove that the third is also finitely generated and that $\chi(X) = \chi(A) + \chi(X,A)$.

2 Let $\{X_1,X_2\}$ be an excisive couple of subsets of X such that $H(X_1)$ and $H(X_2)$ are finitely generated. Prove that $H(X_1 \cup X_2)$ is finitely generated if and only if $H(X_1 \cap X_2)$ is finitely generated, in which case

$$\chi(X_1) + \chi(X_2) = \chi(X_1 \cup X_2) + \chi(X_1 \cap X_2)$$

3 Let γ be an integer-valued function defined on the class of compact polyhedra with base points such that

(*a*) If (X,x_0) is homeomorphic to (Y,y_0), then $\gamma(X,x_0) = \gamma(Y,y_0)$.

(*b*) If (X,A) is a compact polyhedral pair and $x_0 \in A$, then $\gamma(X,x_0) = \gamma(A,x_0) + \gamma(X/A,x_0')$, where X/A denotes the space obtained by collapsing A to a single point x_0'.

Prove that for any X

$$\gamma(X,x_0) = \gamma(S^0,p_0)\chi(X,x_0)$$

[*Hint:*[1] Prove first that if z_0 is a base point of E^n in S^{n-1}, then $\gamma(E^n,z_0) = 0$. Show that the result is true for $X = S^n$, and then use induction on the number of simplexes in a triangulation of X.]

4 If X and Y are compact polyhedra, prove that

$$\chi(X \times Y) = \chi(X)\chi(Y)$$

[1] See C. E. Watts, On the Euler characteristic of polyhedra, *Proceedings of the American Mathematical Society*, vol. 13, pp. 304–306, 1962.

C EXAMPLES

1 Let s be an n-simplex and let $(s)^m$ be its m-dimensional skeleton. Compute $H((s)^m)$.

2 Compute the homology group of an arbitrary surface.

3 Compute the homology group of the lens space $L(p,q)$.

4 Let A be a subspace of S^n which is homeomorphic to the one-point union $S^p \vee S^q$. Compute $H(S^n - A)$.

5 Let X be the space obtained from a closed triangle with vertices v_0, v_1, and v_2 by identifying the edges v_0v_1, v_1v_2, and v_2v_0 linearly with the edges v_1v_2, v_2v_0, and v_0v_1, respectively. Compute $H(X)$.

6 Given an integer $n > 0$ and an integer $m > 1$, prove that there exists a compact polyhedron X such that

$$\tilde{H}_q(X) = \begin{cases} 0 & q \neq n \\ \mathbf{Z}_m & q = n \end{cases}$$

7 Let H be a finitely generated nonnegative graded abelian group such that H_0 is a free abelian group. Prove that there exists a compact polyhedron X such that $\tilde{H}(X) \approx H$.

D JOINS AND PRODUCTS

1 Prove that for any space X there are isomorphisms

$$\tilde{H}_q(X) \approx \tilde{H}_{q+1}(X * S^0)$$

(*Hint:* If Y is contractible, so is $X * Y$.)

2 Prove that for any space X there are isomorphisms

$$H_q(X \times S^n, X \times p_0) \approx H_{q-n}(X)$$

[*Hint:* Use induction on n and the fact that if Y is contractible, $H(X \times Y, X \times y_0) = 0$.]

3 Compute the homology group of the n-dimensional torus $(S^1)^n$.

4 If a space is homeomorphic to a finite product of spheres, prove that the set of spheres which are the factors is unique.

E ORIENTATION

1 Let K be an n-dimensional pseudomanifold. Prove that it is possible to enumerate the n-simplexes of K in a (finite or infinite) sequence $s_0, s_1, \ldots, s_q, \ldots$ and to find a sequence $s'_1, s'_2, \ldots, s'_q, \ldots$ of $(n-1)$-simplexes of K such that for $q \geq 1$, s'_q is a face of s_q and also a face of s_i for some $i < q$.

2 If K is a finite n-dimensional pseudomanifold, prove that exactly one of the following holds:

(*a*) $H_n(K,\dot{K}) \approx \mathbf{Z}$ and $H_{n-1}(K,\dot{K})$ has no torsion.
(*b*) $H_n(K,\dot{K}) = 0$ and $H_{n-1}(K,\dot{K})$ has torsion subgroup isomorphic to $\mathbf{Z}_2$.

3 Let K be a finite simplicial complex which is homogeneously n-dimensional and such that every $(n-1)$-simplex of K is the face of at most two n-simplexes of K. Let $\dot{K}$ be the subcomplex of K generated by the $(n-1)$-simplexes of K which are faces of exactly one n-simplex of K. Prove that if $(K,\dot{K})$ satisfies either (*a*) or (*b*) of exercise 2 above, then K is an n-dimensional pseudomanifold.

A finite n-dimensional pseudomanifold is said to be *orientable* (or *nonorientable*) if it satisfies (*a*) (or (*b*)) of exercise 2. An *orientation* of an orientable n-dimensional pseudomanifold K is a generator of $H_n(K,\dot{K})$, and an *oriented* n-dimensional pseudomanifold is an n-dimensional pseudomanifold together with an orientation of it.

4 Let $z \in H_n(K,\dot{K})$ be an orientation of a finite n-dimensional pseudomanifold. If s is any n-simplex of K, prove that there is a unique orientation of s, denoted by $z \mid s \in H_n(s,\dot{s})$ and called the *induced orientation of* s, characterized by the property that z and $z \mid s$ correspond under the homomorphisms

$$H_n(K,\dot{K}) \to H_n(K, K - s) \xleftarrow{\approx} H_n(s,\dot{s})$$

A collection of orientations $\{\sigma(s) \in H_n(s,\dot{s})\}$ for each n-simplex s of an n-dimensional pseudomanifold is called *compatible* if for any $(n-1)$-simplex s' of $K - \dot{K}$ which is a face of the two n-simplexes s_1 and s_2 of K, $\sigma(s_1)$ and $-\sigma(s_2)$ correspond under the homomorphisms

$$\begin{array}{ccccc} H_n(s_1,\dot{s}_1) \xrightarrow{\partial} H_{n-1}(\dot{s}_1) \to H_{n-1}(\dot{s}_1, \dot{s}_1 - s') & & \\ & \nwarrow{\approx} & \\ & & H_{n-1}(s',\dot{s}') \\ & \swarrow{\approx} & \\ H_n(s_2,\dot{s}_2) \xrightarrow{\partial} H_{n-1}(\dot{s}_2) \to H_{n-1}(\dot{s}_2, \dot{s}_2 - s') & & \end{array}$$

5 If z is an orientation of a finite n-dimensional pseudomanifold, prove that the collection $\{z \mid s\}$ is compatible. Conversely, given a compatible collection $\{\sigma(s)\}$ of orientations of the n-simplexes s of a finite n-dimensional pseudomanifold K, prove that there is a unique orientation z of K such that $z \mid s = \sigma(s)$ for each n-simplex s of K. Use this to define orientability for arbitrary (nonfinite) n-dimensional pseudomanifolds. [*Hint:* Identify $H_n(K,K^{n-1})$ with indexed collections $\{\sigma(s) \in H_n(s,\dot{s})\}$, where s varies over the n-simplexes of K, and show that the image of the homomorphism $H_n(K,\dot{K}) \to H_n(K,K^{n-1})$ consists of the compatible collections.]

F DEGREES OF MAPS

Let K_1 and K_2 be finite n-dimensional pseudomanifolds with orientations z_1 and z_2, respectively. Given a continuous map $f: (|K_1|,|\dot{K}_1|) \to (|K_2|,|\dot{K}_2|)$, its *degree*, denoted by $\deg f$, is the unique integer such that $f_*(z_1) = (\deg f)z_2$ [where we regard $z_1 \in H_n(|K_1|,|\dot{K}_1|)$ and $z_2 \in H_n(|K_2|,|\dot{K}_2|)$].

1 Let $\varphi: (K_1,\dot{K}_1) \to (K_2,\dot{K}_2)$ be a simplicial approximation to f, let s_2 be a fixed n-simplex of K_2, and let $m_+(\varphi)$ (or $m_-(\varphi)$) be the number of n-simplexes s_1 of K_1 such that φ maps the induced orientation $z_1 \mid s_1$ into the induced orientation $z_2 \mid s_2$ (or into $-z_2 \mid s_2$). Prove that $\deg f = m_+(\varphi) - m_-(\varphi)$.

2 In case K is a finite orientable n-dimensional pseudomanifold and $f: (|K|,|\dot{K}|) \to (|K|,|\dot{K}|)$, there is a unique integer $\deg f$ such that $f_*(z) = (\deg f)z$ for any $z \in H_n(|K|,|\dot{K}|)$. Prove that if $f, g: (|K|,|\dot{K}|) \to (|K|,|\dot{K}|)$, then $\deg (g \circ f) = (\deg g)(\deg f)$.

3 Let $f: S^n \to S^n$ be a map such that $f(E^n_+) \subset E^n_+$, $f(E^n_-) \subset E^n_-$ and let $f': S^{n-1} \to S^{n-1}$ be the map defined by f. Prove that $\deg f = \deg f'$.

4 Show that for any $n \geq 1$ and any integer m there is a map $f: S^n \to S^n$ such that $\deg f = m$.

G TOPOLOGICAL INVARIANCE OF PSEUDOMANIFOLDS

1 Let K be a simplicial complex and let $x \in \langle s \rangle$, where s is a simplex of K. Prove that there is an isomorphism

$$H(|K|, |K| - \text{st } s) \approx H(|K|, |K| - x)$$

2 Let K be a simplicial complex and let $x \in \langle s \rangle$, where s is a *principal* n-simplex of K (that is, s is not a proper face of any simplex of K). Prove that

$$H_q(|K|, |K| - x) \approx \begin{cases} 0 & q \neq n \\ \mathbf{Z} & q = n \end{cases}$$

3 Prove that a locally compact polyhedron X has dimension n if and only if n is the largest integer such that there exist points $x \in X$, with $H_n(X, X - x) \neq 0$.

4 Let X be a finite dimensional polyhedron and for each n let X_n be the closure of the set of all $x \in X$ having a neighborhood U such that $H_n(X, X - y) \approx \mathbf{Z}$ for all $y \in U$. If K is any simplicial complex triangulating X and K_n is the subcomplex of K generated by the principal n-simplexes of K, prove that K_n triangulates X_n.

5 Prove that the property of being homogeneously n-dimensional is a topologically invariant property of simplicial complexes (and so we can speak of a homogeneously n-dimensional polyhedron).

6 Let K be an arbitrary simplicial complex triangulating a homogeneously n-dimensional polyhedron X. Prove that every $(n-1)$-simplex of K is the face of at most two n-simplexes of K if and only if $H_q(A, A - x) = 0$ for all $x \in A$ and all $q \geq n - 1$, where A is the closure in X of the set $\{x \in X \mid H_n(X, X - x) \text{ is noncyclic}\}$.

7 Let X be a homogeneously n-dimensional polyhedron satisfying exercise 6 and let $\dot{X} = B_{n-1}$, where B is the closure in X of the set $\{x \in X \mid H_n(X, X - x) = 0\}$ and where B_{n-1} is defined in terms of B, as in exercise 4. If K is any simplicial complex triangulating X, prove that the subcomplex of K generated by the $(n-1)$-simplexes of K which are faces of exactly one n-simplex of K triangulates $\dot{X}$.

8 Prove that the property of being a finite n-dimensional pseudomanifold is a topologically invariant property of simplicial complexes.

H EDGE-PATH GROUPS

1 Let K be a connected simplicial complex with a base vertex $v_0 \in K$. Given an edge $e = (v_0, v_1)$, of K, let $[e]$ be the oriented 1-simplex $[v_0, v_1]$. If $\zeta = e_1 e_2 \cdots e_r$ is a closed edge path of K at v_0, let $\psi(\zeta) = [e_1] + [e_2] + \cdots + [e_r] \in C_1(K)$. Prove that $\psi(\zeta)$ is a cycle and that if ζ and ζ' are equivalent edge paths, then $\psi(\zeta)$ and $\psi(\zeta')$ are homologous.

2 Prove that there is a natural transformation $\psi: E(K, v_0) \to H_1(K)$ (on the category of connected simplicial complexes with a base vertex) defined by $\psi[\zeta] = \{\psi(\zeta)\}$.

3 Prove that the homomorphism ψ is an epimorphism and has kernel equal to the commutator subgroup of $E(K, v_0)$.

I AXIOMATIC HOMOLOGY THEORY

In this group of exercises H will denote an arbitrary homology theory.

1 Let X_1 and X_2 be subspaces of a space X. Prove that the following are equivalent:

(*a*) The excision map $(X_1, X_1 \cap X_2) \subset (X_1 \cup X_2, X_2)$ induces an isomorphism of homology.

(*b*) The excision map $(X_2, X_1 \cap X_2) \subset (X_1 \cup X_2, X_1)$ induces an isomorphism of homology.

(*c*) The inclusion maps

$$i_1\colon (X_1, X_1 \cap X_2) \subset (X_1 \cup X_2, X_1 \cap X_2)$$

and

$$i_2\colon (X_2, X_1 \cap X_2) \subset (X_1 \cup X_2, X_1 \cap X_2)$$

induce monomorphisms on homology and

$$H(X_1 \cup X_2, X_1 \cap X_2) \approx i_{1*}H(X_1, X_1 \cap X_2) \oplus i_{2*}H(X_2, X_1 \cap X_2)$$

(*d*) The inclusion maps

$$j_1\colon (X_1 \cup X_2, X_1 \cap X_2) \subset (X_1 \cup X_2, X_1)$$

and

$$j_2\colon (X_1 \cup X_2, X_1 \cap X_2) \subset (X_1 \cup X_2, X_2)$$

induce epimorphisms on homology and j_{1*} and j_{2*} induce an isomorphism

$$H(X_1 \cup X_2, X_1 \cap X_2) \approx H(X_1 \cup X_2, X_1) \oplus H(X_1 \cup X_2, X_2)$$

(*e*) For any $A \subset X_1 \cap X_2$ there is an exact Mayer-Vietoris sequence

$$\cdots \to H_q(X_1 \cap X_2, A) \to H_q(X_1,A) \oplus H_q(X_2,A)$$
$$\to H_q(X_1 \cup X_2, A) \to H_{q-1}(X_1 \cap X_2, A) \to \cdots$$

(*f*) For any $Y \supset X_1 \cup X_2$ there is an exact Mayer-Vietoris sequence

$$\cdots \to H_q(Y, X_1 \cap X_2) \to H_q(Y,X_1) \oplus H_q(Y,X_2)$$
$$\to H_q(Y, X_1 \cup X_2) \to H_{q-1}(Y, X_1 \cap X_2) \to \cdots$$

2 Let $X_1, \ldots, X_m$ and A be closed subspaces of a space X such that

(*a*) $X = \bigcup X_i$.
(*b*) $X_i \cap X_j = A$ if $i \neq j$.
(*c*) $\overline{X_i - A}$ is disjoint from $\overline{X_j - A}$ if $i \neq j$.

Prove that the homomorphisms $H(X_i,A) \to H(X,A)$ are monomorphisms and $H(X,A)$ is isomorphic to the direct sum of the images.

3 Let $\{X_j\}_{j \in J}$ (with J possibly infinite) be a collection of closed subsets of a space X and let A be a subspace of X such that (*a*), (*b*), and (*c*) of exercise 2 above are satisfied. Assume also that every compact subset of X is contained in a finite union of $\{X_j\}$ and that H is a homology theory with compact supports. Prove that $H(X,A) \approx \bigoplus_{j \in J} H(X_j,A)$.

4 Let (X,A) be a topological pair and let $\{X_s\}$ be a family of subspaces of X indexed by the integers such that

(*a*) $A = X_{-1}$.
(*b*) $X_s \subset X_{s+1}$ for all s.
(*c*) $X = \bigcup X_s$ and every compact subset of X is contained in X_s for some s.
(*d*) $H_q(X_s,X_{s-1}) = 0$ if $q \neq s$ and $s \geq 0$.

Let $C = \{C_q,\partial_q\}$ be the nonnegative chain complex with $C_q = H_q(X_q,X_{q-1})$ for $q \geq 0$ and ∂_q the connecting homomorphism of the triple (X_q, X_{q-1}, X_{q-2}) for $q \geq 1$. If H has compact supports, prove that $H(X,A) \approx H(C)$. [*Hint:* Prove that there are exact sequences

$$H_{q+1}(X_{q+1},X_q) \xrightarrow{\partial} H_q(X_q,A) \to H_q(X,A) \to 0$$

and

$$0 \to H_q(X_q,A) \to H_q(X_q,X_{q-1}) \to H_{q-1}(X_{q-1},A)]$$

5 Let H be a homology theory defined on the category of compact pairs. Prove that there is an extension of H to a homology theory $\bar{H}$ with compact supports such that $\bar{H}(X,A) = \varinjlim \{H(X',A') \mid (X',A') \text{ a compact pair in } (X,A)\}$.

CHAPTER FIVE

PRODUCTS

WE ARE NOW READY TO EXTEND THE DEFINITION OF HOMOLOGY TO MORE GENERAL coefficients. In this framework the homology considered in the last chapter appears as the special case of integral coefficients. The extension is done in a purely algebraic way. Given a chain complex C and an abelian group G, their tensor product is the chain complex $C \otimes G = \{C_q \otimes G, \partial_q \otimes 1\}$, and the homology of $C \otimes G$ is defined to be the homology of C, with coefficients G.

We shall also introduce the concepts of cochain complex and cohomology. These are dual to the concepts of chain complex and homology and arise on replacing the tensor-product functor by the functor Hom.

We shall establish universal-coefficient formulas expressing the homology and cohomology of a space with arbitrary coefficients as functors of the integral homology of the space. Although these new functors do not distinguish between spaces not already distinguished by the integral homology functor, it is nonetheless important to consider them, as it frequently happens that the most natural functor to apply in a given geometrical problem is determined by the problem itself and need not be the integral homology functor. For example, in the obstruction theory developed in Chapter Eight we shall be led to the cohomology of a space with coefficients in the homotopy groups of another space.

A further consideration is that the cohomology of a space has a multiplicative structure in addition to its additive structure, which makes cohomology a more powerful tool than homology. We shall present some applications of this added multiplication structure, the most important of which is the study of the homology properties of fiber bundles, where we establish the exactness of the Thom-Gysin sequence of a sphere bundle.

At the end of the chapter is a brief discussion of cohomology operations. These are natural transformations between two cohomology functors and strengthen even further the applicability of cohomology as a tool. We shall define the particular set of cohomology operations known as the Steenrod squares and establish their basic properties.

Sections 5.1 and 5.2 are devoted to homology with general coefficients and to the universal-coefficient formula for homology. Section 5.3 deals with the tensor product of two chain complexes and contains a proof of the Künneth formula expressing the homology of the tensor product as a functor of the homology of the factor complexes. This is applied geometrically to express the homology of a product space in terms of the homology of its factors.

Sections 5.4 and 5.5 contain the dual concepts of cochain complex and cohomology and the appropriate universal-coefficient formulas for them. In Sec. 5.6 the cup and cap products are defined, the cup product being the multiplicative structure in cohomology mentioned previously, and the cap product being a dual involving cohomology and homology together. These products are used in Sec. 5.7 to study the homology and cohomology of fiber bundles. We establish the Leray-Hirsch theorem, which asserts that certain fiber bundles have homology and cohomology which are additively isomorphic to the homology and cohomology of the corresponding product of the base and the fiber.

Section 5.8 is devoted to a study of the cohomology algebra. The exactness of the Thom-Gysin sequence is used to compute the cohomology algebra of projective spaces, and this, in turn, is used to prove the Borsuk-Ulam theorem. There is also a discussion of the structure of Hopf algebras, which arise in considering the cohomology of an *H* space. In Sec. 5.9 the Steenrod squares are defined and their elementary properties are proved. They will be applied later.

1 HOMOLOGY WITH COEFFICIENTS

In this section we shall extend the concepts dealing with chain complexes to the case where the chain groups are modules over a ring. The tensor product of such a chain complex with a fixed module is another chain complex, and its graded homology module is a functor of the original chain complex and the fixed module. These homology modules have properties analogous to those established in the last chapter for complexes of abelian groups. The section closes with the definition of a homology theory with an arbitrary coefficient module. This is analogous to the concept of homology theory (which has integral coefficients) introduced in the last chapter.

Throughout this section R will denote a commutative ring with a unit. We consider R modules and homomorphisms between them. A *chain complex over* R, $C = \{C_q,\partial_q\}$ consists of a sequence of R modules C_q and homomorphisms $\partial_q: C_q \to C_{q-1}$ such that $\partial_q\partial_{q+1} = 0$ for all q. There is then a graded homology module

$$H(C) = \{H_q(C) = \ker \partial_q/\text{im } \partial_{q+1}\}$$

The concepts of chain maps and chain homotopies can be defined for chain complexes over R, and the results about chain complexes of abelian groups generalize in a straightforward fashion to chain complexes over R. In particular, on the category of short exact sequences of chain complexes over R,

$$0 \to C' \to C \to C'' \to 0$$

there is a functorial connecting homomorphism

$$\partial_*: H_q(C'') \to H_{q-1}(C')$$

and a functorial exact sequence

$$\cdots \xrightarrow{\partial_*} H_q(C') \to H_q(C) \to H_q(C'') \xrightarrow{\partial_*} H_{q-1}(C') \to \cdots$$

If C is a chain complex over R and G' is an R module, an *augmentation of* C *over* G' is an epimorphism $\varepsilon: C_0 \to G'$ such that $\varepsilon \circ \partial_1 = 0$. An *augmented chain complex over* G' consists of a nonnegative chain complex C and an augmentation of C over G'.

If $C = \{C_q,\partial_q\}$ is a chain complex over R and G is an R module, then $C \otimes G = \{C_q \otimes G, \partial_q \otimes 1\}$ is also a chain complex over R, and if C is augmented over G', then $C \otimes G$ is augmented over $G' \otimes G$. The graded homology module $H(C \otimes G)$ is called the *homology module of* C *with coefficients* G and is denoted by $H(C;G)$. If $\tau: C \to C'$ is a chain map, $\tau \otimes 1: C \otimes G \to C' \otimes G$ is also a chain map, and $\tau_*: H(C;G) \to H(C';G)$ denotes the homomorphism induced by $\tau \otimes 1$. Given a homomorphism $\varphi: G \to G'$, there is a chain map $1 \otimes \varphi: C \otimes G \to C \otimes G'$ inducing a homomorphism

$$\varphi_*: H(C;G) \to H(C;G')$$

These remarks are summarized in the following statement.

1 THEOREM *There is a covariant functor of two arguments from the category of chain complexes over R and the category of R modules to the category of graded R modules which assigns to a chain complex C and module G the homology module of C with coefficients G.* ∎

Note that if $c \in C_q$ is a cycle of C and $g \in G$, then $c \otimes g \in C_q \otimes G$ is a cycle of $C \otimes G$, and if c is a boundary, so is $c \otimes g$. Therefore there is a bilinear map

$$H_q(C) \times G \to H_q(C;G)$$

which assigns to $(\{c\},g)$ the homology class $\{c \otimes g\}$. This corresponds to a homomorphism

$$\mu: H(C) \otimes G \to H(C;G)$$

such that $\mu(\{c\} \otimes g) = \{c \otimes g\}$ for $c \in Z(C)$. The homomorphism μ is easily verified to be a natural transformation on the product of the category of chain complexes with the category of modules.

If C is a chain complex over $\mathbf{Z}$ and G is an R module, then $C \underset{\mathbf{Z}}{\otimes} G$ is a chain complex over R. It follows from theorem 4.5 in the Introduction that the homology module over $\mathbf{Z}$ of C with coefficients G is isomorphic, as a graded R module, to the homology module over R of $C \underset{\mathbf{Z}}{\otimes} R$ with coefficients G.

2 EXAMPLE Let $C(K)$ denote the oriented chain complex of the simplicial complex K. Given an abelian group G and a simplicial pair (K,L), the *oriented homology group of* (K,L) *with coefficients* G, denoted by $H(K,L; G)$, is defined to be the graded homology group of $[C(K)/C(L)] \otimes G$ (which is augmented over $\mathbf{Z} \otimes G \approx G$). Then $H(K,L; G)$ is a covariant functor of two arguments from the category of simplicial pairs and the category of abelian groups to the category of graded abelian groups. If G is also an R module, $H(K,L; G)$ is a graded R module. Similar remarks apply to the ordered chain complex $\Delta(K)/\Delta(L)$.

3 EXAMPLE If (X,A) is a topological pair and G is an abelian group, the *singular homology group of* (X,A) *with coefficients* G, denoted by $H(X,A; G)$, is defined to be the graded homology group of $[\Delta(X)/\Delta(A)] \otimes G$ (which is augmented over G). It is a covariant functor of two arguments from the category of topological pairs and the category of abelian groups to the category of graded abelian groups. If G is an R module, $H(X,A; G)$ is a graded R module.

Because the ring R is commutative, there is a canonical isomorphism $G \otimes G' \approx G' \otimes G$ for R modules G and G'. Therefore, if C is a chain complex over R, $G \otimes C$ is canonically isomorphic to $C \otimes G$. Hence no new homology modules are obtained from $G \otimes C$.

We recall some general properties of tensor products which will be important in the next section.

4 LEMMA *The tensor product of two epimorphisms is an epimorphism.*

PROOF Let $\alpha: A \to A''$ and $\beta: B \to B''$ be epimorphisms. $A'' \otimes B''$ is generated by elements of the form $a'' \otimes b''$, where $a'' \in A''$ and $b'' \in B''$. Since α and β are epimorphisms, $A'' \otimes B''$ is generated by elements of the form

$\alpha(a) \otimes \beta(b)$, where $a \in A$ and $b \in B$. Since $(\alpha \otimes \beta)(a \otimes b) = \alpha(a) \otimes \beta(b)$, $A'' \otimes B''$ is generated by $(\alpha \otimes \beta)(A \otimes B)$, showing that $\alpha \otimes \beta$ is an epimorphism. ■

In general, it is not true that the tensor product of two monomorphisms is a monomorphism (see example 7 below). The following lemma shows that something can be said about the kernel of $\alpha \otimes \beta$ when α and β are epimorphisms.

5 LEMMA *If α and β are epimorphisms, the kernel of $\alpha \otimes \beta$ is generated by elements of the form $a \otimes b$, where $a \in \ker \alpha$ or $b \in \ker \beta$.*

PROOF Let $\alpha: A \to A''$ and $\beta: B \to B''$ be epimorphisms and let D be the submodule of $A \otimes B$ generated by elements of the form $a \otimes b$, where $a \in \ker \alpha$ or $b \in \ker \beta$. Let $p: A \otimes B \to (A \otimes B)/D$ be the projection. There is a well-defined bilinear map

$$A'' \times B'' \to (A \otimes B)/D$$

sending (a'',b'') to $p(a \otimes b)$, where $a \in A$ and $b \in B$ are chosen so that $\alpha(a) = a''$ and $\beta(b) = b''$. This bilinear map corresponds to a homomorphism

$$\psi: A'' \otimes B'' \to (A \otimes B)/D$$

such that $\psi(a'' \otimes b'') = p(a \otimes b)$, where $\alpha(a) = a''$ and $\beta(b) = b''$. It is then obvious that p equals the composite

$$A \otimes B \xrightarrow{\alpha \otimes \beta} A'' \otimes B'' \xrightarrow{\psi} (A \otimes B)/D$$

This shows that $\ker (\alpha \otimes \beta) \subset D$. The reverse inclusion is evident, showing that $\ker (\alpha \otimes \beta) = D$. ■

6 COROLLARY *Given an exact sequence*

$$A' \to A \to A'' \to 0$$

and given a module B, there is an exact sequence

$$A' \otimes B \to A \otimes B \to A'' \otimes B \to 0$$

PROOF It follows from lemma 4 that $A \otimes B \to A'' \otimes B$ is an epimorphism, so the sequence is exact at $A'' \otimes B$. If $\bar{A} \subset A$ is the image of $A' \to A$, then, by lemma 4, $A' \otimes B \to \bar{A} \otimes B$ is an epimorphism. Because $\bar{A}$ is also the kernel of $A \to A''$, it follows from lemma 5 that the kernel of $A \otimes B \to A'' \otimes B$ is the image of $\bar{A} \otimes B \to A \otimes B$. Therefore the sequence is exact at $A \otimes B$. ■

If the original sequence is assumed to be a short exact sequence, it need not be true that the tensor-product sequence is a short exact sequence. We present an example to illustrate this.

7 EXAMPLE Over **Z**, consider the short exact sequence

$$0 \to \mathbf{Z} \xrightarrow{\alpha} \mathbf{Z} \xrightarrow{\beta} \mathbf{Z}_2 \to 0$$

where $\alpha(1) = 2$ and $\beta(1)$ is a generator $\bar{1}$ of $\mathbf{Z}_2$. The tensor product of this sequence with $\mathbf{Z}_2$ is not a short exact sequence because $\alpha \otimes 1: \mathbf{Z} \otimes \mathbf{Z}_2 \to \mathbf{Z} \otimes \mathbf{Z}_2$ is not a monomorphism [$\mathbf{Z} \otimes \mathbf{Z}_2 \approx \mathbf{Z}_2 \neq 0$, but $(\alpha \otimes 1)(1 \otimes \bar{1}) = 2 \otimes \bar{1} = 1 \otimes 2 \cdot \bar{1} = 0$].

8 THEOREM *The tensor-product functor commutes with direct sums.*

PROOF Assume $A = \bigoplus A_j$ and consider the bilinear map $A \times B \to \bigoplus (A_j \otimes B)$ sending $(\sum a_j, b)$ to $\sum (a_j \otimes b)$ and the homomorphisms $A_j \otimes B \to A \otimes B$ for all j. By the characteristic properties of tensor product and direct sum, there are commutative triangles

$$\begin{array}{ccc} A \times B & & \\ \downarrow & \searrow & \\ A \otimes B & \xrightarrow{\varphi} & \bigoplus (A_j \otimes B) \end{array} \qquad \begin{array}{ccc} & & A_j \otimes B \\ & \swarrow & \downarrow \\ A \otimes B & \xleftarrow{\psi} & \bigoplus (A_j \otimes B) \end{array}$$

Clearly, the maps φ and ψ are inverses, showing that $A \otimes B \approx \bigoplus (A_j \otimes B)$. If, also, $B = \bigoplus B_k$, then similarly,

$$A \otimes B \approx \bigoplus_{j,k} A_j \otimes B_k \quad \blacksquare$$

9 THEOREM *The tensor-product functor commutes with direct limits.*

PROOF Let $A = \lim_{\to} \{A^\alpha\}$ and consider the bilinear map $A \times B \to \lim_{\to} \{A^\alpha \otimes B\}$ sending $(\{a\},b)$ to $\{a \otimes b\}$ for $a \in A^\alpha$ and the homomorphisms $A^\alpha \otimes B \to A \otimes B$ for all α. By the characteristic properties of tensor product and direct limit, there are commutative triangles

$$\begin{array}{ccc} A \times B & & \\ \downarrow & \searrow & \\ A \otimes B & \xrightarrow{\varphi} & \lim_{\to} \{A^\alpha \otimes B\} \end{array} \qquad \begin{array}{ccc} & & A^\alpha \otimes B \\ & \swarrow & \downarrow \\ A \otimes B & \xleftarrow{\psi} & \lim_{\to} \{A^\alpha \otimes B\} \end{array}$$

Clearly, φ and ψ are inverses, showing that $A \otimes B \approx \lim_{\to} \{A^\alpha \otimes B\}$. If, also, $B = \lim_{\to} \{B^\beta\}$, then similarly, $A \otimes B \approx \lim_{\to} \{A^\alpha \otimes B^\beta\}$. ■

We now consider a special class of short exact sequences. These sequences have the property that their tensor product with any module is again exact. A short exact sequence

$$0 \to A' \xrightarrow{\alpha} A \xrightarrow{\beta} A'' \to 0$$

is said to be *split* if β has a right inverse (that is, if there exists a homomorphism $\beta': A'' \to A$ such that $\beta \circ \beta' = 1_{A''}$). We also say that the sequence *splits*.

10 EXAMPLE Any short exact sequence $0 \to A' \to A \xrightarrow{\beta} A'' \to 0$ with A'' free is split. To see this, let $\{a_j''\}$ be a basis for A'' and for each j choose $a_j \in A$ so that $\beta(a_j) = a_j''$. Let $\beta': A'' \to A$ be the homomorphism such that $\beta'(a_j'') = a_j$ for all j. Then β' is a right inverse of β.

11 LEMMA *Given a short exact sequence*

$$0 \to A' \xrightarrow{\alpha} A \xrightarrow{\beta} A'' \to 0$$

define $A' \xrightarrow{i} A' \oplus A'' \xrightarrow{p} A''$ *by* $i(a') = (a',0)$ *and* $p(a',a'') = a''$. *Then the following are equivalent:*

(*a*) *The sequence is split.*

(*b*) *There is a commutative diagram*

$$\begin{array}{ccccc} & & A & & \\ & \overset{\alpha}{\nearrow} & & \overset{\beta}{\searrow} & \\ A' & & \gamma'\uparrow & & A'' \\ & \underset{i}{\searrow} & & \underset{p}{\nearrow} & \\ & & A' \oplus A'' & & \end{array}$$

(*c*) *There is a commutative diagram*

$$\begin{array}{ccccc} & & A & & \\ & \overset{\alpha}{\nearrow} & & \overset{\beta}{\searrow} & \\ A' & & \gamma\downarrow & & A'' \\ & \underset{i}{\searrow} & & \underset{p}{\nearrow} & \\ & & A' \oplus A'' & & \end{array}$$

(*d*) α *has a left inverse.*

PROOF If $\beta'\colon A'' \to A$ is a right inverse of β, let $\gamma'\colon A' \oplus A'' \to A$ be defined by $\gamma'(a',a'') = \alpha(a') + \beta'(a'')$. Then γ' has the desired properties. Conversely, given γ', define $\beta'\colon A'' \to A$ by $\beta'(a'') = \gamma'(0,a'')$. Then β' is a right inverse of β, so the sequence is split. Therefore (*a*) is equivalent to (*b*). A similar argument shows that (*c*) is equivalent to (*d*). It follows from the five lemma that in the diagram of (*b*) [or (*c*)], γ' [or γ] is necessarily an isomorphism. Therefore (*b*) is equivalent to (*c*) with γ' equal to γ^{-1}. ∎

12 COROLLARY *Given a split short exact sequence*

$$0 \to A' \xrightarrow{\alpha} A \to A'' \to 0$$

and given a module B, the sequence

$$0 \to A' \otimes B \xrightarrow{\alpha \otimes 1} A \otimes B \to A'' \otimes B \to 0$$

is a split short exact sequence.

PROOF By corollary 6 and lemma 11 we need only show that $\alpha \otimes 1$ has a left inverse. By lemma 11, α has a left inverse α'. Then $\alpha' \otimes 1$ is a left inverse of $\alpha \otimes 1$. ∎

In case $0 \to C' \to C \to C'' \to 0$ is a split short exact sequence of chain complexes, it follows from corollary 12 that for any module G the sequence

$$0 \to C' \otimes G \to C \otimes G \to C'' \otimes G \to 0$$

is a short exact sequence of chain complexes. This short exact sequence gives rise to an exact homology sequence, and we obtain the next result.

13 THEOREM *Given a split short exact sequence of chain complexes*

$$0 \to C' \to C \to C'' \to 0$$

and given a module G, there is a functorial exact homology sequence

$$\cdots \to H_q(C';G) \to H_q(C;G) \to H_q(C'';G) \to H_{q-1}(C';G) \to \cdots \quad \blacksquare$$

This implies the exactness of the singular homology sequence (and reduced homology sequence) of a pair with arbitrary coefficients. Similarly, there is an exact sequence of a triple with arbitrary coefficients. All these sequences (except the reduced sequence of a pair) are consequences of the exactness of the relative Mayer-Vietoris sequence, which we now establish. If $\{(X_1,A_1), (X_2,A_2)\}$ is an excisive couple of pairs in a topological space, the short exact sequence of singular chain complexes

$$0 \to \Delta(X_1 \cap X_2)/\Delta(A_1 \cap A_2) \to \Delta(X_1)/\Delta(A_1) \oplus \Delta(X_2)/\Delta(A_2) \to [\Delta(X_1) + \Delta(X_2)]/[\Delta(A_1) + \Delta(A_2)] \to 0$$

is split [by example 10, because $[\Delta(X_1) + \Delta(X_2)]/[\Delta(A_1) + \Delta(A_2)]$ is a free abelian group]. Therefore we obtain the following result.

14 COROLLARY *If $\{(X_1,A_1), (X_2,A_2)\}$ is an excisive couple of pairs in a space and G is an R module, there is an exact relative Mayer-Vietoris sequence of $\{(X_1,A_1), (X_2,A_2)\}$ with coefficients G.* ■

If G is fixed, the singular homology of (X,A) with coefficients G satisfies all the axioms of homology theory except the dimension axiom (all of them are easily seen to hold except exactness, which follows from corollary 14). If P is a one-point space, there is a functorial isomorphism $H_0(P;G) \approx G$. This leads to the following definition.

Let G be an R module. A *homology theory with coefficients* G consists of a covariant functor H from the category of topological pairs to graded R modules and a natural transformation ∂: $H(X,A) \to H(A)$ of degree -1 satisfying the homotopy, exactness, and excision axioms, and satisfying the following form of the dimension axiom: On the category of one-point spaces there is a natural equivalence of H with the constant functor which assigns to every one-point space the graded module which is trivial for degrees other than 0 and equal to G in degree 0. A homology theory with coefficients $\mathbf{Z}$ is called an *integral homology theory*. An integral homology theory is the same as a homology theory as defined in Sec. 4.8.

Singular homology with coefficients G is an example of a homology theory with coefficients G. The uniqueness theorem 4.8.10 is valid for homology theories with coefficients.

In the next section we shall show how the singular homology modules with coefficients are determined by the integral singular homology groups.

2 THE UNIVERSAL-COEFFICIENT THEOREM FOR HOMOLOGY

In order to express $H(C;G)$ in terms of $H(C)$ and G, it is necessary to introduce certain functors of modules that are associated to the tensor-product functor. This section contains a definition of these functors, and a study of them in the special case of a principal ideal domain. This leads to the universal-coefficient theorem. In the next section these new functors will enter in a description of the homology of a product space.

Let A be an R module. A *resolution of A (over R)* is an exact sequence

$$\cdots \to C_n \xrightarrow{\partial_n} \cdots \to C_1 \xrightarrow{\partial_1} C_0 \xrightarrow{\varepsilon} A \to 0$$

If, in addition, each C_q is a free R module, the resolution is said to be *free*. Thus a resolution of A consists of a chain complex $C = \{C_q,\partial_q\}$ over R which is augmented over A and is such that $\tilde{C}$ is acyclic. The resolution is free if and only if the chain complex C is free.

Any R module A has free resolutions. In fact, given an R module B, let $F(B)$ be the free R module generated by the elements of B and let $F(B) \to B$ be the canonical map. The *canonical* free resolution of A is the following resolution (defined inductively):

$$\cdots \to F(\ker \partial_q) \xrightarrow{\partial_{q+1}} F(\ker \partial_{q-1}) \xrightarrow{\partial_q} \cdots \to F(\ker \varepsilon) \xrightarrow{\partial_1} F(A) \xrightarrow{\varepsilon} A \to 0$$

The method of acyclic models applies to chain complexes over R and, when applied to a category consisting of a single object and single morphism, implies the following result.

1 THEOREM *Let C be a free nonnegative chain complex augmented over A and let C' be a resolution of A'. Any homomorphism $\varphi: A \to A'$ extends to a chain map*

$$\begin{array}{ccccccccc} \cdots \to & C_{q+1} & \to & C_q & \to \cdots \to & C_0 & \xrightarrow{\varepsilon} & A & \to 0 \\ & \downarrow & & \downarrow & & \downarrow & & \downarrow \varphi & \\ \cdots \to & C'_{q+1} & \to & C'_q & \to \cdots \to & C'_0 & \xrightarrow{\varepsilon'} & A' & \to 0 \end{array}$$

preserving augmentations, and two such chain maps are chain homotopic. ■

Specializing to the case $\varphi = 1_A: A \subset A$, we obtain the next result.

2 COROLLARY *If C and C' are free resolutions of A, then C and C' are canonically chain-equivalent chain complexes.* ■

For modules A and B and a free resolution C of A, it follows from corollary 2 that the graded module $H(C;B)$ depends only on A and B. Let C be the canonical free resolution of A. For $q \geq 0$ we define the *qth torsion product* $\operatorname{Tor}_q(A,B) = H_q(C;B)$. It is a covariant functor of A and of B. From the short exact sequence

$$0 \to \partial_1 C_1 \to C_0 \xrightarrow{\varepsilon} A \to 0$$

it follows from corollary 5.1.6 that there is an exact sequence

$$\partial_1 C_1 \otimes B \to C_0 \otimes B \xrightarrow{\varepsilon \otimes 1} A \otimes B \to 0$$

By definition, $\operatorname{Tor}_0(A,B)$ is the zeroth homology module of the chain complex

$$\cdots \to C_2 \otimes B \to C_1 \otimes B \xrightarrow{\partial_1 \otimes 1} C_0 \otimes B \to 0$$

Hence $\operatorname{Tor}_0(A,B) = (C_0 \otimes B)/\operatorname{im}(\partial_1 \otimes 1)$. By the above exact sequence,

$$\operatorname{im}(\partial_1 \otimes 1) = \operatorname{im}(\partial_1 C_1 \otimes B \to C_0 \otimes B) = \ker(\varepsilon \otimes 1)$$

Therefore

$$\operatorname{Tor}_0(A,B) = (C_0 \otimes B)/\ker(\varepsilon \otimes 1) \approx A \otimes B$$

and so $\operatorname{Tor}_0(A,B)$ is naturally equivalent to $A \otimes B$.

All the previous remarks are valid for any commutative ring with a unit. For the remainder of this section we specialize to the case where R is a principal ideal domain. Over a principal ideal domain any submodule of a free module is free. Therefore any module A has a short free resolution of the form

$$0 \to C_1 \to C_0 \to A \to 0$$

(simply let $C_0 = F(A)$ and $C_1 = \ker[F(A) \to A]$). Such a short free resolution of A is the same as a free presentation of A. Because there exist short free resolutions, $\operatorname{Tor}_q(A,B) = 0$ if $q > 1$. We define the *torsion product* $A * B$ to equal $\operatorname{Tor}_1(A,B)$. It is characterized by the property that, given any free presentation of A,

$$0 \to C_1 \to C_0 \to A \to 0$$

there is an exact sequence

$$0 \to A * B \to C_1 \otimes B \to C_0 \otimes B \to A \otimes B \to 0$$

In fact, $A * B \approx H_1(C \otimes B) = \ker(C_1 \otimes B \to C_0 \otimes B)$, since $C_2 \otimes B = 0$.

The torsion product is a covariant functor of each of its arguments. Because the tensor product commutes with direct sums and direct limits (by theorems 5.1.8 and 5.1.9) and the direct limit of exact sequences is exact (by theorem 4.5.7), the torsion product also commutes with direct sums and direct limits. Its name derives from the fact that it depends only on the torsion submodules of A and B (see corollary 11 below).

3 EXAMPLE If A is free, it has the free presentation

$$0 \to 0 \to A \to A \to 0$$

from which we see that $A * B = 0$ for any B.

4 EXAMPLE If A is the cyclic R module whose annihilating ideal is generated by an element $v \in R$, then $A \approx R/vR$ and there is a free presentation of A

$$0 \to R \xrightarrow{\alpha} R \to A \to 0$$

in which $\alpha(v') = vv'$ for $v' \in R$. For any module B there is an isomorphism $R \otimes B \approx B$ sending $1 \otimes b$ to b. Under this isomorphism, the map $\alpha \otimes 1: R \otimes B \to R \otimes B$ corresponds to $\alpha': B \to B$, where $\alpha'(b) = vb$ for $b \in B$. Therefore ker α' is the submodule of B annihilated by v, and so

$$(R/vR) * B \approx \{b \in B \mid vb = 0\}$$

The above examples suffice to compute $A * B$ for a finitely generated module A (because of the structure theorem 4.14 in the Introduction). This theoretically determines $A * B$ for arbitrary A, because any A is the direct limit of its finitely generated submodules (see theorem 4.2 in the Introduction) and the torsion product commutes with direct limits.

5 LEMMA *If A or B is torsion free, then $A * B = 0$.*

PROOF Because the torsion product commutes with direct limits, it suffices to consider the case where A and B are finitely generated, in which case being torsion free is equivalent to being free. If A is free, the result follows from example 3. If B is free and finitely generated, it is isomorphic to a direct sum of n copies of R. If

$$0 \to C_1 \to C_0 \to A \to 0$$

is a free presentation of A, then $C_1 \otimes B \to C_0 \otimes B \to A \otimes B \to 0$ is isomorphic to a direct sum of n copies of the sequence $C_1 \otimes R \to C_0 \otimes R \to A \otimes R \to 0$. Since $C_1 \otimes R \to C_0 \otimes R$ is a monomorphism, so is $C_1 \otimes B \to C_0 \otimes B$, and $A * B = 0$. ∎

It follows that if R is a field, then $A * B = 0$ for all modules A and B. The following result is proved similarly by proving it first for finitely generated modules (where being torsion free is equivalent to being free) and taking direct limits to obtain the result for arbitrary modules.

6 LEMMA *Given a short exact sequence of modules*

$$0 \to A' \to A \to A'' \to 0$$

and given a module B, if A'' or B is torsion free, there is a short exact sequence

$$0 \to A' \otimes B \to A \otimes B \to A'' \otimes B \to 0$$

PROOF As remarked above, it suffices to prove the result if A'' or B is free and finitely generated. If A'' is free, the original sequence splits, by example 5.1.10, and the result follows from corollary 5.1.12. If B is free and finitely generated, the map $A' \otimes B \to A \otimes B$ is a finite direct sum of copies of $A' \otimes R \to A \otimes R$, and hence a monomorphism. The result follows from this and corollary 5.1.6. ∎

We use this result to obtain an exact sequence of homology corresponding to a short exact sequence of coefficient modules.

7 THEOREM *On the product category of torsion-free chain complexes C and short exact sequences of modules*

$$0 \to G' \xrightarrow{\varphi} G \xrightarrow{\psi} G'' \to 0$$

there is a natural connecting homomorphism

$$\beta: H(C;G'') \to H(C;G')$$

of degree -1 *and a functorial exact sequence*

$$\cdots \to H_q(C;G') \xrightarrow{\varphi_*} H_q(C;G) \xrightarrow{\psi_*} H_q(C;G'') \xrightarrow{\beta} H_{q-1}(C;G') \to \cdots$$

PROOF By lemma 6, there is a short exact sequence of chain complexes

$$0 \to C \otimes G' \xrightarrow{1 \otimes \varphi} C \otimes G \xrightarrow{1 \otimes \psi} C \otimes G'' \to 0$$

Since this is functorial in C and in the exact coefficient sequence, the result follows from theorem 4.5.4. ∎

The connecting homomorphism β occurring in theorem 7 is called the *Bockstein homology homomorphism* corresponding to the coefficient sequence $0 \to G' \xrightarrow{\varphi} G \xrightarrow{\psi} G' \to 0$. Theorem 7 remains valid over an arbitrary commutative ring R with a unit if C is assumed to be a free chain complex over R.

Let C be a chain complex over R and let G be an R module. Recall the homomorphism μ: $H(C) \otimes G \to H(C;G)$ defined in the last section. This homomorphism enters in the following *universal-coefficient theorem for homology.*

8 THEOREM *Let C be a free chain complex and let G be a module. There is a functorial short exact sequence*

$$0 \to H_q(C) \otimes G \xrightarrow{\mu} H_q(C \otimes G) \to H_{q-1}(C) * G \to 0$$

and this sequence is split.

PROOF Let Z be the subcomplex of C defined by $Z_q = Z_q(C)$ with trivial boundary operator and let B be the complex defined by $B_q = B_{q-1}(C)$ with trivial boundary operator. Both B and Z are free chain complexes and there is a short exact sequence

$$0 \to Z \xrightarrow{\alpha} C \xrightarrow{\beta} B \to 0$$

where $\alpha_q(z) = z$ for $z \in Z_q$ and $\beta_q(c) = \partial_q c$ for $c \in C_q$. Since B is a free complex, this short exact sequence is split. By theorem 5.1.13, there is an exact sequence

$$\cdots \to H_q(Z;G) \xrightarrow{\alpha_*} H_q(C;G) \xrightarrow{\beta_*} H_q(B;G) \xrightarrow{\partial_*} H_{q-1}(Z;G) \to \cdots$$

where $\partial_*\{b\} = \{\alpha_{q-1}^{-1}\partial_q\partial_q^{-1}b\} = \{\alpha_{q-1}^{-1}(b)\}$ for $b \in B_{q-1}$. Since Z and B have trivial boundary operators, so do $Z \otimes G$ and $B \otimes G$. Therefore $H_q(Z;G) = Z_q \otimes G$ and $H_q(B;G) = B_q \otimes G = B_{q-1}(C) \otimes G$, and the above exact sequence becomes

$$\cdots \to B_q(C) \otimes G \xrightarrow{\gamma_q \otimes 1} Z_q(C) \otimes G \to H_q(C;G) \to$$
$$B_{q-1}(C) \otimes G \xrightarrow{\gamma_{q-1} \otimes 1} Z_{q-1}(C) \otimes G \to \cdots$$

where γ_q: $B_q(C) \subset Z_q(C)$. From the exactness of this sequence we obtain a short exact sequence

$$0 \to \text{coker}\,(\gamma_q \otimes 1) \to H_q(C;G) \to \ker\,(\gamma_{q-1} \otimes 1) \to 0$$

and it only remains to interpret the modules on either side of $H_q(C;G)$.

Since $Z_q(C)$ is free, the short exact sequence

$$0 \to B_q(C) \xrightarrow{\gamma_q} Z_q(C) \to H_q(C) \to 0$$

is a free presentation of $H_q(C)$. By the characteristic property of the torsion product, there is an exact sequence

$$0 \to H_q(C) * G \to B_q(C) \otimes G \xrightarrow{\gamma_q \otimes 1} Z_q(C) \otimes G \to H_q(C) \otimes G \to 0$$

Therefore coker $(\gamma_q \otimes 1) \approx H_q(C) \otimes G$ and ker $(\gamma_q \otimes 1) \approx H_q(C) * G$. Substituting these into the short exact sequence above yields the short exact sequence

$$0 \to H_q(C) \otimes G \to H_q(C;G) \to H_{q-1}(C) * G \to 0$$

It is easily verified by checking the definitions that the homomorphism $H_q(C) \otimes G \to H_q(C;G)$ is equal to μ.

If τ: $C \to C'$ is a chain map, τ defines a commutative diagram

$$\begin{array}{ccccccccc} 0 & \to & Z & \xrightarrow{\alpha} & C & \xrightarrow{\beta} & B & \to & 0 \\ & & \downarrow{\scriptstyle \tau'} & & \downarrow{\scriptstyle \tau} & & \downarrow{\scriptstyle \tau''} & & \\ 0 & \to & Z' & \xrightarrow{\alpha'} & C' & \xrightarrow{\beta'} & B' & \to & 0 \end{array}$$

from which we obtain the commutative diagram

$$\begin{array}{ccccccccc} 0 & \to & H_q(C) \otimes G & \xrightarrow{\mu} & H_q(C;G) & \to & H_{q-1}(C) * G & \to & 0 \\ & & \downarrow{\scriptstyle \tau_* \otimes 1} & & \downarrow{\scriptstyle \tau_*} & & \downarrow{\scriptstyle \tau_* * 1} & & \\ 0 & \to & H_q(C') \otimes G & \xrightarrow{\mu} & H_q(C';G) & \to & H_{q-1}(C') * G & \to & 0 \end{array}$$

Therefore the short exact sequence for $H_q(C;G)$ is functorial.

We now prove that the short exact sequence is split (but is not functorially split). Because $B_{q-1}(C)$ is free and $\partial_q C_q = B_{q-1}(C)$, there exist homomorphisms h_q: $B_{q-1}(C) \to C_q$ such that $\partial_q h_q = 1$. Then

$$h_q \otimes 1\colon B_{q-1}(C) \otimes G \to C_q \otimes G$$

maps the kernel of $\gamma_{q-1} \otimes 1$ into cycles of $C_q \otimes G$ and induces a homomorphism $H_{q-1}(C) * G \to H_q(C;G)$ which is a right inverse of the homomorphism $H_q(C;G) \to H_{q-1}(C) * G$ of the short exact sequence in the theorem. ∎

We can use this result to establish some properties of the torsion product, beginning with the following *six-term exact sequence* connecting the tensor and torsion products.

9 COROLLARY *Let* $0 \to B' \xrightarrow{\alpha'} B \xrightarrow{\beta'} B'' \to 0$ *be a short exact sequence of modules and let A be a module. There is an exact sequence*

$$0 \to A * B' \xrightarrow{1 * \alpha'} A * B \xrightarrow{1 * \beta'} A * B'' \to$$
$$A \otimes B' \xrightarrow{1 \otimes \alpha'} A \otimes B \xrightarrow{1 \otimes \beta'} A \otimes B'' \to 0$$

PROOF Let $0 \to C_1 \to C_0 \to A \to 0$ be a free presentation of A and let C be the corresponding free chain complex obtained by adding trivial groups on both sides. Since C is free, it follows from lemma 6 that there is a short exact sequence of chain complexes

$$0 \to C \otimes B' \xrightarrow{1 \otimes \alpha'} C \otimes B \xrightarrow{1 \otimes \beta'} C \otimes B'' \to 0$$

Because $H_q(C) = 0$ if $q \neq 0$ and $H_0(C) = A$, the homology sequence of the above short exact sequence of chain complexes (interpreted by means of theorem 8) gives the desired exact sequence. ∎

This yields the commutativity of the torsion product.

10 COROLLARY *There is a functorial isomorphism*

$$A * B \approx B * A$$

PROOF Let $0 \to C_1 \to C_0 \to B \to 0$ be a free presentation of B. By corollary 9, there is an exact sequence

$$0 \to A * C_1 \to A * C_0 \to A * B \to A \otimes C_1 \to A \otimes C_0 \to A \otimes B \to 0$$

Since C_0 is free, it follows from lemma 5 that $A * C_0 = 0$, and there is an exact sequence

$$0 \to A * B \to A \otimes C_1 \to A \otimes C_0 \to A \otimes B \to 0$$

By the characteristic property of $B * A$, there is an exact sequence

$$0 \to B * A \to C_1 \otimes A \to C_0 \otimes A \to B \otimes A \to 0$$

The functorial isomorphism $A * B \approx B * A$ then results by chasing in the commutative diagram

$$\begin{array}{ccccccccc} 0 \to & A * B & \to & A \otimes C_1 & \to & A \otimes C_0 & \to & A \otimes B & \to 0 \\ & & & \downarrow \approx & & \downarrow \approx & & \downarrow \approx & \\ 0 \to & B * A & \to & C_1 \otimes A & \to & C_0 \otimes A & \to & B \otimes A & \to 0 \end{array}$$

in which the vertical maps are the functorial isomorphisms expressing the commutativity of the tensor product. ■

We can now show that the torsion product of A and B depends only on the torsion submodules of A and B.

11 COROLLARY *Let A and B be modules and let i: Tor $A \subset A$ and j: Tor $B \subset B$. Then $i * j$: Tor $A *$ Tor $B \approx A * B$.*

PROOF There is a short exact sequence

$$0 \to \text{Tor } B \xrightarrow{j} B \to B/\text{Tor } B \to 0$$

where $B/\text{Tor } B$ is without torsion. By lemma 5, $A * (B/\text{Tor } B) = 0$, and, by corollary 9, $1 * j$: $A *$ Tor $B \approx A * B$. By a similar argument, there is an isomorphism $i * 1$: Tor $A *$ Tor $B \approx A *$ Tor B, and the composite of these gives the result. ■

We use these results to extend the universal-coefficient theorem. Given a chain complex C over R, a *free approximation of* C is a chain map $\tau: \bar{C} \to C$ such that

(*a*) $\bar{C}$ is a free chain complex over R.
(*b*) τ is an epimorphism.
(*c*) τ induces an isomorphism $\tau_*: H(\bar{C}) \approx H(C)$.

12 LEMMA *Any chain complex C has a free approximation, uniquely determined up to homotopy equivalence.*

PROOF For each $q \geq 0$ choose a homomorphism α_q: $F_q \to Z_q(C)$ such that F_q is a free R module and α_q is an epimorphism. Let $F'_q = \alpha_q^{-1}(B_q(C))$ and choose a homomorphism β_q: $F'_q \to C_{q+1}$ such that $\partial_{q+1}\beta_q = \alpha_q \mid F'_q$ [such a homomorphism exists because F'_q is free and ∂_{q+1}: $C_{q+1} \to B_q(C)$ is an epimorphism]. Define $\bar{C}_q = F_q \oplus F'_{q-1}$ and define homomorphisms

$$\bar{\partial}_q: \bar{C}_q \to \bar{C}_{q-1} \qquad \text{by} \qquad \bar{\partial}_q(a,b) = (b,0)$$

$$\tau_q: \bar{C}_q \to C_q \qquad \text{by} \qquad \tau_q(a,b) = \alpha_q(a) + \beta_{q-1}(b)$$

Then $\bar{C} = \{\bar{C}_q,\bar{\partial}_q\}$ is a free chain complex and $\tau = \{\tau_q\}$ is a chain map from $\bar{C}$ to C. τ is epimorphic because $\tau_q(\bar{C}_q) \supset \ker \partial_q$ and $\partial_q\tau_q(\bar{C}_q) \supset \operatorname{im} \partial_q$. Since $Z_q(\bar{C}) = F_q$, $B_q(C) = F'_q$, and $\tau_q(Z_q(\bar{C})) = \alpha_q(F_q)$, it follows that

$$\tau_*: Z_q(\bar{C})/B_q(\bar{C}) \approx Z_q(C)/B_q(C)$$

Therefore $\tau: \bar{C} \to C$ is a free approximation of C. The uniqueness will follow from lemma 13 below. ■

If $\tau: \bar{C} \to C$ is a free approximation of C, there is a subcomplex $\bar{\bar{C}} = \{\bar{\bar{C}}_q = \ker \tau_q: \bar{C}_q \to C_q\}$ of $\bar{C}$ and a short exact sequence of chain complexes

$$0 \to \bar{\bar{C}} \xrightarrow{i} \bar{C} \xrightarrow{\tau} C \to 0$$

Because $\tau_*: H(\bar{C}) \approx H(C)$, it follows from the exactness of the homology sequence of the above short exact sequence that $\bar{\bar{C}}$ is acyclic (see corollary 4.5.5*a*). Since $\bar{\bar{C}}$ is a free chain complex (because it is a subcomplex of a free chain complex), it follows from theorem 4.2.5 that $\bar{\bar{C}}$ is contractible. We use this in the following lemma.

13 LEMMA *Given a free approximation $\tau: \bar{C} \to C$ of C and given a free chain complex C' and a chain map $\tau': C' \to C$, there exist chain maps $\bar{\tau}: C' \to \bar{C}$ such that $\tau \circ \bar{\tau} = \tau'$, and any two are chain homotopic.*

PROOF As above, there is a short exact sequence of chain complexes

$$0 \to \bar{\bar{C}} \xrightarrow{i} \bar{C} \xrightarrow{\tau} C \to 0$$

where $\bar{\bar{C}}$ is chain contractible. Let $D = \{D_q: \bar{\bar{C}}_q \to \bar{\bar{C}}_{q+1}\}$ be a contraction of $\bar{\bar{C}}$. Because C'_q is free and $\tau_q: \bar{C}_q \to C_q$ is an epimorphism, there is a homomorphism $\varphi_q: C'_q \to \bar{C}_q$ such that $\tau_q\varphi_q = \tau'_q$. Then

$$h_q = \bar{\partial}_q\varphi_q - \varphi_{q-1}\partial'_q: C'_q \to \bar{C}_{q-1}$$

and

$$\begin{aligned}\tau_{q-1}h_q &= \tau_{q-1}\bar{\partial}_q\varphi_q - \tau_{q-1}\varphi_{q-1}\partial'_q = \partial_q\tau_q\varphi_q - \tau'_{q-1}\partial'_q \\ &= \partial_q\tau'_q - \tau'_{q-1}\partial'_q = 0\end{aligned}$$

Therefore h_q is a homomorphism of C'_q into $i(\bar{\bar{C}}_{q-1})$. It follows immediately that $\bar{\tau} = \{\bar{\tau}_q = \varphi_q - iD_{q-1}i^{-1}h_q\}$ is a chain map $\bar{\tau}: C' \to \bar{C}$ such that $\tau\bar{\tau} = \tau'$.

If $\bar{\tau}, \bar{\tau}': C' \to \bar{C}$ are chain maps such that $\tau\bar{\tau} = \tau\bar{\tau}'$, then $\bar{\tau} - \bar{\tau}' = i\psi$ for some chain map $\psi: C' \to \bar{\bar{C}}$. It follows immediately that

$$\bar{D} = \{\bar{D}_q = iD_q\psi_q: C'_q \to \bar{C}_{q+1}\}$$

is a chain homotopy from $\bar{\tau}$ to $\bar{\tau}'$. ∎

If C is a chain complex over R and G is an R module, let $C * G$ be the chain complex $C * G = \{C_q * G, \partial_q * 1\}$. We use this in the general universal-coefficient theorem.

14 THEOREM *On the subcategory of the product category of chain complexes C and modules G such that $C * G$ is acyclic there is a functorial short exact sequence*

$$0 \to H_q(C) \otimes G \xrightarrow{\mu} H_q(C;G) \to H_{q-1}(C) * G \to 0$$

and this sequence is split.

PROOF Let $\tau: \bar{C} \to C$ be a free approximation to C (which exists, by lemma 12), and consider the short exact sequence

$$0 \to \bar{\bar{C}} \xrightarrow{i} \bar{C} \xrightarrow{\tau} C \to 0$$

in which $\bar{\bar{C}}$ is acyclic. By the characteristic property of the torsion product, there is an exact sequence of chain complexes

$$0 \to C * G \to \bar{\bar{C}} \otimes G \xrightarrow{i \otimes 1} \bar{C} \otimes G \xrightarrow{\tau \otimes 1} C \otimes G \to 0$$

from which we get two short exact sequences

$$0 \to C * G \to \bar{C} \otimes G \to \operatorname{im}(i \otimes 1) \to 0$$

$$0 \to \operatorname{im}(i \otimes 1) \subset \bar{C} \otimes G \xrightarrow{\tau \otimes 1} C \otimes G \to 0$$

In the first of these $C * G$ is acyclic by hypothesis, and $\bar{C} \otimes G$ is also acyclic (by theorem 8, because $\bar{C}$ is free and acyclic). From corollary 4.5.5c it follows that im $(i \otimes 1)$ is also acyclic. In the second exact homology sequence this implies that

$$(\tau \otimes 1)_*: H(\bar{C} \otimes G) \approx H(C \otimes G)$$

The desired short exact sequence is now defined, so that the following diagram is commutative

$$\begin{array}{ccccccccc} 0 & \to & H_q(\bar{C}) \otimes G & \xrightarrow{\mu} & H_q(\bar{C} \otimes G) & \to & H_{q-1}(\bar{C}) * G & \to & 0 \\ & & \downarrow{\scriptstyle \tau_* \otimes 1} & & \downarrow{\scriptstyle (\tau \otimes 1)_*} & & \downarrow{\scriptstyle \tau_* * 1} & & \\ 0 & \to & H_q(C) \otimes G & \xrightarrow{\mu} & H_q(C \otimes G) & \to & H_{q-1}(C) * G & \to & 0 \end{array}$$

where the upper row is the short exact sequence of theorem 8 (it is possible to define the unlabeled homomorphism in the bottom sequence to make the diagram commutative because all the vertical homomorphisms are isomorphisms). Then the bottom sequence splits because the top one does.

The functorial property of the resulting short exact sequence (and the fact that it is independent of the particular free approximation of C) follows from lemma 13. ∎

It should be emphasized again that the sequence of theorem 14 does not split functorially.

15 COROLLARY *Let $\tau: C \to C'$ be a chain map between torsion-free chain complexes such that $\tau_*: H(C) \approx H(C')$. For any R module G, τ induces an isomorphism*

$$\tau_*: H(C;G) \approx H(C';G)$$

PROOF This follows from the functorial exact sequence of theorem 14 and the five lemma. ∎

In corollary 15, if C and C' are free, then τ is a chain equivalence (by theorem 4.6.10), and so is $\tau \otimes 1: C \otimes G \to C' \otimes G$. Therefore $\tau_*: H(C;G) \approx H(C';G)$. Corollary 15 shows that the latter fact remains true (even though τ need not be a chain equivalence) for chain complexes without torsion.

3 THE KÜNNETH FORMULA

In this section we extend the universal-coefficient theorem to obtain the Künneth formula expressing the homology of the tensor product of two chain complexes in terms of the homology of the factors. This is given geometric content by the Eilenberg-Zilber theorem asserting that the singular complex of a product space is chain equivalent to the tensor product of the singular complexes of the factor spaces.

If C and C' are graded R modules, their *tensor product* $C \otimes C'$ is the graded module $\{(C \otimes C')_q\}$, where $(C \otimes C')_q = \bigoplus_{i+j=q} C_i \otimes C'_j$. Similarly, their *torsion product* $C * C'$ is the graded module $\{(C * C')_q = \bigoplus_{i+j=q} C_i * C'_j\}$. If C and C' are chain complexes, their tensor product [and torsion product] are chain complexes $\{(C \otimes C')_q, \partial''_q\}$ [and $\{(C * C')_q, \bar\partial_q\}$], where if $c \in C_i$ and $c' \in C'_j$ with $i + j = q$, then

$$\partial''_q(c \otimes c') = \partial_i c \otimes c' + (-1)^i c \otimes \partial'_j c'$$

[and $\bar\partial_q \mid C_i * C'_j = \partial_i * 1 + (-1)^i 1 * \partial'_j$]. It is easy to verify that $C \otimes C'$ [and $C * C'$] really are chain complexes. We shall see later that the tensor product arises naturally in studying product spaces.

If C' is a chain complex such that $C'_q = 0$ for $q \neq 0$, then $C \otimes C'$ is the same as the tensor product of C with the module C'_0. Therefore the tensor product of two chain complexes is a natural generalization of the tensor product of a chain complex with a module. It is reasonable to expect that there is a generalization of the universal-coefficient theorem to express the homology of $C \otimes C'$ in terms of the homology of C and of C'.

We define a functorial homomorphism of degree 0

$$\mu: H(C) \otimes H(C') \to H(C \otimes C')$$

If $c \in Z_i(C)$ and $c' \in Z_j(C')$, then $c \otimes c' \in Z_{i+j}(C \otimes C')$, and if c or c' is a boundary, so is $c \otimes c'$. Therefore there is a well-defined homomorphism μ such that

$$\mu(\{c\} \otimes \{c'\}) = \{c \otimes c'\}$$

This homomorphism enters in the following *Künneth formula.*

1 LEMMA *Let C and C' be chain complexes, with C' free. Then there is a functorial short exact sequence*

$$0 \to [H(C) \otimes H(C')]_q \xrightarrow{\mu} H_q(C \otimes C') \to [H(C) * H(C')]_{q-1} \to 0$$

If C is also free, this short exact sequence is split.

PROOF As in the proof of theorem 5.2.8, let Z' and B' be the complexes (with trivial boundary operators) defined by $Z'_q = Z_q(C')$ and $B'_q = B_{q-1}(C')$. There is a short exact sequence of chain complexes

$$0 \to Z' \to C' \to B' \to 0$$

Since C' is free, so is B', and there is a short exact sequence

$$0 \to C \otimes Z' \to C \otimes C' \to C \otimes B' \to 0$$

from which we obtain an exact homology sequence

$$\cdots \to H_q(C \otimes Z') \to H_q(C \otimes C') \to H_q(C \otimes B') \xrightarrow{\partial_*} H_{q-1}(C \otimes Z') \to \cdots$$

Note that $C \otimes Z' = \bigoplus C^j$, where $(C^j)_q = C_{q-j} \otimes Z_j(C')$ and $C \otimes B' = \bigoplus \bar{C}^j$, where $(\bar{C}^j)_q = C_{q-j} \otimes B_{j-1}(C')$. Since $Z_j(C')$ and $B_j(C')$ are free, it follows from theorem 5.2.14 that

$$H_q(C \otimes Z') = \bigoplus_j H_q(C^j) = \bigoplus_{i+j=q} H_i(C) \otimes Z_j(C')$$

$$H_q(C \otimes B') = \bigoplus_j H_q(\bar{C}^j) = \bigoplus_{i+j=q-1} H_i(C) \otimes B_j(C')$$

The map ∂_* corresponds under these isomorphisms to the homomorphism $(-1)^i \otimes \gamma_j$, where γ_j is the inclusion map $\gamma_j: B_j(C') \subset Z_j(C')$. Therefore there is a short exact sequence

$$0 \to \bigoplus_{i+j=q} [\text{coker } (-1)^i \otimes \gamma_j] \to H_q(C \otimes C') \to \bigoplus_{i+j=q-1} [\ker (-1)^i \otimes \gamma_j] \to 0$$

To compute the two sides of this sequence, consider the short exact sequence

$$0 \to B_j(C') \xrightarrow{(-1)^i\gamma_j} Z_j(C') \to H_j(C') \to 0$$

Because $Z_j(C')$ is free, it follows from corollary 5.2.9 that there is an exact sequence

$$0 \to H_i(C) * H_j(C') \to H_i(C) \otimes B_j(C') \xrightarrow{(-1)^i \otimes \gamma_j} H_i(C) \otimes Z_j(C') \to H_i(C) \otimes H_j(C') \to 0$$

Hence

$$\bigoplus_{i+j=q} [\text{coker } (-1)^i \otimes \gamma_j] = \bigoplus_{i+j=q} H_i(C) \otimes H_j(C')$$

and

$$\bigoplus_{i+j=q-1} [\ker (-1)^i \otimes \gamma_j] = \bigoplus_{i+j=q-1} H_i(C) * H_j(C')$$

Substituting these into the short exact sequence above gives a short exact sequence

$$0 \to [H(C) \otimes H(C')]_q \xrightarrow{\nu} H_q(C \otimes C') \to [H(C) * H(C')]_{q-1} \to 0$$

We now verify that ν is the map μ. Given $\{c\} \in H(C)$ and $\{c'\} \in H(C')$, then $\{c\} \otimes c' \in H(C) \otimes Z(C')$ and $\{c\} \otimes c' = \{c \otimes c'\}_{C\otimes Z(C')}$. Therefore $\nu(\{c\} \otimes \{c'\}) = \{c \otimes c'\}_{C\otimes C'} = \mu(\{c\} \otimes \{c'\})$. Thus we have the desired short exact sequence, and it is clearly functorial.

Assuming that C is also free, we can show that the sequence splits. By lemma 5.1.11, it suffices to find a left inverse for μ. Because C and C' are free, so are $B(C)$ and $B(C')$, and there are homomorphisms $p: C \to Z(C)$ and $p': C' \to Z(C')$ such that $p(c) = c$ for $c \in Z(C)$ and $p'(c') = c'$ for $c' \in Z(C')$. Then

$$p \otimes p': C \otimes C' \to Z(C) \otimes Z(C')$$

maps $B(C \otimes C')$(which is contained in the union of im $[B(C) \otimes C' \to C \otimes C']$ and im $[C \otimes B(C') \to C \otimes C']$) into the union of im $[B(C) \otimes Z(C') \to Z(C) \otimes Z(C')]$ and im $[Z(C) \otimes B(C') \to Z(C) \otimes Z(C')]$. Therefore the composite

$$Z(C \otimes C') \subset C \otimes C' \xrightarrow{p \otimes p'} Z(C) \otimes Z(C') \to H(C) \otimes H(C')$$

maps $B(C \otimes C')$ into 0 and induces a homomorphism

$$H(C \otimes C') \to H(C) \otimes H(C')$$

which is a left inverse of μ. ∎

A similar functorial short exact sequence can be defined if C (instead of C') is assumed free. The two short exact sequences are identical when C and C' are both free.[1]

2 COROLLARY *If C' is a free chain complex and either C or C' is acyclic, then $C \otimes C'$ is acyclic.* ∎

We now extend lemma 1 to obtain the following general Künneth formula.

3 THEOREM *On the subcategory of the product category of chain complexes C and C' such that $C * C'$ is acyclic there is a functorial short exact sequence*

$$0 \to [H(C) \otimes H(C')]_q \xrightarrow{\mu} H_q(C \otimes C') \to [H(C) * H(C')]_{q-1} \to 0$$

and this sequence is split.

PROOF Let $\tau: \bar{C} \to C$ and $\tau': \bar{C}' \to C'$ be free approximations. Then there is a short exact sequence

$$0 \to \bar{\bar{C}}' \xrightarrow{i'} \bar{C}' \xrightarrow{\tau'} C' \to 0$$

where $\bar{\bar{C}}'$ is acyclic. Since $\bar{C}'$ is free, the six-term exact sequence becomes the exact sequence

$$0 \to C * C' \to C \otimes \bar{\bar{C}}' \to C \otimes \bar{C}' \xrightarrow{1 \otimes \tau'} C \otimes C' \to 0$$

Since $C * C'$ is acyclic by hypothesis and $C \otimes \bar{\bar{C}}'$ is acyclic by corollary 2, it follows (as in the proof of theorem 5.2.14) that there is an isomorphism

$$(1 \otimes \tau')_*: H(C \otimes \bar{C}') \approx H(C \otimes C')$$

There is also a short exact sequence

$$0 \to \bar{\bar{C}} \xrightarrow{i} \bar{C} \xrightarrow{\tau} C \to 0$$

where $\bar{\bar{C}}$ is acyclic. Since $\bar{C}'$ is free, there is a short exact sequence

$$0 \to \bar{\bar{C}} \otimes \bar{C}' \to \bar{C} \otimes \bar{C}' \xrightarrow{\tau \otimes 1} C \otimes \bar{C}' \to 0$$

By corollary 2, $\bar{\bar{C}} \otimes \bar{C}'$ is acyclic, and we have an isomorphism

$$(\tau \otimes 1)_*: H(\bar{C} \otimes \bar{C}') \approx H(C \otimes \bar{C}')$$

[1] This is proved in G. M. Kelley, Observations on the Künneth theorem, *Proceedings of the Cambridge Philosophical Society*, vol. 59, pp. 575–587, 1963.

Hence the composite $(\tau \otimes \tau')_* = (1 \otimes \tau')_*(\tau \otimes 1)_*$ is an isomorphism of $H(\bar{C} \otimes \bar{C}')$ onto $H(C \otimes C')$. The desired short exact sequence is now defined so that the following diagram is commutative

$$\begin{array}{ccccccccc} 0 \to & H(\bar{C}) \otimes H(\bar{C}') & \xrightarrow{\mu} & H(\bar{C} \otimes \bar{C}') & \to & H(\bar{C}) * H(\bar{C}') & \to 0 \\ & \downarrow \tau_* \otimes \tau'_* & & \downarrow (\tau \otimes \tau')_* & & \downarrow \tau_* * \tau'_* & \\ 0 \to & H(C) \otimes H(C') & \xrightarrow{\mu} & H(C \otimes C') & \to & H(C) * H(C') & \to 0 \end{array}$$

where the top row is the short exact sequence of lemma 1 (it is possible to define the homomorphisms in the bottom row to make the diagram commutative because the vertical homomorphisms are isomorphisms). The bottom sequence splits because the top one does.

The functorial property of the sequence (and the fact that it is independent of the free approximations $\bar{C}$ and $\bar{C}'$) follow from the functorial property of the sequence in lemma 1 and from lemma 5.2.13. ∎

If C and C' are chain complexes over R and G and G' are R modules, the composite

$$H(C \otimes G) \otimes H(C' \otimes G') \xrightarrow{\mu} H[(C \otimes G) \otimes (C' \otimes G')] \to H[(C \otimes C') \otimes (G \otimes G')]$$

[where the right-hand homomorphism is induced by the canonical isomorphism $(C \otimes G) \otimes (C' \otimes G') \approx (C \otimes C') \otimes (G \otimes G')$] is a functorial homomorphism

$$\mu' \colon H(C;G) \otimes H(C';G') \to H(C \otimes C'; G \otimes G')$$

called the *cross product*. If $z \in H(C;G)$ and $z' \in H(C';G')$, then

$$z \times z' \in H(C \otimes C'; G \otimes G')$$

denotes the image of $z \otimes z'$ under this homomorphism [that is, $z \times z' = \mu'(z \otimes z')$].

4 COROLLARY *Given torsion-free chain complexes C and C' and modules G and G' such that $G * G' = 0$, there is a functorial short exact sequence*

$$0 \to [H(C;G) \otimes H(C';G')]_q \xrightarrow{\mu'} H_q(C \otimes C'; G \otimes G') \to [H(C;G) * H(C',G')]_{q-1} \to 0$$

and this sequence is split.

PROOF This follows from theorem 3 once we verify that $(C \otimes G) * (C' \otimes G')$ is trivial. To show that $(C \otimes G) * (C' \otimes G') = 0$, let $0 \to F' \to F \to G$ be a free presentation of G. Because $G * G' = 0$, there is an exact sequence

$$0 \to F' \otimes G' \to F \otimes G' \to G \otimes G' \to 0$$

and since C and C' are without torsion, there is an exact sequence

$$0 \to (C \otimes F') \otimes (C' \otimes G') \to (C \otimes F) \otimes (C' \otimes G') \to (C \otimes G) \otimes (C' \otimes G') \to 0$$

Because there is also a short exact sequence

$$0 \to C \otimes F' \to C \otimes F \to C \otimes G \to 0$$

where $C \otimes F$ is without torsion, it follows that $(C \otimes G) * (C' \otimes G')$ is isomorphic to the kernel of the homomorphism

$$(C \otimes F') \otimes (C' \otimes G') \to (C \otimes F) \otimes (C' \otimes G')$$

and hence is 0. ■

The cross product has the following commutativity with connecting homomorphisms.

5 THEOREM *Let $0 \to \bar{\bar{C}} \to \bar{C} \to C \to 0$ be a split short exact sequence of chain complexes and let $z \in H(C;G)$ and $z' \in H(C';G')$. Then*

$$\partial_*(z \times z') = \partial_* z \times z'$$
$$\partial_*(z' \times z) = (-1)^{\deg z'} z' \times \partial_* z$$

PROOF We have a commutative diagram of chain maps

$$\begin{array}{ccccccc}
0 \to & \bar{\bar{C}} \otimes G & \to & \bar{C} \otimes G & \to & C \otimes G & \to 0 \\
& \bar{\bar{\tau}}\downarrow & & \downarrow\bar{\tau} & & \downarrow\tau & \\
0 \to & (\bar{\bar{C}} \otimes G) \otimes (C' \otimes G') & \to & (\bar{C} \otimes G) \otimes (C' \otimes G') & \to & (C \otimes G) \otimes (C' \otimes G') & \to 0
\end{array}$$

with exact rows, with the vertical maps defined by forming the tensor product on the right with $c' \in Z(C' \otimes G')$, where $z' = \{c'\}$ [that is, $\tau(c) = c \otimes c'$ for $c \in C \otimes G$]. Because c' is a cycle, each vertical map is a chain map. Because the connecting homomorphism is functorial, we obtain a commutative diagram

$$\begin{array}{ccccc}
H(C \otimes G) & \xrightarrow{\tau_*} & H((C \otimes G) \otimes (C' \otimes G')) & \xrightarrow{\approx} & H((C \otimes C') \otimes (G \otimes G')) \\
\partial_*\downarrow & & \partial_*\downarrow & & \downarrow\partial_* \\
H(\bar{\bar{C}} \otimes G) & \xrightarrow{\bar{\bar{\tau}}_*} & H((\bar{\bar{C}} \otimes G) \otimes (C' \otimes G')) & \xrightarrow{\approx} & H((\bar{\bar{C}} \otimes C') \otimes (G \otimes G'))
\end{array}$$

in which each vertical map is a suitable connecting homomorphism. The top row sends z into $z \times z'$, and the bottom row sends $\partial_* z$ into $\partial_* z \times z'$. This gives half the result. The second half follows by a similar argument, the only difference being that the tensor product formed on the left with c' is not a chain map but either commutes or anticommutes with the boundary operator, depending on the degree of c'. This accounts for the presence of the factor $(-1)^{\deg z'}$ in the second equation. ■

The following *Eilenberg-Zilber theorem*[1] is the link between the algebra of tensor products and the geometry of product spaces.

6 THEOREM *On the category of ordered pairs of topological spaces X and Y there is a natural chain equivalence of the functor $\Delta(X \times Y)$ with the functor $\Delta(X) \otimes \Delta(Y)$.*

[1] The theorem appears in S. Eilenberg and J. A. Zilber, On products of complexes, *American Journal of Mathematics*, vol. 75, pp. 200–204, 1953.

PROOF We show that both functors are free and acyclic with models $\{\Delta^p,\Delta^q\}_{p,q\geq 0}$. Let $d_n \in \Delta_n(\Delta^n \times \Delta^n)$ be the singular simplex which is the diagonal map $\Delta^n \to \Delta^n \times \Delta^n$. If $\sigma: \Delta^n \to X \times Y$ is any singular n-simplex, then σ is the composite

$$\Delta^n \xrightarrow{d_n} \Delta^n \times \Delta^n \xrightarrow{\sigma' \times \sigma''} X \times Y$$

where $\sigma' = p_1 \circ \sigma$ and $\sigma'' = p_2 \circ \sigma$, and p_1 and p_2 are the projections of $X \times Y$ to X and Y, respectively. Conversely, given $\sigma': \Delta^n \to X$ and $\sigma'': \Delta^n \to Y$, there is a corresponding $\sigma = (\sigma' \times \sigma'')d_n: \Delta^n \to X \times Y$. Therefore the singleton $\{d_n\}$ is a basis for $\Delta_n(X \times Y)$, so $\Delta(X \times Y)$ is free with models $\{\Delta^n,\Delta^n\}$, and hence also free with models $\{\Delta^p,\Delta^q\}$. Since Δ^p and Δ^q are each contractible, so is $\Delta^p \times \Delta^q$. Therefore $\tilde{\Delta}(\Delta^p \times \Delta^q)$ is acyclic, and we have proved that $\Delta(X \times Y)$ is a free acyclic functor with models $\{\Delta^p,\Delta^q\}$.

Since $\Delta_p(X)$ is free with a basis $\xi_p \in \Delta_p(\Delta^p)$ and $\Delta_q(Y)$ is free with basis $\xi_q \in \Delta_q(\Delta^q)$, it follows that $\Delta_p(X) \otimes \Delta_q(Y)$ is free with the basis

$$\xi_p \otimes \xi_q \in \Delta_p(\Delta^p) \otimes \Delta_q(\Delta^q).$$

Then $[\Delta(X) \otimes \Delta(Y)]_n$ is free with the basis $\{\xi_p \otimes \xi_q\}_{p+q=n}$. Hence $\Delta(X) \otimes \Delta(Y)$ is free with models $\{\Delta^p,\Delta^q\}$. Since $\varepsilon: \Delta(\Delta^p) \to \mathbf{Z}$ and $\varepsilon: \Delta(\Delta^q) \to \mathbf{Z}$ are chain equivalences, it follows that

$$\varepsilon \otimes \varepsilon: \Delta(\Delta^p) \otimes \Delta(\Delta^q) \to \mathbf{Z} \otimes \mathbf{Z} = \mathbf{Z}$$

is also a chain equivalence. Hence, by lemma 4.3.2, the reduced complex of $\Delta(\Delta^p) \otimes \Delta(\Delta^q)$ is acyclic, and we have shown that $\Delta(X) \otimes \Delta(Y)$ is also free and acyclic with models $\{\Delta^p,\Delta^q\}$. The theorem now follows by the method of acyclic models. ∎

The same technique based on the method of acyclic models can be used to prove the following results.

7 THEOREM *Given X, Y, and Z, there is a chain homotopy commutative diagram*

$$\begin{array}{ccc} \Delta((X \times Y) \times Z) & \approx & \Delta(X \times (Y \times Z)) \\ \approx\updownarrow & & \updownarrow\approx \\ [\Delta(X) \otimes \Delta(Y)] \otimes \Delta(Z) & \approx & \Delta(X) \otimes [\Delta(Y) \otimes \Delta(Z)] \end{array}$$

where the vertical maps are the natural chain equivalences of theorem 6. ∎

8 THEOREM *For any X and Y there is a chain homotopy commutative diagram*

$$\begin{array}{ccc} \Delta(X \times Y) & \approx & \Delta(Y \times X) \\ \approx\updownarrow & & \updownarrow\approx \\ \Delta(X) \otimes \Delta(Y) & \approx & \Delta(Y) \otimes \Delta(X) \end{array}$$

where the bottom map sends $x \otimes y$ *to* $(-1)^{\deg x \deg y}\, y \otimes x$ *and the vertical maps are the natural chain equivalences of theorem 6.* ∎

The sign in theorem 8 is necessary to make the map a chain map (that is, to make it commute with the boundary operators).

Given topological pairs (X,A) and (Y,B), we define their *product* $(X,A) \times (Y,B)$ to be the pair $(X \times Y, X \times B \cup A \times Y)$. Then we have the following relative form of the Eilenberg-Zilber theorem.

9 THEOREM *On the category of ordered pairs of topological pairs* (X,A) *and* (Y,B) *such that* $\{X \times B, A \times Y\}$ *is an excisive couple in* $X \times Y$ *there is a natural chain equivalence of* $\Delta(X \times Y)/\Delta(X \times B \cup A \times Y)$ *with* $[\Delta(X)/\Delta(A)] \otimes [\Delta(Y)/\Delta(B)]$.

PROOF Because $\{X \times B, A \times Y\}$ is an excisive couple, the natural map

$$\Delta(X \times Y)/[\Delta(X \times B) + \Delta(A \times Y)] \to \Delta(X \times Y)/\Delta(X \times B \cup A \times Y)$$

is a chain equivalence. By theorem 6 there is a functorial equivalence of $\Delta(X) \otimes \Delta(Y)$ with $\Delta(X \times Y)$ taking $\Delta(X) \otimes \Delta(B)$ and $\Delta(A) \otimes \Delta(Y)$ into $\Delta(X \times B)$ and $\Delta(A \times Y)$, respectively. Hence there is a functorial chain equivalence of the quotient

$$\Delta(X) \otimes \Delta(Y)/[\Delta(X) \otimes \Delta(B) + \Delta(A) \otimes \Delta(Y)] \approx [\Delta(X)/\Delta(A)] \otimes [\Delta(Y)/\Delta(B)]$$

with the quotient

$$\Delta(X \times Y)/[\Delta(X \times B) + \Delta(A \times Y)]$$

Combining these two chain equivalences gives the result. ∎

For any two pairs (X,A) and (Y,B) we define the *homology cross product*

$$\mu'\colon H_p(X,A;\, G) \otimes H_q(Y,B;\, G') \to H_{p+q}((X,A) \times (Y,B);\, G \otimes G')$$

to be equal to the cross product

$$H_p([\Delta(X)/\Delta(A)] \otimes G) \otimes H_q([\Delta(Y)/\Delta(B)] \otimes G')$$
$$\downarrow$$
$$H_{p+q}(([\Delta(X)/\Delta(A)] \otimes [\Delta(Y)/\Delta(B)]) \otimes (G \otimes G'))$$

followed by the functorial homomorphism of the bottom module to

$$H_{p+q}(\Delta(X \times Y)/\Delta(X \times B \cup A \times Y);\, G \otimes G')$$

If $z \in H_p(X,A;\, G)$ and $z' \in H_q(Y,B;\, G')$, then we write

$$z \times z' = \mu'(z \otimes z') \in H_{p+q}((X,A) \times (Y,B);\, G \otimes G')$$

Because $\Delta(X)/\Delta(A)$ and $\Delta(Y)/\Delta(B)$ are free, we can combine theorem 9 with corollary 4 to obtain the following Künneth formula for singular homology.

10 THEOREM *If $\{X \times B, A \times Y\}$ is an excisive couple in $X \times Y$ and G and G' are modules over a principal ideal domain such that $G * G' = 0$, there is a functorial short exact sequence*

$$0 \to [H(X,A;\, G) \otimes H(Y,B;\, G')]_q \xrightarrow{\mu'} H_q((X,A) \times (Y,B);\, G \otimes G') \to [H(X,A;\, G) * H(Y,B;\, G')]_{q-1} \to 0$$

and this sequence is split. ∎

In particular, if the right-hand term vanishes (which always happens if R is a field), then the cross product is an isomorphism

$$\mu'\colon H(X,A;\, G) \otimes H(Y,B;\, G') \approx H((X,A) \times (Y,B);\, G \otimes G')$$

The following formulas are consequences of the naturality of μ and of theorems 5, 7, and 8.

11 *Let $f\colon (X,A) \to (X',A')$ and $g\colon (Y,B) \to (Y',B')$ be maps and let $z \in H_p(X,A;\, G)$ and $z' \in H_q(Y,B;\, G')$. Then, in the module*

$$H_{p+q}((X',A') \times (Y',B');\, G \otimes G')$$

we have

$$(f \times g)_*(z \times z') = f_* z \times g_* z' \quad ∎$$

12 *Let $p\colon (X,A) \times Y \to (X,A)$ be the projection to the first factor and let $\varepsilon\colon H(Y;G') \to G'$ be the augmentation map. For $z \in H_q(X,A;\, G)$ and $z' \in H_r(Y;G')$, in $H_{q+r}(X,A;\, G \otimes G')$,*

$$p_*(z \times z') = \mu(z \otimes \varepsilon(z')) \quad ∎$$

13 *For $z \in H_p(X,A;\, G)$, $z' \in H_q(Y,B;\, G')$, and $z'' \in H_r(Z,C;\, G'')$, in*

$$H_{p+q+r}((X,A) \times (Y,B) \times (Z,C);\, G \otimes G' \otimes G''),$$

we have

$$z \times (z' \times z'') = (z \times z') \times z'' \quad ∎$$

14 *Let $T\colon (X,A) \times (Y,B) \to (Y,B) \times (X,A)$ and $\varphi\colon G' \otimes G \to G \otimes G'$ interchange the factors. For $z \in H_p(X,A;\, G)$ and $z' \in H_q(Y,B;\, G')$, in $H_{p+q}((Y,B) \times (X,A);\, G \otimes G')$, we have*

$$T_*(z \times z') = (-1)^{pq}\varphi_*(z' \times z) \quad ∎$$

15 *Let $\{(X_1,A_1), (X_2,A_2)\}$ be an excisive couple of pairs in X and let $z \in H_p(X_1 \cup X_2, A_1 \cup A_2;\, G)$ and $z' \in H_q(Y,B;\, G')$. For the connecting homomorphisms of appropriate Mayer-Vietoris sequences we have*

$$\partial_*(z \times z') = \partial_* z \times z'$$

in $H_{p+q-1}((X_1 \cap X_2, A_1 \cap A_2) \times (Y,B);\, G \otimes G')$ and

$$\partial_*(z' \times z) = (-1)^q z' \times \partial_* z$$

in $H_{p+q-1}((Y,B) \times (X_1 \cap X_2, A_1 \cap A_2);\, G' \otimes G)$ ∎

4 COHOMOLOGY

A chain complex has a differential of degree -1. Related to this is the concept of a cochain complex, which has a differential of degree $+1$. Cochain complexes have many of the properties of chain complexes, and this section is devoted to a discussion of these properties. The functor Hom associates to every chain complex a cochain complex, and vice versa. The cohomology module of a topological pair with coefficients G is the homology module of the cochain complex associated in this way to the singular complex of the pair. The last part of the section is devoted to a brief discussion of axiomatic cohomology theory.

Throughout this section R will denote a commutative ring with a unit. A *cochain complex* (*over* R), denoted by $C^* = \{C^q,\delta^q\}$, is a graded R module together with a homogeneous differential δ of degree $+1$ called the *coboundary operator* (thus $\delta^q\colon C^q \to C^{q+1}$ and $\delta^{q+1}\delta^q = 0$ for all q). The kernel of δ is the module of *cocycles* $Z(C^*)$, and the image of δ is the module of *coboundaries* $B(C^*)$. Then $B(C^*) \subset Z(C^*)$, and the *cohomology module* $H(C^*)$ is defined to be the quotient $Z(C^*)/B(C^*)$.

If C^* is a cochain complex, there is a chain complex C defined by $C_q = C^{-q}$ and $\partial_q\colon C_q \to C_{q-1}$ equal to $\delta^{-q}\colon C^{-q} \to C^{-q+1}$. Then $H_q(C) = H^{-q}(C^*)$, and the cohomology module of C^* corresponds to the homology module of C. In this way results about chain complexes give results about cochain complexes. Thus we have the concepts of *cochain map* and *cochain homotopy*, and there is a category of cochain complexes and cochain homotopy classes of cochain maps. The cohomology module is a covariant functor from this category to the category of graded modules. Furthermore, given a short exact sequence of cochain complexes

$$0 \to \bar{\bar{C}}^* \xrightarrow{\alpha} \bar{C}^* \xrightarrow{\beta} C^* \to 0$$

there is a functorial *connecting homomorphism*

$$\delta^*\colon H(C^*) \to H(\bar{\bar{C}}^*)$$

of degree $+1$ and a functorial exact cohomology sequence

$$\cdots \to H^q(C^*) \xrightarrow{\delta^*} H^{q+1}(\bar{\bar{C}}^*) \xrightarrow{\alpha^*} H^{q+1}(\bar{C}^*) \xrightarrow{\beta^*} H^{q+1}(C^*) \xrightarrow{\delta^*} \cdots$$

Passing from a cochain complex to a chain complex by changing the sign of the degree gives us the following analogues of theorems 5.2.14 and 5.3.3.

1 THEOREM *Given a cochain complex C^* and a module G such that $C^* * G$ is acyclic, there is a functorial short exact sequence*

$$0 \to H^q(C^*) \otimes G \xrightarrow{\mu} H^q(C^* \otimes G) \to H^{q+1}(C^*) * G \to 0$$

and this sequence is split. ∎

2 THEOREM *Given cochain complexes C^* and C'^* such that $C^* * C'^*$ is an acyclic cochain complex, there is a functorial short exact sequence*

$$0 \to [H^*(C^*) \otimes H^*(C'^*)]^q \xrightarrow{\mu} H^q(C^* \otimes C'^*) \to [H^*(C^*) * H^*(C'^*)]^{q+1} \to 0$$

and this sequence is split. ∎

There is also an analogue of corollary 5.3.4 for cochain complexes which we shall not state as a separate theorem. If C^* is a cochain complex over R and G is an R module, an *augmentation of C^* over G* is a monomorphism $\eta: G \to C^0$ such that $\delta^0 \circ \eta = 0$. An *augmented* cochain complex over G consists of a nonnegative cochain complex C^* (that is, $C^q = 0$ for $q < 0$) and an augmentation of C^* over G. Such an augmentation can be regarded as a monomorphic chain map of the cochain complex (also denoted by G) whose only nontrivial cochain module is G in degree 0 to C^*. For this trivial cochain complex G it is clear that $H^q(G) = 0$ for $q \neq 0$ and $H^0(G) = G$. Therefore η induces a monomorphism $\eta^*: G \to H^0(C^*)$. The *reduced cochain complex* $\tilde{C}^*$ of an augmented cochain complex C^* is defined to be the quotient cochain complex $\tilde{C}^q = C^q$ for $q \neq 0$, $\tilde{C}^0 = \operatorname{coker} \eta$, and $\tilde{\delta}^q$ is suitably induced by δ^q. We define $\tilde{H}(C^*) = H(\tilde{C}^*)$. Because there is a short exact sequence of cochain complexes

$$0 \to G \xrightarrow{\eta} C^* \to \tilde{C}^* \to 0$$

we see that $H^q(C^*) \approx \tilde{H}^q(C^*)$ for $q \neq 0$, and there is a short exact sequence

$$0 \to G \to H^0(C^*) \to \tilde{H}^0(C^*) \to 0$$

Our interest in cochain complexes is in their relation to chain complexes. If C is a chain complex over R and G is an R module, there is a cochain complex $\operatorname{Hom}(C,G) = \{\operatorname{Hom}(C_q,G), \delta^q\}$, where, if $f \in \operatorname{Hom}(C_q,G)$, then $\delta^q f \in \operatorname{Hom}(C_{q+1},G)$ is defined by

$$(\delta^q f)(c) = f(\partial_{q+1} c) \qquad c \in C_{q+1}$$

We also write $\langle f,c \rangle$ instead of $f(c)$ and set $\langle f,c \rangle = 0$ if $\deg f \neq \deg c$. In this notation $\langle \delta^q f,c \rangle = \langle f,\partial_{q+1} c \rangle$.

If C is augmented by $\varepsilon: C_0 \to G'$, then $\operatorname{Hom}(C,G)$ is augmented by $\eta: \operatorname{Hom}(G',G) \to \operatorname{Hom}(C_0,G)$, where $\eta(f)(c) = f(\varepsilon(c))$ for $c \in C_0$ and $f \in \operatorname{Hom}(G',G)$. It is easy to verify the following result.

3 THEOREM *There is a functor of two arguments contravariant in chain complexes C and covariant in modules G which assigns to a chain complex C and module G the cochain complex* $\operatorname{Hom}(C,G)$. ∎

For a chain complex C and module G we define the *cohomology module* $H^*(C;G) = \{H^q(C;G)\}$ *of C with coefficients G* by

$$H^q(C;G) = H^q(\text{Hom}\,(C,G))$$

It follows from theorem 3 that $H^*(C;G)$ is a contravariant functor of C and a covariant functor of G to the category of graded modules. For a chain map $\tau: C \to C'$ we use $\tau^*: H^*(C';G) \to H^*(C;G)$ to denote the homomorphism induced by the cochain map Hom $(\tau,1)$: Hom $(C',G) \to$ Hom (C,G), and for a homomorphism $\varphi: G \to G'$ we use $\varphi_*: H^*(C;G) \to H^*(C,G')$ to denote the homomorphism induced by the cochain map Hom $(1,\varphi)$: Hom $(C,G) \to$ Hom (C,G'). To distinguish the homology of C from its cohomology, we shall sometimes denote $H(C;G)$ by $H_*(C;G)$.

4 EXAMPLE Given an abelian group G and a simplicial pair (K,L), the *oriented cohomology group of* (K,L) *with coefficients* G, denoted by $H^*(K,L;\, G)$, is defined to be the graded cohomology group of the cochain complex Hom $(C(K)/C(L),\, G)$ [which is augmented over Hom $(\mathbf{Z},G) \approx G$]. Then $H^*(K,L;\, G)$ is a contravariant functor of (K,L) and a covariant functor of G to the category of graded abelian groups. If G is also an R module, $H^*(K,L;\, G)$ is a graded R module.

5 EXAMPLE If (X,A) is a topological pair and G is an abelian group, the *singular cohomology group of* (X,A) *with coefficients* G, denoted by $H^*(X,A;\, G)$, is defined to be the graded cohomology group of the cochain complex Hom $(\Delta(X)/\Delta(A),\, G)$ (which is augmented over G). It is contravariant in (X,A) and covariant in G, and if G is an R module, $H^*(X,A;\, G)$ is a graded R module. If $(X',A') \subset (X,A)$ and $u \in H^*(X,A;\, G)$, we use $u \mid (X',A')$ to denote the element of $H^*(X',A';\, G)$ equal to i^*u, where $i: (X',A') \subset (X,A)$. We also use $1 \in H^0(X;R)$ to denote the image of $1 \in R$ under the augmentation $\eta: R \to H^0(X;R)$.

We recall some properties of the functor Hom. The following analogue of corollary 5.1.6 is easily established.

6 LEMMA *Given an exact sequence of R modules*

$$A' \to A \to A'' \to 0$$

and an R module B, there is an exact sequence

$$0 \to \text{Hom}\,(A'',B) \to \text{Hom}\,(A,B) \to \text{Hom}\,(A',B) \quad \blacksquare$$

If $A' \to A$ is a monomorphism [that is, if $0 \to A' \to A$ is also exact], it need not be true that Hom $(A,B) \to$ Hom (A',B) is an epimorphism, [that is, that Hom $(A,B) \to$ Hom $(A',B) \to 0$ is exact]. However, there is the following analogue of corollary 5.1.12 (which follows easily by using lemma 5.1.11).

7 LEMMA *Given a split short exact sequence of R modules*

$$0 \to A' \to A \to A'' \to 0$$

and an R module B, the sequence

$$0 \to \text{Hom}(A'',B) \to \text{Hom}(A,B) \to \text{Hom}(A',B) \to 0$$

is a split short exact sequence. ∎

In case $0 \to C' \to C \to C'' \to 0$ is a split short exact sequence of chain complexes, it follows from lemma 7 that for any module G there is a short exact sequence of cochain complexes

$$0 \to \text{Hom}(C'',G) \to \text{Hom}(C,G) \to \text{Hom}(C',G) \to 0$$

This gives the following result.

8 THEOREM *Given a split short exact sequence of chain complexes*

$$0 \to C' \to C \to C'' \to 0$$

and a module G, there is a functorial exact cohomology sequence

$$\cdots \to H^q(C'';G) \to H^q(C;G) \to H^q(C';G) \xrightarrow{\delta^*} H^{q+1}(C'';G) \to \cdots$$ ∎

This leads to an exact *Mayer-Vietoris cohomology sequence* analogous to the exact sequence of corollary 5.1.14.

9 COROLLARY *If $\{(X_1,A_1), (X_2,A_2)\}$ is an excisive couple of pairs in a topological space and G is an R module, there is a functorial exact cohomology sequence*

$$\cdots \xrightarrow{\delta^*} H^q(X_1 \cup X_2, A_1 \cup A_2; G) \xrightarrow{j^*} H^q(X_1, A_1; G) \oplus H^q(X_2,A_2; G) \xrightarrow{i^*} H^q(X_1 \cap X_2, A_1 \cap A_2; G) \xrightarrow{\delta^*} \cdots$$

where $j^*(u) = (j_1^*(u), j_2^*(u))$ and $i^*(u_1 + u_2) = i_1^* u_1 - i_2^* u_2$ *and i_1, i_2, j_1, and j_2 are suitable inclusion maps.* ∎

If $\{X_j\}$ is the set of path components of X, then $\Delta(X) = \bigoplus_j \Delta(X_j)$. Therefore $\text{Hom}(\Delta(X);G) = \times_j \text{Hom}(\Delta(X_j),G)$, and by theorem 4.1.6, we obtain the following result.

10 THEOREM *The singular cohomology module of a space is the direct product of the singular cohomology modules of its path components.* ∎

Because the homology functor does not commute with inverse limits, it is not true that the singular cohomology of a space is isomorphic to the inverse limit of the singular cohomology of its compact subsets (that is, there is no general cohomology analogue of theorem 4.4.6).

There is an exact cohomology sequence corresponding to a short exact sequence of coefficient modules (analogous to theorem 5.2.7).

11 THEOREM *On the category of free chain complexes C over R and short exact sequences of R modules*

$$0 \to G' \xrightarrow{\varphi} G \xrightarrow{\psi} G'' \to 0$$

there is a functorial connecting homomorphism

$$\beta^*: H^*(C;G'') \to H^*(C;G')$$

of degree 1 *and a functorial exact sequence*

$$\cdots \to H^q(C;G') \xrightarrow{\varphi_*} H^q(C;G) \xrightarrow{\psi_*} H^q(C;G'') \xrightarrow{\beta^*} H^{q+1}(C;G') \to \cdots$$

PROOF Because C is free, there is a short exact sequence of cochain complexes

$$0 \to \text{Hom}\,(C,G') \xrightarrow{\bar\varphi} \text{Hom}\,(C,G) \xrightarrow{\bar\psi} \text{Hom}\,(C,G'') \to 0$$

where $\bar\varphi$ and $\bar\psi$ are induced by φ and ψ. The result follows from this. ■

The connecting homomorphism β^* in theorem 11 is called the *Bockstein cohomology homomorphism* corresponding to the coefficient sequence $0 \to G' \xrightarrow{\varphi} G \xrightarrow{\psi} G'' \to 0$.

Let G be an R module. A *cohomology theory with coefficients* G consists of a contravariant functor $H^* = \{H^q\}$ from the category of topological pairs to graded R modules and a natural transformation $\delta^*: H^*(A) \to H^*(X,A)$ of degree 1 such that the following axioms hold.

12 **HOMOTOPY AXIOM** *If* $f_0, f_1: (X,A) \to (Y,B)$ *are homotopic, then*

$$H^*(f_0) = H^*(f_1): H^*(Y,B) \to H^*(X,A)$$

13 **EXACTNESS AXIOM** *For any pair* (X,A) *with inclusion maps* $i: A \subset X$ *and* $j: X \subset (X,A)$, *there is an exact sequence*

$$\cdots \xrightarrow{\delta^*} H^q(X,A) \xrightarrow{H^q(j)} H^q(X) \xrightarrow{H^q(i)} H^q(A) \xrightarrow{\delta^*} H^{q+1}(X,A) \to \cdots$$

14 **EXCISION AXIOM** *For any pair* (X,A) *if* U *is an open subset of* X *such that* $\bar U \subset \text{int}\, A$, *then the excision map* $j: (X - U, A - U) \subset (X,A)$ *induces an isomorphism*

$$H^*(j): H^*(X,A) \approx H^*(X - U, A - U)$$

15 **DIMENSION AXIOM** *On the category of one-point spaces there is a natural equivalence of the constant functor* G *with the functor* H^*.

Singular cohomology theory with coefficients G is an example of a cohomology theory with coefficients G (the exactness axiom following from the application of corollary 9 to a suitable couple). The uniqueness theorem is valid for cohomology theories (that is, a homomorphism from one cohomology theory to another which is an isomorphism for one-point spaces is an isomorphism for compact polyhedral pairs).

The reduced cohomology modules $\tilde H^*$ of a cohomology theory are defined as follows. If X is a nonempty space, let $c: X \to P$ be the unique map from X to some one-point space P. The reduced module $\tilde H^*(X)$ is defined to be the cokernel of the homomorphism

$$H^*(c): H^*(P) \to H^*(X)$$

Because c has a right inverse, $H^*(c)$ has a left inverse. Therefore

$$H^*(X) \approx \tilde{H}^*(X) \oplus H^*(P)$$

and the reduced modules have properties similar to those of the reduced singular cohomology modules.

5 THE UNIVERSAL-COEFFICIENT THEOREM FOR COHOMOLOGY

This section is devoted to relations between cohomology and homology of chain complexes. In order to express the cohomology of a chain complex in terms of its homology it is necessary to introduce certain functors of modules which are associated to the module of homomorphisms of one module to another in much the same way that the torsion products are associated to the tensor product. Over a principal ideal domain there is one associated functor, the module of extensions. We use this to formulate the universal-coefficient theorems and Künneth formulas established in the section.

Let C be a free resolution of the module A and let B be a module. There is a cochain complex $\mathrm{Hom}\,(C,B) = \{[\mathrm{Hom}\,(C,B)]^q = \mathrm{Hom}\,(C_q,B), \delta^q\}$ whose cohomology module depends only on A and B, up to canonical isomorphism (and not on the choice of C). Let C be the canonical free resolution of A and define $\mathrm{Ext}^q\,(A,B) = H^q(\mathrm{Hom}\,(C,B))$. Then $\mathrm{Ext}^q\,(A,B)$ is a functor of two arguments contravariant in A and covariant in B, and $\mathrm{Ext}^0\,(A,B)$ is naturally equivalent to $\mathrm{Hom}\,(A,B)$.

For the rest of this section we assume R is a principal ideal domain. Then, $\mathrm{Ext}^q\,(A,B) = 0$ for $q > 1$, and the *module of extensions* $\mathrm{Ext}\,(A,B)$ is defined to equal $\mathrm{Ext}^1\,(A,B)$. It is characterized by the property that given any free presentation of A

$$0 \to C_1 \xrightarrow{\partial_1} C_0 \to A \to 0$$

there is an exact sequence

$$0 \to \mathrm{Hom}\,(A,B) \to \mathrm{Hom}\,(C_0,B) \xrightarrow{\mathrm{Hom}\,(\partial_1,1)} \mathrm{Hom}\,(C_1,B) \to \mathrm{Ext}\,(A,B) \to 0$$

In fact, because $\mathrm{Hom}\,(C_2,B) = 0$,

$$\begin{aligned}\mathrm{Ext}\,(A,B) = H^1(C;B) &= \mathrm{Hom}\,(C_1,B)/\mathrm{im}\,[\mathrm{Hom}\,(\partial_1,1)]\\ &= \mathrm{coker}\,[\mathrm{Hom}\,(\partial_1,1)]\end{aligned}$$

Clearly, $\mathrm{Ext}\,(A,B)$ is contravariant in A and covariant in B. Its name derives from its connection with the extensions of B by A which we describe briefly after the following examples.

1 If A is free, it has the free presentation

$$0 \to 0 \to A \to A \to 0$$

from which it follows that $\mathrm{Ext}\,(A,B) = 0$ for any B.

2 For $v \in R$, $v \neq 0$ there is a short exact sequence

$$0 \to R \xrightarrow{\alpha} R \to R/vR \to 0$$

where $\alpha(v') = vv'$ for $v' \in R$, which is a free presentation of R/vR. For any R module B, Hom $(R,B) \approx B$ and the homomorphism Hom $(\alpha,1)$: Hom $(R,B) \to$ Hom (R,B) corresponds to $\alpha^*: B \to B$, where $\alpha^*(b) = vb$. Hence there is an isomorphism coker Hom $(\alpha,1) \approx B/vB$, and we have proved

$$\text{Ext}\,(R/vR,B) \approx B/vB \approx (R/vR) \otimes B$$

Since Hom commutes with finite direct sums, it follows that for any finitely generated torsion module A there is an isomorphism (nonfunctorial)

$$\text{Ext}\,(A,B) \approx A \otimes B$$

because such a module A is a finite direct sum of cyclic modules (by theorem 4.14 in the Introduction).

An *extension of B by A* is a short exact sequence

$$0 \to B \to E \to A \to 0$$

With a suitable definition of equivalence of extensions (by a commutative diagram), of the sum of two extensions, and of the product of an extension by an element of R, there is obtained a module whose elements are equivalence classes of extensions of B by A. This module is isomorphic to Ext (A,B). In fact, given an extension $0 \to B \to E \to A \to 0$ and a free presentation of A, $0 \to C_1 \to C_0 \to A \to 0$, there is, by theorem 5.2.1, a commutative diagram

$$\begin{array}{ccccccc} 0 \to & C_1 & \to & C_0 & \searrow & & \\ & \varphi_1\downarrow & & \varphi_0\downarrow & & A \to 0 & \\ 0 \to & B & \to & E & \nearrow & & \end{array}$$

uniquely determined up to chain homotopy. Then $\varphi_1 \in$ Hom (C_1,B) is unique up to im [Hom $(C_0,B) \to$ Hom (C_1,B)], and so determines an element of Ext (A,B). This function from extensions of B by A to Ext (A,B) induces an isomorphism of the module of equivalence classes of extensions with Ext (A,B).

Given a graded module $C = \{C_q\}$, there is a graded module Ext $(C,B) = \{[\text{Ext}\,(C,B)]^q = \text{Ext}\,(C_q,B)\}$. If C is a chain complex, Ext (C,B) is a cochain complex with

$$\delta^q = \text{Ext}\,(\partial_{q+1},1)\colon \text{Ext}\,(C_q,B) \to \text{Ext}\,(C_{q+1},B)$$

A homomorphism

$$h\colon H^q(C;G) \to \text{Hom}\,(H_q(C;G'), G \otimes G')$$

natural in C and G is defined by

$$(h\{f\})\{\Sigma\, c_i \otimes g_i'\} = \Sigma\, f(c_i) \otimes g_i'$$

for $\{f\} \in H^q(C;G)$ and $\{\Sigma\, c_i \otimes g_i'\} \in H_q(C;G')$ [after verification that

$\Sigma f(c_i) \otimes g'_i$ is independent of the choice of f in its cohomology class and $\Sigma\, c_i \otimes g'_i$ in its homology class]. For $u \in H^p(C;G)$ and $z \in H_q(C;G')$ we define $\langle u,z\rangle \in G \otimes G'$ to be 0 if $p \neq q$ and to be $h(u)(z)$ if $p = q$. In this notation

$$\langle \{f\}, \{\Sigma\, c_i \otimes g'_i\}\rangle = \Sigma \langle f,c_i\rangle \otimes g'_i$$

The homomorphism h enters in the following *universal-coefficient theorem for cohomology.*

3 THEOREM *Given a chain complex C and module G such that* Ext (C,G) *is an acyclic cochain complex, there is a functorial short exact sequence*

$$0 \to \operatorname{Ext}(H_{q-1}(C),G) \to H^q(C,G) \xrightarrow{h} \operatorname{Hom}(H_q(C),G) \to 0$$

and this sequence is split.

PROOF We first consider the case in which C is a free chain complex. There is then a short exact sequence of chain complexes

$$0 \to Z \to C \to B \to 0$$

where $Z_q = Z_q(C)$ and $B_q = B_{q-1}(C)$. This sequence is split because B is free, and by theorem 5.4.8, there is an exact cohomology sequence

$$\cdots \to H^{q-1}(Z;G) \xrightarrow{\delta^*} H^q(B;G) \to H^q(C;G) \to H^q(Z;G) \xrightarrow{\delta^*} H^{q+1}(B;G) \to \cdots$$

Since Z and B have trivial boundary operators, $H^q(Z;G) = \operatorname{Hom}(Z_q(C),G)$ and $H^q(B;G) = \operatorname{Hom}(B_{q-1}(C),G)$. Furthermore, the homomorphism

$$\delta^*: H^q(Z;G) \to H^{q+1}(B;G)$$

equals $\operatorname{Hom}(\gamma_q,1)$: $\operatorname{Hom}(Z_q(C),G) \to \operatorname{Hom}(B_q(C),G)$, where $\gamma_q: B_q(C) \subset Z_q(C)$. Hence there is a functorial short exact sequence

$$0 \to \operatorname{coker}[\operatorname{Hom}(\gamma_{q-1},1)] \to H^q(C;G) \to \ker[\operatorname{Hom}(\gamma_q,1)] \to 0$$

To interpret the modules in the above sequence we have the short exact sequence

$$0 \to B_q(C) \xrightarrow{\gamma_q} Z_q(C) \to H_q(C) \to 0$$

which is a free presentation of $H_q(C)$. By the characteristic property of Ext, there is an exact sequence

$$0 \to \operatorname{Hom}(H_q(C),G) \to \operatorname{Hom}(Z_q(C),G) \xrightarrow{\operatorname{Hom}(\gamma_q,1)} \operatorname{Hom}(B_q(C),G) \to \operatorname{Ext}(H_q(C),G) \to 0$$

Therefore, $\ker[\operatorname{Hom}(\gamma_q,1)] \approx \operatorname{Hom}(H_q(C),G)$ and $\operatorname{coker}[\operatorname{Hom}(\gamma_q,1)] \approx \operatorname{Ext}(H_q(C),G)$. Substituting these into the short exact sequence containing $H^q(C;G)$ yields the desired short exact sequence

$$0 \to \operatorname{Ext}(H_{q-1}(C),G) \to H^q(C;G) \to \operatorname{Hom}(H_q(C),G) \to 0$$

with the homomorphism $H^q(C;G) \to \operatorname{Hom}(H_q(C),G)$ easily verified to equal h.

This sequence is functorial and is split (because the sequence of chain complexes

$$0 \to Z \to C \to B \to 0$$

is split).

For arbitrary C such that Ext (C,G) is acyclic, the result follows by using a free approximation to C (as in the proof of theorem 5.2.14) to reduce it to the case of a free complex. ∎

It follows from theorem 3 that if X is a path-connected topological space, then $H^0(X;R)$ is a free R module generated by 1 [or, in other words, the augmentation map is an isomorphism $\eta\colon R \approx H^0(X;R)$]. From theorems 3 and 5.4.10, it follows that for any X, $H^0(X;G)$ is isomorphic to the direct product of as many copies of G as path components of X.

4 COROLLARY *If (X,A) is a topological pair such that $H_q(X,A;R)$ is finitely generated for all q, then the free submodules of $H^q(X,A;R)$ and $H_q(X,A;R)$ are isomorphic and the torsion submodules of $H^q(X,A;R)$ and $H_{q-1}(X,A;R)$ are isomorphic.*

PROOF Let $H_q(X,A;R) = F_q \oplus T_q$, where F_q is free and T_q is the torsion module of H_q. Then

$$\text{Hom}\,(H_q(X,A;R),R) \approx \text{Hom}\,(F_q,R) \oplus \text{Hom}\,(T_q,R) \approx F_q$$

and by example 2,

$$\text{Ext}\,(H_q(X,A;R),R) \approx \text{Ext}\,(F_q,R) \oplus \text{Ext}\,(T_q,R) \approx T_q$$

The result follows from theorem 3. ∎

For many purposes it would be more useful to have a formula expressing $H^*(C;G)$ in terms of $H^*(C;R)$. Such a formula can be proved in the case of C or G finitely generated. We begin by establishing some properties of finitely generated modules.

Let $\mu\colon \text{Hom}\,(A,G) \otimes G' \to \text{Hom}\,(A, G \otimes G')$ be the functorial homomorphism defined by $\mu(f \otimes g')(a) = f(a) \otimes g'$ for $f \in \text{Hom}\,(A,G)$, $g' \in G'$, and $a \in A$.

5 LEMMA *If A is a free module and G' is finitely generated, then for any module G, μ is an isomorphism.*

PROOF The result is trivially true if $G' = R$. Because the tensor product and Hom functors both commute with finite direct sums, it is also true if G' is a finitely generated free module. G' is assumed to be finitely generated, so there is a short exact sequence

$$0 \to \bar{\bar{G}} \to \bar{G} \to G' \to 0$$

where $\bar{G}$ (hence also $\bar{\bar{G}}$) is a finitely generated free module. There is a commutative diagram

$$\begin{array}{ccccccc} \text{Hom}(A,G) \otimes \bar{\bar{G}} & \to & \text{Hom}(A,G) \otimes \bar{G} & \to & \text{Hom}(A,G) \otimes G' & \to & 0 \\ \downarrow \bar{\bar{\mu}} & & \downarrow \bar{\mu} & & \downarrow \mu & & \\ \text{Hom}(A, G \otimes \bar{\bar{G}}) & \to & \text{Hom}(A, G \otimes \bar{G}) & \to & \text{Hom}(A, G \otimes G') & \to & 0 \end{array}$$

with exact rows (exactness follows from corollary 5.1.6 and, for the bottom row, from the fact that A is free). Because $\bar{\bar{\mu}}$ and $\bar{\mu}$ are isomorphisms, it follows from the five lemma that μ is also an isomorphism. ∎

There is also a functorial homomorphism

$$\mu: \text{Hom}(A,G) \otimes \text{Hom}(B,G') \to \text{Hom}(A \otimes B, G \otimes G')$$

defined by $\mu(f \otimes f')(a \otimes b) = f(a) \otimes f'(b)$ for $f \in \text{Hom}(A,G)$, $f' \in \text{Hom}(B,G')$, $a \in A$, and $b \in B$. In case $B = R$, $\text{Hom}(B,G') \approx G'$, and μ corresponds to the homomorphism in lemma 5.

6 LEMMA *If B is a finitely generated free module, for arbitrary modules A and G, μ is an isomorphism*

$$\mu: \text{Hom}(A,G) \otimes \text{Hom}(B,R) \approx \text{Hom}(A \otimes B, G)$$

PROOF The result is trivially true for $B = R$ and follows for a finite sum of copies of R because both sides commute with finite direct sums. ∎

7 COROLLARY *If A and B are free modules and either A and B or B and G' are finitely generated, μ is an isomorphism*

$$\mu: \text{Hom}(A,G) \otimes \text{Hom}(B,G') \approx \text{Hom}(A \otimes B, G \otimes G')$$

PROOF Since A and B are free, so is $A \otimes B$. If A and B are finitely generated, so is $A \otimes B$, and there is a commutative diagram

$$\begin{array}{ccc} [\text{Hom}(R,G) \otimes \text{Hom}(A,R)] \otimes [\text{Hom}(R,G') \otimes \text{Hom}(B,R)] & \xrightarrow{\bar{\mu}} & \text{Hom}(R, G \otimes G') \otimes \text{Hom}(A \otimes B, R) \\ \downarrow \mu \otimes \mu & & \downarrow \mu \\ \text{Hom}(A,G) \otimes \text{Hom}(B,G') & \xrightarrow{\mu} & \text{Hom}(A \otimes B, G \otimes G') \end{array}$$

in which $\bar{\mu}((f_1 \otimes f_2) \otimes (f_3 \otimes f_4)) = \mu(f_1 \otimes f_3) \otimes \mu(f_2 \otimes f_4)$. By lemma 6, $\bar{\mu}$ is an isomorphism and so are both vertical maps. Therefore the bottom map is also an isomorphism.

If B and G' are finitely generated, there is a commutative diagram

$$\begin{array}{ccc} \text{Hom}(A,G) \otimes \text{Hom}(B,R) \otimes G' & \xrightarrow{1 \otimes \mu} & \text{Hom}(A,G) \otimes \text{Hom}(B,G') \\ \downarrow \mu \otimes 1 & & \downarrow \mu \\ \text{Hom}(A \otimes B, G) \otimes G' & \xrightarrow{\mu} & \text{Hom}(A \otimes B, G \otimes G') \end{array}$$

By lemma 5, both horizontal maps are isomorphisms, and by lemma 6, the left-hand vertical map is an isomorphism. Therefore the right-hand map is also an isomorphism. ∎

It follows from lemma 5 that if A is free and finitely generated, μ is an isomorphism

$$\mu\colon \operatorname{Hom}(A,R) \otimes A \approx \operatorname{Hom}(A,A)$$

The following lemma is a partial converse of this result.

8 LEMMA *If A is a module such that*

$$\mu\colon \operatorname{Hom}(A,R) \otimes A \to \operatorname{Hom}(A,A)$$

is an epimorphism, then A is finitely generated.

PROOF By hypothesis, there exist $f_i \in \operatorname{Hom}(A,R)$ and $a_i \in A$ for $1 \leq i \leq n$ such that $\mu(\Sigma f_i \otimes a_i) = 1_A$. Then, for any $a \in A$

$$a = \mu(\Sigma f_i \otimes a_i)(a) = \Sigma f_i(a)a_i$$

showing that A is generated by $\{a_i\}$. ∎

A graded module $\{C_q\}$ is said to be of *finite type* if C_q is finitely generated for every q. Thus a graded module C of finite type is finitely generated (as a graded module) if and only if $C_q = 0$, except for a finite set of integers q. The following lemma asserts that a chain complex whose homology is of finite type can be approximated by a chain complex of finite type.

9 LEMMA *Let C be a free chain complex such that $H(C)$ is of finite type. Then there is a free chain complex C' of finite type chain equivalent to C.*

PROOF For each q let F_q be a finitely generated submodule of $Z_q(C)$ such that F_q maps onto $H_q(C)$ under the epimorphism $Z_q(C) \to H_q(C)$. Let F'_q be the kernel of the epimorphism $F_q \to H_q(C)$. Define a chain complex $C' = \{C'_q,\partial'_q\}$ by $C'_q = F_q \oplus F'_{q-1}$ and $\partial'_q(c,c') = (c',0)$ for $c \in F_q$ and $c' \in F'_{q-1}$. Then C' is a free chain complex of finite type and $H_q(C') = F_q/F'_q \approx H_q(C)$. To define a chain equivalence $\tau\colon C' \to C$, choose for each q a homomorphism $\varphi_q\colon F'_q \to C_{q+1}$ such that $\partial_{q+1}\varphi_q(c') = c'$ for $c' \in F'_q$. Then define τ by $\tau(c,c') = c + \varphi_{q-1}(c')$ for $c \in F_q$ and $c' \in F'_{q-1}$. τ is a chain map and induces an isomorphism $\tau_*\colon H(C') \approx H(C)$. Because C' and C are both free, it follows from theorem 4.6.10 that τ is a chain equivalence. ∎

We are now ready for the universal-coefficient theorems toward which we have been heading.

10 THEOREM *Let C be a free chain complex and G be a module such that either $H(C)$ is of finite type or G is finitely generated. Then there is a functorial short exact sequence*

$$0 \to H^q(C;R) \otimes G \xrightarrow{\mu} H^q(C;G) \to H^{q+1}(C;R) * G \to 0$$

and this sequence is split.

PROOF If G is finitely generated, it follows from lemma 5 that

$$\mu\colon \operatorname{Hom}(C,R) \otimes G \approx \operatorname{Hom}(C,G)$$

Because Hom (C,R) is without torsion, Hom $(C,R) * G = 0$, and the result follows from theorem 5.4.1.

If $H(C)$ is of finite type, we use lemma 9 to replace C by a free chain complex C' of finite type. By corollary 7, μ: Hom $(C',R) \otimes G \approx$ Hom (C',G), and the result again follows for C' (and hence for C) from theorem 5.4.1. ∎

In a similar way we obtain the following *Künneth formula for cohomology*.

11 THEOREM *Let C and C' be nonnegative free chain complexes and G and G' be modules over a principal ideal domain such that $G * G' = 0$ and either $H(C)$ and $H(C')$ are of finite type or $H(C')$ is of finite type and G' is finitely generated. Then there is a functional short exact sequence*

$$0 \to [H^*(C;G) \otimes H^*(C';G')]^q \to H^q(C \otimes C'; G \otimes G') \to [H^*(C;G) * H^*(C';G')]^{q+1} \to 0$$

and this sequence is split.

PROOF If $H(C)$ and $H(C')$ are of finite type, by lemma 9, we can replace C and C' by free chain complexes of finite type. Hence we are reduced to proving the result for the case where C and C' have finite type or where C' has finite type and G' is finitely generated. By corollary 7, there is an isomorphism μ: Hom $(C,G) \otimes$ Hom $(C',G') \approx$ Hom $(C \otimes C', G \otimes G')$. The result will now follow from theorem 5.4.2 as soon as we have verified that Hom $(C,G) *$ Hom (C',G') is acyclic.

We show that Hom $(C,G) *$ Hom $(C',G') = 0$. In case C and C' are both of finite type, Hom (C_p,G) is isomorphic to a finite direct sum of copies of G and Hom (C'_q,G') is isomorphic to a finite direct sum of copies of G'. Because $G * G' = 0$ by hypothesis, Hom $(C_p,G) *$ Hom $(C'_q,G') = 0$, and so Hom $(C,G) *$ Hom $(C',G') = 0$ in this case.

In case C' is of finite type, Hom (C'_q,G') is isomorphic to a finite direct sum of copies of G'. Hence it suffices to show that Hom $(C,G) * G' = 0$ if G' is finitely generated. Let

$$0 \to \bar{\bar{G}} \to \bar{G} \to G' \to 0$$

be a free resolution of G' with $\bar{G}$ finitely generated. Because $G * G' = 0$, there is a short exact sequence

$$0 \to G \otimes \bar{\bar{G}} \to G \otimes \bar{G} \to G \otimes G' \to 0$$

and a short exact sequence of cochain complexes (because C is free)

$$0 \to \text{Hom}(C, G \otimes \bar{\bar{G}}) \to \text{Hom}(C, G \otimes \bar{G}) \to \text{Hom}(C, G \otimes G') \to 0$$

Using lemma 5, this implies the exactness of the sequence

$$0 \to \text{Hom}(C,G) \otimes \bar{\bar{G}} \to \text{Hom}(C,G) \otimes \bar{G} \to \text{Hom}(C,G) \otimes G' \to 0$$

Hence Hom $(C,G) * G' = 0$, and so Hom $(C,G) *$ Hom $(C',G') = 0$ in either case. ∎

If A is a free finitely generated module, then

$$A \approx \text{Hom}(\text{Hom}(A,R), R)$$

Since Hom (A,R) is also free and finitely generated, it follows from corollary 7 that

$$A \otimes G \approx \text{Hom}(\text{Hom}(A,R), R) \otimes \text{Hom}(R,G) \approx \text{Hom}(\text{Hom}(A,R), G)$$

We use this to express homology in terms of cohomology.

12 THEOREM *Let C be a free chain complex such that $H(C)$ is of finite type. For any module G there is a functorial short exact sequence*

$$0 \to \text{Ext}(H^{q+1}(C;R), G) \to H_q(C;G) \xrightarrow{h} \text{Hom}(H^q(C;R), G) \to 0$$

and this sequence is split.

PROOF By lemma 9, we are reduced to the case where C is of finite type. Then $C \otimes G \approx \text{Hom}(\text{Hom}(C,R), G)$, and the result follows, by theorem 3, on changing Hom (C,R) to a chain complex by changing the sign of the degree. ∎

The following result is a version of lemma 8 valid for homology that is a partial converse to theorem 10.

13 THEOREM *Let C be a free chain complex such that for every module G the map μ: Hom $(C,R) \otimes G \to$ Hom (C,G) induces isomorphisms of all cohomology modules. Then $H_*(C)$ is of finite type.*

PROOF Because $\mu: H^q(\text{Hom}(C,R) \otimes H_q(C)) \approx H^q(\text{Hom}(C,H_q(C)))$, it follows from theorem 3 that there exist $f_i \in \text{Hom}(C_q,R)$ and $z_i \in H_q(C)$ such that $h\mu\{\sum f_i \otimes z_i\} = 1_{H_q(C)}$. Then, for any $z \in H_q(C)$ we have

$$z = \langle \mu\{\sum f_i \otimes z_i\}, z\rangle = \sum \langle f_i, z\rangle z_i$$

showing that $H_q(C)$ is generated by z_i. ∎

Note that if the short exact sequence of theorem 10 is valid for a given C for all G, then the hypothesis of theorem 13 is satisfied, and so $H(C)$ is of finite type.

6 CUP AND CAP PRODUCTS

There is a cross product of cohomology classes from the tensor product of the cohomology of two spaces to the cohomology of their product space. By using the diagonal map of a space into its square, the cross product gives rise to a product in the cohomology module of a space. This multiplicative structure provides cohomology with more structure than just the essentially additive module structure. In this section we shall define these products and establish some of their elementary properties.

If $\{X \times B, A \times Y\}$ is an excisive couple in $X \times Y$, there is a *cohomology cross product*

$$\mu' \colon H^p(X,A;\, G) \otimes H^q(Y,B;\, G') \to H^{p+q}((X,A) \times (Y,B);\, G \otimes G')$$

induced by the functorial homomorphism

$$\begin{gathered} \operatorname{Hom}(\Delta(X)/\Delta(A),G) \otimes \operatorname{Hom}(\Delta(Y)/\Delta(B),G') \\ {\scriptstyle\mu}\big\downarrow \\ \operatorname{Hom}([\Delta(X)/\Delta(A)] \otimes [\Delta(Y)/\Delta(B)],\, G \otimes G') \end{gathered}$$

followed by an Eilenberg-Zilber cochain equivalence of the bottom module with $\operatorname{Hom}(\Delta(X \times Y)/\Delta(X \times B \cup A \times Y),\, G \otimes G')$. If $u \in H^p(X,A;\, G)$ and $v \in H^q(Y,B;\, G')$, we define

$$u \times v = \mu'(u \otimes v) \in H^{p+q}((X,A) \times (Y,B);\, G \otimes G')$$

From theorem 5.5.11 we obtain the following *Künneth formula for singular cohomology.*

1 THEOREM *Let $\{X \times B, A \times Y\}$ be an excisive couple in $X \times Y$ and let G and G' be modules such that $G * G' = 0$. If $H_*(X,A;\, R)$ and $H_*(Y,B;\, R)$ are of finite type or if $H_*(Y,B;\, R)$ is of finite type and G' is finitely generated, there is a functorial short exact sequence*

$$0 \to [H^*(X,A;\, G) \otimes H^*(Y,B;\, G')]^q \xrightarrow{\mu'} H^q((X,A) \times (Y,B);\, G \otimes G') \to [H^*(X,A;\, G) * H^*(Y,B;\, G')]^{q+1} \to 0$$

and this sequence is split. ∎

The cohomology cross product satisfies the following analogues of statements 5.3.11 to 5.3.15.

2 *Let $f\colon (X,A) \to (X',A')$ and $g\colon (Y,B) \to (Y',B')$ be maps and let $u' \in H^p(X',A';\, G)$ and $v' \in H^q(Y',B';\, G')$. Then, in $H^{p+q}((X,A) \times (Y,B);\, G \otimes G')$, we have*

$$(f \times g)^*(u' \times v') = f^*u' \times g^*v' \qquad \blacksquare$$

3 *Let $p\colon (X,A) \times Y \to (X,A)$ be the projection to the first factor and let $\eta\colon G' \to H^*(Y;G')$ be the augmentation map. For $u \in H^q(X,A;\, G)$, in $H^q((X,A) \times Y;\, G \otimes G')$, we have*

$$p^*(\mu(u \otimes g')) = u \times \eta(g') \qquad \blacksquare$$

4 *For $u \in H^p(X,A;\, G)$, $v \in H^q(Y,B;\, G')$, and $w \in H^r(Z,C;\, G'')$, in $H^{p+q+r}((X,A) \times (Y,B) \times (Z,C);\, G \otimes G' \otimes G'')$, we have*

$$u \times (v \times w) = (u \times v) \times w \qquad \blacksquare$$

5 *Let $T\colon (X,A) \times (Y,B) \to (Y,B) \times (X,A)$ and $\varphi\colon G \otimes G' \to G' \otimes G$ interchange the factors. For $u \in H^p(X,A;\, G)$ and $v \in H^q(Y,B;\, G')$, in $H^{p+q}((X,A) \times (Y,B);\, G' \otimes G)$, we have*

$$T^*(v \times u) = (-1)^{pq}\varphi_*(u \times v) \quad \blacksquare$$

6 *Let* $\{(X_1,A_1), (X_2,A_2)\}$ *be an excisive couple of pairs in* X *and let* $u \in H^p(X_1 \cap X_2, A_1 \cap A_2; G)$ *and* $v \in H^q(Y,B; G')$. *For the connecting homomorphisms of appropriate Mayer-Vietoris sequences we have*

$$\delta^*(u \times v) = \delta^* u \times v$$

in $H^{p+q+1}((X_1 \cup X_2, A_1 \cup A_2) \times (Y,B); G \otimes G')$ *and*

$$\delta^*(v \times u) = (-1)^q v \times \delta^* u$$

in $H^{p+q+1}((Y,B) \times (X_1 \cup X_2, A_1 \cup A_2); G' \otimes G)$. $\blacksquare$

Consider the two functors $\Delta(X)$ and $\Delta(X) \otimes \Delta(X)$ on the category of topological spaces. Because $\Delta(X)$ is free with models $\{\Delta^q\}_{q\geq 0}$ and $\Delta(X) \otimes \Delta(X)$ is acyclic with models $\{\Delta^q\}_{q\geq 0}$ [that is, the reduced complex of $\Delta(\Delta^q) \otimes \Delta(\Delta^q)$ is acyclic for all q], it follows from the acyclic-model theorem 4.3.3 that there exist functorial chain maps $\tau_*: \Delta(X) \to \Delta(X) \otimes \Delta(X)$ preserving augmentation, and any two are chain homotopic. Such a functorial chain map is called a *diagonal approximation*. The name stems from the fact that if $\tau'_X: \Delta(X \times X) \to \Delta(X) \otimes \Delta(X)$ is a functorial chain equivalence given by the Eilenberg-Zilber theorem and $d: X \to X \times X$ is the diagonal map, then the composite

$$\Delta(X) \xrightarrow{\Delta(d)} \Delta(X \times X) \xrightarrow{\tau'_X} \Delta(X) \otimes \Delta(X)$$

is a diagonal approximation.

We construct a particular diagonal approximation called the *Alexander-Whitney diagonal approximation*. If $\sigma: \Delta^q \to X$ is a singular q-simplex, the *front i-face* ${}_i\sigma$ is defined for $0 \leq i \leq q$ to equal the composite $\sigma \circ \lambda$, where $\lambda: \Delta^i \to \Delta^q$ is the simplicial map defined by $\lambda(p_j) = p_j$ for $0 \leq j \leq i$. Similarly, the *back i-face* σ_i is defined for $0 \leq i \leq q$ to equal the composite $\sigma \circ \lambda'$, where $\lambda': \Delta^i \to \Delta^q$ is the simplicial map defined by $\lambda'(p_j) = p_{j+q-i}$ for $0 \leq j \leq i$. It is easy to verify that

$$\tau(\sigma) = \sum_{i+j=\deg \sigma} {}_i\sigma \otimes \sigma_j$$

defines a functorial chain map $\tau: \Delta(X) \to \Delta(X) \otimes \Delta(X)$, and this chain map is the Alexander-Whitney diagonal approximation.

Let G and G' be R modules. A *pairing* of G and G' to an R module G'' is a homomorphism $\varphi: G \otimes G' \to G''$. For example, G and G' are always paired to $G \otimes G'$. Given such a pairing and given a diagonal approximation τ, there is a functorial cochain map

$$\bar\tau_X: \text{Hom}(\Delta(X),G) \otimes \text{Hom}(\Delta(X),G') \to \text{Hom}(\Delta(X),G'')$$

defined to equal the composite

$$\text{Hom}(\Delta(X),G) \otimes \text{Hom}(\Delta(X),G') \xrightarrow{\mu} \text{Hom}(\Delta(X) \otimes \Delta(X), G \otimes G') \xrightarrow{\text{Hom}(\tau_X,\varphi)} \text{Hom}(\Delta(X),G'')$$

If $A \subset X$, then for $f \in \operatorname{Hom}(\Delta(X),G)$ and $f' \in \operatorname{Hom}(\Delta(X),G')$, we have

$$\bar\tau_X(f \otimes f') \mid \Delta(A) = \bar\tau_A(f \mid \Delta(A) \otimes f' \mid \Delta(A))$$

If $A_1, A_2 \subset X$ and f vanishes on A_1, f' vanishes on A_2, it follows that $\bar\tau_X(f \otimes f')$ vanishes on $\Delta(A_1) + \Delta(A_2)$. If $\{A_1,A_2\}$ is an excisive couple in X, it follows that $\bar\tau_X$ induces a homomorphism

$$H^p(X,A_1;\, G) \otimes H^q(X,A_2;\, G') \to H^{p+q}(X, A_1 \cup A_2;\, G'')$$

which is called the *cup-product homomorphism*. If $u \in H^p(X,A_1;\, G)$ and $v \in H^q(X,A_2;\, G')$, their cup product is denoted by

$$u \smile v \in H^{p+q}(X,A_1 \cup A_2;\, G'')$$

This product is a bilinear function of u and v and depends on the pairing φ but not on the particular diagonal approximation. The Alexander-Whitney diagonal approximation yields a particular map $\bar\tau$ which defines a cup product of cochains $f \smile f'$ for $f \in \operatorname{Hom}(\Delta_p(X),G)$ and $f' \in \operatorname{Hom}(\Delta_q(X),G')$ by

$$(f \smile f')(\sigma) = \varphi(f({}_p\sigma) \otimes f'(\sigma_q))$$

Then $\{f\} \smile \{f'\} = \{f \smile f'\}$ in $H^{p+q}(X, A_1 \cup A_2;\, G'')$.

As pointed out above, there exist diagonal approximations which are factored through $\Delta(d)$. This implies the following relation expressing the cup product in terms of the cross product.

7 THEOREM *If $\{X \times A_2, A_1 \times X\}$ is an excisive couple in $X \times X$, if $\{A_1,A_2\}$ is an excisive couple in X, and $\varphi\colon G \otimes G' \to G''$ is a pairing, then for $u \in H^p(X,A_1;\, G)$ and $v \in H^q(X,A_2;\, G')$, in $H^{p+q}(X, A_1 \cup A_2;\, G'')$, we have*

$$u \smile v = \varphi_*(d^*(u \times v)) \quad ■$$

The cup product has the following properties analogous to the corresponding properties of the cross product.

8 *Let $f\colon X \to Y$ map A_1 into B_1 and A_2 into B_2 and let $u \in H^p(Y,B_1;\, G)$ and $v \in H^q(Y,B_2;\, G')$. Let $f_1\colon (X,A_1) \to (Y,B_1)$, $f_2\colon (X,A_2) \to (Y,B_2)$, and $\bar f\colon (X, A_1 \cup A_2) \to (Y, B_1 \cup B_2)$ be maps defined by f. In $H^{p+q}(X, A_1 \cup A_2;\, G'')$, we have*

$$\bar f^*(u \smile v) = f_1^* u \smile f_2^* v \quad ■$$

9 *For any $u \in H^q(X,A;\, G)$ with the pairings $R \otimes G \approx G \approx G \otimes R$ we have*

$$1 \smile u = u = u \smile 1 \quad ■$$

10 *Given a commutative diagram, where φ, φ', ψ, and ψ' are pairings,*

$$\begin{array}{ccc} G_1 \otimes (G_2 \otimes G_3) \approx (G_1 \otimes G_2) \otimes G_3 & \xrightarrow{\varphi \otimes 1} & G_{12} \otimes G_3 \\ {\scriptstyle 1 \otimes \varphi'}\downarrow & & \downarrow{\scriptstyle \psi} \\ G_1 \otimes G_{23} & \xrightarrow{\psi'} & G_{123} \end{array}$$

and given $u_1 \in H^p(X,A_1;\ G_1)$, $u_2 \in H^q(X,A_2;\ G_2)$, *and* $u_3 \in H^r(X,A_3;\ G_3)$, *then, in* $H^{p+q+r}(X, A_1 \cup A_2 \cup A_3;\ G_{123})$, *we have*

$$u_1 \smile (u_2 \smile u_3) = (u_1 \smile u_2) \smile u_3 \quad \blacksquare$$

11 *Given a commutative diagram of pairings*

$$\begin{array}{c} G \otimes G' \approx G' \otimes G \\ \searrow \quad \swarrow \\ G'' \end{array}$$

and given $u \in H^p(X,A_1;\ G)$ *and* $v \in H^q(X,A_2;\ G')$, *in* $H^{p+q}(X, A_1 \cup A_2;\ G'')$, *we have*

$$u \smile v = (-1)^{pq} v \smile u \quad \blacksquare$$

12 *Let* $\{(X_1,A_1), (X_2,A_2)\}$ *be an excisive couple of pairs in X, let* $A \subset X_1 \cup X_2$, *and let* $i\colon (X_1 \cap X_2,\ A \cap X_1 \cap X_2) \subset (X_1 \cup X_2,\ A)$. *For elements* $u \in H^p(X_1 \cap X_2, A_1 \cap A_2;\ G)$ *and* $v \in H^q(X_1 \cup X_2, A;\ G')$ *and with the connecting homomorphisms of the appropriate Mayer-Vietoris sequences, in* $H^{p+q+1}(X_1 \cup X_2, A_1 \cup A_2 \cup A;\ G'')$, *we have*

$$\delta^*(u \smile i^*v) = \delta^* u \smile v$$
$$\delta^*(i^*v \smile u) = (-1)^q v \smile \delta^* u \quad \blacksquare$$

Let $\tau'\colon \Delta(X \times Y) \to \Delta(X) \otimes \Delta(Y)$ be a functorial chain equivalence given by the Eilenberg-Zilber theorem and let

$$T\colon [\Delta(X) \otimes \Delta(Y)] \otimes [\Delta(X) \otimes \Delta(Y)] \to [\Delta(X) \otimes \Delta(X)] \otimes [\Delta(Y) \otimes \Delta(Y)]$$

be the chain map defined by

$$T((c \otimes d) \otimes (c' \otimes d')) = (-1)^{\deg d \deg c'}(c \otimes c') \otimes (d \otimes d')$$

If τ is any diagonal approximation, it follows by the method of acyclic models that the diagram

$$\begin{array}{ccc} \Delta(X \times Y) & \xrightarrow{\tau_{X\times Y}} & \Delta(X \times Y) \otimes \Delta(X \times Y) \\ \tau' \downarrow & & \downarrow T \circ (\tau' \otimes \tau') \\ \Delta(X) \otimes \Delta(Y) & \xrightarrow{\tau_X \otimes \tau_Y} & [\Delta(X) \otimes \Delta(X)] \otimes [\Delta(Y) \otimes \Delta(Y)] \end{array}$$

is chain homotopy commutative. This implies the following additional relation between cup products and cross products.

13 THEOREM *Let* $\varphi\colon G_1 \otimes G_2 \to G$ *and* $G_1' \otimes G_2' \to G'$ *be pairings and let* $G_1 \otimes G_1'$ *and* $G_2 \otimes G_2'$ *be paired to* $G \otimes G'$ *by the homomorphism*

$$(G_1 \otimes G_1') \otimes (G_2 \otimes G_2') \approx (G_1 \otimes G_2) \otimes (G_1' \otimes G_2') \xrightarrow{\varphi \otimes \varphi'} G \otimes G'$$

Given $u_1 \in H^p(X,A_1;\ G_1)$, $u_2 \in H^q(X,A_2;\ G_2)$, $v_1 \in H^r(Y,B_1\colon\ G_1')$, *and* $v_2 \in H^s(Y,B_2;\ G_2')$ *then with suitable excisiveness assumptions, we have, in* $H^{p+q+r+s}((X, A_1 \cup A_2) \times (Y, B_1 \cup B_2);\ G \otimes G')$,

$$(u_1 \times v_1) \smile (u_2 \times v_2) = (-1)^{qr}(u_1 \smile u_2) \times (v_1 \smile v_2) \quad \blacksquare$$

Combining theorem 13 with statements 3 and 9, we obtain the following result expressing the cross product in terms of the cup products.

14 COROLLARY *Let $\{X \times B, A \times Y\}$ be an excisive couple in $X \times Y$ and let $p_1: (X,A) \times Y \to (X,A)$ and $p_2: X \times (Y,B) \to (Y,B)$ be the projections. Given $u \in H^p(X,A; G)$ and $v \in H^q(Y,B; G')$, then, in $H^{p+q}((X,A) \times (Y,B); G \otimes G')$, we have*

$$u \times v = p_1^*(u) \smile p_2^*(v) \quad \blacksquare$$

With the last result we can give the following example of two polyhedra having isomorphic homology and cohomology modules but not isomorphic cup-product structures.

15 EXAMPLE Let p and q be integers ≥ 1 and let X be the space which is the union of S^p, S^q, and S^{p+q}, all identified at one point. If $i: S^p \subset X$, $j: S^q \subset X$, and $k: S^{p+q} \subset X$, then $i_* \tilde{H}(S^p) \oplus j_* \tilde{H}(S^q) \oplus k_* \tilde{H}(S^{p+q}) \approx \tilde{H}(X)$. Computing $H(S^p \times S^q)$ by the Künneth formula, we see that $H(X) \approx H(S^p \times S^q)$. By the universal-coefficient theorem, X and $S^p \times S^q$ have isomorphic homology and cohomology groups for any coefficient group. Since

$$k^*: H^{p+q}(X;\mathbf{Z}) \approx H^{p+q}(S^{p+q};\mathbf{Z})$$

and k^* commutes with the cup product, it follows that the cup product of integral cohomology classes of degrees p and q, respectively, in X is zero. However, it follows from corollary 14 that there are integral cohomology classes of $S^p \times S^q$ of degrees p and q, respectively, whose cup product is nonzero. Therefore $H^*(X;\mathbf{Z})$ and $H^*(S^p \times S^q; \mathbf{Z})$ are not isomorphic by an isomorphism of graded modules preserving the cup product. Hence X and $S^p \times S^q$ are not homeomorphic, nor even of the same homotopy type.

There is another product closely related to the cup product that multiplies homology and cohomology classes together. We begin with the observation that if C and C' are chain complexes and G and G' are paired to G'' by φ, there is a functorial homomorphism

$$h: \operatorname{Hom}(C',G) \otimes (C \otimes C' \otimes G') \to C \otimes G''$$

such that $h(f \otimes (c \otimes c' \otimes g')) = c \otimes \varphi(\langle f,c' \rangle \otimes g')$. A straightforward calculation shows that for $f \in \operatorname{Hom}(C'_q,G)$ and $\bar{c} \in (C \otimes C')_n \otimes G'$

$$\partial h(f \otimes \bar{c}) = (-1)^{n-q} h(\delta f \otimes \bar{c}) + h(f \otimes \partial \bar{c})$$

If X is a space and $\tau: \Delta(X) \to \Delta(X) \otimes \Delta(X)$ is a diagonal approximation, a functorial map

$$\bar{\tau}: \operatorname{Hom}(\Delta(X),G) \otimes (\Delta(X) \otimes G') \to \Delta(X) \otimes G''$$

is defined by $\bar{\tau}(f \otimes c) = h(f \otimes \tau(c))$. The boundary formula yields

$$\partial \bar{\tau}(f \otimes c) = (-1)^{\deg c - \deg f} \bar{\tau}(\delta f \otimes c) + \bar{\tau}(f \otimes \partial c)$$

Note that if A is a subset of X and $f \in \operatorname{Hom}(\Delta(X),G)$ vanishes on A, then for any $c \in \Delta(A) \otimes G'$, $\bar{\tau}(f \otimes c) = 0$. It follows that if A_1, $A_2 \subset X$,

$f \in \text{Hom}(\Delta(X)/\Delta(A_1),G)$ is a cocycle, and $c \in \Delta(X) \otimes G'$ is a chain such that $\partial c \in [\Delta(A_1) + \Delta(A_2)] \otimes G'$, then $\bar{\bar{\tau}}(f \otimes c)$ is a chain of $\Delta(X) \otimes G''$ whose boundary is in $\Delta(A_2) \otimes G''$ [because $\partial\bar{\bar{\tau}}(f \otimes c) = \tau(\bar{\bar{f}} \otimes \partial c)$]. Furthermore, if f is the coboundary of a cochain which vanishes on $\Delta(A_1)$ or if c equals a boundary modulo $[\Delta(A_1) + \Delta(A_2)] \otimes G'$, then $\bar{\bar{\tau}}(f \otimes c)$ is a boundary modulo $\Delta(A_2) \otimes G''$. Hence $\bar{\bar{\tau}}$ defines a homomorphism [sending $\{f\} \otimes \{c\}$ to $\{\bar{\bar{\tau}}(f \otimes c)\}$]

$$H^q(X,A_1;\, G) \otimes H_n(\Delta(X)/[\Delta(A_1) + \Delta(A_2)];\, G') \to H_{n-q}(X,A_2;\, G'')$$

If $\{A_1,A_2\}$ is an excisive couple in X, this yields a homomorphism

$$H^q(X,A_1;\, G) \otimes H_n(X,\, A_1 \cup A_2;\, G') \to H_{n-q}(X,A_2;\, G'')$$

called the *cap product*. If $u \in H^q(X,A_1;\, G)$ and $z \in H_n(X,\, A_1 \cup A_2;\, G')$, their cap product is denoted by $u \frown z \in H_{n-q}(X,A_2;\, G'')$. It depends on the pairing φ but not on the particular diagonal approximation used to define $\bar{\bar{\tau}}$. The Alexander-Whitney diagonal approximation yields a map $\bar{\bar{\tau}}$ which defines a cap product on cochains and chains, denoted by $f \frown c$, by the formula

$$f \frown c = f \frown \left(\sum_\sigma \sigma \otimes g_\sigma'\right) = \sum_\sigma {}_{n-q}\sigma \otimes \varphi(\langle f,\sigma_q\rangle \otimes g_\sigma')$$

for $f \in \text{Hom}(\Delta_q(X), G)$ and $c = \Sigma_\sigma\, \sigma \otimes g_\sigma' \in \Delta_n(X) \otimes G'$. Then $\{f\} \frown \{c\} = \{f \frown c\}$.

The cap product has the following properties analogous to those of the cup product.

16 *Let $f\colon X \to Y$ map A_1 to B_1 and A_2 to B_2 and let $u \in H^q(Y,B_1;\, G)$ and $z \in H_n(X,\, A_1 \cup A_2;\, G')$. Let $f_1\colon (X,A_1) \to (Y,B_1)$, $f_2\colon (X,A_2) \to (Y,B_2)$, and $\bar{f}\colon (X,\, A_1 \cup A_2) \to (Y,\, B_1 \cup B_2)$ be maps defined by f. Then, in $H_{n-q}(Y,B_2;\, G'')$, we have*

$$f_{2*}(f_1^* u \frown z) = u \frown \bar{f}_* z \quad \blacksquare$$

17 *For any $z \in H_n(X,A;\, G)$ with the pairing $R \otimes G \approx G$*

$$1 \frown z = z \quad \blacksquare$$

18 *Given a commutative diagram, where φ, φ', ψ, and ψ' are pairings,*

$$\begin{array}{ccc} G_1 \otimes (G_2 \otimes G_3) \approx (G_1 \otimes G_2) \otimes G_3 & \xrightarrow{\varphi \otimes 1} & G_{12} \otimes G_3 \\ {\scriptstyle 1 \otimes \varphi'}\downarrow & & \downarrow{\scriptstyle \psi} \\ G_1 \otimes G_{23} & \xrightarrow{\psi'} & G_{123} \end{array}$$

for $u \in H^p(X,A_1;\, G_1)$, $v \in H^q(X,A_2;\, G_2)$, and $z \in H_n(X,\, A_1 \cup A_2 \cup A_3;\, G_3)$, then, in $H_{n-p-q}(X,A_3;\, G_{123})$, we have

$$u \frown (v \frown z) = (u \smile v) \frown z \quad \blacksquare$$

19 *Let $u \in H^q(X,A;\, G)$ and $z \in H_q(X,A;\, G')$ and let $\varepsilon\colon H_0(X;\, G \otimes G') \to G \otimes G'$ be the augmentation. Then, in $G \otimes G'$,*

$$\varepsilon(u \frown z) = \langle u,z \rangle \quad \blacksquare$$

20 *Let* $\{(X_1,A_1), (X_2,A_2)\}$ *be an excisive couple in X and let* $A \subset X_1 \cup X_2$ *and* $i: (X_1 \cap X_2, A \cap X_1 \cap X_2) \subset (X_1 \cup X_2, A)$. *For* $u \in H^q(X_1 \cup X_2, A; G)$ *and* $z \in H_n(X_1 \cup X_2, A_1 \cup A_2 \cup A; G')$, *with the connecting homomorphisms of the appropriate Mayer-Vietoris sequences, in* $H_{n-q-1}(X_1 \cap X_2, A_1 \cap A_2; G'')$, *we have*

$$\partial_*(u \frown z) = i^*u \frown \partial_* z \quad \blacksquare$$

21 *Let* $u_1 \in H^p(X,A_1; G_1)$, $u_2 \in H^q(Y,B_1; G_2)$, $z_1 \in H_m(X, A_1 \cup A_2; G_1')$, *and* $z_2 \in H_n(X, B_1 \cup B_2; G_2')$, *and let* G_1 *and* G_1' *be paired to* G_1'', G_2 *and* G_2' *be paired to* G_2'', *and* $(G_1 \otimes G_2)$ *and* $(G_1' \otimes G_2')$ *be compatibly paired to* $G_1'' \otimes G_2''$. *Then, in* $H_{m+n-p-q}((X,A_2) \times (Y,B_2); G_1'' \otimes G_2'')$, *we have*

$$(u_1 \times u_2) \frown (z_1 \times z_2) = (-1)^{p(n-q)}(u_1 \frown z_1) \times (u_2 \frown z_2) \quad \blacksquare$$

7 HOMOLOGY OF FIBER BUNDLES

Cup and cap products are used in this section to study the homology of fiber bundles. We shall show that in case the cohomology of the total space maps epimorphically onto the cohomology of each fiber, the homology (or cohomology) of the total space is isomorphic to the homology (or cohomology) of the product space of the base and the fiber. For orientable sphere bundles this leads to a proof of the exactness of the Thom-Gysin sequences, which will be applied in the next section to compute the cohomology rings of projective spaces.

We begin with some algebraic considerations. Let $M = \{M_q\}$ be a free finitely generated graded R module and let $M^* = \{M^q = \text{Hom}(M_q,R)\}$. Let (X,A) be a topological pair and $f: X \to Y$ be a continuous map. Given a homomorphism (of degree 0) $\theta: M^* \to H^*(X,A; R)$, there are homomorphisms (of degree 0) for any R module G

$$\Phi: H(X,A; G) \to H(Y;G) \otimes M$$
$$\Phi^*: H^*(Y;G) \otimes M^* \to H^*(X,A; G)$$

defined by $\Phi(z) = \sum_i f_*(\theta(m_i^*) \frown z) \otimes m_i$, where $\{m_i\}$ is a basis of M and $\{m_i^*\}$ is the dual basis of M^* (Φ is uniquely defined by this formula), and $\Phi^*(u \otimes m^*) = f^*u \smile \theta(m^*)$.

1 LEMMA *With the notation above, if* Φ *is an isomorphism for* $G = R$, *then* Φ *and* Φ^* *are isomorphisms for all R modules G.*

PROOF For each i let c_i^* be a cocycle of $\text{Hom}(\Delta(X)/\Delta(A);R)$ representing the class $\theta(m_i^*)$ and assume that m_i (and hence also m_i^* and c_i^*) have degree q_i. Let $\tau: \Delta(X)/\Delta(A) \to \Delta(Y) \otimes M$ be the homomorphism (of degree 0) defined by

$$\tau(c) = \sum_i \Delta(f)(c_i^* \frown c) \otimes m_i$$

An easy computation shows that τ is a chain map and that the induced homomorphisms

$$\tau_*: H_*(X,A;\, G) \to H_*(\Delta(Y) \otimes M;\, G) \approx H_*(Y;G) \otimes M$$
$$\tau^*: H^*(Y;G) \otimes M^* \approx H^*(\operatorname{Hom}(\Delta(Y) \otimes M,\, G)) \to H^*(X,A;\, G)$$

equal Φ and Φ^*, respectively. Since Φ is assumed to be an isomorphism for $G = R$, the chain map τ induces an isomorphism of homology. The universal-coefficient theorems for homology and cohomology then imply that Φ and Φ^* are isomorphisms for all G. ■

A *fiber-bundle pair* with base space B consists of a *total pair* $(E,\dot{E})$, a *fiber pair* $(F,\dot{F})$, and a *projection* $p: E \to B$ such that there exists an open covering $\{V\}$ of B and for each $V \in \{V\}$ a homeomorphism $\varphi_V: V \times (F,\dot{F}) \to (p^{-1}(V), p^{-1}(V) \cap \dot{E})$ such that the composite

$$V \times F \xrightarrow{\varphi_V} p^{-1}(V) \xrightarrow{p} V$$

is the projection to the first factor. If $A \subset B$, we let $E_A = p^{-1}(A)$ and $\dot{E}_A = p^{-1}(A) \cap \dot{E}$, and if $b \in B$, then $(E_b,\dot{E}_b)$ is the *fiber pair over b*.

Following are some examples.

2 For a space B and pair $(F,\dot{F})$ the *product-bundle pair* consists of the total pair $B \times (F,\dot{F})$ with projection to the first factor.

3 Given a bundle projection $\dot{p}: \dot{E} \to B$ with compact fiber $\dot{F}$, let E be the mapping cylinder of $\dot{p}$ and $p: E \to B$ the canonical retraction. Then $(E,\dot{E})$ is the total pair of a fiber-bundle pair over B with fiber $(F,\dot{F})$, where F is the cone over $\dot{F}$, and projection p.

4 If ξ is a q-sphere bundle over B, then $(E_\xi,\dot{E}_\xi)$ is the total pair of a fiber-bundle pair over B with fiber (E^{q+1},S^q) and projection $p_\xi: E_\xi \to B$.

Given a fiber-bundle pair with total pair $(E,\dot{E})$ and fiber pair $(F,\dot{F})$, a *cohomology extension of the fiber* is a homomorphism $\theta: H^*(F,\dot{F};\, R) \to H^*(E,\dot{E};\, R)$ of graded modules (of degree 0) such that for each $b \in B$ the composite

$$H^*(F,\dot{F};\, R) \xrightarrow{\theta} H^*(E,\dot{E};\, R) \to H^*(E_b,\dot{E}_b;\, R)$$

is an isomorphism. The following statements are easily verified.

5 *Let $\bar{p}: B \times (F,\dot{F}) \to (F,\dot{F})$ be the projection to the second factor. Then*

$$\theta = \bar{p}^*: H^*(F,\dot{F};\, R) \to H^*(B \times (F,\dot{F});\, R)$$

is a cohomology extension of the fiber of the product-bundle pair. ■

6 *Let $\theta: H^*(F,\dot{F};\, R) \to H^*(E,\dot{E};\, R)$ be a cohomology extension of the fiber of a fiber-bundle pair over B and let $f: B' \to B$ be a map. There is an induced bundle pair over B', with total pair $(E',\dot{E}')$ and fiber $(F,\dot{F})$, and there is a map*

$\bar{f}$: $(E',\dot{E}') \to (E,\dot{E})$ commuting with projections. Then the composite

$$H^*(F,\dot{F};R) \xrightarrow{\theta} H^*(E,\dot{E};R) \xrightarrow{\bar{f}^*} H^*(E',\dot{E}';R)$$

is a cohomology extension of the fiber in the induced bundle. ■

7 *Given a fiber-bundle pair over B with total pair $(E,\dot{E})$, let the path components of B be $\{B_j\}$ and let $(E_j,\dot{E}_j)$ be the induced total pair over B_j. A cohomology extension θ of the fiber of the bundle pair over B corresponds to a family of cohomology extensions $\{\theta_j\}$ of the induced bundle pairs over B_j.* ■

We now establish the local form of the theorem toward which we are heading. It shows that any cohomology extension of the fiber in a product-bundle pair has homology properties as nice as the one given in statement 5 above.

8 LEMMA *Let $(F,\dot{F})$ be a pair such that $H_*(F,\dot{F};R)$ is free and finitely generated over R and let θ: $H^*(F,\dot{F};R) \to H^*(B \times (F,\dot{F});R)$ be a cohomology extension of the fiber of the product-bundle pair. Then the homomorphisms*

$$\Phi: H_*(B \times (F,\dot{F});G) \to H_*(B;G) \otimes H_*(F,\dot{F};R)$$
$$\Phi^*: H^*(B;G) \otimes H^*(F,\dot{F};R) \to H^*(B \times (F,\dot{F});G)$$

are isomorphisms for all R modules G.

PROOF By lemma 1, it suffices to prove that Φ is an isomorphism for $G = R$. If $\{B_j\}$ is the set of path components of B, then

$$H_*(B \times (F,\dot{F});R) \approx \bigoplus_j H_*(B_j \times (F,\dot{F});R)$$

and

$$H_*(B;R) \otimes H_*(F,\dot{F};R) \approx \bigoplus_j H_*(B_j;R) \otimes H_*(F,\dot{F};R)$$

Therefore it suffices to prove the result for a path-connected space B. For such a B, $R \approx H^0(B;R)$.

By the Künneth formula, $H_*(B \times (F,\dot{F});R) \approx H_*(B;R) \otimes H_*(F,\dot{F};R)$. We define graded submodules N_s of $H_*(B;R) \otimes H_*(F,\dot{F};R)$ by

$$(N_s)_q = \bigoplus_{i+j=q,\, j\geq s} H_i(B;R) \otimes H_j(F,\dot{F};R)$$

Then

$$H_*(B;R) \otimes H_*(F,\dot{F};R) = N_0 \supset N_1 \supset \cdots \supset N_s \supset N_{s+1}$$

and $N_s = 0$ for large enough s. If $u \in H^s(F,\dot{F};R)$, then $\theta(u) = 1 \times \lambda(u) + \bar{u}$, where $\bar{u} \in \bigoplus_{i+j=s,j<s} H^i(B;R) \otimes H^j(F,\dot{F};R)$ and $\theta(u) \mid [b \times (F,\dot{F})] = 1 \times \lambda(u)$. Because θ is a cohomology extension of the fiber, λ is an automorphism of $H^*(F,\dot{F};R)$. Let $z' \in H_s(F,\dot{F};R)$ and consider $z \times z' \in N_s$. Then

$$\Phi(z \times z') = \sum_i p_*(\theta(m_i^*) \frown (z \times z')) \otimes m_i$$

and if $\deg m_i < s$, then $\theta(m_i^*) \frown (z \times z') \in N_1$ and $p_*(N_1) = 0$. Therefore

$\Phi(z \times z') \in N_s$, and so Φ maps N_s into itself for all s. Because of the short exact sequences

$$0 \to N_{s+1} \to N_s \to N_s/N_{s+1} \to 0$$

and the five lemma, it follows by downward induction on s that Φ is an isomorphism if and only if it induces an isomorphism of N_s/N_{s+1} onto itself for all s. For $z' \in H_s(F,\dot{F}; R)$, computing $\Phi(z \times z')$ in N_s/N_{s+1}, we obtain

$$\begin{aligned}\Phi(z \times z') &= \sum_{\deg m_i \geq s} p_*[(1 \times \lambda(m_i^*) + \bar{m}_i^*) \frown (z \times z')] \otimes m_i \\ &= \sum_{\deg m_i = s} p^*[1 \times \lambda(m_i^*) \frown (z \times z')] \otimes m_i\end{aligned}$$

because $\bar{m}_i^* \frown (z \times z') \in N_1$ and $p_*(N_1) = 0$. Now, by properties 5.6.21, 5.6.19, and 5.6.17,

$$\begin{aligned}&\sum_{\deg m_i = s} p_*[1 \times \lambda(m_i^*) \frown (z \times z')] \otimes m_i \\ &\qquad = \sum_{\deg m_i = s} z \otimes \langle \lambda(m_i^*), z' \rangle m_i = z \otimes \lambda_*(z')\end{aligned}$$

where $\lambda_*: H_*(F,\dot{F}; R) \to H_*(F,\dot{F}; R)$ is the automorphism dual to λ. Hence $\Phi(z \times z') = z \times \lambda_*(z')$ in N_s/N_{s+1}, showing that Φ induces an isomorphism of N_s/N_{s+1} for all s. ■

The following *Leray-Hirsch theorem* shows that fiber-bundle pairs with cohomology extensions of the fiber have homology and cohomology modules isomorphic to those of the product of the fiber pair and the base.

9 THEOREM *Let $(E,\dot{E})$ be the total pair of a fiber-bundle pair with base B and fiber pair $(F,\dot{F})$. Assume that $H_*(F,\dot{F}; R)$ is free and finitely generated over R and that θ is a cohomology extension of the fiber. Then the homomorphisms*

$$\Phi: H_*(E,\dot{E}; G) \to H_*(B;G) \otimes H_*(F,\dot{F}; R) \qquad \Phi(z) = \sum_i p_*(\theta(m_i^*) \frown z) \otimes m_i$$

$$\Phi^*: H^*(B;G) \otimes H^*(F,\dot{F}; R) \to H^*(E,\dot{E}; G) \qquad \Phi^*(u \otimes v) = p^*(u) \smile \theta(v)$$

are isomorphisms (of graded modules) for all R modules G.

PROOF By lemma 1, it suffices to prove the result for the map Φ in the case $G = R$. For any subset $A \subset B$ let θ_A be the composite

$$H^*(F,\dot{F}; R) \xrightarrow{\theta} H^*(E,\dot{E}; R) \to H^*(E_A,\dot{E}_A; R)$$

Then θ_A is a cohomology extension of the fiber in the induced bundle over A. It follows from lemma 8 that if the induced bundle over A is homeomorphic to the product-bundle pair $A \times (F,\dot{F})$, then

$$\Phi_A: H_*(E_A,\dot{E}_A; R) \approx H_*(A;R) \otimes H_*(F,\dot{F}; R)$$

Hence Φ_V is an isomorphism for all sufficiently small open sets V.

If V and V' are open sets in B, then $\{(E_V,\dot{E}_V), (E_{V'},\dot{E}_{V'})\}$ is an excisive couple of pairs in E, and it follows from property 5.6.20 that Φ_V, $\Phi_{V'}$, $\Phi_{V\cap V'}$, and $\Phi_{V\cup V'}$ map the exact Mayer-Vietoris sequence of $(E_V,\dot{E}_V)$ and $(E_{V'},\dot{E}_{V'})$ into the tensor product of the exact Mayer-Vietoris sequence of V and V' by $H_*(F,\dot{F};R)$. Since $H_*(F,\dot{F};R)$ is free over R, its tensor product with any exact sequence is exact. Therefore, if Φ_V, $\Phi_{V'}$, and $\Phi_{V\cap V'}$ are isomorphisms, it follows from the five lemma that $\Phi_{V\cup V'}$ is also an isomorphism. By induction, Φ_U is an isomorphism for any U which is a finite union of sufficiently small open sets. Let $\mathfrak{U}$ be the collection of these sets. Since any compact subset of B lies in some element of $\mathfrak{U}$, $H_*(B;R) \approx \lim_{\to} \{H_*(U;R)\}_{U\in\mathfrak{U}}$. Also, any compact subset of E lies in E_U for some $U \in \mathfrak{U}$, so $H_*(E,\dot{E};R) \approx \lim_{\to} \{H_*(E_U,\dot{E}_U;R)\}$. Because the tensor product commutes with direct limits and Φ corresponds to $\lim_{\to} \{\Phi_U\}_{U\in\mathfrak{U}}$ under these isomorphisms, Φ is also an isomorphism. ∎

The above argument proves directly that Φ is an isomorphism for any coefficient module G. A similar argument does not appear possible for Φ^*, because it is not true that $H^*(B;R)$ is isomorphic to the inverse limit $\lim_{\leftarrow} \{H^*(U;R)\}_{U\in\mathfrak{U}}$. It should be noted that in theorem 9 we have said nothing about commutativity of Φ^* with cup products, because it is not true, in general, that Φ^* preserves cup products.

We now specialize to the case of sphere bundles. Because

$$H^r(E^{q+1},S^q;R) \approx \begin{cases} 0 & r \neq q+1 \\ R & r = q+1 \end{cases}$$

if ξ is a q-sphere bundle, a cohomology extension of the fiber in ξ is an element $U \in H^{q+1}(E_\xi,\dot{E}_\xi;R)$ such that for any $b \in B$, the restriction of U to $(p^{-1}(b), p^{-1}(b) \cap \dot{E})$ is a generator of $H^{q+1}(p^{-1}(b), p^{-1}(b) \cap \dot{E};R)$. Such a cohomology class is called an *orientation class* (*over* R) of the bundle. If orientations of the bundle exist, the bundle is called *orientable*. An *oriented sphere bundle* is a pair (ξ,U_ξ) consisting of a sphere bundle ξ and an orientation class of U_ξ of ξ.

If U is an orientation class of ξ over $\mathbf{Z}$ and if 1 is the unit element of R, then $\mu(U \otimes 1)$ is an orientation class of ξ over R. Therefore a sphere bundle orientable over $\mathbf{Z}$ is orientable over any R.

If (ξ,U_ξ) is an oriented sphere bundle over B and $f: B' \to B$, then $(f^*\xi,\bar{f}^*U_\xi)$ is an oriented sphere bundle over B' [where $\bar{f}: (E_{f^*\xi},\dot{E}_{f^*\xi}) \to (E_\xi,\dot{E}_\xi)$ is associated to f].

From theorem 9 we get the following *Thom isomorphism theorem.*

10 THEOREM *Let (ξ,U_ξ) be an oriented q-sphere bundle over B. There are natural isomorphisms for any R module G*

$$\Phi_\xi: H_n(E_\xi,\dot{E}_\xi;G) \xrightarrow{\approx} H_{n-q-1}(B;G) \qquad \Phi_\xi(z) = p_*(U_\xi \frown z)$$

$$\Phi_\xi^*: H^r(B;G) \xrightarrow{\approx} H^{r+q+1}(E_\xi,\dot{E}_\xi;G) \qquad \Phi_\xi^*(v) = p^*v \smile U_\xi$$

PROOF Let m and m^* be dual generators of $H_{q+1}(E^{q+1},S^q;R)$ and $H^{q+1}(E^{q+1},S^q;R)$, respectively, and define a cohomology extension θ by $\theta(m^*) = U_\xi$. Then Φ_ξ is the composite

$$H_n(E_\xi,\dot{E}_\xi;G) \xrightarrow{\Phi} H_{n-q-1}(B;G) \otimes H_{q+1}(E^{q+1},S^q;R) \approx H_{n-q-1}(B;G)$$

where the second map sends $z \otimes m$ to z. By theorem 9, Φ is an isomorphism, and so Φ_ξ is an isomorphism. A similar argument shows that Φ_ξ^* is an isomorphism. These isomorphisms are natural for induced bundles because of naturality properties of the cup and cap products. ∎

This result implies the exactness of the following *Thom-Gysin sequences* of a sphere bundle.

11 THEOREM *Let (ξ, U_ξ) be an oriented q-sphere bundle with base B and projection $\dot{p} = p \mid \dot{E}: \dot{E} \to B$. For any R module G there are natural exact sequences*

$$\cdots \to H_n(\dot{E}_\xi;G) \xrightarrow{\dot{p}_*} H_n(B;G) \xrightarrow{\Psi_\xi} H_{n-q-1}(B;G) \xrightarrow{\rho} H_{n-1}(\dot{E}_\xi;G) \to \cdots$$

$$\cdots \to H^r(B;G) \xrightarrow{\dot{p}^*} H^r(\dot{E}_\xi;G) \xrightarrow{\rho^*} H^{r-q}(B;G) \xrightarrow{\Psi_\xi^*} H^{r+1}(B;G) \to \cdots$$

in which Ψ_ξ and Ψ_ξ^* have properties

$$\Psi_\xi(v \frown z) = (-1)^{(q+1)\deg v}\, \Psi_\xi^*(v) \frown z$$

$$\Psi_\xi^*(v_1 \smile v_2) = v_1 \smile \Psi_\xi^*(v_2)$$

PROOF There is a commutative diagram (with any coefficient module)

$$\begin{array}{ccccccccc}
\cdots \to & H_n(\dot{E}) & \xrightarrow{i_*} & H_n(E) & \xrightarrow{j_*} & H_n(E,\dot{E}) & \xrightarrow{\partial} & H_{n-1}(\dot{E}) & \to \cdots \\
& & \dot{p}_* \searrow & \approx \downarrow p_* & & \approx \downarrow \Phi_\xi & & & \\
& & & H_n(B) & & H_{n-q-1}(B) & & &
\end{array}$$

the top row of which is exact. Since p is a deformation retraction of E onto B, p_* is an isomorphism. By theorem 10, Φ_ξ is an isomorphism. The desired sequence is obtained by defining $\Psi_\xi = \Phi_\xi j_* p_*^{-1}$ and $\rho = \partial \Phi_\xi^{-1}$. Similarly, the cohomology sequence is defined by $\Psi_\xi^* = p^{*-1} j^* \Phi_\xi^*$ and $\rho^* = \Phi_\xi^{*-1}\delta$. We verify the formula for Ψ_ξ.

$$\begin{aligned}
\Psi_\xi(v \frown z) &= \Phi_\xi j_* p_*^{-1}(v \frown z) = \Phi_\xi j_*(p^*(v) \frown p_*^{-1}(z)) \\
&= \Phi_\xi(p^*(v) \frown j_* p_*^{-1}(z)) = p_*(U \frown [p^*(v) \frown j_* p_*^{-1}(z)]) \\
&= p_*(j^*[U \smile p^*(v)] \frown p_*^{-1}(z)) \\
&= (-1)^{(q+1)\deg v} p_*[j^* \Phi_\xi^*(v) \frown p_*^{-1}(z)] \\
&= (-1)^{(q+1)\deg v}\, \Psi_\xi^*(v) \frown z \quad ∎
\end{aligned}$$

Note that the isomorphisms Φ and Φ^* of the Thom isomorphism theorem depend on the choice of the orientation class U of the bundle. Therefore the homomorphisms ρ and Ψ and ρ^* and Ψ^* of the Thom-Gysin sequences also depend on the orientation class. In case B is path connected and U and U' are orientation classes of a sphere bundle over B, it follows from theorem 10 that there is an element $r \in R$ such that

$$U' = p^*(r \times 1) \smile U = r[p^*(1) \smile U]$$

If $b_0 \in B$, then

$$U' \mid (p^{-1}(b_0), p^{-1}(b_0) \cap \dot{E}) = r[U \mid (p^{-1}(b_0), p^{-1}(b_0) \cap \dot{E})]$$

Therefore we have the next result.

12 LEMMA *Two orientation classes U and U' of a sphere bundle over a path-connected base space B are equal if and only if for some $b_0 \in B$*

$$U \mid (p^{-1}(b_0), p^{-1}(b_0) \cap \dot{E}) = U' \mid (p^{-1}(b_0), p^{-1}(b_0) \cap \dot{E}) \quad \blacksquare$$

If B is not path connected, let $\{B_j\}$ be the set of path components of B and let $(E_j, \dot{E}_j)$ be the part of $(E, \dot{E})$ over B_j. Then

$$H^*(E, \dot{E}; R) \approx \times_j H^*(E_j, \dot{E}_j; R)$$

and we also obtain the following result.

13 LEMMA *Two orientation classes U and U' of a sphere bundle with base space B are equal if and only if for all $b \in B$*

$$U \mid (p^{-1}(b), p^{-1}(b) \cap \dot{E}) = U' \mid (p^{-1}(b), p^{-1}(b) \cap \dot{E}) \quad \blacksquare$$

In case $R = \mathbf{Z}_2$, then $H^{q+1}(p^{-1}(b), p^{-1}(b) \cap \dot{E}; \mathbf{Z}_2) \approx \mathbf{Z}_2$ for all $b \in B$. Therefore this module has a unique nonzero element, and we obtain the following consequence of lemma 13.

14 COROLLARY *Any two orientation classes over $\mathbf{Z}_2$ of a sphere bundle are equal.* ■

Thus, for $R = \mathbf{Z}_2$ the homomorphisms Φ, ρ, and Ψ and Φ^*, ρ^*, and Ψ^* are all unique.

The *characteristic class* Ω_ξ of an oriented q-sphere bundle (ξ, U_ξ) is defined to be the element

$$\Omega_\xi = \Psi_\xi^*(1) \in H^{q+1}(B;R)$$

This is functorial (that is, $\Omega_{f^*\xi} = f^*\Omega_\xi$). From the multiplicative properties of Ψ_ξ and Ψ_ξ^* in theorem 11 we obtain the following equations.

15 *For $z \in H_n(B;G)$*

$$\Psi_\xi(z) = \Omega_\xi \frown z$$

and for $v \in H^r(B;G)$

$$\Psi_\xi^*(v) = v \smile \Omega_\xi \quad \blacksquare$$

We now investigate the existence of orientation classes for a sphere bundle. Let (X,X') be a pair and let $\{A_j\}_{j \in J}$ be an indexed collection of subsets $A_j \subset X$. An indexed collection

$$\{u_j \in H^n(A_j, A_j \cap X'; G)\}_{j \in J}$$

is said to be *compatible* if for all $j, j' \in J$

$$u_j \mid (A_j \cap A_{j'}, A_j \cap A_{j'} \cap X') = u_{j'} \mid (A_j \cap A_{j'}, A_j \cap A_{j'} \cap X')$$

The compatible collections $\{u_j\}$ constitute an R module $H^n(\{A_j\}, X'; G)$. Clearly, the restriction maps

$$H^n(X,X'; G) \to H^n(A_j, A_j \cap X'; G)$$

define a natural homomorphism $H^n(X,X'; G) \to H^n(\{A_j\},X'; G)$.

16 LEMMA *Let $(E,\dot{E})$ be a fiber-bundle pair with base B, projection $p: E \to B$, and fiber pair $(F,\dot{F})$. Assume that for some $n > 0$, $H_i(F,\dot{F}; R) = 0$ for $i < n$. Then*

(a) For all $A \subset B$ and all R modules G

$$H_i(p^{-1}(A), p^{-1}(A) \cap \dot{E}; G) = 0 = H^i(p^{-1}(A), p^{-1}(A) \cap \dot{E}; G) \qquad i < n$$

(b) If $\{V\}$ is any open covering of B, then in degree n the natural homomorphism is an isomorphism

$$H^n(E,\dot{E}; G) \approx H^n(\{p^{-1}V\},\dot{E}; G)$$

PROOF By the universal-coefficient formula, it suffices to prove (*a*) for $G = R$. If $A \subset B$ is such that $(p^{-1}(A), p^{-1}(A) \cap \dot{E})$ is homeomorphic to $A \times (F,\dot{F})$, then by the Künneth formula,

$$H_i(p^{-1}(A), p^{-1}(A) \cap \dot{E}; R) \approx H_i(A \times (F,\dot{F}); R) = 0 \qquad i < n$$

From this it follows (as in the proof of theorem 9) by induction on the number of coordinate neighborhoods of the bundle needed to cover A (using the Mayer-Vietoris sequence and the five lemma) that (*a*) holds for all compact $A \subset B$. By taking direct limits, (*a*) holds for any A.

For (*b*), let $\{W\}$ be the collection of finite unions of elements of $\{V\}$. By (*a*) and the universal-coefficient formula for cohomology, there is a commutative diagram

$$\begin{array}{ccc} H^n(E,\dot{E}; G) & \approx & \operatorname{Hom}(H_n(E,\dot{E};R), G) \\ \downarrow & & \downarrow\approx \\ \lim_{\leftarrow}\{H^n(p^{-1}(W), p^{-1}(W) \cap \dot{E}; G)\} & \approx & \lim_{\leftarrow}\{\operatorname{Hom}(H_n(p^{-1}(W), p^{-1}(W) \cap \dot{E}; R),G)\} \end{array}$$

Hence we need only prove that a compatible collection $\{u_V\}_{V\in\{V\}}$ extends to a unique compatible collection $\{u_W\}_{W\in\{W\}}$. This follows by using Mayer-Vietoris sequences again and from the fact that $H^i(p^{-1}(W), p^{-1}(W) \cap \dot{E}; G) = 0$ for $i < n$. ∎

For sphere bundles we have the following immediate consequence.

17 COROLLARY *A sphere bundle ξ with base B is orientable if and only if there is a covering $\{V\}$ of B and a compatible family $\{u_V\}$, where u_V is an orientation class of $\xi \mid V$ for each $V \in \{V\}$.* ∎

Since a trivial sphere bundle is orientable, corollaries 17 and 14 imply the following result.

18 COROLLARY *Any sphere bundle has a unique orientation class over $\mathbf{Z}_2$.* ∎

By theorem 2.8.12, there is a contravariant functor from the fundamental groupoid of the base space B of a sphere bundle ξ to the homotopy category which assigns to $b \in B$ the fiber pair $(E_b,\dot{E}_b)$ over b and to a path class $[\omega]$ in B a homotopy class $h[\omega] \in [E_{\omega(0)},\dot{E}_{\omega(0)}; E_{\omega(1)},\dot{E}_{\omega(1)}]$. For fixed R there is then a covariant functor from the fundamental groupoid of B to the category of R modules which assigns to $b \in B$ the module $H^{q+1}(E_b,\dot{E}_b; R)$ and to a path class $[\omega]$ the homomorphism

$$h[\omega]^*: H^{q+1}(E_{\omega(1)},\dot{E}_{\omega(1)}; R) \to H^{q+1}(E_{\omega(0)},\dot{E}_{\omega(0)}; R)$$

19 THEOREM *A sphere bundle ξ is orientable over R if and only if for every closed path ω in B, $h[\omega]^* = 1$.*

PROOF If ξ is orientable with orientation class $U \in H^{q+1}(E,\dot{E}; R)$, for any small path ω in B (and hence for any path)

$$h[\omega]^*(U \mid (E_{\omega(1)},\dot{E}_{\omega(1)})) = U \mid (E_{\omega(0)},\dot{E}_{\omega(0)})$$

Since $U \mid (E_b,\dot{E}_b)$ is a generator of $H^{q+1}(E_b,\dot{E}_b; R)$, this implies that $h[\omega]^* = 1$ for any closed path ω.

Conversely, if $h[\omega]^* = 1$ for every closed path ω in B, there exist generators $U_b \in H^{q+1}(E_b,\dot{E}_b; R)$ such that for any path class $[\omega]$ in B, $h[\omega]^*(U_{\omega(1)}) = U_{\omega(0)}$. If V is any subset of B such that $\xi \mid V$ is trivial, it is easy to see that there is an orientation class U_V of $\xi \mid V$ such that $U_V \mid (E_b,\dot{E}_b) = U_b$ for all $b \in V$. If $\{V\}$ is an open covering of B by sets such that $\xi \mid V$ is trivial for all V, then $\{U_V\}$ is a compatible family of orientations, and by corollary 17, ξ is orientable. ∎

20 COROLLARY *A sphere bundle with a simply connected base is orientable over any R.* ∎

8 THE COHOMOLOGY ALGEBRA

The cup product in cohomology makes the cohomology (over R) of a topological pair a graded R algebra. In the first part of this section we define the relevant algebraic concepts and compute this algebra over $\mathbf{Z}_2$ for a real projective space and over any R for complex and quaternionic projective space. This is applied to prove the Borsuk-Ulam theorem.

For the case of an H space, there is even more algebraic structure that can be introduced in the cohomology algebra. The cohomology of such a space is a Hopf algebra, and the second part of the section is devoted to its definition and some results about its structure. The section concludes with a proof of the Hopf theorem about the cohomology algebra of a compact connected H space.

A *graded R algebra* consists of a graded R module $A = \{A^q\}$ and a homomorphism of degree 0

$$\mu: A \otimes A \to A$$

called the *product* of the algebra (μ then maps $A^p \otimes A^q$ into A^{p+q} for all p and q). For $a, a' \in A$ we write $aa' = \mu(a \otimes a')$. The product is *associative* if $(aa')a'' = a(a'a'')$ for all $a, a', a'' \in A$ and is *commutative* if $aa' = (-1)^{\deg a \deg a'} a'a$ for all $a, a' \in A$.

1 EXAMPLE If (X,A) is a topological pair, then $H^*(X,A; R)$ is a graded R algebra whose product is the cup product (with respect to the multiplication pairing of R with itself to R). It follows from property 5.6.10 that this product is associative and from property 5.6.11 that it is commutative. If $A = \emptyset$, it follows from property 5.6.9 that 1 is a unit element of the algebra $H^*(X;R)$. $H^*(X,A; R)$ is called the *cohomology algebra* of (X,A) over R.

2 EXAMPLE The *polynomial algebra over R generated by x of degree* $n > 0$, denoted by $S_n(x)$, is defined by

$$[S_n(x)]^q = \begin{cases} 0 & q \not\equiv 0\ (n) \text{ or } q < 0 \\ \text{free } R \text{ module generated by } x_p & q = pn,\ p \geq 0 \end{cases}$$

with the product $(\alpha x_p)(\beta x_q) = (\alpha\beta)x_{p+q}$ for α, $\beta \in R$. It is then clear that x_0 is a unit element and that $x_p = (x_1)^p$. If we denote x_1 by x, then $x_p = x^p$. Thus, disregarding the graded structure, $S_n(x)$ is simply the polynomial algebra over R in one indeterminate x. The *truncated polynomial algebra over R generated by x of degree n and height h*, denoted by $T_{n,h}(x)$, is defined to be the quotient of $S_n(x)$ by the graded ideal generated by x^h. If $h = 2$, this is called the *exterior algebra generated by x of degree n* and is denoted by $E_n(x)$.

If A and B are graded R algebras, their tensor product $A \otimes B$ is also a graded R algebra with product

$$(a \otimes b)(a' \otimes b') = (-1)^{\deg b \deg a'} aa' \otimes bb'$$

If A and B have associative or commutative products, so does $A \otimes B$.

3 EXAMPLE If R is a field and (X,A) and (Y,B) are topological pairs such that either $H_*(X,A; R)$ or $H_*(Y,B; R)$ is of finite type, it follows from theorem 5.5.11 that

$$H^*(X,A; R) \otimes H^*(Y,B; R) \approx H^*((X,A) \times (Y,B); R)$$

We compute the graded $\mathbf{Z}_2$ algebra $H^*(P^n;\mathbf{Z}_2)$ for real projective space P^n. Note that the double covering p; $S^n \to P^n$ is a 0-sphere bundle. We let $w_n \in H^1(P^n;\mathbf{Z}_2)$ be the characteristic class (over $\mathbf{Z}_2$) of this bundle.

4 THEOREM *For* $n \geq 1$, $H^*(P^n;\mathbf{Z}_2)$ *is a truncated polynomial algebra over* $\mathbf{Z}_2$ *generated by* w_n *of degree* 1 *and height* $n + 1$.

PROOF All coefficients in the proof will be $\mathbf{Z}_2$ and will be omitted. By corollary 5.7.18 and theorem 5.7.11, there is an exact Thom-Gysin sequence

$$\cdots \to H^q(S^n) \xrightarrow{p^*} H^q(P^n) \xrightarrow{\Psi^*} H^{q+1}(P^n) \xrightarrow{p^*} H^{q+1}(S^n) \to \cdots$$

starting on the left with $0 \to H^0(P^n) \xrightarrow{p^*} H^0(S^n)$ and terminating on the right with $H^n(S^n) \xrightarrow{p^*} H^n(P^n) \to 0$ [note that $H^q(P^n) = 0$ for $q > n$, because P^n is a polyhedron of dimension n]. Because $H^q(S^n) = 0$ for $0 < q < n$, it follows that

$$\Psi^*: H^q(P^n) \to H^{q+1}(P^n)$$

is an epimorphism for $0 \le q < n-1$ and is a monomorphism for $0 < q \le n-1$. Because P^n and S^n are connected for $n \ge 1$, $p^* H^0(P^n) = H^0(S^n)$, which implies that $\Psi^*: H^0(P^n) \to H^1(P^n)$ is also a monomorphism. Therefore $H^q(P^n) \ne 0$ for $0 \le q \le n$, and because $\rho^* H^n(S^n) = H^n(P^n)$ and $H^n(S^n) \approx \mathbf{Z}_2$, it follows that ρ^* is a monomorphism and that $\Psi^*: H^{n-1}(P^n) \to H^n(P^n)$ is also an epimorphism.

We have shown that for $0 \le q \le n-1$

$$\Psi^*: H^q(P^n) \approx H^{q+1}(P^n)$$

Then $w_n = \Psi^*(1)$ is the nonzero element of $H^1(P^n)$, and by equation 5.7.15, $\Psi^*(w_n{}^q) = w_n{}^{q+1}$. Therefore, for $1 \le q \le n$, $w_n{}^q$ is the nonzero element of $H^q(P^n)$. ∎

By corollary 3.8.9, $P_n(\mathbf{C})$ and $P_n(\mathbf{Q})$ are simply connected. It follows from corollary 5.7.20 that the Hopf bundles $S^{2n+1} \to P_n(\mathbf{C})$ with fiber S^1 and $S^{4n+3} \to P_n(\mathbf{Q})$ with fiber S^3 are orientable over any R. Let $x_n \in H^2(P_n(\mathbf{C});R)$ and $y_n \in H^4(P_n(\mathbf{Q});R)$ be the characteristic classses of these Hopf bundles (based on some orientation class of each bundle). An argument analogous to that of theorem 4, using the Thom-Gysin sequences of the Hopf bundles, establishes the following result.

5 THEOREM *For $n \ge 1$, $H^*(P_n(\mathbf{C});R)$ is a truncated polynomial algebra over R generated by x_n of degree 2 and height $n+1$, and $H^*(P_n(\mathbf{Q});R)$ is a truncated polynomial algebra over R generated by y_n of degree 4 and height $n+1$.* ∎

6 COROLLARY *Let $n > m \ge 1$ and let $i: P^m \subset P^n$ be a linear imbedding. Then for $q \le m$*

$$i^*: H^q(P^n;\mathbf{Z}_2) \approx H^q(P^m,\mathbf{Z}_2)$$

PROOF The hypothesis that i is a linear imbedding implies that the 0-sphere bundle over P^m induced by i from the double covering $S^n \to P^n$ is the double covering $S^m \to P^m$. By the naturality of the characteristic class, $i^* w_n = w_m$. The result now follows from theorem 4 and the fact that $i^*(w_n{}^q) = (i^* w_n)^q$. ∎

7 COROLLARY *Let $n > m \ge 1$ and let $f: P^n \to P^m$ be a map. There exists a map $f': P^n \to S^m$ such that $p \circ f' = f$, where $p: S^m \to P^m$ is the double covering.*

PROOF By the lifting theorem 2.4.5, it suffices to prove $f_\#(\pi(P^n)) = 0$. If $m = 1$, this follows from the fact that $\pi(P^n) = \mathbf{Z}_2$ and $\pi(P^1) = \mathbf{Z}$. Assume that $m > 1$ and observe that because $H^1(P^n)$ has just the two elements 0 and w_n, either $f^*(w_m) = 0$ or $f^*(w_m) = w_n$. Because f^* is an algebra homomorphism, the latter is impossible [since $0 \ne w_n{}^{m+1}$ and $f^*(w_m{}^{m+1}) = 0$]. Therefore $f^*(w_m) = 0$.

We know that $\pi(P^n) = \mathbf{Z}_2$, and a generator for this group is the homotopy class of the linear inclusion map $i: P^1 \subset P^n$. Because $f^*(w_m) = 0$, it follows that $i^* f^*(w_m) = 0$. If $j: P^1 \subset P^m$ is the linear inclusion map, by corollary 6, $j^*(w_m) \neq 0$. Since $(f \circ i)^*(w_m) \neq j^*(w_m)$, $f \circ i$ is not homotopic to j. Since $\pi(P^m) = \mathbf{Z}_2$, $f \circ i$ is null homotopic. Hence $f_\#[i] = [f \circ i] = 0$, and so $f_\#(\pi(P^n)) = 0$ in this case also. ■

8 COROLLARY *For $n > m \geq 1$ there is no continuous map $g\colon S^n \to S^m$ such that $g(-x) = -g(x)$ for all $x \in S^n$.*

PROOF If there were such a map, it would define a map $f\colon P^n \to P^m$ such that the following square (where p and p' are the double coverings) is commutative

$$\begin{array}{ccc} S^n & \xrightarrow{g} & S^m \\ {\scriptstyle p'}\downarrow & & \downarrow{\scriptstyle p} \\ P^n & \xrightarrow{f} & P^m \end{array}$$

By corollary 7, f can be lifted to a map $f'\colon P^n \to S^m$. Then

$$pf'p' = fp' = pg$$

Therefore $f'p'$ and g are liftings of the same map. For any $x \in S^n$ either $g(x) = f'p'(x)$ or $g(-x) = f'p'(x) = f'p'(-x)$. In any event, $f'p'$ and g must agree at some point of S^n. By the unique-lifting property 2.2.2, $f'p' = g$. This is a contradiction, because for any $x \in S^n$, p' maps x and $-x$ into the same point, while g maps them into separate points. ■

This last result is equivalent to the *Borsuk-Ulam theorem,* which is next.

9 THEOREM *Given a continuous map $f\colon S^n \to R^n$ for $n \geq 1$, there exists $x \in S^n$ such that $f(x) = f(-x)$.*

PROOF Assume there is no such x and let $g\colon S^n \to S^{n-1}$ be the map defined by

$$g(x) = \frac{f(x) - f(-x)}{\| f(x) - f(-x) \|}$$

Then $g(-x) = -g(x)$, which would contradict corollary 8. ■

Dual to the concept of graded R algebra is that of graded R coalgebra, which is defined by dualizing the concept of product. A *graded R coalgebra* consists of a graded R module $A = \{A^q\}$ and a homomorphism of degree 0

$$d\colon A \to A \otimes A$$

called the *coproduct* of the coalgebra (so d maps A^q into $\bigoplus_{i+j=q} A^i \otimes A^j$ for all q). The coproduct is said to be *associative* if

$$(d \otimes 1)d = (1 \otimes d)d\colon A \to A \otimes A \otimes A$$

and is said to be *commutative* if $Td = d$, where $T\colon A \otimes A \to A \otimes A$ is the homomorphism $T(a \otimes a') = (-1)^{\deg a \deg a'} a' \otimes a$. A *counit* for the coalgebra is a homomorphism $\varepsilon\colon A \to R$ (where R is regarded as a graded R module consisting of R in degree 0) such that each of the composites

$$\begin{array}{ccccc} & & & R \otimes A & \\ & & \overset{\varepsilon \otimes 1}{\nearrow} & & \searrow^{\approx} \\ A \xrightarrow{d} A \otimes A & & & & A \\ & & \underset{1 \otimes \varepsilon}{\searrow} & & \nearrow_{\approx} \\ & & & A \otimes R & \end{array}$$

is the identity map.

A *Hopf algebra over* R is a graded R algebra B which is also a coalgebra whose coproduct

$$d: B \to B \otimes B$$

is a homomorphism of graded R algebras. A Hopf algebra B is said to be *connected* if B^0 is the free R module generated by a unit element 1 for the algebra and the homomorphism $\varepsilon: B \to R$ defined by $\varepsilon(\alpha 1) = \alpha$ for $\alpha \in R$ is a counit for the coalgebra.

10 EXAMPLE If X is a connected H space whose homology over a field R is of finite type, then the multiplication map $\mu: X \times X \to X$ defines a coproduct

$$d = \mu^*: H^*(X;R) \to H^*(X;R) \otimes H^*(X;R)$$

$H^*(X;R)$ with this coproduct is a connected Hopf algebra of finite type whose product is associative and commutative (the fact that X has a homotopy unit x_0 implies that the map $H^*(X;R) \to H^*(x_0;R) \approx R$ is a counit).

We shall study connected Hopf algebras having an associative and commutative product and describe the algebra structure of those which are of finite type over a field of characteristic 0. The following is the inductive step of the structure theorem toward which we are heading.

11 LEMMA *Let B be a connected Hopf algebra with an associative and commutative product over a field R of characteristic 0. Let B' be a connected sub Hopf algebra of B such that B is generated as an algebra by B' and some element $x \in B - B'$. If x has odd degree n, then as a graded algebra $B \approx B' \otimes E_n(x)$ and if x has even degree n, then as a graded algebra $B \approx B' \otimes S_n(x)$.*

PROOF Because B' is a sub Hopf algebra of B, the unit element of B belongs to B'. Since $x \in B - B'$, x has positive degree n. Let A be the ideal in B generated by the elements of positive degree in B', and if $\eta: B \to B/A$ is the projection, let

$$d' = (1 \otimes \eta)d: B \to B \otimes B \to B \otimes (B/A)$$

Then d' is an algebra homomorphism, $d'(\beta) = \beta \otimes 1$ for $\beta \in B'$, and $d'(x) = x \otimes 1 + 1 \otimes \eta(x)$. Note that $x \notin A$, because A consists of finite sums $\sum_{i \geq 0} \beta_i x^i$, where $\beta_i \in B'$ is of positive degree, so $\beta_i x^i$ is of degree larger than n unless $i = 0$. Therefore $\eta(x) \neq 0$ in B/A.

Assume that x is of odd degree. Because B has a commutative product and R has characteristic different from 2, $x^2 = 0$. We show that there is no relation of the form $\beta_0 + \beta_1 x = 0$ with $\beta_0, \beta_1 \in B'$ and $\beta_1 \neq 0$. If there were such a relation, then

$$0 = d'(\beta_0 + \beta_1 x) = \beta_0 \otimes 1 + (\beta_1 \otimes 1)[x \otimes 1 + 1 \otimes \eta(x)]$$
$$= \beta_1 \otimes \eta(x)$$

Since $\eta(x) \neq 0$, this implies $\beta_1 = 0$, which is a contradiction. Therefore the homomorphism $B' \otimes E_n(x) \to B$ sending $\beta \otimes 1$ to β and $\beta \otimes x$ to βx is an isomorphism of graded algebras.

Assume that x is of even degree. We shall show that there is no relation of the form $\sum_{0 \le i \le r} \beta_i x^i = 0$ with $\beta_i \in B'$, $r \geq 1$, and $\beta_r \neq 0$. If there were such a relation, consider one of minimal degree in x. Then

$$0 = d'(\sum \beta_i x^i) = \sum (\beta_i \otimes 1)[x \otimes 1 + 1 \otimes \eta(x)]^i$$
$$= (\sum i\beta_i x^{i-1}) \otimes \eta(x) + \cdots + \beta_r \otimes (\eta(x))^r$$

The only term on the right in $B \otimes (B/A)^n$ is the term $(\sum i\beta_i x^{i-1}) \otimes \eta(x)$. It must be 0, and because $\eta(x) \neq 0$, $\sum i\beta_i x^{i-1} = 0$. If $r > 1$, this is a relation of smaller degree in x (note that $r\beta_r \neq 0$ because R has characteristic 0), and this is a contradiction. If $r = 1$, we get $\beta_1 = 0$, which is also a contradiction. Therefore there is no relation, and the homomorphism $B' \otimes S_n(x) \to B$ sending $\beta \otimes x^q$ to βx^q for $\beta \in B'$ and $q \geq 0$ is an isomorphism of graded algebras. ∎

We use this result to establish the following *Leray structure theorem* for Hopf algebras over a field of characteristic 0[1].

12 THEOREM *Let B be a connected Hopf algebra with an associative and commutative product and of finite type over a field R of characteristic 0. As a graded R algebra either $B \approx R$ or B is the tensor product of a countable number of exterior algebras with generators of odd degree and a countable number of polynomial algebras with generators of even degree.*

PROOF Because B is of finite type, there is a countable sequence $1 = x_0, x_1, x_2, \ldots$ of elements of B such that $i < j$ implies that $\deg x_i \leq \deg x_j$ and such that as an algebra B is generated by the set $\{x_j\}_{j \geq 0}$. For $n \geq 0$ let B_n be the subalgebra of B generated by $x_0, x_1, \ldots, x_n$. We can also assume that x_{n+1} does not belong to B_n. Because of the condition that $\deg x_j$ is a nondecreasing function of j, each B_n is a connected sub Hopf algebra of B (that is, d maps B_n into $B_n \otimes B_n$). Since B_{n+1} is generated as an algebra by B_n and x_{n+1}, lemma 11 applies. Since $B_0 \approx R$, $B_1 \approx R \otimes E(x_1)$ or $B_1 \approx R \otimes S(x_1)$. Therefore $B = B_0 \approx R$ or B_1 is either an exterior algebra on an odd-degree generator or a polynomial algebra on an even-degree generator. By induction on n, using lemma 11, each B_{n+1} is a tensor product of the desired form. Since B has finite type, $B \approx \lim_{\to} B_n$, and B has the desired form. ∎

[1] A structure theorem valid over a perfect field of arbitrary characteristic can be found in A. Borel, Sur la cohomologie des espaces fibrés principaux et des espaces homogenes de groupes de Lie compacts, *Annals of Mathematics*, vol. 57, pp. 115–207, 1953.

For a connected H space whose homology is finitely generated over a field F no polynomial algebra factors can occur in the above structure theorem, and we obtain the following *Hopf theorem on H spaces.*

13 COROLLARY *Let X be a connected H space whose homology over a field R of characteristic 0 is finitely generated. Then the cohomology algebra of X over R is isomorphic to the cohomology algebra over R of a product of a finite number of odd-dimensional spheres.* ■

In particular, we obtain the following result about spheres that can be H spaces.

14 COROLLARY *No even-dimensional sphere of positive dimension is an H space.* ■

9 THE STEENROD SQUARING OPERATIONS

In the last section the cup product in cohomology was used to prove the Borsuk-Ulam theorem, a geometric result. Any other algebraic structure which can be introduced into cohomology (or homology) and which is functorial can be similarly applied. A particular example of such an additional algebraic structure is a natural transformation from one cohomology functor to another. These natural transformations are called cohomology operations. In this section we introduce the concept of cohomology operation and define the particular set of cohomology operations called the Steenrod squares.

Let p and q be fixed integers and G and G' fixed R modules. A *cohomology operation θ of type $(p,q;\, G,G')$* is a natural transformation from the functor $H^p(\ ;G)$ to the functor $H^q(\ ;G')$ (both functors being contravariant singular cohomology functors defined on the category of topological pairs). Thus θ assigns to a pair (X,A) a function (which is not assumed to be a homomorphism)

$$\theta_{(X,A)}\colon H^p(X,A;\, G) \to H^q(X,A;\, G')$$

such that if $f\colon (X,A) \to (Y,B)$ is a map, there is a commutative square

$$\begin{array}{ccc} H^p(Y,B;\, G) & \xrightarrow{\theta_{(Y,B)}} & H^q(Y,B;\, G') \\ f^*\downarrow & & \downarrow f^* \\ H^p(X,A;\, G) & \xrightarrow{\theta_{(X,A)}} & H^q(X,A;\, G') \end{array}$$

A *homology operation* is defined similarly, but we shall not discuss homology operations.

Following are some examples.

1 If $\varphi\colon G \to G'$ is a homomorphism, φ_* is a cohomology operation of type $(q,q;\, G,G')$ for every q, where

$$\varphi_*\colon H^q(X,A;\, G) \to H^q(X,A;\, G')$$

is defined as in Sec. 5.4. φ_* is called the *operation induced by the coefficient homomorphism* φ.

2 Given a short exact sequence of R modules $0 \to G' \to G \to G'' \to 0$, the *Bockstein cohomology operation* β^* of type $(q, q + 1; G'',G')$ for every q is defined to equal the Bockstein homomorphism

$$\beta^*: H^q(X,A; G'') \to H^{q+1}(X,A; G')$$

corresponding to the coefficient sequence $0 \to G' \to G \to G'' \to 0$ as defined in theorem 5.4.11.

3 For any p and q there is an operation θ_p of type $(q,pq; R,R)$, called the *pth-power operation*, defined by

$$\theta_p(u) = u^p \qquad u \in H^q(X,A; R)$$

An operation θ is said to be *additive* if $\theta_{(X,A)}$ is a homomorphism for every (X,A). The operations in examples 1 and 2 are additive; however, the operation θ_p of example 3 is not additive, in general.

Any cohomology operation provides a necessary condition for a homomorphism between the cohomology modules of two pairs to be the induced homomorphism of some continuous map between the pairs. For example, if θ is of type $(p,q; G,G)$, a necessary condition that a homomorphism

$$\psi: H^*(Y,B; G) \to H^*(X,A; G)$$

be induced by some map $f: (X,A) \to (Y,B)$ is that

$$\psi\theta_{(Y,B)} = \theta_{(X,A)}\psi: H^p(Y,B; G) \to H^q(X,A; G)$$

In these terms the algebraic idea underlying corollaries 5.8.7 and 5.8.8 is that for $n > m \geq 1$ there is no homomorphism

$$\psi: H^*(P^m;\mathbf{Z}_2) \to H^*(P^n;\mathbf{Z}_2)$$

such that ψ sends the nonzero element of $H^1(P^m;\mathbf{Z}_2)$ to the nonzero element of $H^1(P^n;\mathbf{Z}_2)$ and commutes with the $(m + 1)$st-power operation θ_{m+1} of type $(1, m + 1; \mathbf{Z}_2,\mathbf{Z}_2)$.

We shall now define a sequence of operations Sq^i called the Steenrod squares, each Sq^i being a cohomology operation of type $(q, q + i; \mathbf{Z}_2,\mathbf{Z}_2)$ for every q. These operations include the squaring operation θ_2 and are related to it by "reducing" the value of $\theta_2(u)$ in a certain way. For this reason, the operations Sq^i are also called the reduced squares.

For the remainder of this section we make the assumption that all modules are over $\mathbf{Z}_2$ and all homology and cohomology modules have coefficients $\mathbf{Z}_2$. The *Steenrod squares,* or *reduced squares,* $\{Sq^i\}_{i\geq 0}$ are additive cohomology operations

$$Sq^i: H^q(X,A) \to H^{q+i}(X,A)$$

defined for all q such that

(*a*) $Sq^0 = 1$.
(*b*) If $\deg u = q$, then $Sq^q u = u \smile u$.
(*c*) If $q > \deg u$, then $Sq^q u = 0$.
(*d*) If $u \in H^*(X,A)$ and $v \in H^*(Y,B)$ and $\{X \times B, A \times Y\}$ is an excisive couple in $X \times Y$, the following *Cartan formula* is valid:

$$Sq^k(u \times v) = \sum_{i+j=k} Sq^i u \times Sq^j v$$

The above properties characterize the cohomology operations Sq^i. We shall not prove the uniqueness[1], but shall content ourselves with their construction. First we establish a formula equivalent to the Cartan formula.

4 LEMMA *If $u, v \in H^*(X,A)$, then*

$$Sq^k(u \smile v) = \sum_{i+j=k} Sq^i u \smile Sq^j v$$

PROOF Since $u \smile v = d^*(u \times v)$, where $d: (X,A) \to (X,A) \times (X,A)$ is the diagonal map, this follows from the Cartan formula and functorial properties of Sq^i. ∎

For any chain complex C let $T: C \otimes C \to C \otimes C$ be the chain map interchanging the factors [$T(c_1 \otimes c_2) = c_2 \otimes c_1$ is a chain map over $\mathbf{Z}_2$].

5 LEMMA *There exists a sequence $\{D_j\}_{j \geq 0}$ of functorial homomorphisms $D_j: \Delta(X) \to \Delta(X) \otimes \Delta(X)$ of degree j such that*

(*a*) *D_0 is a chain map commuting with augmentation.*
(*b*) *For $j > 0$, $\partial D_j + D_j \partial + D_{j-1} + TD_{j-1} = 0$.*

If $\{D_j\}$ and $\{D'_j\}$ are two such sequences, there exists a sequence $\{E_j\}_{j \geq 0}$ of functorial homomorphisms $E_j: \Delta(X) \to \Delta(X) \otimes \Delta(X)$ of degree j such that

(*c*) $E_0 = 0$.
(*d*) *For $j \geq 0$, $\partial E_{j+1} + E_{j+1}\partial + E_j + TE_j + D_j + D'_j = 0$.*

PROOF We use the method of acyclic models. Let R be the group ring of $\mathbf{Z}_2$ over the field $\mathbf{Z}_2$. We regard R as the quotient ring of the polynomial ring $\mathbf{Z}_2(t)$ modulo the ideal generated by the polynomial $t^2 + 1 = 0$. Thus the elements of R have the form $a + bt$, where a and $b \in \mathbf{Z}_2$.

Let $\mathbf{Z}_2$ be regarded as a trivial R module (that is, the element t of R induces the identity map of $\mathbf{Z}_2$) and let C be the free resolution of $\mathbf{Z}_2$ over R in which C_q is free with one generator d_q for all $q \geq 0$ and which has boundary operator $\partial(d_q) = (1 + t)d_{q-1}$ for $q \geq 1$ and augmentation $\varepsilon(d_0) = 1$. The functor which assigns to a space X the chain complex $\Delta(X) \otimes_{\mathbf{Z}_2} C$ is augmented and free over R with models $\{\Delta_q\}_{q \geq 0}$ and basis $\{\xi_q \otimes d_j\}$. We regard

[1] For a proof see N. Steenrod and D. Epstein, Cohomology operations, *Annals of Mathematics Studies No.* 50, Princeton University Press, Princeton, N.J., 1962.

$\Delta(X) \underset{\mathbf{Z}_2}{\otimes} \Delta(X)$ as a chain complex over R, with t acting on $\Delta(X) \otimes \Delta(X)$ in the same way T does. Then $\Delta(X) \otimes \Delta(X)$ is augmented and acyclic, with models $\{\Delta_q\}_{q\geq 0}$. It follows from theorem 4.3.3 (which is valid for chain complexes over R) that there exist natural chain maps $\tau\colon \Delta(X) \otimes C \to \Delta(X) \otimes \Delta(X)$ preserving augmentation, and any two are naturally chain homotopic.

A map $\tau\colon \Delta(X) \otimes C \to \Delta(X) \otimes \Delta(X)$ of degree 0 corresponds bijectively to a sequence of maps

$$D_j\colon \Delta(X) \to \Delta(X) \otimes \Delta(X) \qquad j \geq 0$$

of degree j such that $D_j(c) = \tau(c \otimes d_j)$. Then τ is a chain map preserving augmentation if and only if $\{D_j\}$ satisfies (*a*) and (*b*). Thus there exist families $\{D_j\}$ satisfying (*a*) and (*b*), and any such family corresponds to some τ.

Similarly, a map $H\colon \Delta(X) \otimes C \to \Delta(X) \otimes \Delta(X)$ of degree 1 corresponds bijectively to a sequence of maps

$$E_j\colon \Delta(X) \to \Delta(X) \otimes \Delta(X) \qquad j \geq 0$$

of degree j such that $E_0 = 0$ and $E_j(c) = H(c \otimes d_{j-1})$ for $j \geq 1$. Then H is a chain homotopy from τ to τ' if and only if $\{E_j\}$ satisfies (*c*) and (*d*) for the sequences $\{D_j\}$ and $\{D'_j\}$ corresponding to τ and τ', respectively. Thus, if $\{D_j\}$ and $\{D'_j\}$ are two sequences satisfying (*a*) and (*b*), there is a sequence $\{E_j\}$ satisfying (*c*) and (*d*). ∎

Given a sequence $\{D_j\}_{j\geq 0}$ as in lemma 5, we define homomorphisms

$$D_j^*\colon \text{Hom}\,(\Delta(X) \otimes \Delta(X), \mathbf{Z}_2) \to \text{Hom}\,(\Delta(X), \mathbf{Z}_2)$$

of degree $-j$ by $(D_j^* f)(\sigma) = f(D_j\sigma)$ for $\sigma \in \Delta_q(X)$ and $f \in \text{Hom}\,(\Delta(X) \otimes \Delta(X), \mathbf{Z}_2)$. If $c^* \in \text{Hom}\,(\Delta_q(X), \mathbf{Z}_2)$ is a q-cochain of $\Delta(X)$, then

$$c^* \otimes c^* \in \text{Hom}\,(\Delta(X) \otimes \Delta(X), \mathbf{Z}_2),$$

and we define a $(q + i)$-cochain $\text{Sq}^i c^* \in \text{Hom}\,(\Delta(X), \mathbf{Z}_2)$ by

$$\text{Sq}^i c^* = \begin{cases} 0 & i > q \\ D_{q-i}^*(c^* \otimes c^*) & i \leq q \end{cases}$$

Let us now establish some properties of these cochain maps. It will be convenient to understand $D_j = 0$ for $j < 0$. Then lemma 5*b* holds for all j.

6 *If c^* is zero on $\Delta(A)$ for some $A \subset X$, then $Sq^i c^*$ is zero on $\Delta(A)$.*

PROOF This follows from the naturality of $\{D_j\}$, and hence of $\{Sq^i\}$. ∎

7 *If $\delta c^* = 0$, then $\delta(Sq^i c^*) = 0$.*

PROOF This is trivial if $i > q$. If $i \leq q$, we have

$$\begin{aligned}\delta(Sq^i c^*)(\sigma) &= D_{q-i}^*(c^* \otimes c^*)(\partial\sigma) = (c^* \otimes c^*)(D_{q-i}\partial\sigma) \\ &= (c^* \otimes c^*)(\partial D_{q-i}\sigma) + (c^* \otimes c^*)(D_{q-i-1}\sigma + TD_{q-i-1}\sigma) \\ &= (c^* \otimes c^*)(\partial D_{q-i}\sigma)\end{aligned}$$

the last equality because $(c^* \otimes c^*)(Tc) = (c^* \otimes c^*)c$ for any $c \in \Delta(X) \otimes \Delta(X)$. Then we have

$$(c^* \otimes c^*)(\partial D_{q-i}\sigma) = \delta(c^* \otimes c^*)(D_{q-i}\sigma) = 0$$

because $\delta c^* = 0$. ∎

8 *If $c^* = \delta \bar{c}^*$, then $Sq^i c^* = \delta[D^*_{q-i}(\bar{c}^* \otimes c^*) + D^*_{q-i-1}(\bar{c}^* \otimes \bar{c}^*)]$.*

PROOF If $i > q$, both sides are zero. If $i \leq q$, we have

$$\begin{aligned}(Sq^i c^*)(\sigma) &= D^*_{q-i}(\delta\bar{c}^* \otimes \delta\bar{c}^*)(\sigma) = \delta(\bar{c}^* \otimes \delta\bar{c}^*)(D_{q-i}(\sigma)) \\ &= (\bar{c}^* \otimes \delta\bar{c}^*)(D_{q-i}\partial\sigma + D_{q-i-1}\sigma + TD_{q-i-1}\sigma) \\ &= D^*_{q-i}(\bar{c}^* \otimes c^*)(\partial\sigma) + \delta(\bar{c}^* \otimes \bar{c}^*)(D_{q-i-1}\sigma)\end{aligned}$$

the last equality because

$$(\bar{c}^* \otimes \delta\bar{c}^*)(D_{q-i-1}\sigma + TD_{q-i-1}\sigma) = (\bar{c}^* \otimes \delta\bar{c}^* + \delta\bar{c}^* \otimes \bar{c}^*)(D_{q-i-1}\sigma)$$

We also have

$$\begin{aligned}\delta(\bar{c}^* \otimes \bar{c}^*)(D_{q-i-1}\sigma) &= (\bar{c}^* \otimes \bar{c}^*)(D_{q-i-1}\partial\sigma + D_{q-i-2}\sigma + TD_{q-i-2}\sigma) \\ &= D^*_{q-i-1}(\bar{c}^* \otimes \bar{c}^*)(\partial\sigma)\end{aligned}$$

The result follows by substituting this into the right-hand side of the other equation. ∎

9 *If c_1^* and c_2^* are cocycles, then*

$$Sq^i(c_1^* + c_2^*) = Sq^i c_1^* + Sq^i c_2^* + \delta D^*_{q-i+1}(c_1^* \otimes c_2^*)$$

PROOF If $i > q$, both sides are zero. If $i \leq q$, we have

$$\begin{aligned}Sq^i(c_1^* + c_2^*)(\sigma) &= [(c_1^* + c_2^*) \otimes (c_1^* + c_2^*)](D_{q-i}\sigma) \\ &= (c_1^* \otimes c_1^* + c_2^* \otimes c_2^*)(D_{q-i}\sigma) + (c_1^* \otimes c_2^*)(D_{q-i}\sigma + TD_{q-i}\sigma) \\ &= (Sq^i c_1^* + Sq^i c_2^*)(\sigma) + (c_1^* \otimes c_2^*)(D_{q-i+1}\partial\sigma + \partial D_{q-i+1}\sigma) \\ &= [Sq^i c_1^* + Sq^i c_2^* + \delta D^*_{q-i+1}(c_1^* \otimes c_2^*)](\sigma)\end{aligned}$$

the last equality because $\delta(c_1^* \otimes c_2^*) = 0$. ∎

It follows that there is a well-defined functorial homomorphism

$$Sq^i\colon H^q(X,A) \to H^{q+i}(X,A)$$

defined by $Sq^i\{c^*\} = \{Sq^i c^*\}$. If $\{D'_j\}$ is another system satisfying lemma 5*a* and 5*b*, and Sq'^i is defined using this system, let $\{E_j\}$ satisfy 5*c* and 5*d*. If c^* is a q-cocycle of $\Delta(X)/\Delta(A)$, then

$$(c^* \otimes c^*)(D_{q-i}\sigma + D'_{q-i}\sigma + E_{q+1-i}\partial\sigma) = 0$$

Therefore

$$Sq^i c^* + Sq'^i c^* + \delta E^*_{q+1-i}(c^* \otimes c^*) = 0$$

showing that $Sq^i\{c^*\} = Sq'^i\{c^*\}$. Hence Sq^i is uniquely defined independent of the particular choice of $\{D_j\}$. We shall now verify that these cohomology operations $\{Sq^i\}$ satisfy the axioms characterizing the Steenrod squares.

10 THEOREM *The additive cohomology operations $\{Sq^i\}$ defined above satisfy conditions (a) to (d), inclusive, on page 271.*

PROOF Let $C(\Delta^q)$ denote the oriented chain complex of the simplex. Over $\mathbf{Z}_2$ there is a unique orientation for each simplex, and $C(\Delta^q)$ is isomorphic to the subcomplex of $\Delta(\Delta^q)$ generated by the singular simplexes which are the faces of Δ^q. We regard $C(\Delta^q)$ as imbedded in $\Delta(\Delta^q)$ in this way. $\tilde{C}(\Delta^q)$ is acyclic, and if $\lambda: \Delta^p \to \Delta^q$ is a p-face of Δ^q, then $\Delta(\lambda)(C(\Delta^p)) \subset C(\Delta^q)$. It follows that a sequence $\{D_j\}$ can be found satisfying lemma 5*a* and 5*b* such that $D_j(\xi_q) \in C(\Delta^q) \otimes C(\Delta^q)$ for all q and j. For such a sequence, $D_j(\xi_q) = 0$ if $j > q$ (because $[C(\Delta^q) \otimes C(\Delta^q)]_s = 0$ if $s > 2q$), whence $D_j(\sigma) = 0$ for any $\sigma \in \Delta_q(X)$ with $q < j$.

We now shall prove $D_q(\xi_q) = \xi_q \otimes \xi_q$ for all q by induction on q. If $q = 0$, then $D_0(\xi_0)$ must have nonzero augmentation, by lemma 5*a*. The only element of $C(\Delta^0) \otimes C(\Delta^0)$ with nonzero augmentation is $\xi_0 \otimes \xi_0$. Therefore $D_0(\xi_0) = \xi_0 \otimes \xi_0$. Assume that $q > 0$ and $D_{q-1}(\xi_{q-1}) = \xi_{q-1} \otimes \xi_{q-1}$. Either $D_q(\xi_q) = \xi_q \otimes \xi_q$ or $D_q(\xi_q) = 0$. In the latter case, by lemma 5*b*, we have [because $D_q(\partial\xi_q) = 0$]

$$D_{q-1}(\xi_q) + TD_{q-1}(\xi_q) = 0$$

From this it follows that $D_{q-1}(\xi_q) = \sum a_i(\xi_q \otimes \xi_q^{(i)} + \xi_q^{(i)} \otimes \xi_q)$, where $a_i = 0$ or $a_i = 1$. This is a contradiction, because

$$D_{q-2}(\xi_q) + TD_{q-2}(\xi_q) = \partial D_{q-1}(\xi_q) + D_{q-1}(\partial\xi_q)$$

and $\xi_q^{(i)} \otimes \xi_q^{(i)}$ has a coefficient of $2a_i + 1 = 1$ on the right and a coefficient of 0 on the left.

Therefore, with this choice of $\{D_j\}$ we have $D_q(\sigma) = \sigma \otimes \sigma$ if σ has degree q. Then

$$(Sq^0 c^*)(\sigma) = (c^* \otimes c^*)(D_q(\sigma)) = [c^*(\sigma)]^2$$

Because $a^2 = a$ for $a \in \mathbf{Z}_2$, we see that $Sq^0 c^* = c^*$, and so $Sq^0 = 1$, showing that condition (*q*) is satisfied.

By definition, D_0 is a chain approximation to the diagonal. Therefore $\{D_0^*(c^* \otimes c^*)\} = \{c^*\} \smile \{c^*\}$ for any cocycle c^*, and so $Sq^q u = u \smile u$ if $\deg u = q$. Hence condition (*b*) is satisfied. From the definition of Sq^i condition (*c*) is trivially satisfied.

It merely remains to verify the Cartan formula. Let $\{D_j\}$ be a system satisfying lemma 5*a* and 5*b* and let $\{D_j^X\}$ be the collection of homomorphisms for $\Delta(X)$. On the category of pairs of topological spaces X and Y the system $\{D_k^{X\times Y}\}$ and the system $\{\bar{T} \sum_{i+j=k} T^k D_i^X \otimes D_j^Y\}$, where

$$\bar{T}: [\Delta(X) \otimes \Delta(X)] \otimes [\Delta(Y) \otimes \Delta(Y)] \to [\Delta(X) \otimes \Delta(Y)] \otimes [\Delta(X) \otimes \Delta(Y)]$$

interchanges the second and third factors, both satisfy lemma 5*a* and 5*b*.

Then a system $\{E_k^{X \times Y}\}$ satisfying 5c and 5d with respect to them can be defined by the method of acyclic models. Therefore the system

$$\{\bar{T} \sum_{i+j=k} T^k D_i^X \otimes D_j^Y\}$$

can be used to define $Sq^k(u \times v)$ for $u \in H^*(X,A)$ and $v \in H^*(Y,B)$. Let c_1^* be a p-cochain of X, c_2^* a q-cochain of Y, σ_1 a singular p'-simplex of X with $p \leq p' \leq 2p$, and σ_2 a singular q'-simplex of Y with $q \leq q' \leq 2q$, where $p' + q' = p + q + k$. Then

$$\begin{aligned}
Sq^k(c_1^* \otimes c_2^*)(\sigma_1 \otimes \sigma_2) \\
&= [(c_1^* \otimes c_2^*) \otimes (c_1^* \otimes c_2^*)](D_{p+q-k}^{X \times Y}(\sigma_1 \otimes \sigma_2)) \\
&= [(c_1^* \otimes c_1^*) \otimes (c_2^* \otimes c_2^*)]\Big(\sum_{i+j=p+q-k} T^{p+q-k} D_i^X \sigma_1 \otimes D_j^Y \sigma_2\Big) \\
&= [(c_1^* \otimes c_1^*)(D_{2p-p'}^X \sigma_1)]\,[(c_2^* \otimes c_2^*)(D_{2q-q'}^Y \sigma_2)] \\
&= (Sq^{p'-p} c_1^* \otimes Sq^{q'-q} c_2^*)(\sigma_1 \otimes \sigma_2)
\end{aligned}$$

Letting σ_1 and σ_2 vary, we see that $Sq^k(c_1^* \otimes c_2^*) = \sum_{i+j=k} Sq^i c_1^* \otimes Sq^j c_2^*$. Passing to cohomology and using the natural homomorphism

$$H^*(X,A) \otimes H^*(Y,B) \to H^*([\Delta(X)/\Delta(A)] \otimes [\Delta(Y)/\Delta(B)]) \approx H^*((X,A) \times (Y,B))$$

sending the tensor product to the cross product, we obtain

$$Sq^k(u \times v) = \sum_{i+j=k} Sq^i u \times Sq^j v$$

showing that condition (*d*) is satisfied. ∎

11 EXAMPLE Observe that, by condition (*b*) on page 271 and theorem 5.8.5,

$$Sq^2\colon H^2(P_2(\mathbf{C})) \to H^4(P_2(\mathbf{C}))$$

is nontrivial. If $u \in H^2(P_2(\mathbf{C}))$ is such that $Sq^2 u \neq 0$ and $v \in H^1(I,\dot{I})$ is the nontrivial element, it follows from condition (*d*) that

$$Sq^2(u \times v) = Sq^2 u \times v$$

and $Sq^2\colon H^3(P_2(\mathbf{C}) \times (I,\dot{I})) \to H^5(P_2(\mathbf{C}) \times (I,\dot{I}))$ is nontrivial. Let X be the unreduced suspension of $P_2(\mathbf{C})$ obtained from $P_2(\mathbf{C}) \times I$ by identifying $P_2(\mathbf{C}) \times 0$ to one point x_0 and $P_2(\mathbf{C}) \times 1$ to another point x_1. There is then a continuous map

$$f\colon P_2(\mathbf{C}) \times (I,\dot{I}) \to (X, x_0 \cup x_1)$$

inducing an isomorphism

$$f^*\colon H^q(X, x_0 \cup x_1) \approx H^q(P_2(\mathbf{C}) \times (I,\dot{I}))$$

for all q. Therefore $Sq^2\colon H^3(X) \to H^5(X)$ is nontrivial. Let Y be the one-point union of S^3 and S^5. An easy computation shows that X and Y have isomorphic homology and cohomology for any coefficient group, and even isomorphic cup and cap products. However, because $Sq^2\colon H^3(X) \to H^5(X)$ is nontrivial and $Sq^2\colon H^3(Y) \to H^5(Y)$ is trivial, X and Y are not of the same homotopy type.

Further applications of the Steenrod squares will be given in the next chapter and in Chap. 8.

It is obvious that cohomology operations of the same type can be added and that the sum is again a cohomology operation of the same type. Given cohomology operations θ of type $(p,q;\, G,G')$ and θ' of type $(q,r;\, G',G'')$, their composite $\theta'\theta$ (of natural transformations) is a cohomology operation of type $(p,r;\, G,G'')$. In this way the Steenrod squares can be added and multiplied, and they generate an algebra of cohomology operations called the *modulo 2 Steenrod algebra.*

In this algebra the following *Adem relations*[1] hold:

$$Sq^iSq^j = \sum_{0\le k\le[i/2]} \binom{j-k-1}{i-2k} Sq^{i+j-k}Sq^k \qquad 0 < i < 2j$$

where $[i/2]$ denotes as usual the largest integer $\le i/2$ and the binomial coefficient $\binom{j-k-1}{i-2k}$ is reduced modulo 2. Using these relations, it is easily shown that the algebra of cohomology operations generated by Sq^i, where i is a power of 2, contains all the Steenrod squares. This implies that the only spheres that can be H spaces have dimension $2^n - 1$ for some n. By using deeper properties of the algebra of cohomology operations Adams[2] has shown that the only spheres that can be H spaces are the spheres S^0, S^1, S^3, and S^7. Each of these is, in fact, an H space, with multiplication defined to be the multiplication of the reals, complex numbers, quaternions, or Cayley numbers, respectively, of norm 1.

EXERCISES

A DISSECTIONS

Let C be a graded module over R. A *filtration (increasing)* of C is a sequence $\{F_sC\}$ of graded submodules of C such that $F_sC \subset F_{s+1}C$ for all s. It is said to be *bounded below* if for any t there is $s(t)$ such that $F_{s(t)}C_t = 0$, and it is *convergent above* if $\cup F_sC = C$.

1 If $\{F_sC\}$ is a filtration of a chain complex C by subcomplexes, there is an increasing filtration of $H_*(C)$ defined by $F_sH_*(C) = \operatorname{im}[H_*(F_sC) \to H_*(C)]$. If the original filtration on C is bounded below or convergent above, prove that the same is true of the induced filtration on $H_*(C)$.

An increasing filtration $\{F_sC\}$ of a chain complex C by subcomplexes is called a *dissection* if it is bounded below, convergent above, and if

$$H_q(F_{s+1}C,F_sC) = 0 \qquad q \ne s+1$$

[1] See J. Adem, The iteration of the Steenrod squares in algebraic topology, *Proceedings of the National Academy of Sciences, USA*, vol. 38, pp. 720–726, 1952, or H. Cartan, Sur l'iteration des operations de Steenrod, *Commentarii Mathematici Helvetici*, vol. 29, pp. 40–58, 1955.

[2] See J. F. Adams, On the non-existence of elements of Hopf invariant one, *Annals of Mathematics*, vol. 72, pp. 20–104, 1960.

Given a dissection $\{F_sC\}$ of a chain complex C, the sequence

$$\cdots \to H_{q+1}(F_{q+1}C,F_qC) \xrightarrow{\partial} H_q(F_qC,F_{q-1}C) \xrightarrow{\partial} H_{q-1}(F_{q-1}C,F_{q-2}C) \to \cdots$$

is a chain complex $\bar{C}$, called the *chain complex associated to the dissection.*

2 If $\bar{C}$ is the chain complex associated to a dissection of C, prove that $H_*(\bar{C}) \approx H_*(C)$.

3 Let $\{F_sC\}$ be a dissection of a free chain complex C by free subcomplexes such that $F_{s+1}C/F_sC$ is free for all s. If $\bar{C}$ is the chain complex associated to the dissection, prove that $\bar{C}$ and C have isomorphic homology and cohomology for all coefficient modules. [*Hint:* The freeness hypotheses ensure that the universal-coefficient theorems hold for both homology and cohomology. Then $\{F_sC \otimes G\}$ is a dissection of $C \otimes G$ whose associated chain complex is isomorphic to $\bar{C} \otimes G$. Dual considerations apply to $\{\text{Hom}\,(F_sC,G)\}$ and $\text{Hom}\,(C,G)$.]

A ***block dissection*** of a chain complex C is a collection of subcomplexes $\{E_j{}^q\}$, called *blocks*, where q varies over the set of integers and for each q, j varies over a set J_q, such that if F_sC is the subcomplex of C generated by $\{E_j{}^q\}_{q\le s}$ and if $\dot{E}_j{}^q = E_j{}^q \cap F_{s-1}C$, then

$$E_j{}^q \cap E_k{}^q \subset F_{q-1}C \qquad j \ne k$$
$$E_j{}^q = 0 \qquad q \text{ sufficiently small}$$
$$\cup F_sC = C$$
$$H_i(E_j{}^q,\dot{E}_j{}^q) \approx \begin{cases} 0 & i \ne q \\ R & i = q \end{cases}$$

4 If $\{E_j{}^q\}$ is a block dissection of a chain complex C, prove that the corresponding collection $\{F_sC\}$ is a dissection of C whose associated chain complex $\bar{C}$ is free with generators for $\bar{C}_q$ in one-to-one correspondence with the set J_q.

A ***block dissection*** of a simplicial complex K is a collection of subcomplexes $\{K_j{}^q\}$, where q varies over the set of integers and for each q, j varies over some indexing set J_q, such that if $F_sK = \bigcup_{j\le s} K_j{}^q$ and $\dot{K}_j{}^q = F_{s-1}K \cap K_j{}^q$, then

$$K_j{}^q \cap K_k{}^q \subset F_{q-1}K \qquad j \ne k$$
$$K_j{}^q = 0 \qquad q \text{ sufficiently small}$$
$$\cup F_sK = K$$
$$H_i(K_j{}^q,\dot{K}_j{}^q) \approx \begin{cases} 0 & i \ne q \\ Z & i = q \end{cases}$$

5 If $\{K_j{}^q\}$ is a block dissection of K, prove that $\{C(K_j{}^q)\}$ is a block dissection of the chain complex $C(K)$ by free subcomplexes. If $\bar{C}$ is the chain complex associated to the dissection, prove that $\bar{C}$ and $C(K)$ have isomorphic homology and cohomology with any coefficient group.

B HOMOLOGY MANIFOLDS

A *homology n-manifold* is a locally compact Hausdorff space X such that for all $x \in X$, $H_q(X, X - x) = 0$ for $q \ne n$ and either $H_n(X, X - x) = 0$ or $H_n(X, X - x) \approx \mathbf{Z}$. Furthermore, if the *boundary* $\dot{X}$ of X is defined to be the subset

$$\dot{X} = \{x \in X \mid H_n(X, X - x) = 0\}$$

then we also assume that $X - \dot{X}$ is a nonempty connected set. If $\dot{X} = \varnothing$, X is said to be *without boundary.*

1 If X is a homology n-manifold and Y is a homology m-manifold, prove that $X \times Y$ is a homology $(n+m)$-manifold whose boundary equals $\dot{X} \times Y \cup X \times \dot{Y}$.

2 Prove that if a polyhedron is a homology n-manifold, its boundary is a subpolyhedron.

3 If K is a simplicial complex triangulating a homology n-manifold X, prove that K is an n-dimensional pseudomanifold and $\dot{K}$ triangulates $\dot{X}$. (A polyhedral homology n-manifold is said to be *orientable* or *nonorientable*, according to whether any triangulation of it is orientable or nonorientable as a pseudomanifold.)

4 Let $(K,\dot{K})$ be a simplicial pair triangulating a polyhedral homology n-manifold $(X,\dot{X})$ and let L be the subcomplex of the barycentric subdivision K' consisting of all simplexes disjoint from $\dot{K}'$. If s^q is a q-simplex of $K - \dot{K}$, let $E^{n-q}(s^q)$ be the subcomplex of L generated by the star of the barycenter $b(s^q)$. Prove that $\{E^{n-q}(s^q)\}_{s^q \in K-\dot{K}}$ is a block dissection of L and that if $\bar{C}$ is the chain complex associated to this block dissection, then $\bar{C}$ has homology and cohomology isomorphic to that of $X - \dot{X}$. (*Hint:* let st $s^q = s^q * B(s^q)$, where $B(s^q)$ is a subcomplex of K. Then $E^{n-q}(s^q) = b(s^q) * [B(s^q)]'$ and $\dot{E}^{n-q}(s^q) = [B(s^q)]'$. Also note that $|L|$ is a strong deformation retract of $|K| - |\dot{K}|$.)

5 *Lefschetz duality theorem.* Let $(K,\dot{K})$ be a simplicial pair triangulating a compact homology n-manifold $(X,\dot{X})$ and assume that $z \in H_n(K,\dot{K})$ is an orientation of K. For each q-simplex s^q of $K - \dot{K}$ let $z(s^q) \in H_n(K, K - \text{st } s^q)$ be the image of z, and assume an orientation σ^q of s^q chosen once and for all. Then $z(s^q) = \sigma^q * \bar{z}(\sigma^q)$, where $\bar{z}(\sigma^q) \in H_{n-q-1}(B(s^q))$. Define $z'(\sigma^q) \in H_{n-q}(E^{n-q}(s^q),\dot{E}^{n-q}(s^q))$ to correspond to $\bar{z}(\sigma^q)$ under the isomorphisms

$$H_{n-q-1}(B(s^q)) \approx H_{n-q-1}(\dot{E}^{n-q}(s^q)) \approx H_{n-q}(E^{n-q}(s^q),\dot{E}^{n-q}(s^q))$$

Let φ: Hom $(C_q(K,\dot{K}), G) \to \bar{C}_{n-q} \otimes G$ be the homomorphism defined by

$$\varphi(u) = \sum_{\sigma^q} z'(\sigma^q) \otimes u(\sigma^q) \qquad u \in \text{Hom } (C_q(K,\dot{K}), G)$$

Prove that φ is an isomorphism and that it commutes up to sign with the respective coboundary and boundary operators. Deduce isomorphisms

$$H^q(X,\dot{X}; G) \approx H_{n-q}(X - \dot{X}; G) \qquad \text{and} \qquad H_q(X,\dot{X}; G) \approx H^{n-q}(X - \dot{X}; G)$$

C PROPERTIES OF THE TORSION PRODUCT AND EXT

In this group of exercises all modules will be over a principal ideal domain R.

1 Prove that the torsion product is associative.

2 If A, B, and C are modules, prove that

$$A \otimes (B * C) \oplus A * (B \otimes C)$$

is symmetric in A, B, and C.

3 Given a module A and a short exact sequence of modules

$$0 \to B' \to B \to B'' \to 0$$

prove there is an exact sequence

$$0 \to \text{Hom } (A,B') \to \text{Hom } (A,B) \to \text{Hom } (A,B'') \to \text{Ext } (A,B') \to \text{Ext } (A,B) \to \text{Ext } (A,B'') \to 0$$

4 Given a short exact sequence of modules

$$0 \to A' \to A \to A'' \to 0$$

and given a module B, prove there is an exact sequence

$$0 \to \text{Hom}(A'',B) \to \text{Hom}(A,B) \to \text{Hom}(A',B) \to \text{Ext}(A'',B) \to \text{Ext}(A,B) \to \text{Ext}(A',B) \to 0$$

If $C = \{C_i\}$ and $C^* = \{C^j\}$ are graded modules, there is a graded module $\text{Hom}(C,C^*) = \{\text{Hom}^q(C,C^*)\}$, where $\text{Hom}^q(C,C^*) = \times_{i+j=q} \text{Hom}(C_i,C^j)$ [thus an element of $\text{Hom}^q(C,C^*)$ is an indexed family $\{\varphi_i: C_i \to C^{q-i}\}_i$]. Similarly, there is a graded module $\text{Ext}(C,C^*) = \{\text{Ext}^q(C,C^*)\}$, where $\text{Ext}^q(C,C^*) = \times_{i+j=q} \text{Ext}(C_i,C^j)$.

5 If C is a chain complex and C^* is a cochain complex, prove that $\text{Hom}(C,C^*)$ is a cochain complex, with

$$(\delta\varphi)_{i,j} = \varphi_{i-1,j} \circ \partial_i + (-1)^i \delta^{j-1} \circ \varphi_{i,j-1} \qquad \varphi = \{\varphi_{i,j}\} \in \text{Hom}^q(C,C^*)$$

and that $\text{Ext}(C,C^*)$ is a cochain complex with

$$(\delta\psi)_{i,j} = \text{Ext}(\partial_i,1)(\psi_{i-1,j}) + (-1)^i \text{Ext}(1,\delta^{j-1})(\psi_{i,j-1}) \qquad \psi = \{\psi_{i,j}\} \in \text{Ext}^q(C,C^*)$$

6 If C is a chain complex and C^* is a cochain complex such that $\text{Ext}(C,C^*)$ is acyclic, prove that there is a split short exact sequence

$$0 \to \text{Ext}^{q-1}(H_*(C),H^*(C^*)) \to H^q(\text{Hom}(C,C^*)) \to \text{Hom}^q(H_*(C),H^*(C^*)) \to 0$$

7 If C and C' are chain complexes and C^* is a cochain complex, prove that the exponential correspondence is an isomorphism

$$\text{Hom}(C, \text{Hom}(C',C^*)) \approx \text{Hom}(C \otimes C', C^*)$$

8 Let (X,A) and (Y,B) be topological pairs such that $\{X \times B, A \times Y\}$ is an excisive couple in $X \times Y$. For any module G prove that there is a split short exact sequence

$$0 \to \text{Ext}^{q-1}(H_*,H^*) \to H^q((X,A) \times (Y,B); G) \to \text{Hom}^q(H_*,H^*) \to 0$$

where $H_* = H_*(X,A; R)$ and $H^* = H^*(Y,B; G)$.

D CATEGORY

A topological space X is said to have *category* $\leq n$, denoted as cat $X \leq n$, if X is the union of n closed sets, each deformable to a point in X.

1 If X is a connected polyhedron of dimension n, prove that cat $X \leq n + 1$.

2 If X is any space, prove that cat $(SX) \leq 2$.

3 If cat $X \leq n$, prove that all n-fold cup products of positive-dimensional cohomology classes of X vanish.

4 Prove that cat $P^n = n + 1$ and cat $(P^{n_1} \times \cdots \times P^{n_k}) = n_1 + \cdots + n_k + 1$.

E HOMOLOGY OF FIBER BUNDLES

1 Let $p: E \to B$ be a fiber-bundle pair, with total pair $(E,\dot{E})$ and fiber pair $(F,\dot{F})$, such that $H_*(F,\dot{F}) = 0$. Prove that $H_*(E,\dot{E}) = 0$.

2 If $p: E \to B$ is a fiber-bundle pair over a path-connected base space B, prove that a homomorphism $\theta: H^*(F,\dot{F}; R) \to H^*(E,\dot{E}; R)$ is a cohomology extension of the fiber if and only if for some $b \in B$ the composite

$$H^*(F,\dot{F}; R) \xrightarrow{\theta} H^*(E,\dot{E}; R) \to H^*(E_b,\dot{E}_b; R)$$

is an isomorphism.

3 Let $p: E \to B$ be a fiber-bundle pair over a path-connected base space. If for some $b \in B$ the pair $(E_b,\dot{E}_b)$ is a weak retract of $(E,\dot{E})$, prove there exists a cohomology extension of the fiber.

4 Prove that a q-sphere bundle ξ with base space B is orientable over R if and only if for every map $\alpha: S^1 \to B$ the bundle $\alpha^*(\xi)$ is orientable over R.

5 Prove that a q-sphere bundle ξ is orientable over $\mathbf{Z}$ if and only if there is an element $U \in H^{q+1}(E_\xi,\dot{E}_\xi; \mathbf{Z}_4)$ whose image in $H^{q+1}(E_\xi,\dot{E}_\xi; \mathbf{Z}_2)$ is the unique orientation class of ξ over $\mathbf{Z}_2$. (*Hint:* Show that there is such an element U if and only if for every closed path ω in the base space, $h[\omega]^*$ is the identity map of $H^{q+1}(E_{\omega(1)},\dot{E}_{\omega(1)}; \mathbf{Z}_4)$, and this, in turn, is equivalent to the condition that $h[\omega]^*$ is the identity map of $H^{q+1}(E_{\omega(1)},\dot{E}_{\omega(1)}; \mathbf{Z})$.)

6 Let ξ be a q-sphere bundle with base space B and with orientation class $U_\xi \in H^{q+1}(E_\xi,\dot{E}_\xi; R)$ and let $\Omega_\xi \in H^{q+1}(B;R)$ be the corresponding characteristic class. Prove that $\Phi_\xi^*(\Omega_\xi) = U_\xi \smile U_\xi$.

7 Prove that the characteristic class Ω_ξ of an even-dimensional sphere bundle ξ oriented over $\mathbf{Z}$ has order 2.

8 Let ξ be a sphere bundle oriented over R, with base space B. If ξ has a section in $\dot{E}_\xi$, (that is, if the map $\dot{p}_\xi: \dot{E}_\xi \to B$ has a right inverse), prove that its characteristic class $\Omega_\xi = 0$. [*Hint:* Any two sections $B \to E_\xi$ are homotopic in E_ξ. Since E_ξ is the mapping cylinder of $\dot{p}_\xi: \dot{E}_\xi \to B$, there is an inclusion map $k: B \subset E_\xi$ which is a section. There is a section in $\dot{E}_\xi$ if and only if k is homotopic to a map $B \to \dot{E}_\xi$, in which case the composite

$$H^{q+1}(E_\xi,\dot{E}_\xi; R) \xrightarrow{i^*} H^{q+1}(E_\xi;R) \xrightarrow{p^{*-1}} H^{q+1}(B;R)$$

is trivial, because $p^{*-1} = k^*$.]

F HOPF ALGEBRAS

1 Prove that the tensor product of connected Hopf algebras is a connected Hopf algebra.

2 If B is a connected Hopf algebra of finite type over a field R, prove that $B^* = \text{Hom}(B;R)$ is a connected Hopf algebra over R whose product and coproduct are dual, respectively, to the coproduct and product of B.

3 Let B be a connected Hopf algebra over a field of characteristic $p \neq 0$ and assume that B has an associative and commutative product and is generated as an algebra by a single element x of positive degree. Prove that if $\deg x$ is odd and $p \neq 2$, then $B = E(x)$, and if $\deg x$ is even or $p = 2$, then either $B = S_{\deg x}(x)$ or $B = T_{\deg x, h}(x)$, where $h = p^k$ for some $k \geq 1$.

4 Let B be a connected Hopf algebra of finite type over a field of finite characteristic $p \neq 0$ and assume that B has an associative and commutative product. If the pth power of every element of positive degree of B is 0, prove that B is the tensor product of exterior algebras (with generators of odd degree if $p \neq 2$) and truncated polynomial algebras of height p (with generators of even degree if $p \neq 2$).

G THE BOCKSTEIN HOMOMORPHISM

1 Show that the Bockstein homomorphism in homology (or cohomology) anticommutes with the boundary homomorphism (or coboundary homomorphism) of a pair.

For any prime p let β_p be the Bockstein homomorphism in either homology or cohomology for the short exact sequence of abelian groups

$$0 \to \mathbf{Z}_p \to \mathbf{Z}_{p^2} \to \mathbf{Z}_p \to 0$$

Let $\tilde{\beta}_p$ be the Bockstein homomorphism for the short exact sequence

$$0 \to \mathbf{Z} \xrightarrow{\lambda_p} \mathbf{Z} \xrightarrow{\mu_p} \mathbf{Z}_p \to 0$$

where $\lambda_p(n) = pn$ and μ_p is reduction modulo p.

2 Prove that $\beta_p = (\mu_p)_* \circ \bar{\beta}_p$.

3 Prove that $\beta_p \circ \beta_p = 0$.

4 Prove that $\beta_p(u \smile v) = \beta_p(u) \smile v + (-1)^{\deg u} u \smile \beta_p(v)$.

5 Prove that $Sq^{2i+1} = \beta_2 \circ Sq^{2i}$ for $i \geq 0$. [*Hint:* Show that there exist functorial homomorphisms $\{D_j\}_{j \geq 0}$, with D_j of degree j from the integral singular chain complex $\Delta(X)$ to $\Delta(X) \otimes \Delta(X)$, such that D_0 is a chain map commuting with augmentation and

$$\begin{aligned} \partial D_{2j-1} + D_{2j-1}\partial &= D_{2j} - TD_{2j} \qquad j \geq 0 \\ \partial D_{2j} - D_{2j}\partial &= D_{2j-1} + TD_{2j-1} \qquad j > 0 \end{aligned}$$

where $T(\sigma_1 \otimes \sigma_2) = (-1)^{\deg \sigma_1 \deg \sigma_2} \sigma_2 \otimes \sigma_1$.]

6 Let ξ be a q-sphere bundle and let $U_\xi \in H^{q+1}(E_\xi, \dot{E}_\xi; \mathbf{Z}_2)$ be its unique orientation over $\mathbf{Z}_2$. Prove that ξ is orientable over $\mathbf{Z}$ if and only if $\beta_2(U_\xi) = 0$.

H STIEFEL-WHITNEY CHARACTERISTIC CLASSES

Let ξ be a q-sphere bundle, with base space B, and let $U_\xi \in H^{q+1}(E_\xi, \dot{E}_\xi; \mathbf{Z}_2)$ be its orientation class over $\mathbf{Z}_2$. The *ith Stiefel-Whitney characteristic class* $w_i(\xi) \in H^i(B; \mathbf{Z}_2)$ for $i \geq 0$ is defined by

$$\Phi_\xi^*(w_i(\xi)) = Sq^i(U_\xi)$$

1 Let $f: B' \to B$ be continuous. Prove that $f^*(w_i(\xi)) = w_i(f^*\xi)$.

2 If ξ is a product bundle, prove that $w_i(\xi) = 0$ for $i > 0$.

3 Prove the following:

(*a*) $w_0(\xi)$ is the unit class of $H^0(B; \mathbf{Z}_2)$.

(*b*) $\beta_2(w_{2i}(\xi)) = w_{2i+1}(\xi) + w_1(\xi) \smile w_{2i}(\xi)$ for $i \geq 0$.

(*c*) If ξ is a q-sphere bundle, then $w_i(\xi) = 0$ for $i > q + 1$, and $w_{q+1}(\xi)$ is the characteristic class of ξ over $\mathbf{Z}_2$.

(*d*) ξ is orientable over $\mathbf{Z}$ if and only if $w_1(\xi) = 0$.

If ξ is a q-sphere bundle over B and ξ' is a q'-sphere bundle over B', their *cross product* $\xi \times \xi'$ is a $(q + q' + 1)$-sphere bundle with $E_{\xi \times \xi'} = E_\xi \times E_{\xi'}$, $\dot{E}_{\xi \times \xi'} = E_\xi \times \dot{E}_{\xi'} \cup \dot{E}_\xi \times E_{\xi'}$ and $p_{\xi \times \xi'} = p_\xi \times p_{\xi'}$.

4 If $U_\xi \in H^{q+1}(E_\xi, \dot{E}_\xi; \mathbf{Z}_2)$ and $U_{\xi'} \in H^{q'+1}(E_{\xi'}, \dot{E}_{\xi'}; \mathbf{Z}_2)$ are respective orientation classes, prove that

$$U_\xi \times U_{\xi'} \in H^{q+q'+2}(E_{\xi \times \xi'}, \dot{E}_{\xi \times \xi'}; \mathbf{Z}_2)$$

is the orientation class of $\xi \times \xi'$.

5 Prove that $w_k(\xi \times \xi') = \sum_{i+j=k} w_i(\xi) \times w_j(\xi')$.

If ξ and ξ' are sphere bundles with the same base space B, their *Whitney sum* $\xi \oplus \xi'$ is the sphere bundle over B induced from $\xi \times \xi'$ by the diagonal map $B \to B \times B$.

6 *Whitney duality theorem.* Prove that

$$w_k(\xi \oplus \xi') = \sum_{i+j=k} w_i(\xi) \smile w_j(\xi')$$

I HOMOLOGY WITH LOCAL COEFFICIENTS

If $\sigma: \Delta^q \to X$ is a singular q-simplex of X, with $q \geq 1$, let ω_σ be the path in X obtained by composing the linear path in Δ^q from v_0 to v_1 with σ. Given a local system Γ of

R modules on X, define $\Delta_q(X;\Gamma)$ to be the R module of finitely nonzero formal sums $\sum \alpha_\sigma \sigma$ in which σ varies over the set of singular q-simplexes of X and $\alpha_\sigma \in \Gamma(\sigma(v_0))$ is zero except for a finite set of σ. For $q > 0$ define a homomorphism $\partial: \Delta_q(X;\Gamma) \to \Delta_{q-1}(X;\Gamma)$ by

$$\partial(\alpha\sigma) = \sum_{0<i\leq q} (-1)^i \alpha\sigma^{(i)} + \Gamma(\omega_\sigma)(\alpha)\sigma^{(0)}$$

1 Prove that $\Delta(X;\Gamma) = \{\Delta_q(X;\Gamma), \partial\}$ is a chain complex which is free (or torsion free) if Γ is a local system of free (or torsion free) R modules, and if $A \subset X$, show that $\Delta(A; \Gamma \mid A)$ is a subcomplex of $\Delta(X;\Gamma)$.

The *homology of* (X,A) *with local coefficients* Γ, denoted by $H_*(X,A;\Gamma)$, is defined to be the graded homology module of $\Delta(X,A;\Gamma) = \Delta(X;\Gamma)/\Delta(A;\Gamma \mid A)$.

2 For a fixed ring R let $\mathcal{C}$ be the category whose objects are topological pairs (X,A), together with local systems Γ of R modules on X, and whose morphisms from (X,A) and Γ to (Y,B) and Γ' are continuous maps $f: (X,A) \to (Y,B)$, together with indexed families of homomorphisms $\{f_x: \Gamma(x) \to \Gamma'(f(x))\}_{x\in X}$ such that $f_{\omega(0)} \circ \Gamma(\omega) = \Gamma'(f \circ \omega) \circ f_{\omega(1)}$ for any path ω in X. Prove that $H_*(X,A;\Gamma)$ is a covariant functor from $\mathcal{C}$ to the category of graded R modules.

3 *Exactness.* Given $A \subset B \subset X$ and a local system Γ of R modules on X, prove that there is an exact sequence

$$\cdots \to H_q(B,A;\Gamma \mid B) \to H_q(X,A;\Gamma) \to H_q(X,B;\Gamma) \to H_{q-1}(B,A;\Gamma \mid B) \to \cdots$$

4 *Excision.* Let X_1 and X_2 be subsets of a space X such that $X_1 \cup X_2 = \text{int } X_1 \cup \text{int } X_2$. For any local system Γ of R modules on X prove that the excision map j_1 from $(X_1, X_1 \cap X_2)$ and $\Gamma \mid X_1$ to $(X_1 \cup X_2, X_2)$ and $\Gamma \mid (X_1 \cup X_2)$ induces an isomorphism

$$j_{1*}: H_*(X_1, X_1 \cap X_2; \Gamma \mid X_1) \approx H_*(X_1 \cup X_2, X_2; \Gamma \mid (X_1 \cup X_2))$$

5 Two morphisms f and g in $\mathcal{C}$ from (X,A) and Γ to (Y,B) and Γ' are said to be *homotopic* in $\mathcal{C}$ if there is a homotopy $F: (X,A) \times I \to (Y,B)$ from f to g and an indexed family of homomorphisms $\{F_{(x,t)}: \Gamma(x) \to \Gamma'(F(x,t))\}_{(x,t)\in X\times I}$ such that $F_{(x,0)} = f_x$ and $F_{(x,1)} = g_x$. Prove that homotopy is an equivalence relation in the set of morphisms from (X,A) and Γ to (Y,B) and Γ' and that the composites of homotopic morphisms are homotopic (so that the homotopy category of $\mathcal{C}$ can be defined).

6 *Homotopy.* If f and g are morphisms from (X,A) and Γ to (Y,B) and Γ' and f is homotopic to g in $\mathcal{C}$, prove that $f_* = g_*: H_*(X,A;\Gamma) \to H_*(Y,B;\Gamma')$.

7 If Γ and Γ' are local systems of R modules on X, there is a local system $\Gamma \otimes \Gamma'$ on X with $(\Gamma \otimes \Gamma')(x) = \Gamma(x) \otimes \Gamma'(x)$ and $(\Gamma \otimes \Gamma')(\omega) = \Gamma(\omega) \otimes \Gamma'(\omega)$. In case Γ' is the constant local system equal to G, then prove that

$$\Delta(X,A;\Gamma \otimes G) \approx \Delta(X,A;\Gamma) \otimes G$$

Deduce a universal-coefficient formula for homology with local coefficients.

8 If Γ and Γ' are local systems of R modules on X and Y, respectively, let $\Gamma \times \Gamma' = p^*(\Gamma) \otimes p'^*(\Gamma')$ be the local system on $X \times Y$, where $p^*(\Gamma)$ and $p'^*(\Gamma')$ are induced from Γ and Γ', respectively, by the projections $p: X \times Y \to X$ and $p': X \times Y \to Y$. Prove that there is a natural chain equivalence of $\Delta(X;\Gamma) \otimes \Delta(Y;\Gamma')$ with $\Delta(X \times Y; \Gamma \times \Gamma')$. Deduce a Künneth formula for homology with local coefficients.

J COHOMOLOGY WITH LOCAL COEFFICIENTS

If Γ is a local system of R modules on X, define $\Delta^q(X;\Gamma)$ to be the module of functions φ assigning to every singular q-simplex σ of X an element $\varphi(\sigma) \in \Gamma(\sigma(v_0))$. Define a homomorphism $\delta: \Delta^q(X;\Gamma) \to \Delta^{q+1}(X;\Gamma)$ by

$$(\delta\varphi)(\sigma) = \sum_{0<i\leq q+1} (-1)^i \varphi(\sigma^{(i)}) + \Gamma(\omega_\sigma^{-1})(\varphi(\sigma^{(0)}))$$

1 Prove that $\Delta^*(X;\Gamma) = \{\Delta^q(X;\Gamma), \delta\}$ is a cochain complex and that if $A \subset X$, the restriction map $\Delta^*(X;\Gamma) \to \Delta^*(A; \Gamma \mid A)$ is an epimorphism.

The *cohomology of* (X,A) *with local coefficients* Γ, denoted by $H^*(X,A; \Gamma)$, is defined to be the graded cohomology module of

$$\Delta^*(X,A; \Gamma) = \ker [\Delta^*(X;\Gamma) \to \Delta^*(A; \Gamma \mid A)]$$

2 For a fixed ring R let $\mathcal{C}$ be the category whose objects are topological pairs (X,A), together with local systems Γ of R modules on X, and whose morphisms from (X,A) and Γ to (Y,B) and Γ' are continuous maps $f: (X,A) \to (Y,B)$, together with indexed families of homomorphisms $\{f^x: \Gamma'(f(x)) \to \Gamma(x)\}_{x\in X}$ such that $\Gamma(\omega) \circ f^{\omega(1)} = f^{\omega(0)} \circ \Gamma'(f \circ \omega)$ for any path ω in X. Prove that $H^*(X,A; \Gamma)$ is a contravariant functor from $\mathcal{C}$ to the category of graded R modules.

3 Prove that the cohomology with local coefficients has exactness, excision, and homotopy properties analogous to those of the homology with local coefficients.

4 If Γ is a local system of R modules on X and G is an R module, there is a local system Hom (Γ,G) of R modules on X which assigns to $x \in X$ the module Hom $(\Gamma(x),G)$. Prove that

$$\Delta^*(X,A; \text{Hom}(\Gamma,G)) \approx \text{Hom}(\Delta(X,A; \Gamma), G)$$

Deduce a universal-coefficient formula for cohomology with local coefficients.

Let ξ be a q-sphere bundle with base space B and let Γ_ξ be the local system on B such that $\Gamma_\xi(b) = H_{q+1}(E_b,\dot{E}_b)$. Let $p_\xi^*(\Gamma_\xi)$ be the local system on E_ξ induced from Γ_ξ by $p_\xi: E_\xi \to B$. A *Thom class* of ξ is an element $U_\xi \in H^{q+1}(E_\xi,\dot{E}_\xi; p_\xi^*(\Gamma_\xi))$ such that for every $b \in B$ the element

$$U_\xi \mid (E_b,\dot{E}_b) \in H^{q+1}(E_b,\dot{E}_b; p_\xi^*(\Gamma_\xi) \mid E_b) = H^{q+1}(E_b,\dot{E}_b; H_{q+1}(E_b,\dot{E}_b))$$

corresponds to the identity map of $H_{q+1}(E_b,\dot{E}_b)$ under the universal-coefficient isomorphism

$$H^{q+1}(E_b,\dot{E}_b; H_{q+1}(E_b,\dot{E}_b)) \approx \text{Hom}(H_{q+1}(E_b,\dot{E}_b), H_{q+1}(E_b,\dot{E}_b))$$

5 Prove that every q-sphere bundle has a unique Thom class. (*Hint:* Prove the result first for a product bundle, and then use Mayer-Vietoris sequences to extend the result to arbitrary bundles.)

6 Let ξ be a q-sphere bundle with a base space B and let U_ξ be its Thom class. If Γ is any local system of abelian groups on X, prove that the homomorphism

$$\Phi_\xi: H_n(E_\xi,\dot{E}_\xi; p^*(\Gamma)) \to H_{n-q-1}(B; \Gamma_\xi \otimes \Gamma)$$

such that $\Phi_\xi(z) = p_*(U_\xi \frown z)$, where $U_\xi \frown z$ is an element of $H_{n-q-1}(E; p^*(\Gamma_\xi \otimes \Gamma))$, is an isomorphism. If B is compact, prove that the homomorphism

$$\Phi_\xi^*: H^r(B;\Gamma) \to H^{r+q+1}(E_\xi,\dot{E}_\xi; p^*(\Gamma \otimes \Gamma_\xi))$$

such that $\Phi_\xi^*(v) = p^*(v) \smile U_\xi$ is an isomorphism.

CHAPTER SIX
GENERAL COHOMOLOGY THEORY AND DUALITY

IN THIS CHAPTER WE CONTINUE THE STUDY OF HOMOLOGY AND COHOMOLOGY functors, with particular emphasis on the homological properties of topological manifolds. For this important class of spaces we shall establish the duality theorem equating the cohomology of a compact pair in an orientable manifold with the homology, in complementary dimensions, of the complementary pair.

The cohomology which enters in the duality theorem is the direct limit of the singular cohomology of neighborhoods of the pair, with the family of neighborhoods directed downward by inclusion. For the case of a closed pair in a manifold, the resulting direct limit depends only on the pair itself. In fact, it is isomorphic to the Alexander cohomology of the pair, Alexander cohomology being another cohomology theory distinct from the singular cohomology.

Thus we are led to consider Alexander cohomology. We define it and prove that it is a cohomology theory in the sense that it satisfies the axioms of cohomology theory. We also establish the special properties of tautness and continuity possessed by this theory and not generally valid for singular cohomology. For deeper properties of the Alexander theory we introduce the cohomology of a space with coefficients in a presheaf. The definition of this cohomology involves a Čech construction, using nerves of open coverings. We use general properties of this cohomology to prove that for paracompact spaces the Alexander and Čech cohomologies are isomorphic, and with this result establish universal-coefficient formulas for the Alexander cohomology of compact pairs and for the Alexander cohomology with compact supports of locally compact pairs.

The cohomology of presheaves is also applied to compare the singular and Alexander cohomology theories, and we prove that they are isomorphic for manifolds. Another application of the cohomology of presheaves is in the proof of the Vietoris-Begle mapping theorem. The final topic is a discussion of homological properties of one manifold imbedded in another.

In Sec. 6.1 we define the slant product as a pairing from the cohomology of a product space and the homology of one of its factors to the cohomology of the other factor. This furnishes the map that is the isomorphism in the duality theorem for manifolds, and the duality theorem itself is proved in Sec. 6.2. In Sec. 6.3 we consider various formulations of orientability for manifolds.

The Alexander cohomology theory is defined in Secs. 6.4 and 6.5, and the axioms of cohomology theory are verified for it. Section 6.6 contains a proof of the tautness property for Alexander cohomology, that the Alexander cohomology of a closed pair in a paracompact space is isomorphic to the direct limit of the Alexander cohomology of its neighborhoods. We deduce the continuity property of Alexander cohomology and show that the continuity property characterizes Alexander cohomology on compact pairs. We also define the Alexander cohomology with compact supports.

Sections 6.7, 6.8, and 6.9 develop the theory of the cohomology of spaces with coefficients in a presheaf and illustrate its application to the Alexander theory. In this way we equate the Alexander and singular cohomology for paracompact spaces that are homologically locally connected in all dimensions.

Section 6.10 contains definitions of the characteristic classes of a manifold and the normal characteristic classes of one manifold imbedded in another. These are related in the Whitney duality theorem, which is a useful tool for establishing non-imbeddability results.

1 THE SLANT PRODUCT

We are ready now to introduce a new product which pairs cohomology of a product space and homology of one of the factors to the cohomology of the other factor. This product will be used in the next section to prove the duality theorem for topological manifolds. In this section we shall establish some of its properties. We shall also introduce new cohomology modules of a pair (A,B) in a space X which appear to depend on the imbedding of (A,B) in X. These will be used in the proof of the duality theorem in the next section. Later in the chapter, we shall introduce the Alexander cohomology modules and prove that these are isomorphic to the abovementioned ones in all relevant cases.

Given chain complexes C and C' over R and a cochain

$$c^* \in \text{Hom}\,((C \otimes C')_n, G)$$

and chain $c' \in C'_q \otimes G'$, their *slant product* $c^*/c' \in \text{Hom}\,(C_{n-q}, G \otimes G')$ is the $(n-q)$-cochain such that if $c' = \sum_i c'_i \otimes g'_i$ with $c'_i \in C'_q$ and $g'_i \in G'$, then

$$\langle c^*/c', c\rangle = \sum_i \langle c^*, c \otimes c'_i\rangle \otimes g'_i \qquad c \in C_{n-q}$$

It is easily verified that

$$\delta(c^*/c') = [(\delta c^*)/c'] + (-1)^{n-q} c^*/\partial c'$$

Therefore the slant product of a cocycle and a cycle is a cocycle, and if the cocycle is a coboundary or the cycle is a boundary, the slant product is a coboundary. Hence there is a slant product of $H^n(C \otimes C'; G)$ and $H_q(C';G')$ to $H^{n-q}(C; G \otimes G')$ such that $\{c^*\}/\{c'\} = \{c^*/c'\}$ for $\{c^*\} \in H^n(C \otimes C'; G)$ and $\{c'\} \in H_q(C';G')$.

For topological pairs (X,A) and (Y,B) let

$$\tau\colon [\Delta(X)/\Delta(A)] \otimes [\Delta(Y)/\Delta(B)] \to [\Delta(X \times Y)]/[\Delta(X \times B \cup A \times Y)]$$

be a functorial chain map given by the Eilenberg-Zilber theorem. For $u \in H^n((X,A) \times (Y,B); G)$ and $z \in H_q(Y,B; G')$, their *slant product*

$$u/z \in H^{n-q}(X,A; G \otimes G')$$

is defined to equal the slant product $(\tau^* u)/z$. The following properties of this slant product are easy consequences of the definitions.

1 *Given $f\colon (X,A) \to (X',A')$, $g\colon (Y,B) \to (Y',B')$, $u \in H^n((X',A') \times (Y',B'); G)$, and $z \in H_q(Y,B; G')$, then, in $H^{n-q}(X,A; G \otimes G')$,*

$$[(f \times g)^* u]/z = f^*(u/g_* z) \quad \blacksquare$$

2 *Given $u \in H^p(X,A; G)$, $v \in H^q(Y,B; G')$, and $z \in H_q(Y,B; G'')$, if $\{X \times B, A \times Y\}$ is an excisive couple in $X \times Y$, then, in $H^p(X,A; G \otimes G' \otimes G'')$,*

$$(u \times v)/z = \mu(u \otimes \langle v,z\rangle) \quad \blacksquare$$

3 *Let $\{(X_1,A_1), (X_2,A_2)\}$ and $\{(Y_1,B_1), (Y_2,B_2)\}$ be excisive couples in X and Y, respectively. Given*

$$u \in H^n((X_1 \cup X_2) \times (Y_1 \cup Y_2), X_1 \times B_1 \cup X_2 \times B_2 \cup A_1 \times Y_1 \cup A_2 \times Y_2; G)$$

and

$$z \in H_q(Y_1 \cup Y_2, B_1 \cup B_2; G')$$

then, in $H^{n-q+1}(X_1 \cup X_2, A_1 \cup A_2; G \otimes G')$,

$$\begin{aligned}&[u \mid (X_1 \cup X_2, A_1 \cup A_2) \times (Y_1 \cap Y_2, B_1 \cap B_2)]/\partial_* z \\ &\quad = (-1)^{n-q-1}\delta^*([u \mid (X_1 \cap X_2, A_1 \cap A_2) \times (Y_1 \cup Y_2, B_1 \cup B_2)]/z) \quad \blacksquare\end{aligned}$$

The following formulas express relations between the slant product and the cup and cap products. We sketch proofs in which the Alexander-Whitney diagonal approximation $\sigma \to \sum_{i+j=\deg\sigma} {}_i\sigma \otimes \sigma_j$ is used in $\Delta(X)$ and its tensor product with itself

$$\sigma \otimes \sigma' \to \sum_{i,j} (-1)^{j(p-i)}({}_i\sigma \otimes {}_j\sigma') \otimes (\sigma_{p-i} \otimes \sigma'_{q-j}) \qquad \deg \sigma = p, \deg \sigma' = q$$

is used in $\Delta(X) \otimes \Delta(Y)$.

4 *Given* $v \in H^p(X,A; G)$, $u \in H^n((X,A') \times (Y,B); G')$, *and* $z \in H_q(Y,B; G'')$, *then, in* $H^{p+n-q}(X, A \cup A';\ G \otimes G' \otimes G'')$,

$$v \cup (u/z) = [(v \times 1) \cup u]/z$$

PROOF Let c_1^* be a p-cochain of $\Delta(X)$, c_2^* an n-cochain of $\Delta(X) \otimes \Delta(Y)$, and $\sigma' \in \Delta_q(Y)$. It suffices to prove that

$$c_1^* \cup (c_2^*/\sigma') = [(c_1^* \otimes 1) \cup c_2^*]/\sigma'$$

If $\sigma \in \Delta_{p+n-q}(X)$, then

$$\begin{aligned}\langle c_1^* \cup (c_2^*/\sigma'), \sigma\rangle &= \langle c_1^*, {}_p\sigma\rangle \otimes \langle c_2^*/\sigma', \sigma_{n-q}\rangle\\ &= \langle c_1^*, {}_p\sigma\rangle \otimes \langle c_2^*, \sigma_{n-q} \otimes \sigma'\rangle\\ &= \langle c_1^* \otimes 1, {}_p\sigma \otimes {}_0\sigma'\rangle \otimes \langle c_2^*, \sigma_{n-q} \otimes \sigma'\rangle\\ &= \langle (c_1^* \otimes 1) \cup c_2^*, \sigma \otimes \sigma'\rangle = \langle [(c_1^* \otimes 1) \cup c_2^*]/\sigma', \sigma\rangle \quad \blacksquare\end{aligned}$$

5 *If* $u \in H^n((X,A) \times (Y,B); G)$, $v \in H^p(Y,B'; G')$, *and* $z \in H_q(Y,B \cup B'; G'')$, *then, in* $H^{n-(q-p)}(X,A; G \otimes G' \otimes G'')$,

$$u/(v \cap z) = [u \cup (1 \times v)]/z$$

PROOF Let c_1^* be an n-cochain of $\Delta(X) \otimes \Delta(Y)$, c_2^* be a p-cochain of $\Delta(Y)$, and $\sigma' \in \Delta_q(Y)$. It suffices to prove that

$$c_1^*/(c_2^* \cap \sigma') = [c_1^* \cup (1 \otimes c_2^*)]/\sigma'$$

If $\sigma \in \Delta_{n-(q-p)}(X)$, then

$$\begin{aligned}\langle c_1^*/(c_2^* \cap \sigma'), \sigma\rangle &= \langle c_1^*, \sigma \otimes (c_2^* \cap \sigma')\rangle\\ &= \langle c_1^*, (1 \otimes c_2^*) \cap (\sigma \otimes \sigma')\rangle\\ &= \langle c_1^* \cup (1 \otimes c_2^*), \sigma \otimes \sigma'\rangle\\ &= \langle [c_1^* \cup (1 \otimes c_2^*)]/\sigma', \sigma\rangle \quad \blacksquare\end{aligned}$$

6 *Given* $u \in H^n((X,A) \times (Y,B); G)$, $w \in H_r(X,A; G')$, *and* $z \in H_q(Y,B; G'')$, *let* $p: X \times Y \to X$ *be the projection to the first factor and let*

$$T: G \otimes G'' \otimes G' \to G \otimes G' \otimes G''$$

interchange the last two factors. Then, in $H_{r-(n-q)}(X; G \otimes G' \otimes G'')$,

$$p_*(u \cap (w \times z)) = T_*[(u/z) \cap w]$$

PROOF Let c^* be an n-cochain of $\Delta(X) \otimes \Delta(Y)$, $\sigma \in \Delta_r(X)$, and $\sigma' \in \Delta_q(Y)$.

Then

$$\begin{aligned}\Delta(p)(c^* \frown (\sigma \otimes \sigma')) &= \Delta(p)[\sum_{i+j=n} (-1)^{i(q-j)}({}_{r-i}\sigma \otimes {}_{q-j}\sigma') \otimes \langle c^*, \sigma_i \otimes \sigma'_j\rangle] \\ &= {}_{r-(n-q)}\sigma \otimes \langle c^*, \sigma_{n-q} \otimes \sigma'\rangle \\ &= {}_{r-(n-q)}\sigma \otimes \langle c^*/\sigma', \sigma_{n-q}\rangle \\ &= (c^*/\sigma') \frown \sigma \quad \blacksquare\end{aligned}$$

For a topological space X let $\delta(X)$ be the *diagonal* of X defined by $\delta(X) = \{(x,x') \in X \times X \mid x = x'\}$. Given $u \in H^n(X \times X, X \times X - \delta(X); R)$ and a pair (A,B) in X, define

$$\gamma_u: H_q(X - B, X - A; G) \to H^{n-q}(A,B; G)$$

by $\gamma_u(z) = [u \mid (A,B) \times (X - B, X - A)]/z$ (with $R \otimes G$ identified with G). If $i: (A,B) \subset (A',B')$ and $j: (X - B', X - A') \subset (X - B, X - A)$, it follows from property 1 that there is a commutative diagram (all coefficients G)

$$\begin{array}{ccc} H_q(X - B', X - A') & \xrightarrow{\gamma_u} & H^{n-q}(A',B') \\ {\scriptstyle j_*}\downarrow & & \downarrow{\scriptstyle i^*} \\ H_q(X - B, X - A) & \xrightarrow{\gamma_u} & H^{n-q}(A,B) \end{array}$$

Thus γ_u is a natural transformation from $H_q(X - B, X - A)$ to $H^{n-q}(A,B)$ on the category of pairs of subspaces and inclusion maps in X. It follows from property 3 that γ_u commutes up to sign with the connecting homomorphisms of relative Mayer-Vietoris sequences.

For a pair (A,B) in a topological space X we define a *neighborhood* (U,V) *of* (A,B) to be a pair in X such that U is a neighborhood of A and V is a neighborhood of B. The family of all neighborhoods of (A,B) in X is directed downward by inclusion. Hence

$$\{H^q(U,V; G) \mid (U,V) \text{ a neighborhood of } (A,B)\}$$

is a direct system, and we define

$$\bar{H}^q(A,B; G) = \lim_{\rightarrow} \{H^q(U,V; G)\}$$

where (U,V) varies over neighborhoods of (A,B) [or over the cofinal family of open neighborhoods of (A,B)]. The restriction maps $H^q(U,V; G) \to H^q(A,B; G)$ define a natural homomorphism

$$i: \bar{H}^q(A,B; G) \to H^q(A,B; G)$$

The pair (A,B) is said to be *tautly imbedded* in X, or to be a *taut pair in* X (with respect to singular cohomology), if i is an isomorphism for all q and G. The definition of tautness can be formulated for any cohomology theory (or any contravariant functor). We shall see examples later of a subspace taut with respect to one cohomology theory but not with respect to another.

Following are some examples.

7 If (A,B) is an open pair, or, more generally, if it has arbitrarily small neighborhoods which are homotopy equivalent to (A,B), then (A,B) is a taut pair in X.

8 Let $A' = \{(x,y) \in \mathbf{R}^2 \mid x > 0,\ y = \sin 1/x\}$, let $A'' = \{(x,y) \in \mathbf{R}^2 \mid x = 0, |y| \leq 1\}$, and let $A = A' \cup A'' \subset \mathbf{R}^2$. Then A' and A'' are the path components of A, and so $H^0(A;\mathbf{Z}) \approx \mathbf{Z} \oplus \mathbf{Z}$. Since A is connected, in any open neighborhood U of A in $\mathbf{R}^2$, A' and A'' must be in the same path component of U (the path components of U are the same as the components of U because U is locally path connected). It follows that $\bar{H}^0(A;\mathbf{Z}) = \lim_{\to} \{H^0(U;\mathbf{Z})\}$, where U varies over the connected open neighborhoods of A in $\mathbf{R}^2$. Therefore $\bar{H}^0(A;\mathbf{Z}) \approx \mathbf{Z}$ and $i: \bar{H}^0(A;\mathbf{Z}) \to H^0(A;\mathbf{Z})$ is not an epimorphism. Thus A is not a taut subspace of $\mathbf{R}^2$ with respect to singular cohomology.

9 LEMMA *Let (A,B) be a pair in X. Then, if two of the three pairs $(B,\varnothing)$, $(A,\varnothing)$, and (A,B) are taut in X, so is the third.*

PROOF This follows from the exactness of the cohomology sequence of a triple, from the fact that a direct limit of exact sequences is exact, and from the five lemma. ■

Recall (exercise set 1.C) that a normal space X is an absolute neighborhood retract if it has the property that whenever it is imbedded as a closed subset of a normal space, it is a retract of some neighborhood. Also recall that a space X is binormal if $X \times I$ (hence also X) is normal.

10 THEOREM *Any imbedding of an absolute neighborhood retract as a closed subspace of a binormal absolute neighborhood retract is taut.*

PROOF Assume $A \subset X$, where A and X are absolute neighborhood retracts and A is closed in the binormal space X. There is a neighborhood U of A in X such that A is a retract in U. Then $H^*(U) \to H^*(A)$ is an epimorphism, and this implies that

$$i: \bar{H}^*(A) \to H^*(A)$$

is an epimorphism.

To show that it is also a monomorphism, let U be an open neighborhood of A in X. There is a closed neighborhood U' of A in U of which A is a retract. Let $r: U' \to A$ be a retraction and define a map

$$F: (U' \times 0) \cup (A \times I) \cup (U' \times 1) \to U$$

by $F(x,0) = x$ and $F(x,1) = r(x)$ for $x \in U'$ and $F(x,t) = x$ for $x \in A$ and $t \in I$. Because A is closed in X, $(U' \times 0) \cup (A \times I) \cup (U' \times 1)$ is closed in $U' \times I$, the latter being a normal space because it is a closed subset of the normal space $X \times I$. Since U is an open subset of the absolute neighborhood retract X, it follows (see exercise 1.C.4) that U is an absolute neighborhood retract and F can be extended to a map $F': N \to U$, where N is a neighborhood of $(U' \times 0) \cup (A \times I) \cup (U' \times 1)$ in $U' \times I$. N contains a set of the form $V \times I$, where V is a neighborhood of A in U', and $F' \mid V \times I$ is a homotopy from the inclusion map $j: V \subset U$ to kr', where $r' = r \mid V: V \to A$ and $k: A \subset U$. Therefore there is a commutative triangle

$$\begin{array}{ccc} H^*(U) & \xrightarrow{k^*} & H^*(A) \\ & {}_{j^*}\searrow \quad \swarrow_{r'^*} & \\ & H^*(V) & \end{array}$$

which shows that $\ker k^* \subset \ker j^*$. Thus, if an element in $H^*(U)$ restricts to 0 in $H^*(A)$, it restricts to 0 in $H^*(V)$ for some smaller neighborhood V, hence it represents 0 in $\lim_\to \{H^*(U)\} = \bar{H}^*(A)$. Therefore $i: \bar{H}^*(A) \to H^*(A)$ is a monomorphism and A is taut in X. ∎

11 COROLLARY *If A, B, and X are compact polyhedra, any imbedding of (A,B) in X is taut.*

PROOF This follows from the fact (exercise 3.A.1) that a compact polyhedron is an absolute neighborhood retract and from theorem 10 and lemma 9. ∎

One reason for introducing the modules $\bar{H}^q(A,B; G)$ is the following result, which asserts that any pair (A,B) in X is taut with respect to the functor $\bar{H}^q$.

12 THEOREM *As U varies over the neighborhoods of A, there is an isomorphism*

$$\lim_\to \{\bar{H}^q(U;G)\} \approx \bar{H}^q(A;G)$$

PROOF Restricting U to the cofinal family of open neighborhoods, we have $\bar{H}^q(U;G) = H^q(U;G)$, and the limit on the left is, by definition, equal to the module on the right. ∎

If (A,B) and (A',B') are pairs in X and (U,V) and (U',V') are respective open neighborhoods, there is a relative Mayer-Vietoris sequence of $\{(U,V), (U',V')\}$. As (U,V) and (U',V') vary over open neighborhoods of (A,B) and (A',B'), respectively, $(U \cup U', V \cup V')$ varies over a cofinal family of neighborhoods of $(A \cup A', B \cup B')$. If (A,B) and (A',B') are closed pairs in a normal space X, it is also true that $(U \cap U', V \cap V')$ varies over a cofinal family of neighborhoods of $(A \cap A', B \cap B')$. Because the direct limit of exact sequences is exact, we obtain the following result, which is another reason for our interest in the modules $\bar{H}^*(A,B)$.

13 THEOREM *If (A,B) and (A',B') are closed pairs in a normal space X, there is an exact relative Mayer-Vietoris sequence (for any coefficient module G)*

$$\cdots \to \bar{H}^q(A \cup A', B \cup B') \to \bar{H}^q(A,B) \oplus \bar{H}^q(A',B') \to \bar{H}^q(A \cap A', B \cap B') \to \cdots$$ ∎

Given $u \in H^n(X \times X, X \times X - \delta(X); R)$, as (U,V) varies over neighborhoods of (A,B), the homomorphisms

$$\gamma_u: H_q(X - V, X - U; G) \to H^{n-q}(U,V; G)$$

define a homomorphism

$$\lim_{\rightarrow} \{H_q(X - V, X - U; G)\} \to \lim_{\rightarrow} \{H^{n-q}(U,V; G)\}$$

Because singular homology has compact supports, if X is a Hausdorff space the limit on the left is isomorphic to $H_q(X - B, X - A; G)$. Therefore we obtain a natural homomorphism

$$\bar{\gamma}_u\colon H_q(X - B, X - A; G) \to \bar{H}^{n-q}(A,B; G)$$

such that if (U,V) is a neighborhood of (A,B), there is a commutative diagram (all coefficients G)

$$\begin{array}{ccccc} H_q(X - V, X - U) & \to & H_q(X - B, X - A) & & \\ \gamma_u \downarrow & & \downarrow \bar{\gamma}_u & \searrow^{\gamma_u} & \\ H^{n-q}(U,V) & \to & \bar{H}^{n-q}(A,B) & \xrightarrow{i} & H^{n-q}(A,B) \end{array}$$

If (A,B) and (A',B') are closed pairs in a normal space X, then $\bar{\gamma}_u$ maps the exact Mayer-Vietoris sequence of the couple of open pairs

$$\{(X - B, X - A), (X - B', X - A')\}$$

into the exact Mayer-Vietoris sequence of theorem 13 in such a way that each square is commutative up to sign.

2 DUALITY IN TOPOLOGICAL MANIFOLDS

This section is devoted to a study of homology properties of topological manifolds. Over a connected manifold as base space there is a fiber-bundle pair called the homology tangent bundle. An orientation class of this bundle gives rise to a duality in the manifold asserting that the cohomology of a compact pair in the manifold is isomorphic to the homology of its complement. This duality theorem is proved by using the orientation class and the slant product to define a natural homomorphism from homology to cohomology. The resulting homomorphism is shown to be an isomorphism by proving it first in euclidean space and then in an arbitrary manifold using the piecing-together technique based on Mayer-Vietoris sequences.

A topological *n-manifold* (*without boundary*) is a paracompact Hausdorff space in which each point has an open neighborhood homeomorphic to $\mathbf{R}^n$ (called a *coordinate neighborhood* in the manifold). Following are some examples of n-manifolds.

1 $\mathbf{R}^n$ and S^n are n-manifolds.

2 An open subset of an n-manifold is an n-manifold.

3 The product of an n-manifold and an m-manifold is an $(n + m)$-manifold.

4 P^n is an n-manifold, $P_n(\mathbf{C})$ a $2n$-manifold, and $P_n(\mathbf{Q})$ a $4n$-manifold for all n. In fact, if X denotes one of these spaces and is coordinatized by homogeneous coordinates $[t_0,t_1, \ldots ,t_n]$, then for each $0 \le i \le n$ the subset $A_i \subset X$ of points having ith coordinate 0 is a projective space of dimension $n-1$ and $X - A_i$ is homeomorphic to $\mathbf{R}$, $\mathbf{R}^2$, or $\mathbf{R}^4$, respectively. Hence, $X - A_i$ is a coordinate neighborhood of X, and X is covered by these $n+1$ coordinate neighborhoods.

5 LEMMA *In an n-manifold X each point x has an open neighborhood V such that $(V \times X, V \times X - \delta(V))$ is homeomorphic to $V \times (X, X - x)$ by a homeomorphism preserving first coordinates.*

PROOF Let U be a coordinate neighborhood containing x. Without loss of generality, we can suppose that there is a homeomorphism $\varphi: U \approx \mathbf{R}^n$ such that $\varphi(x) = 0$. Let $D' = \{z \in \mathbf{R}^n \mid \|z\| \le 2\}$ and $V' = \{z \in \mathbf{R}^n \mid \|z\| < 1\}$ and define $D = \varphi^{-1}(D')$ and $V = \varphi^{-1}(V')$. Then V is an open neighborhood of x contained in the compact set D. If $(x',x'') \in V \times D - \delta(V)$, there is a unique point $z''' \in \mathbf{R}^n$ such that $\|z'''\| = 2$ and $\varphi(x'')$ belongs to the closed segment from $\varphi(x')$ to z'''. If $\varphi(x'') = t\varphi(x') + (1-t)z'''$, with $t \in I$, let $h(x',x'') \in D - x$ be the point such that $\varphi h(x',x'') = (1-t)z'''$, as illustrated

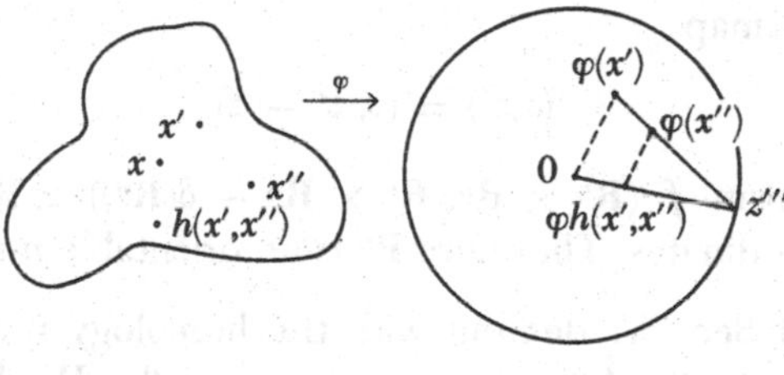

and define $h(x',x') = x$. A homeomorphism

$$\psi: (V \times X, V \times X - \delta(X)) \approx V \times (X, X - x)$$

having the desired properties is defined by

$$\psi(x',x'') = \begin{cases} (x',x'') & x'' \notin D \\ (x', h(x',x'')) & x'' \in D \end{cases} \quad \blacksquare$$

It follows from lemma 5 that if $x' \in V$ then $(X, X - x')$ is homeomorphic to $(X, X - x)$. Hence we obtain the following result.

6 COROLLARY *In a connected n-manifold X the group of homeomorphisms acts transitively; in particular, the topological type of $(X, X - x)$ is independent of x. Furthermore, projection to the first factor $p: X \times X \to X$ is the projection of a fiber-bundle pair $(X \times X, X \times X - \delta(X))$ with fiber pair $(X, X - x)$.* $\blacksquare$

If V is a coordinate neighborhood of x in an n-manifold X, the couple $\{V, X - x\}$ is excisive, and so there is an excision isomorphism

$$H_*(V, V - x; G) \approx H_*(X, X - x; G)$$

Since $H_*(V, V - x; G) \approx H_*(\mathbf{R}^n, \mathbf{R}^n - 0; G)$, it follows that

$$H_q(X, X - x; G) \approx \begin{cases} 0 & q \neq n \\ G & q = n \end{cases}$$

and so the fiber pair $(X, X - x)$ of the fiber-bundle pair of corollary 6 has the same homology as $(\mathbf{R}^n, \mathbf{R}^n - 0)$. For this reason the fiber-bundle pair of corollary 6 will be called the *homology tangent bundle* of X (the tangent bundle itself is an n-plane bundle defined if X is a differentiable manifold and having homology properties isomorphic to those of the homology tangent bundle).

A connected n-manifold X is said to be *orientable* (over R) if its homology tangent bundle is orientable [that is, if there exists an element $U \in H^n(X \times X, X \times X - \delta(X); R)$ such that for all $x \in X$, $U \mid x \times (X, X - x)$ is a generator of $H^n(x \times (X, X - x); R)$]. Such a cohomology class U is called an *orientation* of X. An n-manifold X (which is not assumed to be connected) is said to be *orientable* if each component is orientable, and an *orientation of* X is defined to be a cohomology class $U \in H^n(X \times X, X \times X - \delta(X); R)$ whose restriction to each component is an orientation of that component.

7 EXAMPLE For $\mathbf{R}^n$ the fiber-bundle pair $(\mathbf{R}^n \times \mathbf{R}^n, \mathbf{R}^n \times \mathbf{R}^n - \delta(\mathbf{R}^n))$ is trivial, because the map

$$f(z,z') = (z, z' - z)$$

is a homeomorphism $f\colon (\mathbf{R}^n \times \mathbf{R}^n, \mathbf{R}^n \times \mathbf{R}^n - \delta(\mathbf{R}^n)) \approx \mathbf{R}^n \times (\mathbf{R}^n, \mathbf{R}^n - 0)$ preserving first coordinates. Therefore $\mathbf{R}^n$ is an orientable n-manifold.

The results of Sec. 5.7 dealing with the homology properties of sphere bundles carry over to the homology tangent bundle. We list some of these explicitly.

8 *Two orientations U and U' of a connected manifold X are equal if and only if for some $x_0 \in X$*

$$U \mid x_0 \times (X, X - x_0) = U' \mid x_0 \times (X, X - x_0) \quad ■$$

9 *Any manifold has a unique orientation over Z_2.* ■

10 *A simply connected manifold is orientable over any R.* ■

11 *An n-manifold X is orientable if and only if there is an open covering $\{V\}$ of X and a compatible family $\{U_V \in H^n(V \times X, V \times X - \delta(V); R)\}$, where U_V corresponds to an orientation of V under the excision isomorphism*

$$H^n(V \times X, V \times X - \delta(V); R) \approx H^n(V \times V, V \times V - \delta(V); R) \quad ■$$

The duality theorem asserts that if $U \in H^n(X \times X, X \times X - \delta(X); R)$ is an orientation of X, then for any compact pair (A,B) in X, $\bar{\gamma}_U$ is an isomorphism of $H_q(X - B, X - A; G)$ onto $\bar{H}^{n-q}(A,B; G)$. We prove this first for $\mathbf{R}^n$ by a sequence of lemmas.

12 LEMMA *Let $A \subset \mathbf{R}^n$ be homeomorphic to a simplex and let $a_0 \in A$. Then $H_q(\mathbf{R}^n - a_0, \mathbf{R}^n - A; G) = 0$ for all q and G.*

PROOF Regarding $\mathbf{R}^n$ as an open subset of S^n, there is an excision isomorphism $H_q(\mathbf{R}^n - a_0, \mathbf{R}^n - A; G) \approx H_q(S^n - a_0, S^n - A; G)$. Because $S^n - a_0$ is homeomorphic to $\mathbf{R}^n$, $\tilde{H}_q(S^n - a_0; G) = 0$. From lemma 4.7.13 and the universal-coefficient formula, $\tilde{H}_q(S^n - A; G) = 0$. The lemma now follows from exactness of the reduced homology sequence of the pair $(S^n - a_0, S^n - A)$. ■

13 COROLLARY *If $A \subset \mathbf{R}^n$ is homeomorphic to a simplex and U is an orientation of $\mathbf{R}^n$ over R, then for all q and R modules G*

$$\gamma_U\colon H_q(\mathbf{R}^n, \mathbf{R}^n - A; G) \approx H^{n-q}(A;G)$$

PROOF Let $a_0 \in A$ and consider the diagram (all coefficients G)

$$\begin{array}{ccccccccc}
\cdots \to & H_q(\mathbf{R}^n - a_0, \mathbf{R}^n - A) & \to & H_q(\mathbf{R}^n, \mathbf{R}^n - A) & \to & H_q(\mathbf{R}^n, \mathbf{R}^n - a_0) & \to & H_{q-1}(\mathbf{R}^n - a_0, \mathbf{R}^n - A) & \to \cdots \\
& \gamma_U\downarrow & & \gamma_U\downarrow & & \gamma_U\downarrow & & \gamma_U\downarrow & \\
\cdots \to & H^{n-q}(A,a_0) & \to & H^{n-q}(A) & \to & H^{n-q}(a_0) & \to & H^{n-q+1}(A,a_0) & \to \cdots
\end{array}$$

The rows are exact, and each square either commutes or anticommutes. Since A is contractible, $H^*(A,a_0) = 0$. Using lemma 12, we see that trivially $\gamma_U\colon H_q(\mathbf{R}^n - a_0, \mathbf{R}^n - A) \approx H^{n-q}(A,a_0)$. By the five lemma, to complete the proof we need only verify that $\gamma_U\colon H_q(\mathbf{R}^n, \mathbf{R}^n - a_0) \approx H^{n-q}(a_0)$. Because U is an orientation, $U \mid [a_0 \times (\mathbf{R}^n, \mathbf{R}^n - a_0)] = 1 \times u$, where $u \in H^n(\mathbf{R}^n, \mathbf{R}^n - a_0; R)$ is a generator. By property 6.1.2,

$$\gamma_U(z) = \langle u,z\rangle 1$$

Since u is a generator of $H^n(\mathbf{R}^n, \mathbf{R}^n - a_0; R) \approx \operatorname{Hom}(H_n(\mathbf{R}^n, \mathbf{R}^n - a_0; R), R)$, it follows that the map $z \to \langle u,z\rangle$ of $H_n(\mathbf{R}^n, \mathbf{R}^n - a_0; R)$ to R is an isomorphism; and hence so is $\gamma_U\colon H_n(\mathbf{R}^n, \mathbf{R}^n - a_0; R) \approx H^0(a_0;R)$. If $q \neq n$, it is trivially true that $\gamma_U\colon H_q(\mathbf{R}^n, \mathbf{R}^n - a_0; R) \approx H^{n-q}(a_0;R)$, since both modules are trivial. ■

14 THEOREM *If U is an orientation of $\mathbf{R}^n$ over R and (A,B) is a compact polyhedral pair in $\mathbf{R}^n$, then for all q and all R modules G there is an isomorphism*

$$\gamma_U\colon H_q(\mathbf{R}^n - B, \mathbf{R}^n - A; G) \approx H^{n-q}(A,B; G)$$

PROOF Because of the naturality properties of γ_U, it suffices to prove this for the case where B is empty. The theorem follows for A from corollary 13 by induction on the number of simplexes in a triangulation of A, using Mayer-Vietoris sequences and the five lemma. ■

15 COROLLARY *If U is an orientation of $\mathbf{R}^n$ over R and (A,B) is a compact pair in $\mathbf{R}^n$, then for all q and R modules G there is an isomorphism*

$$\bar{\gamma}_U\colon H_q(\mathbf{R}^n - B, \mathbf{R}^n - A; G) \approx \bar{H}^{n-q}(A,B; G)$$

PROOF Since the family of compact polyhedral pairs is cofinal in the family of all neighborhoods of a compact pair (A,B) in $\mathbf{R}^n$, the corollary follows from theorem 14 by taking direct limits. ∎

Because of the commutativity of the triangle

$$\begin{array}{ccc} & H_q(\mathbf{R}^n - B, \mathbf{R}^n - A; G) & \\ \bar{\gamma}_U \swarrow & & \searrow \gamma_U \\ \bar{H}^{n-q}(A,B; G) & \xrightarrow{i} & H^{n-q}(A,B; G) \end{array}$$

it follows from theorem 14 and corollary 15 that any imbedding of a compact polyhedral pair in $\mathbf{R}^n$ is taut (which is also a consequence of corollary 6.1.11).

As an immediate result of corollary 15, we obtain the following *Alexander duality theorem.*

16 THEOREM *If A is a compact subset of $\mathbf{R}^n$, then for all q and R modules G*

$$\tilde{H}_q(\mathbf{R}^n - A; G) \approx \bar{H}^{n-q-1}(A;G)$$

PROOF Because $\tilde{H}_*(\mathbf{R}^n;G) = 0$, there is an isomorphism

$$\partial_*: H_{q+1}(\mathbf{R}^n, \mathbf{R}^n - A; G) \approx \tilde{H}_q(\mathbf{R}^n - A; G)$$

The result is obtained by composing the inverse of this isomorphism with the isomorphism of corollary 15. ∎

For general orientable manifolds there is the following *duality theorem.*

17 THEOREM *Let U be an orientation over R of an n-manifold X and let (A,B) be a compact pair in X. Then for all q and R modules G there is an isomorphism*

$$\bar{\gamma}_U: H_q(X - B, X - A; G) \approx \bar{H}^{n-q}(A,B; G)$$

PROOF Because of the naturality properties of $\bar{\gamma}_U$, it suffices to prove the theorem for the case where B is empty. If A is contained in some coordinate neighborhood V of X and $U' = U|(V \times V, V \times V - \delta(V))$ is the induced orientation of V, there is a commutative triangle (all coefficients G)

$$\begin{array}{ccc} H_q(V, V - A) & \underset{\approx}{\to} & H_q(X, X - A) \\ \bar{\gamma}_{U'} \searrow & & \swarrow \bar{\gamma}_U \\ & \bar{H}^{n-q}(A) & \end{array}$$

By corollary 15, $\bar{\gamma}_{U'}$ is an isomorphism, hence $\bar{\gamma}_U$ is also an isomorphism. The result for arbitrary compact A follows by induction on the finite number of coordinate neighborhoods needed to cover A, using naturality of $\bar{\gamma}_U$, the usual Mayer-Vietoris technique, and the five lemma. ∎

In case X is compact, by applying theorem 17 to the pair $(X, \varnothing)$ and observing that $i: \bar{H}^q(X;G) \approx H^q(X;G)$, we obtain the following *Poincaré duality theorem.*

18 THEOREM *If U is an orientation over R of a compact n-manifold X, then for all q and R modules G there is an isomorphism*

$$\gamma_U\colon H_q(X;G) \approx \bar{H}^{n-q}(X;G) \quad \blacksquare$$

A pair (X,A) is called a *relative n-manifold* if X is a Hausdorff space, A is closed in X (A may be empty), and $X - A$ is an n-manifold. For relative manifolds there is the following *Lefschetz duality theorem.*

19 THEOREM *Let (X,A) be a compact relative n-manifold such that $X - A$ is orientable over R. For all q and R modules G there is an isomorphism*

$$H_q(X - A;\, G) \approx \bar{H}^{n-q}(X,A;\, G)$$

PROOF Let $\{N\}$ be the family of closed neighborhoods of A directed downward by inclusion. There are isomorphisms

$$\lim_{\to} \{H_q(X - N;\, G)\} \approx H_q(X - A;\, G)$$
$$\lim_{\to} \{\bar{H}^{n-q}(X,N;\, G)\} \approx \bar{H}^{n-q}(X,A;\, G)$$

the first because singular homology has compact supports and the second as a consequence of theorem 6.1.12. Let V be an open neighborhood of A with $\bar{V}$ contained in the interior of N and let U be an orientation of $X - A$ over R. By theorem 17 and standard excision properties, there are isomorphisms (all coefficients G)

$$\begin{array}{ccc} H_q(X - N) & \xrightarrow{\approx} & H_q((X - A) - (N - V), (X - A) - (X - V)) \\ & & \approx\downarrow \gamma_U \\ \bar{H}^{n-q}(X,N) & \xrightarrow{\approx} & \bar{H}^{n-q}(X - V, N - V) \end{array}$$

which yield the result on passing to the limit. $\blacksquare$

An *n-manifold X with boundary $\dot{X}$* is a paracompact Hausdorff space such that $(X,\dot{X})$ is a relative n-manifold and every point $x \in \dot{X}$ has a neighborhood V such that $(V, V \cap \dot{X})$ is homeomorphic to $\mathbf{R}^{n-1} \times (I,0)$. Since $\dot{X}$ may be empty, the concept of manifold with boundary encompasses that of manifold without boundary.

If X is an n-manifold with boundary $\dot{X}$, then $\dot{X}$ has neighborhoods N such that $(N,\dot{X})$ is homeomorphic to $\dot{X} \times (I,0)$.[1] Such a neighborhood N is called a *collaring* of $\dot{X}$, and its interior is called an *open collaring* of $\dot{X}$. (In case $\dot{X}$ is compact, any neighborhood of $\dot{X}$ contains a collaring of $\dot{X}$.) Because of the existence of such collarings, $X - \dot{X}$ is a weak deformation retract of X, and the pair $((X - \dot{X}) \times (X - \dot{X}), (X - \dot{X}) \times (X - \dot{X}) - \delta(X - \dot{X}))$ is a weak deformation retract of $(X \times X, X \times X - \delta(X))$.

An n-manifold X with boundary $\dot{X}$ is said to be *orientable* over R if $X - \dot{X}$ is orientable over R. An *orientation* over R of X is a class

[1] See M. Brown, Locally flat imbeddings of topological manifolds, *Annals of Mathematics*, vol. 75, pp. 331–341, 1962.

$U \in H^n(X \times X, X \times X - \delta(X); R)$ whose restriction to $((X - \dot{X}) \times (X - \dot{X}), (X - \dot{X}) \times (X - \dot{X}) - \delta(X - \dot{X}))$ is an orientation of $X - \dot{X}$ over R. For manifolds with boundary the *Lefschetz duality theorem* takes the following form.

20 THEOREM *Let X be a compact n-manifold with boundary $\dot{X}$ and orientation U over R. For all q and R modules G there are isomorphisms (where $j: X - \dot{X} \subset X$)*

$$H_q(X;G) \overset{j_*}{\underset{\approx}{\leftarrow}} H_q(X - \dot{X}; G) \overset{\gamma_U}{\underset{\approx}{\rightarrow}} H^{n-q}(X,\dot{X}; G)$$

$$H_q(X,\dot{X}; G) \overset{\gamma_U}{\underset{\approx}{\rightarrow}} H^{n-q}(X - \dot{X}; G) \overset{j^*}{\underset{\approx}{\leftarrow}} H^{n-q}(X;G)$$

PROOF Because j is a homotopy equivalence, j_* and j^* are isomorphisms. Let N be a collaring of $\dot{X}$ with interior $\mathring{N}$. Let U' be the orientation of $X - \dot{X}$ obtained by restricting U. In the following commutative diagram each horizontal map is induced by inclusion and is an isomorphism because it is an excision (labelled e) or a homotopy equivalence (labelled h) (all coefficients G):

$$\begin{array}{ccccc}
H_q(X - \dot{X}) & \overset{h}{\underset{\approx}{\leftarrow}} & H_q(X - N) & \overset{e}{\underset{\approx}{\rightarrow}} & H_q((X - \dot{X}) - (N - \mathring{N}), (X - \dot{X}) - (X - \mathring{N})) \\
\gamma_U \downarrow & & \gamma_U \downarrow & & \downarrow \gamma_{U'} \\
H^{n-q}(X,\dot{X}) & \overset{h}{\underset{\approx}{\leftarrow}} & H^{n-q}(X,N) & \overset{e}{\underset{\approx}{\rightarrow}} & H^{n-q}(X - \mathring{N}, N - \mathring{N})
\end{array}$$

Because $(X - \mathring{N}, N - \mathring{N})$ has arbitrarily small neighborhoods of which it is a deformation retract $i: \bar{H}^{n-q}(X - \mathring{N}, N - \mathring{N}) \approx H^{n-q}(X - \mathring{N}, N - \mathring{N})$, and it follows from theorem 17 that the right-hand vertical map is an isomorphism (because it corresponds to the isomorphism $\bar{\gamma}_{U'}$). Therefore the left-hand vertical map is also an isomorphism proving the first part of the theorem.

Similarly, there is a commutative diagram

$$\begin{array}{ccccc}
H_q(X,\dot{X}) & \overset{h}{\underset{\approx}{\rightarrow}} & H_q(X,\mathring{N}) & \overset{e}{\underset{\approx}{\leftarrow}} & H_q(X - \dot{X}, (X - \dot{X}) - (X - \mathring{N})) \\
\gamma_U \downarrow & & \gamma_U \downarrow & & \downarrow \gamma_{U'} \\
H^{n-q}(X - \dot{X}) & \overset{h}{\underset{\approx}{\rightarrow}} & H^{n-q}(X - \mathring{N}) & \overset{e}{\underset{\approx}{\leftarrow}} & H^{n-q}(X - \mathring{N})
\end{array}$$

Because $X - \mathring{N}$ has arbitrarily small neighborhoods of which it is a deformation retract, it follows from theorem 17 that the right-hand vertical map is an isomorphism. Therefore the left-hand vertical map is also an isomorphism, proving the second part of the theorem. ∎

From the isomorphisms of theorem 20 and the universal-coefficient theorem for homology, we obtain a short exact sequence

$$0 \to H^q(X;R) \otimes G \xrightarrow{\mu} H^q(X;G) \to H^{q+1}(X;R) * G \to 0$$

and a similar short exact sequence for $H^q(X,\dot{X}; G)$. Since this is so for every R module G, from theorem 5.5.13 we have the following result.

21 COROLLARY *If X is a compact n-manifold with boundary $\dot{X}$ orientable over R, then $H_*(X;R)$ and $H_*(X,\dot{X}; R)$ are finitely generated.* ∎

Later in the chapter (see theorem 6.9.11) we shall prove that corollary 21 is also valid for nonorientable manifolds.

3 THE FUNDAMENTAL CLASS OF A MANIFOLD

In view of the importance of the concept of orientability of manifolds, we shall now investigate some equivalent formulations. We shall show that a compact connected n-manifold is orientable if and only if its n-dimensional homology module is nonzero. In fact, any orientation class of the manifold will be shown to correspond to a generator of the n-dimensional homology module. Moreover, if z is the element of H_n corresponding to the orientation, then the cap product of z and a cohomology class defines a homomorphism which equals, up to sign, the inverse of the duality isomorphism. The methods in this section rely heavily on the technique of piecing together homology classes,[1] analogous to the piecing together of cohomology classes in lemma 5.7.16.

Let X be a space, X' a subspace of X, and $\mathcal{A} = \{A\}$ a collection of subsets of $X - X'$. A *compatible $\mathcal{A}$ family* is a family $\{z_A \in H_q(X, X - A; G)\}$ (for some fixed q and G) indexed by $\mathcal{A}$ such that if $A, A' \in \mathcal{A}$, then z_A and $z_{A'}$ map to the same element of $H_q(X, X - A \cap A'; G)$ under the homomorphisms

$$H_q(X, X - A; G) \to H_q(X, X - A \cap A'; G) \leftarrow H_q(X, X - A'; G)$$

The compatible $\mathcal{A}$ families form a module with respect to componentwise operations that will be denoted by $H_q^{\mathcal{A}}(X,X'; G)$. For the collection $\mathcal{A}$ of all compact subsets of $X - X'$ we use $H_q{}^c(X,X'; G)$ to denote the corresponding module.

We are interested in the module $H_n{}^c(X,\dot{X}; R)$ for an n-manifold X with boundary $\dot{X}$. The following lemma is important in this connection.

1 LEMMA *Let X be an n-manifold with boundary $\dot{X}$ and let A be a compact subset of $X - \dot{X}$. For all R modules G*

$$H_q(X, X - A; G) = 0 \qquad q > n$$

PROOF Assume first that A is contained in some coordinate neighborhood V in $X - \dot{X}$. By excision, $H_q(V, V - A) \approx H_q(X, X - A)$, and since V is homeomorphic to $\mathbf{R}^n$, we can use corollary 6.2.15 to obtain

$$H_q(V, V - A) \approx \bar{H}^{n-q}(A) = 0 \qquad q > n$$

For arbitrary compact A the result follows by induction on the number of coordinate neighborhoods needed to cover A, using Mayer-Vietoris sequences. ∎

In an n-manifold X with boundary $\dot{X}$ a *small cell* in $X - \dot{X}$ is defined to be a compact subset A having an open neighborhood $V \subset X - \dot{X}$ such that

[1] This technique can be found in H. Cartan, Méthodes modernes en topologie algébrique, *Commentarii Mathematici Helvetici*, vol. 18, pp. 1–15, 1945.

(V,A) is homeomorphic to $(\mathbf{R}^n,E^n)$. Every point of $X - \dot{X}$ has arbitrarily small neighborhoods which are small cells. If A and V are as above, there is an excision isomorphism

$$H_q(X, X - A; G) \approx H_q(V, V - A; G) \approx \begin{cases} 0 & q \neq n \\ G & q = n \end{cases}$$

If $x_0 \in A$, then the inclusion map induces isomorphisms

$$H_q(X, X - A; G) \approx H_q(X, X - x_0; G)$$

We use $H_q{}^{sc}(X,\dot{X}; G)$ to denote the module of compatible $\mathcal{A}$ families, where $\mathcal{A}$ consists of the collection of small cells of $X - \dot{X}$. Since the collection of small cells is contained in the collection of compact subsets of $X - \dot{X}$, there is a natural homomorphism

$$H_q{}^c(X,\dot{X}; G) \to H_q{}^{sc}(X,\dot{X}; G)$$

which assigns to a compatible family $\{z_A\}$ indexed by all compact A the compatible subfamily of elements indexed by small cells.

2 LEMMA *Let X be an n-manifold with boundary $\dot{X}$. Then, for all G*

$$H_n{}^c(X,\dot{X}; G) \approx H_n{}^{sc}(X,\dot{X}; G)$$

PROOF For each positive integer i let $\mathcal{A}_i$ be the collection of compact subsets of $X - \dot{X}$ contained in the union of i small cells. Then $\mathcal{A}_i \subset \mathcal{A}_{i+1}$ and $\bigcup \mathcal{A}_i$ is the collection of all compact subsets of $X - \dot{X}$. There are homomorphisms

$$\cdots \to H_n^{\mathcal{A}_{i+1}} \to H_n^{\mathcal{A}_i} \to \cdots \to H_n^{\mathcal{A}_1} \to H_n{}^{sc}$$

and an isomorphism $H_n{}^c \approx \lim\ \{H_n^{\mathcal{A}_i}\}$.

Since every element of $\mathcal{A}_1$ is contained in some small cell, it is obvious that $H_n^{\mathcal{A}_1} \approx H_n{}^{sc}$. By the usual Mayer-Vietoris technique and lemma 1, it follows that for any $i \geq 1$ $H_n^{\mathcal{A}_{i+1}} \approx H_n^{\mathcal{A}_i}$. Combining these isomorphisms yields the result. ∎

This gives the following important result.

3 THEOREM *Let X be an n-manifold with boundary $\dot{X}$ and let*

$$\{z_A\} \in H_n{}^c(X,\dot{X}; G)$$

(*a*) *$\{z_A\} = 0$ if and only if $z_x = 0$ for all $x \in X - \dot{X}$.*

(*b*) *If X is connected, $\{z_A\} = 0$ if and only if $z_x = 0$ for some $x \in X - \dot{X}$.*

PROOF (*a*) follows from lemma 2 and the observation that if A is a small cell and $x \in A$, then

$$H_n(X, X - A; G) \approx H_n(X, X - x; G)$$

and so $z_A = 0$ if and only if $z_x = 0$.

To prove (*b*), assume $z_{x_0} = 0$ for some $x_0 \in X - \dot{X}$. Because X is connected, so is its weak deformation retract $X - \dot{X}$. This implies that if

$x \in X - \dot{X}$, there is a finite sequence of small cells $A_1, \ldots, A_m$ in $X - \dot{X}$ such that $x_0 \in A_1$ and $x \in A_m$, and A_i meets A_{i+1} for $1 \le i < m$. Choose a point $x_i \in A_i \cap A_{i+1}$ for $1 \le i < m$. There are isomorphisms

$$H_n(X, X - x_0) \underset{\approx}{\leftarrow} H_n(X, X - A_1) \underset{\approx}{\rightarrow} H_n(X, X - x_1) \underset{\approx}{\leftarrow} \cdots$$
$$\underset{\approx}{\leftarrow} H_n(X, X - A_m) \underset{\approx}{\rightarrow} H_n(X, X - x)$$

from which it follows that if $z_{x_0} = 0$, then $z_x = 0$. Since this is so for all $x \in X - \dot{X}$, the result follows from *(a)*. ∎

If X is an n-manifold with boundary $\dot{X}$, a *fundamental family of* X over R is an element $\{z_A\} \in H_n^c(X,\dot{X}; R)$ such that for all $x \in X - \dot{X}$, z_x is a generator of $H_n(X, X - x; R)$. The relation between fundamental families and orientations is made precise in the next result.

4 THEOREM *Let X be an n-manifold with boundary $\dot{X}$. There is a one-to-one correspondence between orientations U (over R) of X and fundamental families $\{z_A\}$ (over R) of X such that U and $\{z_A\}$ correspond if and only if $\gamma_U(z_A) = 1 \in H^0(A;R)$ for all compact A in $X - \dot{X}$.*

PROOF If U is an orientation of X, let U' be the induced orientation of $X - \dot{X}$. For any compact $A \subset X - \dot{X}$ we have the commutative diagram (all coefficients R)

$$\begin{array}{ccc} H_n(X,X - A) & \overset{j_*}{\underset{\approx}{\leftarrow}} & H_n(X - \dot{X}, (X - \dot{X}) - A) \\ \gamma_U \downarrow & \overset{\gamma_{U'}}{\swarrow} & \downarrow \bar{\gamma}_{U'} \\ H^0(A) & \overset{i}{\leftarrow} & \bar{H}^0(A) \end{array}$$

By theorem 6.2.17, the right-hand vertical map is an isomorphism, and since $1 \in H^0(A)$ is the image of $1 \in \bar{H}^0(A)$, there is a unique $z_A \in H_n(X, X - A)$ such that $\bar{\gamma}_{U'} j_*^{-1}(z_A) = 1 \in \bar{H}^0(A)$. Because of the uniqueness of z_A and the naturality of γ_U and $\bar{\gamma}_{U'}$, the collection $\{z_A\}$ is a compatible family. From the commutativity of the above diagram, $\gamma_U(z_A) = 1 \in H^0(A)$ for all compact A in $X - \dot{X}$. Hence we need only verify that $\{z_A\}$ is a fundamental family. In case $A = x$, it follows from the commutativity of the above square and the fact that i: $\bar{H}^0(x) \approx H^0(x)$ that γ_U: $H_n(X, X - x) \approx H^0(x)$. Therefore $z_x = \gamma_U^{-1}(1)$ is a generator of $H_n(X, X - x)$. Hence $\{z_A\}$ is a fundamental family with the desired property, and the collection $\{z_x\}_{x \in X - \dot{x}}$ (and hence, by theorem 3*a*, $\{z_A\}$) is uniquely characterized by the property $\gamma_U(z_x) = 1 \in H^0(x)$.

Conversely, given a fundamental family $\{z_A\}$, let V be any open subset of $X - \dot{X}$ homeomorphic to $\mathbf{R}^n$. If $x_0 \in V$, then $H^*(V;R) \approx H^*(x_0;R)$, which implies that

$$H^*(V \times X, V \times X - \delta(V); R) \approx H^*(x_0 \times (X, X - x_0); R)$$

If $u \in H^n(V \times X, V \times X - \delta(V); R)$, it follows from the Künneth formula for cohomology (theorem 5.6.1) that $u \mid x_0 \times (X, X - x_0) = 1 \times u'$ for a unique $u' \in H^n(X, X - x_0; R) \approx \text{Hom}\,(H_n(X, X - x_0; R), R)$. By property 6.1.2,

$$[u \mid x_0 \times (X, X - x_0)]/z_{x_0} = \langle u', z_{x_0} \rangle 1$$

Since z_{x_0} is a generator of $H_n(X, X - x_0; R)$, $\langle u', z_{x_0} \rangle$ completely determines u'. Therefore there is a unique element $U \in H^n(V \times X, V \times X - \delta(V); R)$ such that $[U \mid x_0 \times (X, X - x_0)]/z_{x_0} = 1 \in H^0(x_0;R)$.

We now show that for any $x \in V$, $[U|x \times (X, X - x)]/z_x = 1 \in H^0(x;R)$. If x and x' belong to a small cell $A \subset V$, then z_A maps to z_x and to $z_{x'}$. Therefore $[U \mid A \times (X, X - A)]/z_A \in H^0(A;R)$ maps to $[U \mid x \times (X, X - x)]/z_x$ and to $[U \mid x' \times (X, X - x')]/z_{x'}$ by naturality of γ_U. Since $H^0(A;R) \approx H^0(x;R)$ and $H^0(A;R) \approx H^0(x';R)$, it follows that both $[U \mid x \times (X, X - x)]/z_x = 1 \in H^0(x;R)$ and $[U \mid x' \times (X, X - x')]/z_{x'} = 1 \in H^0(x';R)$ or neither equation is true. Hence the set of $x \in V$ for which $[U \mid x \times (X, X - x)]/z_x = 1 \in H^0(x;R)$ is open and its complement in V is open. Since V is connected and $[U \mid x_0 \times (X, X - x_0)]/z_{x_0} = 1$, it follows that $[U \mid x \times (X, X - x)]/z_x = 1$ for all $x \in V$.

This means that U is an orientation of V, and if U' is a similarly defined orientation for another coordinate neighborhood V' in $X - \dot{X}$, then for any $x \in V \cap V'$, $U \mid x \times (X, X - x) = U' \mid x \times (X, X - x)$. This implies that U and U' induce the same orientation of $V \cap V'$. Hence the collection $\{U_V\}$ for coordinate neighborhoods V in $X - \dot{X}$ is compatible. Therefore there is an orientation U of X such that $U \mid (V \times X, V \times X - \delta(V)) = U_V$. From the construction of U_V we see that $\gamma_U(z_x) = 1 \in H^0(x;R)$ for all $x \in X - \dot{X}$. By the first half of the proof, there is a fundamental family $\{z'_A\}$ such that $\gamma_U(z'_A) = 1 \in H^0(A;R)$. Then $z'_x = z_x$ for all $x \in X - \dot{X}$, and by theorem 3*a*, $z'_A = z_A$ for all compact $A \subset X - \dot{X}$. Therefore $\gamma_U(z_A) = 1 \in H^0(A;R)$ for all A, proving that every fundamental family $\{z_A\}$ corresponds to some orientation U.

The orientation U is uniquely characterized by the fundamental family $\{z_A\}$, for if U and U' are two orientations of X such that $\gamma_U(z_x) = \gamma_{U'}(z_x)$ for all $x \in X - \dot{X}$, then $U \mid x \times (X, X - x) = U' \mid x \times (X, X - x)$ for all $x \in X - \dot{X}$. Therefore, by lemma 5.7.13, $U = U'$. ∎

This last result gives the following useful characterization of orientability for connected manifolds.

5 THEOREM *Let X be a connected n-manifold with boundary $\dot{X}$. If $H_n{}^c(X,\dot{X}; R) \neq 0$, then $H_n{}^c(X,\dot{X}; R) \approx R$ and any generator is a fundamental family of X.*

PROOF From theorem 3*b* it follows that, given $x_0 \in X - \dot{X}$, the homomorphism

$$H_n{}^c(X,\dot{X}; R) \to H_n(X, X - x_0; R)$$

sending $\{z_A\}$ to z_{x_0} is a monomorphism. Since $H_n(X, X - x_0; R) \approx R$, either $H_n{}^c(X,\dot{X}; R) = 0$ or $H_n{}^c(X,\dot{X}; R) \approx R$. Assume $H_n{}^c(X,\dot{X}; R) \approx R$ and let $\{z_A\}$ be a generator of $H_n{}^c(X,\dot{X}; R)$. Assume that for some $x \in X - \dot{X}$, z_x is not a generator of $H_n(X, X - x; R)$. There is then a noninvertible element $r \in R$ such that $z_x = rz'_x$ for some $z'_x \in H_n(X, X - x; R)$. It follows that for any small cell A containing x, $z_A = rz'_A$ for some $z'_A \in H_n(X, X - A; R)$. Because X

is connected, it follows, as in the proof of theorem 3*b*, that for any small cell A in $X - \dot{X}$, $z_A = rz'_A$ for some $z'_A \in H_n(X, X - A; R)$. If A' is a small cell in A, then rz'_A maps to $rz'_{A'}$ in $H_n(X, X - A'; R)$. Because $H_n(X, X - A'; R)$ is torsion free, by lemma 1, z'_A maps to $z'_{A'}$. Therefore $\{z'_A\} \in H_n{}^{sc}(X,\dot{X}; R)$. By lemma 2, it follows that the original element $\{z_A\} \in H_n{}^c(X,\dot{X}; R)$ is divisible by the element $r \in R$. Since r is not invertible, this contradicts the hypothesis that $\{z_A\}$ is a generator of $H_n{}^c(X,\dot{X}; R)$. ∎

6 COROLLARY *If X is a connected n-manifold with boundary $\dot{X}$, then X is orientable over R if and only if $H_n{}^c(X,\dot{X}; R) \neq 0$.*

PROOF This is immediate from theorems 4 and 5. ∎

We now specialize to the case of a compact manifold.

7 LEMMA *If X is a compact n-manifold with boundary $\dot{X}$, there is an isomorphism*

$$H_n(X,\dot{X}; G) \approx H_n{}^c(X,\dot{X}; G)$$

sending $z \in H_n(X,\dot{X}; G)$ to $\{z_A = \text{image of } z \text{ in } H_n(X, X - A; G)\}$.

PROOF Let V be an open collaring of $\dot{X}$ and let $B = X - V$. Then B is compact and there is a homomorphism

$$H_n{}^c(X,\dot{X}; G) \to H_n(X, X - B; G)$$

sending $\{z_A\}$ to z_B. Since $X - B = V$ and $(X,\dot{X}) \subset (X,V)$ is a homotopy equivalence, the composite

$$H_n(X,\dot{X}; G) \to H_n{}^c(X,\dot{X}; G) \to H_n(X, X - B; G)$$

is an isomorphism. To complete the proof we need only show that the right-hand map is a monomorphism. Assume that $\{z_A\}$ is a compatible family such that $z_B = 0$ and let A be any compact set in $X - \dot{X}$. There is then an open collaring V' of $\dot{X}$ such that $V' \subset V$ and V' is disjoint from A. Let $B' = X - V'$. Then $A, B \subset B'$, and we have homomorphisms (all coefficients G)

$$H_n(X, X - A) \leftarrow H_n(X, X - B') \xrightarrow{\approx} H_n(X, X - B)$$

the second map being an isomorphism because $(X,V') \subset (X,V)$ is a homotopy equivalence. Since $z_B = 0$, $z_{B'} = 0$ and $z_A = 0$. Therefore $\{z_A\} = 0$ in $H_n{}^c(X,\dot{X}; G)$. ∎

8 COROLLARY *A compact connected n-manifold X with boundary $\dot{X}$ is orientable over R if and only if $H_n(X,\dot{X}; R) \neq 0$.*

PROOF This is immediate from corollary 6 and lemma 7. ∎

If X is a compact n-manifold with boundary $\dot{X}$, a *fundamental class* over R of X is an element $z \in H_n(X,\dot{X}; R)$ whose image in $H_n{}^c(X,\dot{X}; R)$ under the isomorphism of lemma 7 is a fundamental family [that is, for every $x \in X - \dot{X}$ the image of z in $H_n(X, X - x; R)$ is a generator of the latter].

9 THEOREM *If X is a compact n-manifold with boundary $\dot{X}$, there is a one-to-one correspondence between orientations U over R and fundamental classes z over R such that U corresponds to z if and only if $\gamma_U(z) = 1 \in H^0(X;R)$.*

PROOF This follows from theorem 4 and lemma 7 on observing that an element $v \in H^0(X;R)$ equals 1 if and only if $v \mid x = 1 \in H^0(x;R)$ for all $x \in X - \dot{X}$. ∎

10 COROLLARY *If X is a compact n-manifold with boundary $\dot{X}$, then if X is orientable, so is $\dot{X}$, and any fundamental class of X maps to a fundamental class of $\dot{X}$ under the connecting homomorphism*

$$\partial_*: H_n(X,\dot{X};\, R) \to H_{n-1}(\dot{X};R)$$

PROOF Let N be a collaring of $\dot{X}$ with interior $\mathring{N}$. Then N is an n-manifold with boundary $\dot{X} \cup (N - \mathring{N})$, and there is a commutative diagram (all coefficients R)

$$\begin{array}{ccccc} H_n(X,\dot{X}) & & \xrightarrow{i_*} & & H_n(X, \dot{X} \cup (X - \mathring{N})) \\ \partial_* \downarrow & & & & j_* \uparrow \approx \\ H_{n-1}(\dot{X}) & \xrightarrow[\approx]{k_*} & H_{n-1}(\dot{X} \cup (N - \mathring{N}), N - \mathring{N}) & \xleftarrow{\partial_*} & H_n(N, \dot{X} \cup (N - \mathring{N})) \end{array}$$

It is clear from the definition of fundamental class that if $z \in H_n(X,\dot{X})$ is a fundamental class of X, then $j_*{}^{-1} i_* z = z'$ is a fundamental class of N. Because N is homeomorphic to $\dot{X} \times I$ in such a way that $\dot{X}$ and $N - \mathring{N}$ correspond to $\dot{X} \times 0$ and $\dot{X} \times 1$, respectively, the Künneth formula implies

$$H_n(N, \dot{X} \cup (N - \mathring{N})) \approx H_{n-1}(\dot{X}) \otimes H_1(I,\dot{I})$$

Let $w \in H_1(I,\dot{I})$ be a generator and let $\{\dot{X}_j\}$ be the components of $\dot{X}$. Then z' corresponds to $\sum z_j' \times w$ for some $z_j' \in H_{n-1}(\dot{X}_j)$, and $k_*{}^{-1}\partial_* z' = \pm\sum z_j'$. Hence $\partial_* z = \pm\sum z_j'$, and since z is a fundamental class of X, $z_j' \times w$ corresponds to a fundamental class of $\dot{X}_j \times I$. Therefore z_j' is nonzero and is a generator of $H_{n-1}(\dot{X}_j)$. Then z_j' is a fundamental class of $\dot{X}_j$, whence $\pm\sum z_j' = \partial_* z$ is a fundamental class of $\dot{X}$. ∎

We are now heading toward a proof that cap product with a fundamental class is an isomorphism which, up to sign, is inverse to the duality isomorphism in a compact manifold. First we need a lemma.

11 LEMMA *Let X be a compact orientable n-manifold with boundary $\dot{X}$ and let $p_1, p_2: X \times X \to X$ be the projections. Given*

$$u \in H^q(X \times X, X \times X - \delta(X);\, R),\ z \in H_m(X \times X, X \times X - \delta(X);\, G),$$

and $v \in H^r(X;G)$, then

$$\begin{array}{lll} p_{1*}(u \frown z) = p_{2*}(u \frown z) & \text{in} & H_{m-q}(X;G) \\ u \smile p_1^* v = u \smile p_2^* v & \text{in} & H^{q+r}(X \times X, X \times X - \delta(X);\, G) \end{array}$$

PROOF Let $T: (X \times X, X \times X - \delta(X)) \to (X \times X, X \times X - \delta(X))$ be the map interchanging the factors. If $w \in H_n(X,\dot{X}; R)$ is a fundamental class of X, then $w \times w \in H_{2n}((X,\dot{X}) \times (X,\dot{X}); R)$ is a fundamental class of $X \times X$ (whence $X \times X$ is orientable), and $T_*(w \times w) = (-1)^n w \times w$. By theorem 9, T maps the orientation of $X \times X$ corresponding to $w \times w$ into $(-1)^n$ times itself. Let

$$\gamma\colon H_m(X \times X, X \times X - \delta(X); G) \approx \bar{H}^{2n-m}(\delta(X),\delta(\dot{X});G)$$

be the duality map associated to this orientation. Then we have a commutative diagram (all coefficients G)

$$\begin{array}{ccc} H_m(X \times X, X \times X - \delta(X); G) & \xrightarrow{T_*} & H_m(X \times X, X \times X - \delta(X); G) \\ & {\scriptstyle \gamma}\searrow\approx \qquad \approx\swarrow{\scriptstyle (-1)^n\gamma} & \\ & \bar{H}^{2n-m}(\delta(X),\delta(\dot{X});G) & \end{array}$$

Therefore $T_*(z) = (-1)^n z$ for any $z \in H_*(X \times X, X \times X - \delta(X); G)$ (which implies $T^*(u) = (-1)^n u$ for any $u \in H^*(X \times X, X \times X - \delta(X); G)$). Then

$$p_{2*}(u \frown z) = p_{1*}T_*(u \frown z) = p_{1*}(T^*u \frown T_*z) = p_{1*}(u \frown z)$$

and $u \smile p_2^* v = (-1)^n T^*(u \smile p_2^* v) = u \smile T^* p_2^* v = u \smile p_1^* v$ ∎

12 THEOREM *Let z be a fundamental class over R of a compact n-manifold X with boundary $\dot{X}$. For all q and R modules G the homomorphism $\kappa_z(v) = v \frown z$ defines isomorphisms*

$$\kappa_z\colon H^q(X;G) \approx H_{n-q}(X,\dot{X}; G)$$
$$\kappa_z\colon H^q(X,\dot{X}; G) \approx H_{n-q}(X;G)$$

which are, up to sign, the inverse of the duality isomorphisms of theorem 6.2.20 *defined by the orientation corresponding to z.*

PROOF Let U be the orientation of X corresponding to z as in theorem 9, and let $j\colon X - \dot{X} \subset X$. We prove commutativity up to sign in the triangle (all coefficients G)

$$\begin{array}{ccc} H_q(X - \dot{X}) & \xrightarrow{\gamma_U} & H^{n-q}(X,\dot{X}) \\ & {\scriptstyle j_*}\searrow \qquad \swarrow{\scriptstyle \kappa_z} & \\ & H_q(X) & \end{array}$$

For $w \in H_q(X - \dot{X})$, by property 6.1.6,

$$\begin{aligned} k_z\gamma_U(w) &= \{[U \mid (X,\dot{X}) \times (X - \dot{X})]/w\} \frown z \\ &= p_{1*}\{[U \mid (X,\dot{X}) \times (X - \dot{X})] \frown (z \times w)\} \end{aligned}$$

By lemma 11, this equals

$$\begin{aligned} &p_{2*}\{[U \mid (X,\dot{X}) \times (X - \dot{X})] \frown (z \times w)\} \\ &\qquad = p_{1*}T_*\{[U \mid (X,\dot{X}) \times (X - \dot{X})] \frown (z \times w)\} \\ &\qquad = \pm j_*\bar{p}_{1*}\{[U \mid (X - \dot{X}) \times (X,\dot{X})] \frown (w \times z)\} \end{aligned}$$

where $\bar{p}_1\colon (X - \dot{X}) \times X \to X - \dot{X}$ is projection to the first factor. Again by property 6.1.6,

$$\bar{p}_{1*}\{[U \mid (X - \dot{X}) \times (X,\dot{X})] \frown (w \times z)\} = \gamma_U(z) \frown w = w$$

Therefore

$$\kappa_z \gamma_U(w) = \pm j_*(w)$$

Similarly, we prove commutativity up to sign in the triangle

$$\begin{array}{ccc} H^q(X) & \xrightarrow{\kappa_z} & H_{n-q}(X,\dot{X}) \\ & {}^{j^*}\searrow \quad \swarrow{}^{\gamma_U} & \\ & H^q(X - \dot{X}) & \end{array}$$

For $v \in H^q(X)$, by property 6.1.5,

$$\begin{aligned} \gamma_U \kappa_z(v) &= [U \mid (X - \dot{X}) \times (X,\dot{X})]/(v \frown z) \\ &= \{[U \smile p_2^*(v)] \mid (X - \dot{X}) \times (X,\dot{X})\}/z \end{aligned}$$

By lemma 11 and property 6.1.4, this equals

$$\pm\{[\bar{p}_1^* j^*(v) \smile U] \mid (X - \dot{X}) \times (X,\dot{X})\}/z = \pm j^*(v) \smile \gamma_U(z) = \pm j^*(v)$$

Therefore

$$\gamma_U \kappa_z(v) = \pm j^* v \quad \blacksquare$$

4 THE ALEXANDER COHOMOLOGY THEORY

We shall now describe a cohomology theory particularly suited for applications in which a space is mapped into polyhedra (the singular theory is more suitable for applications where polyhedra are mapped into a space). One approach to the theory, called the Čech construction, is based on approximating a space by nerves of open coverings; another approach, called the Alexander-Kolmogoroff construction, is based on complexes built of "small" simplexes consisting of finite sets of points. We shall begin with the Alexander construction, and show later in the chapter (see corollary 6.9.9 and the following paragraph) that if (A,B) is a closed pair in a manifold X, then $\bar{H}^q(A,B;\, G)$ as defined in Sec. 6.1 is the Alexander cohomology of (A,B) with coefficients G.

Let G be an R module and let X be a topological space. For $q \geq 0$ let $C^q(X;G)$ be the module of all functions φ from X^{q+1} to G with addition and scalar multiplication defined pointwise. Thus, if $x_0, x_1, \ldots, x_q \in X$, then $\varphi(x_0,x_1, \ldots ,x_q) \in G$, and if $\varphi_1, \varphi_2 \in C^q(X;G)$ and $r \in R$, then

$$\begin{aligned} r\varphi_1(x_0, \ldots ,x_q) &= r(\varphi_1(x_0, \ldots ,x_q)) \\ (\varphi_1 + \varphi_2)(x_0, \ldots ,x_q) &= \varphi_1(x_0, \ldots ,x_q) + \varphi_2(x_0, \ldots ,x_q) \end{aligned}$$

We shall omit the symbol G from $C^q(X;G)$ where its absence will not cause confusion.

A coboundary homomorphism $\delta\colon C^q(X) \to C^{q+1}(X)$ is defined by the formula

$$(\delta\varphi)(x_0, \ldots, x_{q+1}) = \sum_{0 \le i \le q+1} (-1)^i \varphi(x_0, \ldots, \hat{x}_i, \ldots, x_{q+1})$$

Then $\delta\delta = 0$ and $C^*(X) = \{C^q(X),\delta\}$ is a cochain complex over R. If X is nonempty, it is augmented over G by η: $G \to C^0(X)$, where $(\eta(g))(x) = g$ for $g \in G$ and all $x \in X$. So far the topology of X has played no role, and the following result shows that $C^*(X)$ has uninteresting cohomology.

1 LEMMA *If X is a nonempty space, $\eta^*: G \approx H^*(C^*(X;G))$.*

PROOF Let $\bar{x}$ be a fixed point of X and define a cochain homotopy $D: C^*(X) \to C^*(X)$ by

$$(D\varphi)(x_0, \ldots, x_q) = \varphi(\bar{x}, x_0, \ldots, x_q) \qquad q \ge 0$$

Then
$$\delta D\varphi + D\delta\varphi = \begin{cases} \varphi & \deg \varphi > 0 \\ \varphi - \eta(\varphi(\bar{x})) & \deg \varphi = 0 \end{cases}$$

Therefore, if τ: $C(X;G) \to G$ is the cochain map defined by

$$\tau(\varphi) = \begin{cases} 0 & \deg \varphi > 0 \\ \varphi(\bar{x}) & \deg \varphi = 0 \end{cases}$$

then $\tau\eta = 1_G$ and D is a cochain homotopy from $1_{C^*(X)}$ to $\eta\tau$. Therefore η is a cochain equivalence, whence the result. ∎

We now use the topology of X to pass to a more interesting quotient complex. An element $\varphi \in C^q(X)$ is said to be *locally zero* if there is a covering $\mathcal{U}$ of X by open sets such that φ vanishes on any $(q + 1)$-tuple of X which lies in some element of $\mathcal{U}$. Thus, if we define $\mathcal{U}^{q+1} = \bigcup_{U \in \mathcal{U}} U^{q+1} \subset X^{q+1}$, then φ vanishes on $\mathcal{U}^{q+1}$. The subset of $C^q(X)$ consisting of locally zero functions is a submodule, denoted by $C_0{}^q(X)$, and if φ vanishes on $\mathcal{U}^{q+1}$, then $\delta\varphi$ vanishes on $\mathcal{U}^{q+2}$, whence $C_0^*(X) = \{C_0{}^q(X),\delta\}$ is a cochain subcomplex of $C^*(X)$. We define $\bar{C}^*(X)$ to be the quotient cochain complex of $C^*(X)$ by $C_0^*(X)$. If X is nonempty, the composite

$$G \xrightarrow{\eta} C^*(X) \to \bar{C}^*(X)$$

is an augmentation of $\bar{C}^*(X)$, also denoted by η. The cohomology module of $\bar{C}^*(X)$ of degree q is denoted by $\bar{H}^q(X;G)$.

Given a function f: $X \to Y$ (not necessarily continuous), there is an induced cochain map

$$f^\#: C^*(Y;G) \to C^*(X;G)$$

defined by the formula

$$(f^\#\varphi)(x_0, \ldots, x_q) = \varphi(f(x_0), \ldots, f(x_q)) \qquad \varphi \in C^q(Y);\ x_0, \ldots, x_q \in X$$

If φ vanishes on $\mathcal{V}^{q+1}$, where $\mathcal{V}$ is an open covering of Y, and if there is an open covering $\mathcal{U}$ of X such that f maps each element of $\mathcal{U}$ into some element of $\mathcal{V}$, then $f^\#\varphi$ vanishes on $\mathcal{U}^{q+1}$. In particular, if f is continuous, $f^{-1}\mathcal{V}$ is an open covering of X which can be taken as $\mathcal{U}$, and therefore $f^\#$

maps $C_0^*(Y)$ into $C_0^*(X)$. It follows that if f is continuous, there is an induced cochain map

$$f^\#: \bar{C}^*(Y;G) \to \bar{C}^*(X;G)$$

Let A be a subspace of X and let $i: A \subset X$. Then $i^\#: \bar{C}^*(X;G) \to \bar{C}^*(A;G)$ is an epimorphism. Therefore the kernel of $i^\#$ is a cochain subcomplex of $\bar{C}^*(X;G)$, denoted by $\bar{C}^*(X,A;G)$. The relative module $\bar{H}^q(X,A;G)$ is defined to be the cohomology module of $\bar{C}^*(X,A;G)$ of degree q.

Since there is a short exact sequence of cochain complexes

$$0 \to \bar{C}^*(X,A;G) \xrightarrow{j^\#} \bar{C}^*(X;G) \xrightarrow{i^\#} \bar{C}^*(A;G) \to 0$$

it follows that there is an exact sequence

2 $$\cdots \to \bar{H}^q(X,A;G) \xrightarrow{j^*} \bar{H}^q(X;G) \xrightarrow{i^*} \bar{H}^q(A;G) \xrightarrow{\delta^*} \bar{H}^{q+1}(X,A;G) \to \cdots$$

The graded module $\bar{H}^*(X,A) = \{\bar{H}^q(X,A;G)\}$ is the module function of the cohomology theory we are constructing, and the homomorphism $\delta^*: \bar{H}^q(A;G) \to \bar{H}^{q+1}(X,A;G)$ is the connecting homomorphism of the theory. Given a continuous map $f: (X,A) \to (Y,B)$, there is induced by f a commutative diagram of cochain maps

$$\begin{array}{ccccccccc} 0 & \to & \bar{C}^*(Y,B;G) & \to & \bar{C}^*(Y;G) & \to & \bar{C}^*(B;G) & \to & 0 \\ & & \downarrow f^\# & & \downarrow (f|X)^\# & & \downarrow (f|A)^\# & & \\ 0 & \to & \bar{C}^*(X,A;G) & \to & \bar{C}^*(X;G) & \to & \bar{C}^*(A;G) & \to & 0 \end{array}$$

The homomorphism $f^*: \bar{H}^*(Y,B;G) \to \bar{H}^*(X,A;G)$ is defined to be the homomorphism induced by the cochain map $f^\#$ in the above diagram. It is then clear that for fixed G, $\bar{H}^*(X,A;G)$ and f^* constitute a contravariant functor from the category of topological pairs to the category of graded R modules. Furthermore, the connecting homomorphism δ^* is a natural transformation of degree 1 from $\bar{H}^*(A;G)$ to $\bar{H}^*(X,A;G)$. Therefore we have the constituents of a cohomology theory, and we shall verify that the axioms are satisfied. The resulting cohomology theory is called the *Alexander* (or *Alexander-Spanier*[1]) *cohomology theory*, and $\bar{H}^q(X,A;G)$ is called the *Alexander cohomology module of* (X,A) *of degree* q *with coefficients* G.

The exactness axiom is a consequence of the exactness of the sequence 2. The dimension axiom will follow from the next result.

3 LEMMA *If X is a one-point space, $\eta^*: G \approx \bar{H}^*(X;G)$.*

PROOF Because X is a one-point space, a locally zero function on X is zero. Therefore $\bar{C}^*(X;G) = C^*(X;G)$ and the result follows from lemma 1. ■

Before proving the excision axiom it will be useful to introduce another cochain complex for the relative theory. If $A \subset X$, let $C^*(X,A)$ be the sub-

[1] See E. Spanier, Cohomology theory for general spaces, *Annals of Mathematics*, vol. 49 pp. 407–427, 1948.

complex of $C^*(X)$ of functions φ which are locally zero on A. Thus there is a short exact sequence

$$0 \to C^*(X,A) \to C^*(X) \to \bar{C}^*(A) \to 0$$

and $C_0^*(X) \subset C^*(X,A)$. It follows that $\bar{C}^*(X,A) = C^*(X,A)/C_0^*(X)$. The excision axiom follows from the next result.

4 LEMMA *Let U be a subset of $A \subset X$ such that U has an open neighborhood W with $\bar{W} \subset \operatorname{int} A$. Then the inclusion map $j: (X - U, A - U) \subset (X,A)$ induces an isomorphism*

$$j^{\#}: \bar{C}^*(X,A) \approx \bar{C}^*(X - U, A - U)$$

PROOF There is a commutative diagram with exact rows

$$\begin{array}{ccccccccc} 0 \to & C_0^*(X) & \to & C^*(X,A) & \to & \bar{C}^*(X,A) & \to 0 \\ & \downarrow & & \downarrow k^{\#} & & \downarrow j^{\#} & \\ 0 \to & C_0^*(X - U) & \to & C^*(X - U, A - U) & \xrightarrow{\lambda} & \bar{C}^*(X - U, A - U) & \to 0 \end{array}$$

It suffices to prove that $\lambda k^{\#}$ is an epimorphism and that $(k^{\#})^{-1}(C_0^*(X - U)) = C_0^*(X)$. If $\varphi \in C^q(X - U, A - U)$, let $\bar{\varphi} \in C^q(X)$ be defined by

$$\bar{\varphi}(x_0, \ldots, x_q) = \begin{cases} 0 & x_i \in W \text{ for some } 0 \le i \le q \\ \varphi(x_0, \ldots, x_q) & x_0, \ldots, x_q \in X - W \end{cases}$$

If $\mathcal{V}$ is an open covering of $A - U$ such that φ vanishes on $\mathcal{V}^{q+1}$, then $\mathcal{U} = \{V \cup W \mid V \in \mathcal{V}\}$ is an open covering of A such that $\bar{\varphi}$ vanishes on $\mathcal{U}^{q+1}$. Therefore $\bar{\varphi} \in C^q(X,A)$, and from the definition of $\bar{\varphi}$, $k^{\#}\bar{\varphi} - \varphi$ vanishes on $\mathscr{W}^{q+1}$ where $\mathscr{W} = \{V \cap \operatorname{int} A \mid V \in \mathscr{V}\} \cup \{X - \overline{W}\}$, which is an open covering of $X - U$. Therefore $\lambda k^{\#}\bar{\varphi} = \lambda\varphi$, and because λ is an epimorphism, so is $\lambda k^{\#}$.

Assume that $\varphi \in C^q(X,A)$ is such that $k^{\#}\varphi \in C_0{}^q(X - U)$. Because φ is locally zero on A, there is an open covering $\mathcal{U}_1$ of A such that φ vanishes on $\mathcal{U}_1{}^{q+1}$. Because $k^{\#}\varphi \in C_0{}^q(X - U)$, there is an open covering $\mathcal{U}_2$ of $X - U$ such that φ vanishes on $\mathcal{U}_2{}^{q+1}$. Let

$$\mathcal{V}_1 = \{U_1 \cap \operatorname{int} A \mid U_1 \in \mathcal{U}_1\} \qquad \mathcal{V}_2 = \{U_2 \cap (X - \bar{U}) \mid U_2 \in \mathcal{U}_2\}$$

Then $\mathcal{V} = \mathcal{V}_1 \cup \mathcal{V}_2$ is an open covering of X such that φ vanishes on $\mathcal{V}^{q+1}$. Therefore $\varphi \in C_0{}^q(X)$ and so

$$(k^{\#})^{-1}(C_0^*(X - U)) = C_0^*(X) \quad \blacksquare$$

The homotopy axiom will be proved in the next section. We conclude this section with a study of $\bar{H}^0$. A function φ from a topological space X to a set is said to be *locally constant* if there is an open covering $\mathcal{U}$ of X such that φ is constant on each element of $\mathcal{U}$.

5 THEOREM *If $A \subset X$, then $\bar{H}^0(X,A; G)$ is isomorphic to the module of locally constant functions from X to G which vanish on A.*

PROOF A locally zero function from X to G is zero. Therefore $C_0^0(X) = 0$, and so

$$\bar{C}^0(X,A) = C^0(X,A)/C_0^0(X) = C^0(X,A)$$

Therefore $\bar{H}^0(X,A;\,G)$ is the kernel of the composite

$$C^0(X,A) \xrightarrow{\delta} C^1(X,A) \to \bar{C}^1(X,A)$$

$C^0(X,A)$ is the module of functions from X to G which vanish on A. If $\varphi \in C^0(X,A)$, then φ is in the kernel of the above composite if and only if there is some open covering $\mathfrak{U}$ of X such that $\delta\varphi$ vanishes on $\mathfrak{U}^2$. Since $(\delta\varphi)(x,y) = \varphi(y) - \varphi(x)$, this is equivalent to the condition that there is an open covering $\mathfrak{U}$ such that φ is constant on each element of $\mathfrak{U}$. Hence the kernel of the above composite equals the module of functions vanishing on A that are locally constant on X. ■

6 COROLLARY *Let X be a topological space in which every quasi-component is open and let $A \subset X$. Then $\bar{H}^0(X,A;\,G)$ is isomorphic to the module of functions from the set of those quasi-components of X which do not intersect A to G.*

PROOF This follows from theorem 5 and the fact that a locally constant function on X is constant on every quasi-component of X. ■

7 COROLLARY *A nonempty space X is connected if and only if*

$$\eta^*\colon G \approx \bar{H}^0(X;G)$$

PROOF This follows from theorem 5 and the trivial observation that every locally constant function on X is constant if and only if X is connected. ■

It follows that there exist spaces for which the singular cohomology and Alexander cohomology differ. In fact, for any connected space which is not path connected, corollary 7 and theorem 5.4.10 show that they differ in degree 0.

We now present a version of theorem 5.4.10 valid for the Alexander theory.

8 THEOREM *Let $\{U_j\}$ be an open covering of X by pairwise disjoint sets. Then there is a canonical isomorphism*

$$\bar{H}^q(X;G) \approx \times\, \bar{H}^q(U_j;G)$$

PROOF Because $\{U_j\}$ consists of pairwise disjoint sets, the map induced by restriction

$$i\#\colon C^*(X) \to \times\, C^*(U_j)$$

is an epimorphism. Because $\{U_j\}$ is an open covering of X, it follows that

$$(i\#)^{-1}(\times\, C_0^*(U_j)) = C_0^*(X)$$

Therefore $i\#$ induces an isomorphism $\bar{C}^*(X) \approx \times\, \bar{C}^*(U_j)$. ■

9 COROLLARY *Let $\{C_j\}$ be the collection of components of a locally connected space X. Then there is a canonical isomorphism*

$$\bar{H}^q(X;G) \approx \times \bar{H}^q(C_j;G)$$

PROOF Because X is locally connected, its components are open, and the result follows from theorem 8. ∎

5 THE HOMOTOPY AXIOM FOR THE ALEXANDER THEORY

In this section we shall prove the homotopy axiom for the Alexander cohomology theory. The proof will be based on a description of the Alexander cochain complex as the limit of cochain complexes of abstract simplicial complexes. We shall also use this description to construct a homomorphism of the Alexander cohomology theory into the singular cohomology theory. Because the Alexander theory satisfies all the axioms, this homomorphism is an isomorphism from the Alexander theory to the singular cohomology theory on the category of compact polyhedral pairs.

We shall be considering a fixed R module G as coefficient module for cohomology and will usually not mention G explicitly. Let $\mathfrak{U}$ be a collection of subsets covering a set X. Let $X(\mathfrak{U})$ be the abstract simplicial complex whose vertices are the points of X and whose simplexes are finite subsets F of X such that there is some $U \in \mathfrak{U}$ containing F. Let $C(\mathfrak{U})$ be the ordered chain complex of $X(\mathfrak{U})$ over R. Given a subset $A \subset X$ and a subcollection $\mathfrak{U}' \subset \mathfrak{U}$ which covers A, we let $A(\mathfrak{U}')$ be the subcomplex of $X(\mathfrak{U})$ whose vertices are the points of A and whose simplexes are finite subsets of A lying in some element of $\mathfrak{U}'$. Then $C'(\mathfrak{U}')$ will denote the chain subcomplex of $C(\mathfrak{U})$ corresponding to $A(\mathfrak{U}')$.

Let $(\mathcal{V},\mathcal{V}')$ be another pair consisting of a covering $\mathcal{V}$ of X and a subset $\mathcal{V}' \subset \mathcal{V}$ which is a covering of A. Assume that $(\mathcal{V},\mathcal{V}')$ is a refinement of $(\mathfrak{U},\mathfrak{U}')$ in the sense that every element of $\mathcal{V}$ is contained in some element of $\mathfrak{U}$ and every element of $\mathcal{V}'$ is contained in some element of $\mathfrak{U}'$. Then the pair $(C(\mathcal{V}),C'(\mathcal{V}'))$ is mapped injectively into the pair $(C(\mathfrak{U}),C'(\mathfrak{U}'))$ by the identity map of (X,A) to itself.

Let X be a topological space and A a subspace of X. Consider pairs $(\mathfrak{U},\mathfrak{U}')$, where $\mathfrak{U}$ is an open covering of X and $\mathfrak{U}'$ is a subset of $\mathfrak{U}$ which covers A. Such a pair is called an *open covering* of (X,A). Let $C^*(\mathfrak{U},\mathfrak{U}')$ be the cochain complex of the pair $(C(\mathfrak{U}),C'(\mathfrak{U}'))$ (with coefficients in G). An element u of $C^q(\mathfrak{U},\mathfrak{U}')$ is a function defined on $(q+1)$-tuples of X which lie in some element of $\mathfrak{U}$, taking values in G, and vanishing on $(q+1)$-tuples of A which lie in some element of $\mathfrak{U}'$. If $(\mathcal{V},\mathcal{V}')$ is a refinement of $(\mathfrak{U},\mathfrak{U}')$, the restriction map is a cochain map

$$C^*(\mathfrak{U},\mathfrak{U}') \to C^*(\mathcal{V},\mathcal{V}')$$

If $(\mathcal{U},\mathcal{U}')$ and $(\mathcal{V},\mathcal{V}')$ are two open coverings of (X,A) as above, let $\mathcal{W} = \{U \cap V \mid U \in \mathcal{U}, V \in \mathcal{V}\}$ and let $\mathcal{W}' = \{U' \cap V' \mid U' \in \mathcal{U}', V' \in \mathcal{V}'\}$. Then $(\mathcal{W},\mathcal{W}')$ is another open covering of (X,A) and $(\mathcal{W},\mathcal{W}')$ is a refinement of $(\mathcal{U},\mathcal{U}')$ and of $(\mathcal{V},\mathcal{V}')$. Therefore the cochain complexes $\{C^*(\mathcal{U},\mathcal{U}')\}$ form a direct system, and we have a limit cochain complex

$$\lim_{\to} \{C^*(\mathcal{U},\mathcal{U}')\}$$

We shall show that this limit cochain complex is canonically isomorphic to $\bar{C}^*(X,A)$. If $\varphi \in C^q(X,A)$, let $\mathcal{U}'$ be a collection of open subsets of X covering A such that φ vanishes on $(\mathcal{U}')^{q+1} \cap A^{q+1}$ (such a $\mathcal{U}'$ exists because φ is locally zero) and let $\mathcal{U} = \mathcal{U}' \cup \{X\}$. Then $(\mathcal{U},\mathcal{U}')$ is an open covering of (X,A) and φ determines by restriction an element $\varphi \mid (\mathcal{U},\mathcal{U}') \in C^q(\mathcal{U},\mathcal{U}')$. Passing to the limit, we obtain a homomorphism (by restriction)

$$\lambda: C^*(X,A) \to \lim_{\to} \{C^*(\mathcal{U},\mathcal{U}')\}$$

which is a canonical cochain map. The following result explains our interest in the cochain complexes $C^*(\mathcal{U},\mathcal{U}')$.

1 THEOREM *The canonical cochain map*

$$\lambda: C^*(X,A) \to \lim_{\to} \{C^*(\mathcal{U},\mathcal{U}')\}$$

is an epimorphism and has kernel equal to $C_0^*(X)$.

PROOF To prove that λ is an epimorphism, let $u \in C^q(\mathcal{U},\mathcal{U}')$. Define $\varphi_u \in C^q(X)$ by

$$\varphi_u(x_0, \ldots, x_q) = \begin{cases} u(x_0, \ldots, x_q) & \text{if } x_0, \ldots, x_q \in U, \text{ where } U \in \mathcal{U} \\ 0 & \text{otherwise} \end{cases}$$

Then φ_u vanishes on $(\mathcal{U}')^{q+1} \cap A^{q+1}$, and therefore $\varphi_u \in C^q(X,A)$. By definition, $\varphi_u \mid (\mathcal{U},\mathcal{U}') = u$, and λ is an epimorphism.

An element $\varphi \in C^q(X,A)$ is in the kernel of λ if and only if there is some $(\mathcal{U},\mathcal{U}')$ such that $\varphi \mid (\mathcal{U},\mathcal{U}') = 0$. Thus $\lambda(\varphi) = 0$ if and only if there is some open covering $\mathcal{U}$ such that φ vanishes on $\mathcal{U}^{q+1}$. By the definition of $C_0^*(X)$, $\lambda(\varphi) = 0$ if and only if $\varphi \in C_0^*(X)$. ∎

From theorem 1 and the analogue of theorem 4.1.7 for cochain complexes, we have the following corollary.

2 COROLLARY *For the Alexander cohomology theory there is a canonical isomorphism*

$$\bar{H}^q(X,A; G) \approx \lim_{\to} \{H^q(C^*(\mathcal{U},\mathcal{U}'; G))\} \quad ∎$$

We are now ready for the proof of the homotopy axiom for the Alexander cohomology theory. In the presence of the other axioms, it suffices to prove it for the case of the two mappings

$$h_0, h_1: (X,A) \to (X \times I, A \times I)$$

where $h_0(x) = (x,0)$, $h_1(x) = (x,1)$. The proof consists in showing that if $(\mathcal{U},\mathcal{U}')$ is any open covering of $(X \times I, A \times I)$, there is an open covering $(\mathcal{V},\mathcal{V}')$ of (X,A) such that h_0 and h_1 induce chain-homotopic chain maps from $(C(\mathcal{V}),C(\mathcal{V}'))$ to $(C(\mathcal{U}),C(\mathcal{U}'))$. This is a result about free chain complexes, and the technique of acyclic models is available for obtaining the desired chain homotopy.

Let Y be an arbitrary set and n a nonnegative integer. Let $C(Y,n)$ be the chain complex over R of the abstract simplicial complex $(Y \times I)(\mathcal{U}(Y,n))$, where $\mathcal{U}(Y,n)$ is the covering of $Y \times I$ defined by

$$\mathcal{U}(Y,n) = \left\{ Y \times \left[\frac{m}{2^n}, \frac{m+1}{2^n}\right] \middle|\, 0 \le m < 2^n \right\}$$

3 LEMMA *If Y is nonempty, the chain complex $\tilde{C}(Y,n)$ is acyclic.*

PROOF For $0 \le m < 2^n$ let K_m be the subcomplex of $(Y \times I)(\mathcal{U}(Y,n))$ consisting of all the finite subsets of $Y \times [m/2^n, (m+1)/2^n]$. For $0 \le m \le 2^n$ let L_m be the subcomplex of $(Y \times I)(\mathcal{U}(Y,n))$ consisting of all the finite subsets of $Y \times (m/2^n)$. Then $(Y \times I)(\mathcal{U}(Y,n)) = \bigcup_m K_m$ and $K_i \cap K_j = 0$ if $|i - j| > 1$ and $K_i \cap K_{i+1} = L_{i+1}$. Because Y is nonempty, each K_m (and L_m) is nonempty and is the join of K_m (or L_m) with any vertex in it. Therefore, by theorem 4.3.6, $\tilde{C}(K_m)$ and $\tilde{C}(L_m)$ are acyclic. Let $N_q = \bigcup_{m \le q} K_m$. Then $N_{q+1} = N_q \cup K_{q+1}$ and $N_q \cap K_{q+1} = L_{q+1}$. By induction on q, using the exactness of the reduced Mayer-Vietoris sequence, it follows that $\tilde{C}(N_q)$ is acyclic for all q. Therefore $\tilde{C}(Y,n) = \tilde{C}(N_{2^n-1})$ is acyclic. ∎

From this we have our next result, which will provide the acyclic model for the homotopy axiom.

4 LEMMA *Let $Y_1, \ldots, Y_q$ be subsets of a nonempty set Y, where $Y = Y_1$, and for each i let n_i be a nonnegative integer. Let K be the simplicial complex defined by*

$$K = \bigcup_i (Y_i \times I)(\mathcal{U}(Y_i,n_i))$$

Then $\tilde{C}(K)$ is acyclic.

PROOF We prove the lemma by induction on q. If $q = 1$, it follows from lemma 3. Assume that $q > 1$, and the result is valid for fewer than q sets Y_i. Let $\bar{K} = \bigcup_{i \le q-1} (Y_i \times I)(\mathcal{U}(Y_i,n_i))$. Then $\bar{K} \cup (Y_q \times I)(\mathcal{U}(Y_q,n_q)) = K$. If Y_q is empty, $\tilde{C}(K) = \tilde{C}(\bar{K})$ is acyclic, by the inductive assumption. If Y_q is nonempty, $\tilde{C}(Y_q,n_q)$ is acyclic, by lemma 3, and $\tilde{C}(\bar{K})$ is acyclic, by the inductive assumption. To prove that $\tilde{C}(K)$ is acyclic, from the exactness of the reduced Mayer-Vietoris sequence it suffices to prove that $\tilde{C}(\bar{K} \cap (Y_q \times I)(\mathcal{U}(Y_q,n_q)))$ is acyclic. However, $\bar{K} \cap (Y_q \times I)(\mathcal{U}(Y_q,n_q)) = \bigcup_{1 \le i < q} (Y'_i \times I)(\mathcal{U}(Y'_i,n'_i))$, where $Y'_i = Y_i \cap Y_q$ are subsets of Y_q (and $Y'_1 = Y_q$) and $n'_i = \max(n_i,n_q)$. Therefore, by the inductive assumption, $\tilde{C}(\bar{K} \cap (Y_q \times I)(\mathcal{U}(Y_q,n_q)))$ is acyclic. ∎

We now come to the following main step in the proof of the homotopy axiom.

5 LEMMA *Let $(\mathcal{U},\mathcal{U}')$ be any open covering of $(X \times I, A \times I)$. There is an open covering $(\mathcal{V},\mathcal{V}')$ of (X,A) such that h_0 and h_1 induce chain-homotopic chain maps from $(C(\mathcal{V}),C'(\mathcal{V}'))$ to $(C(\mathcal{U}),C'(\mathcal{U}'))$.*

PROOF For each $x \in X$ it follows from the compactness of $x \times I$ that there is an open set V_x about x and an integer $n \geq 0$ such that for $0 \leq m < 2^n$ the set $V_x \times [m/2^n, (m+1)/2^n]$ is contained in some element of $\mathcal{U}$. Furthermore, if $x \in A$, we can choose V_x and n so that $V_x \times [m/2^n, (m+1)/2^n]$ is contained in some element of $\mathcal{U}'$. Let $\mathcal{V}$ be the collection $\{V_x\}_{x \in X}$ and $\mathcal{V}'$ the subcollection $\{V_x\}_{x \in A}$. To show that $(\mathcal{V},\mathcal{V}')$ has the desired property, let $\mathcal{C}$ be the category consisting of the subcomplexes of $X(\mathcal{V})$ partially ordered by inclusion. For each subcomplex K of $X(\mathcal{V})$ let $G(K)$ be the ordered chain complex of K. For each simplex s of $X(\mathcal{V})$ [or $A(\mathcal{V}')$] define $n(s)$ to be the smallest nonnegative integer such that for $0 \leq m < n(s)$ each set $s \times [m/2^{n(s)}, (m+1)/2^{n(s)}]$ is contained in some element of $\mathcal{U}$ [or $\mathcal{U}'$]. Such an integer exists because of the way $(\mathcal{V},\mathcal{V}')$ was chosen. For a subcomplex K of $X(\mathcal{V})$ let $\hat{K}$ be the subcomplex of $(X \times I)(\mathcal{U})$ defined by

$$\hat{K} = \cup \{(s \times I)(\mathcal{U}(s,n(s)) \mid s \in K\}$$

and let $G'(K)$ be the ordered chain complex of $\hat{K}$. Then G and G' are covariant functors from $\mathcal{C}$ to the category of augmented chain complexes.

Let $\mathfrak{M}$ be the set of subcomplexes $\{\bar{s} \subset X(\mathcal{V}) \mid s \in X(\mathcal{V})\}$. Then G is free on $\mathcal{C}$ with models $\mathfrak{M}$, and by lemma 4, G' is acyclic on $\mathcal{C}$ with models $\mathfrak{M}$. If $\sigma = (x_0,x_1, \ldots ,x_q)$ is an ordered q-simplex of $X(\mathcal{V})$, then

$$h_0(\sigma) = ((x_0,0), \ldots ,(x_q,0)) \qquad \text{and} \qquad h_1(\sigma) = ((x_0,1), \ldots ,(x_q,1))$$

are natural chain maps preserving augmentation from G to G'. It follows from theorem 4.3.3 that there is a natural chain homotopy from h_0 to h_1. ■

If $u \in H^q(C^*(\mathcal{U},\mathcal{U}'))$, where $(\mathcal{U},\mathcal{U}')$ is an open covering of $(X \times I, A \times I)$, it follows from lemma 5 that there is an open covering $(\mathcal{V},\mathcal{V}')$ of (X,A) such that $h_0(\mathcal{V},\mathcal{V}') \subset (\mathcal{U},\mathcal{U}')$, $h_1(\mathcal{V},\mathcal{V}') \subset (\mathcal{U},\mathcal{U}')$, and $h_0^\# u = h_1^\# u$ in $H^q(C^*(\mathcal{V},\mathcal{V}'))$. Passing to the limit and using corollary 2 gives us the final result.

6 THEOREM *The Alexander cohomology theory satisfies the homotopy axiom.* ■

We have now verified all the axioms of cohomology theory for the Alexander cohomology theory. We construct a homomorphism μ from the Alexander cohomology theory to the singular cohomology theory. Let $(\mathcal{U},\mathcal{U}')$ be an open covering of (X,A). There is a canonical chain transformation

$$(\Delta(\mathcal{U}),\Delta(\mathcal{U}' \cap A)) \to (C(\mathcal{U}),C'(\mathcal{U}'))$$

which assigns to a singular q-simplex $\sigma\colon \Delta^q \to X$ the ordered simplex $(\sigma(v_0),\sigma(v_1), \ldots ,\sigma(v_q))$ of $C(\mathcal{U})$. This induces a homomorphism

$$C^*(\mathfrak{U},\mathfrak{U}'; G) \to C^*(\Delta(\mathfrak{U}), \Delta(\mathfrak{U}' \cap A); G)$$

Passing to the limit and using corollary 2, we obtain a canonical homomorphism

$$\mu': \bar{H}^q(X,A; G) \to \lim_{\to} \{H^q(\Delta(\mathfrak{U}), \Delta(\mathfrak{U}' \cap A); G)\}$$

By theorem 4.4.14, there is a canonical isomorphism

$$\mu'': H^q(\Delta(X), \Delta(A); G) \approx \lim_{\to} \{H^q(\Delta(\mathfrak{U}), \Delta(\mathfrak{U}' \cap A); G)\}$$

and the homomorphism

$$\mu: \bar{H}^q(X,A; G) \to H^q(\Delta(X), \Delta(A); G)$$

is defined to equal the composite $\mu''^{-1}\mu'$. It can be verified that this homomorphism has the commutativity properties necessary to be a natural transformation of cohomology theories.

We now introduce a cup product in the Alexander theory, which will have the usual properties of a cup product (as in Sec. 5.6) and will be compatible with the singular cup product by the homomorphism μ.

Let G and G' be R modules paired to an R module G''. Given $\varphi_1 \in C^p(X;G)$ and $\varphi_2 \in C^q(X;G')$, we define $\varphi_1 \smile \varphi_2 \in C^{p+q}(X;G'')$ by

$$(\varphi_1 \smile \varphi_2)(x_0, \ldots ,x_{p+q}) = \varphi_1(x_0, \ldots ,x_p)\varphi_2(x_p, \ldots ,x_{p+q})$$

If φ_1 is locally zero on A_1, so is $\varphi_1 \smile \varphi_2$, and if φ_2 is locally zero on A_2, so is $\varphi_1 \smile \varphi_2$. Therefore $\varphi_1 \smile \varphi_2$ induces a cup product from $\bar{C}^p(X;G)$ and $\bar{C}^q(X;G')$ to $\bar{C}^{p+q}(X;G'')$. An easy verification shows that

$$\delta(\varphi_1 \smile \varphi_2) = \delta\varphi_1 \smile \varphi_2 + (-1)^p\varphi_1 \smile \delta\varphi_2$$

Therefore the cup product induces a cup product on cohomology classes, and this cup product is clearly mapped by μ to the singular cup product.

In order to get a cup product from $C^p(X,A_1; G)$ and $C^q(X,A_2; G')$ to $C^{p+q}(X, A_1 \cup A_2; G'')$, we need to ensure that an element of $C^{p+q}(X;G'')$ which is locally zero on A_1 and locally zero on A_2 will be locally zero on $A_1 \cup A_2$. If $A_1 \cup A_2 = \operatorname{int}_{A_1 \cup A_2} A_1 \cup \operatorname{int}_{A_1 \cup A_2} A_2$, this is so. With this modification properties 5.6.8 to 5.6.12 are all valid for the resulting cohomology product.

6 TAUTNESS AND CONTINUITY

In this section we shall consider tautness for the Alexander theory and establish the strong result that any paracompact space imbedded as a closed subspace of a paracompact space is tautly imbedded. This implies a strong excision property for paracompact pairs (X,A) with A closed in X. It also implies the continuity property (that the Alexander cohomology theory commutes with limits of compact Hausdorff spaces directed by inclusion). This continuity property, together with the other axioms of cohomology theory, characterizes the Alexander theory on the category of compact Hausdorff pairs (that is, pairs with X compact Hausdorff and A closed in X). The section closes with a brief discussion of the Alexander cohomology with compact supports. Our proof of the special tautness properties of the Alexander cohomology theory is based on techniques of Wallace.[1]

Let $\mathcal{U}$ be a collection of subsets of a set X. Let $\mathcal{U}^* = \{U^*\}_{U \in \mathcal{U}}$, where

$$U^* = \cup\{U' \in \mathcal{U} \mid U' \cap U \neq \emptyset\}$$

A collection $\mathcal{V}$ is said to be a *star refinement* of $\mathcal{U}$ if $\mathcal{V}^*$ is a refinement of $\mathcal{U}$. A topological space X is said to be *fully normal* if every open covering of X has an open star refinement. It is known that for Hausdorff spaces paracompactness is equivalent to full normality.

1 LEMMA *Let A be a subset of a topological space X and let $\mathcal{V}$ be an open covering of X. There exist a neighborhood N of A and a function $f\colon N \to A$ (not necessarily continuous) such that*

(a) $f(x) = x$ for $x \in A$.
(b) If $V \in \mathcal{V}$, then $f(V \cap N) \subset V^*$.

PROOF If A is empty, let $N = A$ and f be the identity map. If A is nonempty, let $N = \cup\{V \in \mathcal{V} \mid V \cap A \neq \emptyset\}$ and define $f\colon N \to A$ by $f(x) = x$ for $x \in A$, or if $x \notin A$, choose $f(x) \in A$ so that there is $V \in \mathcal{V}$ with $x, f(x) \in V$. Such a choice of $f(x)$ is always possible because of the way N was defined. Clearly, if $x \in V \cap N$, there is $V' \in \mathcal{V}$ with $x, f(x) \in V'$. Therefore $x \in V \cap V'$ and $V' \subset V^*$. Hence, $f(V \cap N) \subset V^*$ and (a) and (b) are satisfied. ■

This last result may be interpreted as asserting that A is a discontinuous neighborhood retract of X with a retraction that is not too discontinuous. If A is a closed subset of a paracompact space, it is similar enough to an absolute neighborhood retract so that we have the following generalization of theorem 6.1.10 for the Alexander theory.

2 THEOREM *A closed subspace of a paracompact Hausdorff space is a taut subspace relative to the Alexander cohomology theory.*

PROOF Let A be a closed subspace of a paracompact space X and let $\varphi \in C^q(A)$ be a cochain such that $\delta\varphi$ vanishes on $\mathcal{W}^{q+2}$, where $\mathcal{W}$ is an open covering of A. Let $\mathcal{U} = \{W \cup (X - A) \mid W \in \mathcal{W}\}$ and observe that $\mathcal{U}$ is an open covering of X because A is closed in X. Let $\mathcal{V}$ be an open star refinement of $\mathcal{U}$ and let N be a neighborhood of A and $f\colon N \to A$ a function (not necessarily continuous) satisfying lemma 1 relative to $\mathcal{V}$. Then $f^\#\varphi \in C^q(N)$, and we show that $\delta f^\#\varphi = f^\#\delta\varphi$ vanishes on $\mathcal{V}^{q+2} \cap N^{q+2}$. By lemma 1b, for any $V \in \mathcal{V}$ there is $U \in \mathcal{U}$ such that $f(V \cap N) \subset U$. Then $f(V \cap N) \subset U \cap A \subset W$ for some $W \in \mathcal{W}$. Therefore $\delta f^\#\varphi$ vanishes on $(V \cap N)^{q+2}$. This means that $f^\#\varphi$ represents a cocycle of $\bar{C}^q(N)$ and, by lemma 1a, $(f^\#\varphi) \mid A = \varphi$. Hence

[1] See A. D. Wallace, The map excision theorem, *Duke Mathematical Journal*, vol. 19, pp. 177–182, 1952.

the cohomology class $\{\varphi\} \in \bar{H}^q(A)$ is the image under restriction of the cohomology class $\{f^\#\varphi\} \in \bar{H}^q(N)$, showing that $\lim_\rightarrow \{\bar{H}^q(N)\} \to \bar{H}^q(A)$ is an epimorphism.

To prove that it is a monomorphism, let N' be a paracompact neighborhood of A and assume that $\varphi \in C^q(N')$ is such that $\delta\varphi$ vanishes on $\mathcal{W}^{q+2}$ and $\varphi \mid A = \delta\varphi'$ on $(\mathcal{W}')^{q+1}$, where $\mathcal{W}$ is an open covering of N' and $\mathcal{W}'$ is an open covering of A. Let $\mathcal{U} = \{W' \cup (N' - A) \mid W' \in \mathcal{W}'\}$ and observe that $\mathcal{U}$ is an open covering of N' (because A is closed.) Let $\mathcal{V}$ be an open star refinement of both $\mathcal{W}$ and $\mathcal{U}$ ($\mathcal{V}$ is a covering of N') and let N be a neighborhood of A in N' and $f: N \to A$ a function (not necessarily continuous) defined with respect to $\mathcal{V}$ to satisfy lemma 1. If $V \in \mathcal{V}$, then $f(V \cap N) \subset W'$ for some $W' \in \mathcal{W}'$. Therefore $f^\#(\varphi \mid A) = \delta f^\#\varphi'$ on $V^{q+1} \cap N^{q+1}$.

To show that $f^\#(\varphi \mid A)$ is cohomologous in $C^q(N)$ to $\varphi \mid N$, for $\psi \in C^p(N)$ define $D\psi \in C^{p-1}(N)$ by

$$(D\psi)(x_0, \ldots, x_{p-1}) = \sum_{0 \le j \le p-1} (-1)^j \psi(x_0, \ldots, x_j, f(x_j), \ldots, f(x_{p-1}))$$

An easy computation establishes the formula

$$\delta D\psi + D\delta\psi = f^\#(\psi \mid A) - \psi$$

For every $V \in \mathcal{V}$, $(V \cap N) \cup f(V \cap N) \subset W$ for some $W \in \mathcal{W}$ (by lemma 1*b*), and because $\delta\varphi$ vanishes on $\mathcal{W}^{q+2}$, $\delta D(\varphi \mid N) = f^\#(\varphi \mid A) - \varphi \mid N$ on $\mathcal{V}^{q+1} \cap N^{q+1}$. Therefore the cohomology class $\{\varphi\} \in \bar{H}^q(N')$ maps to zero in $\bar{H}^q(N)$. This suffices to show that $\lim_\rightarrow \{\bar{H}^q(N)\} \to \bar{H}^q(A)$ is a monomorphism, and so A is a taut subspace of X. ■

3 COROLLARY *Let $X \supset A \supset B$, where X is a paracompact Hausdorff space and A and B are closed subspaces of X. Then, relative to the Alexander cohomology theory, (A,B) is a taut pair in X.*

PROOF This is an immediate consequence of theorem 2 and lemma 6.1.9. ■

4 EXAMPLE Let X be the subspace of $\mathbf{R}^2 \subset S^2$ defined in example 2.4.8. The space $\tilde{X}$ obtained by retopologizing X by the topology generated by the path components of open sets in X is a half-open interval. Since X has the same singular homology as $\tilde{X}$, $H^1(X;\mathbf{Z}) = 0$. Since $S^2 - X$ has two components, it follows from the Alexander duality theorem that $\lim_\rightarrow \{H^1(U;\mathbf{Z})\} = \mathbf{Z}$ as U varies over neighborhoods of X. Therefore $\lim_\rightarrow \{H^1(U;\mathbf{Z})\} \to H^1(X;\mathbf{Z})$ is not a monomorphism, and so X is not a taut subspace of $\mathbf{R}^2$ with respect to singular cohomology. Since X is closed in $\mathbf{R}^2$, it is taut with respect to Alexander cohomology.

Note that the above example is one in which $\lim_\rightarrow \{H^1(U;\mathbf{Z})\} \to H^1(X;\mathbf{Z})$ is not a monomorphism, whereas in example 6.1.8 a subspace $A \subset \mathbf{R}^2$ was given such that $\lim_\rightarrow \{H^0(U;\mathbf{Z})\} \to H^0(A;\mathbf{Z})$ was not an epimorphism.

The tautness property 3 implies that the Alexander cohomology theory satisfies the following *strong excision property.*

5 THEOREM *Let (X,A) and (Y,B) be pairs, with X and Y paracompact Hausdorff and A and B closed. Let f: (X,A) → (Y,B) be a closed continuous map such that f induces a one-to-one map of X − A onto Y − B. Then, for all q and all G*

$$f^*: \bar{H}^q(Y,B;\, G) \approx \bar{H}^q(X,A;\, G)$$

PROOF Because f is closed, continuous, and one-to-one from $X - A$ onto $Y - B$, it follows that f is a homeomorphism of $X - A$ onto $Y - B$. Let $\{U_\alpha\}$ be the family of open neighborhoods of B in Y and let $V_\alpha = f^{-1}(U_\alpha)$. Then V_α is an open neighborhood of A in X, and because f is a closed map, the collection $\{V_\alpha\}$ is cofinal in the family of all neighborhoods of A in X. We have a commutative diagram

$$\begin{array}{ccccc} \bar{H}^q(Y,B) & \leftarrow & \varinjlim \{\bar{H}^q(Y,U_\alpha)\} & \rightarrow & \varinjlim \{\bar{H}^q(Y - B,\, U_\alpha - B)\} \\ f^*\downarrow & & f_1^*\downarrow & & \downarrow f_2^* \\ \bar{H}^q(X,A) & \leftarrow & \varinjlim \{\bar{H}^q(X,V_\alpha)\} & \rightarrow & \varinjlim \{\bar{H}^q(X - A,\, V_\alpha - A)\} \end{array}$$

in which the vertical maps are induced by f and the horizontal maps are induced by inclusions. By corollary 3 and lemma 6.4.4, the horizontal maps are isomorphisms. Because $f \mid X - A$ is a homeomorphism of $X - A$ onto $Y - B$, it follows that for each α, $f \mid (X - A,\, V_\alpha - A)$ is a homeomorphism of $(X - A,\, V_\alpha - A)$ onto $(Y - B,\, U_\alpha - B)$. Therefore f_2^* is an isomorphism, and by commutativity of the diagram, f^* is also an isomorphism. ∎

The following *weak continuity property* of the Alexander cohomology theory is another consequence of its tautness properties.

6 THEOREM *Let $\{(X_\alpha,A_\alpha)\}_\alpha$ be a family of compact Hausdorff pairs in some space, directed downward by inclusion, and let $(X,A) = (\bigcap X_\alpha, \bigcap A_\alpha)$. The inclusion maps i_α: $(X,A) \subset (X_\alpha,A_\alpha)$ induce an isomorphism*

$$\{i_\alpha^*\}: \varinjlim \bar{H}^q(X_\alpha,A_\alpha;\, M) \approx \bar{H}^q(X,A;\, M)$$

PROOF If F is a closed subset of X_β for some β, the collection $\{X_\alpha \cap F\}_\alpha$ consists of compact sets directed downward by inclusion, and $X \cap F = \bigcap (X_\alpha \cap F)$. It follows that if $X \cap F = \varnothing$, there is some α such that $X_\alpha \cap F = \varnothing$. Therefore, if U is any neighborhood of X in X_β, there exists α such that $X_\alpha \subset U$. Similarly, if (U,V) is any neighborhood of (X,A) in X_β, there is α such that $(X_\alpha,A_\alpha) \subset (U,V)$.

To show that $\{i_\alpha^*\}$ is an epimorphism, let $u \in \bar{H}^q(X,A)$ be arbitrary. For any β, (X,A) is a taut pair in X_β, by corollary 3. Therefore there is a neighborhood (U,V) of (X,A) in X_β and an element $v \in \bar{H}^q(U,V)$ such that $v \mid (X,A) = u$. Let α be such that $(X_\alpha,A_\alpha) \subset (U,V)$ and $v_\alpha = v \mid (X_\alpha,A_\alpha)$. Then $v_\alpha \in \bar{H}^q(X_\alpha,A_\alpha)$ and $i_\alpha^* v_\alpha = u$, which proves that $\{i_\alpha^*\}$ is an epimorphism.

To prove that $\{i_\alpha^*\}$ is a monomorphism, let $u \in \bar{H}^q(X_\beta,A_\beta)$ be such that $i_\beta^* u = 0$. By corollary 3, (X,A) is a taut pair in X_β. Therefore there is a neighborhood (U,V) of (X,A) in X_β such that $u \mid (U,\, V \cap A_\beta) = 0$. Choose α

so that $(X_\alpha,A_\alpha) \subset (U, V \cap A_\beta)$. Then $u \mid (X_\alpha,A_\alpha) = 0$, and $\{i_\alpha^*\}$ is an isomorphism. ∎

The *continuity property* involves an assertion analogous to that of theorem 6 for an arbitrary inverse system $\{(X_\alpha,A_\alpha)\}$ of compact Hausdorff pairs, where $(X,A) = \lim_{\leftarrow} \{(X_\alpha,A_\alpha)\}$. It is not hard to prove that the continuity property is equivalent to the weak continuity property.[1] A cohomology theory having the weak continuity property is called *weakly continuous.* Such theories are characterized on the category of compact Hausdorff spaces in view of the following result.

7 LEMMA *Any compact Hausdorff pair can be imbedded in a space in which it is the intersection of a family of pairs directed downward by inclusion, each pair of the family being a compact Hausdorff space of the same homotopy type as a compact polyhedral pair.*

PROOF It is a standard fact that any compact Hausdorff space can be imbedded in a cube I^J; hence we assume (X,A) imbedded in I^J. For each finite subset $\alpha \subset J$ let $p_\alpha: I^J \to I^\alpha$ be the projection map and let (U,V) be a compact polyhedral neighborhood of $(p_\alpha(X),p_\alpha(A))$ in I^α. It can be verified that the collection of pairs $\{(p_\alpha^{-1}(U),p_\alpha^{-1}(V))\}$ corresponding to all finite $\alpha \subset J$ and compact polyhedral neighborhoods of $(p_\alpha(X),p_\alpha(A))$ in I^α is directed downward by inclusion and has (X,A) as intersection. Furthermore, $(p_\alpha^{-1}(U),p_\alpha^{-1}(V))$ is a compact pair in I^J homeomorphic to $(U,V) \times I^{J-\alpha}$, and the projection map

$$p_\alpha: (p_\alpha^{-1}(U),p_\alpha^{-1}(V)) \to (U,V)$$

is a homotopy equivalence. Therefore the family $\{(p_\alpha^{-1}(U),p_\alpha^{-1}(V))\}$ has the desired properties. ∎

This yields the following extension of the uniqueness theorem for weakly continuous cohomology theories.

8 THEOREM *Given two weakly continuous cohomology theories, any homomorphism between them which is an isomorphism for some one-point space is an isomorphism for all compact Hausdorff pairs.* ∎

We now describe the Alexander cohomology with compact supports. This is a cohomology theory on a suitable category of topological pairs and maps, and we shall discuss the category first.

A subset A of a topological space X is said to be *bounded* if $\bar{A}$ is compact. A subset $B \subset X$ is said to be *cobounded* if $X - B$ is bounded. A function f from a space X to a space Y is said to be *proper* if it is continuous and if for every bounded set A of Y, $f^{-1}(A)$ is a bounded set of X (or, equivalently, for every cobounded set B of Y, $f^{-1}(B)$ is a cobounded set of X). Clearly, the composite of proper maps is proper, and there is a category of topological spaces and proper maps. There is also a category of topological pairs and

[1] See S. Eilenberg and N. E. Steenrod, "Foundations of Algebraic Topology," Princeton University Press, Princeton, N.J., 1952, or exercise 6.C.2 at the end of this chapter.

proper maps, a proper map from (X,A) to (Y,B) being a proper map from X to Y which maps A to B. This is the category on which the Alexander cohomology theory with compact supports will be defined.

Given a topological pair (X,A), let $C_c^q(X,A;\,G)$ be the submodule of $C^q(X,A;\,G)$ consisting of all $\varphi \in C^q(X,A;\,G)$ such that φ is locally zero on some cobounded subset of X. If φ is locally zero on B, so is $\delta\varphi$, and therefore there is a cochain complex $C_c^*(X,A;\,G) = \{C_c^q(X,A;\,G), \delta\}$ which is a subcomplex of $C^*(X,A;\,G)$. Clearly, $C_0^*(X;G) \subset C_c^*(X,A;\,G)$, and we define

$$\bar{C}_c^*(X,A;\,G) = C_c^*(X,A;\,G)/C_0^*(X;G)$$

The *Alexander cohomology of* (X,A) *with compact supports,* denoted by $\bar{H}_c^*(X,A;\,G)$, is the cohomology module of $\bar{C}_c^*(X,A;\,G)$. If $f: (X,A) \to (Y,B)$ is a proper map, $f^\#$ maps $C_c^*(Y,B;\,G)$ to $C_c^*(X,A;\,G)$ and induces a homomorphism

$$f^*: \bar{H}_c^*(Y,B;\,G) \to \bar{H}_c^*(X,A;\,G)$$

The Alexander cohomology with compact supports satisfies suitable modifications of all the axioms of cohomology theory.

The homotopy axiom holds for proper homotopies, a proper homotopy being a proper map $(X,A) \times I \to (Y,B)$. In general, an inclusion map $(X',A') \subset (X,A)$ is not a proper map. It is a proper map, however, if X' is closed in X. Because of this, the coboundary homomorphism

$$\delta^*: \bar{H}_c^q(A;G) \to \bar{H}_c^{q+1}(X,A;\,G)$$

is defined only when A is a closed subset of X. When A is a closed subset of X, there are proper inclusion maps $i: A \subset X$ and $j: X \subset (X,A)$ and there is a short exact sequence of cochain complexes (for any coefficient module G)

$$0 \to \bar{C}_c^*(X,A) \xrightarrow{j^\#} \bar{C}_c^*(X) \xrightarrow{i^\#} \bar{C}_c^*(A) \to 0$$

The connecting homomorphism of this short exact sequence is a natural transformation from $\bar{H}_c^*(A)$ to $\bar{H}_c^*(X,A)$, of degree 1 on the category of pairs (X,A), with A closed in X and proper maps between such pairs. The exactness axiom then holds for pairs (X,A) with A closed in X.

The excision axiom holds for proper excisions, a proper excision map being an inclusion map $j: (X - U, A - U) \subset (X,A)$ such that U is an open subset of X with $\bar{U} \subset \operatorname{int} A$, in which case it can be shown (analogous to the proof of lemma 6.4.4) that

$$j^\#: \bar{C}_c^*(X,A) \approx \bar{C}_c^*(X - U, A - U)$$

The dimension axiom is obviously satisfied.

We now consider relations between the Alexander cohomology with compact supports and the Alexander cohomology theory previously defined. The following is one case in which they agree.

9 LEMMA *If A is a cobounded subset of X, then*

$$\bar{H}_c^*(X,A;\,G) = \bar{H}^*(X,A;\,G)$$

PROOF Because A is cobounded in X,

$$C_c^*(X,A) = C^*(X,A)$$

and so $\bar{C}_c^*(X,A) = \bar{C}^*(X,A)$. ∎

10 LEMMA *Let B be a closed subset of a Hausdorff space A. Then a subset U of $A - B$ is cobounded in $A - B$ if and only if $U \cup B$ is a neighborhood of B cobounded in A.*

PROOF If U' is a neighborhood of B in A, then the closure of $A - U'$ in A equals the closure of $(A - B) - (U' - B)$ in $A - B$. Hence one is compact if and only if the other is. Therefore the result will follow once we have verified that if U is a cobounded subset of $A - B$, then $U \cup B$ is a neighborhood of B in A. However, if C is the compact set which equals the closure of $(A - B) - U$ in $A - B$, then C is closed in A (because A is Hausdorff). Therefore $A - C$ is an open subset of A containing B. Since $(A - B) - C \subset U$, it follows that $(A - C) \subset U \cup B$, and $U \cup B$ is a neighborhood of B in A. ∎

Let B be a closed subset of a normal space A. If U is a neighborhood of B in A which is a cobounded subset of A, then $\bar{C}^*(A,U) \subset \bar{C}_c^*(A,B)$. Therefore $\lim_{\to} \{\bar{C}^*(A,U)\} = \cup \bar{C}^*(A,U)$ is imbedded as a subcomplex of $\bar{C}_c^*(A,B)$. By the excision property 6.4.4,

$$\cup \bar{C}^*(A,U) \approx \cup \bar{C}^*(A - B, U - B)$$

As U varies over cobounded neighborhoods of B in A, it follows from lemma 10 that $U - B$ varies over cobounded subsets of $A - B$. Therefore

$$\cup \bar{C}^*(A - B, U - B) = \bar{C}_c^*(A - B)$$

and we have defined a functorial imbedding

$$j\colon \bar{C}_c^*(A - B) \subset \bar{C}_c^*(A,B)$$

such that $j(\bar{C}_c^*(A - B)) = \lim_{\to} \{\bar{C}^*(A,U)\}$, where U varies over cobounded neighborhoods of B in A. Hence j induces an isomorphism of cohomology if and only if

$$\lim_{\to} \{\bar{H}^*(A,U)\} \approx \bar{H}_c^*(A,B)$$

We shall now consider cases in which j induces an isomorphism of cohomology.

11 LEMMA *If A is a compact Hausdorff space and B is closed in A, for all q and all G there is an isomorphism*

$$\bar{H}_c^q(A - B; G) \approx \bar{H}^q(A,B; G)$$

PROOF By lemma 9 and the above remarks, it suffices to prove that as U varies over neighborhoods of B in A (any such neighborhood being cobounded because A is compact), there is an isomorphism

$$\lim_{\to} \{\bar{H}^q(A,U; G)\} \approx \bar{H}^q(A,B; G)$$

Since A is paracompact, this is a consequence of the tautness property 3 of Alexander cohomology. ■

This result allows the following interpretation of the cohomology with compact supports of a locally compact space.

12 COROLLARY *If X is a locally compact Hausdorff space and X^+ is the one-point compactification of X, there is an isomorphism*

$$\bar{H}_c^q(X;G) \approx \tilde{\bar{H}}^q(X^+;G)$$

PROOF By lemma 11, $\bar{H}_c^q(X;G) \approx \bar{H}^q(X^+, X^+ - X; G)$ and because $\tilde{\bar{H}}^*(X^+ - X; G) = 0$, there is an isomorphism

$$\bar{H}^q(X^+, X^+ - X; G) \approx \tilde{\bar{H}}^q(X^+;G) \quad ■$$

13 EXAMPLE It follows from corollary 12 that

$$\bar{H}_c^q(\mathbf{R}^n;G) \approx \begin{cases} 0 & q \neq n \\ G & q = n \end{cases}$$

because $(\mathbf{R}^n)^+$ is homeomorphic to S^n. Hence, if $n \neq m$, $\mathbf{R}^n$ and $\mathbf{R}^m$ are not of the same proper homotopy type.

14 EXAMPLE Regarding $\mathbf{R}^1$ as a linear subspace of $\mathbf{R}^2$, then

$$\bar{H}_c^q(\mathbf{R}^2,\mathbf{R}^1; G) \approx \begin{cases} 0 & q \neq 2 \\ G \oplus G & q = 2 \end{cases}$$

15 THEOREM *Let B be a closed subset of a locally compact Hausdorff space A. For all q and all G there is an isomorphism*

$$\lim_{\rightarrow} \{\bar{H}^q(A,U; G)\} \approx \bar{H}_c^q(A,B; G)$$

where U varies over cobounded neighborhoods of B in A.

PROOF If A is compact, this follows from lemmas 9 and 11. If A is not compact, let A^+ be the one-point compactification of A. Set $p^+ = A^+ - A$ and $B^+ = B \cup p^+ \subset A^+$. Then B^+ is closed in the compact space A^+. There is a commutative diagram of chain maps

$$\begin{array}{ccccccccc} & & \bar{C}_c^*(A - B) & \rightarrow & \bar{C}_c^*(A) & \rightarrow & \bar{C}_c^*(B) & & \\ & & \downarrow & & \downarrow & & \downarrow & & \\ 0 & \rightarrow & \bar{C}^*(A^+,B^+) & \rightarrow & \bar{C}^*(A^+,p^+) & \rightarrow & \bar{C}^*(B^+,p^+) & \rightarrow & 0 \end{array}$$

and, by corollary 12 and lemma 11, each vertical map induces an isomorphism on cohomology. Since the bottom row is exact and $\bar{C}_c^*(A - B) \subset \bar{C}_c^*(A)$, it follows that $\bar{C}_c^*(A)/\bar{C}_c^*(A - B) \rightarrow \bar{C}_c^*(B)$ induces isomorphisms of cohomology. Since there is a short exact sequence of cochain complexes

$$0 \rightarrow \bar{C}_c^*(A,B)/\bar{C}_c^*(A - B) \rightarrow \bar{C}_c^*(A)/\bar{C}_c^*(A - B) \rightarrow \bar{C}_c^*(B) \rightarrow 0$$

it follows that $\bar{C}_c^*(A,B)/\bar{C}_c^*(A - B)$ has trivial cohomology. Therefore

$\bar{H}_c^*(A - B) \approx \bar{H}_c^*(A,B)$, and this is equivalent to the statement of the theorem. ■

The last result is a form of tautness for Alexander cohomology with compact supports. This and the five lemma easily imply the next result.

16 THEOREM *Let (A,B) be a pair of closed subsets of a locally compact Hausdorff space X. For all q and all G there is an isomorphism*

$$\lim_{\rightarrow} \{\bar{H}^q(U,V;G)\} \approx \bar{H}_c^q(A,B;G)$$

where (U,V) varies over neighborhoods of (A,B) in X, both U and V being cobounded subsets of X. ■

In a similar fashion, we may consider the singular cohomology with compact supports. A singular cochain $c^* \in \operatorname{Hom}(\Delta_q(X)/\Delta_q(A),G)$ is said to have *compact support* if there is some cobounded set $U \subset X$ such that for every $x \in U$ there is a neighborhood V of x such that $c^*(\sigma) = 0$ for every singular q-simplex σ in V. The singular cochains with compact support form a subcomplex of the singular cochain complex, whose cohomology module is denoted by $H_c^*(X,A;G)$.

7 PRESHEAVES

In this section the Čech construction will be introduced. Because of the ultimate applications, we define the Čech cohomology of a space not merely for coefficients in a module, but, more generally, for coefficient modules which may vary from one point of X to another. This leads to the concepts of presheaf and sheaf. We shall introduce these and give the definition of the Čech cohomology of a space with coefficients in a presheaf. Applications will be given in the next two sections.

A *presheaf* Γ of R modules on a topological space X is a contravariant functor from the category of open subsets U of X and inclusion maps $U \subset V$ to the category of R modules such that $\Gamma(\varnothing) = 0$. Thus Γ assigns to every open subset $U \subset X$ an R module $\Gamma(U)$ and to every inclusion map $U \subset V$ a homomorphism $\rho_{UV}\colon \Gamma(V) \to \Gamma(U)$, called the *restriction map*, such that

$$\rho_{UU} = 1_{\Gamma(U)}$$
$$\rho_{UW} = \rho_{UV} \circ \rho_{VW}\colon \Gamma(W) \to \Gamma(U) \qquad U \subset V \subset W$$

Given $\gamma \in \Gamma(V)$ and $U \subset V$, we use $\gamma \mid U$ to denote the image $\rho_{UV}(\gamma) \in \Gamma(U)$.

In a similar manner, we define presheaves on X with values in any category. We are interested primarily in the case of a presheaf of modules or of cochain complexes. Following are some examples.

1 Given an R module G, the *constant presheaf* G on X assigns to every nonempty open $U \subset X$, the module G (and to $\varnothing$ the trivial module).

2 Given a subset $A \subset X$, the *relative Alexander presheaf of* (X,A) *with* *coefficients* G, denoted by $C^*(\cdot, \cdot \cap A; G)$, assigns to an open $U \subset X$ the cochain complex $C^*(U, U \cap A; G)$.

3 The *relative singular presheaf of* (X,A) *with coefficients* G, denoted by $\Delta^*(\cdot, \cdot \cap A; G)$, assigns to an open $U \subset X$ the cochain complex $\Delta^*(U, U \cap A; G)$ equal to the subcomplex of Hom $(\Delta_*(U), G)$ of cochains locally zero on $U \cap A$ (i.e. cochains that are zero on $\Delta_*(\mathfrak{U})$ for some open covering $\mathfrak{U}$ of $U \cap A$).

Given two presheaves Γ and Γ' on X taking values in the same category, a *homomorphism* α: $\Gamma \to \Gamma'$ is defined to be a natural transformation from Γ to Γ'. It is then clear that there is a category of presheaves on X with values in any fixed category and homomorphisms between them. In particular, there is a category of presheaves of modules and a category of presheaves of cochain complexes. If α: $\Gamma \to \Gamma'$ is a homomorphism of presheaves of modules (or cochain complexes), it is clear how to define ker α, im α, and coker α so as to be presheaves of modules (or cochain complexes) on X. Therefore it is meaningful to consider exact sequences of presheaves of modules (or cochain complexes) on X.

If Γ and Γ' are presheaves of modules (or cochain complexes) on X, their *tensor product* $\Gamma \otimes \Gamma'$ is the presheaf of modules (or cochain complexes) on X such that for open $U \subset X$

$$(\Gamma \otimes \Gamma')(U) = \Gamma(U) \otimes \Gamma'(U)$$

Consider two examples.

4 There is a homomorphism

$$\tau: C^*(\cdot, \cdot \cap A; G) \to \Delta^*(\cdot, \cdot \cap A; G)$$

such that if $\varphi \in C^q(U, U \cap A; G)$ and σ: $\Delta^q \to U$, then $\tau(\varphi)(\sigma) = \varphi(\sigma(p_0), \ldots, \sigma(p_q))$, where $p_0, \ldots, p_q$ are the vertices of Δ^q.

5 There is a homomorphism

$$\tau: C^*(\cdot, \cdot \cap A; R) \otimes G \to C^*(\cdot, \cdot \cap A; G)$$

such that if $\varphi \in C^q(U, U \cap A; R)$ and $g \in G$, then

$$\tau(\varphi \otimes g)(x_0, \ldots, x_q) = \varphi(x_0, \ldots, x_q)g \qquad x_i \in U$$

Similar to the concept of presheaf on X with values in a category is the concept of sheaf on X with values in a category. We are interested only in sheaves of modules, and for this case the following formulation will do.

Let Γ be a presheaf of modules on X. If $\mathfrak{U} = \{U\}$ is a collection of open subsets of X, a *compatible* $\mathfrak{U}$ *family of* Γ is an indexed family $\{\gamma_U \in \Gamma(U)\}_{U \in \mathfrak{U}}$ such that

$$\gamma_U \mid U \cap U' = \gamma_{U'} \mid U \cap U' \qquad U, U' \in \mathfrak{U}$$

The presheaf Γ is said to be a *sheaf* if both the following conditions hold:

(*a*) Given a collection $\mathcal{U}$ of open subsets of X with $V = \bigcup_{U \in \mathcal{U}} U$ and given $\gamma \in \Gamma(V)$ such that $\gamma \mid U = 0$ for all $U \in \mathcal{U}$, then $\gamma = 0$.
(*b*) Given a collection $\mathcal{U}$ of open subsets of X with $V = \bigcup_{U \in \mathcal{U}} U$ and given a compatible $\mathcal{U}$ family $\{\gamma_U\}_{U \in \mathcal{U}}$, there is an element $\gamma \in \Gamma(V)$ such that $\gamma \mid U = \gamma_U$ for all $U \in \mathcal{U}$.

It follows from (*a*) that the element γ in (*b*) is unique.

We now associate to every presheaf Γ of modules another presheaf $\hat{\Gamma}$, called its completion, whose elements are compatible families of Γ. Given a collection of open sets $\mathcal{U} = \{U\}$, let $\Gamma(\mathcal{U})$ be the module of compatible $\mathcal{U}$ families of Γ. If $\mathcal{V}$ is another collection of open sets which refines $\mathcal{U}$, there is a homomorphism $\Gamma(\mathcal{U}) \to \Gamma(\mathcal{V})$ which assigns to a compatible $\mathcal{U}$ family $\{\gamma_U\}$ the compatible $\mathcal{V}$ family $\{\gamma'_V\}$ such that if $V \in \mathcal{V}$ is contained in $U \in \mathcal{U}$, then $\gamma'_V = \gamma_U \mid V$ (γ'_V is uniquely defined by this condition because of the compatibility of $\{\gamma_U\}$). As $\mathcal{U}$ varies over the family of open coverings of a fixed open set $W \subset X$, the collection $\{\Gamma(\mathcal{U})\}$ is a direct system of modules, and we define

$$\hat{\Gamma}(W) = \lim_{\rightarrow} \{\Gamma(\mathcal{U})\}$$

If $W' \subset W$ and $\mathcal{U}$ is an open covering of W, then $\mathcal{U}' = \{U \cap W' \mid U \in \mathcal{U}\}$ is an open covering of W' which refines $\mathcal{U}$. Hence there is a homomorphism $\Gamma(\mathcal{U}) \to \Gamma(\mathcal{U}')$ which defines (by passage to the limit) a homomorphism $\hat{\Gamma}(W) \to \hat{\Gamma}(W')$. A trivial verification shows that $\hat{\Gamma}$ is a presheaf [if $\mathcal{U} = \{\varnothing\}$, then trivially $\Gamma(\mathcal{U}) = 0$, and so $\hat{\Gamma}(\varnothing) = 0$]. There is a natural homomorphism $\alpha\colon \Gamma \to \hat{\Gamma}$ such that α assigns to $\gamma \in \Gamma(V)$ the element of $\hat{\Gamma}(V)$ represented by the compatible $\mathcal{V}$ family $\{\gamma\}$, where $\mathcal{V}$ consists solely of V. The presheaf $\hat{\Gamma}$ is called the *completion* of Γ. It depends only on the values $\Gamma(U)$ for small open sets $U \subset X$.

6 LEMMA *A presheaf Γ is a sheaf if and only if*

$$\alpha\colon \Gamma \approx \hat{\Gamma}$$

PROOF In fact, condition (*a*) above is satisfied if and only if α is a monomorphism. If condition (*b*) is satisfied, α is an epimorphism. If α is an isomorphism, then (*b*) is satisfied. ∎

7 EXAMPLE The constant presheaf G defined by a module G is not generally a sheaf [if U is a disconnected open set, $G(U) \not\approx \hat{G}(U)$].

8 EXAMPLE If C^* is the relative Alexander presheaf of (X,A) (with some coefficient module G), the kernel of $\alpha\colon C^* \to \hat{C}^*$ is C_0^* (the locally zero functions). To show that α satisfies condition (*b*) (and hence induces an isomorphism $\bar{C}^* \approx \hat{C}^*$), let $\varphi' \in \hat{C}^q(V, V \cap A)$ and assume φ' represented by a compatible $\mathcal{U}$ family $\{\varphi_U\}_{U \in \mathcal{U}}$, where $\mathcal{U}$ is an open covering of V. Then $\varphi_U\colon U^{q+1} \to G$ for $U \in \mathcal{U}$ is locally zero on $U \cap A$ and

$$\varphi_U \mid (U \cap U')^{q+1} = \varphi_{U'} \mid (U \cap U')^{q+1} \qquad U, U' \in \mathcal{U}$$

Therefore there is a function $\varphi: V^{q+1} \to G$ such that $\varphi \mid U^{q+1} = \varphi_U$ for $U \in \mathcal{U}$ and $\varphi(x_0, \ldots, x_q) = 0$ if $x_0, \ldots, x_q$ do not all lie in some element of $\mathcal{U}$. Then φ is locally zero on A, whence $\varphi \in C^q(V, V \cap A)$ and $\alpha(\varphi) = \varphi'$.

This example shows that, in general, $H^*(C^*) \neq H^*(\hat{C}^*)$, so it is not generally true that a presheaf of cochain complexes and its completion have isomorphic cohomology.

9 EXAMPLE If Δ^* is the relative singular presheaf of (X,A) (with some coefficient module G), the kernel of $\alpha: \Delta^* \to \hat{\Delta}^*$ is the subcomplex of locally zero cochains [that is, $c^* \in \operatorname{Hom}(\Delta_q(V),G)$ is in the kernel of α if and only if there is some open covering $\mathcal{U}$ of V such that c^* is zero on $\Delta_q(\mathcal{U}) \subset \Delta_q(V)$]. Also α satisfies condition (*b*) (as can be shown by an argument similar to that of example 8). If $\mathcal{U}$ is an open covering of X, it is clear that $\Delta^*(\mathcal{U}) = \bigcup \operatorname{Hom}(\Delta_*(\mathcal{U})/\Delta_*(\mathcal{U}'), G)$ where the union is over all open coverings $\mathcal{U}'$ of A that refine $\mathcal{U} \cap A$. As $\mathcal{U}$ and $\mathcal{U}'$ vary over open coverings, respectively, of X and A such that $\mathcal{U}'$ refines $\mathcal{U} \cap A$, there is an inverse system of chain complexes $\{\Delta_*(\mathcal{U})/\Delta_*(\mathcal{U}')\}$ and a direct system of cochain complexes

$$\{\operatorname{Hom}(\Delta_*(\mathcal{U})/\Delta_*(\mathcal{U}'), G)\}$$

Therefore there is an isomorphism

$$\lim_{\to} \{\operatorname{Hom}(\Delta_*(\mathcal{U})/\Delta_*(\mathcal{U}'), G)\} \approx \hat{\Delta}^*(\cdot, \cdot \cap A; G)(X)$$

It follows from theorem 4.4.14 that

$$\operatorname{Hom}(\Delta_*(X)/\Delta_*(A), G) \to \lim_{\to} \{\operatorname{Hom}(\Delta_*(\mathcal{U})/\Delta_*(\mathcal{U}'), G)\}$$

induces isomorphisms of the cohomology modules. Therefore α induces an isomorphism

$$H^*(\Delta^*(\cdot, \cdot \cap A; G)(X)) \approx H^*(\hat{\Delta}^*(\cdot, \cdot \cap A; G)(X))$$

10 EXAMPLE Let ξ be an n-sphere bundle with base space B and let R be fixed. A presheaf Γ on B is defined by $\Gamma(V) = H^{n+1}(p_\xi^{-1}(V), p_\xi^{-1}(V) \cap \dot{E}_\xi; R)$ for an open $V \subset B$. Γ is called the *orientation presheaf of* ξ *over* R. It can be verified that if B is connected, ξ is orientable over R if and only if $\hat{\Gamma}(B) \neq 0$.

11 EXAMPLE Let X be an n-manifold with boundary $\dot{X}$ and let R be fixed. Define a presheaf Γ on $X - \dot{X}$ such that $\Gamma(V) = H_n(X, X - V; R)$ for open $V \subset X - \dot{X}$. Γ is called the *fundamental presheaf of* X *over* R. It can be verified (using lemma 6.3.2) that $\hat{\Gamma}(X) \approx H_n{}^c(X,\dot{X}; R)$. By theorem 6.3.5, it follows that if X is connected, it is orientable over R if and only if $\hat{\Gamma}(X) \neq 0$.

There are cohomology modules of X with coefficients in sheaves,[1] and cohomology modules with coefficients in presheaves. For paracompact spaces

[1] See R. Godement, "Théorie des faisceaux," Hermann et Cie, Paris, 1958.

these theories are equivalent. We now define the Čech cohomology with coefficients in a presheaf of modules.

Let Γ be a presheaf of modules on a space X and let $\mathcal{U}$ be an open covering of X. For $q \geq 0$ define $C^q(\mathcal{U};\Gamma)$ to be the module of functions ψ which assign to an ordered $(q+1)$-tuple $U_0, U_1, \ldots, U_q$ of elements of $\mathcal{U}$ an element $\psi(U_0, \ldots, U_q) \in \Gamma(U_0 \cap \cdots \cap U_q)$. A coboundary operator

$$\delta: C^q(\mathcal{U};\Gamma) \to C^{q+1}(\mathcal{U};\Gamma)$$

is defined by

$$(\delta\psi)(U_0, \ldots, U_{q+1}) = \sum_{0 \leq i \leq q+1} (-1)^i \psi(U_0, \ldots, \hat{U}_i, \ldots, U_{q+1}) \mid (U_0 \cap \cdots \cap U_{q+1})$$

Then $\delta\delta = 0$ and $C^*(\mathcal{U};\Gamma) = \{C^q(\mathcal{U};\Gamma),\delta\}$ is a cochain complex. Its cohomology module is denoted by $H^*(\mathcal{U};\Gamma)$.

12 EXAMPLE It is an immediate consequence of the definition that $H^0(\mathcal{U};\Gamma) = \Gamma(\mathcal{U})$ (the module of compatible $\mathcal{U}$ families).

Let $\mathcal{V}$ be a refinement of $\mathcal{U}$ and let $\lambda: \mathcal{V} \to \mathcal{U}$ be a function such that $V \subset \lambda(V)$ for all $V \in \mathcal{V}$. There is a cochain map $\lambda^*: C^*(\mathcal{U};\Gamma) \to C^*(\mathcal{V};\Gamma)$ defined by

$$(\lambda^*\psi)(V_0, \ldots, V_q) = \psi(\lambda(V_0), \ldots, \lambda(V_q)) \mid (V_0 \cap \cdots \cap V_q)$$

If $\mu: \mathcal{V} \to \mathcal{U}$ is another function such that $V \subset \mu(V)$ for all $V \in \mathcal{V}$, a cochain homotopy $D: C^q(\mathcal{U};\Gamma) \to C^{q-1}(\mathcal{V};\Gamma)$ from λ^* to μ^* is defined by

$$(D\psi)(V_0, \ldots, V_{q-1}) = \sum_{0 \leq j \leq q-1} (-1)^j \psi(\lambda(V_0), \ldots, \lambda(V_j), \mu(V_j), \ldots, \mu(V_{q-1})) \mid (V_0 \cap \cdots \cap V_{q-1})$$

It follows that there is a well defined homomorphism

$$\lambda^*: H^*(\mathcal{U};\Gamma) \to H^*(\mathcal{V};\Gamma)$$

such that $\lambda^*\{\varphi\} = \{\lambda^*\varphi\}$ that is independent of the particular choice of λ. As $\mathcal{U}$ varies over open coverings of X, the collection $\{H^*(\mathcal{U};\Gamma)\}$ is a direct system, and the *Čech cohomology of X with coefficients* Γ is defined by

$$\check{H}^*(X;\Gamma) = \lim_{\to} \{H^*(\mathcal{U};\Gamma)\}$$

13 EXAMPLE For any presheaf Γ, $\check{H}^0(X;\Gamma) = \hat{\Gamma}(X)$.

14 EXAMPLE The *Čech cohomology of X with coefficients G*, denoted by $\check{H}^*(X;G)$, is defined to be the cohomology of X with coefficients the constant presheaf G.

We now establish some basic properties of the cohomology with coefficients in a presheaf.

15 THEOREM *There is a covariant functor from the category of short exact sequences of presheaves on X to the category of exact sequences which assigns to a short exact sequence* $0 \to \Gamma' \to \Gamma \to \Gamma'' \to 0$ *of presheaves on X an exact sequence*

$$\cdots \to \check{H}^q(X;\Gamma') \to \check{H}^q(X;\Gamma) \to \check{H}^q(X;\Gamma'') \to \check{H}^{q+1}(X;\Gamma') \to \cdots$$

PROOF For any open covering $\mathfrak{U}$ there is a short exact sequence of cochain complexes

$$0 \to C^*(\mathfrak{U};\Gamma') \to C^*(\mathfrak{U};\Gamma) \to C^*(\mathfrak{U};\Gamma'') \to 0$$

This yields an exact cohomology sequence, and the result follows from this on passing to the direct limit. ∎

Given a short exact sequence of modules

$$0 \to G' \to G \to G'' \to 0$$

the corresponding constant presheaves on X constitute a short exact sequence of presheaves. The corresponding exact cohomology sequence of Čech cohomology modules given by theorem 15 is an analogue for Čech theory of the exact sequence of theorem 5.4.11.

Given a presheaf Γ on X and given a subspace $A \subset X$, define a presheaf Γ_A on X by

$$\Gamma_A(U) = \begin{cases} \Gamma(U) & U \cap A \neq \varnothing \\ 0 & U \cap A = \varnothing \end{cases}$$

Also define a presheaf Γ^A on X by

$$\Gamma^A(U) = \begin{cases} \Gamma(U) & U \cap A = \varnothing \\ 0 & U \cap A \neq \varnothing \end{cases}$$

Then Γ^A is a sub-presheaf of Γ, and there is a short exact sequence of presheaves

$$0 \to \Gamma^A \to \Gamma \to \Gamma_A \to 0$$

The corresponding exact cohomology sequence given by theorem 15 is an exact Čech cohomology sequence of the pair (X,A) with coefficients Γ when we define $\check{H}^q(A;\Gamma) = \check{H}^q(X,\Gamma_A)$ and $\check{H}^q(X,A;\Gamma) = \check{H}^q(X;\Gamma^A)$. Thus the exact sequence of theorem 15 gives rise to exact sequences corresponding to a change of coefficients or to a change of space.

A presheaf Γ of modules on X is said to be *locally zero* if, given $\gamma \in \Gamma(V)$, there is an open covering $\mathfrak{U}$ of V such that $\gamma \mid U = 0$ for all $U \in \mathfrak{U}$. This is so if and only if the completion $\hat{\Gamma}$ of Γ is the zero presheaf and is equivalent to the condition that for all $x \in X$, $\lim_{\rightarrow} \{\Gamma(U)\} = 0$ as U varies over open neighborhoods of x.

16 THEOREM *If X is a paracompact Hausdorff space and* Γ *is a locally zero presheaf on X, then* $\check{H}^*(X;\Gamma) = 0$.

PROOF Let $\mathfrak{U}$ be a locally finite open covering of X and φ a q-cochain of

$C^*(\mathscr{U};\Gamma)$. Let $\mathscr{W}$ be a locally finite open star refinement of $\mathscr{U}$. For $x \in X$, because Γ is locally zero, there is an open neighborhood V_x contained in some element of $\mathscr{W}$ such that $x \in U_0 \cap \cdots \cap U_q$ with $U_0, \ldots, U_q \in \mathscr{U}$ implies that $V_x \subset U_0 \cap \cdots \cap U_q$ and $\varphi(U_0, \ldots, U_q) \mid V_x = 0$ (only a finite number of conditions because $\mathscr{U}$ is locally finite). Let $\mathscr{V} = \{V_x\}_{x \in X}$ and define $\lambda: \mathscr{V} \to \mathscr{U}$ so that for each $x \in X$ there is $W_x \in \mathscr{W}$ with $V_x \subset W_x \subset W_x^* \subset \lambda(V_x)$. Then if $V_{x_0} \cap \cdots \cap V_{x_q} \neq 0$, $V_{x_0} \subset W_{x_j}$ for each j so that $V_{x_0} \subset \lambda(V_{x_j})$ for each j. Therefore, $\phi(\lambda(V_{x_0}), \ldots, \lambda(V_{x_q})) \mid V_{x_0} = 0$, so $\lambda^*\phi = 0$ in $C^*(\mathscr{V};\Gamma)$. Therefore, $\check{H}^q(X;\Gamma) = 0$ for all q.

A homomorphism $\alpha: \Gamma \to \Gamma'$ between presheaves on X is called a *local isomorphism* if ker α and coker α are both locally zero. This is equivalent to the condition that for all $x \in X$, α induces an isomorphism

$$\lim_{\to} \{\Gamma(U)\} \approx \lim_{\to} \{\Gamma'(U)\}$$

where U varies over open neighborhoods of x. There are short exact sequences of presheaves

$$0 \to \ker \alpha \to \Gamma \xrightarrow{\alpha'} \operatorname{im} \alpha \to 0$$

$$0 \to \operatorname{im} \alpha \xrightarrow{\alpha''} \Gamma' \to \operatorname{coker} \alpha \to 0$$

with $\alpha = \alpha''\alpha'$. Combining theorems 15 and 16, we obtain the following result.

17 COROLLARY *If $\alpha: \Gamma \to \Gamma'$ is a local isomorphism of presheaves on a paracompact Hausdorff space X, then*

$$\alpha_*: \check{H}^*(X;\Gamma) \approx \check{H}^*(X;\Gamma') \quad ■$$

18 COROLLARY *If X is a paracompact Hausdorff space, the natural homomorphism $\alpha: \Gamma \to \hat{\Gamma}$ induces isomorphisms*

$$\alpha_*: \check{H}^*(X;\Gamma) \approx \check{H}^*(X;\hat{\Gamma})$$

PROOF It suffices to prove that $\alpha: \Gamma \to \hat{\Gamma}$ is a local isomorphism. Let $\gamma \in (\ker \alpha)(V)$. Then $\gamma \in \Gamma(V)$, and there is an open covering $\mathfrak{U}$ of V such that $\gamma \mid U = 0$ for all $U \in \mathfrak{U}$. Hence ker α is locally zero.

If $\gamma' \in (\operatorname{coker} \alpha)(V)$, there is an open covering $\mathfrak{U}$ of V and a compatible $\mathfrak{U}$ family $\{\gamma_U\}$ which represents γ'. For each $U \in \mathfrak{U}$, $\gamma' \mid U$ is represented by $\gamma_U \in \alpha(\Gamma(U))$. Therefore $\gamma' \mid U = 0$, and coker α is locally zero. ■

8 FINE PRESHEAVES

In this section we shall introduce the concept of fine presheaf and show that the positive dimensional cohomology of a paracompact space with coefficients in a fine presheaf is zero. This leads to uniqueness theorems for cohomology of cochain complexes of fine presheaves on a paracompact space, which we apply to compare the Alexander and Čech cohomology. Further applications will be given in the next section.

A presheaf Γ on X is said to be *fine* if, given any locally finite open covering $\mathcal{U}$ of X, there exists an indexed family $\{e_U\}_{U\in\mathcal{U}}$ of endomorphisms of Γ such that (for every open set V in X)

(*a*) For $\gamma \in \Gamma(V)$, $e_U(\gamma) \mid (V - \bar{U}) = 0$.
(*b*) If V meets only finitely many elements of $\{\bar{U}\}$, then for $\gamma \in \Gamma(V)$, $\gamma = \sum_{U\in\mathcal{U}} e_U(\gamma)$.

Note that the sum in condition (*b*) is finite because, by (*a*), $e_U(\gamma) = 0$ if $\bar{U} \cap V = \varnothing$.

1 EXAMPLE The relative Alexander presheaf of (X,A) of degree q with coefficients G is fine. In fact, if $\mathcal{U}$ is a locally finite open covering of X, for each $x \in X$ choose an element $U_x \in \mathcal{U}$ containing x and for $\varphi \in C^q(V, V \cap A; G)$ define $e_U\varphi \in C^q(V, V \cap A; G)$ by

$$(e_U\varphi)(x_0, \ldots ,x_q) = \begin{cases} \varphi(x_0, \ldots ,x_q) & U = U_{x_0} \\ 0 & U \neq U_{x_0} \end{cases}$$

If $V' \subset V$, there is a commutative square

$$\begin{array}{ccc} C^q(V,V \cap A; G) & \xrightarrow{e_U} & C^q(V, V \cap A; G) \\ \downarrow & & \downarrow \\ C^q(V', V' \cap A; G) & \xrightarrow{e_U} & C^q(V', V' \cap A; G) \end{array}$$

showing that e_U is an endomorphism of C^q. If

$$(x_0, \ldots , x_q) \in (V^{q+1} - \bar{U}^{q+1}) \subset (V^{q+1} - U^{q+1}),$$

then $U_{x_0} \neq U$ and $(e_U\varphi)(x_0, \ldots ,x_q) = 0$. Hence $e_U\varphi \mid V - \bar{U} = 0$, and condition (*a*) is satisfied. To show that (*b*) is also satisfied, observe that, given $x_0, \ldots , x_q$, there is a unique U, namely U_{x_0}, such that $(e_U\varphi)(x_0, \ldots ,x_q) \neq 0$. Then

$$(\textstyle\sum e_U\varphi)(x_0, \ldots ,x_q) = (e_{U_{x_0}}\varphi)(x_0, \ldots ,x_q) = \varphi(x_0, \ldots ,x_q)$$

It should be noted that e_U does not commute with the coboundary operator in $C^*(V, V \cap A; G)$. Therefore e_U is not an endomorphism of the Alexander presheaf $C^*(\cdot, \cdot \cap A; G)$ of cochain complexes.

2 EXAMPLE The relative singular presheaf of (X,A) of degree q with coefficients G is also fine. If $\mathcal{U}$ is a locally finite open covering of X and U_x is chosen so that $x \in U_x \in \mathcal{U}$, then

$$e_U\colon \operatorname{Hom}(\Delta_q(V)/\Delta_q(V \cap A), G) \to \operatorname{Hom}(\Delta_q(V)/\Delta_q(V \cap A), G)$$

is defined by

$$(e_Uc^*)(\sigma) = \begin{cases} c^*(\sigma) & U = U_{\sigma(p_0)} \\ 0 & U \neq U_{\sigma(p_0)} \end{cases}$$

Then the family $\{e_U\}_{U\in\mathcal{U}}$ satisfies conditions (*a*) and (*b*) of the definition of fineness [but e_U is not an endomorphism of $\Delta^*(\cdot, \cdot \cap A; G)$ so $\Delta^*(\cdot, \cdot \cap A; G)$ has not been shown to be a fine presheaf of cochain complexes].

Given a presheaf Γ on X and a continuous map $f: X \to Y$, there is a presheaf $f_* \Gamma$ on Y defined by $(f_* \Gamma)(V) = \Gamma(f^{-1}V)$ for an open $V \subset Y$. Clearly, f defines a covariant functor from the category of presheaves of any type on X to the category of presheaves of the same type on Y. Some of the nice properties of fine presheaves are made explicit in the following result.

3 THEOREM *Let Γ be a fine presheaf of modules on X.*

(a) For any presheaf Γ' of modules on X, $\Gamma \otimes \Gamma'$ is fine.

(b) If $f: X \to Y$ is continuous, $f_* \Gamma$ is fine on Y.

(c) $\hat{\Gamma}$ is a fine presheaf on X.

PROOF For (a), observe that if $\mathfrak{U}$ is a locally finite open covering of X and $\{e_U\}_{U \in \mathfrak{U}}$ are the corresponding endomorphisms of Γ, then $\{e_U \otimes 1\}_{U \in \mathfrak{U}}$ is a family of endomorphisms of $\Gamma \otimes \Gamma'$, showing that $\Gamma \otimes \Gamma'$ is fine.

For (b), observe that if $\mathfrak{U}$ is a locally finite open covering of Y, then $f^{-1}\mathfrak{U} = \{f^{-1}U \mid U \in \mathfrak{U}\}$ is a locally finite open covering of X. If $\{e_U\}_{U \in \mathfrak{U}}$ is a family of endomorphisms of Γ corresponding to the covering $f^{-1}\mathfrak{U}$, they induce endomorphisms of $f_* \Gamma$, showing that $f_* \Gamma$ is fine.

(c) follows easily on observing that any endomorphism of Γ induces an endomorphism of $\hat{\Gamma}$. ■

Given an open covering $\mathfrak{U}$ of a space X, a *shrinking* of $\mathfrak{U}$ is an open covering $\mathfrak{V}$ of X in one-to-one correspondence with $\mathfrak{U}$ such that if $U \in \mathfrak{U}$ corresponds to $V_U \in \mathfrak{V}$, then $\bar{V}_U \subset U$. Any locally finite open covering of a normal Hausdorff space has shrinkings. Any shrinking of a locally finite open covering is clearly locally finite.

The following theorem is the main result on fine presheaves.

4 THEOREM *If Γ is a fine presheaf on a paracompact Hausdorff space X, then $\check{H}^q(X;\Gamma) = 0$ for $q > 0$.*

PROOF Let $\mathfrak{U} = \{U\}$ be a locally finite open covering of X and let $\mathfrak{U}' = \{U'\}$ be a shrinking of $\mathfrak{U}$. Let $\{e_U\}_{U \in \mathfrak{U}}$ be fineness endomorphisms of Γ corresponding to the covering $\mathfrak{U}'$ (but indexed by the covering $\mathfrak{U}$). Let $\mathfrak{V} = \{V\}$ be an open refinement of $\mathfrak{U}$ covering X such that each $V \in \mathfrak{V}$ meets only a finite number of elements of $\mathfrak{U}$ and for any $U \in \mathfrak{U}$ either $V \subset U$ or $V \subset X - \bar{U}'$. Let $\lambda: \mathfrak{V} \to \mathfrak{U}$ be a function such that $V \subset \lambda(V)$ for all $V \in \mathfrak{V}$.

Since each e_U is an endomorphism of Γ, e_U induces a cochain map, denoted by $e_U: C^*(\mathfrak{U};\Gamma) \to C^*(\mathfrak{U};\Gamma)$ such that for $\psi \in C^q(\mathfrak{U};\Gamma)$ and $U_0, \ldots, U_q \in \mathfrak{U}$

$$(e_U\psi)(U_0, \ldots, U_q) = e_U(\psi(U_0, \ldots, U_q))$$

Then e_U acts similarly as a cochain map on $C^*(\mathfrak{V};\Gamma)$ and commutes with the cochain map $\lambda^*: C^*(\mathfrak{U};\Gamma) \to C^*(\mathfrak{V};\Gamma)$.

Let $q > 0$ and $\psi \in C^q(\mathfrak{U};\Gamma)$ be a cocycle. Define $\psi_U \in C^q(\mathfrak{V};\Gamma)$ by

$\psi_U = e_U(\lambda^* \psi)$. Then ψ_U is a cocycle for each $U \in \mathcal{U}$, and if $V_0, \ldots, V_q \in \mathcal{V}$, then $\psi_U(V_0, \ldots, V_q) = 0$, except for a finite number of $U \in \mathcal{U}$. Therefore $\Sigma\, \psi_U$ exists, and $\Sigma\, \psi_U = \lambda^* \psi$.

Define $\psi'_U \in C^{q-1}(\mathcal{V};\Gamma)$ by

$$\psi'_U(V_0, \ldots, V_{q-1}) = \begin{cases} e_U(\psi(U,\lambda(V_0), \ldots, \lambda(V_{q-1})) \mid (V_0 \cap \cdots \cap V_{q-1})) & V_0 \cap \cdots \cap V_{q-1} \subset U \\ 0 & V_0 \cap \cdots \cap V_{q-1} \subset X - \bar{U}' \end{cases}$$

Then $\delta\psi'_U = \psi_U$ for all U, and because $\Sigma\, \psi'_U$ can be formed [for given $V_0, \ldots, V_{q-1}$, $\psi'_U(V_0, \ldots, V_{q-1}) = 0$, except for a finite number of $U \in \mathcal{U}$], we see that

$$\lambda^* \psi = \Sigma\, \psi_U = \delta(\Sigma\, \psi'_U)$$

Therefore $\lambda^* \psi$ is a coboundary, and $\check{H}^q(X;\Gamma) = 0$. ∎

Our next results are technical lemmas about cochain complexes of presheaves. If Γ^* is a cochain complex of presheaves of modules on X, we use Z^q and B^{q+1} to denote the kernel and image, respectively, of $\delta: \Gamma^q \to \Gamma^{q+1}$ and H^q to denote Z^q/B^q, all of these being presheaves of modules on X. (Note that a fine presheaf of cochain complexes is a cochain complex of fine presheaves, but the converse is not generally true.)

5 LEMMA *Let Γ^* be a cochain complex of presheaves of modules on X. For every q there is an exact sequence, functorial in Γ^*,*

$$0 \to \ker(\check{H}^0(X;B^q) \to \check{H}^1(X;Z^{q-1})) \to \check{H}^0(X;Z^q) \to H^q(\hat{\Gamma}^*(X)) \to 0$$

PROOF By example 6.7.13, $\hat{\Gamma}^q(X) = \check{H}^0(X;\Gamma^q)$. From the short exact sequence of presheaves

$$0 \to Z^q \to \Gamma^q \to B^{q+1} \to 0$$

there follows, by theorem 6.7.15, an exact sequence

$$0 \to \check{H}^0(X;Z^q) \to \check{H}^0(X;\Gamma^q) \to \check{H}^0(X;B^{q+1}) \to \check{H}^1(X;Z^q)$$

Because $B^{q+1} \subset \Gamma^{q+1}$, it follows from a similar exactness property that $\check{H}^0(X;B^{q+1}) \subset \check{H}^0(X;\Gamma^{q+1})$. Combining these, we see that

$$\begin{aligned}\check{H}^0(X;Z^q) &\approx \ker[\check{H}^0(X;\Gamma^q) \to \check{H}^0(X;B^{q+1})] \\ &\approx \ker[\check{H}^0(X;\Gamma^q) \to \check{H}^0(X;\Gamma^{q+1})]\end{aligned}$$

and also that

$$\operatorname{im}[\check{H}^0(X;\Gamma^q) \to \check{H}^0(X;\Gamma^{q+1})] \approx \ker[\check{H}^0(X;B^{q+1}) \to \check{H}^1(X;Z^q)]$$

Since

$$H^q(\hat{\Gamma}^*(X)) = \ker[\check{H}^0(X;\Gamma^q) \to \check{H}^0(X;\Gamma^{q+1})]/\operatorname{im}[\check{H}^0(X;\Gamma^{q-1}) \to \check{H}^0(X;\Gamma^q)]$$

the result follows. ∎

6 COROLLARY *Let Γ^* be a cochain complex of presheaves of modules on a paracompact Hausdorff space X. For any q there is a short exact sequence, functorial in Γ^*,*

$$0 \to \operatorname{im} [\check{H}^0(X;B^q) \to \check{H}^1(X;Z^{q-1})] \to H^q(\hat{\Gamma}^*(X)) \to \ker [\check{H}^0(X;H^q) \to \check{H}^1(X;B^q)] \to 0$$

If Γ^{q-1} is fine, this becomes

$$0 \to \check{H}^1(X;Z^{q-1}) \to H^q(\hat{\Gamma}^*(X)) \to \ker [\check{H}^0(X;H^q) \to \check{H}^1(X;B^q)] \to 0$$

PROOF From the short exact sequence of presheaves

$$0 \to B^q \to Z^q \to H^q \to 0$$

it follows, by theorem 6.7.15, that there is an isomorphism

$$\check{H}^0(X;Z^q)/\check{H}^0(X;B^q) \approx \ker [\check{H}^0(X;H^q) \to \check{H}^1(X;B^q)]$$

From lemma 5, there is an isomorphism

$$\check{H}^0(X;Z^q)/\ker [\check{H}^0(X;B^q) \to \check{H}^1(X;Z^{q-1})] \approx H^q(\hat{\Gamma}^*(X))$$

It follows that $H^q(\hat{\Gamma}^*(X))$ maps epimorphically to $\ker [\check{H}^0(X;H^q) \to \check{H}^1(X;B^q)]$ with kernel isomorphic to

$$\check{H}^0(X;B^q)/\ker [\check{H}^0(X;B^q) \to \check{H}^1(X;Z^{q-1})] \approx \operatorname{im} [\check{H}^0(X;B^q) \to \check{H}^1(X;Z^{q-1})]$$

This gives the first short exact sequence. For the second, there is a short exact sequence of presheaves

$$0 \to Z^{q-1} \to \Gamma^{q-1} \to B^q \to 0$$

and if Γ^{q-1} is fine, it follows from theorems 6.7.15 and 4 that

$$\operatorname{im} [\check{H}^0(X;B^q) \to \check{H}^1(X;Z^{q-1})] = \check{H}^1(X;Z^{q-1}) \quad \blacksquare$$

7 THEOREM *Let Γ^* be a nonnegative cochain complex of fine presheaves of modules on a paracompact Hausdorff space X. Assume that for some integers $0 \le m < n$, $H^q(\Gamma^*)$ is locally zero for $q < m$ and $m < q < n$. Then there are functorial isomorphisms*

$$\check{H}^{q-m}(X;H^m(\Gamma^*)) \approx H^q(\hat{\Gamma}^*(X)) \qquad q < n$$

and a functorial monomorphism

$$\check{H}^{n-m}(X;H^m(\Gamma^*)) \to H^n(\hat{\Gamma}^*(X))$$

PROOF For each q there is a short exact sequence of presheaves

$$0 \to Z^q \to \Gamma^q \to B^{q+1} \to 0$$

Because Γ^q is fine, it follows from theorems 6.7.15 and 4 that

(a) $$\check{H}^p(X;B^{q+1}) \approx \check{H}^{p+1}(X;Z^q) \qquad p \ge 1$$

For each q there is also a short exact sequence of presheaves

$$0 \to B^q \to Z^q \to H^q \to 0$$

Because H^q is locally zero for $q < m$ and $m < q < n$, it follows from theorems 6.7.15 and 6.7.16 that

$$\text{(b)} \qquad \check{H}^p(X;B^q) \approx \check{H}^p(X;Z^q) \qquad q < m \text{ or } m < q < n, \text{ all } p$$

Since B^0 is the zero presheaf, it follows by induction on q from equations (b) and (a) that for $q < m$

$$\text{(c)} \qquad \check{H}^p(X;Z^q) = 0 = \check{H}^p(X;B^{q+1}) \qquad p \geq 1$$

From this and corollary 6, it follows that $H^i(\hat{\Gamma}^*(X)) = 0$ for $i < m$. Hence the theorem holds for $q < m$ (both modules being trivial). For $q = m$ we have [by corollary 6 and equation (c)]

$$H^m(\hat{\Gamma}^*(X)) \approx \check{H}^0(X;H^m)$$

and the theorem holds in this case too.

To obtain the result for $m < q \leq n$, note that, by equation (c), $\check{H}^p(X;B^m) = 0$, if $p \geq 1$. From the short exact sequence of presheaves

$$0 \to B^m \to Z^m \to H^m \to 0$$

it follows that

$$\check{H}^p(X;Z^m) \approx \check{H}^p(X;H^m) \qquad p \geq 1$$

For $m < i < n$ it follows from corollary 6 that

$$\check{H}^1(X;Z^{i-1}) \approx H^i(\hat{\Gamma}^*(X))$$

and for $i = n$ there is a monomorphism

$$\check{H}^1(X;Z^{n-1}) \to H^n(\hat{\Gamma}^*(X))$$

Using equations (b) and (a), we see that for $m < i \leq n$

$$\check{H}^1(X;Z^{i-1}) \approx \check{H}^1(X;B^{i-1}) \approx \check{H}^2(X;Z^{i-2}) \approx \cdots \approx \check{H}^{i-m}(X;Z^m) \approx \check{H}^{i-m}(X;H^m)$$

and this gives the result for $m < q \leq n$. ∎

This last result has as an immediate consequence the following isomorphism between the Čech and Alexander cohomologies with coefficients G.

8 COROLLARY *For any paracompact Hausdorff space and module G there is a functorial isomorphism*

$$\check{H}^*(X;G) \approx \bar{H}^*(X;G)$$

of the Čech and Alexander cohomology modules.

PROOF Let C^* be the Alexander presheaf of X with coefficients G. Since C^q is fine for all q (by example 1), this is a nonnegative cochain complex of fine sheaves. Furthermore, for any nonempty U, by lemma 6.4.1,

$$H^q(C^*(U;G)) \approx \begin{cases} 0 & q \neq 0 \\ G & q = 0 \end{cases}$$

Therefore $H^q(C^*)$ is locally zero for $q > 0$ and $H^0(C^*)$ is isomorphic to the constant presheaf G. The hypotheses of theorem 7 are satisfied with $m = 0$ and any n, and there is a functorial isomorphism

$$\check{H}^q(X;G) \approx H^q(\hat{C}^*)$$

for all q. As pointed out in example 6.7.8, there is a canonical isomorphism $\bar{C}^* \approx \hat{C}^*$, and so $\bar{H}^q(X;G) \approx H^q(\hat{C}^*)$. Combining these isomorphisms yields the result. ∎

The last result is also true without the assumption of paracompactness (see exercise 6.D.3). The next result is the main uniqueness theorem of the cohomology of presheaves.

9 THEOREM *Let X be a paracompact Hausdorff space and let τ: $\Gamma^* \to \Gamma'^*$ be a cochain map between nonnegative cochain complexes of fine presheaves of modules on X. Assume that for some $n \geq 0$, $\tau_*: H^q(\Gamma^*) \to H^q(\Gamma'^*)$ is a local isomorphism for $q < n$ and a local monomorphism for $q = n$. Then the induced map*

$$\hat{\tau}_*: H^q(\hat{\Gamma}^*(X)) \to H^q(\hat{\Gamma}'^*(X))$$

is an isomorphism for $q < n$ and a monomorphism for $q = n$.

PROOF Let Γ_τ^* be the mapping cone of τ (defined for cochain complexes analogous to the definition in Sec. 4.2 for chain complexes). Then $\Gamma_\tau{}^q = \Gamma^{q+1} \oplus \Gamma'^q$, and for $\gamma \in \Gamma^{q+1}(U)$ and $\gamma' \in \Gamma'^q(U)$, $\delta(\gamma,\gamma') = (-\delta(\gamma), \tau(\gamma) + \delta(\gamma'))$. Γ_τ^* is a nonnegative cochain complex of fine presheaves on X, and for any open $U \subset X$ there is an exact sequence

$$\cdots \to H^q(\Gamma'^*(U)) \to H^q(\Gamma_\tau^*(U)) \to H^{q+1}(\Gamma^*(U)) \xrightarrow{\tau_*} H^{q+1}(\Gamma'^*(U)) \to \cdots$$

Taking the direct limit as U varies over open neighborhoods of $x \in X$, we see that $\tau_*: H^q(\Gamma^*) \to H^q(\Gamma'^*)$ is a local isomorphism for $q < n$ and a local monomorphism for $q = n$ if and only if $H^q(\Gamma_\tau^*)$ is locally zero for $q < n$. By theorem 7, it follows that $H^q(\hat{\Gamma}_\tau^*(X)) = 0$ for $q < n$ (if $n = 0$ this is trivially true, and if $n > 0$ it follows from theorem 7 with $m = 0$).

It is obvious that $\hat{\Gamma}_\tau^*$ is the mapping cone $\Gamma_{\hat{\tau}}^*$ of the induced map $\hat{\tau}$: $\hat{\Gamma}^* \to \hat{\Gamma}'^*$ between the completions. Therefore

$$\cdots \to H^q(\hat{\Gamma}'^*(X)) \to H^q(\hat{\Gamma}_\tau^*(X)) \to H^{q+1}(\hat{\Gamma}^*(X)) \xrightarrow{\hat{\tau}_*} H^{q+1}(\hat{\Gamma}'^*(X)) \to \cdots$$

Since $H^q(\hat{\Gamma}_\tau^*(X))$ was shown to be zero for $q < n$ in the first paragraph above, the result follows from the exactness of this sequence. ∎

For compact spaces there is the following *universal-coefficient formula for Čech cohomology.*

10 THEOREM *Let X be a compact Hausdorff space. On the product category* *of presheaves* Γ *on* X *consisting of torsion free* R *modules and the category of* R *modules* G *there is a functorial short exact sequence*

$$0 \to \check{H}^q(X;\Gamma) \otimes G \to \check{H}^q(X; \Gamma \otimes G) \to \check{H}^{q+1}(X;\Gamma) * G \to 0$$

PROOF Let $\mathcal{U}$ be a finite open covering of X. The cochain map

$$\tau: C^*(\mathcal{U};\Gamma) \otimes G \to C^*(\mathcal{U}; \Gamma \otimes G)$$

defined by $\tau(\psi \otimes g)(U_0, \ldots ,U_q) = \psi(U_0, \ldots ,U_q) \otimes g$ is an isomorphism (this is a consequence of the finiteness of $\mathcal{U}$ analogous to lemma 5.5.6). From the universal-coefficient formula for cochain complexes (theorem 5.4.1), there is a functorial short exact sequence

$$0 \to H^q(\mathcal{U};\Gamma) \otimes G \to H^q(\mathcal{U}; \Gamma \otimes G) \to H^{q+1}(\mathcal{U};\Gamma) * G \to 0$$

The result follows by taking direct limits over the cofinal family of finite open coverings of X (because the tensor product and the torsion product both commute with direct limits). ■

From corollary 8, this gives a universal-coefficient formula for Alexander cohomology of compact spaces. The following theorem generalizes this result to compact pairs and includes the statement that the short exact sequence in question is split.

11 THEOREM *On the product category of pairs* (X,A), *where* A *is a closed subset of a compact Hausdorff space* X, *and the category of* R *modules* G, *there is a functorial short exact sequence*

$$0 \to \bar{H}^q(X,A; R) \otimes G \to \bar{H}^q(X,A; G) \to \bar{H}^{q+1}(X,A; R) * G \to 0$$

and this sequence is split.

PROOF Let $\tau: C^*(\cdot, \cdot \cap A; R) \otimes G \to C^*(\cdot, \cdot \cap A; G)$ be the homomorphism of presheaves defined as in example 6.7.5 [that is, $\tau(\varphi \otimes g)(x_0, \ldots ,x_q) = \varphi(x_0, \ldots ,x_q)g$]. Both $C^*(\cdot, \cdot \cap A; R) \otimes G$ and $C^*(\cdot, \cdot \cap A; G)$ are nonnegative cochain complexes of fine presheaves. First we prove that

$$\tau_*: H^*(C^*(\cdot, \cdot \cap A; R) \otimes G) \to H^*(C^*(\cdot, \cdot \cap A; G))$$

is a local isomorphism. If $U \subset X - A$, $C^*(U, U \cap A; R) = C^*(U;R)$, and $C^*(U, U \cap A; G) = C^*(U;G)$, it follows from lemma 6.4.1 and theorem 5.4.1 that

$$\tau_*: H^*(C^*(U, U \cap A; R) \otimes G) \approx H^*(C^*(U, U \cap A; G))$$

Since A is closed in X, for any $x \in X - A$, τ_* is an isomorphism of $\lim_{\to} \{H^*(C^*(U, U \cap A; R) \otimes G)\}$ onto $\lim_{\to} \{H^*(C^*(U, U \cap A; G))\}$, both limits as U varies over open neighborhoods of x in X.

For any U intersecting A there is a commutative diagram with exact rows

$$\begin{array}{ccccccccc} 0 \to & C^*(U, U \cap A; R) \otimes G & \to & C^*(U;R) \otimes G & \to & \bar{C}^*(U \cap A; R) \otimes G & \to 0 \\ & \downarrow & & \downarrow & & \downarrow & \\ 0 \to & C^*(U, U \cap A; G) & \to & C^*(U;G) & \to & \bar{C}^*(U \cap A; G) & \to 0 \end{array}$$

By lemma 6.4.1, the middle cochain complexes have trivial reduced modules. Therefore there is a commutative square

$$\begin{array}{ccc} \tilde{H}^q(\bar{C}^*(U \cap A; R) \otimes G) & \xrightarrow{\approx} & H^{q+1}(C^*(U, U \cap A; R) \otimes G) \\ \downarrow & & \downarrow \\ \tilde{H}^q(\bar{C}^*(U \cap A; G)) & \xrightarrow{\approx} & H^{q+1}(C^*(U, U \cap A; G)) \end{array}$$

To complete the proof that τ_* is a local isomorphism, therefore, we need only prove that for $x \in A$

$$\lim_{\to} \{\tilde{H}^q(\bar{C}^*(U \cap A; R) \otimes G)\} \approx \lim_{\to} \{\tilde{H}^q(\bar{C}^*(U \cap A; G))\}$$

as U varies over neighborhoods of x in X. This is equivalent to the condition that

$$\tilde{H}^q(\lim_{\to} \{\bar{C}^*(U \cap A; R)\}) \otimes G \approx \tilde{H}^q(\lim_{\to} \{\bar{C}^*(U \cap A; G)\})$$

where $U \cap A$ varies over neighborhoods of x in A. This is trivially true because both sides are zero for all q (this follows from the tautness property of x in the paracompact space A but can be proved without assuming the paracompactness of A, because any one-point subspace is taut in any space with respect to Alexander cohomology).

We have verified that τ satisfies the hypotheses of theorem 9 for all n. Therefore τ induces an isomorphism

$$\bar{\tau}_*: H^*([\hat{C}^*(\cdot, \cdot \cap A; R) \otimes G](X)) \approx H^*(\hat{C}^*(\cdot, \cdot \cap A; G)(X))$$

By example 6.7.8, the right-hand side is isomorphic to $H^*(\bar{C}^*(X,A; G))$. By example 6.7.13, the left hand side is the qth cohomology module of $\check{H}^0(X;C^*(\cdot, \cdot \cap A; R) \otimes G)$. By theorem 10 and the fineness of $C^*(\cdot, \cdot \cap A; R) \otimes G$, this is isomorphic to

$$\begin{aligned} \check{H}^0(X;C^*(\cdot, \cdot \cap A; R)) \otimes G &\approx (\hat{C}^*(\cdot, \cdot \cap A; R)(X)) \otimes G \\ &\approx \bar{C}^*(X,A; R) \otimes G \end{aligned}$$

It follows that the map

$$\bar{\tau}: \bar{C}^*(X,A; R) \otimes G \to \bar{C}^*(X,A; G)$$

induced by τ induces an isomorphism of cohomology. The result now follows from the universal-coefficient formula for cochain complexes (theorem 5.4.1). ■

This implies the following *universal-coefficient formula for Alexander cohomology with compact supports.*

12 COROLLARY *On the product category of pairs (X,A), where A is a closed subset of a locally compact Hausdorff space X, and the category of R modules G, there is a functorial short exact sequence*

$$0 \to \bar{H}_c^q(X,A;R) \otimes G \to \bar{H}_c^q(X,A;G) \to \bar{H}_c^{q+1}(X,A;R) * G \to 0$$

and this sequence is split.

PROOF Let N be a closed cobounded neighborhood of A in X. There is a commutative square of cochain maps

$$\begin{array}{ccc} \bar{C}^*(X,N;R) \otimes G & \to & \bar{C}^*(X,N;G) \\ \downarrow & & \downarrow \\ \bar{C}^*(\overline{X-N}, \overline{X-N} \cap N; R) \otimes G & \to & \bar{C}^*(\overline{X-N}, \overline{X-N} \cap N; G) \end{array}$$

in which, by theorem 6.6.5, each vertical map induces an isomorphism of cohomology. By theorem 11, the bottom horizontal map induces an isomorphism of cohomology. Therefore the top horizontal map also induces an isomorphism of cohomology.

There is also a commutative square (in which the limit is over closed cobounded neighborhoods N of A in X)

$$\begin{array}{ccc} \varinjlim \{\bar{C}^*(X,N;R)\} \otimes G & \to & \varinjlim \{\bar{C}^*(X,N;G)\} \\ \downarrow & & \downarrow \\ \bar{C}_c^*(X,A;R) \otimes G & \to & \bar{C}_c^*(X,A;G) \end{array}$$

It follows from the first paragraph above that the top horizontal map induces an isomorphism of cohomology. Since the closed cobounded neighborhoods of A in X are cofinal in the family of all cobounded neighborhoods of A in X, it follows from theorem 6.6.15 that each vertical map induces an isomorphism in cohomology. Therefore the bottom horizontal map induces an isomorphism in cohomology. The result follows from this and theorem 5.4.1. ∎

9 APPLICATIONS OF THE COHOMOLOGY OF PRESHEAVES

This section is devoted to two main applications of the theory developed in the last two sections. One is the study of the relation between Alexander and singular cohomology. We shall prove that in a homologically locally connected space (for example, a manifold) the two are isomorphic. The other application is to a study of the relation between the Alexander cohomology of two spaces connected by a continuous map. We conclude with a proof of the Vietoris-Begle mapping theorem.

Let (X,A) be a pair and let G be an R module. Recall the homomorphism

$$\tau: C^*(\cdot, \cdot \cap A; G) \to \Delta^*(\cdot, \cdot \cap A; G)$$

defined in example 6.7.4. This induces a homomorphism

$$\hat{\tau}: \hat{C}^*(\cdot, \cdot \cap A; G) \to \hat{\Delta}^*(\cdot, \cdot \cap A; G)$$

such that the following square is commutative

$$\begin{array}{ccc} C^*(\cdot, \cdot \cap A; G) & \xrightarrow{\tau} & \Delta^*(\cdot, \cdot \cap A; G) \\ {\scriptstyle\alpha}\downarrow & & \downarrow{\scriptstyle\alpha} \\ \hat{C}^*(\cdot, \cdot \cap A; G) & \xrightarrow{\hat{\tau}} & \hat{\Delta}^*(\cdot, \cdot \cap A; G) \end{array}$$

By examples 6.7.8 and 6.7.9, there are isomorphisms

$$\bar{C}^*(\cdot, \cdot \cap A; G) \approx \hat{C}^*(\cdot, \cdot \cap A; G)$$
$$\alpha_*: H^*(\Delta^*(\cdot, \cdot \cap A; G)) \approx H^*(\hat{\Delta}^*(\cdot, \cdot \cap A; G))$$

In Sec. 6.5 a natural homomorphism

$$\mu: \bar{H}^*(X,A; G) \to H^*(X,A; G)$$

was defined, and it is a simple matter to check that commutativity holds in the diagram

$$\begin{array}{ccc} H^*(\bar{C}^*(X,A; G)) & \xrightarrow{\mu} & H^*(\Delta^*(X,A; G)) \\ \approx\downarrow & & \approx\downarrow{\scriptstyle\alpha_*} \\ H^*(\hat{C}^*(\cdot, \cdot \cap A; G)(X)) & \xrightarrow{\hat{\tau}_*} & H^*(\hat{\Delta}^*(\cdot, \cdot \cap A; G)(X)) \end{array}$$

Therefore μ is an isomorphism if and only if $\hat{\tau}_*$ is.

1 THEOREM *Let X be a paracompact Hausdorff space and suppose there is $n \geq 0$ such that each $x \in X$ is taut with respect to singular cohomology with coefficients G in degrees $< n$. Then*

$$\mu: \bar{H}^q(X;G) \to H^q(X;G)$$

is an isomorphism for $q < n$ and a monomorphism for $q = n$.

PROOF Both $C^*(\cdot;G)$ and $\Delta^*(\cdot;G)$ are nonnegative cochain complexes of fine presheaves. The tautness assumption of the points of X with respect to singular cohomology implies that $\tau_*: H^q(C^*(\cdot;G)) \to H^q(\Delta^*(\cdot;G))$ is a local isomorphism for $q < n$ and a local monomorphism for $q = n$ (in fact, it is always a local monomorphism for all q). By theorem 6.8.9,

$$\hat{\tau}_*: H^q(\hat{C}^*(X;G)) \to H^q(\hat{\Delta}^*(X;G))$$

is an isomorphism for $q < n$ and a monomorphism for $q = n$. ∎

There is a partial converse of theorem 1 which asserts that if $\mu: \bar{H}^q(U;G) \to H^q(U;G)$ is an isomorphism for $q < n$ and every open $U \subset X$, then each point $x \in X$ is taut with respect to singular cohomology in degrees $< n$. This follows from commutativity of the following diagram (where U varies over open neighborhoods of $x \in X$):

$$\begin{array}{ccc} \lim_{\rightarrow} \{\bar{H}^q(U;G)\} & \xrightarrow{\approx} & \bar{H}^q(x;G) \\ \downarrow^{\mu} & & \approx\downarrow^{\mu} \\ \lim_{\rightarrow} \{H^q(U;G)\} & \rightarrow & H^q(x;G) \end{array}$$

In case X is a Hausdorff space in which every open subset is paracompact (for example, X is metrizable), we see that each point $x \in X$ is taut with respect to singular cohomology in degrees $< n$ if and only if μ: $\bar{H}^q(U;G) \rightarrow H^q(U;G)$ is an isomorphism for all $q < n$ and all open $U \subset X$.

A space X is said to be *homologically locally connected in dimension n* if for every $x \in X$ and neighborhood U of x there exists a neighborhood V of x in U such that $\tilde{H}_q(V) \rightarrow \tilde{H}_q(U)$ is trivial for $q \leq n$. It is said to be *homologically locally connected* if it is homologically locally connected in dimension n for all n.

2 EXAMPLE Any locally contractible space, in particular any polyhedron or any manifold, is homologically locally connected in dimension n for all n.

3 EXAMPLE Let $X_q = S^q$ for $q \geq 1$ and let x_q be a base point of X_q. The subspace of $\times X_q$ consisting of all points having at most one coordinate different from the corresponding base point is homologically locally connected in dimension n for all n but is not locally contractible.

4 LEMMA *If X is homologically locally connected in dimension n, then $\bar{H}^q(\Delta^*(\cdot\,;G))$ is locally zero for $q \leq n$ and all G.*

PROOF Let $c^* \in \operatorname{Hom}(\tilde{\Delta}_q(U),G)$ be a cocycle $(0 \leq q \leq n)$ and let $x \in U$. If $q = 0$, let V be a neighborhood of x in U such that $\tilde{H}_0(V) \rightarrow \tilde{H}_0(U)$ is trivial. If $c \in \tilde{\Delta}_0(V)$, there is $c' \in \Delta_1(U)$ such that $c = \partial c'$. Then $c^*(c) = c^*(\partial c') = (\delta c^*)(c') = 0$. Therefore $c^* \mid \tilde{\Delta}_0(V) = 0$, proving that $\bar{H}^0(\Delta^*(\cdot\,;G))$ is locally trivial.

If $0 < q$, let V and V' be neighborhoods of x in U, with $V \subset V'$ and such that $\tilde{H}_{q-1}(V) \rightarrow \tilde{H}_{q-1}(V')$ and $H_q(V') \rightarrow H_q(U)$ are both trivial. If c is a reduced singular $(q-1)$-cycle of V, let c' be a q-chain of V' such that $\partial c' = c$. Then $c^*(c') \in G$ is independent of the choice of c'; if c'' is another q-chain in V' such that $\partial c'' = c$, then $c' - c'' = \partial d$ for some $(q+1)$-chain d in U and

$$c^*(c' - c'') = c^*(\partial d) = (\delta c^*)(d) = 0$$

Hence there is a homomorphism $\bar{c}^*: \tilde{Z}_{q-1}(V) \rightarrow G$ such that $\bar{c}^*(c) = c^*(c')$ if $\partial c' = c$. Because $\Delta_{q-1}(V)/\tilde{Z}_{q-1}(V)$ is free (since it is isomorphic to a subgroup of $\Delta_{q-2}(V)$ if $q > 1$ or to $\mathbf{Z}$ if $q = 1$), there is a homomorphism $d^*: \Delta_{q-1}(V) \rightarrow G$ which is an extension of $\bar{c}^*$. Then $c^* \mid \Delta_q(V) = \delta d^*$, proving that $H^q(\Delta^*(\cdot\,;G))$ is locally trivial. ■

5 COROLLARY *If X is a paracompact Hausdorff space homologically locally connected in dimension n, then μ: $\bar{H}^q(X;G) \rightarrow H^q(X;G)$ is an isomorphism for $q \leq n$ and a monomorphism for $q = n + 1$.* ■

6 COROLLARY *Let A be a closed subset, homologically locally connected in dimension n, of a Hausdorff space X, homologically locally connected in dimension n. If X has the property that every open subset is paracompact,* $\mu: \bar{H}_c^q(X,A;G) \to H_c^q(X,A;G)$ *is an isomorphism for* $q \le n$ *and a monomorphism for* $q = n + 1$.

PROOF From the definitions, there is a commutative square (where U varies over open cobounded subsets of X)

$$\begin{array}{ccc} \lim_{\rightarrow} \{\bar{H}^q(X,U;G)\} & \xrightarrow{\approx} & \bar{H}_c^q(X;G) \\ \mu\downarrow & & \downarrow\mu \\ \lim_{\rightarrow} \{H^q(X,U;G)\} & \xrightarrow{\approx} & H_c^q(X;G) \end{array}$$

Since an open subset of a space homologically locally connected in dimension n is again a space homologically locally connected in dimension n corollary 5 applies to X and to every open $U \subset X$. By the five lemma,

$$\mu: \bar{H}^q(X,U;G) \to H^q(X,U;G)$$

is an isomorphism for $q \le n$ and a monomorphism for $q = n + 1$. Passing to the limit, $\mu: \bar{H}_c^q(X;G) \to H_c^q(X;G)$ is an isomorphism for $q \le n$ and a monomorphism for $q = n + 1$. Since A has the same properties as X,

$$\mu: \bar{H}_c^q(A;G) \to H_c^q(A;G)$$

is an isomorphism for $q \le n$ and a monomorphism for $q = n + 1$. The result now follows from the five lemma. ∎

Since a manifold is homologically locally connected in dimension n for all n, and every open subset is paracompact, this implies the next result.

7 COROLLARY *If X is a manifold,* $\mu: \bar{H}^*(X;G) \approx H^*(X;G)$. *If A is a closed homologically locally connected subset of X,*

$$\mu: \bar{H}_c^*(X,A;G) \approx H_c^*(X,A;G). \quad ∎$$

8 COROLLARY *If X is a homologically locally connected space imbedded as a closed subset of a manifold Y, then X is taut in Y with respect to singular cohomology.*

PROOF By corollary 5, $\bar{H}^*(X;G) \approx H^*(X;G)$, and for an open set U in Y, by corollary 7, $\bar{H}^*(U;G) \approx H^*(U;G)$. Since X is taut in Y with respect to Alexander cohomology, these isomorphisms imply that it is also taut with respect to singular cohomology. ∎

9 COROLLARY *If A is any closed subset of a manifold X, then as U varies over neighborhoods of A in X,*

$$\lim_{\rightarrow} \{H^*(U;G)\} \approx \bar{H}^*(A;G)$$

where the right-hand side is Alexander cohomology.

PROOF By corollary 7, $\lim_{\rightarrow} \{\bar{H}^*(U;G)\} \approx \lim_{\rightarrow} \{H^*(U;G)\}$, so the result follows from the tautness of A with respect to the Alexander cohomology theory. ∎

This shows that the modules $\bar{H}^*(A;G)$ and $\bar{H}^*(A,B;G)$ introduced in Sec. 6.1 are the Alexander cohomology modules if A [or (A,B)] is a closed subset [or pair] of a manifold. The next result generalizes the duality theorem 6.2.17 to arbitrary closed pairs.

10 THEOREM *Let X be an n-manifold orientable over R. For any closed pair (A,B) in X and any R module G there is an isomorphism*

$$H_q(X - B, X - A; G) \approx \bar{H}_c^{n-q}(A,B; G)$$

PROOF Let N be a closed cobounded neighborhood of B in A. By theorem 6.6.5, there is an isomorphism

$$\bar{H}^{n-q}(A,N; G) \approx \bar{H}^{n-q}(\overline{A - N}, \overline{A - N} \cap N; G)$$

Since $(\overline{A - N}, \overline{A - N} \cap N)$ is a compact pair in X, by theorem 6.2.17,

$$H_q(X - (\overline{A - N} \cap N), X - (\overline{A - N}); G) \approx \bar{H}^{n-q}(\overline{A - N}, \overline{A - N} \cap N; G)$$

Since $X - (\overline{A - N})$ and $X - N$ are open, there is an excision isomorphism

$$H_q(X - N, X - A; G) \approx H_q(X - (\overline{A - N} \cap N), X - (\overline{A - N}); G)$$

Combining these gives an isomorphism

$$H_q(X - N, X - A; G) \approx \bar{H}^{n-q}(A,N; G)$$

As N varies over closed cobounded neighborhoods of B in A, the limit of the modules on the left is $H_q(X - B, X - A; G)$ and the limit of the modules on the right is $\bar{H}_c^{n-q}(A,B; G)$, whence the result. ∎

11 THEOREM *If X is a compact Hausdorff space which is homologically locally connected in dimension n, then $H_q(X)$ is finitely generated for $q \leq n$.*

PROOF This follows from corollary 5, theorem 6.8.11, and theorem 5.5.13. ∎

The last result gives a generalization of corollary 6.2.21 to arbitrary compact manifolds (orientable or not). We now work toward a proof of the Vietoris-Begle mapping theorem.

12 LEMMA *Let (X,A) be a pair and let Γ be the presheaf on X defined by $\Gamma(V) = \bar{C}^q(V, V \cap A; G)$ for open $V \subset X$ (q and G being fixed).*

(a) For any open covering $\mathcal{U}$ of X the map $\Gamma(X) \to \Gamma(\mathcal{U})$ sending $\gamma \in \Gamma(X)$ to the compatible $\mathcal{U}$ family $\{\gamma \mid U\}_{U \in \mathcal{U}}$ is a monomorphism.

(b) If $\mathcal{U}$ is a locally finite open covering of X and $\mathcal{V}$ is a shrinking of $\mathcal{U}$, the image of $\Gamma(\mathcal{U}) \to \Gamma(\mathcal{V})$ equals the image of the composite

$$\Gamma(X) \to \Gamma(\mathcal{U}) \to \Gamma(\mathcal{V})$$

PROOF For (*a*), assume that $\gamma \in \bar{C}^q(X,A; G)$ is in the kernel of $\Gamma(X) \to \Gamma(\mathcal{U})$

(that is, $\gamma \mid U = 0$ for all $U \in \mathfrak{U}$). Let $\varphi \in C^q(X,A; G)$ be a representative of γ. Then $\gamma \mid U = 0$ implies that $\varphi \mid U$ is locally zero on U. Since this is so for all $U \in \mathfrak{U}$, φ is locally zero on X and $\gamma = 0$, proving (*a*).

To prove (*b*), let $\{\gamma_U\}_{U\in\mathfrak{U}}$ be a compatible $\mathfrak{U}$ family and suppose that $\varphi_U \in C^q(U, U \cap A; G)$ is a representative of γ_U for $U \in \mathfrak{U}$. Then, for $U, U' \in \mathfrak{U}$, $\varphi_U \mid U \cap U' - \varphi_{U'} \mid U \cap U'$ is locally zero on $U \cap U'$. If $x \in X$, some neighborhood of x meets only finitely many elements of $\mathfrak{U}$, and there is a smaller neighborhood W_x of x such that

$$\begin{array}{ll}
\text{(i)} & W_x \text{ intersects } \bar{V}_U \Leftrightarrow x \in \bar{V}_U \\
\text{(ii)} & x \in U \Rightarrow W_x \subset U \\
\text{(iii)} & x \in V_U \Rightarrow W_x \subset V_U \\
\text{(iv)} & x \in \bar{V}_U \cap \bar{V}_{U'} \Rightarrow \varphi_U \mid W_x = \varphi_{U'} \mid W_x
\end{array}$$

The first three conditions are clearly satisfied by taking W_x small enough (because there are only a finite number of conditions to be satisfied) and (iv) can also be satisfied, because for $x \in \bar{V}_U \cap \bar{V}_{U'}$, $\varphi_U \mid U \cap U' - \varphi_{U'} \mid U \cap U'$ is locally zero.

For $x \in X$ choose U so that $x \in \bar{V}_U$ and set $\varphi_x = \varphi_U \mid W_x \in C^q(W_x, W_x \cap A; G)$. By (iv), this is independent of the choice of U. If $x'' \in W_x \cap W_{x'}$, then $x'' \in \bar{V}_U$ for some $U \in \mathfrak{U}$. Then W_x and $W_{x'}$ meet $\bar{V}_U$, and by (i), x, $x' \in \bar{V}_U$. Therefore $\varphi_x = \varphi_U \mid W_x$ and $\varphi_{x'} = \varphi_U \mid W_{x'}$, whence $\varphi_x \mid W_x \cap W_{x'} = \varphi_{x'} \mid W_x \cap W_{x'}$. Hence the collection $\{\varphi_x \in C^q(W_x, W_x \cap A; G)\}$ is a compatible $\{W_x\}$ family [of $C^q(\cdot, \cdot \cap A; G)$]. By example 6.7.8, there is an element $\varphi \in C^q(X,A; G)$ such that $\varphi \mid W_x = \varphi_x$ for all $x \in X$. To complete the proof of (*b*) it suffices to prove that for each $U \in \mathfrak{U}$, $\varphi \mid V_U - \varphi_U \mid V_U$ is locally zero on V_U. However, if $x \in V_U$, then, by (iii), $W_x \subset V_U$ and $\varphi \mid W_x = \varphi_x = \varphi_U \mid W_x$. Hence $\{W_x\}_{x\in V_U}$ is an open covering of V_U on which $\varphi \mid V_U$ and $\varphi_U \mid V_U$ agree. ∎

13 THEOREM *Let $f: X' \to X$ be a closed continuous map between paracompact Hausdorff spaces. Let A' be a closed subset of X' and suppose there are integers $0 \le m < n$ such that $\bar{H}^q(f^{-1}x, f^{-1}x \cap A'; G) = 0$ for all $x \in X$ and for $q < m$ or $m < q < n$. Let Γ be the presheaf on X defined by $\Gamma(U) = \bar{H}^m(f^{-1}(U), f^{-1}(U) \cap A'; G)$. Then there are isomorphisms*

$$\check{H}^{q-m}(X;\Gamma) \approx \bar{H}^q(X',A'; G) \qquad q < n$$

and a monomorphism

$$\check{H}^{n-m}(X;\Gamma) \to \bar{H}^n(X',A'; G)$$

PROOF Let Γ^* be the nonnegative cochain complex of presheaves on X defined by $\Gamma^*(U) = \bar{C}^*(f^{-1}(U), f^{-1}(U) \cap A'; G)$. Thus Γ^q is the image under f_* of the fine presheaf on X' which assigns $\bar{C}^q(U', U' \cap A'; G)$ to $U' \subset X'$. By theorem 6.8.3*c*, the latter is a fine presheaf on X' [being the completion of the fine presheaf $C^q(\cdot, \cdot \cap A'; G)$; see example 6.8.1], and by theorem 6.8.3*b*, Γ^q is fine on X. As U varies over neighborhoods of x in X, $(f^{-1}(U), f^{-1}(U) \cap A')$

varies over a cofinal family of neighborhoods of $(f^{-1}x, f^{-1}x \cap A')$ in (X',A') (because f is closed and continuous). From the standard tautness properties and the hypothesis about $\bar{H}^*(f^{-1}x, f^{-1}x \cap A'; G)$, it follows that $H^q(\Gamma^*)$ is locally zero for $q < m$ and $m < q < n$. By theorem 6.8.7, there are functorial isomorphisms

$$\check{H}^{q-m}(X;H^m(\Gamma^*)) \approx H^q(\hat{\Gamma}^*(X)) \qquad q < n$$

and a monomorphism

$$\check{H}^{n-m}(X;H^m(\Gamma^*)) \to H^n(\hat{\Gamma}^*(X))$$

Since $\Gamma = H^m(\Gamma^*)$, it merely remains to verify that

$$H^p(\hat{\Gamma}^*(X)) \approx \bar{H}^p(X',A'; G) \qquad \text{all } p$$

As $\mathfrak{U}$ varies over the cofinal family of locally finite open coverings of X it follows from lemma 12 that

$$\hat{\Gamma}^*(X) = \lim_{\to} \{\Gamma^*(\mathfrak{U})\} = \lim_{\to} \{\bar{C}^*(\cdot, \cdot \cap A'; G)(f^{-1}\mathfrak{U})\} \approx \bar{C}^*(X',A'; G)$$

and this yields the result. ∎

If ξ is an m-sphere bundle over a paracompact Hausdorff base space B, then $\bar{H}^q(p_\xi^{-1}(x), p_\xi^{-1}(x) \cap \dot{E}) = 0$ if $q \neq m + 1$. Therefore the hypotheses of theorem 13 are satisfied for all n. Since the presheaf Γ that occurs in theorem 13 is the tensor product of the orientation presheaf of ξ and G, we obtain the following generalization of the Thom isomorphism theorem to nonorientable sphere bundles.

14 THEOREM *Let ξ be an m-sphere bundle over a paracompact Hausdorff base space B and let Γ be the orientation presheaf of ξ over R. For all R modules G and all q there is an isomorphism*

$$\check{H}^q(B; \Gamma \otimes G) \approx \bar{H}^{q+m+1}(E_\xi,\dot{E}_\xi; G) \quad ∎$$

Another interesting consequence of theorem 13 is the following *Vietoris-Begle mapping theorem.*

15 THEOREM *Let $f: X' \to X$ be a closed continuous surjective map between paracompact Hausdorff spaces. Assume that there is $n \geq 0$ such that $\tilde{\bar{H}}^q(f^{-1}x;G) = 0$ for all $x \in X$ and for $q < n$. Then*

$$f^*: \bar{H}^q(X;G) \to \bar{H}^q(X';G)$$

is an isomorphism for $q < n$ and a monomorphism for $q = n$.

PROOF Let Z be the mapping cylinder of f and regard X' as imbedded in Z. Then Z is a paracompact Hausdorff space, X' is closed in Z, and the retraction $r: Z \to X$ is a closed continuous map. For $x \in X$, $r^{-1}(x)$ is contractible [since it is homeomorphic to the join of x with $f^{-1}(x)$], and so $\tilde{\bar{H}}^*(r^{-1}(x)) = 0$. Because $r^{-1}(x) \cap X' = f^{-1}(x)$ is nonempty, we have

$$\bar{H}^{q+1}(r^{-1}(x), r^{-1}(x) \cap X'; G) \approx \tilde{\bar{H}}^q(f^{-1}(x); G) = 0 \qquad q < n$$

It follows from theorem 13 that $\bar{H}^q(Z,X'; G) = 0$ for $q \leq n$. Since there is a commutative diagram with an exact row

$$\begin{array}{ccccccc} \cdots \to \bar{H}^q(Z,X') \to & \bar{H}^q(Z) & \to & \bar{H}^q(X') & \to \bar{H}^{q+1}(Z,X') \to \cdots \\ & r^*\uparrow\approx & \nearrow_{f^*} & & \\ & \bar{H}^q(X) & & & \end{array}$$

the result follows. ∎

There is a partial converse of theorem 15 asserting that if $f\colon X' \to X$ is a closed continuous surjective map between paracompact Hausdorff spaces and there is $n \geq 0$ such that for every open $U \subset X$, $f^*\colon H^q(U;G) \to H^q(f^{-1}(U);G)$ is an isomorphism for $q < n$, then $\tilde{\bar{H}}^q(f^{-1}(x);G) = 0$ for all $x \in X$ and for $q < n$. This follows from commutativity of the following diagram (where U varies over open neighborhoods of $x \in X$):

$$\begin{array}{ccc} \varinjlim \{\tilde{\bar{H}}^q(U;G)\} & \xrightarrow{\approx} & \tilde{\bar{H}}^q(x;G) \\ f^*\downarrow & & \downarrow f^* \\ \varinjlim \{\tilde{\bar{H}}^q(f^{-1}(U);G)\} & \xrightarrow{\approx} & \tilde{\bar{H}}^q(f^{-1}(x);G) \end{array}$$

In particular, if X and X' are metrizable (or have the property that every open subset is paracompact), then for $n \geq 0, f^*\colon \bar{H}^q(U;G) \to \bar{H}^q(f^{-1}(U);G)$ is an isomorphism for all open $U \subset X$ and all $q < n$ if and only if $\tilde{\bar{H}}^q(f^{-1}(x);G) = 0$ for all $x \in X$ and all $q < n$.

We present an example to show that the condition that f be a closed map is necessary in theorem 15.

16 EXAMPLE Let $X' = \{(x,y) \in \mathbf{R}^2 \mid x^2 + y^2 = 1 \text{ or } x^2 + y^2 < 1, x > 0\}$ and let $X = [0,1]$. Define $f\colon X' \to X$ by

$$f(x,y) = \begin{cases} 0 & x \leq 0 \\ x & x \geq 0 \end{cases}$$

Then f is a continuous surjective map but not a closed map. Furthermore,

$$f^{-1}(t) = \begin{cases} \text{closed semicircle} & t = 0 \\ \text{closed interval} & 0 < t < 1 \\ \text{single point} & t = 1 \end{cases}$$

Because the unit circle S^1 is a strong deformation retract of X',

$$\bar{H}^1(X';G) \approx \bar{H}^1(S^1;G) \approx G.$$

Since $\bar{H}^1(X;G) = 0$, the homomorphism $f^*\colon \bar{H}^1(X;G) \to \bar{H}^1(X';G)$ is not an isomorphism.

17 EXAMPLE Let $X \subset \mathbf{R}^2$ be the space of example 2.4.8, illustrated below:

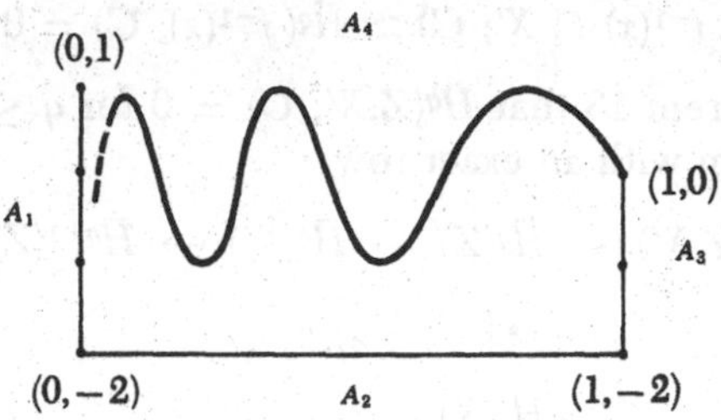

There is a closed continuous surjective map f of X onto the space Y consisting of the four sides of the rectangle

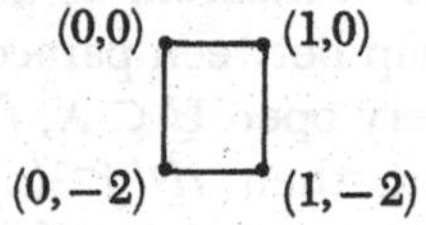

such that

$$f^{-1}(y) = \begin{cases} \text{single point} & y \neq (0,0) \\ \text{closed interval} & y = (0,0) \end{cases}$$

It follows from theorem 15 that $f^*: \bar{H}^*(Y;G) \approx \bar{H}^*(X;G)$ for any G, and therefore the map f is not null homotopic.

18 THEOREM *Let $f: X' \to X$ be a proper surjective map between locally compact Hausdorff spaces and assume that for some $n > 0$, $\tilde{\bar{H}}^q(f^{-1}(x);G) = 0$ for all $x \in X$ and all $q < n$. Then*

$$f^*: \bar{H}_c^q(X;G) \to \bar{H}_c^q(X';G)$$

is an isomorphism for $q < n$ and a monomorphism for $q = n$.

PROOF If either X or X' is compact, the other one is also compact, and the result follows from lemma 6.6.9 and theorem 15. If neither X nor X' is compact, let X^+ and X'^+ be their one-point compactifications and extend f to a map $f^+: X'^+ \to X^+$ mapping the point at infinity of X'^+ to the point at infinity of X^+. Then f^+ satisfies the hypotheses of theorem 15, and the result follows from corollary 6.6.12 and theorem 15. ∎

10 CHARACTERISTIC CLASSES

This section is a culmination of our general work on homology theory. We use the cup product and Steenrod squares to define characteristic classes of a manifold and of one manifold imbedded in another. These characteristic classes are important invariants of the manifold and have interesting applications to nonimbedding problems.

Let X be an n-manifold with boundary $\dot{X}$ and $U \in H^n(X \times X, X \times X - \delta(X))$

be an orientation (over R) of X. Let $j: X - \dot{X} \subset X$ be the inclusion map. Then the maps

$$j \times 1: (X - \dot{X}) \times (X,\dot{X}) \subset X \times (X,\dot{X})$$
$$1 \times j: (X,\dot{X}) \times (X - \dot{X}) \subset (X,\dot{X}) \times X$$

are both homotopy equivalences. Therefore there are elements

$$U_1 \in H^n(X \times (X,\dot{X})) \qquad U_2 \in H^n((X,\dot{X}) \times X)$$

such that

$$(j \times 1)^* U_1 = U \mid (X - \dot{X}) \times (X,\dot{X}) \qquad (1 \times j)^* U_2 = U \mid (X,\dot{X}) \times (X - \dot{X})$$

If X is compact, let $z \in H_n(X,\dot{X})$ be the fundamental class of X corresponding to U, as in theorem 6.3.9. The *Euler class* of a compact oriented manifold X, denoted by $\chi \in H^n(X,\dot{X})$, is defined by

$$\chi = (U_1 \cup U_2)/z$$

The reason for the name is furnished by theorem 2 below.

Assume that R is a field and that X is a compact n-manifold with boundary $\dot{X}$. By theorem 6.9.11, $H_*(X)$ and $H_*(X,\dot{X})$ are finitely generated. If $\{u_i\}$ is a basis of $H^*(X)$ and $\{v_j\}$ is a basis of $H^*(X,\dot{X})$, then by the Künneth formula for cohomology, $\{u_i \times v_j\}$ is a basis of $H^*(X \times (X,\dot{X}))$. Hence

$$U_1 = \sum_{i,j} a_{ij} u_i \times v_j$$

for some scalars a_{ij}. Let $b_{jk} = \langle v_j \cup u_k, z\rangle$, where z is the fundamental class corresponding to U. Then we have matrices $A = (a_{ij})$ and $B = (b_{jk})$, and the following expresses their relation to each other.

1 LEMMA *With the above notation,*

$$(AB)_{ik} = (-1)^{n \deg u_k} \delta_{ik}$$

PROOF The proof is essentially the same as that for theorem 6.3.12. Because z is the fundamental class corresponding to U, it follows that

$$U_1/z = 1 \in H^0(X)$$

By property 6.1.4, for any k

$$u_k = u_k \cup 1 = u_k \cup U_1/z = [(u_k \times 1) \cup U_1]/z$$

From lemma 6.3.11 it follows readily that

$$(u_k \times 1) \cup U_1 = (1 \times u_k) \cup U_1 = (-1)^{n \deg u_k} U_1 \cup (1 \times u_k)$$
$$= \sum_{i,j} (-1)^{n \deg u_k} a_{ij} u_i \times (v_j \cup u_k)$$

Hence by property 6.1.2

$$u_k = \sum_{i,j} (-1)^{n \deg u_k} a_{ij} b_{jk} u_i$$

Since $\{u_i\}$ is a basis, this implies the result. ∎

2 THEOREM *If χ is the Euler class of a compact n-manifold X oriented over a field, then $\langle \chi, z\rangle$ is the Euler characteristic of X.*

PROOF We first compute U_2. Let $T: X \times X \to X \times X$ be the map interchanging the factors. There is a commutative diagram, with all vertical maps induced by maps defined by T and all horizontal maps induced by inclusions,

$$\begin{array}{ccccc} H^n(X \times X, X \times X - \delta(X)) & \to & H^n((X - \dot{X}) \times (X,\dot{X})) & \xleftarrow[\approx]{(j \times 1)^*} & H^n(X \times (X,\dot{X})) \\ T_1^* \downarrow & & T_2^* \downarrow & & \downarrow T_3^* \\ H^n(X \times X, X \times X - \delta(X)) & \to & H^n((X,\dot{X}) \times (X - \dot{X})) & \xleftarrow[\approx]{(1 \times j)^*} & H^n((X,\dot{X}) \times X) \end{array}$$

In the proof of lemma 6.3.11 it was shown that $T_1^* U = (-1)^n U$. Therefore $T_3^* U_1 = (-1)^n U_2$, and so

$$\begin{aligned} U_2 &= (-1)^n T_3^* \Big(\sum_{k,l} a_{kl} u_k \times v_l \Big) \\ &= (-1)^n \sum_{k,l} (-1)^{\deg u_k \deg v_l} a_{kl} v_l \times u_k \end{aligned}$$

Therefore

$$\begin{aligned} U_1 \smile U_2 &= (-1)^n \sum (-1)^{\deg v_l \deg v_j + \deg u_k \deg v_l} a_{ij} a_{kl} (u_i \smile v_l) \times (v_j \smile u_k) \\ &= (-1)^n \sum (-1)^{\deg v_l \deg v_j + \deg u_k \deg v_l + \deg u_i \deg v_l} a_{ij} a_{kl} (v_l \smile u_i) \\ &\qquad \times (v_j \smile u_k) \end{aligned}$$

where the summation is over all i, j, k, and l such that

$$\deg u_i + \deg v_j = n = \deg u_k + \deg v_l$$

It follows that

$$U_1 \smile U_2 = \sum (-1)^{\deg u_k} a_{ij} a_{kl} (v_l \smile u_i) \times (v_j \smile u_k)$$

Using lemma 1,

$$\begin{aligned} \langle \chi, z\rangle &= \langle U_1 \smile U_2, z \times z\rangle \\ &= \sum_{i,j,k,l} (-1)^{\deg u_k} a_{ij} b_{jk} a_{kl} b_{li} \\ &= \sum_{i,k} (-1)^{\deg u_k} (AB)_{ik} (AB)_{ki} \\ &= \sum_k (-1)^{\deg u_k} \end{aligned}$$

and the last sum is the Euler characteristic of X. ∎

Classically, the Euler class is usually taken to be the Euler class (in our sense) over **Z**. For any pair (Y,B) whose homology is of finite type, it follows from the universal-coefficient formula for cohomology (theorem 5.5.10) that

$$H^q(Y,B; \mathbf{R}) \approx H^q(Y,B; \mathbf{Z}) \otimes \mathbf{R}$$

Therefore the monomorphism $\mathbf{Z} \to \mathbf{R}$ induces a monomorphism

$$H^q(Y,B; \mathbf{Z}) \to H^q(Y,B; \mathbf{R})$$

In particular, the monomorphism $H^n(X,\dot X;\, \mathbf{Z}) \to H^n(X,\dot X;\, \mathbf{R})$ maps Euler class to Euler class, and therefore theorem 2 remains valid for the integral Euler class of X.

We now specialize to the case where the coefficient field is $\mathbf{Z}_2$, in which case U, hence also U_1, and (if X is compact) z, are all unique. There is the Thom isomorphism

$$\Phi^*\colon H^q(X - \dot X) \approx H^{q+n}((X - \dot X) \times (X - \dot X), (X - \dot X) \times (X - \dot X) - \delta(X - \dot X))$$

defined by $\Phi^*(v) = (v \times 1) \smile U'$, where

$$U' = U \,|\, ((X - \dot X) \times (X - \dot X), (X - \dot X) \times (X - \dot X) - \delta(X - \dot X))$$

Φ^* can be extended to

$$\Phi^*\colon H^q(X) \to H^{q+n}(X \times X, X \times X - \delta(X))$$

by $\Phi^*(v) = (v \times 1) \smile U$. There is a commutative diagram whose vertical maps are isomorphisms

$$\begin{array}{ccc} H^q(X) & \xrightarrow{\Phi^*} & H^{q+n}(X \times X, X \times X - \delta(X)) \\ \approx\downarrow & & \approx\downarrow \\ H^q(X - \dot X) & \xrightarrow{\Phi^*} & H^{q+n}((X - \dot X) \times (X - \dot X), (X - \dot X) \times (X - \dot X) - \delta(X - \dot X)) \end{array}$$

from which it follows that Φ^* is also an isomorphism on $H^q(X)$. For $i \geq 0$ the *ith Stiefel-Whitney class of X*, $w_i \in H^i(X;\mathbf{Z}_2)$, is defined by the formula

$$\Phi^*(w_i) = Sq^i U$$

[that is, $Sq^i U = (w_i \times 1) \smile U$]. Following are some examples.

3 By condition (*a*) on page 271, $w_0 = 1$.

4 By condition (*b*) on page 271, if X is a compact n-manifold without boundary, w_n is the Euler class of X over $\mathbf{Z}_2$.

5 By condition (*c*) on page 271, $w_i = 0$ for $i > \dim X$.

6 A manifold X is orientable over $\mathbf{Z}$ if and only if $w_1 = 0$ (see exercise 5.H.3*d*).

If X is compact and $z \in H_n(X,\dot X)$ is the fundamental class of X over $\mathbf{Z}_2$, then, by property 6.1.4,

$$w_i = [(w_i \times 1) \smile U_1]/z = Sq^i U_1/z$$

where $U_1 \in H^n(X \times (X,\dot X))$ corresponds to U. We use this to determine the Stiefel-Whitney classes of a compact X in terms of cohomology operations in X. For $i \geq 0$ the homomorphism $Sq^i\colon H^{n-i}(X,\dot X) \to H^n(X,\dot X)$ has a transpose homomorphism $\overline{Sq^i}\colon H_n(X,\dot X) \to H_{n-i}(X,\dot X)$ such that

$$\langle Sq^i u, z\rangle = \langle u, \overline{Sq^i} z\rangle \qquad u \in H^{n-i}(X,\dot X)$$

where z is the fundamental class of X. By the isomorphism of theorem 6.3.12,

$$\kappa_z: H^i(X) \approx H_{n-i}(X,\dot{X})$$

and there is a unique $V_i \in H^i(X)$ such that $\kappa_z(V_i) = \overline{Sq^i}(z)$. Then for $u \in H^{n-i}(X,\dot{X};\mathbf{Z}_2)$

$$\begin{aligned}\langle Sq^i u,z\rangle &= \langle u,\overline{Sq^i}z\rangle = \langle u,\kappa_z(V_i)\rangle \\ &= \langle u, V_i \frown z\rangle = \langle u \smile V_i, z\rangle\end{aligned}$$

This equation holds trivially if $\deg u \neq n - i$. The following *Wu formula* shows that the classes V_i and the Stiefel-Whitney classes w_i determine each other.

7 THEOREM *In a compact n-manifold, for* $q \geq 0$

$$w_q = \sum_{0 \leq i \leq q} Sq^{q-i} V_i$$

PROOF We have $U_1 = \Sigma\, a_{ij} u_i \times v_j$, where $\{u_i\}$ is a basis of $H^*(X,\mathbf{Z}_2)$ and $\{v_j\}$ is a basis of $H^*(X,\dot{X};\mathbf{Z}_2)$. By the Cartan formula, condition (*d*) on page 271

$$Sq^q U_1 = \sum_{k+l=q} a_{ij} Sq^k u_i \times Sq^l v_j$$

Let $V_l = \Sigma\, c_{lm} u_m$. Then we have

$$\begin{aligned}w_q = (Sq^q U_1)/z &= \sum_{k+l=q} a_{ij} \langle Sq^l v_j, z\rangle Sq^k u_i \\ &= \sum_{k+l=q} a_{ij} \langle v_j \smile V_l, z\rangle Sq^k u_i \\ &= \sum_{k+l=q} a_{ij} c_{lm} \langle v_j \smile u_m, z\rangle Sq^k u_i \\ &= \sum_{k+l=q} a_{ij} b_{jm} c_{lm} Sq^k u_i\end{aligned}$$

Using lemma 1, we find that

$$w_q = \sum_{k+l=q} c_{li} Sq^k u_i = \sum_{k+l=q} Sq^k V_l \quad \blacksquare$$

Let P^n be the real projective n-space and let w be a generator of $H^1(P^n)$ for any $n \geq 1$. We use lemma 5.9.4 to compute $Sq^i(w^j)$ in the following examples.

8 For the real projective plane P^2, $Sq^1(w) = w^2$; therefore $V_1(P^2) = w$, $w_1(P^2) = w$, and $w_2(P^2) = w^2$.

9 For P^3, $Sq^2(w) = 0$ and $Sq^1(w^2) = 0$, so $V_i(P^3) = 0$ for $i > 0$ and $w_i(P^3) = 0$ for $i > 0$.

10 For P^4, $Sq^2(w^2) = w^4$ and $Sq^1(w^3) = w^4$, so $V_1(P^4) = w$, $V_2(P^4) = w^2$, $w_1(P^4) = w$, $w_2(P^4) = 0$, $w_3(P^4) = 0$, and $w_4(P^4) = w^4$.

11 For P^5, $Sq^2(w^3) = w^5$ and $Sq^1(w^4) = 0$, and $V_2(P^5) = w^2$ is the only nonzero $V_i(P^5)$, where $i > 0$. Hence $w_1(P^5) = 0$, $w_2(P^5) = w^2$, $w_3(P^5) = 0$, $w_4(P^5) = w^4$, and $w_5(P^5) = 0$.

The Euler class and Stiefel-Whitney classes of a manifold X are topological invariants associated to X. We shall now define characteristic classes for a manifold X imbedded in a manifold Y. These will be topological invariants of the imbedding. First, however, we need an algebraic digression.

In our consideration of the slant product we limited ourselves to one of the two possible slant products. We now introduce the other one. Given chain complexes C and C', a cochain $c^* \in \operatorname{Hom}((C \otimes C')_n, G)$, and chain $c \in C_q \otimes G'$, there is a slant product $c\backslash c^* \in \operatorname{Hom}(C'_{n-q}, G \otimes G')$ which is the cochain such that if $c = \sum c_i \otimes g'_i$, with $c_i \in C_q$ and $g'_i \in G'$, then

$$\langle c\backslash c^*, c'\rangle = \sum \langle c^*, c_i \otimes c'\rangle \otimes g'_i \qquad c' \in C'_{n-q}$$

Then
$$\delta(c\backslash c^*) = (-1)^q(c\backslash \delta c^* - \partial c\backslash c^*)$$

from which it follows that there is an induced slant product of $H^n(C \otimes C'; G)$ and $H_q(C;G')$ to $H^{n-q}(C'; G \otimes G')$. This gives rise to a topological slant product of $H^n((X,A) \times (Y,B); G)$ and $H_q(X,A; G')$ to $H^{n-q}(Y,B; G \otimes G')$ having properties analogous to 6.1.1 to 6.1.6. We list without proof two of these, to which we shall have occasion to refer.

12 *Given $u \in H^n((X,A) \times (Y,B); G)$, $z \in H_q(X,A; G'')$, and $v \in H^p(Y,B; G')$, let $T: G \otimes G'' \otimes G' \to G \otimes G' \otimes G''$ interchange the last two factors. In $H^{n-q+p}(Y,B; G \otimes G' \otimes G'')$ we have*

$$T_*((z\backslash u) \smile v) = z\backslash[u \smile (1 \times v)] \quad ■$$

13 *Given $u \in H^n((X,A) \times (Y,B); G)$, $v \in H^p(X,A; G')$, and $z \in H_q(X,A; G'')$, then, in $H^{n+p-q}(Y,B; G \otimes G' \otimes G'')$,*

$$(v \frown z)\backslash u = z\backslash[u \smile (v \times 1)] \quad ■$$

Let Y be an m-manifold without boundary and

$$U \in H^m(Y \times Y, Y \times Y - \delta(Y); R)$$

an orientation of Y over R. Given a pair (A,B) in Y, we define

$$\gamma'_U: H_q(A,B; G) \to H^{n-q}(Y - B, Y - A; G)$$

by

$$\gamma'_U(z) = z\backslash[U \mid (A,B) \times (Y - B, Y - A)] \qquad z \in H_q(A,B; G)$$

Then we have the following complement to the duality theorem.

14 LEMMA *Let X be a compact homologically locally connected space in an m-manifold Y with orientation class U. Then we have an isomorphism for all q and all G*

$$\gamma'_U: H_q(X;G) \approx H^{m-q}(Y, Y - X; G)$$

PROOF Since X is compact and homologically locally connected, it follows from theorem 6.9.11 that $H(\Delta(X))$ is of finite type. By lemma 5.5.9, there is a free chain complex C of finite type which is chain equivalent to $\Delta(X)$. Let $\lambda: C \to \Delta(X)$ be a chain equivalence. Let Δ' and C' be the chain complexes obtained by reindexing the cochain complexes $\operatorname{Hom}(\Delta(X),R)$ and $\operatorname{Hom}(C,R)$, respectively, so that $\Delta'_q = \operatorname{Hom}(\Delta_{m-q}(X),R)$ and $C'_q = \operatorname{Hom}(C_{m-q},R)$. The chain equivalence λ defines a chain equivalence $\lambda': \Delta' \to C'$. Because C is free and of finite type, so is C' [Δ' will not be free, in general, because $\Delta(X)$ need not be of finite type].

Let $c^* \in \operatorname{Hom}([\Delta(X) \otimes (\Delta(Y)/\Delta(Y - X))]_m, R)$ be an m-cocycle corresponding to $U \mid X \times (Y, Y - X)$ under the Eilenberg-Zilber isomorphism and define a map

$$\tau: \Delta(Y)/\Delta(Y - X) \to \Delta'$$

by $\tau(c) = c^*/c$ for $c \in \Delta(Y)/\Delta(Y - X)$. If deg $c = q$,

$$\partial(\tau(c)) = \delta(c^*/c) = (-1)^{m-q} c^*/\partial c = (-1)^{m-q}\tau(\partial c)$$

so τ either commutes or anticommutes with ∂, depending on degree. Hence τ induces homomorphisms τ_* on homology and τ^* on cohomology for any coefficient module. Clearly,

$$\tau_* = \gamma_U: H_q(Y, Y - X; R) \to H^{m-q}(X;R)$$

Because X is homologically locally connected, by corollary 6.9.8, X is taut in Y, and by the duality theorem, γ_U, and hence τ_*, is an isomorphism. Therefore the composite $\lambda' \circ \tau$ induces an isomorphism $\lambda'_* \circ \tau_*$ of $H_q(Y, Y - X; R)$ with $H_q(C') = H^{m-q}(C;R)$. Since $\Delta(Y)/\Delta(Y - X)$ and C' are both free, it follows from the universal-coefficient formula for cohomology (theorem 5.5.3) that for any G

$$(\lambda' \circ \tau)^* = \tau^* \circ \lambda'^*: H^*(\operatorname{Hom}(C',G)) \approx H^*(\operatorname{Hom}(\Delta(Y)/\Delta(Y - X), G))$$

There is also a commutative diagram

$$\begin{array}{ccc} H_q(\Delta(X) \otimes G) & \xleftarrow[\approx]{(\lambda\otimes 1)_*} & H_q(C \otimes G) \\ \downarrow & & \downarrow\approx \\ H^{m-q}(\operatorname{Hom}(\Delta',G)) & \xleftarrow[\approx]{\lambda'^*} & H^{m-q}(\operatorname{Hom}(C', G)) \end{array}$$

where the vertical maps are induced by the canonical map

$$A \otimes G \to \operatorname{Hom}(\operatorname{Hom}(A,R), G)$$

for any module A (the right-hand vertical map being an isomorphism because C is of finite type). Hence there are isomorphisms

$$H_q(X;G) \underset{\approx}{\to} H^{m-q}(\operatorname{Hom}(\Delta',G)) \underset{\approx}{\xrightarrow{\tau^*}} H^{m-q}(Y, Y - X; G)$$

It only remains to verify that this composite is γ'_U. If $\sigma \in \Delta_q(X)$ and $g \in G$, the composite

$$\Delta_q(X) \otimes G \to \operatorname{Hom}(\Delta'_{m-q},G) \xrightarrow{\operatorname{Hom}(\tau,1)} \operatorname{Hom}(\Delta_{m-q}(Y)/\Delta_{m-q}(Y - X),G)$$

maps $\sigma \otimes g$ to the homomorphism h such that if $\sigma' \in \Delta_{m-q}(Y)$,

$$\begin{aligned} h(\sigma') &= \tau(\sigma')(\sigma \otimes g) = (c^*/\sigma')(\sigma \otimes g) \\ &= \langle c^*, \sigma \otimes \sigma'\rangle g = [(\sigma \otimes g)\backslash c^*](\sigma') \end{aligned}$$

Therefore $h = (\sigma \otimes g)\backslash c^*$, and this gives the result on passing to homology. ∎

Let X be a closed subset of a space Y tautly imbedded with respect to singular cohomology and let $A \subset X$. Assuming $X - A$ taut in $Y - A$, we define

$$H^p(Y, Y - X; G) \smile H^q(X, X - A; G') \to H^{p+q}(Y, Y - A; G'')$$

where G and G' are paired to G''. If V is any neighborhood of X in Y and V' is a neighborhood of $X - A$ in $V - A$, there is a cup product

$$H^p(V, V - X; G) \smile H^q(V,V'; G') \to H^{p+q}(V, (V - X) \cup V'; G'')$$

There are excision isomorphisms (for all coefficients)

$$H^p(Y, Y - X) \approx H^p(V, V - X)$$
$$H^{p+q}(Y, Y - A) \approx H^{p+q}(V, V - A)$$

Since $V - A = (V - X) \cup V'$ we have a cup product

$$H^p(Y, Y - X; G) \smile H^q(V,V'; G') \to H^{p+q}(Y, Y - A; G'')$$

As V varies over neighborhoods of X in Y and V' varies over neighborhoods of $X - A$ in $V - A$, it follows from the tautness assumptions and the five lemma that $\lim_{\to} \{H^*(V,V'; G')\} \approx H^*(X, X - A; G')$. The desired cup product is thus obtained by passing to the direct limit with the above cup product.

Let X be a compact n-manifold without boundary imbedded in an m-manifold Y without boundary. Assume that U and U' are orientations of X and Y, respectively, over R. There is then an isomorphism (for any R module G)

$$\theta: H^q(X;G) \approx H^{m-n+q}(Y, Y - X; G)$$

characterized by commutativity in the triangle of isomorphisms (note that X is homologically locally connected, and so lemma 14 applies to $X \subset Y$)

$$\begin{array}{ccc} & H_{n-q}(X;G) & \\ {\scriptstyle\gamma_U}\swarrow & & \searrow{\scriptstyle\gamma_{U'}} \\ H^q(X;G) & \xrightarrow{\theta} & H^{m-n+q}(Y, Y - X; G) \end{array}$$

This map θ is similar to a Thom isomorphism and has the following multiplicative property.

15 LEMMA *The isomorphism θ: $H^q(X,G) \approx H^{m-n+q}(Y, Y - X; G)$ has the property that for $v \in H^q(X;G)$*

$$\theta(v) = \pm\theta(1) \smile v$$

where $\theta(1) \in H^{m-n}(Y, Y - X; R)$

PROOF Let $z \in H_n(X;R)$ be the fundamental class of X corresponding to U and suppose $v = i^*v'$ for $v' \in H^q(V;G)$ and i: $X \subset V$, where V is a neighborhood of X in Y. By theorem 6.3.12, $\gamma_U^{-1}(v) = \pm v \frown z = \pm i^*v' \frown z$. Then, using properties 12 and 13 (with all equations holding up to sign),

$$\begin{aligned}\theta(v) \mid (V, V - X) &= \pm(i^* v' \frown z)\backslash[U' \mid X \times (V, V - X)] \\ &= \pm i_*(i^* v' \frown z)\backslash[U' \mid V \times (V, V - X)] \\ &= \pm(v' \frown i_* z)\backslash[U' \mid V \times (V, V - X)] \\ &= \pm i_* z\backslash\{[U' \mid V \times (V, V - X)] \smile (v' \times 1_V)\} \\ &= \pm i_* z\backslash\{[U' \mid V \times (V, V - X)] \smile (1_V \times v')\} \\ &= \pm z\backslash\{[U' \mid X \times (V, V - X)] \smile (1_X \times v')\} \\ &= \pm[\theta(1) \mid (V, V - X)] \smile v' \\ &= \pm[\theta(1) \smile v] \mid (V, V - X)\end{aligned}$$

Since $H^*(Y, Y - X) \approx H^*(V, V - X)$, this gives the result. ∎

Our next result, a consequence of lemma 15, follows immediately from the definition of the cup product, $H^p(Y, Y - X) \smile H^q(X) \to H^{p+q}(Y, Y - X)$.

16 COROLLARY *Let X be a compact oriented n-manifold imbedded in an oriented m-manifold Y, both without boundary. For any element $v \in H^q(Y;G)$ we have*

$$\theta(v \mid X) = \pm\theta(1) \smile v \quad ∎$$

The *normal Euler class of X in Y*, denoted by $\chi_{X,Y} \in H^{m-n}(X;R)$, is defined by the equation

$$\theta(\chi_{X,Y}) = \theta(1) \smile \theta(1) \in H^{2(m-n)}(Y, Y - X; R)$$

Since $\theta(1) \smile \theta(1) = \theta(1) \smile [\theta(1) \mid Y]$, we obtain from corollary 16 the following characterization of the normal Euler class.

17 THEOREM *If a compact n-manifold X is imbedded in an m-manifold Y, both without boundary and oriented over R, the normal Euler class $\chi_{X,Y} = \theta(1) \mid X$.* ∎

In particular, if $H^{m-n}(Y;R) \to H^{m-n}(X;R)$ is trivial, it follows that the normal Euler class is zero. Thus, if Y is Euclidean space, the normal Euler class of any compact X imbedded in Y is zero.

For $i \geq 0$ the *ith normal Stiefel-Whitney class of X in Y*, $\bar{w}_i \in H^i(X;\mathbf{Z}_2)$, is defined by

$$\theta(\bar{w}_i) = Sq^i\theta(1)$$

Here are some examples.

18 By condition (*a*) on page 271, $\bar{w}_0 = 1$.

19 By condition (*b*) on page 271, if $k = \dim Y - \dim X$ then $\bar{\omega}_k$ is the normal Euler class of X in Y over $\mathbf{Z}_2$.

20 By condition (*c*) on page 271, $\bar{w}_i = 0$ for $i > \dim Y - \dim X$.

There is an important relation between the Stiefel-Whitney classes of X and Y and the normal Stiefel-Whitney classes of X in Y toward which we are heading.

21 LEMMA *Let X be a compact n-manifold imbedded in an m-manifold Y, both without boundary. Let U and U' be the orientation classes of X and Y, respectively, over $\mathbf{Z}_2$ and let $\theta(1) \in H^{m-n}(Y, Y - X; \mathbf{Z}_2)$. Then*

$$U' \mid (X \times Y, X \times Y - \delta(X)) = [1 \times \theta(1)] \smile U$$

PROOF If X' is a component of X, it suffices to prove that

$$U' \mid (X' \times Y, X' \times Y - \delta(X')) = ([1 \times \theta(1)] \smile U) \mid (X' \times Y, X' \times Y - \delta(X'))$$

Hence we may assume X connected, in which case $(X \times Y, X \times Y - \delta(X))$ is a fiber-bundle pair over X with fiber pair $(Y, Y - x_0)$, where $x_0 \in X$. Since $U' \mid (X \times Y, X \times Y - \delta(X))$ is an orientation over $\mathbf{Z}_2$ of this bundle pair, and there is a unique orientation over $\mathbf{Z}_2$, it suffices to prove that $[1 \times \theta(1)] \smile U$ is also an orientation over $\mathbf{Z}_2$ of this bundle pair. That is, we need only show that for $x \in X$, $([1 \times \theta(1)] \smile U) \mid x \times (Y, Y - x)$ is nonzero. This will be so if its image in $x \times (Y, Y - X)$, which equals $([1 \times \theta(1)] \smile U) \mid x \times (Y, Y - X)$, is nonzero. Because $U \in H^n(X \times X, X \times X - \delta(X))$ is an orientation, $U \mid x \times (X, X - x) = 1_x \times u$, where $u \in H^n(X, X - x)$ is nonzero. Because $H^n(X, X - x) \to H^n(X)$ is a monomorphism [dual to the monomorphism $H_0(x) \to H_0(X)$], $u \mid X$ is nonzero. We have

$$\begin{aligned}([1 \times \theta(1)] \smile U) \mid x \times (Y, Y - X) &= [1_x \times \theta(1)] \smile (1_x \times u \mid X) \\ &= 1_x \times [\theta(1) \smile u \mid X] = 1_x \times \theta(u \mid X)\end{aligned}$$

Since θ is an isomorphism, this implies that $([1 \times \theta(1)] \smile U) \mid x \times (Y, Y - X)$ is nonzero. ■

From this result we have the following *Whitney duality theorem.*

22 THEOREM *Let X be a compact n-manifold imbedded in an m-manifold Y, both without boundary. For $k \geq 0$*

$$w_k(Y) \mid X = \sum_{i+j=k} \bar{w}_i \smile w_j(X)$$

where $w_k(Y)$, $w_j(X)$, and $\bar{w}_i$ denote the Stiefel-Whitney classes of Y, X, and X in Y, respectively.

PROOF The result follows easily from lemma 21 and the Cartan formula (rather, the equivalent form of lemma 5.9.4):

$$\begin{aligned}
&([w_k(Y) \mid X] \times 1_Y) \smile U' \mid (X \times Y, X \times Y - \delta(X)) \\
&\quad = ([w_k(Y) \times 1_Y] \smile U') \mid (X \times Y, X \times Y - \delta(X)) \\
&\quad = Sq^k U' \mid (X \times Y, X \times Y - \delta(X)) = Sq^k(U' \mid (X \times Y, X \times Y - \delta(X))) \\
&\quad = Sq^k([1_X \times \theta(1)] \smile U) = \sum_{i+j=k} [1_X \times Sq^i\theta(1)] \smile Sq^j U \\
&\quad = \sum_{i+j=k} (1_X \times [\theta(1) \smile \bar{w}_i]) \smile [w_j(X) \times 1_X] \smile U \\
&\quad = \sum_{i+j=k} (\bar{w}_i \times 1_X) \smile [w_j(X) \times 1_X] \smile [1_X \times \theta(1)] \smile U \\
&\quad = (([\sum_{i+j=k} \bar{w}_i \smile w_j(X)] \times 1_Y) \smile U') \mid (X \times Y, X \times Y - \delta(X))
\end{aligned}$$

By the Thom isomorphism theorem, this implies the result. ■

In case Y is Euclidean space, $w_k(Y) = 0$ for $k > 0$, and theorem 22 shows that $\bar{w}_i$ and $w_j(X)$ determine each other recursively. In particular, the classes $\bar{w}_i$ are independent of the imbedding of X in the Euclidean space. If X is a compact n-manifold imbedded in $\mathbf{R}^{n+d}$, it follows from example 19 and 20 and from the fact that the Euler class of X in $\mathbf{R}^{n+d}$ is zero that $\bar{w}_i = 0$ for $i \geq d$. This gives the following necessary condition for imbeddability of X in $\mathbf{R}^{n+d}$.

23 COROLLARY *Let X be a compact n-manifold imbedded in $\mathbf{R}^{n+d}$ and let $\bar{w}_i \in H^i(X;\mathbf{Z}_2)$ be defined by*

$$\sum_{i+j=k} \bar{w}_i \cup w_j(X) = \begin{cases} 1 & k = 0 \\ 0 & k > 0 \end{cases}$$

Then $\bar{w}_i = 0$ for $i \geq d$. ■

We present some examples.

24 For P^2, $\bar{w}_1(P^2) = w$ and $\bar{w}_2(P^2) = 0$, so P^2 cannot be imbedded in $\mathbf{R}^3$.

25 For P^3, $\bar{w}_i(P^3) = 0$ for $i > 0$.

26 For P^4, $\bar{w}_1(P^4) = w$, $\bar{w}_2(P^4) = w^2$, $\bar{w}_3(P^4) = w^3$, and $\bar{w}_4(P^4) = 0$. Therefore P^4 cannot be imbedded in $\mathbf{R}^7$.

27 For P^5, $\bar{w}_1(P^5) = 0$, $\bar{w}_2(P^5) = w^2$, $\bar{w}_3(P^5) = 0$, $\bar{w}_4(P^5) = 0$, and $\bar{w}_5(P^5) = 0$. Hence P^5 cannot be imbedded in $\mathbf{R}^7$ (which is also a consequence of example 26).

The last examples show the importance of calculating $w_i(P^n)$, which we now do.

28 THEOREM *Let $\binom{n}{i}_2$ be the binomial coefficient $\binom{n}{i} = n!/i!(n-i)!$ reduced modulo 2. Then*

$$w_i(P^n) = \binom{n+1}{i}_2 w^i$$

PROOF Since $\binom{n+1}{n}_2 \equiv n + 1 = \chi(P^n)$, the result is true for $i = n$. For $i < n$, where $n > 1$, we suppose P^{n-1} linearly imbedded in P^n. Then $P^n - P^{n-1}$ is an affine space, hence $\tilde{H}^*(P^n - P^{n-1}) = 0$ and $H^q(P^n, P^n - P^{n-1}) \approx \tilde{H}^q(P^n)$. Then the normal Thom class $\theta(1) \in H^1(P^n, P^n - P^{n-1})$ maps to w in $H^q(P^n)$, so $\bar{w}_1 = w$. By theorem 22, $w_i(P^n) \mid P^{n-1} = w_i(P^{n-1}) + w \cup w_{i-1}(P^{n-1})$. Since $H^q(P^n) \approx H^q(P^{n-1})$ for $q < n$, it follows by induction on n that

$$w_i(P^n) = [\binom{n}{i-1}_2 + \binom{n}{i}_2]w^i = \binom{n+1}{i}_2 w^i \quad ■$$

EXERCISES

A MANIFOLDS

1 If X is an n-manifold with boundary $\dot{X}$, prove that X is a homology n-manifold whose boundary, as a homology manifold, equals $\dot{X}$.

In the rest of the exercises of this group, X will be an n-manifold without boundary and R will be a fixed principal ideal domain.

2 If Γ is a local system of R modules on X, prove that for any $A \subset X$

$$H_q(A \times X, A \times X - \delta(A); R \times \Gamma) = 0 \qquad q < n$$

(*Hint:* Prove this first for $\bar{A}$ contained in a coordinate neighborhood of X. Prove it next for compact $\bar{A}$ by using the Mayer-Vietoris technique. Then prove it for arbitrary A by taking direct limits over the family of compact subsets of A.)

3 Prove that there is a local system Γ_X of R modules on X such that $\Gamma_X(x) = H^n(X, X - x; R)$ for $x \in X$.

For $x \in X$ let $z_x \in H_n(X, X - x; \Gamma_X)$ be the generator corresponding under the isomorphism

$$H_n(X, X - x; \Gamma_X) \approx \text{Hom}\,(H^n(X, X - x; R), H^n(X, X - x; R))$$

to the identity homomorphism of $H^n(X, X - x; R)$. A *Thom class* of X is an element

$$U \in H^n(X \times X, X \times X - \delta(X); R \times \text{Hom}\,(\Gamma_X, R))$$

such that $(U \mid [x \times (X, X - x)])/z_x = 1 \in H^0(x;R)$ for all $x \in X$.

4 If V is an open subset of X and U is a Thom class of X, prove that $U \mid (V \times V, V \times V - \delta(V))$ is a Thom class of V.

5 Prove that $\mathbf{R}^n$ has a unique Thom class.

6 Prove that X has a unique Thom class. [*Hint:* Use exercise 2 to show that

$$H^n(X \times X, X \times X - \delta(X); R \times \text{Hom}\,(\Gamma_X, R)) \approx \varprojlim \{H^n(V \times X, V \times X - \delta(V); R \times \text{Hom}\,(\Gamma_X, R))\}$$

where V varies over finite unions of coordinate neighborhoods. Then the result follows from exercises 4 and 5 by Mayer-Vietoris techniques.]

If (A,B) is a pair in X and G is an R module, define

$$\gamma\colon H_q(X - B, X - A; \Gamma_X \otimes G) \to H^{n-q}(A,B; G)$$

by $\gamma(z) = [U \mid (A,B) \times (X - B, X - A)]/z$, where U is the Thom class of X. As (V,W) varies over neighborhoods of a closed pair (A,B) in X, there are isomorphisms

$$\varinjlim \{H_q(X - W, X - V; \Gamma_X \otimes G)\} \approx H_q(X - B, X - A; \Gamma_X \otimes G)$$

and

$$\varinjlim \{H^{n-q}(V,W; G)\} \approx \bar{H}^{n-q}(A,B; G)$$

and a homomorphism

$$\bar{\gamma}\colon H_q(X - B, X - A; \Gamma_X \otimes G) \to \bar{H}^{n-q}(A,B; G)$$

is defined by passing to the limit with γ.

7 *Duality theorem.* Prove that for a compact pair (A,B) in X, $\bar{\gamma}$ is an isomorphism.

B THE INDEX OF A MANIFOLD

1 Let X be a compact n-manifold, with boundary $\dot{X}$ oriented over a field R, and let $[X] \in H_n(X,\dot{X}; R)$ be the corresponding fundamental class. For $u \in H^q(X,\dot{X}; R)$ and $v \in H^{n-q}(X;R)$ prove that $\varphi_X(u,v) = \langle u \smile v, [X]\rangle \in R$ is a nonsingular bilinear form from $H^q(X,\dot{X}) \times H^{n-q}(X)$ to R [that is, $u = 0$ if and only if $\varphi_X(u,v) = 0$ for all v].

2 With the same hypotheses as above, let $[\dot{X}] = \partial[X] \in H_{n-1}(\dot{X};R)$ and let $\varphi_{\dot{X}}$ be the corresponding bilinear form from $H^{q-1}(\dot{X};R) \times H^{n-q}(\dot{X};R)$ to R. Let $j\colon \dot{X} \subset X$, and if $u \in H^{q-1}(\dot{X};R)$ and $v \in H^{n-q}(X;R)$, prove that

$$\varphi_{\dot X}(u,j^*(v)) = \varphi_X(\delta(u),v)$$

3 Prove that the Euler characteristic of any odd-dimensional compact manifold is 0 and the Euler characteristic of an even-dimensional compact manifold which is a boundary is even. (*Hint:* If $\dot X$ is the boundary of a $(2n+1)$-manifold X, then, with $\mathbf{Z}_2$ coefficients,

$$\dim \operatorname{im} [j^*: H^n(X) \to H^n(\dot X)] = \dim \operatorname{im} [\delta: H^n(\dot X) \to H^{n+1}(X,\dot X)]$$

and their sum equals $\dim H^n(\dot X)$.)

Let Y be a compact $4m$-manifold, without boundary oriented over $\mathbf{R}$, and define the *index* of Y to be the index of the nonsingular bilinear form φ_Y from $H^{2m}(Y;\mathbf{R}) \times H^{2m}(Y;\mathbf{R})$ to $\mathbf{R}$ (when φ_Y is represented as a sum of k squares minus a sum of j squares, the index of φ_Y is $k - j$).

4 If Y is oppositely oriented, prove that its index changes sign. Show that the index of the product of oriented manifolds is the product of their indices.

5 If X is a compact $(4m+1)$-manifold, with boundary $\dot X$ oriented over $\mathbf{R}$, prove that The index of $\dot X$ is 0. [*Hint:* Prove that $j^*(H^{2m}(X;\mathbf{R}))$ is a subspace of $H^{2m}(\dot X;\mathbf{R})$ whose dimension equals one-half the dimension of $H^{2m}(\dot X;\mathbf{R})$ and on which $\varphi_{\dot X}$ is identically zero. This implies the result.]

C CONTINUITY

1 Let $\{(X_j,A_j), \pi_j^k\}_{j\in J}$ be an inverse system of compact Hausdorff pairs and let $(X,A) = \lim_{\leftarrow} \{(X_j,A_j)\}$. Prove that (X,A) can be imbedded in a space in which it is a directed intersection of compact Hausdorff pairs $\{(X_j',A_j')\}_{j\in J}$, where (X_j',A_j') has the same homotopy type as (X_j,A_j). [*Hint:* For each $j \in J$ imbed X_j in a contractible compact Hausdorff space Y_j, for example, a cube, and let $(X_k',A_k') \subset \times_{j\in J} Y_j$ be defined as the pair of all points (y_j) with y_k in X_k or in A_k, respectively, such that if $j \le k$, then $y_j = \pi_j^k(y_k)$, and if $j \not\le k$, then y_j is arbitrary.]

2 Prove that a cohomology theory has the continuity property if and only if it has the weak continuity property.

3 The *p-adic solenoid* is defined to be the inverse limit of the sequence

$$S^1 \xleftarrow{f} S^1 \leftarrow \cdots \leftarrow S^1 \xleftarrow{f} S^1 \leftarrow \cdots$$

where $f(z) = z^p$. Compute the Alexander cohomology groups of the p-adic solenoid for coefficients $\mathbf{Z}$, $\mathbf{Z}_p$, and $\mathbf{R}$.

4 Generalize the solenoid of the preceding example to the case where there is a sequence of integers $n_1, n_2, \ldots$ such that the mth map of S^1 to S^1 sends z to z^{n_m}. Compute the integral Alexander cohomology groups of the resulting space.

5 Find a compact Hausdorff space X such that $\bar{\bar H}^q(X;\mathbf{Z}) = 0$ if $q \ne 1$ and $\bar H^1(X;\mathbf{Z}) \approx \mathbf{R}$.

D ČECH COHOMOLOGY THEORY

1 Let $(\mathcal{U},\mathcal{U}')$ be an open covering of (X,A) ($\mathcal{U}$ is an open covering of X and $\mathcal{U}' \subset \mathcal{U}$ is a covering of A) and let $K(\mathcal{U})$ be the nerve of $\mathcal{U}$ and $K'(\mathcal{U}')$ the subcomplex of $K(\mathcal{U})$ which is the nerve of $\mathcal{U}' \cap A = \{U' \cap A \mid U' \in \mathcal{U}'\}$. Prove that the chain complexes $(C(K(\mathcal{U})),C(K'(\mathcal{U}')))$ and $(C(X(\mathcal{U})),C(A(\mathcal{U}')))$ are canonically chain equivalent. (*Hint:* If $s = \{U_0, \ldots, U_q\}$ is a simplex of $K(\mathcal{U})$ [or of $K'(\mathcal{U}')$], let $\lambda(s)$ be the subcomplex of $X(\mathcal{U})$ [or of $A(\mathcal{U}')$] generated by all simplexes of $X(\mathcal{U})$ [or of $A(\mathcal{U}')$] in $\cap U_i$. If $s' = \{x_0, \ldots, x_q\}$ is a simplex of $X(\mathcal{U})$ [or of $A'(\mathcal{U}')$], let $\mu(s')$ be the subcomplex of $K(\mathcal{U})$ [or of $K'(\mathcal{U}')$] generated by all simplexes $\{U_0, \ldots, U_r\}$ of $K(\mathcal{U})$ [or of $K'(\mathcal{U}')$] such that U_i contains s' for $0 \le i \le r$. Then $C(\lambda(s))$ and $C(\mu(s'))$ are acyclic, and the method of acyclic models can be applied to prove the existence of chain maps

$$\tau: (C(K(\mathcal{U})), C(K'(\mathcal{U}'))) \to (C(X(\mathcal{U})), C(A(\mathcal{U}')))$$
$$\tau': (C(X(\mathcal{U})), C(A(\mathcal{U}))) \to (C(K(\mathcal{U})), C(K'(\mathcal{U}')))$$

such that $\tau(C(s)) \subset C(\lambda(s))$ and $\tau'(C(s')) \subset C(\mu(s'))$. Similarly, the method of acyclic models shows that τ and τ' are chain homotopy inverses of each other.[1])

2 Let $(\mathcal{V},\mathcal{V}')$ be a refinement of $(\mathcal{U},\mathcal{U}')$, let $\pi: (K(\mathcal{V}),K'(\mathcal{V}')) \to (K(\mathcal{U}),K'(\mathcal{U}'))$ be a projection map, and let $j: (X(\mathcal{V}),A(\mathcal{V}')) \subset (X(\mathcal{U}),A(\mathcal{U}'))$. For any abelian group G prove that there is a commutative diagram

$$\begin{array}{ccc} H^*(K(\mathcal{U}),K'(\mathcal{U}'); G) & \approx & H^*(X(\mathcal{U}),A(\mathcal{U}'); G) \\ \pi^*\downarrow & & \downarrow j^* \\ H^*(K(\mathcal{V}),K'(\mathcal{V}'); G) & \approx & H^*(X(\mathcal{V}),A(\mathcal{V}'); G) \end{array}$$

where the horizontal maps are induced by the canonical chain equivalences of exercise 1 above.

3 The *Čech cohomology group* of (X,A) with coefficients G is defined by $\check{H}^*(X,A; G) = \lim_{\rightarrow} \{H^*(K(\mathcal{U}),K'(\mathcal{U}'); G)\}$. Prove that there is a natural isomorphism

$$\check{H}^*(X,A; G) \approx \bar{H}^*(X,A; G).$$

4 If $\dim (X - A) \le n$, prove that $\bar{H}^q(X,A; G) = 0$ for all $q > n$ and all G.

E THE KÜNNETH FORMULA FOR ČECH COHOMOLOGY

If K_1 and K_2 are simplicial complexes, their *simplicial product* $K_1 \Delta K_2$ is the simplicial complex whose vertex set is the cartesian product of the vertex sets of K_1 and of K_2 and whose simplexes are sets $\{(v_0,w_0), \ldots, (v_q,w_q)\}$, where $v_0, \ldots, v_q$ are vertices of some simplex of K_1 and $w_0, \ldots, w_q$ are vertices of some simplex of K_2.

1 Prove that $K_1 \Delta K_2$ is a simplicial complex, and if $L_1 \subset K_1$ and $L_2 \subset K_2$, then $L_1 \Delta L_2 \subset K_1 \Delta K_2$.

2 For simplicial pairs (K_1,L_1) and (K_2,L_2) define

$$(K_1,L_1) \Delta (K_2,L_2) = (K_1 \Delta K_2, K_1 \Delta L_2 \cup L_1 \Delta K_2)$$

Prove that $C((K_1,L_1) \Delta (K_2,L_2))$ is canonically chain equivalent to $C(K_1,L_1) \otimes C(K_2,L_2)$. (*Hint:* Use the method of acyclic models.)

3 Call an open covering $(\mathcal{U},\mathcal{U}')$ of (X,A) *special* if $\mathcal{U}' = \{U \in \mathcal{U} \mid U \cap A \neq \phi\}$. If $(\mathcal{U},\mathcal{U}')$ is a special open covering of (X,A) and $(\mathcal{V},\mathcal{V}')$ is a special open covering of (Y,B), let $(\mathcal{U},\mathcal{U}') \times (\mathcal{V},\mathcal{V}') = (\mathcal{W},\mathcal{W}')$ be the special open covering of $(X,A) \times (Y,B)$ where $\mathcal{W} = \{U \times V \mid U \in \mathcal{U}, V \in \mathcal{V}\}$ and $\mathcal{W}' = \{U \times V \mid U \in \mathcal{U}' \text{ or } V \in \mathcal{V}'\}$. Prove that $(K(\mathcal{W}), K'(\mathcal{W}')) = (K(\mathcal{U}), K'(\mathcal{U}')) \Delta (K(\mathcal{V}), K'(\mathcal{V}'))$.

4 If A is closed in X, prove that the family of special open coverings of (X,A) is cofinal in the family of all open coverings of (X,A). If (X,A) and (Y,B) are compact Hausdorff pairs, prove that the family of coverings of $(X,A) \times (Y,B)$ of the form $(\mathcal{U},\mathcal{U}') \times (\mathcal{V},\mathcal{V}')$ where $(\mathcal{U},\mathcal{U}')$ is a special open covering of (X,A) and $(\mathcal{V},\mathcal{V}')$ is a special open covering of (Y,B) is cofinal in the family of all open coverings of $(X,A) \times (Y,B)$.

[1] For details see C. H. Dowker, Homology groups of relations, *Annals of Mathematics*, vol. 56, pp. 84–95, 1952.

5 If (X,A) and (Y,B) are compact Hausdorff pairs and G and G' are modules such that $G * G' = 0$, prove that there is a short exact sequence

$$0 \to (\check{H}_1^* \otimes \check{H}_2^*)^q \to \check{H}^q((X,A) \times (Y,B); G \otimes G') \to (\check{H}_1^* * \check{H}_2^*)^{q+1} \to 0$$

where $\check{H}_1^* = \check{H}^*(X,A; G)$ and $\check{H}_2^* = \check{H}^*(Y,B; G')$.

6 Let (X,A) and (Y,B) be locally compact Hausdorff pairs with A and B closed in X and Y, respectively. If G and G' are modules such that $G * G' = 0$, prove that there is a short exact sequence

$$0 \to (\bar{H}_{c,1}^* \otimes \bar{H}_{c,2}^*)^q \to \bar{H}_c^q((X,A) \times (Y,B); G * G') \to (\bar{H}_{c,1}^* * \bar{H}_{c,2}^*)^{q+1} \to 0$$

where $\bar{H}_{c,1}^* = \bar{H}_c^*(X,A; G)$ and $\bar{H}_{c,2}^* = \bar{H}_c^*(Y,B; G')$.

F LOCAL SYSTEMS AND SHEAVES

Throughout this group of exercises we assume X to be a paracompact Hausdorff space.

1 If Γ is a local system on X, let $\hat{\Gamma}$ be the presheaf on X such that for an open set $V \subset X$, $\hat{\Gamma}(V)$ is the set of all functions f assigning to each $x \in X$ an element $f(x) \in \Gamma(x)$ with the property that for any path ω in V, $f(\omega(1)) = \Gamma(\omega)(f(\omega(0)))$. Prove that $\hat{\Gamma}$ is a sheaf on X and the association of $\hat{\Gamma}$ to Γ is a natural transformation from local systems to sheaves.

2 A presheaf Γ on X is said to be *locally constant* if there is an open covering $\mathcal{U} = \{U\}$ of X such that if $U \in \mathcal{U}$ and $x \in U$, then $\Gamma(U) \approx \lim_{\rightarrow} \{\Gamma(V)\}$, where V varies over open neighborhoods of x. If $U \in \mathcal{U}$ and U' is a connected open subset of U, prove that the composite

$$\Gamma(U) \to \Gamma(U') \to \hat{\Gamma}(U')$$

is an isomorphism. Deduce that if Γ is a locally constant sheaf and U' is a connected open subset of $U \in \mathcal{U}$, then $\Gamma(U) \approx \Gamma(U')$.

3 If X is locally path connected and Γ' is a locally constant sheaf on X, prove that there is a local system Γ on X such that $\hat{\Gamma} \approx \Gamma'$.

4 If X is locally path connected and semilocally 1-connected, prove that there is a one-to-one correspondence between equivalence classes of local systems on X and equivalence classes of locally constant sheaves on X.

5 If Γ is a local system of R modules on X, let $\Delta^q(\cdot\,;\Gamma)$ be the presheaf on X such that $\Delta^q(\cdot\,;\Gamma)(V) = \Delta^q(V;\Gamma \mid V)$ for V open in X. Prove that $\Delta^q(\cdot\,;\Gamma)$ is fine.

6 If Γ is a local system of R modules on X, let $\Delta^*(\cdot\,;\Gamma)$ be the cochain complex of presheaves $\Delta^q(\cdot\,;\Gamma)$ on X and let $\hat{\Delta}^*(\cdot\,;\Gamma)$ be the cochain complex of completions $\hat{\Delta}^q(\cdot\,;\Gamma)$. Prove that there is an isomorphism

$$H^*(\Delta^*(\cdot\,;\Gamma)(X)) \approx H^*(\hat{\Delta}^*(\cdot\,;\Gamma)(X))$$

7 Let Γ be a local system of R modules on X and assume that $H^q(\Delta^*(\cdot\,;\Gamma))$ is locally zero on X for all $q > 0$. Prove that there is an isomorphism

$$\check{H}^*(X;\hat{\Gamma}) \approx H^*(X;\Gamma)$$

(*Hint:* Note that $\hat{\Gamma} = H^0(\Delta^*(\cdot\,;\Gamma))$ and apply theorem 6.8.7.)

G SOME PROPERTIES OF EUCLIDEAN SPACE

1 Find a compact subset X of $\mathbf{R}^2$ that is n-connected for all n and such that $\bar{H}^1(X;\mathbf{Z}) \approx \mathbf{Z}$.

2 If X is a compact subset of $\mathbf{R}^n$ and $\dim X < n - 1$, prove that $\mathbf{R}^n - X$ is connected.

Let A_1 and A_2 be disjoint closed subsets of $\mathbf{R}^n$ and let $z_1 \in H_p(A_1;R)$ and $z_2 \in H_q(A_2;R)$, with $p + q = n - 1$. If $z_1 \in \tilde{H}_p(A_1;R)$, let $z_1' \in H_{p+1}(\mathbf{R}^n,\mathbf{R}^n - A_2;R)$ be the image of z_1 under the composite

$$\tilde{H}_p(A_1) \to \tilde{H}_p(\mathbf{R}^n - A_2) \xrightarrow[\approx]{\partial^{-1}} H_{p+1}(\mathbf{R}^n, \mathbf{R}^n - A_2)$$

The *linking number* $\text{Lk}(z_1,z_2) \in R$ is defined by

$$\text{Lk}(z_1,z_2) = \langle \gamma_U(z_1'),z_2 \rangle$$

where U is an orientation class of $\mathbf{R}^n$ over R fixed once and for all.

3 Prove that $\text{Lk}(z_1,z_2) = \langle U, i_*(z_2 \times z_1') \rangle$, where

$$i: A_2 \times (\mathbf{R}^n, \mathbf{R}^n - A_2) \subset (\mathbf{R}^n \times \mathbf{R}^n, \mathbf{R}^n \times \mathbf{R}^n - \delta(\mathbf{R}^n))$$

4 Assume that $\text{Lk}(z_2,z_1)$ is also defined [that is, $z_2 \in \tilde{H}_q(A_2)$]. Prove that $\text{Lk}(z_1,z_2) = (-1)^{pq+1}\,\text{Lk}(z_2,z_1)$.

5 Let A_1 be a p-sphere and A_2 a q-sphere imbedded as disjoint subsets of $\mathbf{R}^n$, where $p + q = n - 1$. Prove that $H_p(A_1) \to H_p(\mathbf{R}^n - A_2)$ is trivial if and only if $H_q(A_2) \to H_q(\mathbf{R}^n - A_1)$ is trivial.

H IMBEDDINGS OF MANIFOLDS IN EUCLIDEAN SPACE

1 Prove that a compact n-manifold which is nonorientable over $\mathbf{Z}$ cannot be imbedded in $\mathbf{R}^{n+1}$.

2 Let X be a compact connected n-manifold imbedded in $\mathbf{R}^{n+1}$ and let U and V be the components of $\mathbf{R}^{n+1} - X$. Let $i: X \subset \mathbf{R}^{n+1} - U$ and $j: X \subset \mathbf{R}^{n+1} - V$ and prove that over any R, $i^*(\tilde{H}^*(\mathbf{R}^{n+1} - U))$ and $j^*(\tilde{H}^*(\mathbf{R}^{n+1} - V))$ are subalgebras of $\tilde{H}^*(X)$

$$\{i^*,j^*\}: \tilde{H}^q(\mathbf{R}^{n+1} - U) \oplus \tilde{H}^q(\mathbf{R}^{n+1} - V) \approx \tilde{H}^q(X) \qquad 0 < q < n$$

3 Prove that for $n \geq 2$ the real projective n-space P^n cannot be imbedded in $\mathbf{R}^{n+1}$.

CHAPTER SEVEN

HOMOTOPY THEORY

WITH THIS CHAPTER WE RETURN TO THE CONSIDERATION OF GENERAL HOMOTOPY theory. Now that we have homology theory available as a tool, we are able to obtain deeper results about homotopy than we could without it. We shall consider the higher homotopy groups in some detail and prove they satisfy analogues of all the axioms of homology theory except the excision axiom. We introduce the Hurewicz homomorphism as a natural transformation from the homotopy groups to the integral singular homology groups. It leads us to the Hurewicz isomorphism theorem, which states roughly that the lowest-dimensional nontrivial homotopy group is isomorphic to the corresponding integral homology group.

We discuss next the concept of CW complex. The class of CW complexes is particularly suited for homotopy theory because it is the smallest class of spaces containing the empty space and, up to homotopy type, is closed with respect to the operation of attaching cells (even an infinite number).

The last main result is the Brown representability theorem. It characterizes by means of simple properties those contravariant functors from the homotopy category of path-connected pointed CW complexes to the category of pointed sets that are naturally equivalent to the functor assigning to a *CW* complex the set of homotopy classes of maps from it to some fixed pointed space.

Section 7.1 contains a general exactness property for sets of homotopy classes. Section 7.2 contains definitions of the absolute and relative homotopy groups and proofs of the exactness of the homotopy sequences of a pair, a triple, and a fibration. In Sec. 7.3 we consider the extent to which the homotopy groups depend on the choice of the base point used in their definition and prove analogues for the higher homotopy groups of properties established in Chapter One for the fundamental group.

The Hurewicz homomorphism is defined in Sec. 7.4 and the Hurewicz isomorphism theorem is proved in Sec. 7.5. The proof establishes the absolute and relative Hurewicz theorems, as well as a homotopy addition theorem, by simultaneous induction. The Hurewicz theorem implies the Whitehead theorem, which asserts that a continuous map between simply connected spaces induces isomorphisms of all homotopy groups if and only if it induces isomorphisms of all integral singular homology groups.

Section 7.6 introduces the concept of *CW* complex. Among the elementary properties established is the cellular-approximation theorem, which is an analogue for *CW* complexes of the simplicial-approximation theorem. Section 7.7 deals with contravariant functors on the homotopy category of path-connected pointed spaces. We prove the representability theorem cited above, and apply it in Sec. 7.8 to obtain *CW* approximations to a space or a pair and to discuss the related concept of weak homotopy type. The representability theorem will be used again in Chapter Eight.

1 EXACT SEQUENCES OF SETS OF HOMOTOPY CLASSES

One of the most important properties of the homology functor is the exactness property relating the homology of the pair and the homology of each of the spaces in the pair. A similar exactness property is valid for functors defined by homotopy classes. This section is devoted to preliminaries about homotopy classes and a proof of this exactness property. Throughout the section we shall work in the category of pointed spaces, and unless stated to the contrary, (X,A) will be understood as a pair of pointed spaces (that is, A has the same base point as X) in which the subspace A and the base point are closed in X. Homotopies in this category are understood to preserve base points. If $A \subset X$, we use X/A to denote the space obtained from X by collapsing A to a single point (this point serving as the base point of X/A). If X' and A are closed subsets of X, then $A/(A \cap X')$ is a closed subset of X/X'. Hence, if (X,A) is a pair and X' is closed in X, there is a pair $(X/X', A/(A \cap X'))$, which will also be denoted by $(X,A)/X'$.

The unit interval I will be a pointed space with 0 as base point. The *reduced cone* CX over X is defined to be the space obtained from $X \times I$ by collapsing $X \times 0 \cup x_0 \times I$ to a point (so $CX = X \times I/(X \times 0 \cup x_0 \times I)$). We shall use $[x,t]$ to denote the point of CX corresponding to the point $(x,t) \in X \times I$ under the collapsing map $X \times I \to CX$. X is imbedded as a closed subset of CX by the map $x \to [x,1]$. If (X,A) is a pair, then CA is a subspace of CX and $C(X,A)$ is defined to be the pair (CX,CA).

1 LEMMA *A map f: $(X,A) \to (Y,B)$ is null homotopic if and only if there is a map F: $C(X,A) \to (Y,B)$ such that $F[x,1] = f(x)$ for all $x \in X$.*

PROOF There is a one-to-one correspondence between null homotopies H: $(X,A) \times I \to (Y,B)$ of f and maps F:$C(X,A) \to (Y,B)$ such that $F[x,1] = f(x)$, given by the formula

$$F[x,t] = H(x, 1 - t) \quad \blacksquare$$

The following relative homotopy extension property can also be deduced from the relative form of theorem 1.4.12.

2 LEMMA *Given f: $C(X,A) \to (Y,B)$ and a homotopy G: $(X,A) \times I \to (Y,B)$ of $f \mid (X,A)$, there is a homotopy F: $C(X,A) \times I \to (Y,B)$ of f such that $F \mid (X,A) \times I = G$.*

PROOF An explicit formula for F is

$$F([x,t], t') = \begin{cases} f[x, t(1 + t')] & t(1 + t') \le 1 \\ G(x, t(1 + t') - 1) & 1 \le t(1 + t') \end{cases} \quad \blacksquare$$

The homotopy class of the unique constant map $(X,A) \to (Y,B)$ is denoted by $0 \in [X,A; Y,B]$ [it consists of the null-homotopic maps $(X,A) \to (Y,B)$]. Because the composite, on either side, of a null-homotopic map and an arbitrary map is null homotopic, the element 0 is a distinguished element of $[X,A; Y,B]$, and we regard $[X,A; Y,B]$ as a pointed set with this distinguished element. Given a map f: $(X',A') \to (X,A)$, the *kernel* of the induced map

$$f^\#: [X,A; Y,B] \to [X',A'; Y,B]$$

is defined to be the pointed set $f^{\#-1}(0)$ and is denoted by $\ker f^\#$.

We now show how to map another set of homotopy classes into $[X,A; Y,B]$ so that its image equals $\ker f^\#$. This will be the basis for the exactness property we seek. The *mapping cone* C_f of a map f: $X' \to X$ is defined to be the quotient space of $CX' \vee X$ by the identifications $[x',1] = f(x')$ for all $x' \in X'$. Given a map f: $(X',A') \to (X,A)$, let f': $X' \to X$ and f'': $A' \to A$ be maps defined by f. Then $C_{f''}$ is a closed subspace of $C_{f'}$ and there is a pair $(C_{f'},C_{f''})$. There is a functorial imbedding i of (X,A) as a closed subpair of $(C_{f'},C_{f''})$.

A three-term sequence of pairs and maps

$$(X',A') \xrightarrow{f} (X,A) \xrightarrow{g} (X'',A'')$$

is said to be *exact* if for any pair (Y,B) (where B is not necessarily closed in Y) the associated sequence of pointed sets

$$[Y,B;\ X',A'] \xrightarrow{f_\#} [Y,B;\ X,A] \xrightarrow{g_\#} [Y,B;\ X'',A'']$$

is exact (that is, $\ker g_\# = \operatorname{im} f_\#$). Similarly, it is said to be *coexact* if the sequence of pointed sets

$$[X'',A'';\ Y,B] \xrightarrow{g^\#} [X,A;\ Y,B] \xrightarrow{f^\#} [X',A';\ Y,B]$$

is exact (that is, $\ker f^\# = \operatorname{im} g^\#$). A sequence of pairs and maps (which may terminate at either or both ends)

$$\cdots \to (X_{n+1},A_{n+1}) \xrightarrow{f_n} (X_n,A_n) \xrightarrow{f_{n-1}} (X_{n-1},A_{n-1}) \to \cdots$$

is said to be an *exact sequence* (or a *coexact sequence*) if every three-term sequence of consecutive pairs is exact (or coexact).

3 THEOREM *For any map $f\colon (X',A') \to (X,A)$ the sequence*

$$(X',A') \xrightarrow{f} (X,A) \xrightarrow{i} (C_{f'},C_{f''})$$

is coexact.

PROOF Let (Y,B) be arbitrary (with B not necessarily closed in Y) and consider the sequence

$$[C_{f'},C_{f''};\ Y,B] \xrightarrow{i^\#} [X,A;\ Y,B] \xrightarrow{f^\#} [X',A';\ Y,B]$$

We now show that $\operatorname{im} i^\# \subset \ker f^\#$. The composite $i \circ f\colon (X',A') \to (C_{f'},C_{f''})$ equals the composite

$$(X',A') \subset C(X',A') \subset C(X',A') \vee (X,A) \xrightarrow{k} (C_{f'},C_{f''})$$

where k is the canonical map to the quotient. However, the inclusion map $(X',A') \subset C(X',A')$ is null homotopic [by lemma 1, because this inclusion map can be extended to the identity map of $C(X',A')$]. Therefore $i \circ f$ is null homotopic, and so $\operatorname{im}(f^\# \circ i^\#) = 0$, proving that $\operatorname{im} i^\# \subset \ker f^\#$.

Assume that $g\colon (X,A) \to (Y,B)$ is such that $f^\#[g] = 0$ (that is, $g \circ f$ is null homotopic). By lemma 1, there is a map $G\colon C(X',A') \to (Y,B)$ which extends $g \circ f$. Then G and g define a map $G'\colon C(X',A') \vee (X,A) \to (Y,B)$ such that $G' \mid C(X',A') = G$ and $G' \mid (X,A) = g$. Since

$$G'[x',1] = G[x',1] = g(f(x')) = G'(f(x')) \qquad x' \in X'$$

there is a map $h\colon (C_{f'},C_{f''}) \to (Y,B)$ such that $G' = h \circ k$. Then $h \mid (X,A) = g$, showing that $h \circ i = g$ or $[g] = i^\#[h]$. Therefore $\ker f^\# \subset \operatorname{im} i^\#$. ∎

For a map $f\colon (X',A') \to (X,A)$ we have a sequence

4 $$(X',A') \xrightarrow{f} (X,A) \xrightarrow{i} (C_{f'},C_{f''}) \xrightarrow{j} (C_{i'},C_{i''}) \xrightarrow{l} (C_{j'},C_{j''})$$

and by theorem 3, it follows that this sequence is coexact.

Thus we have succeeded in imbedding the map

$$f^\#\colon [X,A;\ Y,B] \to [X',A';\ Y,B]$$

in an exact sequence. We shall show that the pairs $(C_{i'},C_{i''})$ and $(C_{j'},C_{j''})$ in sequence 4 can be replaced by other pairs more explicitly expressed in terms of (X',A'), (X,A), and f.

5 LEMMA *Let (Y,B) be a pair and let Y' be a closed subset of Y. Assume that there is a homotopy $H: (Y,B) \times I \to (Y,B)$ such that*

(a) $H(y,0) = y$, for $y \in Y$.
(b) $H(Y' \times I) \subset Y'$.
(c) $H(Y' \times 1) = y_0$.

Then the collapsing map $k: (Y,B) \to (Y,B)/Y'$ is a homotopy equivalence.

PROOF Define a map $f: (Y,B)/Y' \to (Y,B)$ by the equation

$$f(k(y)) = H(y,1) \qquad y \in Y$$

[this is well-defined, because $H(Y' \times 1) = y_0$]. We show that f is a homotopy inverse of k. By definition of f, we see that H is a homotopy from $1_{(Y,B)}$ to $f \circ k$. On the other hand, because $H(Y' \times 1) \subset Y'$, there is a homotopy

$$H': ((Y,B)/Y') \times I \to (Y,B)/Y'$$

such that $H'(k(y),t) = k(H(y,t))$ for $y \in Y$ and $t \in I$. Then

$$k(f(k(y))) = k(H(y,1)) = H'(k(y),1) \qquad y \in Y$$

Therefore H' is a homotopy from the identity map of $(Y,B)/Y'$ to $k \circ f$, and f is a homotopy inverse of k. ∎

6 COROLLARY *Let $f: (X',A') \to (X,A)$ be a map and let $i: (X,A) \subset (C_{f'},C_{f''})$. Then $CX \subset C_{i'}$, $(C_{i'},C_{i''})/CX = (C_{f'},C_{f''})/X$, and the collapsing map*

$$k: (C_{i'},C_{i''}) \to (C_{i'},C_{i''})/CX$$

is a homotopy equivalence.

PROOF $C_{i'}$ is the quotient space of $CX' \vee CX$ with the identifications $[x',1] = [f(x'),1]$ for all $x' \in X'$, hence $CX \subset C_{i'}$. Since $C_{i'}$ is the union of the closed subspaces CX and $C_{f'}$, it follows that

$$C_{i'}/CX = C_{f'}/(C_{f'} \cap CX) = C_{f'}/X$$

Similarly, $C_{i''}/CA = C_{f''}/A$, and because $C_{i''} \cap CX = CA$,

$$(C_{i'},C_{i''})/CX = (C_{f'},C_{f''})/X$$

This proves the first two parts of the corollary.

Let $F: C(X,A) \times I \to C(X,A)$ be the contraction defined by $F([x,t], t') = [x, (1 - t')t]$ and let $g: C(X',A') \to (C_{i'},C_{i''})$ be the composite

$$C(X',A') \subset C(X',A') \vee C(X,A) \to (C_{i'},C_{i''})$$

where the second map is the canonical map. The composite

$$(X',A') \times I \xrightarrow{f \times 1} (X,A) \times I \subset C(X,A) \times I \xrightarrow{F} C(X,A) \subset (C_{i'},C_{i''})$$

is a homotopy $G: (X',A') \times I \to (C_{i'},C_{i''})$ such that $G(x',0) = [f(x'),1] = g[x',1]$. By lemma 2, there is a homotopy $F': C(X',A') \times I \to (C_{i'}C_{i''})$ such that $F' \mid (X',A') \times I = G$ and $F'([x',t], 0) = g([x',t])$. Then a homotopy

$$H: (C_{i'},C_{i''}) \times I \to (C_{i'},C_{i''})$$

is defined by the equations

$$\begin{aligned} H([x',t], t') &= F'([x',t], t') \qquad & x' \in X';\ t, t' \in I \\ H([x,t], t') &= F([x,t], t') \qquad & x \in X;\ t, t' \in I \end{aligned}$$

[this is well-defined because $F'([x',1], t') = G(x',t') = F([f(x'),1], t')]$. Then H satisfies a, b, and c of lemma 5 with $(Y,B) = (C_{i'},C_{i''})$ and $Y' = CX$. Therefore $k: (C_{i'},C_{i''}) \to (C_{i'},C_{i''})/CX$ is a homotopy equivalence. ∎

Recall from Sec. 1.6 that the suspension SX is defined as the space $X \times I/(X \times 0 \cup x_0 \times I \cup X \times 1)$ (therefore $SX = CX/X$). For a pair (X,A) we define $S(X,A) = (SX,SA)$. Then, for any map $f: (X',A') \to (X,A)$, we have $(C_{f'},C_{f''})/X = S(X',A')$, and we let $k: (C_{f'},C_{f''}) \to S(X',A')$ be the collapsing map.

7 LEMMA *For any map $f: (X',A') \to (X,A)$ the sequence*

$$(X',A') \xrightarrow{f} (X,A) \xrightarrow{i} (C_{f'},C_{f''}) \xrightarrow{k} S(X',A') \xrightarrow{Sf} S(X,A)$$

is coexact.

PROOF We shall use the coexact sequence 4,

$$(X',A') \xrightarrow{f} (X,A) \xrightarrow{i} (C_{f'},C_{f''}) \xrightarrow{j} (C_{i'},C_{i''}) \xrightarrow{l} (C_{j'},C_{j''})$$

By corollary 6, there is a homotopy equivalence

$$(C_{i'},C_{i''}) \xrightarrow{k'} (C_{f'},C_{f''})/X = S(X',A')$$

and the composite $(C_{f'},C_{f''}) \xrightarrow{j} (C_{i'},C_{i''}) \xrightarrow{k'} S(X',A')$ is seen to be the collapsing map $k: (C_{f'},C_{f''}) \to S(X',A')$. Also by corollary 6, there is a homotopy equivalence

$$(C_{j'},C_{j''}) \xrightarrow{k''} (C_{j'},C_{j''})/CC_{f'} = (C_{i'},C_{i''})/C_{f'} = S(X,A)$$

and the composite $(C_{i'},C_{i''}) \xrightarrow{l} (C_{j'},C_{j''}) \xrightarrow{k''} S(X,A)$ is easily seen to be the collapsing map $\bar{k}: (C_{i'},C_{i''}) \to (C_{i'},C_{i''})/C_{f'} = S(X,A)$. Let $g: S(X',A') \to S(X,A)$ be the map defined by $g([x',t]) = [f(x'), 1 - t]$. The triangle

$$\begin{array}{ccc} & (C_{i'},C_{i''}) & \\ {\scriptstyle k'}\swarrow & & \searrow{\scriptstyle \bar{k}} \\ S(X',A') & \xrightarrow{g} & S(X,A) \end{array}$$

is homotopy commutative because a homotopy

$$H: (C_{i'},C_{i''}) \times I \to S(X,A)$$

from $\bar{k}$ to $g \circ k'$ is defined by

$$\begin{aligned} H([x',t], t') &= [f(x'), 1 - tt'] \qquad & x' \in X';\ t, t' \in I \\ H([x,t], t') &= [x, (1 - t')t] \qquad & x \in X;\ t, t' \in I \end{aligned}$$

[this is well-defined because $H([x',1], t') = [f(x'), 1 - t'] = H([f(x'),1], t')$]. Therefore there is a homotopy-commutative diagram

$$\begin{array}{ccccc} (C_{f'},C_{f''}) & \xrightarrow{j} & (C_{i'},C_{i''}) & \xrightarrow{l} & (C_{j'},C_{j''}) \\ & \searrow^{k} & \downarrow^{k'} & & \downarrow^{k''} \\ & & S(X',A') & \xrightarrow{g} & S(X,A) \end{array}$$

in which k' and k'' are homotopy equivalences. From the coexactness of the sequence 4, the coexactness of the sequence

$$(X',A') \xrightarrow{f} (X,A) \xrightarrow{i} (C_{f'},C_{f''}) \xrightarrow{k} S(X',A') \xrightarrow{g} S(X,A)$$

follows. Let $h\colon S(X,A) \to S(X,A)$ be the homeomorphism defined by $h([x,t]) = [x, 1 - t]$. The coexactness of the above sequence implies the coexactness of the sequence

$$(X',A') \xrightarrow{f} (X,A) \xrightarrow{i} (C_{f'},C_{f''}) \xrightarrow{k} S(X',A') \xrightarrow{h \circ g} S(X,A)$$

Because $h \circ g = Sf$, this is the desired result. ■

8 LEMMA *If the sequence*

$$(X',A') \xrightarrow{f} (X,A) \xrightarrow{g} (X'',A'')$$

is coexact, so is the suspended sequence

$$S(X',A') \xrightarrow{Sf} S(X,A) \xrightarrow{Sg} S(X'',A'')$$

PROOF For any pair (Y,B) let $\Omega(Y,B) = (\Omega Y,\Omega B)$. By theorem 2.8 in the Introduction, there is a commutative diagram (in which the vertical maps are equivalences of pointed sets)

$$\begin{array}{ccccc} [S(X'',A'');\, Y,B] & \xrightarrow{(Sg)\#} & [S(X,A);\, Y,B] & \xrightarrow{(Sf)\#} & [S(X',A');\, Y,B] \\ \updownarrow & & \updownarrow & & \updownarrow \\ [X'',A'';\, \Omega(Y,B)] & \xrightarrow{g\#} & [X,A;\, \Omega(Y,B)] & \xrightarrow{f\#} & [X',A';\, \Omega(Y,B)] \end{array}$$

Hence $\operatorname{im} (Sg)\# = \ker (Sf)\#$ in the top sequence is equivalent to $\operatorname{im} g\# = \ker f\#$ in the bottom sequence. ■

We define $S^n(X,A)$ inductively for $n \geq 0$ so that

$$S^0(X,A) = (X,A)$$
$$S^n(X,A) = S(S^{n-1}(X,A)) \qquad n \geq 1$$

9 THEOREM *For any map $f\colon (X',A') \to (X,A)$ the sequence*

$$(X',A') \xrightarrow{f} (X,A) \xrightarrow{i} \cdots \xrightarrow{S^nf} S^n(X,A) \xrightarrow{S^ni} S^n(C_{f'},C_{f''}) \xrightarrow{S^nk} S^{n+1}(X',A') \xrightarrow{S^{n+1}f} \cdots$$

is coexact.

PROOF From lemmas 7 and 8, for $n \geq 0$ there is a coexact sequence

$$S^n(X',A') \xrightarrow{S^nf} S^n(X,A) \xrightarrow{S^ni} S^n(C_{f'},C_{f''}) \xrightarrow{S^nk} S^{n+1}(X',A') \xrightarrow{S^{n+1}f} S^{n+1}(X,A)$$

Since every three-term subsequence of the sequence in the theorem is contained in one of these five-term coexact sequences, the result follows. ∎

In the coexact sequence of theorem 9 all but the first three pairs are H cogroup pairs, and all but the first three of these are abelian. Furthermore, all maps between H cogroup pairs are homomorphisms. Thus, for any (Y,B) the coexact sequence of homotopy classes of maps of the sequence of theorem 9 into the fixed pair (Y,B) (with B not necessarily closed in Y) consist of groups and homomorphisms except for the last three pointed sets, and all but three of the groups are abelian.

We now show how the last group in the sequence, namely $[S(X',A'); Y,B]$, acts as a group of operators on the left on the next set in the sequence, namely $[C_{f'},C_{f''}; Y,B]$, in such a way that the orbits are mapped injectively by $i\#$ into $[X,A; Y,B]$. If α: $S(X',A') \to (Y,B)$ and β: $(C_{f'},C_{f''}) \to (Y,B)$, we define

$$\alpha \top \beta\colon (C_{f'},C_{f''}) \to (Y,B)$$

by
$$(\alpha \top \beta)[x',t] = \begin{cases} \alpha[x',2t] & 0 \le t \le \tfrac{1}{2}, x' \in X', t \in I \\ \beta[x', 2t-1] & \tfrac{1}{2} \le t \le 1, x' \in X', t \in I \end{cases}$$

and
$$(\alpha \top \beta)(x) = \beta(x) \qquad x \in X$$

It is then clear that $(\alpha \top \beta) \mid (X,A) = \beta \mid (X,A)$, and the following statements are easily verified.

10 $\alpha \simeq \alpha'$ *and* $\beta \simeq \beta'$ *(or* $\beta \simeq \beta'$ rel X*) implies* $\alpha \top \beta \simeq \alpha' \top \beta'$ *(or* $\alpha \top \beta \simeq \alpha' \top \beta'$ rel X*).* ∎

11 *If* α_0 *is the constant map, then* $\alpha_0 \top \beta \simeq \beta$ rel X. ∎

12 $(\alpha_1 * \alpha_2) \top \beta \simeq \alpha_1 \top (\alpha_2 \top \beta)$ rel X. ∎

13 $\alpha_1 \top (\alpha_2 \circ k) \simeq (\alpha_1 * \alpha_2) \circ k$ rel X. ∎

Given maps β_1,β_2: $(C_{f'},C_{f''}) \to (Y,B)$ such that $\beta_1 \mid (X,A) = \beta_2 \mid (X,A)$, we define $d(\beta_1,\beta_2)$: $S(X',A') \to (Y,B)$ by

$$d(\beta_1,\beta_2)[x',t] = \begin{cases} \beta_1[x',2t] & 0 \le t \le \tfrac{1}{2}, x' \in X', t \in I \\ \beta_2[x', 2-2t] & \tfrac{1}{2} \le t \le 1, x' \in X', t \in I \end{cases}$$

The following results are easily verified.

14 $\beta_1 \simeq \beta_1'$ rel X *and* $\beta_2 \simeq \beta_2'$ rel X *imply* $d(\beta_1,\beta_2) \simeq d(\beta_1',\beta_2')$. ∎

15 $d(\beta_1,\beta_3) \simeq d(\beta_1,\beta_2) * d(\beta_2,\beta_3)$. ∎

16 $d(\alpha \top \beta,\beta) \simeq \alpha$. ∎

17 $\beta_1 \simeq d(\beta_1,\beta_2) \top \beta_2$ rel X. ∎

From statements 17, 10, and 11, it follows that if $d(\beta_1,\beta_2)$ is null homotopic, then $\beta_1 \simeq \beta_2$ rel X. Conversely, if $\beta_1 \simeq \beta_2$ rel X, it follows from statements 11, 14, and 16 that

$$d(\beta_1,\beta_2) \simeq d(\alpha_0 \top \beta_1,\beta_1) \simeq \alpha_0$$

Therefore we have $\beta_1 \simeq \beta_2$ rel X if and only if $d(\beta_1,\beta_2)$ is null homotopic.

By statements 10, 11, and 12, there is an action of $[S(X',A'); Y,B]$ on the left on $[C_{f'},C_{f''}; Y,B]$ defined by $[\alpha] \top [\beta] = [\alpha \top \beta]$.

18 THEOREM *Given $[\beta_1], [\beta_2] \in [C_{f'},C_{f''}; Y,B]$, then $i\#[\beta_1] = i\#[\beta_2]$ if and only if there is $[\alpha] \in [S(X',A'); Y,B]$ such that $[\beta_1] = [\alpha] \top [\beta_2]$.*

PROOF By the definition of $\alpha \top \beta_2$ we see that

$$i\#[\alpha \top \beta_2] = [(\alpha \top \beta_2) \mid (X,A)] = [\beta_2 \mid (X,A)] = i\#[\beta_2]$$

showing that $i\#([\alpha] \top [\beta_2]) = i\#[\beta_2]$. Conversely, if $i\#[\beta_1] = i\#[\beta_2]$, we can choose representatives β_1 and β_2 such that $\beta_1 \mid (X,A) = \beta_2 \mid (X,A)$ [because the map i: $(X,A) \subset (C_{f'},C_{f''})$ is a cofibration]. Then, by statement 17,

$$[\beta_1] = [d(\beta_1,\beta_2) \top \beta_2] = [d(\beta_1,\beta_2)] \top [\beta_2] \quad \blacksquare$$

19 THEOREM *Given $[\alpha_1], [\alpha_2] \in [S(X',A'); Y,B]$, then $k\#[\alpha_1] = k\#[\alpha_2]$ if and only if there is $[\gamma] \in [S(X,A); Y,B]$ such that $[\alpha_2] = [\alpha_1] + (Sf)\#[\gamma]$.*

PROOF By statement 13, if β_0: $(C_{f'},C_{f''}) \to (Y,B)$ is the constant map

$$\begin{aligned} k\#[\alpha_1 * (\gamma \circ Sf)] &= [\alpha_1] \top (k\#Sf\#[\gamma]) = [\alpha_1] \top [\beta_0] \\ &= [\alpha_1] \top k\#[\alpha_0] = k\#[\alpha_1 * \alpha_0] \end{aligned}$$

Therefore $k\#([\alpha_1] + (Sf)\#[\gamma]) = k\#[\alpha_1]$. Conversely, if $k\#[\alpha_1] = k\#[\alpha_2]$, then by statements 10 and 13,

$$0 = k\#[\alpha_1{}^{-1} * \alpha_1] = [\alpha_1{}^{-1}] \top k\#[\alpha_1] = [\alpha_1{}^{-1}] \top k\#[\alpha_2] = k\#[\alpha_1{}^{-1} * \alpha_2]$$

Therefore there is $[\gamma] \in [S(X,A); Y,B]$ such that $[\alpha_1{}^{-1} * \alpha_2] = (Sf)\#[\gamma]$, and so

$$[\alpha_2] = [\alpha_1] + [\alpha_1{}^{-1} * \alpha_2] = [\alpha_1] + (Sf)\#[\gamma] \quad \blacksquare$$

2 HIGHER HOMOTOPY GROUPS

The higher homotopy groups of a space or pair are covariant functors defined to be sets of homotopy classes of maps of fixed spaces or pairs into the given one. In the absolute case these are the functors already defined in Sec. 1.6. The exactness property established in the last section implies an important exactness property relating relative and absolute homotopy groups. This section contains definitions of the homotopy groups, some of their elementary properties, and a proof of the exactness of the homotopy sequence of a fibration.

We shall use 0 as base point for I and for the subspace $\dot{I} \subset I$. Let X be a space with base point x_0. For $n \geq 1$ the homotopy group $\pi_n(X)$ [or $\pi_n(X,x_0)$, when it is important to indicate the base point] is the group $[S^n(\dot{I});X]$ [this being equivalent to the definition given in Sec. 1.6, because S^n is homeomorphic to $S^n(S^0) \approx S^n(\dot{I})$]. For $n = 0$ the homotopy set $\pi_0(X)$ is defined to be the pointed set $[\dot{I};X]$ (that is, the set of path components of X with the path component of x_0 as distinguished element). Then π_n is a covariant functor from the category of pointed spaces to the category of abelian groups if $n \geq 2$, the category of groups if $n = 1$, and the category of pointed sets if $n = 0$.

Let (X,A) be a pair with base point $x_0 \in A$. For $n \geq 1$ the *nth relative homotopy group* (or *homotopy set* for $n = 1$), denoted by $\pi_n(X,A)$ or $\pi_n(X,A,x_0)$, is defined to equal $[S^{n-1}(I,\dot{I}); X,A]$. Then π_n is a covariant functor from the category of pairs of pointed spaces to the category of abelian groups if $n \geq 3$, the category of groups if $n = 2$, and the category of pointed sets if $n = 1$.

There is a homeomorphism of $S(\dot{I})$ with $I/\dot{I}$ which sends $[0,t] \in S(\dot{I})$ to the base point of $I/\dot{I}$ and $[1,t] \in S(\dot{I})$ to that point of $I/\dot{I}$ determined by the point $t \in I$. Therefore, for $n \geq 1$, $S^n(\dot{I})$ and $S^{n-1}(I/\dot{I}) = S^{n-1}(I)/S^{n-1}(\dot{I})$ are homeomorphic. This induces a natural one-to-one correspondence between $[S^{n-1}(I,\dot{I}); X,\{x_0\}]$ and $[S^n(\dot{I});X]$. By means of this correspondence we identify the relative homotopy group $\pi_n(X,\{x_0\})$ for $n \geq 1$ with the absolute homotopy group $\pi_n(X)$. Then the inclusion map j: $(X,\{x_0\}) \subset (X,A)$ induces a homomorphism

$$j_\#: \pi_n(X) \to \pi_n(X,A) \qquad n \geq 1$$

Because $S^n(\dot{I})$ is path connected for $n \geq 1$, it follows that if X' is the path component of X containing x_0, the inclusion map $X' \subset X$ induces isomorphisms $\pi_n(X') \approx \pi_n(X)$ for $n \geq 1$. Similarly, if A' is the path component of A containing x_0, the inclusion map $(X',A') \subset (X,A)$ induces isomorphisms $\pi_n(X',A') \approx \pi_n(X,A)$ for $n \geq 1$.

We present an alternative description of the relative homotopy groups. For $n \geq 1$ there is a homeomorphism of $S^{n-1}(I,\dot{I})$ with $(I \times I^{n-1}, \dot{I} \times I^{n-1})/(I \times \dot{I}^{n-1} \cup 0 \times I^{n-1})$ sending $[\cdots[t,t_1], \ldots ,t_{n-1}]$ to $[t,t_1, \ldots ,t_{n-1}]$ (I^0 is a single point and $\dot{I}^0$ is empty). Therefore, for $n \geq 1$, $\pi_n(X,A,x_0)$ is in one-to-one correspondence with the set of homotopy classes of maps

$$(I^n, \dot{I}^n, I \times \dot{I}^{n-1} \cup 0 \times I^{n-1}) \to (X,A,x_0)$$

Since $I \times \dot{I}^{n-1} \cup 0 \times I^{n-1}$ is contractible, if $z_0 = (0,0, \ldots ,0)$, the inclusion map

$$(I^n,\dot{I}^n,z_0) \subset (I^n, \dot{I}^n, I \times \dot{I}^{n-1} \cup 0 \times I^{n-1})$$

is a homotopy equivalence. Hence, for $n \geq 1$, $\pi_n(X,A,x_0)$ is in one-to-one correspondence with the set of homotopy classes of maps

$$(I^n,\dot{I}^n,z_0) \to (X,A,x_0)$$

Since $(I^n,\dot{I}^n,z_0)$ is homeomorphic to (E^n,S^{n-1},p_0) for $n \geq 1$, $\pi_n(X,A,x_0)$ is in one-to-one correspondence with the set of homotopy classes of maps

$$(E^n,S^{n-1},p_0) \to (X,A,x_0)$$

The following condition for a map $(E^n,S^{n-1},p_0) \to (X,A,x_0)$ to represent the trivial element of $\pi_n(X,A,x_0)$ is a relative version of theorem 1.6.7.

1 THEOREM *Given a map* α: $(E^n,S^{n-1},p_0) \to (X,A,x_0)$, *then* $[\alpha] = 0$ *in*

$\pi_n(X,A,x_0)$ *if and only if* α *is homotopic relative to* S^{n-1} *to some map of* E^n *to* A.

PROOF Assume $[\alpha] = 0$ in $\pi_n(X,A,x_0)$. Then there is a homotopy

$$H: (E^n,S^{n-1},p_0) \times I \to (X,A,x_0)$$

from α to the constant map $E^n \to x_0$. A homotopy H' relative to S^{n-1} from α to some map E^n to A is defined by

$$H'(z,t) = \begin{cases} H\left(\dfrac{z}{1-t/2}, t\right) & 0 \le \|z\| \le 1 - \dfrac{t}{2} \\ H\left(\dfrac{z}{\|z\|}, 2 - 2\|z\|\right) & 1 - \dfrac{t}{2} \le \|z\| \le 1 \end{cases}$$

Conversely, if α is homotopic relative to S^{n-1} to some map α' such that $\alpha'(E^n) \subset A$, then $[\alpha] = [\alpha']$ in $\pi_n(X,A,x_0)$, and it suffices to show that $[\alpha'] = 0$ in $\pi_n(X,A,x_0)$. A homotopy $H: (E^n,S^{n-1},p_0) \times I \to (X,A,x_0)$ from α' to the constant map $E^n \to x_0$ is defined by

$$H(z,t) = \alpha'((1-t)z + tp_0) \quad \blacksquare$$

A pair (X,A) is said to be *n-connected* for $n \ge 0$ if for $0 \le k \le n$ every map $\alpha: (E^k,S^{k-1}) \to (X,A)$ is homotopic relative to S^{k-1} to some map of E^k to A. For $k = 0$, (E^0,S^{-1}) is a pair consisting of a single point and the empty set, and the condition that every map $\alpha: (E^0,S^{-1}) \to (X,A)$ be homotopic to a map $E^0 \to A$ is equivalent to the condition that every point of X be joined by a path to some point of A. From theorem 1 we obtain the following relation between n-connectedness of (X,A) and the vanishing of relative homotopy groups of (X,A).

2 COROLLARY *A pair* (X,A) *is n-connected for* $n \ge 0$ *if and only if every path component of* X *intersects* A *and for every point* $a \in A$ *and every* $1 \le k \le n$, $\pi_k(X,A,a) = 0$. $\blacksquare$

For $n \ge 1$ there is a map (which is a homomorphism for $n \ge 2$)

$$\partial: \pi_n(X,A,x_0) \to \pi_{n-1}(A,x_0)$$

defined by restriction. That is, given $\alpha: S^{n-1}(I,\dot{I}) \to (X,A)$, then

$$\partial[\alpha] = [\alpha \mid S^{n-1}(\dot{I})]$$

It is trivial that if $f: (X',A',x_0') \to (X,A,x_0)$ is a map, there is a commutative square

$$\begin{array}{ccc} \pi_n(X',A',x_0') & \xrightarrow{\partial} & \pi_{n-1}(A',x_0') \\ f_\# \downarrow & & \downarrow (f|A')_\# \\ \pi_n(X,A,x_0) & \xrightarrow{\partial} & \pi_n(A,x_0) \end{array}$$

In other words, ∂ is a natural transformation between covariant functors $\pi_n(X,A)$ and $\pi_{n-1}(A)$ on the category of pairs (X,A) of pointed spaces. Thus the homotopy-group functors and the natural transformation ∂ are in analogy with the constituents of a homology theory. We shall show that they also satisfy many of the axioms of homology theory. It is obvious that the homotopy axiom and the dimension axiom are satisfied for the homotopy-group functors.

We shall now investigate the exactness property. Given a pair (X,A) of pointed spaces, let $i: A \subset X$ and $j: (X,\{x_0\}) \subset (X,A)$. The *homotopy sequence of* (X,A) [or of (X,A,x_0)] is the sequence of pointed sets (all but the last three being groups)

$$\cdots \to \pi_{n+1}(X,A) \xrightarrow{\partial} \pi_n(A) \xrightarrow{i_\#} \pi_n(X) \xrightarrow{j_\#} \pi_n(X,A) \xrightarrow{\partial} \cdots \xrightarrow{i_\#} \pi_0(X)$$

3 THEOREM *The homotopy sequence of a pair is exact.*

PROOF Let $f: (\dot{I},\{0\}) \subset (I,\dot{I})$ and let $f': \dot{I} \subset \dot{I}$ and $f'': \{0\} \subset \dot{I}$. By theorem 7.1.9, there is a coexact sequence

$$(\dot{I},\{0\}) \xrightarrow{f} (I,\dot{I}) \xrightarrow{i} (C_{f'},C_{f''}) \xrightarrow{k} S(\dot{I},\{0\}) \xrightarrow{Sf} S(I,\dot{I}) \to \cdots$$

Let $g: (C_{f'},C_{f''}) \to (I,\dot{I})$ be the homeomorphism defined by $g([0,t]) = 0$ and $g([1,t]) = t$. Then the composite $g \circ i$ is the inclusion map $i': (\dot{I},\dot{I}) \subset (I,\dot{I})$, and the composite $k \circ g^{-1}$ equals the composite

$$(I,\dot{I}) \xrightarrow{k'} (I/\dot{I},\{0\}) \xrightarrow{h} (S(\dot{I}),\{0\})$$

where k' is the collapsing map and h is the homeomorphism used in identifying $\pi_n(X,\{x_0\})$ with $\pi_n(X)$. Therefore there is a coexact sequence

$$(\dot{I},\{0\}) \xrightarrow{f} (\dot{I},\dot{I}) \xrightarrow{i'} (I,\dot{I}) \xrightarrow{h \circ k'} S(\dot{I},\{0\}) \xrightarrow{Sf} \cdots$$

This yields an exact sequence

$$\cdots \to \pi_{n+1}(X,A) \xrightarrow{(S^n i')^\#} \pi_n(A) \xrightarrow{(S^n f)^\#} \pi_n(X) \xrightarrow{(S^{n-1}(h \circ k'))^\#} \pi_n(X,A) \to \cdots \to \pi_0(X)$$

The proof is completed by the trivial verification that

$$(S^n i')^\# = \partial, \; (S^n f)^\# = i_\#, \text{ and } \; (S^{n-1}(h \circ k'))^\# = j_\# \quad ■$$

4 COROLLARY *For $n \geq 0$, (E^{n+1},S^n) is n-connected.*

PROOF For $n \geq 0$, E^{n+1} is path connected and S^n is nonempty; therefore every path component of E^{n+1} meets S^n. If $x \in S^n$, then $\pi_k(E^{n+1},x) = 0$ for $0 \leq k$, because E^{n+1} is contractible. By theorem 3.4.11, $\pi_k(S^n,x) = 0$ if $0 \leq k < n$. It follows from theorem 3 that $\pi_k(E^{n+1},S^n,x) = 0$ for $1 \leq k \leq n$. The result follows from corollary 2. ■

We shall see that the excision property fails to hold for the homotopy group functors. There is, however, a different property possessed by the homotopy group functors but not by homology functors. This property is the existence of an isomorphism between the absolute homotopy groups of the base space of a fibration and the corresponding relative homotopy groups of the total space modulo the fiber. This is true for a more general class of maps than fibrations, and we now present the relevant definition.

A map $p: E \to B$ is called a *weak fibration* (or *Serre fiber space* in the literature) if p has the homotopy lifting property with respect to the collection of cubes $\{I^n\}_{n\geq 0}$. E is called the *total space* and B the *base space* of the weak fibration. For $b \in B$, $p^{-1}(b)$ is called the *fiber over b*.

If s is a simplex, $|s|$ is homeomorphic to some cube, and so a map $p: E \to B$ is a weak fibration if and only if it has the homotopy lifting property with respect to the space of any simplex. We shall show that, in fact, a weak fibration has the homotopy lifting property with respect to any polyhedron.

It is clear that a fibration is a weak fibration. If $p: E \to B$ is a weak fibration and $f: B' \to B$ is a map, let E' be the fibered product of B' and E. Then there is a weak fibration $p': E' \to B'$, called the *weak fibration induced from p by f*.

5 LEMMA *Let $p: E \to B$ be a weak fibration and let $g: I^n \times 0 \cup \dot{I}^n \times I \to E$ and $H: I^n \times I \to B$ be maps, with $n \geq 0$, such that H is an extension of $p \circ g$. Then there is a map $G: I^n \times I \to E$ such that $p \circ G = H$ and G is an extension of g.*

PROOF The lemma asserts that the dotted arrow in the diagram

$$\begin{array}{ccc} I^n \times 0 \cup \dot{I}^n \times I & \xrightarrow{g} & E \\ \cap\downarrow & \nearrow & \downarrow p \\ I^n \times I & \xrightarrow{H} & B \end{array}$$

represents a map making the diagram commutative. This follows from the homotopy lifting property of p since the pair $(I^n \times I, I^n \times 0 \cup \dot{I}^n \times I)$ is homeomorphic to the pair $(I^n \times I, I^n \times 0)$. ■

6 THEOREM *Let (X,A) be a polyhedral pair and let $p: E \to B$ be a weak fibration. Given maps $g: X \times 0 \cup A \times I \to E$ and $H: X \times I \to B$ such that H is an extension of $p \circ g$, there is a map $G: X \times I \to E$ such that $p \circ G = H$ and G is an extension of g.*

PROOF The method of obtaining G involves a standard stepwise-extension procedure over the successive skeleta of a triangulation of X. Let (K,L) be a triangulation of (X,A) and identify (X,A) with $(|K|,|L|)$. For $q \geq -1$ set $X_q = |K| \times 0 \cup (|K^q \cup L| \times I)$, so that $X_{-1} = X \times 0 \cup A \times I$ and $X_{q-1} \subset X_q$ for $q \geq 0$. By induction on q, we shall define a sequence of maps $G_q: X_q \to E$ such that

(a) $G_{-1} = g$
(b) $G_q \mid X_{q-1} = G_{q-1}$ for $q \geq 0$
(c) $p \circ G_q = H \mid X_q$ for $q \geq -1$

Once a sequence $\{G_q\}$ with these properties is obtained, a map $G: X \times I \to E$ with the desired properties is defined by the conditions $G \mid X_q = G_q$, for $q \geq -1$. Thus the problem is reduced to the construction of such a sequence $\{G_q\}$.

Condition (*a*) defines G_{-1}. Assume G_q defined for $q < n$, where $n \geq 0$. To define G_n to satisfy conditions (*b*) and (*c*), for every n-simplex $s \in K - L$ let $g_s\colon |s| \times 0 \cup |\dot{s}| \times I \to E$ and $H_s\colon |s| \times I \to B$ be the maps defined by $g_s = G_{n-1} \mid (|s| \times 0 \cup |\dot{s}| \times I)$ and $H_s = H \mid (|s| \times I)$. Because $(|s|,|\dot{s}|)$ is homeomorphic to $(I^n,\dot{I}^n)$, it follows from lemma 5 that there is a map $G_s\colon |s| \times I \to E$ such that $G_s \mid (|s| \times 0 \cup |\dot{s}| \times I) = g_s$ and $p \circ G_s = H_s$. Then a map $G_n\colon X_n \to E$ satisfying conditions (*b*) and (*c*) is defined by the conditions $G_n \mid X_{n-1} = G_{n-1}$ and $G_n \mid (|s| \times I) = G_s$ for s an n-simplex of $K - L$. ■

Taking A to be empty in theorem 6, we see that a weak fibration has the homotopy lifting property with respect to any polyhedron.

7 COROLLARY *Let (X',A') be a polyhedral pair such that A' is a strong deformation retract of X' and let $p\colon E \to B$ be a weak fibration. Given maps $g'\colon A' \to E$ and $H'\colon X' \to B$ such that $H' \mid A' = p \circ g'$, there is a map $G'\colon X' \to E$ such that $p \circ G' = H'$ and $G' \mid A' = g'$.*

PROOF Let $D\colon X' \times I \to X'$ be a strong deformation retraction of X' to A'. Then $D(X' \times 1 \cup A' \times I) \subset A'$, and we define $g\colon X' \times 1 \cup A' \times I \to E$ to be the composite

$$X' \times 1 \cup A' \times I \xrightarrow{D} A' \xrightarrow{g'} E'$$

Let $H\colon X' \times I \to B$ be the composite

$$X' \times I \xrightarrow{D} X' \xrightarrow{H'} B$$

Then H is an extension of $p \circ g$, and it follows from theorem 6 that there is a map $G\colon X' \times I \to E$ such that $p \circ G = H$ and G is an extension of g. Let $G'\colon X' \to E$ be defined by $G'(x') = G(x',0)$. Then G' has the desired properties. ■

The following theorem is the main result relating the homotopy groups of the base and total space of a weak fibration.

8 THEOREM *Let $p\colon E \to B$ be a weak fibration and suppose $b_0 \in B' \subset B$. Let $E' = p^{-1}(B')$ and let $e_0 \in p^{-1}(b_0)$. Then p induces a bijection*

$$p_\#\colon \pi_n(E,E',e_0) \approx \pi_n(B,B',b_0) \qquad n \geq 1$$

PROOF To show that $p_\#$ is surjective, let $\alpha\colon (I^n,\dot{I}^n,z_0) \to (B,B',b_0)$ represent an element of $\pi_n(B,B',b_0)$. Because z_0 is a strong deformation retract of I^n, we can apply corollary 7 to the pair $(I^n,\{z_0\})$ and to maps $g'\colon \{z_0\} \to E$ and $H'\colon I^n \to B$, where $g'(z_0) = e_0$ and $H' = \alpha \mid I^n$. We then obtain a map $G'\colon I^n \to E$ such that $p \circ G' = H'$ and $G'(z_0) = e_0$. Then

$$G'(\dot{I}^n) \subset p^{-1}(H'(\dot{I}^n)) \subset p^{-1}(B') = E'$$

Therefore G' defines a map $\alpha'\colon (I^n,\dot{I}^n,z_0) \to (E,E',e_0)$ such that $p \circ \alpha' = \alpha$. Then α' represents an element $[\alpha'] \in \pi_n(E,E',e_0)$ and $p_\#[\alpha'] = [\alpha]$.

To show that $p_\#$ is injective, let $\alpha_0,\alpha_1\colon (I^n,\dot{I}^n,z_0) \to (E,E',e_0)$ be such that $p \circ \alpha_0 \simeq p \circ \alpha_1$. Let $X' = I^n \times I$ and $A' = (I^n \times 0) \cup (z_0 \times I) \cup (I^n \times 1)$.

Then (X',A') is a polyhedral pair, and because X' and A' are both contractible, A' is a strong deformation retract of X'. Let $g'\colon A' \to E$ be defined by $g'(z,0) = \alpha_0(z)$, $g'(z,1) = \alpha_1(z)$, and $g'(z_0,t) = e_0$ and let $H'\colon X' \to B$ be a map which is a homotopy from $p \circ \alpha_0$ to $p \circ \alpha_1$ in (B,B',b_0). By corollary 7, there is a map $G'\colon X' \to E$ such that $p \circ G' = H'$ and $G' \mid A' = g'$. Since

$$G'(\dot{I}^n \times I) \subset p^{-1}(H'(\dot{I}^n \times I)) \subset p^{-1}(B') = E'$$

G' is a homotopy from α_0 to α_1 in (E,E',e_0); hence $[\alpha_0] = [\alpha_1]$ in $\pi_n(E,E',e_0)$. ■

9 COROLLARY *Let $p\colon E \to B$ be a weak fibration, $b_0 \in B$, and $e_0 \in F = p^{-1}(b_0)$. Then p induces a bijection*

$$p_\#\colon \pi_n(E,F,e_0) \approx \pi_n(B,b_0) \qquad n \geq 1$$

PROOF This follows from theorem 8 on taking $B' = \{b_0\}$ and using the canonical identification $\pi_n(B,\{b_0\},b_0) = \pi_n(B,b_0)$. ■

If $p\colon E \to B$ is a weak fibration with $F = p^{-1}(b_0)$ and $e_0 \in F$, we define

$$\bar{\partial}\colon \pi_n(B,b_0) \to \pi_{n-1}(F,e_0) \qquad n \geq 1$$

to be the composite

$$\pi_n(B,b_0) \xrightarrow{p_\#^{-1}} \pi_n(E,F,e_0) \xrightarrow{\partial} \pi_{n-1}(F,e_0)$$

The *homotopy sequence of the weak fibration* is the sequence

$$\cdots \to \pi_n(F,e_0) \xrightarrow{i_\#} \pi_n(E,e_0) \xrightarrow{p_\#} \pi_n(B,b_0) \xrightarrow{\bar{\partial}} \pi_{n-1}(F,e_0) \to \cdots \xrightarrow{p_\#} \pi_0(B,b_0)$$

where $i\colon (F,e_0) \subset (E,e_0)$.

10 THEOREM *The homotopy sequence of a weak fibration is exact.*

PROOF Exactness at $\pi_0(E,e_0)$ is an easy consequence of the homotopy lifting property. Exactness at any set to the left of $\pi_0(E,e_0)$ is a consequence of the exactness of the homotopy sequence of the pair (E,F). ■

11 COROLLARY *Let $p\colon E \to B$ be a weak fibration with unique path lifting. Then p induces an isomorphism*

$$p_\#\colon \pi_q(E,e_0) \approx \pi_q(B,p(e_0)) \qquad q \geq 2$$

PROOF Because $F = p^{-1}(p(e_0))$ has no nonconstant paths (by theorem 2.2.5), $\pi_q(F,e_0) = 0$ for $q \geq 1$. The result then follows from theorem 10. ■

12 COROLLARY *For $q \geq 2$, $\pi_q(S^1) = 0$.*

PROOF This follows from application of corollary 11 to the covering projection $p\colon \mathbf{R} \to S^1$ and the fact that because $\mathbf{R}$ is contractible, $\pi_q(\mathbf{R}) = 0$ for all $q \geq 0$. ■

13 COROLLARY *Let $p\colon S^{2n+1} \to P_n(\mathbf{C})$ be the Hopf fibration. Then p induces an isomorphism*

$$p_\#\colon \pi_q(S^{2n+1}) \approx \pi_q(P_n(\mathbf{C})) \qquad q \geq 3$$

PROOF Because $F = S^1$ for the Hopf fibration, this follows from corollary 12 and theorem 10. ∎

14 COROLLARY $\pi_3(S^2) \neq 0$.

PROOF Because the identity map $(S^3,p_0) \subset (S^3,p_0)$ induces a nontrivial homomorphism of $H_3(S^3,p_0)$, it is not homotopic to the constant map. Therefore $\pi_3(S^3) \neq 0$, and the result follows from corollary 13, with $n = 1$ (since $P_1(\mathbf{C}) \approx S^2$). ∎

This last result shows that, unlike the homology groups, the homotopy groups of a polyhedron need not vanish in degrees larger than the dimension of the polyhedron.

If H is a closed hemisphere of S^2 and a is the pole in H, then the pair $(S^2 - a, H - a)$ has the same homotopy type as (E^2,S^1). Therefore

$$\pi_3(S^2 - a, H - a) \approx \pi_3(E^2,S^1) \overset{\partial}{\approx} \pi_2(S^1) = 0$$

On the other hand, (S^2,H) has the same homotopy type as $(S^2,\{a\})$. Therefore

$$\pi_3(S^2,H) \approx \pi_3(S^2,\{a\}) = \pi_3(S^2) \neq 0$$

Hence we see that the excision map j: $(S^2 - a, H - a) \subset (S^2,H)$ does not induce an isomorphism of $\pi_3(S^2 - a, H - a)$ with $\pi_3(S^2,H)$. Therefore the excision property does not hold for homotopy groups.

Recall the path fibration p: $P(X,x_0) \to X$ with fiber $p^{-1}(x_0) = \Omega X$ (see corollary 2.8.8). Since $P(X,x_0)$ is contractible (by lemma 2.4.3), $\pi_n(P(X,x_0)) = 0$ for $n \geq 0$, and by theorem 10, there is an isomorphism

$$\bar{\partial}: \pi_n(X) \approx \pi_{n-1}(\Omega X) \qquad n \geq 1$$

This result can also be deduced directly from the canonical one-to-one correspondence $[S^n(\dot{I});X] \approx [S^{n-1}(\dot{I});\Omega X]$ given by the exponential law. We shall use the path space to prove the exactness of the homotopy sequence of a triple.

Given a triple (X,A,B) with base point $x_0 \in B$, let i: $(A,B) \subset (X,B)$ and j: $(X,B) \subset (X,A)$ and let j': $(A,\{x_0\}) \subset (A,B)$. Define

$$\partial': \pi_n(X,A,x_0) \to \pi_{n-1}(A,B,x_0) \qquad n \geq 2$$

to equal the composite

$$\pi_n(X,A,x_0) \xrightarrow{\partial} \pi_{n-1}(A,x_0) \xrightarrow{j'_\#} \pi_{n-1}(A,B,x_0)$$

The *homotopy sequence of the triple* (X,A,B) is defined to be the sequence

$$\cdots \to \pi_{n+1}(X,A) \xrightarrow{\partial'} \pi_n(A,B) \xrightarrow{i_\#} \pi_n(X,B) \xrightarrow{j_\#} \pi_n(X,A) \to \cdots \to \pi_1(X,A)$$

15 THEOREM *The homotopy sequence of a triple is exact.*

PROOF Let p: $P(X,x_0) \to X$ be the path fibration and let $X' = P(X,x_0)$, $A' = p^{-1}(A)$, and $B' = p^{-1}(B)$. Then (X',A',B') is a triple, and it follows from theorem 8 that $p_\#$ maps the homotopy sequence of (X',A',B') bijectively to the homotopy sequence of (X,A,B). Therefore it suffices to prove that the homotopy sequence of the triple (X',A',B') is exact.

Let $i: (A',B') \subset (X',B')$, $j: (X',B') \subset (X',A')$, $i': B' \subset A'$, and $j': A' \subset (A',B')$. There is a commutative diagram

$$\begin{array}{ccccccccc} \cdots \to & \pi_{n+1}(X',A') & \xrightarrow{\partial'} & \pi_n(A',B') & \xrightarrow{i_\#} & \pi_n(X',B') & \xrightarrow{j_\#} & \pi_n(X',A') & \to \cdots \\ & \partial\downarrow & & \downarrow= & & \partial\downarrow & & \downarrow\partial & \\ \cdots \to & \pi_n(A') & \xrightarrow{j'_\#} & \pi_n(A',B') & \xrightarrow{\partial} & \pi_{n-1}(B') & \xrightarrow{i'_\#} & \pi_{n-1}(A') & \to \cdots \end{array}$$

in which each vertical map is a bijection (because X' is contractible). Therefore the exactness of the homotopy sequence of the triple (X',A',B') follows from the exactness of the homotopy sequence of the pair (A',B'). ■

This result can also be derived from the exactness of the homotopy sequence of a pair and functorial properties of the homotopy groups (as was the case with the corresponding exactness property for homology, theorem 4.8.5).

3 CHANGE OF BASE POINTS

The absolute and relative homotopy groups are defined for pointed spaces and pairs. This section is devoted to a study of the extent to which these groups depend on the choice of base point. By generalizing the methods of Sec. 1.8, we shall see that these groups based at different base points in the same path component are isomorphic, but the isomorphism between them is not usually unique. Much of these considerations apply to more general homotopy sets, and we begin with this.

Let (X,A) be a pair with base point $x_0 \in A$ and let (Y,B) be a pair. Two maps $\alpha_0, \alpha_1: (X,A) \to (Y,B)$ are said to be *freely homotopic* if they are homotopic as maps of (X,A) to (Y,B) (that is, no restriction is placed on the base point during the homotopy). If ω is a path in B from $\alpha_0(x_0)$ to $\alpha_1(x_0)$, an *ω-homotopy from α_0 to α_1* is a homotopy

$$H: (X,A) \times I \to (Y,B)$$

such that $H(x,0) = \alpha_0(x)$, $H(x,1) = \alpha_1(x)$, and $H(x_0,t) = \omega(t)$. If such a homotopy exists, we say that α_0 is *ω-homotopic* to α_1. It is clear that α_0 and α_1 are freely homotopic if and only if there is some path ω in B such that α_0 and α_1 are ω-homotopic. In particular, two maps $\alpha_0, \alpha_1: (X,A,x_0) \to (Y,B,y_0)$ are freely homotopic if and only if there is some closed path ω in B at y_0 such that α_0 is ω-homotopic to α_1.

Although the relation of free homotopy is an equivalence relation in the set of maps from (X,A) to (Y,B), for a fixed ω the relation of ω-homotopy is not generally an equivalence relation. For example, if ω is not a closed path, it is impossible for any map α_0 to be ω-homotopic to itself.

1 LEMMA *(a) Given a map $f: (X',A',x_0') \to (X,A,x_0)$, maps $\alpha_0, \alpha_1: (X,A) \to (Y,B)$, and a path ω in B such that α_0 is ω-homotopic to α_1, then $\alpha_0 \circ f$ is ω-homotopic to $\alpha_1 \circ f$.*

(*b*) *Given a map* $g: (Y,B) \to (Y',B')$, *maps* $\alpha_0, \alpha_1: (X,A) \to (Y,B)$, *and a path* ω *in* B *such that* α_0 *is* ω*-homotopic to* α_1, *then* $g \circ \alpha_0$ *is* $(g \circ \omega)$*-homotopic to* $g \circ \alpha_1$.

(*c*) *Given maps* $\alpha_0, \alpha_0': (SX,SA,x_0) \to (Y,B,\omega(0))$ *and* $\alpha_1, \alpha_1': (SX,SA,x_0) \to (Y,B,\omega(1))$ *such that* α_0 *is* ω*-homotopic to* α_1 *and* α_0' *is* ω*-homotopic to* α_1', *then* $\alpha_0 * \alpha_0'$ *is* ω*-homotopic to* $\alpha_1 * \alpha_1'$.

PROOF If $H: (X,A) \times I \to (Y,B)$ is an ω-homotopy from α_0 to α_1, then for (*a*) the composite

$$(X',A') \times I \xrightarrow{f \times 1} (X,A) \times I \xrightarrow{H} (Y,B)$$

is an ω-homotopy from $\alpha_0 \circ f$ to $\alpha_1 \circ f$, and for (*b*) the composite

$$(X,A) \times I \xrightarrow{H} (Y,B) \xrightarrow{g} (Y',B')$$

is a $(g \circ \omega)$-homotopy from $g \circ \alpha_0$ to $g \circ \alpha_1$.

In (*c*), if $H, H': (SX,SA) \times I \to (Y,B)$ are ω-homotopies from α_0 and α_0' to α_1 and α_1', respectively, the map

$$H * H': (SX,SA) \times I \to (Y,B)$$

defined by

$$(H * H')([x,t], t') = \begin{cases} H([x,2t], t') & 0 \le t \le \frac{1}{2} \\ H'([x, 2t-1], t') & \frac{1}{2} \le t \le 1 \end{cases}$$

is an ω-homotopy from $\alpha_0 * \alpha_0'$ to $\alpha_1 * \alpha_1'$. ∎

The base point x_0 for a pair (X,A) is said to be a *nondegenerate base point* if the inclusion map $(x_0,x_0) \subset (X,A)$ is a cofibration [that is, if, given a map $\alpha_0: (X,A) \to (Y,B)$ and a homotopy $\omega: x_0 \times I \to B$, there is a homotopy $H: (X,A) \times I \to (Y,B)$ such that $H(x,0) = \alpha_0(x)$ and $H(x_0,t) = \omega(t)$]. It follows from lemma 3.8.1 and corollary 3.2.4 that any point of a polyhedral pair is a nondegenerate base point.

2 LEMMA *Let* (X,A) *be a pair with nondegenerate base point and let* (Y,B) *be an arbitrary pair.*

(*a*) *Given a path* ω *in* B *and a map* $\alpha_1: (X,A,x_0) \to (Y,B,\omega(1))$, *there is a map* $\alpha_0: (X,A,x_0) \to (Y,B,\omega(0))$ *such that* α_0 *is* ω*-homotopic to* α_1.

(*b*) *If* $\alpha_0, \alpha_0': (X,A,x_0) \to (Y,B,\omega(0))$ *are both* ω*-homotopic to* α_1, *then* $[\alpha_0] = [\alpha_0']$ *in* $[X,A,x_0; Y,B,\omega(0)]$.

(*c*) *If* α_0 *is* ω*-homotopic to* α_1 *and* $\alpha_0 \simeq \alpha_0'$ *as maps from* (X,A,x_0) *to* $(Y,B,\omega(0))$, $\alpha_1 \simeq \alpha_1'$ *as maps from* (X,A,x_0) *to* $(Y,B,\omega(1))$, *and* $\omega \simeq \omega'$ *as paths in* B, *then* α_0' *is* ω'*-homotopic to* α_1'.

PROOF (*a*) Given α_1 and ω, it follows from the nondegeneracy of x_0 that there is a map $H': (X,A) \times I \to (Y,B)$ such that $H'(x,0) = \alpha_1(x)$ and $H'(x_0,t) = \omega(1-t)$. Define $\alpha_0: (X,A,x_0) \to (Y,B,\omega(0))$ by $\alpha_0(x) = H'(x,1)$. Then $H: (X,A) \times I \to (Y,B)$ defined by $H(x,t) = H'(x, 1-t)$ is an ω-homotopy from α_0 to α_1.

(*b*) Because x_0 is a nondegenerate base point, there is a retraction

$r: (X,A) \times I \to (x_0 \times I \cup X \times 1, x_0 \times I \cup A \times 1)$ (by theorem 2.8.1), and we let $r_t: (X,A) \to (x_0 \times I \cup X \times 1, x_0 \times I \cup A \times 1)$ be defined by $r_t(x) = r(x,t)$. Let $G: (x_0 \times I \cup X \times 1, x_0 \times I \cup A \times 1) \times I \to (X,A) \times I$ be the homotopy relative to $(x_0,0)$ defined by $G(x,t,t') = (x,tt')$ and define

$$G_{t'}: (x_0 \times I \cup X \times 1, x_0 \times I \cup A \times 1) \to (X,A) \times I$$

by $G_{t'}(x,t) = G(x,t,t')$. Then $G_0 \circ r_0 \simeq G_1 \circ r_0$ rel x_0, and because $G_0(x_0 \times I) = (x_0,0)$, $G_0 \circ r_0 \simeq G_0 \circ r_1$ rel x_0. Let $H: (X,A) \times I \to (Y,B)$ be an ω-homotopy from α_0 to α_1. Then $H \circ G_1 \circ r_0 \simeq H \circ G_0 \circ r_1$ rel x_0. Clearly, $H \circ G_0 \circ r_1 = \alpha_0$, and so $\alpha_0 \simeq H \circ G_1 \circ r_0$ rel x_0. Similarly, if $H': (X,A) \times I \to (Y,B)$ is an ω-homotopy from α_0' to α_1, then $\alpha_0' \simeq H' \circ G_1 \circ r_0$ rel x_0. Because

$$H \mid (x_0 \times I \cup X \times 1) = H' \mid (x_0 \times I \cup X \times 1)$$

$H \circ G_1 \circ r_0 = H' \circ G_1 \circ r_0$, and so $\alpha_0 \simeq \alpha_0'$ rel x_0.

(*c*) First we observe that the inclusion map

$$(X \times \dot{I} \cup x_0 \times I, A \times \dot{I} \cup x_0 \times I) \subset (X,A) \times I$$

is a cofibration. In fact, let $h: (I \times I, I \times 0 \cup \dot{I} \times I) \approx (I \times I, 0 \times I)$ be a homeomorphism. Then there is a homeomorphism

$$1 \times h: (X \times I \times I, A \times I \times I) \approx (X \times I \times I, A \times I \times I)$$

which maps

$$X \times I \times 0 \cup X \times \dot{I} \times I \cup x_0 \times I \times I \quad \text{to} \quad X \times 0 \times I \cup x_0 \times I \times I$$

and

$$A \times I \times 0 \cup A \times \dot{I} \times I \cup x_0 \times I \times I \quad \text{to} \quad A \times 0 \times I \cup x_0 \times I \times I.$$

Thus we need only show that $(X \times 0 \cup x_0 \times I, A \times 0 \cup x_0 \times I) \times I$ is a retract of $(X \times I, A \times I) \times I$, which follows from thc fact that $(X \times 0 \cup x_0 \times I, A \times 0 \cup x_0 \times I)$ is a retract of $(X \times I, A \times I)$.

Now let $F, F': (X \times \dot{I} \cup x_0 \times I, A \times \dot{I} \cup x_0 \times I) \to (Y,B)$ be defined by

$$F(x,0) = \alpha_0(x) \qquad F(x,1) = \alpha_1(x) \qquad F(x_0,t) = \omega(t)$$
$$F'(x,0) = \alpha_0'(x) \qquad F'(x,1) = \alpha_1'(x) \qquad F'(x_0,t) = \omega'(t)$$

Because $\alpha_0 \simeq \alpha_0'$, $\alpha_1 \simeq \alpha_1'$, and $\omega \simeq \omega'$, it follows that $F \simeq F'$. Because α_0 is ω-homotopic to α_1, F can be extended to a map $H: (X,A) \times I \to (Y,B)$. By the cofibration property established above, F' can be extended to a map $H': (X,A) \times I \to (Y,B)$. Then H' is an ω'-homotopy from α_0' to α_1'. ∎

It follows from lemmas 2*a* and 2*b* that, given ω and $\alpha_1: (X,A,x_0) \to (Y,B,\omega(1))$, the set of all maps $\alpha_0: (X,A,x_0) \to (Y,B,\omega(0))$ which are ω-homotopic to α_1 belong to a single homotopy class of maps $(X,A,x_0) \to (Y,B,\omega(0))$. It follows from lemma 2*c* that this set of maps equals a homotopy class of maps $(X,A,x_0) \to (Y,B,\omega(0))$ which depends only on the homotopy class $[\alpha_1] \in [X,A,x_0; Y,B,\omega(1)]$ and the path class $[\omega]$. Therefore, if (X,A) has a nondegenerate base point, there is a map

$$h_{[\omega]}: [X,A,x_0;\ Y,B,\omega(1)] \to [X,A,x_0;\ Y,B,\omega(0)]$$

characterized by the property $h_{[\omega]}[\alpha_1] = [\alpha_0]$ if and only if α_0 is ω-homotopic to α_1. It follows from lemmas 1*a* and 1*b* that this map is functorial in (X,A) and in (Y,B) and from lemma 1*c* that if (X,A) is a suspension, the map is a homomorphism.

3 THEOREM *Let (X,A) be a pair with nondegenerate base point x_0. For any pair (Y,B) there is a covariant functor from the fundamental groupoid of B to the category of pointed sets which assigns to a point $y_0 \in B$ the set $[X,A,x_0;\ Y,B,y_0]$ and to a path class $[\omega]$ in B the map $h_{[\omega]}$. If (X,A) is a suspension, this functor takes values in the category of groups and homomorphisms.*

PROOF We need only verify the two functorial properties. If $\alpha: (X,A,x_0) \to (Y,B,y_0)$ is arbitrary and ε is the constant path at y_0, the constant homotopy is an ε-homotopy from α to α proving that $h_{[\varepsilon]} = 1$.

Given paths ω and ω' in B such that $\omega(1) = \omega'(0)$, an ω-homotopy H from α_0 to α_1, and an ω'-homotopy H' from α_1 to α_2 [where α_0, α_1, α_2 are maps of (X,A) to (Y,B)], an $(\omega * \omega')$-homotopy $H * H'$ from α_0 to α_2 is defined by

$$(H * H')(x,t) = \begin{cases} H(x,2t) & 0 \le t \le \tfrac{1}{2} \\ H'(x,\ 2t-1) & \tfrac{1}{2} \le t \le 1 \end{cases}$$

This shows that $h_{[\omega * \omega']} = h_{[\omega]} \circ h_{[\omega']}$. ∎

4 COROLLARY *If $B \subset Y$ is path connected and (X,A) has a nondegenerate base point x_0, then for any y_0, $y_1 \in B$ the pointed sets $[X,A,x_0;\ Y,B,y_0]$ and $[X,A,x_0;\ Y,B,y_1]$ are in one-to-one correspondence. Furthermore, $\pi_1(B,y_0)$ acts as a group of operators on the left on $[X,A,x_0;\ Y,B,y_0]$, and the one-to-one correspondence above is determined up to this action of $\pi_1(B,y_0)$.*

PROOF If $[\omega]$ is any path class in B, $h_{[\omega]}$ is a one-to-one correspondence. If $[\omega] \in \pi_1(B,y_0)$, then $h_{[\omega]}$ is a permutation of $[X,A,x_0;\ Y,B,y_0]$, and in this way $\pi_1(B,y_0)$ acts as a group of operators. If y_0 and y_1 are points in B, the set of one-to-one correspondence $h_{[\omega]}$ determined by path classes $[\omega]$ in Y from y_0 to y_1 is the same as the set of maps $h_{[\omega_0]} \circ h_{[\omega']}$, where $[\omega_0]$ is a fixed path class from y_0 to y_1 and $[\omega'] \in \pi_1(B,y_0)$. ∎

In all of the above, by taking $B = Y$, we get the corresponding results for the absolute case. Thus, if X is a space with nondegenerate base point x_0 and $y_0 \in Y$, then $\pi_1(Y,y_0)$ acts as a group of operators on $[X,x_0;\ Y,y_0]$. If Y is path connected and y_0, $y_1 \in Y$, then $[X,x_0;\ Y,y_0]$ and $[X,x_0;\ Y,y_1]$ are in one-to-one correspondence by a bijection determined up to the action of $\pi_1(Y,y_0)$.

In case Y is an H space and $B \subset Y$ is a sub-H-space, there is the following result, which can be regarded as a generalization of theorem 1.8.4.

5 THEOREM *Let (X,A) have a nondegenerate base point x_0 and let (Y,B) be a pair of H spaces. If $y_0 \in B$ is the base point, $\pi_1(B,y_0)$ acts trivially on $[X,A,x_0;\ Y,B,y_0]$.*

PROOF Let $\mu\colon (Y \times Y,\ B \times B) \to (Y,B)$ be the multiplication. Given $\alpha\colon (X,A,x_0) \to (Y,B,y_0)$ and a closed path $\omega\colon (I,\dot{I}) \to (B,y_0)$, define an ω'-homotopy $H\colon (X,A) \times I \to (Y,B)$ from α' to α' (where $\omega' \simeq \omega$ and $\alpha' \simeq \alpha$) by

$$H(x,t) = \mu(\alpha(x),\omega(t))$$

Therefore $h_{[\omega]}[\alpha] = [\alpha]$ for all $[\alpha] \in [X,A,x_0;\ Y,B,y_0]$ and all $[\omega] \in \pi_1(B,y_0)$. ■

There is an interesting relation between the action of $\pi_1(B,y_0)$ on $[X,A,x_0;\ Y,B,y_0]$ and the action of $\pi_1(B,y_0)$ as covering transformations on a universal covering space of B. We assume that B and Y are path connected and locally path connected, that $\pi_1(B,y_0) \approx \pi_1(Y,y_0)$, and that $\tilde{Y}$ is a simply connected covering space of Y with covering projection $p\colon \tilde{Y} \to Y$. Then $\tilde{B} = p^{-1}(B)$ is a simply connected covering space of B [because $\pi_1(B,y_0) \approx \pi_1(Y,y_0)$]. Let $\tilde{y}_0 \in p^{-1}(y_0)$. There is a canonical map

$$\theta\colon [X,A,x_0;\ \tilde{Y},\tilde{B},\tilde{y}_0] \to [X,A;\ \tilde{Y},\tilde{B}]$$

from base-point-preserving homotopy classes to free homotopy classes. Because $\tilde{B}$ is simply connected, this map is a bijection [recall that two maps α_0, $\alpha_1\colon (X,A) \to (Y,B)$ are freely homotopic if and only if there is a path ω in B from $\alpha_0(x_0)$ to $\alpha_1(x_0)$ such that α_0 is ω-homotopic to α_1].

6 LEMMA *With the notation above, let $g\colon (\tilde{Y},\tilde{B},\tilde{y}_0) \to (\tilde{Y},\tilde{B},\tilde{y}_1)$ be a covering transformation and let $\tilde{\omega}$ be a path in $\tilde{B}$ from $\tilde{y}_0$ to $\tilde{y}_1$. There is a commutative diagram*

$$\begin{array}{ccccc}
[X,A,x_0;\ Y,B,y_0] & \xleftarrow{p_\#} & [X,A,x_0;\ \tilde{Y},\tilde{B},\tilde{y}_0] & \xrightarrow[\approx]{\theta} & [X,A;\ \tilde{Y},\tilde{B}] \\
h_{p_\#[\tilde{\omega}]}\downarrow\approx & & \approx\downarrow h_{[\tilde{\omega}]} \circ g_\# & & \approx\downarrow g_\# \\
[X,A,x_0;\ Y,B,y_0] & \xleftarrow{p_\#} & [X,A,x_0;\ \tilde{Y},\tilde{B},\tilde{y}_0] & \xrightarrow[\approx]{\theta} & [X,A;\ \tilde{Y},\tilde{B}]
\end{array}$$

PROOF Because g is a covering transformation, $p = p \circ g$ and $p_\# = p_\# \circ g_\#$. The commutativity of the left-hand square follows from this and from lemma 1*b*. Since $\theta \circ h_{[\tilde{\omega}]} = \theta$, the commutativity of the right-hand side follows from the trivial verification that $\theta \circ g_\# = g_\# \circ \theta$. ■

Recall the isomorphism $\psi\colon G(\tilde{B}\,|\,B) \approx \pi_1(B,y_0)$ of corollary 2.6.4, which assigns to g the element $[p \circ \tilde{\omega}] \in \pi_1(B,y_0)$. Therefore lemma 6 expresses a relation between the action of $G(\tilde{B}\,|\,B) \approx G(\tilde{Y}\,|\,Y)$ on the free homotopy classes $[X,A;\ \tilde{Y},B]$ and the action of $\pi_1(B,y_0)$ on $[X,A,x_0;\ Y,B,y_0]$.

7 COROLLARY *Let X be a simply connected locally path-connected space with nondegenerate base point and let $\tilde{Y}$ be a simply connected covering space of a locally path-connected space Y. There is a bijection from the free homotopy classes $[X;\tilde{Y}]$ to the pointed homotopy classes $[X,x_0;\ Y,y_0]$ compatible with the action of $G(\tilde{Y}\,|\,Y)$ on the former, the action of $\pi_1(Y,y_0)$ on the latter, and the isomorphism $\psi\colon G(\tilde{Y}\,|\,Y) \approx \pi_1(Y,y_0)$.*

PROOF This follows from lemma 6, with $B = Y$ and $A = X$, and from the observation that because X is simply connected, it follows from the lifting theorem 2.4.5, the homotopy lifting property of $p: \tilde{Y} \to Y$, theorem 2.2.3, and the unique-lifting property, theorem 2.2.2, that $p_\#: [X,x_0; \tilde{Y},\tilde{y}_0] \to [X,x_0; Y,y_0]$ is a bijection. ∎

We now specialize to the homotopy groups. Because

$$\pi_n(X,x_0) = [S^n(\dot{I}),0; X,x_0] = [S^n(\dot{I}),S^n(\dot{I}),0; X,X,x_0]$$

we obtain the following result.

8 THEOREM *For any space X and any $n \geq 1$, there is a covariant functor from the fundamental groupoid of X to the category of groups and homomorphisms which assigns to $x \in X$ the group $\pi_n(X,x)$ and to a path class $[\omega]$ in X the map $h_{[\omega]}: \pi_n(X,\omega(1)) \to \pi_n(X,\omega(0))$. In this way, $\pi_1(X,x_0)$ acts as a group of operators on the left on $\pi_n(X,x_0)$, by conjugation if $n = 1$, and if X is path connected and $x_0, x_1 \in X$, then $\pi_n(X,x_0)$ and $\pi_n(X,x_1)$ are isomorphic by an isomorphism determined up to the action of $\pi_1(X,x_0)$.*

PROOF Everything follows from theorem 3 and corollary 4 except for the statement that $\pi_1(X,x_0)$ acts on $\pi_1(X,x_0)$ by conjugation. For this let $H: S(\dot{I}) \times I \to X$ be on ω-homotopy from α_0 to α_1, where ω, α_0, and α_1 are closed paths in X at x_0. Define $H': I \times I \to X$ by

$$H'(t,t') = H([1,t], t')$$

Then $H' \mid 0 \times I = H' \mid 1 \times I = \omega$ and $H' \mid I \times 0 = \alpha_0$ and $H' \mid I \times 1 = \alpha_1$. It follows from lemma 1.8.6 that $(\omega * \alpha_1) * (\omega^{-1} * \alpha_0^{-1})$ is null homotopic. Therefore $[\alpha_0] = [\omega][\alpha_1][\omega]^{-1}$, and so $h_{[\omega]}[\alpha_1] = [\omega][\alpha_1][\omega]^{-1}$. ∎

Theorem 8 shows that the action of $\pi_1(X,x_0)$ on itself by conjugation, as in theorem 1.8.3, is extended to an action of $\pi_1(X,x_0)$ on $\pi_n(X,x_0)$ for every $n \geq 1$.

A path-connected space X is said to be *n-simple* (for $n \geq 1$) if for some $x_0 \in X$ (and hence all base points $x \in X$) $\pi_1(X,x_0)$ acts trivially on $\pi_n(X,x_0)$. Thus a simply connected space is n-simple for every $n \geq 1$, and a path-connected space X is 1-simple if and only if $\pi_1(X,x_0)$ is abelian. For n-simple spaces there is a unique canonical isomorphism $\pi_n(X,x_0) \approx \pi_n(X,x_1)$, any map $\alpha: S^n \to X$ determines a unique element of $\pi_n(X,x_0)$ (whether or not α maps the base point $p_0 \in S^n$ to x_0), and $\pi_n(X,x_0)$, is in one-to-one correspondence with the free homotopy classes of maps $S^n \to X$. The latter is a useful property, and for n-simple spaces X we shall usually omit the base point and merely write $\pi_n(X)$. From theorem 5 we obtain the following generalization of theorem 1.8.4.

9 THEOREM *A path-connected H space is n-simple for every $n \geq 1$.* ∎

Similar consideration apply to the relative homotopy groups.

10 THEOREM *For any pair (X,A) and any $n \geq 1$ there is a covariant functor from the fundamental groupoid of A to the category of pointed sets if $n = 1$ and the category of groups if $n > 1$ which assigns $\pi_n(X,A,x)$ to $x \in A$ and to a path class $[\omega]$ in A the map*

$$h_{[\omega]}: \pi_n(X,A,\omega(1)) \to \pi_n(X,A,\omega(0))$$

In this way, $\pi_1(A,x_0)$ acts as a group of operators on the left on $\pi_n(X,A,x_0)$, and if A is path connected and $x_0, x_1 \in A$, then $\pi_n(X,A,x_0)$ and $\pi_n(X,A,x_1)$ are isomorphic by an isomorphism determined up to the action of $\pi_1(A,x_0)$. ∎

If ω is a path in A, it follows from lemma 1*a* that there is a commutative square for $n > 1$,

$$\begin{array}{ccc} \pi_n(X,A,\omega(1)) & \xrightarrow{\partial} & \pi_{n-1}(A,\omega(1)) \\ {\scriptstyle h_{[\omega]}}\downarrow & & \downarrow{\scriptstyle h_{[\omega]}} \\ \pi_n(X,A,\omega(0)) & \xrightarrow{\partial} & \pi_{n-1}(A,\omega(0)) \end{array}$$

Thus there is also a covariant functor from the fundamental groupoid of A to the category of exact sequences which assigns to $x \in A$ the homotopy sequence of (X,A,x).

A pair (X,A) with A path connected is said to be *n-simple* (for $n \geq 1$) if $\pi_1(A,x_0)$ acts trivially on $\pi_n(X,A,x_0)$ for some (and hence all) base points $x_0 \in A$. If A is simply connected, (X,A) is n-simple for every $n \geq 1$.

11 THEOREM *Let (X,A) be a pair of H spaces with A path connected. Then (X,A) is n-simple for all $n \geq 1$.*

PROOF This is immediate from theorem 5. ∎

If (X,A) is n-simple and $x_0, x_1 \in A$, then $\pi_n(X,A,x_0)$ and $\pi_n(X,A,x_1)$ are canonically isomorphic. Therefore any map $\alpha: (E^n,S^{n-1}) \to (X,A)$ determines a unique element of $\pi_n(X,A,x_0)$ (whether or not α maps the base point $p_0 \in S^{n-1}$ to x^0), and $\pi_n(X,A,x_0)$ is in one-to-one correspondence with the free homotopy classes $[E^n,S^{n-1}; X,A]$. If (X,A) is n-simple, we shall frequently omit the base point and write $\pi_n(X,A)$.

The action of $\pi_1(A,x_0)$ on $\pi_2(X,A,x_0)$ is closely related to conjugation, as shown by the next result.

12 THEOREM *If $a, b \in \pi_2(X,A,x_0)$, then*

$$aba^{-1} = h_{\partial a}(b)$$

PROOF Let $X' = P(X,x_0)$ and let $p: X' \to X$ be the path fibration. Let $A' = p^{-1}(A)$ and let $x_0' \in A'$ be the constant path at x_0. By theorem 7.2.8, there is an isomorphism

$$p_\#: \pi_2(X',A',x_0') \approx \pi_2(X,A,x_0)$$

Let $a' = p_\#^{-1}(a)$ and $b' = p_\#^{-1}(b)$ and observe that, by lemma 1*b*,

$$h_{\partial a}(b) = p_\#(h_{\partial a'}(b'))$$

Hence it suffices to prove that $a'b'a'^{-1} = h_{\partial a'}(b')$. Because X' is contractible, it follows from the exactness of the homotopy sequence of (X',A',x_0') that

$$\partial: \pi_2(X',A',x_0') \approx \pi_1(A',x_0')$$

So to complete the proof we need only prove that

$$\partial(a'b'a'^{-1}) = \partial(h_{\partial a'}(b'))$$

The left-hand side equals $(\partial a')(\partial b')(\partial a')^{-1}$, and because ∂ commutes with $h_{\partial a'}$, the right-hand side equals $h_{\partial a'}(\partial b')$. The result now follows from the fact that the action of $\pi_1(A',x_0')$ on itself given by h is the same as conjugation. ∎

This again implies that $\pi_2(X,x_0) \approx \pi_2(X,\{x_0\},x_0)$ is abelian. Together with the exactness of the homotopy sequence, it yields the next result.

13 COROLLARY *The inclusion map* $j\colon (X,x_0) \subset (X,A)$ *induces a homomorphism*

$$j_\#\colon \pi_2(X,x_0) \to \pi_2(X,A,x_0)$$

whose image is in the center of $\pi_2(X,A,x_0)$. ∎

The following result is a generalization of theorem 1.8.7 to the higher relative homotopy groups.

14 THEOREM *Let* $f\colon (X,A,x_0) \to (Y,B,y_0)$ *and* $g\colon (X,A,x_0) \to (Y,B,y_1)$ *be freely homotopic. Then there is a path* ω *in* B *from* y_0 *to* y_1 *such that*

$$f_\# = h_{[\omega]} \circ g_\#\colon \pi_n(X,A,x_0) \to \pi_n(Y,B,y_0) \qquad n \geq 2$$

PROOF Let $F\colon (X,A) \times I \to (Y,B)$ be a homotopy from $f \mid (X,A)$ to $g \mid (X,A)$ and let $\omega(t) = F(x_0,t)$. Then ω is a path in B from y_0 to y_1, and if $\alpha\colon (I^n,\dot I^n,p_0) \to (X,A,x_0)$ represents an element of $\pi_n(X,A,x_0)$, then the composite

$$(I^n,\dot I^n) \times I \xrightarrow{\alpha \times 1} (X,A) \times I \xrightarrow{F} (Y,B)$$

is an ω-homotopy from $f \circ \alpha$ to $g \circ \alpha$. Therefore

$$f_\#[\alpha] = [f \circ \alpha] = h_{[\omega]}([g \circ \alpha]) = (h_{[\omega]} \circ g_\#)[\alpha] \quad ∎$$

This yields the following analogue of theorem 1.8.8.

15 COROLLARY *Let* $f\colon (X,A) \to (Y,B)$ *be a homotopy equivalence. For any* $x \in A$, f *induces isomorphisms*

$$f_\#\colon \pi_n(X,A,x) \approx \pi_n(Y,B,f(x))$$

PROOF Let $g\colon (Y,B) \to (X,A)$ be a homotopy inverse of f. By theorem 14, there are paths ω in A from $gf(x)$ to x and ω' in B from $fgf(x)$ to $f(x)$ such that the following diagram is commutative

$$\begin{array}{ccc} \pi_n(X,A,x) & \xrightarrow{h_{[\omega]}} & \pi_n(X,A,gf(x)) \\ f_\# \downarrow & \overset{g_\#}{\nearrow} & \downarrow f_\# \\ \pi_n(Y,B,f(x)) & \xrightarrow{h_{[\omega']}} & \pi_n(Y,B,fgf(x)) \end{array}$$

Since the maps $h_{[\omega]}$ and $h_{[\omega']}$ are isomorphisms, all the maps in the diagram are isomorphisms. ∎

4 THE HUREWICZ HOMOMORPHISM

There are no algorithms for computing the absolute or relative homotopy groups of a topological space (even when the space is given with a triangulation). One of the few main tools available for the general study of homotopy groups is their comparison with the corresponding integral singular homology groups. Such a comparison is effected by means of a canonical homomorphism from homotopy groups to homology groups. The definition and functorial properties of this homomorphism are our concern in this section. A theorem asserting that in the lowest nontrivial dimension for the homotopy group this homomorphism is an isomorphism will be established in the next section.

We shall be working with the integral singular homology theory throughout this section. Let $n \geq 1$ and recall that $H_q(I^n,\dot{I}^n) = 0$ for $q \neq n$ and $H_n(I^n,\dot{I}^n)$ is infinite cyclic. To consider relations among the homology groups of certain pairs in I^n, for $n \geq 1$ we define

$$\begin{aligned} I_1{}^n &= \{(t_1, \ldots ,t_n) \in I^n \mid t_n \leq ½\} \\ \dot{I}_1{}^n &= (I_1{}^n \cap \dot{I}^n) \cup \{(t_1, \ldots ,t_n) \in I^n \mid t_n = ½\} \\ I_2{}^n &= \{(t_1, \ldots ,t_n) \in I^n \mid t_n \geq ½\} \\ \dot{I}_2{}^n &= (I_2{}^n \cap \dot{I}^n) \cup \{(t_1, \ldots ,t_n) \in I^n \mid t_n = ½\} \end{aligned}$$

Then $I_1{}^n \cup I_2{}^n = I^n$ and $(I_1{}^n \cup \dot{I}_2{}^n) \cap (\dot{I}_1{}^n \cup I_2{}^n) = \dot{I}_1{}^n \cup \dot{I}_2{}^n$. By the exactness of the Mayer-Vietoris sequence of the excisive couple $\{I_1{}^n \cup \dot{I}_2{}^n, \dot{I}_1{}^n \cup I_2{}^n\}$, we have

$$H_q(I_1{}^n \cup \dot{I}_2{}^n, \dot{I}_1{}^n \cup \dot{I}_2{}^n) \oplus H_q(\dot{I}_1{}^n \cup I_2{}^n, \dot{I}_1{}^n \cup \dot{I}_2{}^n) \approx H_q(I^n, \dot{I}_1{}^n \cup \dot{I}_2{}^n)$$

By excision, we also have isomorphisms

$$\begin{aligned} H_q(I_1{}^n,\dot{I}_1{}^n) &\approx H_q(I_1{}^n \cup \dot{I}_2{}^n, \dot{I}_1{}^n \cup \dot{I}_2{}^n) \\ H_q(I_2{}^n,\dot{I}_2{}^n) &\approx H_q(\dot{I}_1{}^n \cup I_2{}^n, \dot{I}_1{}^n \cup \dot{I}_2{}^n) \end{aligned}$$

Combining these, we see that if we let $i_1\colon (I_1{}^n,\dot{I}_1{}^n) \subset (I^n, \dot{I}_1{}^n \cup \dot{I}_2{}^n)$ and we let $i_2\colon (I_2{}^n,\dot{I}_2{}^n) \subset (I^n, \dot{I}_1{}^n \cup \dot{I}_2{}^n)$, then we have the following result.

1 LEMMA *The inclusion maps i_1 and i_2 define a direct-sum representation*

$$i_{1*} \oplus i_{2*}\colon H_q(I_1{}^n,\dot{I}_1{}^n) \oplus H_q(I_2{}^n,\dot{I}_2{}^n) \approx H_q(I^n, \dot{I}_1{}^n \cup \dot{I}_2{}^n) \quad ■$$

Let $\nu_1\colon (I^n,\dot{I}^n) \to (I_1{}^n,\dot{I}_1{}^n)$ be defined by $\nu_1(t_1, \ldots ,t_n) = (t_1, \ldots ,t_{n-1},t_n/2)$ and define $\nu_2\colon (I^n,\dot{I}^n) \to (I_2{}^n,\dot{I}_2{}^n)$ by $\nu_2(t_1, \ldots ,t_n) = (t_1, \ldots ,t_{n-1}, (t_n + 1)/2)$. Let $i\colon (I^n,\dot{I}^n) \subset (I^n, \dot{I}_1{}^n \cup \dot{I}_2{}^n)$.

2 COROLLARY *For any $z \in H_n(I^n,\dot{I}^n)$*

$$i_* z = i_{1*} \nu_{1*} z + i_{2*} \nu_{2*} z$$

PROOF Let $j_1\colon (I^n, \dot{I}_1{}^n \cup \dot{I}_2{}^n) \subset (I^n, I_1{}^n \cup \dot{I}_2{}^n)$ and $j_2\colon (I^n, \dot{I}_1{}^n \cup \dot{I}_2{}^n) \subset (I^n, \dot{I}_1{}^n \cup I_2{}^n)$. Then $j_{1*} i_{1*} = 0$ and $j_{1*} i_{2*}$ is an isomorphism of $H_q(I_2{}^n,\dot{I}_2{}^n)$

onto $H_q(I^n, I_1{}^n \cup \dot{I}_2{}^n)$ (induced by the inclusion map, which is an excision). Similarly, $j_{2*} i_{2*} = 0$ and $j_{2*} i_{1*}$ is an isomorphism of $H_q(I_1{}^n, \dot{I}_1{}^n)$ onto $H_q(I^n, \dot{I}_1{}^n \cup I_2{}^n)$. It follows from lemma 1 that

$$\ker j_{1*} \cap \ker j_{2*} = 0$$

Therefore, to prove the corollary it suffices to prove that

$$i_* z - i_{1*} \nu_{1*} z - i_{2*} \nu_{2*} z$$

is in the kernel of j_{1*} and in the kernel of j_{2*}.

We first prove that $j_{1*}(i_* z - i_{1*} \nu_{1*} z - i_{2*} \nu_{2*} z) = 0$. Because $j_{1*} i_{1*} = 0$, we must show that $j_{1*} i_* z = j_{1*} i_{2*} \nu_{2*} z$. Clearly $j_1 i$ is the inclusion map $(I^n, \dot{I}^n) \subset (I^n, I_1{}^n \cup \dot{I}_2{}^n)$ and $j_1 i_2 \nu_2$ is the map $f\colon (I^n, \dot{I}^n) \to (I^n, I_1{}^n \cup \dot{I}_2{}^n)$ defined by $f(t_1, \ldots, t_n) = (t_1, \ldots, t_{n-1}, (t_n + 1)/2)$. A homotopy H from $j_1 i$ to f is defined by

$$H((t_1, \ldots, t_n), t) = (t_1, \ldots, t_{n-1}, (t_n + t)/(1 + t))$$

Therefore $j_{1*} i_* = f_* = j_{1*} i_{2*} \nu_{2*}$. A similar argument shows that

$$j_{2*}(i_* z - i_{1*} \nu_{1*} z - i_{2*} \nu_{2*} z) = 0 \quad \blacksquare$$

For $n \geq 1$ the subset $I \times \dot{I}^{n-1} \cup 0 \times I^{n-1} \subset \dot{I}^n$ is contractible. Therefore $H_q(I^n, I \times \dot{I}^{n-1} \cup 0 \times I^{n-1}) = 0$ for all q. By exactness of the homology sequence of the triple $(I^n, \dot{I}^n, I \times \dot{I}^{n-1} \cup 0 \times I^{n-1})$, it follows that the map

$$\partial\colon H_q(I^n, \dot{I}^n) \to H_{q-1}(\dot{I}^n, I \times \dot{I}^{n-1} \cup 0 \times I^{n-1})$$

is an isomorphism for all q. For $n \geq 2$ let

$$j\colon (I^{n-1}, \dot{I}^{n-1}) \to (\dot{I}^n, I \times \dot{I}^{n-1} \cup 0 \times I^{n-1})$$

be defined by $j(t_1, \ldots, t_{n-1}) = (1, t_1, \ldots, t_{n-1})$. Then j is the composite of a homeomorphism from $(I^{n-1}, \dot{I}^{n-1})$ to $(1 \times I^{n-1}, 1 \times \dot{I}^{n-1})$ and the excision map

$$(1 \times I^{n-1}, 1 \times \dot{I}^{n-1}) \subset (\dot{I}^n, I \times \dot{I}^{n-1} \cup 0 \times I^{n-1})$$

Therefore the homomorphism

$$j_*\colon H_q(I^{n-1}, \dot{I}^{n-1}) \to H_q(\dot{I}^n, I \times \dot{I}^{n-1} \cup 0 \times I^{n-1})$$

is an isomorphism for all q.

We define canonical generators $Z_n \in H_n(I^n, \dot{I}^n)$ for $n \geq 1$ by induction on n as follows:

(a) $Z_1 \in H_1(I, \dot{I})$ is the unique element with $\partial Z_1 = (1) - (0)$ in $H_0(\dot{I})$.
(b) For $n \geq 2$, $Z_n \in H_n(I^n, \dot{I}^n)$ is the unique element such that $\partial Z_n = j_* Z_{n-1}$ in $H_{n-1}(\dot{I}^n, I \times \dot{I}^{n-1} \cup 0 \times I^{n-1})$.

Given a map $\alpha\colon (I^n, \dot{I}^n) \to (X,A)$, then $\alpha_* Z_n \in H_n(X,A)$. If $\alpha \simeq \beta$, then $\alpha_* Z_n = \beta_* Z_n$. Therefore there is for $n \geq 1$ a well-defined map

$$\varphi\colon \pi_n(X,A,x_0) \to H_n(X,A)$$

such that $\varphi[\alpha] = \alpha_* Z_n$, where $\alpha: (I^n,\dot{I}^n) \to (X,A)$ maps z_0 to x_0 and represents an element of $\pi_n(X,A,x_0)$. By identifying $\pi_n(X,x_0)$ with $\pi_n(X,\{x_0\},x_0)$, we also have a map $\varphi: \pi_n(X,x_0) \to H_n(X,x_0)$. Some of the basic properties of φ are summarized in the next result.

3 THEOREM *If $n \geq 2$ or if $n = 1$ and $A = \{x_0\}$, the map φ is a homomorphism. It has the following functorial properties:*

(*a*) *For $n \geq 2$ commutativity holds in the square*

$$\begin{array}{ccc} \pi_n(X,A,x_0) & \xrightarrow{\partial} & \pi_{n-1}(A,x_0) \\ \varphi\downarrow & & \downarrow\varphi \\ H_n(X,A) & \xrightarrow{\partial} & H_{n-1}(A,x_0) \end{array}$$

(*b*) *Given $f: (X,A,x_0) \to (Y,B,y_0)$, commutativity holds in the square*

$$\begin{array}{ccc} \pi_n(X,A,x_0) & \xrightarrow{f_\#} & \pi_n(Y,B,y_0) \\ \varphi\downarrow & & \downarrow\varphi \\ H_n(X,A) & \xrightarrow{f_*} & H_n(Y,B) \end{array}$$

PROOF Let $\alpha_1, \alpha_2: (I^n,\dot{I}^n) \to (X,A)$ be such that

$$\alpha_1(t_1, \ldots ,t_{n-1}, 1) = \alpha_2(t_1, \ldots ,t_{n-1}, 0)$$

[any two maps of $(I^n,\dot{I}^n)$ to (X,A) are homotopic to such maps if $n \geq 2$ or if $n = 1$ and $A = \{x_0\}$]. Then $\alpha_1 * \alpha_2 = \beta \circ i$, where $i: (I^n,\dot{I}^n) \subset (I^n, \dot{I}_1^n \cup \dot{I}_2^n)$ and $\beta: (I^n, \dot{I}_1^n \cup \dot{I}_2^n) \to (X,A)$ is defined by

$$\beta(t_1, \ldots ,t_n) = \begin{cases} \alpha_1(t_1, \ldots ,t_{n-1}, 2t_n) & t_n \leq ½ \\ \alpha_2(t_1, \ldots ,t_{n-1}, 2t_n - 1) & t_n \geq ½ \end{cases}$$

Then $\varphi[\alpha_1 * \alpha_2] = \beta_* i_* Z_n = \beta_*(i_{1*}\nu_{1*}Z_n + i_{2*}\nu_{2*}Z_n)$, the last equality by corollary 2. Since $\beta i_1\nu_1 = \alpha_1$ and $\beta i_2\nu_2 = \alpha_2$, we see that

$$\varphi[\alpha_1 * \alpha_2] = \alpha_{1*} Z_n + \alpha_{2*} Z_n = \varphi[\alpha_1] + \varphi[\alpha_2]$$

which shows that φ is a homomorphism whenever $\pi_n(X,A,x_0)$ is a group.

To prove (*a*), let $\alpha: (I^n,\dot{I}^n) \to (X,A)$ represent an element of $\pi_n(X,A)$ for $n \geq 2$ and suppose that $\alpha(I \times \dot{I}^{n-1} \cup 0 \times I^{n-1}) = x_0$. Then $\partial[\alpha] = [\alpha']$, where $\alpha': (I^{n-1},\dot{I}^{n-1}) \to (A,x_0)$ is defined by $\alpha' = (\alpha \mid (\dot{I}^n, I \times \dot{I}^{n-1} \cup 0 \times I^{n-1})) \circ j$. Then

$$\begin{aligned} \varphi\partial[\alpha] &= \alpha'_* Z_{n-1} = (\alpha|(\dot{I}^n, I\times \dot{I}^{n-1} \cup 0 \times I^{n-1}))_* j_* Z_{n-1} \\ &= (\alpha \mid (\dot{I}^n, I \times \dot{I}^{n-1} \cup 0 \times I^{n-1}))_* \partial Z_n \\ &= \partial\alpha_* Z_n = \partial\varphi[\alpha] \end{aligned}$$

Finally, (*b*) follows from the fact that $(f\alpha)_* = f_*\alpha_*$. ■

The map φ is called the *Hurewicz homomorphism*. The next result follows from theorem 3.

4 COROLLARY *The Hurewicz homomorphism maps the homotopy sequence of (X,A,x_0) into the homology sequence of (X,A,x_0).* ∎

Our next objective is to show that the Hurewicz homomorphism commutes with the actions of the appropriate fundamental group on the homotopy set. We consider the relative case first.

5 LEMMA *Let $[\alpha] \in \pi_n(X,A,x_0)$ for $n \geq 2$ and let $[\omega] \in \pi_1(A,x_0)$. Then*

$$\varphi(h_{[\omega]}[\alpha]) = \varphi[\alpha]$$

PROOF Let $[\alpha]$ be represented by $\alpha\colon (I^n,\dot{I}^n) \to (X,A)$ and let $h_{[\omega]}[\alpha]$ be represented by $\alpha'\colon (I^n,\dot{I}^n) \to (X,A)$. Then α and α' are freely homotopic [that is, α and α' are homotopic as maps of $(I^n,\dot{I}^n)$ to (X,A)]. Therefore

$$\varphi[\alpha] = \alpha_* Z_n = \alpha'_* Z_n = \varphi[\alpha'] = \varphi(h_{[\omega]}[\alpha]) \quad ∎$$

Next we prove the corresponding result for the absolute case.

6 LEMMA *Let $[\alpha] \in \pi_n(X,x_0)$ and $[\omega] \in \pi_1(X,x_0)$. Then*

$$\varphi(h_{[\omega]}[\alpha]) = \varphi[\alpha]$$

PROOF Let Y be the space obtained from I^n by collapsing $\dot{I}^n$ to a single point, this point to be the base point of Y, denoted by y_0. The collapsing map $g\colon (I^n,\dot{I}^n) \to (Y,y_0)$ induces a one-to-one correspondence between $[Y,y_0; X,x_0]$ and $[I^n,\dot{I}^n; X,x_0]$. Therefore $\pi_n(X,x_0)$ can be identified with $[Y,y_0; X,x_0]$. Furthermore, $g_*\colon H_n(I^n,\dot{I}^n) \approx H_n(Y,y_0)$, and we let $g_* Z_n = Z'_n \in H_n(Y,y_0)$. In these terms, if an element of $\pi_n(X,x_0)$ is represented by $\alpha\colon (Y,y_0) \to (X,x_0)$, then $\varphi[\alpha] = \alpha_* Z'_n$. Let $h_{[\omega]}[\alpha]$ be represented by $\alpha'\colon (Y,y_0) \to (X,x_0)$. Then α and α' are homotopic as maps of Y to X. Therefore, if $Z''_n \in H_n(Y)$ is the unique element such that $i'_* Z''_n = Z'_n$ [where $i'\colon Y \subset (Y,y_0)$], then

$$(\alpha \mid Y)_* Z''_n = (\alpha' \mid Y)_* Z''_n$$

Let $j'\colon X \subset (X,x_0)$. Then

$$\varphi[\alpha] = \alpha_* Z'_n = \alpha_* i'_* Z''_n = j'_* (\alpha \mid Y)_* Z'_n$$

Similarly, $\varphi[\alpha'] = j'_* (\alpha' \mid Y)_* Z''_n$, and

$$\varphi[\alpha] = \varphi[\alpha'] = \varphi(h_{[\omega]}[\alpha]) \quad ∎$$

We define $\pi'_n(X,A,x_0)$ for $n \geq 2$ to be the quotient group of $\pi_n(X,A,x_0)$ by the normal subgroup G generated by

$$\{(h_{[\omega]}[\alpha])[\alpha]^{-1} \mid [\alpha] \in \pi_n(X,A,x_0),\ [\omega] \in \pi_1(A,x_0)\}$$

By lemma 5, φ maps G to 0 and there is a homomorphism

$$\varphi'\colon \pi'_n(X,A,x_0) \to H_n(X,A)$$

whose composite with the canonical map $\eta\colon \pi_n(X,A,x_0) \to \pi'_n(X,A,x_0)$ is φ. Note that, by theorem 7.3.12, $\pi'_n(X,A,x_0)$ is abelian for all $n \geq 2$.

Similarly, we define $\pi'_n(X,x_0)$ for $n \geq 1$ to be the quotient group of

$\pi_n(X,x_0)$ by the normal subgroup H generated by

$$\{(h_{[\omega]}[\alpha])[\alpha]^{-1} \mid [\alpha] \in \pi_n(X,x_0), [\omega] \in \pi_1(X,x_0)\}$$

By lemma 6, φ maps H to 0, and there is a homomorphism

$$\varphi': \pi'_n(X,x_0) \to H_n(X,x_0)$$

whose composite with the canonical map $\eta: \pi_n(X,x_0) \to \pi'_n(X,x_0)$ is φ. Note that $\pi'_1(X,x_0)$ is the quotient group of $\pi_1(X,x_0)$ by its commutator subgroup. In particular, $\pi'_n(X,x_0)$ is abelian for all $n \geq 1$.

Because the groups $\pi'_n(X,A,x_0)$ and $\pi'_n(X,x_0)$ are abelian, we shall find them easier to compare with the homology groups (which are abelian) than the homotopy groups themselves. For the comparison it will be convenient to replace the triple $(I^n,\dot{I}^n,z_0)$, which is the antecedent triple used to define $\pi_n(X,A,x_0)$, by the homeomorphic triple $(\Delta^n,\dot{\Delta}^n,v_0)$, where Δ^n is the standard n-simplex used in Sec. 4.1 to define the singular complex (vertices of Δ^n will be denoted by $v_0, v_1, \ldots, v_n$). To achieve this replacement we need only choose a homeomorphism of $(\Delta^n,\dot{\Delta}^n,v_0)$ onto $(I^n,\dot{I}^n,z_0)$. Any homeomorphism $h: (\Delta^n,\dot{\Delta}^n) \to (I^n,\dot{I}^n)$ will induce an isomorphism

$$h_*: H_n(\Delta^n,\dot{\Delta}^n) \approx H_n(I^n,\dot{I}^n)$$

The identity map $\xi_n: \Delta^n \subset \Delta^n$ is a singular simplex which is a cycle modulo $\dot{\Delta}^n$ and whose homology class $\{\xi_n\}$ is a generator of the infinite cyclic group $H_n(\Delta^n,\dot{\Delta}^n)$. Since Z_n is a generator of $H_n(I^n,\dot{I}^n)$ and h_* is an isomorphism, either $h_*\{\xi_n\} = Z_n$ or $h_*\{\xi_n\} = -Z_n$. We want to choose h so that the former holds. If $n = 1$, the choice of Z_1 is such that the simplicial homeomorphism $h: \Delta^1 \to I$ with $h(v_0) = 0$ and $h(v_1) = 1$ will have the desired property (that is, $h_*\{\xi_1\} = Z_1$). If $n > 1$, we choose an arbitrary homeomorphism $h: (\Delta^n,\dot{\Delta}^n) \to (I^n,\dot{I}^n)$ such that $h(v_0) = z_0$. If $h_*\{\xi_n\} = -Z_n$, we replace h by $h\lambda$, where λ is a simplicial homeomorphism of Δ^n to itself such that $\lambda(v_0) = v_0$ and $\lambda_*\{\xi_n\} = -\{\xi_n\}$ (for example, λ is the simplicial map which interchanges v_1 and v_2 and leaves all other vertices of Δ^n fixed). Therefore, in any event, we can find a homeomorphism $h: (\Delta^n,\dot{\Delta}^n,v_0) \to (I^n,\dot{I}^n,z_0)$ such that $h_*\{\xi_n\} = Z_n$. Using such a homeomorphism to represent elements of $\pi_n(X,A,x_0)$ by maps $\alpha: (\Delta^n,\dot{\Delta}^n) \to (X,A)$ such that $\alpha(v_0) = x_0$, we see that $\varphi[\alpha] = \alpha_*\{\xi_n\} = \{\alpha\}$, the latter being the homology class in (X,A) of the singular simplex α.

For any pair (X,A) with base point $x_0 \in A$ and any $n \geq 0$, let $\Delta(X,A,x_0)^n$ be the subcomplex of $\Delta(X)$ generated by singular simplexes $\sigma: \Delta^q \to X$ having the property that σ maps each vertex of Δ^q to x_0 and maps the n-dimensional skeleton $(\Delta^q)^n$ of Δ^q into A. Then $\Delta(X,A,x_0)^{n+1} \subset \Delta(X,A,x_0)^n$, and these two chain complexes agree in degrees $\leq n$. Thus we have a decreasing sequence of subcomplexes $\Delta(X,A,x_0)^n$ (where $n \geq 0$) of $\Delta(X)$ whose intersection is contained in $\Delta(A)$. If X is path connected and (X,A) is n-connected for some $n \geq 0$, we shall see that the inclusion map $\Delta(X,A,x_0)^n \subset \Delta(X)$ is a chain equivalence. The following lemma will be used for this purpose.

7 LEMMA *Let C be a subcomplex of the free chain complex $\Delta(X)$ such that C is generated by the singular simplexes of X in it. Assume that to every singular simplex $\sigma: \Delta^q \to X$ there is assigned a map $P(\sigma): \Delta^q \times I \to X$ such that*

(*a*) $P(\sigma)(z,0) = \sigma(z)$ *for* $z \in \Delta^q$.
(*b*) *Define* $\bar\sigma: \Delta^q \to X$ *by* $\bar\sigma(z) = P(\sigma)(z,1)$. *Then* $\bar\sigma$ *is a singular simplex in C, and if* σ *is in C,* $\bar\sigma = \sigma$.
(*c*) *If* $e_q^i: \Delta^{q-1} \to \Delta^q$ *omits the ith vertex, then* $P(\sigma) \circ (e_q^i \times 1) = P(\sigma^{(i)})$.

Then the inclusion map $C \subset \Delta(X)$ *is a chain equivalence.*

PROOF Let $j: C \subset \Delta(X)$ be the inclusion chain map and let $\tau: \Delta(X) \to C$ be the chain map defined by $\tau(\sigma) = \bar\sigma$ [(*c*) implies that τ is a chain map]. By (*b*), $\tau \circ j = 1_C$, hence to complete the proof we need only verify that $j \circ \tau \simeq 1_{\Delta(X)}$.

For any space Y let $h_0, h_1: Y \to Y \times I$ be the maps $h_0(y) = (y,0)$ and $h_1(y) = (y,1)$. In the proof of theorem 4.4.3 it was shown (by the method of acyclic models) that there exists a natural chain homotopy $D: \Delta(Y) \to \Delta(Y \times I)$ from $\Delta(h_0)$ to $\Delta(h_1)$. Define a chain homotopy

$$D': \Delta(X) \to \Delta(X)$$

by $D'(\sigma) = \Delta(P(\sigma))(D(\xi_q))$, where $\sigma: \Delta^q \to X$ and $\xi_q: \Delta^q \subset \Delta^q$. By (*c*), D' is a chain homotopy, and by (*a*) and the definition of $\bar\sigma$, D' is a chain homotopy from $1_{\Delta(X)}$ to $j \circ \tau$. ■

8 THEOREM *Let* $x_0 \in A \subset X$ *and assume that X is path connected and (X,A) is n-connected for some* $n \geq 0$. *Then the inclusion map* $\Delta(X,A,x_0)^n \subset \Delta(X)$ *is a chain equivalence.*

PROOF For $\sigma: \Delta^q \to X$ we define $P(\sigma)$ by induction on q to satisfy the properties of lemma 7, and to have the additional property that if σ is in $\Delta(X,A,x_0)^n$, then $P(\sigma)$ is the composite

$$\Delta^q \times I \xrightarrow{p} \Delta^q \xrightarrow{\sigma} X$$

where p is projection to the first factor.

If $q = 0$, then $\sigma: \Delta^0 \to X$ is a point of X, and because X is path connected, there is a map $P(\sigma): \Delta^0 \times I \to X$ such that $P(\sigma)(\Delta^0 \times 0) = \sigma(\Delta^0)$ and $P(\sigma)(\Delta^0 \times 1) = x_0$ [and if $\sigma(\Delta^0) = x_0$, we take $P(\sigma)$ to be the constant map to x_0]. This defines $P(\sigma)$ for all σ of degree 0 to have the desired properties.

Assume $0 < q \leq n$ and that $P(\sigma)$ has been defined for all σ of degree $< q$ to have the properties stated above. Given a singular simplex $\sigma: \Delta^q \to X$, if σ is in $\Delta(X,A,x_0)^n$, define $P(\sigma) = \sigma \circ p$. If σ is not in $\Delta(X,A,x_0)^n$, (*a*) and (*c*) of lemma 7 define $P(\sigma)$ on $\Delta^q \times 0 \cup \dot\Delta^q \times I$, and we let $f: \Delta^q \times 0 \cup \dot\Delta^q \times I \to X$ be this map. There is a homeomorphism $h: E^q \times I \approx \Delta^q \times I$ such that

$$h(E^q \times 0) = \Delta^q \times 0 \cup \dot\Delta^q \times I, \quad h(S^{q-1} \times 0) = \dot\Delta^q \times 1$$

and

$$h(S^{q-1} \times I \cup E^q \times 1) = \Delta^q \times 1$$

Let $f'\colon (E^q,S^{q-1}) \to (X,A)$ be defined by $f'(z) = f(h(z,0))$. Because $q \le n$ and (X,A) is n-connected, there is a homotopy $H\colon (E^q,S^{q-1}) \times I \to (X,A)$ from f' to some map of E^q into A (in fact, by the definition of n-connectedness, there is even such a homotopy relative to S^{q-1}). Then the composite

$$\Delta^q \times I \xrightarrow{h^{-1}} E^q \times I \xrightarrow{H} X$$

can be taken as $P(\sigma)$.

In this way $P(\sigma)$ is defined for all degrees $q \le n$. Note that a singular simplex of degree $> n$ is in $\Delta(X,A,x_0)^n$ if and only if every proper face is in $\Delta(X,A,x_0)^n$. Therefore, if $P(\sigma)$ has been defined for all degrees $< q$, where $q > n$, and if $\sigma\colon \Delta^q \to X$, then we define $P(\sigma) = \sigma \circ p$ if σ is in $\Delta(X,A,x_0)^n$ and to be any map $\Delta^q \times I \to X$ satisfying (*a*) and (*c*) of lemma 7 (such maps exist by the homotopy extension property). Then $P(\sigma)$ will necessarily satisfy (*b*) of lemma 7, and we have shown that $P(\sigma)$ can be defined for all σ to satisfy lemma 7. ■

For $n \ge 0$ we define

$$H_q^{(n)}(X,A,x_0) = H_q(\Delta(X,A,x_0)^n, \Delta(X,A,x_0)^n \cap \Delta(A))$$

There are canonical homomorphisms

$$\cdots \to H_q^{(n)}(X,A,x_0) \to H_q^{(n-1)}(X,A,x_0) \to \cdots \to H_q^{(0)}(X,A,x_0) \to H_q(X,A)$$

9 COROLLARY *Assume that A is path connected and for some $n \ge 0$, (X,A) is n-connected. Then the canonical map is an isomorphism for all q*

$$H_q^{(n)}(X,A,x_0) \approx H_q(X,A)$$

PROOF For any $n \ge 0$, $\Delta(X,A,x_0)^n \cap \Delta(A)$ is generated by the set of singular simplexes of A all of whose vertices are at x_0. This is independent of n, and because A is path connected, $(A,\{x_0\})$ is 0-connected, and it follows from theorem 8 that the inclusion map $\Delta(X,A,x_0)^n \cap \Delta(A) \subset \Delta(A)$ is a chain equivalence for all $n \ge 0$.

Since (X,A) is n-connected, where $n \ge 0$, and A is path connected, X is also path connected, and by theorem 8, the inclusion map $\Delta(X,A,x_0)^n \subset \Delta(X)$ is a chain equivalence. The result follows from these facts, using exactness and the five lemma. ■

5 THE HUREWICZ ISOMORPHISM THEOREM

The main result of this section asserts that if X and A are path connected and for some $n \ge 1$, (X,A) is n-connected, then the Hurewicz homomorphism φ induces an isomorphism φ' of $\pi'_{n+1}(X,A,x_0)$ with $H_{n+1}(X,A)$. This result is equivalent to a homotopy addition theorem which asserts that the sum of the $(n+1)$-dimensional faces of an $(n+2)$-simplex is the homotopy boundary of the identity map of the simplex. We prove both these theorems simultaneously by induction on n.

In the proof we shall make essential use of the complexes $\Delta(X,A,x_0)^n$ and of corollary 7.4.9. Let α: $(\Delta^n,\dot\Delta^n,(\Delta^n)^0) \to (X,A,x_0)$ represent an element of $\pi_n(X,A,x_0)$. Then α is a singular simplex in $\Delta(X,A,x_0)^{n-1}$ and represents a homology class $\{\alpha\} \in H_n^{(n-1)}(X,A,x_0)$. Since any element of $\pi_n(X,A,x_0)$ can be represented by such a map α, the Hurewicz homomorphism φ': $\pi'_n(X,A,x_0) \to H_n(X,A)$ factors into the composite

$$\pi'_n(X,A,x_0) \xrightarrow{\varphi''} H_n^{(n-1)}(X,A,x_0) \to H_n(X,A)$$

and there is a commutative diagram

$$\begin{array}{ccc} \pi_n(X,A,x_0) & \xrightarrow{\eta} & \pi'_n(X,A,x_0) \\ \varphi\downarrow & \swarrow\varphi' & \downarrow\varphi'' \\ H_n(X,A) & \leftarrow & H_n^{(n-1)}(X,A,x_0) \end{array}$$

We now formulate the propositions corresponding to the relative and absolute Hurewicz isomorphism theorems.

1 PROPOSITION Φ_n $(n \geq 2)$. *Let A be path connected and let (X,A) be $(n-1)$-connected. Then φ' is an isomorphism*

$$\varphi'\colon \pi'_n(X,A,x_0) \approx H_n(X,A)$$

2 PROPOSITION $\bar\Phi_n$ $(n \geq 1)$. *Let X be $(n-1)$-connected. Then φ' is an isomorphism*

$$\varphi'\colon \pi'_n(X,x_0) \approx H_n(X,x_0)$$

We shall prove both these propositions simultaneously by induction on n, together with a third proposition, which we now formulate. For $n \geq 2$, each face map e^i_{n+1} is a map of triples

$$e^0_{n+1}\colon (\Delta^n,\dot\Delta^n,v_0) \to (\Delta^{n+1},(\Delta^{n+1})^{n-1},v_1)$$
$$e^i_{n+1}\colon (\Delta^n,\dot\Delta^n,v_0) \to (\Delta^{n+1},(\Delta^{n+1})^{n-1},v_0) \qquad 0 < i \leq n+1$$

For vertices v and v' of Δ^{n+1} we use $[vv']$ to denote the path class of the linear path in Δ^{n+1} from v to v'. We define an element $b_1 \in \pi_1(\dot\Delta^2,v_0)$ and, for $n \geq 2$, an element $b_n \in \pi_n(\dot\Delta^{n+1},(\Delta^{n+1})^{n-1},v_0)$ by

$$b_1 = [v_0v_1] * [v_1v_2] * [v_2v_0]$$
$$b_2 = (h_{[v_0v_1]}[e_3{}^0])[e_3{}^2][e_3{}^1]^{-1}[e_3{}^3]^{-1}$$
$$b_n = h_{[v_0v_1]}[e^0_{n+1}] + \sum_{0<i\leq n+1} (-1)^i[e^i_{n+1}] \qquad n \geq 3$$

For $n = 1$ let j: $(\dot\Delta^2,v_0) \subset (\Delta^2,v_0)$ and for $n \geq 2$ let j: $(\dot\Delta^{n+1},(\Delta^{n+1})^{n-1},v_0) \subset (\Delta^{n+1},(\Delta^{n+1})^{n-1},v_0)$. The following proposition corresponds to the homotopy addition theorem.

3 PROPOSITION B_n $(n \geq 1)$. $j_\# b_n = 0$.

The simultaneous proof of propositions 1, 2, and 3 will consist of the following five parts:

(*a*) Proof of B_1
(*b*) Proof that $B_1 \Rightarrow \bar{\Phi}_1$
(*c*) Proof that $\bar{\Phi}_1, \bar{\Phi}_2, \ldots, \bar{\Phi}_{n-1} \Rightarrow B_n$ for $n \geq 2$
(*d*) Proof that $B_n \Rightarrow \Phi_n$ for $n \geq 2$
(*e*) Proof that $\Phi_n \Rightarrow \bar{\Phi}_n$ for $n \geq 2$

(*a*) **PROOF OF** B_1 We must prove that $j_\# b_1 = 0$. But $j_\# b_1 \in \pi_1(\Delta^2, v_0)$, and $\pi_1(\Delta^2, v_0) = 0$ because Δ^2 is contractible. ∎

(*b*) **PROOF THAT** $B_1 \Rightarrow \bar{\Phi}_1$ Let X be path connected. We must prove that $\varphi'\colon \pi_1'(X,x_0) \approx H_1(X,x_0)$. Because X is path connected, the inclusion map $\Delta(X,\{x_0\},x_0)^0 \subset \Delta(X)$ is a chain equivalence, and we need only show that

$$\varphi''\colon \pi_1'(X,x_0) \approx H_1^{(0)}(X,\{x_0\},x_0)$$

If $\alpha\colon (\Delta^1,\dot{\Delta}^1) \to (X,x_0)$ represents an element $[\alpha]' \in \pi_1'(X,x_0)$, then $\varphi''[\alpha]' = \{\alpha\}$, where $\{\alpha\}$ is the homology class in $H_1^{(0)}(X,\{x_0\},x_0)$ of the singular cycle α. Given a singular 1-simplex $\sigma\colon (\Delta^1,\dot{\Delta}^1) \to (X,x_0)$ in $\Delta(X,\{x_0\},x_0)^0$, it determines an element $[\sigma] \in \pi_1(X,x_0)$, and therefore an element $[\sigma]' \in \pi_1'(X,x_0)$. If σ is the constant singular 1-simplex at x_0, then clearly, $[\sigma]' = 0$. Because $\pi_1'(X,x_0)$ is abelian and $\Delta_1(X,\{x_0\},x_0)^0$ is the free abelian group generated by the singular simplexes in it, there is a homomorphism

$$\psi\colon \Delta_1(X,\{x_0\},x_0)^0/\Delta_1(x_0) \to \pi_1'(X,x_0)$$

such that $\psi(\sigma) = [\sigma]'$. We shall show, by using B_1, that the composite

$$\Delta_2(X,\{x_0\},x_0)^0/\Delta_2(x_0) \xrightarrow{\partial} \Delta_1(X,\{x_0\},x_0)^0/\Delta_1(x_0) \xrightarrow{\psi} \pi_1'(X,x_0)$$

is trivial. Given $\sigma\colon (\Delta^2,(\Delta^2)^0) \to (X,x_0)$, let $\sigma^{(0)}$, $\sigma^{(1)}$, and $\sigma^{(2)}$ be the faces of σ, as usual. Then

$$\begin{aligned}\psi\partial[\sigma] &= [\sigma^{(2)}]' + [\sigma^{(0)}]' - [\sigma^{(1)}]' = [(\sigma^{(2)} * \sigma^{(0)}) * (\sigma^{(1)})^{-1}]' \\ &= \eta(\sigma \mid \dot{\Delta}^2)_\#([v_0v_1] * [v_1v_2] * [v_2v_0]) = \eta\sigma_\# j_\# b_1 = 0\end{aligned}$$

Therefore ψ defines a homomorphism

$$\psi'\colon H_1^{(0)}(X,\{x_0\},x_0) \to \pi_1'(X,x_0)$$

and this is easily seen to be an inverse of φ''. ∎

(*c*) **PROOF THAT** $\bar{\Phi}_1, \ldots, \bar{\Phi}_{n-1} \Rightarrow B_n$ **FOR** $n \geq 2$ Consider the commutative diagram

$$\begin{array}{ccccc}
\pi_{n+1}(\Delta^{n+1},\dot{\Delta}^{n+1},v_0) & & & & \pi_n(\Delta^{n+1},(\Delta^{n+1})^{n-1},v_0) \\
& \searrow{\scriptstyle\partial'} & & \nearrow{\scriptstyle j_\#} & \\
\partial\downarrow & & \pi_n(\dot{\Delta}^{n+1},(\Delta^{n+1})^{n-1},v_0) & & \\
& \nearrow{\scriptstyle i_\#} & & \searrow{\scriptstyle\partial''} & \\
\pi_n(\dot{\Delta}^{n+1},v_0) & & & & \pi_{n-1}((\Delta^{n+1})^{n-1},v_0)
\end{array}$$

The top row, being part of the homotopy sequence of the triple $(\Delta^{n+1},\dot{\Delta}^{n+1},(\Delta^{n+1})^{n-1})$, is exact. The bottom row, being part of the homotopy sequence of the pair $(\dot{\Delta}^{n+1},(\Delta^{n+1})^{n-1})$, is also exact. From the exactness of the homotopy sequence of the pair $(\Delta^{n+1},\dot{\Delta}^{n+1})$ and the fact that Δ^{n+1} is contractible, it follows that ∂ is an isomorphism. Therefore

$$\ker j_\# = \operatorname{im} \partial' = \operatorname{im} (i_\# \circ \partial) = \operatorname{im} i_\# = \ker \partial''$$

Thus B_n is equivalent to the equation $\partial''(b_n) = 0$. We prove the latter, giving one proof for $n = 2$ and another for $n > 2$.

If $n = 2$, we have

$$\partial''(b_2) = (h_{[v_0v_1]}\partial''[e_3{}^0])\partial''[e_3{}^2]\partial''[e_3{}^1]^{-1}\partial''[e_3{}^3]^{-1}$$

To calculate $\partial''[e_3{}^i]$, let ξ: $(\Delta^2,\dot{\Delta}^2,v_0) \subset (\Delta^2,\dot{\Delta}^2,v_0)$ be the identity map. Then $[\xi] \in \pi_2(\Delta^2,\dot{\Delta}^2,v_0)$, and because $\pi_1(\dot{\Delta}^2,v_0)$ is infinite cyclic (since $\dot{\Delta}^2$ is homeomorphic to S^1), it follows from $\bar{\Phi}_1$ that φ: $\pi_1(\dot{\Delta}^2,v_0) \approx H_1(\dot{\Delta}^2,v_0)$. There is a commutative square

$$\begin{array}{ccc} \pi_2(\Delta^2,\dot{\Delta}^2,v_0) & \xrightarrow{\partial} & \pi_1(\dot{\Delta}^2,v_0) \\ \varphi\downarrow & & \approx\downarrow\varphi \\ H_2(\Delta^2,\dot{\Delta}^2) & \xrightarrow{\partial} & H_1(\dot{\Delta}^2,v_0) \end{array}$$

and $$\partial\varphi[\xi] = \partial\{\xi\} = \{\xi^{(2)} + \xi^{(0)} - \xi^{(1)}\} = \{\omega\} = \varphi[\omega]$$

where ω: $(\Delta^1,\dot{\Delta}^1) \to (\dot{\Delta}^2,v_0)$ is the path $\omega = (\xi^{(2)} * \xi^{(0)}) * (\xi^{(1)})^{-1}$. (The 2-chain $\xi^{(2)} + \xi^{(0)} - \xi^{(1)}$ is homologous to ω because it is easy to find singular 2-simplexes σ_1 and σ_2 in $\dot{\Delta}^2$ such that

$$\begin{array}{lll} \sigma_1{}^{(0)} = \xi^{(0)} & \sigma_1{}^{(1)} = \xi^{(2)} * \xi^{(0)} & \sigma_1{}^{(2)} = \xi^{(2)} \\ \sigma_2{}^{(0)} = \xi^{(1)} & \sigma_2{}^{(1)} = \xi^{(2)} * \xi^{(0)} & \sigma_2{}^{(2)} = (\xi^{(2)} * \xi^{(0)}) * (\xi^{(1)})^{-1} \end{array}$$

Then $\partial(\sigma_1 - \sigma_2) = \xi^{(2)} + \xi^{(0)} - \xi^{(1)} - \omega$.) Because φ is an isomorphism, it follows that

$$\partial[\xi] = [\omega] = [v_0v_1] * [v_1v_2] * [v_2v_0]$$

To return to the calculation of $\partial''[e_3{}^i]$, we have

$$\begin{aligned} \partial''[e_3{}^i] &= \partial''(e_3{}^i)_\#[\xi] = (e_3{}^i \mid \dot{\Delta}^2)_\#\partial[\xi] \\ &= [e_3{}^i(v_0)e_3{}^i(v_1)] * [e_3{}^i(v_1)\, e_3{}^i(v_2)] * [e_3{}^i(v_2)\, e_3{}^i(v_0)] \end{aligned}$$

Using this, direct substitution into the right-hand side of the equation for $\partial''(b_2)$ shows that $\partial''(b_2) = 0$.

For $n > 2$ note that $(\Delta^{n+1})^{n-1}$ contains the two-dimensional skeleton of Δ^{n+1}. Therefore $(\Delta^{n+1})^{n-1}$ is simply connected (because Δ^{n+1} is simply connected). Similarly, for $q \le n - 2$, $H_q((\Delta^{n+1})^{n-1},v_0) \approx H_q(\Delta^{n+1},v_0) = 0$. By $\bar{\Phi}_1, \ldots, \bar{\Phi}_{n-2}$, it follows that $(\Delta^{n+1})^{n-1}$ is $(n - 2)$-connected, and by $\bar{\Phi}_{n-1}$, there is an isomorphism

$$\varphi\colon \pi_{n-1}((\Delta^{n+1})^{n-1},v_0) \approx H_{n-1}((\Delta^{n+1})^{n-1},v_0)$$

Hence, to complete the proof it suffices to show that $\varphi\partial''(b_n) = 0$. This follows from the equalities

$$\varphi\partial''(b_n) = \partial''\varphi(b_n) = \partial''\{\Sigma\,(-1)^i e^i_{n+1}\} = \partial''\partial'\{\xi_{n+1}\} = \partial'' i_* \partial\{\xi_{n+1}\} = 0 \quad \blacksquare$$

(*d*) PROOF THAT $B_n \Rightarrow \Phi_n$ FOR $n \geq 2$ The argument is similar to the proof of part (*b*) above. The map φ' factors into the composite

$$\pi'_n(X,A,x_0) \xrightarrow{\varphi''} H_n^{(n-1)}(X,A,x_0) \underset{\approx}{\to} H_n(X,A)$$

If α: $(\Delta^n,\dot{\Delta}^n,v_0) \to (X,A,x_0)$ is a map such that α maps all the vertices to x_0, then $\varphi''[\alpha]' = \{\alpha\} \in H_n^{(n-1)}(X,A,x_0)$. To define an inverse of φ'', if σ: $(\Delta^n,\dot{\Delta}^n,(\Delta^n)^0) \to (X,A,x_0)$ is a singular simplex in $\Delta_n(X,A,x_0)^{n-1}$, then $[\sigma] \in \pi_n(X,A,x_0)$ and $\eta[\sigma] = [\sigma]' \in \pi'_n(X,A,x_0)$. If $\sigma(\Delta^n) \subset A$, then $[\sigma]' = 0$, and because $\pi'_n(X,A,x_0)$ is abelian, there is a homomorphism

$$\psi\colon \Delta_n(X,A,x_0)^{n-1}/(\Delta_n(X,A,x_0)^{n-1} \cap \Delta_n(A)) \to \pi'_n(X,A,x_0)$$

such that $\psi(\sigma) = [\sigma]'$.

We show that the composite

$$\psi \circ \partial\colon \Delta_{n+1}(X,A,x_0)^{n-1}/(\Delta_{n+1}(X,A,x_0)^{n-1} \cap \Delta_{n+1}(A)) \to \pi'_n(X,A,x_0)$$

is trivial. This follows from B_n, because if

$$\sigma\colon (\Delta^{n+1},(\Delta^{n+1})^{n-1},(\Delta^{n+1})^0) \to (X,A,x_0)$$

then

$$\begin{aligned}\psi\partial(\sigma) &= \Sigma\,(-1)^i[\sigma^{(i)}]' = \eta(\sigma \mid (\dot{\Delta}^{n+1},(\Delta^{n+1})^{n-1})_\#(b_n))\\ &= \eta\sigma_\# j_\#(b_n) = 0\end{aligned}$$

Therefore ψ defines a homomorphism

$$\psi'\colon H_n^{(n-1)}(X,A,x_0) \to \pi'_n(X,A,x_0)$$

such that $\psi'\{\sigma\} = [\sigma]'$, and ψ' is easily seen to be an inverse of φ''. ■

(*e*) PROOF THAT $\Phi_n \Rightarrow \bar{\Phi}_n$ FOR $n \geq 2$ For $n \geq 2$, if X is $(n-1)$-connected, then the pair $(X,\{x_0\})$ is $(n-1)$-connected and $\pi'_n(X,\{x_0\},x_0)$ is canonically isomorphic to $\pi'_n(X,x_0) = \pi_n(X,x_0)$. Then $\bar{\Phi}_n$ results from Φ_n applied to the pair $(X,\{x_0\})$. ■

This completes the proof of propositions 1, 2, and 3. From proposition 1 we obtain the following *relative Hurewicz isomorphism theorem.*

4 THEOREM *Let $x_0 \in A \subset X$ and assume that A and X are path connected. If there is an $n \geq 2$ such that $\pi_q(X,A,x_0) = 0$ for $q < n$, then $H_q(X,A) = 0$ for $q < n$ and φ' is an isomorphism*

$$\varphi'\colon \pi'_n(X,A,x_0) \approx H_n(X,A)$$

Conversely, if A and X are simply connected and there is an $n \geq 2$ such that $H_q(X,A) = 0$ for $q < n$, then $\pi_q(X,A,x_0) = 0$ for $q < n$ and φ is an isomorphism

$$\varphi\colon \pi_n(X,A,x_0) \approx H_n(X,A) \quad \blacksquare$$

Similarly, from proposition 2 we obtain the following *absolute Hurewicz isomorphism theorem.*

5 THEOREM *Let $x_0 \in X$ and assume that there is $n \geq 1$ such that $\pi_q(X,x_0) = 0$ for $q < n$. Then $H_q(X,x_0) = 0$ for $q < n$ and φ' is an isomorphism*

$$\varphi' \colon \pi'_n(X,x_0) \approx H_n(X,x_0)$$

Conversely, if X is simply connected and there is $n \geq 2$ such that $H_q(X,x_0) = 0$ for $q < n$, then $\pi_q(X,x_0) = 0$ for $q < n$ and φ is an isomorphism

$$\varphi \colon \pi_n(X,x_0) \approx H_n(X,x_0) \quad \blacksquare$$

In the absolute case when X is simply connected and in the relative case when X and A are simply connected, each of these theorems asserts that the first nonvanishing homotopy group is isomorphic to the first nonvanishing homology group.

6 COROLLARY *For $n \geq 1$ there is a commutative diagram of isomorphisms*

$$\begin{array}{ccc} \pi_{n+1}(E^{n+1},S^n,p_0) & \xrightarrow{\partial} & \pi_n(S^n,p_0) \\ \varphi\downarrow & & \downarrow\varphi \\ H_{n+1}(E^{n+1},S^n) & \xrightarrow{\partial} & H_n(S^n,p_0) \end{array}$$

PROOF The diagram is commutative, by theorem 7.4.3*a*, and both horizontal maps are isomorphisms because E^{n+1} is contractible [and because the homotopy and homology sequences of (E^{n+1},S^n,p_0) are exact]. The right-hand vertical map is an isomorphism, by proposition 2 and the fact that (in the case $n = 1$) $\pi_1(S^1,p_0)$ is abelian. $\blacksquare$

The following useful consequence of corollary 6 is called the *Brouwer degree theorem.*

7 COROLLARY *For $n \geq 1$ two maps $f, g \colon S^n \to S^n$ are homotopic if and only if $f_* = g_* \colon H_n(S^n) \to H_n(S^n)$. Similarly, two maps $f, g \colon (E^{n+1},S^n) \to (E^{n+1},S^n)$ are homotopic if and only if $f_* = g_* \colon H_{n+1}(E^{n+1},S^n) \to H_{n+1}(E^{n+1},S^n)$.*

PROOF We consider the absolute case first. Given maps $f, g \colon S^n \to S^n$, there exist homotopic maps f' and g', respectively, such that $f'(p_0) = g'(p_0) = p_0$ (because S^n is path connected). Because S^n is n-simple, f' and g' are freely homotopic if and only if they are homotopic as maps from (S^n,p_0) to (S^n,p_0). Therefore $f \simeq g$ if and only if $[f'] = [g']$ in $\pi_n(S^n,p_0)$. By corollary 6, $[f'] = [g']$ if and only if $\varphi[f'] = \varphi[g']$, and from the definition of φ, $\varphi[f'] = \varphi[g']$ if and only if

$$f'_* = g'_* \colon H_n(S^n,p_0) \to H_n(S^n,p_0)$$

Since there are commutative squares

$$\begin{array}{ccc} H_n(S^n) & \xrightarrow{\approx} & H_n(S^n,p_0) \\ f_*\downarrow & & \downarrow f'_* \\ H_n(S^n) & \xrightarrow{\approx} & H_n(S^n,p_0) \end{array} \qquad \begin{array}{ccc} H_n(S^n) & \xrightarrow{\approx} & H_n(S^n,p_0) \\ g_*\downarrow & & \downarrow g'_* \\ H_n(S^n) & \xrightarrow{\approx} & H_n(S^n,p_0) \end{array}$$

the result follows.

For the relative case note that because E^{n+1} is contractible, it follows from the homotopy extension property of (E^{n+1},S^n) that two maps $f, g\colon (E^{n+1},S^n) \to (E^{n+1},S^n)$ are homotopic if and only if $f \mid S^n, g \mid S^n\colon S^n \to S^n$ are homotopic. Since there are commutative squares

$$\begin{array}{ccc} H_{n+1}(E^{n+1},S^n) & \underset{\approx}{\overset{\partial}{\to}} & H_n(S^n) \\ f_*\downarrow & & \downarrow (f\mid S^n)_* \\ H_{n+1}(E^{n+1},S^n) & \underset{\approx}{\overset{\partial}{\to}} & H_n(S^n) \end{array} \qquad \begin{array}{ccc} H_{n+1}(E^{n+1},S^n) & \underset{\approx}{\overset{\partial}{\to}} & H_n(S^n) \\ g_*\downarrow & & \downarrow (g\mid S^n)_* \\ H_{n+1}(E^{n+1},S^n) & \underset{\approx}{\overset{\partial}{\to}} & H_n(S^n) \end{array}$$

the relative case follows from the absolute case. ∎

8 COROLLARY *For $x_0 \in X$ the map*

$$\psi\colon [S^n,p_0; X,x_0] \to \text{Hom}\,(\pi_n(S^n,p_0), \pi_n(X,x_0))$$

sending $[\alpha]$ to $\alpha_\#$ is an isomorphism.

PROOF This follows from corollary 6, because the fact that $\pi_n(S^n,p_0)$ is infinite cyclic implies that there is an isomorphism

$$\beta\colon \text{Hom}\,(\pi_n(S^n,p_0), \pi_n(X,x_0)) \approx \pi_n(X,x_0)$$

sending a homomorphism λ to $\lambda(a)$, where $a \in \pi_n(S^n,p_0)$ is the homotopy class of the identity map. Then, $(\beta \circ \psi)[\alpha] = \alpha_\#(a) = [\alpha]$, and so ψ is an isomorphism. ∎

The following useful consequence of the relative Hurewicz isomorphism theorem is known as the *Whitehead theorem.*

9 THEOREM *Let X and Y be path-connected pointed spaces and let $f\colon (X,x_0) \to (Y,y_0)$ be a map. If there is $n \geq 1$ such that*

$$f_\#\colon \pi_q(X,x_0) \to \pi_q(Y,y_0)$$

is an isomorphism for $q < n$ and an epimorphism for $q = n$, then

$$f_*\colon H_q(X,x_0) \to H_q(Y,y_0)$$

is an isomorphism for $q < n$ and an epimorphism for $q = n$. Conversely, if X and Y are simply connected and f_* is an isomorphism for $q < n$ and an epimorphism for $q = n$, then $f_\#$ is an isomorphism for $q < n$ and an epimorphism for $q = n$.

PROOF Let Z be the mapping cylinder of f. There are inclusion maps $i\colon X \subset Z$ and $j\colon Y \subset Z$ and a deformation retraction $r\colon Z \to Y$ such that $f = r \circ i$. Then $r\colon (Z,y_0) \to (Y,y_0)$ induces isomorphisms $r_\#\colon \pi_q(Z,y_0) \approx \pi_q(Y,y_0)$ and $r_*\colon H_q(Z,y_0) \approx H_q(Y,y_0)$ for all q. Because X and Y are path connected, so is Z, and $\pi_q(Z,x_0) \approx \pi_q(Z,y_0)$. Therefore $r\colon (Z,x_0) \to (Y,y_0)$ also induces isomorphisms $r_\#\colon \pi_q(Z,x_0) \approx \pi_q(Y,y_0)$ and $r_*\colon H_q(Z,x_0) \approx H_q(Y,y_0)$ for all q. It follows that we can replace (Y,y_0) in the theorem by (Z,x_0) and the conditions on $f_\#$ and f_* by the corresponding conditions on $i_\#$ and i_*. From the exactness of the homotopy sequence of (Z,X,x_0), it follows that $i_\#$ is an isomorphism for $q < n$ and an epimorphism for $q = n$ if and only if $\pi_q(Z,X,x_0) = 0$ for $q \leq n$. Similarly, from the exactness of the homology sequence of the triple (Z,X,x_0), it follows that i_* is an isomorphism for $q < n$ and an epimorphism for $q = n$ if and only if $H_q(Z,X) = 0$ for $q \leq n$. The result now follows from the relative Hurewicz isomorphism theorem 4. ∎

6 CW COMPLEXES

For homotopy theory the most tractable family of topological spaces seems to be the family of *CW* complexes (or the family of spaces each having the same homotopy type as a *CW* complex). *CW* complexes are built in stages, each stage being obtained from the preceding by adjoining cells of a given dimension. The cellular structure of such a complex bears a direct connection with its homotopy properties. Even for such nice spaces as polyhedra it is useful to consider representations of them as *CW* complexes, because such complexes will frequently require fewer cells than a simplicial triangulation.

In this section we shall investigate *CW* complexes and related concepts. In Sec. 7.8 we shall show that any topological space can be approximated by a *CW* complex which is unique up to homotopy. We begin with some results about a space X obtained from a subspace A by adjoining n-cells (defined in Sec. 3.8).

1 LEMMA *If X is obtained from A by adjoining n-cells, then $X \times 0 \cup A \times I$ is a strong deformation retract of $X \times I$.*

PROOF For each n-cell e_j^n of $X - A$ let

$$f_j\colon (E^n,S^{n-1}) \to (e_j^n,\dot{e}_j^n)$$

be a characteristic map. Let $D\colon (E^n \times I) \times I \to E^n \times I$ be a strong deformation retraction of $E^n \times I$ to $E^n \times 0 \cup S^{n-1} \times I$ (which exists, by corollary 3.2.4). There is a well-defined map $D_j\colon (e_j^n \times I) \times I \to e_j^n \times I$ characterized by the equation

$$D_j((f_j(z),t), t') = (f_j \times 1_I)(D(z,t,t')) \qquad z \in E^n;\ t, t' \in I$$

Then there is a map $D'\colon (X \times I) \times I \to X \times I$ such that $D' \mid (e_j \times I) \times I = D_j$ and $D'(a,t,t') = (a,t)$ for $a \in A$, and $t, t' \in I$, and D' is a strong deformation retraction of $X \times I$ to $X \times 0 \cup A \times I$. ∎

2 COROLLARY *If X is obtained from A by adjoining n-cells, then the inclusion map $A \subset X$ is a cofibration.* ∎

3 LEMMA *Let X be obtained from A by adjoining n-cells and let (Y,B) be a pair such that $\pi_n(Y,B,b) = 0$ for all $b \in B$ if $n \geq 1$ and such that every point of Y can be joined to B by a path if $n = 0$. Then any map from (X,A) to (Y,B) is homotopic relative to A to a map from X to B.*

PROOF This follows from theorem 7.2.1 by a technique similar to that in lemma 1 above. ■

A *relative CW complex* (X,A) consists of a topological space X, a closed subspace A, and a sequence of closed subspaces $(X,A)^k$ for $k \geq 0$ such that

(*a*) $(X,A)^0$ is obtained from A by adjoining 0-cells.
(*b*) For $k \geq 1$, $(X,A)^k$ is obtained from $(X,A)^{k-1}$ by adjoining k-cells.
(*c*) $X = \cup (X,A)^k$.
(*d*) X has a topology coherent with $\{(X,A)^k\}_k$.

In this case $(X,A)^k$ is called the *k-skeleton of X relative to A*. If $X = (X,A)^n$ for some n, then we say *dimension* $(X - A) \leq n$. An *absolute CW complex X* is a relative CW complex $(X, \varnothing)$, and its k-skeleton is denoted by X^k.

Following are a number of examples.

4 If (K,L) is a simplicial pair, there is a relative CW complex $(|K|,|L|)$, with $(|K|,|L|)^k = |K^k \cup L|$.

5 If (X,A) is a relative CW complex, for any k the pair $(X, (X,A)^k)$ is a relative CW complex, with

$$(X, (X,A)^k)^q = \begin{cases} (X,A)^k & q \leq k \\ (X,A)^q & q > k \end{cases}$$

Similarly, the pair $((X,A)^k, A)$ is a relative CW complex, with

$$((X,A)^k, A)^q = \begin{cases} (X,A)^q & q \leq k \\ (X,A)^k & q > k \end{cases}$$

6 As in example 3.8.7, for $i = 1, 2,$ or 4 let F_i be **R**, **C**, or **Q**, respectively, and for $q \geq 0$ let $P_q(F_i)$ be the corresponding projective space of dimension q over F_i. Then $P_q(F_i)$ is a CW complex, with

$$(P_q(F_i))^k = \begin{cases} P_{[k/i]}(F_i) & k \leq iq \\ P_q(F_i) & k > iq \end{cases}$$

7 E^n is a CW complex, with $(E^n)^k = p_0$ for $k < n - 1$, $(E^n)^{n-1} = S^{n-1}$, and $(E^n)^k = E^n$ for $k \geq n$.

8 I is a CW complex, with $(I)^0 = \dot{I}$ and $(I)^k = I$ for $k \geq 1$.

9 If (X,A) and (Y,B) are relative CW complexes and either X or Y is locally compact, then $(X,A) \times (Y,B)$ is also a CW complex,[1] with

$$((X,A) \times (Y,B))^k = \cup_{i+j=k} (X,A)^i \times (Y,B)^j \cup X \times B \cup A \times Y$$

10 If (X,A) is a relative CW complex, so is $(X,A) \times I$, with

$$(X \times I, A \times I)^k = (X,A)^k \times \dot{I} \cup (X,A)^{k-1} \times I \cup A \times I$$

[1] It is not true that the product of two CW complexes is always a CW complex. For a counterexample, see C. H. Dowker, Topology of metric complexes, *American Journal of Mathematics*, vol. 74, pp. 555–577, 1952.

11 If (X,A) is a relative CW complex, then X/A is a CW complex, with $(X/A)^k = (X,A)^k/A$.

A *subcomplex* (Y,B) of a relative CW complex (X,A) is a relative CW complex such that Y is a closed subset of X and $(Y,B)^k = Y \cap (X,A)^k$ for all k. If (Y,B) is a subcomplex of (X,A), then $(X, A \cup Y)$ is a relative CW complex, with $(X, A \cup Y)^k = (X,A)^k \cup Y$ for all k. In particular, if X is a CW complex and A is a subcomplex of X, then (X,A) is a relative CW complex. A *CW pair* (X,A) consists of a CW complex X and subcomplex A (hence a CW pair is a relative CW complex).

The definition of relative CW complex suggests its inductive construction. We start with a space A, attach 0-cells to A to obtain a space A_0, attach 1-cells to A_0 to obtain A_1, and continue in this way to define A_k for all $k \geq 0$. Letting X be the space obtained by topologizing $\cup A_k$ with the topology coherent with $\{A_k\}_{k\geq 0}$, then (X,A) is a relative CW complex, with $(X,A)^k = A_k$.

12 THEOREM *If (X,A) is a relative CW complex, then the inclusion map $A \subset X$ is a cofibration.*

PROOF This follows from corollary 2, using induction and the fact that $X \times I$ has the topology coherent with $\{(X,A)^k \times I\}_k$. ∎

13 THEOREM *Let (X,A) be a relative CW complex, with dimension $(X - A) \leq n$, and let (Y,B) be n-connected. Then any map from (X,A) to (Y,B) is homotopic relative to A to a map from X to B.*

PROOF This follows, using induction, from corollary 7.2.2, lemma 3, and theorem 12. ∎

14 COROLLARY *Let (X,A) be a relative CW complex and let (Y,B) be n-connected for all n. Then any map from (X,A) to (Y,B) is homotopic relative to A to a map from X to B.*

PROOF Let $f: (X,A) \to (Y,B)$ be a map. It follows from theorems 12 and 13 that there is a sequence of homotopies

$$H_k: (X,A) \times I \to (Y,B) \qquad k \geq 0$$

constructed by induction on k such that

(*a*) $H_0(x,0) = f(x)$ for $x \in X$.
(*b*) $H_k(x,1) = H_{k+1}(x,0)$ for $x \in X$.
(*c*) H_k is a homotopy relative to $(X,A)^{k-1}$.
(*d*) $H_k((X,A)^k \times 1) \subset B$.

Then a homotopy $H: (X,A) \times I \to (Y,B)$ with the required properties is defined by

$$H(x,t) = H_{k-1}\left(x, \frac{t - (1 - 1/k)}{(1/k) - 1/(k+1)}\right) \qquad 1 - \frac{1}{k} \leq t \leq 1 - \frac{1}{k+1}$$

$$H(x,1) = H_k(x,1) \qquad x \in (X,A)^k \quad ∎$$

15 LEMMA *If X is obtained from A by adjoining n-cells, then for $n \geq 1$, (X,A) is $(n-1)$-connected.*

PROOF For $k \le n - 1$ let $f: (E^k, S^{k-1}) \to (X,A)$ be a map. Because $f(E^k)$ is compact, there exist a finite number, say, $e_1, \ldots, e_m$, of n-cells of $X - A$ such that $f(E^k) \subset e_1 \cup \cdots \cup e_m \cup A$. For $1 \le i \le m$ let x_i be a point of $e_i - \dot{e}_i$. Each of the sets $Y = A \cup (e_1 - x_1) \cup \cdots \cup (e_m - x_m)$ and $e_i - \dot{e}_i$ for $1 \le i \le m$ intersects $f(E^k)$ in a set open in $f(E^k)$. There is a simplicial triangulation of E^k, say K, such that (identifying $|K|$ with E^k) for every simplex $s \in K$ either $f(|s|) \subset Y$ or for some $1 \le i \le m$, $f(|s|) \subset e_i - \dot{e}_i$. Let A' be the subpolyhedron of E^k which is the space of all simplexes $s \in K$ such that $f(|s|) \subset Y$, and for $1 \le i \le m$ let B_i be the subpolyhedron which is the space of all simplexes s of K such that $f(|s|) \subset e_i - \dot{e}_i$. Then $S^{k-1} \subset A'$, $E^k = A' \cup B_1 \cup \cdots \cup B_m$, and if $i \ne j$, then $B_i - A'$ is disjoint from $B_j - A'$. Let $\dot{B}_i = B_i \cap A'$ and observe that $(B_i, \dot{B}_i)$ is a relative CW complex, with $\dim (B_i - \dot{B}_i) \le k \le n - 1$.

For $1 \le i \le m$ the pair $((e_i - \dot{e}_i), (e_i - \dot{e}_i) - x_i)$ is homeomorphic to $(E^n - S^{n-1}, (E^n - S^{n-1}) - 0)$ and has the same homotopy groups as (E^n, S^{n-1}). By corollary 7.2.4, (E^n, S^{n-1}) is $(n - 1)$-connected. It follows from theorem 13 that $f \mid (B_i, \dot{B}_i)$ is homotopic relative to $\dot{B}_i$ to a map from B_i to $(e_i - \dot{e}_i) - x_i$. Because $B_i - \dot{B}_i$ is disjoint from $B_j - \dot{B}_j$ for $i \ne j$, these homotopies fit together to define a homotopy relative to A' of f to some map f' such that $f'(E^k) \subset Y$. Clearly, A is a strong deformation retract of Y. Therefore f' is homotopic relative to S^{k-1} to a map f'' such that $f''(E^k) \subset A$. Then $f \simeq f' \simeq f''$, all homotopies relative to S^{k-1}. Therefore (X,A) is $(n - 1)$-connected. ∎

16 COROLLARY *If (X,A) is a relative CW complex, then for any $n \ge 0$, $(X, (X,A)^n)$ is n-connected.*

PROOF We prove by induction on m that $((X,A)^m, (X,A)^n)$ is n-connected for $m > n$. Since $(X,A)^{n+1}$ is obtained from $(X,A)^n$ by adjoining $(n + 1)$-cells, it follows from lemma 15 that $((X,A)^{n+1}, (X,A)^n)$ is n-connected. Assume $m > n + 1$ and that $((X,A)^{m-1}, (X,A)^n)$ is n-connected. By lemma 15, the pair $((X,A)^m, (X,A)^{m-1})$ is $(m - 1)$-connected, and since $n < m - 1$, it is also n-connected. Then $\pi_0((X,A)^n) \to \pi_0((X,A)^{m-1})$ and $\pi_0((X,A)^{m-1}) \to \pi_0((X,A)^m)$ are both surjective, whence $\pi_0((X,A)^n) \to \pi_0((X,A)^m)$ is also surjective. Furthermore, for any $x \in (X,A)^n$, it follows from the exactness of the homotopy sequence of the triple $((X,A)^m, (X,A)^{m-1}, (X,A)^n)$, with base point x, that $\pi_k((X,A)^m, (X,A)^n, x) = 0$ for $1 \le k \le n$. By corollary 7.2.2, $((X,A)^m, (X,A)^n)$ is n-connected.

To show that $(X, (X,A)^n)$ is n-connected, if $0 \le k \le n$ and $\alpha: (E^k, S^{k-1}) \to (X, (X,A)^n)$, then because $\alpha(E^k)$ is compact and X has a topology coherent with the subspaces $(X,A)^m$, there is $m > n$ such that $\alpha(E^k) \subset (X,A)^m$. Hence α can be regarded as a map from (E^k, S^{k-1}) to $((X,A)^m, (X,A)^n)$ for some $m > n$. Because $((X,A)^m, (X,A)^n)$ is n-connected, α is homotopic relative to S^{k-1} to some map of E^k to $(X,A)^n$. ∎

Given relative CW complexes (X,A) and (X',A'), a map $f: (X,A) \to (X',A')$ is said to be *cellular* if $f((X,A)^k) \subset (X',A')^k$ for all k. Similarly, a homotopy $F: (X,A) \times I \to (X',A')$ is said to be *cellular* if $F((X,A) \times I)^k \subset (X',A')^k$ for all k. Analogous to the simplicial-approximation theorem is the following *cellular-approximation theorem.*

17 THEOREM *Given a map $f: (X,A) \to (X',A')$ between relative CW complexes which is cellular on a subcomplex (Y,B) of (X,A), there is a cellular map $g: (X,A) \to (X',A')$ homotopic to f relative to Y.*

PROOF It follows from corollary 16, theorem 13, and theorem 12 that there is a sequence of homotopies $H_k: (X,A) \times I \to (X',A')$ relative to Y, for $k \geq 0$, such that

(a) $H_0(x,0) = f(x)$ for $x \in X$.
(b) $H_k(x,1) = H_{k+1}(x,0)$ for $x \in X$.
(c) H_k is a homotopy relative to $(X,A)^{k-1}$.
(d) $H_k((X,A)^k \times 1) \subset (X',A')^k$.

Then a homotopy $H: (X,A) \times I \to (X',A')$ with the desired properties is defined by

$$H(x,t) = H_{k-1}\left(x, \frac{t - (1 - 1/k)}{(1/k) - 1/(k+1)}\right) \qquad 1 - \frac{1}{k} \leq t \leq 1 - \frac{1}{k+1}$$

$$H(x,1) = H_k(x,1) \qquad x \in (X,A)^k \quad \blacksquare$$

18 COROLLARY *Any map between relative CW complexes is homotopic to a cellular map. If two cellular maps between relative CW complexes are homotopic, there is a cellular homotopy between them.* $\blacksquare$

A continuous map $f: X \to Y$ is called an *n-equivalence* for $n \geq 1$ if f induces a one-to-one correspondence between the path components of X and of Y and if for every $x \in X$, $f_\#: \pi_q(X,x) \to \pi_q(Y,f(x))$ is an isomorphism for $0 < q < n$ and an epimorphism for $q = n$ (the condition concerning the case $q = n$ is sometimes omitted in the definitions occurring in the literature). A map $f: X \to Y$ is called a *weak homotopy equivalence* or *∞-equivalence* if f is an n-equivalence for all $n \geq 1$. The following results are immediate from the definition and from corollary 7.3.15.

19 *A composite of n-equivalences is an n-equivalence.* $\blacksquare$

20 *Any map homotopic to an n-equivalence is an n-equivalence.* $\blacksquare$

21 *A homotopy equivalence is a weak homotopy equivalence.* $\blacksquare$

Let $f: X \to Y$ be a map and let Z_f be the mapping cylinder of f. Then $f = r \circ i$, where $r: Z_f \to Y$ is a homotopy equivalence. Therefore f is an n-equivalence if and only if $i: X \subset Z_f$ is an n-equivalence. It follows from the exactness of the homotopy sequence of (Z_f,X) and from corollary 7.2.2 that i is an n-equivalence if and only if (Z_f,X) is n-connected.

22 THEOREM *Let $f: X \to Y$ be an n-equivalence (n finite or infinite) and let (P,Q) be a relative CW complex, with $\dim (P - Q) \leq n$. Given maps $g: Q \to X$ and $h: P \to Y$ such that $h \mid Q = f \circ g$, there exists a map $g': P \to X$ such that $g' \mid Q = g$ and $f \circ g' \simeq h$ relative to Q.*

PROOF Let Z_f be the mapping cylinder of f, with inclusion maps $i: X \subset Z_f$

and $j\colon Y \subset Z_f$, and retraction $r\colon Z_f \to Y$ a homotopy inverse of j. Then in

$$\begin{array}{ccc} Q & \subset & P \\ {\scriptstyle g}\downarrow & & \downarrow{\scriptstyle j\circ h} \\ X & \xrightarrow{i} & Z_f \end{array}$$

a homotopy $i \circ g \simeq j \circ h \mid Q$ can be found whose composite with r is constant. By theorem 12, there is a map $h'\colon P \to Z_f$ such that $h' \mid Q = i \circ g$ and such that $r \circ h' \simeq r \circ j \circ h$ relative to Q. We regard h' as a map from (P,Q) to (Z_f,X). Since (Z_f,X) is n-connected and $\dim (P - Q) \le n$, it follows from theorem 13 that h' is homotopic relative to Q to some map $g'\colon P \to X$. Then $g' \mid Q = g$ and

$$f \circ g' = r \circ i \circ g' \simeq r \circ h' \simeq r \circ j \circ h = h$$

all the homotopies being relative to Q. Hence g' has the desired properties. ■

23 COROLLARY *Let $f\colon X \to Y$ be an n-equivalence (n finite or infinite) and consider the map*

$$f_\#\colon [P;X] \to [P;Y]$$

If P is a CW complex of dimension $\le n$, this map is surjective, and if $\dim P \le n - 1$, *it is injective.*

PROOF The first part follows from theorem 22 applied to the relative CW complex $(P, \varnothing)$.

For the second part, we apply theorem 22 to the relative CW complex $(P \times I, P \times \dot{I})$. Given $g_0, g_1\colon P \to X$ such that $f \circ g_0 \simeq f \circ g_1$, there is a map $g\colon P \times \dot{I} \to X$ such that $g(z,0) = g_0(z)$ and $g(z,1) = g_1(z)$ for $z \in P$ and a map $h\colon P \times I \to Y$ such that $h \mid P \times \dot{I} = f \circ g$. Since $\dim (P \times I) \le n$, by theorem 22 there is a mapping $g'\colon P \times I \to X$ such that $g' \mid P \times \dot{I} = g$. Then g' is a homotopy from g_0 to g_1, showing that $[g_0] = [g_1]$. ■

24 COROLLARY *A map between CW complexes is a weak homotopy equivalence if and only if it is a homotopy equivalence.*

PROOF It follows from statement 21 that a map which is a homotopy equivalence is always a weak homotopy equivalence. Conversely, if $f\colon X \to Y$ is a weak homotopy equivalence between CW complexes, it follows from corollary 23 that f induces bijections

$$f_\#\colon [Y;X] \to [Y;Y] \qquad f_\#\colon [X;X] \to [X;Y]$$

If $g\colon Y \to X$ is any map such that $f_\#[g] = [1_Y]$, then $f \circ g \simeq 1_Y$, and also

$$f_\#[g \circ f] = [f \circ g \circ f] = [1_Y \circ f] = [f \circ 1_X] = f_\#[1_X]$$

Therefore $[g \circ f] = [1_X]$ or $g \circ f \simeq 1_X$, and so f is a homotopy equivalence. ■

Thus, for CW complexes the concepts of homotopy equivalence and weak homotopy equivalence coincide. The following theorem is a direct consequence of the Whitehead theorem 7.5.9.

25 THEOREM *A weak homotopy equivalence induces isomorphisms of the corresponding integral singular homology groups. Conversely, a map between simply connected spaces which induces isomorphisms of the corresponding integral singular homology groups is a weak homotopy equivalence.* ∎

7 HOMOTOPY FUNCTORS

In this section we shall study a general class of functors on the homotopy category of path-connected pointed spaces. The main result characterizes, on the subcategory of *CW* complexes, those functors of the form π^Y for some Y in terms of simple properties. In the next section we shall apply this result to prove the existence of approximations to any space by a *CW* complex.[1]

In a category $\mathcal{C}$, given objects A and X and morphisms $f_0: A \to X$ and $f_1: A \to X$, an *equalizer* of f_0 and f_1 is a morphism $j: X \to Z$ such that

(*a*) $j \circ f_0 = j \circ f_1$.
(*b*) If $j': X \to Z'$ is a morphism in $\mathcal{C}$ such that $j' \circ f_0 = j' \circ f_1$, there is a morphism $g: Z \to Z'$ such that $j' = g \circ j$.

Note that it is not asserted in condition (*b*) that g is unique.

We define $\mathcal{C}_0$ to be the homotopy category of path-connected pointed spaces having nondegenerate base points.

1 LEMMA *The category $\mathcal{C}_0$ has equalizers.*

PROOF Let A and X be arbitrary objects of $\mathcal{C}_0$ and let $f_0: A \to X$ and $f_1: A \to X$ be maps preserving base points. Let Z be the space obtained from the topological sum $X \vee (A \times I)$ by identifying $(a,0) \in A \times I$ with $f_0(a) \in X$, $(a,1) \in A \times I$ with $f_1(a) \in X$ for all $a \in A$, and $(a_0,t) \in A \times I$ with $(a_0,0)$ (a_0 the base point of A) for all $t \in I$. Then Z is an object of $\mathcal{C}_0$ and the inclusion map $j: X \subset Z$ has the property that $j \circ f_0 \simeq j \circ f_1$ [in fact, the composite $A \times I \subset X \vee (A \times I) \to Z$ is a homotopy from $j \circ f_0$ to $j \circ f_1$]. Furthermore, if $j': X \to Z'$ is a map such that $j' \circ f_0 \simeq j' \circ f_1$, there is a map $G: X \vee (A \times I) \to Z'$ such that $G|X = j'$ and $G|A \times I$ is a homotopy from $j' \circ f_0$ to $j' \circ f_1$. Then G is compatible with the collapsing map $k: X \vee (A \times I) \to Z$, so there is a map $g: Z \to Z'$ such that $G = g \circ k$. Then $j' = g \circ j$, and therefore $[j]: X \to Z$ is an equalizer of $[f_0]$ and $[f_1]$ in $\mathcal{C}_0$. ∎

2 LEMMA *Let $\{Y_n\}_{n \geq 0}$ be objects of $\mathcal{C}_0$ that are subspaces of a space Y in $\mathcal{C}_0$ such that $Y_n \subset Y_{n+1}$ is a cofibration for all $n \geq 0$, $Y = \cup_n Y_n$, and Y has the topology coherent with $\{Y_n\}$. Let $i_n: Y_n \subset Y_{n+1}$, $1_n: Y_n \subset Y_n$, and $j_n: Y_n \subset Y$ be the inclusion maps. Then the homotopy class $[\{j_n\}]: \bigvee Y_n \to Y$ is an equalizer in $\mathcal{C}_0$ of the homotopy classes*

[1] The techniques of this section are based on E. Brown, Cohomology theories, *Annals of Mathematics,* vol. 75, pp. 467–484, 1962.

$$[\vee i_n]: \vee Y_n \to \vee Y_n \quad \text{and} \quad [\vee 1_n]: \vee Y_n \to \vee Y_n$$

PROOF Since $j_{n+1} \circ i_n = j_n \circ 1_n$, it follows that $\{j_n\} \circ \vee i_n = \{j_n\} \circ \vee 1_n$. Given a map $j': \vee Y_n \to Z'$ such that $j' \circ \vee i_n \simeq j' \circ \vee 1_n$, let $j'_n: Y_n \to Z'$ be defined by $j'_n = j' \mid Y_n$. Then $j'_{n+1} \circ i_n \simeq j'_n$, and using the fact that $Y_n \subset Y_{n+1}$ is a cofibration and by induction on n, there is a sequence of maps $g_n: Y_n \to Z'$ such that $g_n \simeq j'_n$ and $g_{n+1} \circ i_n = g_n$. Let $g: Y \to Z'$ be the map such that $g \mid Y_n = g_n$. If $j = \{j_n\}: \vee Y_n \to Y$, then $g \circ j \simeq j'$ completing the proof. ■

A *homotopy functor* is a contravariant functor H from $\mathcal{C}_0$ to the category of pointed sets such that both of the following hold:

(*a*) If $[j]: X \to Z$ is an equalizer of $[f_0], [f_1]: A \to X$ and if $u \in H(X)$ is such that $H([f_0])u = H([f_1])u$, there is $v \in H(Z)$ such that $H([j])v = u$.
(*b*) If $\{X_\lambda\}_\lambda$ is an indexed family of objects in $\mathcal{C}_0$ and $i_\lambda: X_\lambda \subset \vee X_\lambda$, there is an equivalence

$$\{H[i_\lambda]\}_\lambda: H(\vee X_\lambda) \approx \times H(X_\lambda)$$

If $f: X \to Y$ is a base-point-preserving map and H is a homotopy functor, we shall also use $H(f)$ for $H([f])$. If $X \subset X'$ and $u \in H(X')$, we use $u \mid X$ for $H(i)u$, where $i: X \subset X'$.

If X is a one-point space, and X_1 and X_2 are both equal to X, then $X_1 \vee X_2$ is also equal to X, and the equivalence of condition (*b*)

$$\{H(i_1), H(i_2)\}: H(X_1 \vee X_2) \approx H(X_1) \times H(X_2)$$

corresponds to the diagonal map of $H(X)$ to $H(X) \times H(X)$. Because this is a bijection, $H(X)$ consists of a single element.

Following are some examples.

3 Let Y be a pointed space. Then the functor π^Y on $\mathcal{C}_0$ defined as in Sec. 1.3 (that is, $\pi^Y(X) = [X;Y]$ for an object X in $\mathcal{C}_0$) is a homotopy functor.

4 Fix an integer $n > 0$ and an abelian group G. Then the functor $H(X) = H^n(X,x_0; G)$ (singular cohomology) on $\mathcal{C}_0$ is a homotopy functor called the *nth cohomology functor with coefficients G*.

5 Let G be an arbitrary group (possibly nonabelian). There is a homotopy functor H such that $H(X)$ is the set of all homomorphisms $\pi_1(X,x_0) \to G$ with the trivial homomorphism as base point.

An important result of this section is that on the subcategory of pointed path-connected CW complexes every homotopy functor is naturally equivalent to π^Y for a suitable pointed space Y.

6 **LEMMA** *Let $\nu: SX \to SX \vee SX$ be the comultiplication map. If X is in $\mathcal{C}_0$ and H is a homotopy functor, the composite*

$$H(SX) \times H(SX) \xrightarrow{\{H(i_1),H(i_2)\}^{-1}} H(SX \vee SX) \xrightarrow{H(\nu)} H(SX)$$

is a group multiplication on $H(SX)$, which is abelian if X is a suspension. If H

is a homotopy functor taking values in the category of groups, the two group structures on $H(SX)$ agree.

PROOF Each of the group properties for this multiplication follows from the corresponding H cogroup property of ν. The final statement of the lemma follows from theorem 1.6.8, because the two multiplications in $H(SX)$ are mutually distributive. ■

In particular, for any homotopy functor H, $H(S^q)$ is a group for $q \geq 1$ and abelian for $q \geq 2$ and is called the *qth coefficient group of H*. Thus the qth coefficient group of the functor π^Y of example 3 is $\pi_q(Y)$. The qth coefficient group of the nth cohomology functor with coefficients G of example 4 is 0 if $q \neq n$ and isomorphic to G if $q = n$. The qth coefficient group of the functor of example 5 is G if $q = 1$ and 0 if $q > 1$.

If Y is an object of $\mathcal{C}_0$ and H is a homotopy functor, any element $u \in H(Y)$ determines a natural transformation

$$T_u\colon \pi^Y \to H$$

defined by $T_u([f]) = H([f])(u)$ for $[f] \in [X;Y]$. For a suspension SX, T_u is a homomorphism from $\pi^Y(SX) = [SX;Y]$ to the group $H(SX)$, with the multiplication of lemma 6 (because both group multiplications are induced by the comultiplication $\nu\colon SX \to SX \vee SX$). An element $u \in H(Y)$ is said to be *n-universal for H*, where $n \geq 1$, if the homomorphism

$$T_u\colon \pi^Y(S^q) \to H(S^q)$$

is an isomorphism for $1 \leq q < n$ and an epimorphism for $q = n$. An element $u \in H(Y)$ is said to be *universal* for H if it is n-universal for all $n \geq 1$, in which case Y is called a *classifying space for H*.

7 THEOREM *Assume that H is a homotopy functor with universal elements $u \in H(Y)$ and $u' \in H(Y')$ and let $f\colon Y \to Y'$ be a map such that $H(f)u' = u$. Then f is a weak homotopy equivalence.*

PROOF Since Y and Y' are path connected, this is a consequence of the commutativity of the diagram (for $q \geq 1$)

$$\begin{array}{ccc} [S^q;Y] & \xrightarrow{f_\#} & [S^q;Y'] \\ & {\scriptstyle T_u}\searrow\approx \quad \approx\swarrow{\scriptstyle T_{u'}} & \\ & H(S^q) & \end{array}$$

■

The same kind of argument establishes the next result.

8 LEMMA *Let Y be an object of $\mathcal{C}_0$ and let Y′ be an arbitrary path-connected space. A map $f\colon Y \to Y'$ is a weak homotopy equivalence if and only if $[f] \in [Y;Y'] = \pi^{Y'}(Y)$ is universal for $\pi^{Y'}$.* ■

We are heading toward a proof of the existence of universal elements for any homotopy functor. The following two lemmas will be used in this proof.

9 LEMMA *Let H be a homotopy functor, Y an object in $\mathcal{C}_0$, and $u \in H(Y)$. There exist an object Y' in $\mathcal{C}_0$, obtained from Y by attaching 1-cells, and a 1-universal element $u' \in H(Y')$ such that $u' \mid Y = u$.*

PROOF For each $\lambda \in H(S^1)$ let $S_\lambda{}^1$ be a 1-sphere and define $Y' = Y \vee \bigvee_\lambda S_\lambda{}^1$. Then Y' is an object of $\mathcal{C}_0$ obtained from Y by attaching 1-cells. If g_λ is the composite $S^1 \approx S_\lambda{}^1 \subset Y'$, it follows from condition (*b*) on page 407 that there is an element $u' \in H(Y')$ such that $u' \mid Y = u$ and $H(g_\lambda)u' = \lambda$ for $\lambda \in H(S^1)$. Since $T_{u'}([g_\lambda]) = \lambda$, $T_{u'}([S^1;Y']) = H(S^1)$, and u' is 1-universal. ∎

10 LEMMA *Let H be a homotopy functor and $u \in H(Y)$ an n-universal element for H, with $n \geq 1$. There exist an object Y' in $\mathcal{C}_0$, obtained from Y by attaching $(n+1)$-cells, and an $(n+1)$-universal element $u' \in H(Y')$ such that $u' \mid Y = u$.*

PROOF For each $\lambda \in H(S^{n+1})$ let $S_\lambda{}^{n+1}$ be an $(n+1)$-sphere, and for each map $\alpha: S^n \to Y$ such that $H(\alpha)u = 0$ attach an $(n+1)$-cell $e_\alpha{}^{n+1}$ to Y by α. Let Y' be the space obtained from $Y \vee \bigvee_\lambda S_\lambda{}^{n+1}$ by attaching the $(n+1)$-cells $\{e_\alpha{}^{n+1}\}$. Then Y' is an object of $\mathcal{C}_0$ obtained from Y by attaching $(n+1)$-cells. If $g_\lambda: S^{n+1} \to Y \vee \bigvee_\lambda S_\lambda{}^{n+1}$ is the composite $S^{n+1} \approx S_\lambda{}^{n+1} \subset Y \vee \bigvee_\lambda S_\lambda{}^{n+1}$, it follows from condition (*b*) on page 407 that there is an element $\bar{u} \in H(Y \vee \bigvee_\lambda S_\lambda{}^{n+1})$ such that $\bar{u} \mid Y = u$ and $H(g_\lambda)\bar{u} = \lambda$ for $\lambda \in H(S^{n+1})$.

For each map $\alpha: S^n \to Y$ such that $H(\alpha)u = 0$ let $S_\alpha{}^n$ be an n-sphere and let $f_0: \bigvee_\alpha S_\alpha{}^n \to Y \vee \bigvee_\lambda S_\lambda{}^{n+1}$ be the constant map and let $f_1: \bigvee_\alpha S_\alpha{}^n \to Y \vee \bigvee_\lambda S_\lambda{}^{n+1}$ be the map such that $S_\alpha{}^n$ is mapped by α. Then

$$j: Y \vee \bigvee_\lambda S_\lambda{}^{n+1} \subset Y'$$

is a map such that $[j]$ is an equalizer of $[f_0]$ and $[f_1]$. Since $H(f_0)\bar{u} = 0 = H(f_1)\bar{u}$, by condition (*a*) on page 407 there is an element $u' \in H(Y')$ such that $H(j)u' = \bar{u}$. Then $u' \mid Y = u$ and to complete the proof we need only show that u' is $(n+1)$-universal.

There is a commutative diagram

$$\begin{array}{ccccccc} \pi_{q+1}(Y',Y) & \xrightarrow{\partial} & \pi_q(Y) & \xrightarrow{i_\#} & \pi_q(Y') & \to & \pi_q(Y',Y) \\ & & {\scriptstyle T_u}\searrow & & \swarrow{\scriptstyle T_{u'}} & & \\ & & & H(S^q) & & & \end{array}$$

with the top row exact. Since Y' is obtained from Y by attaching $(n+1)$-cells, it follows from lemma 7.6.15 that $\pi_q(Y',Y) = 0$ for $q \leq n$. Therefore $i_\#$ is an isomorphism for $q < n$ and an epimorphism for $q = n$. Since u is n-universal, T_u is an isomorphism for $q < n$ and an epimorphism for $q = n$. It follows that $T_{u'}$ is also an isomorphism for $q < n$ and an epimorphism for $q = n$. Furthermore, if $a \in [S^n;Y]$ is in the kernel of T_u, then a is represented by a map $\alpha: S^n \to Y$ and

$$a = [\alpha] \in \partial(\pi_{n+1}(e_\alpha{}^{n+1},\dot{e}_\alpha{}^{n+1})) \subset \partial(\pi_{n+1}(Y',Y)) = \ker i_\#$$

Therefore, for $q = n$, ker T_u = ker $i_\#$, and so $T_{u'}$ is an isomorphism from $\pi_n(Y')$ to $H(S^n)$. For any $\lambda \in H(S^{n+1})$ the map $j \circ g_\lambda: S^{n+1} \to Y'$ has the property that

$$T_{u'}([j \circ g_\lambda]) = H(g_\lambda)\bar{u} = \lambda$$

showing that $T_{u'}$ is an epimorphism for $q = n + 1$, and so u' is $(n + 1)$-universal. ∎

11 THEOREM *Let H be a homotopy functor, let Y be an object in $\mathcal{C}_0$, and let $u \in H(Y)$. Then there are a classifying space Y′ for H containing Y such that (Y′,Y) is a relative CW complex and a universal element $u' \in H(Y')$ such that $u' \mid Y = u$.*

PROOF Using lemmas 9 and 10, we construct, by induction on n, a sequence of objects $\{Y_n\}_{n \geq 0}$ in $\mathcal{C}_0$ and elements $u_n \in H(Y_n)$ such that

(a) $Y_0 = Y$ and $u_0 = u$.
(b) Y_{n+1} is obtained from Y_n by attaching $(n + 1)$-cells, where $n \geq 0$.
(c) $u_{n+1} \mid Y_n = u_n$.
(d) u_n is n-universal for $n \geq 1$.

It follows from (b) above that $Y' = \cup Y_n$ topologized with the topology coherent with $\{Y_n\}$ is a path-connected pointed space containing Y such that (Y',Y) is a relative CW complex. By lemma 2, the homotopy class $[\{j_n\}]: \vee Y_n \to Y'$ is an equalizer of the homotopy classes $[\vee i_n]: \vee Y_n \to \vee Y_n$ and $[\vee 1_n]: \vee Y_n \to \vee Y_n$. By condition (b) on page 407 there is an element $\bar{u} \in H(\vee Y_n)$ such that $\bar{u} \mid Y_n = u_n$. It follows from (c) above that $H(\vee i_n)\bar{u} = H(\vee 1_n)\bar{u}$, and by condition (a) on page 407 there is an element $u' \in H(Y')$ such that $H(\{j_n\})u' = \bar{u}$ (that is, $u' \mid Y_n = u_n$ for $n \geq 0$). Then $u' \mid Y = u$, and it remains to show that u' is universal.

By the definition of Y' and u', there is a commutative diagram for $q \geq 1$

$$\begin{array}{ccc} \lim_{\to} \{\pi_q(Y_n)\} & \xrightarrow{\approx} & \pi_q(Y') \\ & {\scriptstyle \{T_{u_n}\}} \searrow \quad \swarrow {\scriptstyle T_{u'}} & \\ & H(S^q) & \end{array}$$

Since u_n is n-universal, T_{u_n} is an isomorphism for $n > q$, and so the left-hand map is an isomorphism. Therefore $T_{u'}$ is also an isomorphism, and u' is universal. ∎

12 COROLLARY *For any homotopy functor there exist classifying spaces which are CW complexes.*

PROOF Apply theorem 11 to a one-point space Y, with u the unique element of $H(Y)$. ∎

13 COROLLARY *Let $u \in H(Y)$ be a universal element for a homotopy functor H. Let (X,A) be a relative CW complex, where A and X are objects* *in $\mathcal{C}_0$. Given a map $g: A \to Y$ and an element $v \in H(X)$ such that $v \mid A = H(g)u$, there exists a map $g': X \to Y$ such that $g = g' \mid A$ and $v = H(g')u$.*

PROOF Let $i: X \subset X \vee Y$ and $i': Y \subset X \vee Y$ and let $j: X \vee Y \to Z$ be a map such that $[j]$ is an equalizer of $[i \circ f]$ (where $f: A \subset X$) and $[i' \circ g]$. By condition (*b*) on page 407, there is an element $\bar{v} \in H(X \vee Y)$ such that $\bar{v} \mid X = v$ and $\bar{v} \mid Y = u$. Since $H(f)v = H(g)u$, it follows that $H(i \circ f)\bar{v} = H(i' \circ g)\bar{v}$, and by condition (*a*) on page 407, there is an element $\bar{u} \in H(Z)$ such that $H(j)\bar{u} = \bar{v}$. We now apply theorem 11 to $\bar{u}$ to obtain a Y' containing Z and a universal element $u' \in H(Y')$ such that $\bar{u} = u' \mid Z$. Let $j': Y \to Y'$ be the composite

$$Y \overset{i'}{\subset} X \vee Y \xrightarrow{j} Z \overset{h}{\subset} Y'$$

Then $H(j')u' = u$, and by theorem 7, j' is a weak homotopy equivalence. Since the composite

$$A \overset{f}{\subset} X \overset{i}{\subset} X \vee Y \xrightarrow{j} Z \overset{h}{\subset} Y'$$

is homotopic to $j' \circ g$, it follows from the fact that f is a cofibration that there is a map $\bar{g}: X \to Y'$ such that $\bar{g} \mid A = j' \circ g$ and $\bar{g}$ is homotopic to $h \circ j \circ i$. Since j' is a weak homotopy equivalence, by theorem 7.6.22, there is a map $g': X \to Y$ such that $g' \mid A = g$ and $j' \circ g' \simeq \bar{g}$. Then

$$H(g')u = H(g')H(j')u' = H(i)H(j)H(h)u' = \bar{v} \mid X = v$$

showing that g' has the requisite properties. ∎

14 THEOREM *If Y is a classifying space and $u \in H(Y)$ is a universal element for a homotopy functor H, then for any CW complex X in $\mathcal{C}_0$, T_u is a natural equivalence of $\pi^Y(X)$ with $H(X)$.*

PROOF Given $v \in H(X)$, apply corollary 13, with $A = x_0$ and g the constant map, to obtain a map $g': X \to Y$ such that $H(g')u = v$. Then $T_u[g'] = v$, proving that T_u is surjective.

If $g_0, g_1: X \to Y$ are maps such that $T_u[g_0] = T_u[g_1]$, let X' be the CW complex $X \times I/x_0 \times I$, with $(X')^q = [(X^q \times \dot{I}) \cup (X^{q-1} \times I)]/(x_0 \times I)$ for $q \geq 0$. Let $v \in H(X')$ be defined by $v = H(h)H(g_0)u$, where $h: X' \to X$ is the map $h([x,t]) = x$. Let $A = X \times \dot{I}/x_0 \times \dot{I}$ and let $g: A \to Y$ be the map such that $g([x,0]) = g_0(x)$ and $g([x,1]) = g_1(x)$. Then $H(g)u = v \mid A$, and by corollary 13, there is a map $g': X' \to Y$ such that $g' \mid A = g$. Then the composite

$$X \times I \to X \times I/x_0 \times I \xrightarrow{g'} Y$$

is a homotopy relative to x_0 from g_0 to g_1, showing that T_u is injective. ∎

15 COROLLARY *If Y and Y' are classifying spaces which are CW complexes and $u \in H(Y)$ and $u' \in H(Y')$ are universal elements for a homotopy functor H, there is a homotopy equivalence $h: Y \to Y'$, unique up to homotopy, such that $H(h)u' = u$.*

PROOF By theorem 14, there exists a unique homotopy class $[g]: Y \to Y'$ such that $H(g)u' = u$. By theorem 7, g is a weak homotopy equivalence. By corollary 7.6.24, g is a homotopy equivalence. ∎

8 WEAK HOMOTOPY TYPE

In this section we shall show that any space can be approximated by CW complexes. This leads to an equivalence relation based on weak homotopy equivalence which is weaker than homotopy equivalence. We shall also consider the same equivalence relation in the category of maps. This will be used in defining and analyzing the general relative-lifting problem.

A *relative CW approximation* to a pair (X,A) consists of a relative CW complex (Y,A) and a weak homotopy equivalence $f: Y \to X$ such that $f(a) = a$ for all $a \in A$. A *CW approximation* to a space X is a relative CW approximation to $(X, \varnothing)$.

1 THEOREM *Any pair has relative CW approximations, and two relative CW approximations to the same pair have the same homotopy type.*

PROOF First we consider the case where X is path connected. Let $x_0 \in X$ and let $\{A_j\}_{j \in J}$ be the set of path components of A, and for each $j \in J$ choose a point $a_j \in A_j$. There is a relative CW complex (A',A) with $(A',A)^0 = A \cup e^0$, where e^0 is a single point and

$$A' = (A',A)^1 = (A',A)^0 \cup \bigcup_{j \in J} e_j^1$$

where e_j^1 is a 1-cell such that $\dot{e}_j^1 = e^0 \cup a_j$ for $j \in J$. Let $g: A' \to X$ be a map such that $g(a) = a$ for $a \in A$, $g(e^0) = x_0$, and $g \mid e_j^1$ is a path in X with end points x_0 and a_j for each $j \in J$. Then A' is a path-connected space with nondegenerate base point e^0 and $[g] \in \pi^X(A')$. It follows from theorem 7.7.11 that there is a relative CW complex (Y,A') [which can be chosen such that $(Y,A')^1 = A' \vee \bigvee S_\alpha^1$] and a universal element $[g'] \in \pi^X(Y)$ for π^X such that $g' \mid A' \simeq g$. Since $A' \subset Y$ is a cofibration, there is a map $f: Y \to X$ such that $[f] \in \pi^X(Y)$ is universal for π^X and $f \mid A' = g$. By lemma 7.7.8, f is a weak homotopy equivalence. Since (Y,A) is a relative CW complex [with $(Y,A)^0 = (A',A)^0$ and $(Y,A)^q = (Y,A')^q$ for $q \geq 1$] and since $f(a) = a$ for $a \in A$, (Y,A) and f constitute a relative CW approximation to (X,A).

Next we consider the case where X is not path connected and we let $\{X_\alpha\}$ be the set of path components of X. By the case already considered, for each α there is a relative CW approximation $f_\alpha: (Y_\alpha, X_\alpha \cap A) \to (X_\alpha, X_\alpha \cap A)$. Let Y be the space obtained from the disjoint union $A \cup \bigcup Y_\alpha$ by identifying $x \in X_\alpha \cap A \subset Y_\alpha$ with $x \in A$ for each α and let $k: A \cup \bigcup Y_\alpha \to Y$ be the collapsing map. Then $k \mid A: A \to Y$ is an imbedding and (Y,A) is a relative CW complex with $(Y,A)^q = k(A \cup \bigcup (Y_\alpha, X_\alpha \cap A)^q)$ for all $q \geq 0$. There is a map $f: Y \to X$ such that $fk(a) = a$ for $a \in A$ and $f \circ (k \mid Y_\alpha) = f_\alpha$ for all α.

Since $\{k(Y_\alpha)\}$ is the set of path components of Y and f induces a weak homotopy equivalence of each of these with the corresponding path component X_α of X, f is a weak homotopy equivalence from Y to X. Identifying A with $k(A)$, we see that (Y,A) and f constitute a CW approximation to (X,A).

Given two relative CW approximations to (X,A), say $f_1: (Y_1,A) \to (X,A)$ and $f_2: (Y_2,A) \to (X,A)$, it follows from theorem 7.6.22 that there are maps $g_1: (Y_1,A) \to (Y_2,A)$ and $g_2: (Y_2,A) \to (Y_1,A)$ such that $f_2 \circ g_1 \simeq f_1$ and $f_1 \circ g_2 \simeq f_2$, both homotopies relative to A. Then $f_2 \circ (g_1 \circ g_2) \simeq f_2 \circ 1$ rel A, and by theorem 7.6.22 again, $g_1 \circ g_2 \simeq 1$ rel A. Similarly, $g_2 \circ g_1 \simeq 1$ rel A, and so (Y_1,A) and (Y_2,A) have the same homotopy type. ∎

Two spaces X_1 and X_2 will be said to have the same *weak homotopy type* if there exists a space Y and weak homotopy equivalences $f_1: Y \to X_1$ and $f_2: Y \to X_2$. By replacing such a space Y with a CW approximation to it, we see that X_1 and X_2 have the same weak homotopy type if and only if they have CW approximations by the same CW complex.

2 LEMMA *The relation of having the same weak homotopy type is an equivalence relation.*

PROOF The relation is reflexive and symmetric by its definition. To prove it transitive, let X_1, X_2, and X_3 be spaces and let Y_1 and Y_2 be CW complexes such that there exist weak homotopy equivalences

$$\begin{array}{ccccc} & Y_1 & & Y_2 & \\ {}^{f_1}\swarrow & & \searrow^{f_2} \quad {}^{g_2}\swarrow & & \searrow^{g_3} \\ X_1 & & X_2 & & X_3 \end{array}$$

Then $f_2: Y_1 \to X_2$ and $g_2: Y_2 \to X_2$ are both CW approximations to X_2, and by theorem 1, there is a homotopy equivalence $h: Y_1 \to Y_2$ such that $f_2 \simeq g_2 \circ h$. Then $g_3 \circ h: Y_1 \to X_3$, being the composite of weak homotopy equivalences, is a weak homotopy equivalence. Therefore X_1 and X_3 have the same weak homotopy type. ∎

We are interested in applying these ideas to weak fibrations. The main result is that any two fibers of a weak fibration with path-connected base space have the same weak homotopy type.

3 LEMMA *Let $p: E \to B$ be a weak fibration with contractible base space B. For any $b_0 \in B$ the inclusion map $i: p^{-1}(b_0) \subset E$ is a weak homotopy equivalence.*

PROOF Let $F = p^{-1}(b_0)$. Since B is contractible, $\pi_q(B,b_0) = 0$ for $q \geq 0$. From the exactness of the homotopy sequence of p, it follows that for any $e \in F$, i induces an isomorphism $i_\#: \pi_q(F,e) \approx \pi_q(E,e)$ for $q \geq 1$ and $i_\#(\pi_0(F,e)) = \pi_0(E,e)$.

It only remains to verify that $i_\#$ maps $\pi_0(F,e)$ injectively into $\pi_0(E,e)$. Assume that $e, e' \in F$ are such that there is a path ω in E from e to e'. Since B is simply connected and $p \circ \omega$ is a closed path in B at b_0, there is a map

$H: I \times I \to B$ such that $H(t,0) = p\omega(t)$ and $H(0,t') = H(1,t') = H(t,1) = b_0$. Let $g: I \times 0 \cup \dot{I} \times I \to E$ be the map defined by $g(t,0) = \omega(t)$, $g(0,t') = e$, and $g(1,t') = e'$. By lemma 7.2.5, there is a map $G: I \times I \to E$ such that $p \circ G = H$ and $G | I \times 0 \cup \dot{I} \times I = g$. Let $\omega': I \to E$ be the path defined by $\omega'(t) = G(1,t)$. Then ω' is a path in F from e to e' [because $p\omega'(t) = b_0$], showing that $i_\#: \pi_0(F,e) \to \pi_0(E,e)$ is injective. ∎

4 COROLLARY *Let $p: E \to B$ be a weak fibration and let ω be a path in B. Then $p^{-1}(\omega(0))$ and $p^{-1}(\omega(1))$ have the same weak homotopy type.*

PROOF Let $p': E' \to I$ be the weak fibration induced from p by $\omega: I \to B$. Then $p^{-1}(\omega(0))$ and $p^{-1}(\omega(1))$ are homeomorphic to $p'^{-1}(0)$ and $p'^{-1}(1)$, respectively. By lemma 3, each of the inclusion maps $p'^{-1}(0) \subset E'$ and $p'^{-1}(1) \subset E'$ is a weak homotopy equivalence. The corollary follows from this and lemma 2. ∎

This result implies the following analogue of corollary 2.8.13 for weak fibrations.

5 COROLLARY *If $p: E \to B$ is a weak fibration with path-connected base space, any two fibers have the same weak homotopy type.* ∎

We now consider the category whose objects are continuous maps $\alpha: P'' \to P'$ between topological spaces and whose morphisms (also called *map pairs*) $f: \alpha \to \beta$ are commutative squares

$$\begin{array}{ccc} P'' & \xrightarrow{f''} & Q'' \\ {\scriptstyle\alpha}\downarrow & & \downarrow{\scriptstyle\beta} \\ P' & \xrightarrow{f'} & Q' \end{array}$$

In this category a *homotopy pair* $H: f_0 \simeq f_1$, where $f_0, f_1: \alpha \to \beta$, is a commutative square

$$\begin{array}{ccc} P'' \times I & \xrightarrow{H''} & Q'' \\ {\scriptstyle\alpha \times 1}\downarrow & & \downarrow{\scriptstyle\beta} \\ P' \times I & \xrightarrow{H'} & Q' \end{array}$$

such that $H'': f_0'' \simeq f_1''$ and $H': f_0' \simeq f_1'$ (note that H is a map pair from $\alpha \times 1_I$ to β). If such a homotopy pair exists, f_0 is said to be *homotopic* to f_1. This is an equivalence relation in the set of map pairs from α to β, and the corresponding equivalence classes are called *homotopy classes*. We use $[\alpha;\beta]$ to denote the set of homotopy classes of map pairs from α to β, and if $f: \alpha \to \beta$ is a map pair, its homotopy class is denoted by $[f]$. It is trivial to verify that the composites of homotopic map pairs are homotopic, so there is a *homotopy category of maps* whose objects are maps $\alpha: P'' \to P'$ and whose morphisms $\alpha \to \beta$ are homotopy classes $[f]$, where $f: \alpha \to \beta$ is a map pair. A map pair $f: \alpha \to \beta$ is called a *homotopy equivalence from α to β* if $[f]$ is an equivalence in the homotopy category of maps. Two maps α and β are said to have the *same homotopy type* if they are equivalent in the homotopy category of maps.

Given a map pair $g: \alpha' \to \alpha$ (or a map pair $h: \beta \to \beta'$) there is an induced map $g^\#: [\alpha;\beta] \to [\alpha';\beta]$ (or $h_\#: [\alpha;\beta] \to [\alpha;\beta']$) such that $g^\#[f] = [f \circ g]$ (or $h_\#[f] = [h \circ f]$). Since $g^\# \circ h_\# = h_\# \circ g^\#$, the function which assigns $[\alpha;\beta]$ to α and β and $g^\#$ and $h_\#$ to $[g]$ and $[h]$, respectively, is a functor of two variables from the product of the homotopy category of maps by itself to the category of sets that is contravariant in α and covariant in β.

If $\alpha: P'' \to P'$ and $\beta: Q'' \to Q'$ are maps, given a map $\tilde{f}: P' \to Q''$, there is a map pair $\rho(\tilde{f}): \alpha \to \beta$ consisting of the commutative square

$$\begin{array}{ccc} P'' & \xrightarrow{\tilde{f} \circ \alpha} & Q'' \\ {\scriptstyle \alpha}\downarrow & & \downarrow{\scriptstyle \beta} \\ P' & \xrightarrow{\beta \circ \tilde{f}} & Q' \end{array}$$

[that is, $(\rho(\tilde{f}))'' = \tilde{f} \circ \alpha$ and $(\rho(\tilde{f}))' = \beta \circ \tilde{f}$]. Given a map pair $f: \alpha \to \beta$, a *lifting* of f is a map $\tilde{f}: P' \to Q''$ such that $\rho(\tilde{f}) = f$. Two liftings $\tilde{f}_0, \tilde{f}_1: P' \to Q''$ of $f: \alpha \to \beta$ are *homotopic relative to* f if there is a homotopy $\tilde{H}: P' \times I \to Q''$ from $\tilde{f}_0$ to $\tilde{f}_1$ such that $\tilde{H} \circ (\alpha \times 1_I)$ and $\beta \circ \tilde{H}$ are both constant homotopies [that is, $\rho(\tilde{H})$ is the constant homotopy pair from f to f]. Such a map $\tilde{H}$ is called a *homotopy relative to f*, and we write $\tilde{H}: \tilde{f}_0 \simeq \tilde{f}_1$ rel f. Homotopy relative to f is an equivalence relation in the set of liftings of f, and the set of equivalence classes is denoted by $[P';Q'']_f$. The *relative-lifting problem* is the study of $[P';Q'']_f$ (for example, do liftings of f exist, and if so, how many homotopy classes relative to f of liftings of f are there?).

6 EXAMPLE If P'' is empty, then a map pair $f: \alpha \to \beta$ consists of a map $f': P' \to Q'$, and a lifting $\tilde{f}: P' \to Q''$ of f is a lifting of f' to Q'' in the sense defined in Sec. 2.2. In this case, if β is a fibration, two liftings $\tilde{f}_0, \tilde{f}_1: P' \to Q''$ of f' are homotopic relative to f if and only if they are fiber homotopic in the sense of Sec. 2.8. Thus the absolute-lifting problem is a special case of a relative-lifting problem.

7 EXAMPLE If α is an inclusion map and Q' is a one-point space, then a map pair $f: \alpha \to \beta$ corresponds bijectively to a map $f'': P'' \to Q''$ and a lifting $\tilde{f}: P' \to Q''$ of f corresponds bijectively to an extension of f'' to P'. In this case two extensions $\tilde{f}_0, \tilde{f}_1: P' \to Q''$ are homotopic relative to f (as liftings) if and only if they are homotopic relative to P''. Thus the extension problem is a special case of a relative-lifting problem.

8 EXAMPLE Let $\tilde{f}_0, \tilde{f}_1: P' \to Q''$ be liftings of a map pair $f: \alpha \to \beta$. Let $R' = P' \times I$ and let R'' be the quotient space of the disjoint union of $P' \times \dot{I}$ and $P'' \times I$ by the identifications $(z'',0) \in P'' \times I$ equals $(\alpha(z''),0) \in P' \times \dot{I}$ and $(z'',1) \in P'' \times I$ equals $(\alpha(z''),1)$. Define a map $\gamma: R'' \to R'$ by $\gamma(z'',t) = (\alpha(z''),t)$ for $(z'',t) \in P'' \times I$ and $\gamma(z',t) = (z',t)$ for $(z',t) \in P' \times \dot{I}$. There is a map pair $g: \gamma \to \beta$ consisting of the maps $g'': R'' \to Q''$ and $g': R' \to Q'$ such that

$g''(z'',t) = f''(z'')$ for $(z'',t) \in P'' \times I$, $g''(z',0) = \tilde{f}_0(z')$ and $g''(z',1) = \tilde{f}_1(z')$ for $z' \in P'$, and $g'(z',t) = f'(z')$ for $(z',t) \in P' \times I$. Then $\tilde{f}_0$ and $\tilde{f}_1$ are homotopic relative to f if and only if there exists a lifting of g.

We are particularly interested in the relative-lifting problem in case α is the inclusion map of a relative CW complex and β is a weak fibration. Thus, if $i: A \subset X$ is an inclusion map and $p: E \to B$ is a weak fibration, a map pair $f: i \to p$ consists of a map $f': X \to B$ and a lifting $f'': A \to E$ of $f'|A$. A lifting $\tilde{f}$ of f is a lifting of f' to E, which is an extension of f''. Two liftings of f are homotopic relative to f if and only if there is a fiber homotopy relative to A between them. The following relative homotopy extension theorem is the main reason for giving particular attention to this case.

9 THEOREM *Let (X,A) be a relative CW complex, with inclusion map $i: A \subset X$, and let $p: E \to B$ be a weak fibration. Given a map $\tilde{f}: X \to E$ and a homotopy pair $H: i \times 1_I \to p$ consisting of a homotopy $H': X \times I \to B$ starting at $p \circ \tilde{f}$ and a homotopy $H'': A \times I \to E$ starting at $\tilde{f} \circ i$, there is a homotopy $\tilde{H}: X \times I \to E$ starting at $\tilde{f}$ such that $H' = p \circ \tilde{H}$ and $H'' = \tilde{H} \circ (i \times 1_I)$.*

PROOF Let $g: X \times 0 \cup A \times I \to E$ be the map defined by $g(x,0) = \tilde{f}(x)$ for $x \in X$ and $g(a,t) = H''(a,t)$ for $a \in A$ and $t \in I$. Then H' is an extension of $p \circ g$, and by the standard stepwise-extension procedure over the successive skeleta of (X,A) (applied to polyhedral pairs in the proof of theorem 7.2.6 and equally applicable to any relative CW complex), there is a map $\tilde{H}: X \times I \to E$ such that $p \circ \tilde{H} = H'$ and $\tilde{H}|X \times 0 \cup A \times I = g$. Then $\tilde{H}$ has the desired properties. ■

Let us reinterpret this last result. A map pair $f: i \to p$ is a commutative square

$$\begin{array}{ccc} A & \xrightarrow{f''} & E \\ {\scriptstyle i}\downarrow & & \downarrow{\scriptstyle p} \\ X & \xrightarrow{f'} & B \end{array}$$

Therefore, if we let $B^X \times' E^A$ denote the fibered product of the map $B^X \to B^A$ induced by restriction and the map $E^A \to B^A$ induced by p, the pair (f',f'') is a point of $B^X \times' E^A$. In this way the set of map pairs $f: i \to p$ is identified with the fibered product $B^X \times' E^A$. The map ρ corresponds to a map $\rho: E^X \to B^X \times' E^A$, and $[X;E]_f$ is the set of path components of $\rho^{-1}(f)$.

10 COROLLARY *Let (X,A) be a relative CW complex with X locally compact Hausdorff, with inclusion map $i: A \subset X$, and let $p: E \to B$ be a weak fibration. Then $\rho: E^X \to B^X \times' E^A$ is a weak fibration.*

PROOF Given a map $g: I^n \to E^X$ and a homotopy $H: I^n \times I \to B^X \times' E^A$ starting with $\rho(g)$, the exponential correspondence assigns to g a map $\bar{g}: X \times I^n \to E$ and to H a homotopy pair H_1 from $(i \times 1_{I^n}) \times 1_I$ to p, starting with $\rho(\bar{g})$. By theorem 9, there is a homotopy $\bar{H}_1: X \times I^n \times I \to E$ starting with $\bar{g}$ such that $\rho(\bar{H}_1) = H_1$. Then the exponential correspondence associates to $\bar{H}_1$ a map $G: I^n \times I \to E^X$ starting with g such that $\rho \circ G = H$. ∎

It follows from corollaries 10 and 4 that if $f_0, f_1: i \to p$ are homotopic map pairs with X locally compact Hausdorff, then $[X;E]_{f_0}$ and $[X;E]_{f_1}$ are in one-to-one correspondence. Thus the relative-lifting problem for f_0 is equivalent to the relative-lifting problem for f_1.

Given weak fibrations $p_1: E_1 \to B_1$ and $p_2: E_2 \to B_2$, a map pair $g: p_1 \to p_2$ is called a *weak homotopy equivalence* if $g'': E_1 \to E_2$ and $g': B_1 \to B_2$ are weak homotopy equivalences. We shall show that a weak homotopy equivalence in the category of maps has much the same properties as a weak homotopy equivalence in the category of spaces. The following analogue of theorem 7.6.22 is our starting point.

11 LEMMA *Let (X,A) be a relative CW complex, with inclusion map $i: A \subset X$, and let $g: p_1 \to p_2$ be a weak homotopy equivalence between weak fibrations. Given a map pair $f: i \to p_1$ and a lifting $\bar{h}: X \to E_2$ of the map pair $g \circ f$, there is a lifting $\bar{f}: X \to E_1$ of f such that $g'' \circ \bar{f}$ and $\bar{h}$ are homotopic relative to $g \circ f$.*

PROOF The proof involves two applications of theorem 7.6.22 and then two applications of theorem 9. We shall not make specific reference to these when they are invoked.

We have a commutative diagram

$$\begin{array}{ccccc} A & \xrightarrow{f''} & E_1 & \xrightarrow{g''} & E_2 \\ i\downarrow & & p_1\downarrow & & \downarrow p_2 \\ X & \xrightarrow{f'} & B_1 & \xrightarrow{g'} & B_2 \end{array}$$

in which g'' and g' are weak homotopy equivalences, and we are given a map $\bar{h}: X \to E_2$ such that $\bar{h} \circ i = g'' \circ f''$ and $p_2 \circ \bar{h} = g' \circ f'$. Then there is a map $\bar{\bar{f}}: X \to E_1$ such that $\bar{\bar{f}} \circ i = f''$ and a homotopy $G'': g'' \circ \bar{\bar{f}} \simeq \bar{h}$ rel A. The maps $p_1 \circ \bar{\bar{f}}$ and f' agree on A and $p_2 \circ G''$ is a homotopy relative to A from $g' \circ p_1 \circ \bar{\bar{f}} = p_2 \circ g'' \circ \bar{\bar{f}}$ to $g' \circ f' = p_2 \circ \bar{h}$. Therefore there is a homotopy $F': p_1 \circ \bar{\bar{f}} \simeq f'$ rel A and a homotopy $H': g' \circ F' \simeq p_2 \circ G''$ rel $A \times I \cup X \times \dot{I}$.

Let $F'': X \times I \to E_1$ be a lifting of F' such that $F''(x,0) = \bar{\bar{f}}(x)$ for $x \in X$ and $F''(a,t) = f''(a)$ for $a \in A$ and $t \in I$. Define $\bar{f}: X \to E_1$ by $\bar{f}(x) = F''(x,1)$. We show that $\bar{f}$ has the desired properties. It is clearly a lifting of f.

The maps $g'' \circ F''$ and G'' are homotopies relative to A from $g'' \circ \bar{\bar{f}}$ to $g'' \circ \bar{f}$ and to $\bar{h}$, respectively, and H' is a homotopy from $p_2 \circ g'' \circ F''$ to $p_2 \circ G''$ rel $A \times I \cup X \times \dot{I}$. Since there is a homeomorphism of $(X \times I \times I, A \times I \times I)$ onto itself taking $X \times (I \times \dot{I} \cup 0 \times I)$ onto $X \times I \times 0$, there is a lifting H'' of H' which is a homotopy from $g'' \circ F''$ to G'' rel $X \times 0 \cup A \times I$. Then the map $H: X \times I \to E_2$ defined by $H(x,t) = H''(x,1,t)$ is a homotopy from $g'' \circ \bar{f}$ to h relative to $g \circ f$. ∎

This gives us the following important result.

12 THEOREM *Let (X,A) be a relative CW complex, with inclusion map $i: A \subset X$, and let $g: p_1 \to p_2$ be a weak homotopy equivalence between weak fibrations. Given a map pair $f: i \to p_1$, the map pair g induces a bijection*

$$g''_\#: [X;E_1]_f \approx [X;E_2]_{g \circ f}$$

PROOF The fact that $g''_\#$ is surjective follows immediately from lemma 11. The fact that $g''_\#$ is injective follows from application of lemma 11 to the relative CW complex $(X,A) \times (I,\dot{I})$. ∎

EXERCISES

A EXACTNESS OF HOMOTOPY SETS

1 Assume that $j: (X',A') \subset (X,A)$ is a cofibration, where A and X' are closed subsets of X and $A' = A \cap X'$. Prove that the collapsing map

$$(C_{j'},C_{j''}) \to (C_{j'},C_{j''})/CX' = (X,A)/X' = (X/X', A/A')$$

is a homotopy equivalence.

2 With the same hypotheses as in exercise 1, let $g': (X,A) \to C(X',A')$ be any map such that $g'(x') = x'$ for $x' \in X'$ and let $g: (X/X',A/A') \to S(X',A')$ be the map such that the following square is commutative, where k' and k'' are the collapsing maps:

$$\begin{array}{ccc} (X,A) & \xrightarrow{g'} & C(X',A') \\ {\scriptstyle k'}\downarrow & & \downarrow{\scriptstyle k''} \\ (X/X',A/A') & \xrightarrow{g} & S(X',A') \end{array}$$

Prove that there is a coexact sequence

$$(X',A') \to \cdots \to S^n(X',A') \xrightarrow{S^n j} S^n(X,A) \xrightarrow{S^n k'} S^n(X/X', A/A') \xrightarrow{S^n g} \cdots$$

3 If (X,A) is a relative CW complex, prove that there is a coexact sequence

$$A \subset X \to X/A \to SA \subset SX \to \cdots \to S^n A \subset S^n X \to \cdots$$

B HOMOTOPY GROUPS

1 If A is a retract of X, prove that there is an isomorphism

$$\pi_n(X,x_0) \approx \pi_n(A,x_0) \oplus \pi_n(X,A,x_0) \qquad n \geq 2$$

2 If X is deformable into A relative to $x_0 \in A$, prove that there is an isomorphism

$$\pi_n(A,x_0) \approx \pi_n(X,x_0) \oplus \pi_{n+1}(X,A,x_0) \qquad n \geq 2$$

3 If $p: E \to B$ is a weak fibration such that the fiber $F = p^{-1}(b_0)$ is contractible in E relative to $e_0 \in F$, prove that there is an isomorphism

$$\pi_n(B,b_0) \approx \pi_n(E,e_0) \oplus \pi_{n-1}(F,e_0) \qquad n \geq 2$$

4 If $p: E \to B$ is a weak fibration which admits a section, prove that there is an isomorphism for $e_0 \in F = p^{-1}(b_0)$

$$\pi_n(E,e_0) \approx \pi_n(B,b_0) \oplus \pi_n(F,e_0) \qquad n \geq 2$$

5 Let $\{X_j\}$ be an indexed family of spaces with base points $x_j \in X_j$. Prove that there is an isomorphism

$$\pi_n(\times X_j,(x_j)) \approx \times \pi_n(X_j,x_j) \qquad n \geq 0$$

6 Given $X \vee Y = X \times y_0 \cup x_0 \times Y \subset X \times Y$, prove that there is an isomorphism

$$\pi_n(X \vee Y, (x_0,y_0)) \approx \pi_n(X,x_0) \oplus \pi_n(Y,y_0) \oplus \pi_{n+1}(X \times Y, X \vee Y, (x_0,y_0))$$

C BASE POINTS[1]

1 Give an example of a degenerate base point.

2 If X and Y have nondegenerate base points, prove that also $X \vee Y$, $X \times Y$, and $X \times Y/X \vee Y$ have nondegenerate base points.

3 If (X,x_0) and (Y,y_0) have the same homotopy type, prove that x_0 is a nondegenerate base point of X if and only if y_0 is a nondegenerate base point of Y.

4 Prove that any space has the same homotopy type as some space with a nondegenerate base point.

5 Let X and Y be path-connected spaces with nondegenerate base points x_0 and y_0, respectively. Prove that X and Y have the same homotopy type if and only if (X,x_0) and (Y,y_0) have the same homotopy type.

D THE WHITEHEAD PRODUCT

Let $p \geq 1$ and $q \geq 1$ and let $h: (I^{p+q},\dot{I}^{p+q}) \to (I^p,\dot{I}^p) \times (I^q,\dot{I}^q)$ be the homeomorphism $h(t_1, \ldots ,t_{p+q}) = ((t_1, \ldots ,t_p),(t_{p+1}, \ldots ,t_{p+q}))$. Then h determines an element $[h] \in \pi_{p+q}((I^p,\dot{I}^p) \times (I^q,\dot{I}^q), (0,0))$ and an element

$$\eta_{p,q} = \partial[h] \in \pi_{p+q-1}(I^p \times \dot{I}^q \cup \dot{I}^p \times I^q, (0,0))$$

Given maps $\alpha: (I^p,\dot{I}^p) \to (X,x_0)$ and $\beta: (I^q,\dot{I}^q) \to (X,x_0)$, define a map $\gamma: (I^p \times \dot{I}^q \cup \dot{I}^p \times I^q, (0,0)) \to (X,x_0)$ by

$$\gamma(z,z') = \begin{cases} \alpha(z) & z' \in \dot{I}^q,\ (z,z') \in I^p \times I^q \\ \beta(z') & z \in \dot{I}^p,\ (z,z') \in I^p \times I^q \end{cases}$$

1 Prove that $\gamma_\#(\eta_{p,q}) \in \pi_{p+q-1}(X,x_0)$ depends only on $[\alpha]$ and $[\beta]$. It is called the *Whitehead product* of $[\alpha]$ and $[\beta]$ and is denoted by $[[\alpha],[\beta]] \in \pi_{p+q-1}(X,x_0)$.

2 Prove that if $p = q = 1$, then $[[\alpha],[\beta]] = [\alpha][\beta][\alpha]^{-1}[\beta]^{-1}$.

3 If $p > 1$ and $q = 1$, prove that $[[\alpha],[\beta]] = [\alpha]h_{[\beta]}([\alpha]^{-1})$.

4 If $p + q > 2$, prove that $[[\alpha],[\beta]] = (-1)^{pq}[[\beta],[\alpha]]$.

5 If $f: (X,x_0) \to (Y,y_0)$, prove that $f_\#[[\alpha],[\beta]] = [f_\#[\alpha],f_\#[\beta]]$.

6 If ω is a path in X, prove that $h_{[\omega]}[[\alpha],[\beta]] = [h_{[\omega]}[\alpha],h_{[\omega]}[\beta]]$.

7 Prove that $[[\alpha],[\beta]] = 0$ if and only if there is a map $f: I^p \times I^q \to X$ such that

$$f(t_1, \ldots ,f_{p+q}) = \begin{cases} \alpha(t_1, \ldots ,t_p) \text{ if } t_i = 0 \text{ or } 1 \text{ for some } p+1 \leq i \leq p+q \\ \beta(t_{p+1}, \ldots ,t_{p+q}) \text{ if } t_i = 0 \text{ or } 1 \text{ for some } 1 \leq i \leq p \end{cases}$$

8 If X is an H space, prove that $[[\alpha],[\beta]] = 0$ for all $[\alpha]$ and $[\beta]$.

[1] See D. Puppe, Homotopiemengen und ihre induzierten Abbildungen. I, *Mathematische Zeitschriften*, vol. 69, pp. 299–344, 1958.

9 Prove that S^n is an H space if and only if $[[\alpha],[\beta]] = 0$ for all $[\alpha]$, $[\beta] \in \pi_n(S^n)$.

E CW COMPLEXES

1 If (X,A) is a relative CW complex, prove that X has a topology coherent with the collection $\{A\} \cup \{e \mid e \text{ a cell of } X - A\}$.

2 If (X,A) is a relative CW complex, prove that X is compactly generated if and only if A is compactly generated.

3 If (X,A) is a relative CW complex and A is paracompact, prove that X is paracompact.

4 If (X,A) is a relative CW complex and A has the same homotopy type as a CW complex, prove that X has the same homotopy type as a CW complex.

5 Prove that a CW complex is locally contractible.

6 Prove that a CW complex has the same homotopy type as a polyhedron.

F ACTION OF THE FUNDAMENTAL GROUP

1 Prove that the real projective n-space P^n is simple if and only if n is odd.

2 For $1 < n < m$ show that $P^{2n+1} \times S^{2m+1}$ and $P^{2m+1} \times S^{2n+1}$ are simple compact polyhedra having isomorphic homotopy groups in all dimensions, but are not of the same homotopy type.

3 Let $(Z,\dot{Z})$ be an $(n-1)$-connected CW pair, with $n \geq 2$, such that $\dot{Z}$ is simply connected. Let (X^*,X) be the adjunction space obtained by adjoining Z to a CW complex X by a map $f: (\dot{Z},z_0) \to (X,x_0)$ and let $g: (Z,\dot{Z},z_0) \to (X^*,X,x_0)$ be the canonical map. Prove that (X^*,X) is $(n-1)$-connected and that the map

$$\bigoplus_{[\omega] \in \pi_1(X,x_0)} [\pi_n(Z,\dot{Z},z_0)]_{[\omega]} \to \pi_n(X^*,X,x_0)$$

sending $[\alpha]_{[\omega]}$ to $h_{[\omega]}(g_\#[\alpha])$ for $[\alpha] \in \pi_n(Z,\dot{Z},z_0)$ is an isomorphism. [*Hint:* Let $\tilde{X}$ be the universal covering space of X and let $\{f_{[\omega]}: \dot{Z} \to \tilde{X}\}_{[\omega] \in \pi_1(X,x_0)}$ be the set of liftings of f. Show that the space $\tilde{X}^*$ obtained by attaching a copy of Z to $\tilde{X}$ for each map $f_{[\omega]}$ is the universal covering space of X^*. Then use the fact that $\pi_q(\tilde{X}^*,\tilde{X}) \approx \pi_q(X^*,X)$ and compute $\pi_n(\tilde{X}^*,\tilde{X})$ by the Hurewicz theorem.]

4 Let X be the CW complex obtained from $S^1 \vee S^2$ by attaching a 3-cell by a map representing $2[\alpha] - h_{[\omega]}[\alpha]$, where $[\alpha]$ is a generator of $\pi_2(S^2)$ and $[\omega]$ is a generator of $\pi_1(S^1)$. Prove that the inclusion map $S^1 \subset X$ induces an isomorphism of the fundamental groups and all homology groups but not of the two-dimensional homotopy groups.

G CW APPROXIMATIONS

1 If (X,A) is an arbitrary pair, prove that there is a CW pair (X',A') and a map $f: (X',A') \to (X,A)$ such that $f \mid X': X' \to X$ and $f \mid A': A' \to A$ are both weak homotopy equivalences.

2 If $f_1: X_1 \to Y_1$ and $f_2: X_2 \to Y_2$ are weak homotopy equivalences, prove that $f_1 \times f_2: X_1 \times X_2 \to Y_1 \times Y_2$ is also a weak homotopy equivalence.

3 If $f_1: X_1 \to Y_1$ and $f_2: X_2 \to Y_2$ are weak homotopy equivalences, show by an example that $f_1 \vee f_2: X_1 \vee X_2 \to Y_1 \vee Y_2$ need not be a weak homotopy equivalence.

4 Show by an example that a weak homotopy equivalence need not induce isomorphisms of the corresponding Alexander cohomology groups.

5 If X is simply connected and $H_*(X)$ is finitely generated, prove that X has the same weak homotopy type as some finite CW complex.

6 A space X is said to be *dominated* by a space Y if there exist maps $f: X \to Y$ and $g: Y \to X$ such that $g \circ f \simeq 1_X$. Prove that a space is dominated by a CW complex if and only if it has the same homotopy type as some CW complex.

H GROUPS OF HOMOTOPY CLASSES

Throughout this group of exercises it is assumed that Y is $(n-1)$-connected, where $n \geq 2$, with base point y_0, and that X is a CW complex of dimension $\leq 2n-2$.

1 Prove that any map $X \to Y$ is homotopic to a map sending X^{n-1} to y_0 and that if $f, g: (X,X^{n-1}) \to (Y,y_0)$ are homotopic as maps from X to Y, they are homotopic relative to X^{n-2}.

2 Prove that the diagonal map $d: X \to X \times X$ is homotopic to a map d' such that $d'(X) \subset (X \times X^{n-2}) \cup (X^{n-2} \times X)$. Prove that maps $d', d'': X \to (X \times X^{n-2}) \cup (X^{n-2} \times X)$ which are homotopic in $X \times X$ are homotopic in $(X \times X^{n-1}) \cup (X^{n-1} \times X)$. (*Hint:* Use the cellular-approximation theorem.)

Let $d': X \to (X \times X^{n-2}) \cup (X^{n-2} \times X)$ be homotopic in $X \times X$ to the diagonal map. Given $f, g: X \to Y$, let $f', g': (X,X^{n-1}) \to (Y,y_0)$ be homotopic to f and g, respectively. Then $(f' \times g') \circ d': X \to Y \times Y$ maps X into $Y \vee Y$. Let $\gamma: Y \vee Y \to Y$ be defined by $\gamma(y,y_0) = y = \gamma(y_0,y)$.

3 Prove that $[\gamma \circ (f' \times g') \circ d']$ depends only on $[f]$ and $[g]$ and that the operation $[f] + [g] = [\gamma \circ (f' \times g') \circ d']$ is associative, commutative, and has a unit element, making $[X;Y]$ into a commutative semigroup with unit.

4 Prove that if $g: Y \to Y'$, where Y' is also $(n-1)$-connected (or if $h: X' \to X$, where X' is a CW complex of dimension $\leq 2n-2$), then $g_\#: [X;Y] \to [X;Y']$ is a homomorphism (or $h^\#: [X;Y] \to [X';Y]$ is a homomorphism).

5 The semigroup $[X;Y]$ is a group. (*Hint:* Use induction on the dimension of X, the fact that $[X^{k+1}/X^k;Y]$ is a group for any k and any Y, because X^{k+1}/X^k, being a wedge of $(k+1)$-spheres, is a suspension, and the exactness of the sequence of homomorphisms

$$[X^{k+1}/X^k;Y] \to [X^{k+1};Y] \to [X^k;Y]\ [X';Y]$$

where X' is a disjoint union of k spheres, one for each $(k+1)$-cell of X.)

In case $Y = S^n$ and dimension $X \leq 2n-2$, the group $[X;S^n]$ is called the *nth cohomotopy group of* X,[1] denoted by $\pi^n(X)$.

I MISCELLANEOUS

1 Let $\partial': \pi_{n+1}(\Delta^{n+1},\dot{\Delta}^{n+1},v_0) \to \pi_n(\dot{\Delta}^{n+1},(\Delta^{n+1})^{n-1},v_0)$ if $n \geq 2$ and let

$$\partial': \pi_2(\Delta^2,\dot{\Delta}^2,v_0) \to \pi_1(\dot{\Delta}^2,v_0)$$

if $n = 1$. Prove that $\partial'[\xi_{n+1}] = b_n$ for $n \geq 1$ (see page 394 for definition of b_n).

2 Let H be a homotopy functor and let $f: X \to Y$ be a base-point-preserving map between path-connected spaces, with nondegenerate base points. Prove that the sequence

$$H(C_f) \to H(Y) \to H(X)$$

is exact.

3 If H is a homotopy functor and (X,A) is a CW pair, prove that there is an exact sequence

$$H(A) \leftarrow H(X) \leftarrow H(X/A) \leftarrow H(SA) \leftarrow \cdots \leftarrow H(S^nA) \leftarrow \cdots$$

[1] For more details see E. Spanier, Borsuk's cohomotopy groups, *Annals of Mathematics*, vol. 50, pp. 203–245, 1949.

CHAPTER EIGHT

OBSTRUCTION THEORY

IN THIS CHAPTER WE DEVELOP OBSTRUCTION THEORY FOR THE GENERAL LIFTING problem. A sequence of obstructions is defined whose vanishing is necessary and sufficient for the existence of a lifting. The kth obstruction in the sequence is defined if and only if all the lower obstructions are defined and vanish, in which case the vanishing of the kth obstruction is a necessary condition for definition of the $(k + 1)$st obstruction.

We begin by applying the general theory of homotopy functors to study the set of homotopy classes of maps from a CW complex to a space with exactly one nonzero homotopy group and we show that a suitable cohomology functor serves to classify maps up to homotopy in this case. This result is then used to obtain a solution, in terms of cohomology, of the lifting problem for a fibration whose fiber has exactly one nonzero homotopy group.

With this in mind, we then consider the problem of factorizing an arbitrary fibration into simpler ones each of which has a fiber with exactly one nonzero homotopy group. We show that such factorizations do exist for a large class of fibrations, and that when they exist, a sequence of obstructions can be associated to the factorization. These obstructions are subsets of cohomology groups, and we apply the general machinery to some special cases where, because of dimension restrictions, the only obstructions which enter are either the first one or the first two. For the case of only one obstruction we obtain the Hopf classification theorem.

Finally, we prove the suspension theorem, which we use to compute the $(n + 1)$st homotopy group of the n-sphere. Combining this with the technique of obstruction theory, we obtain a proof of the Steenrod classification theorem.

Section 8.1 is devoted to spaces with exactly one nonzero homotopy group. We prove that a suitable cohomology functor serves both to classify maps from a CW complex to such a space and to provide a solution for the extension problem for maps involving a relative CW complex and such a space. We use this result to derive the Hopf extension and classification theorems for maps of an n-dimensional CW complex to S^n. Section 8.2 deals with fibrations whose fiber has exactly one nonzero homotopy group, and again it is shown that a suitable cohomology functor serves to provide a solution for the lifting problem and to classify liftings of a given map.

In Sec. 8.3 we prove that many fibrations can be factored as infinite composites of fibrations each of which has a fiber with exactly one nonzero homotopy group. The corresponding lifting problem is then represented as an infinite sequence of simpler lifting problems. In Sec. 8.4 we show how to define obstructions inductively for such a sequence of fibrations, and how to apply the resulting machinery.

In Sec. 8.5 we shall study the suspension map and prove the exactness of the Wang sequence of a fibration with base space a sphere. This result is used to prove the suspension theorem, which is applied to compute $\pi_{n+1}(S^n)$ for all n. We then prove the Steenrod classification theorem for maps of an $(n + 1)$-dimensional CW complex to S^n.

1 EILENBERG-MACLANE SPACES

This section is devoted to a study of spaces with exactly one nonzero homotopy group. Such spaces are classifying spaces for the cohomology functors, and because of this, there is an important relation between the cohomology of these spaces and cohomology operations. At the end of the section we shall apply the results to derive the Hopf classification and extension theorems. Then, later in the chapter, we shall study arbitrary spaces by representing them as iterated fibrations whose fibers are spaces with exactly one nonzero homotopy group. Thus, these homotopically simple spaces serve as building blocks for more complicated spaces.

Let π be a group and let n be an integer ≥ 1. A *space of type* (π,n) is a path-connected pointed space Y such that $\pi_q(Y,y_0) = 0$ for $q \neq n$ and $\pi_n(Y,y_0)$ is isomorphic to π. An *Eilenberg-MacLane space*[1] is a path-connected pointed space all of whose homotopy groups vanish, except possibly for a

[1] See S. Eilenberg and S. MacLane, On the groups $H(\pi,n)$, I, *Annals of Mathematics*, vol. 58, pp. 55–106, 1953.

single dimension. Thus a space of type (π,n) is an Eilenberg-MacLane space. Conversely, if Y is an Eilenberg-MacLane space and $\pi_q(Y,y_0) = 0$ for $q \neq n$, then Y is a space of type $(\pi_n(Y,y_0), n)$. Let us consider a few examples.

1 It follows from corollary 7.2.12 that S^1 is a space of type $(\mathbf{Z},1)$.

2 Let P^∞ be the *CW* complex which is the union of the sequence $P^1 \subset P^2 \subset \cdots$ topologized by the topology coherent with the collection $\{P^j\}_{j\geq 1}$. Then $\pi_q(P^\infty) \approx \lim_\to \{\pi_q(P^j)\}$, and it follows from application of corollary 7.2.11 to the covering $S^n \to P^n$ that P^∞ is a space of type $(\mathbf{Z}_2,1)$.

3 Let $P_\infty(\mathbf{C})$ be the *CW* complex which is the union of the sequence $P_1(\mathbf{C}) \subset P_2(\mathbf{C}) \subset \cdots$ topologized by the topology coherent with the collection $\{P_j(\mathbf{C})\}_{j\geq 1}$. Then $\pi_q(P_\infty(\mathbf{C})) \approx \lim_\to \{\pi_q(P_j(\mathbf{C}))\}$, and it follows from corollary 7.2.13 that $P_\infty(\mathbf{C})$ is a space of type $(\mathbf{Z},2)$.

Let π be an abelian group and Y a path-connected pointed space. An element $v \in H^n(Y,y_0;\pi)$ is said to be *n-characteristic* for Y if the composite

$$\pi_n(Y,y_0) \xrightarrow{\varphi} H_n(Y,y_0) \xrightarrow{h(v)} \pi$$

is an isomorphism (where φ is the Hurewicz homomorphism and h is the homomorphism defined in Sec. 5.5). If Y is $(n-1)$-connected, it follows from the absolute Hurewicz isomorphism theorem and the universal-coefficient theorem for cohomology that there is an n-characteristic element $v \in H^n(Y,y_0;\pi)$ if and only if $\pi \approx \pi_n(Y,y_0)$. Such an element is unique up to automorphisms of π. In particular, a space Y of type (π,n) with π abelian has n-characteristic elements $v \in H^n(Y,y_0;\pi)$.

4 LEMMA *Let $u \in H^n(Y,y_0;G)$ be a universal element for the nth cohomology functor with coefficients G, where $n \geq 1$. Then Y is a space of type (G,n) and u is n-characteristic for Y.*

PROOF By theorem 7.7.14, there are isomorphisms

$$T_u\colon \pi_q(Y,y_0) \approx H^n(S^q,p_0;G) \qquad q \geq 1$$

Therefore $\pi_q(Y,y_0) = 0$ if $q \neq n$, and $T_u\colon \pi_n(Y,y_0) \approx H^n(S^n,p_0;G)$. If $\alpha\colon (S^n,p_0) \to (Y,y_0)$, then $T_u([\alpha]) = \alpha^*(u)$, and there is a commutative diagram

$$\begin{array}{ccccc}
\pi_n(S^n,p_0) & \underset{\approx}{\xrightarrow{\varphi}} & H_n(S^n,p_0) & & \\
 & & & \searrow^{h(\alpha^*(u)) = h(T_u[\alpha])} & \\
\alpha_\#\downarrow & & \alpha_*\downarrow & & G \\
 & & & \nearrow_{h(u)} & \\
\pi_n(Y,y_0) & \underset{\approx}{\xrightarrow{\varphi}} & H_n(Y,y_0) & &
\end{array}$$

Let $\nu\colon H^n(S^n,p_0;G) \approx G$ be the isomorphism defined by

$$\nu(v) = h(v)(\varphi[1_{S^n}]) \qquad v \in H^n(S^n,p_0;G)$$

From the commutativity of the diagram above,

$$(h(u) \circ \varphi)[\alpha] = (h(u) \circ \varphi \circ \alpha_\#)[1_{S^n}] = h(T_u[\alpha])\,(\varphi[1_{S^n}]) = (\nu \circ T_u)[\alpha]$$

It follows that $h(u) \circ \varphi$ equals the composite

$$\pi_n(Y,y_0) \xrightarrow[\approx]{T_u} H^n(S^n,p_0;\, G) \xrightarrow[\approx]{\nu} G$$

and so is an isomorphism. Therefore Y is a space of type (G,n) and u is n-characteristic for Y. ∎

5 COROLLARY *Given $n \geq 1$ and a group π (abelian if $n > 1$), there exists a space of type (π,n).*

PROOF If π is abelian, it follows from lemma 4 that any classifying space for the nth cohomology functor with coefficients π is a space of type (π,n). If $n = 1$ and π is arbitrary, it is easy to see that a classifying space for the homotopy functor of example 7.7.5 which assigns to a pointed path-connected space X the set of homomorphisms $\pi_1(X,x_0) \to \pi$ is a space of type $(\pi,1)$. In either case, since any homotopy functor has a classifying space by corollary 7.7.12, the result follows. ∎

6 COROLLARY *Let $\{\pi_n\}_{n\geq 1}$ be a sequence of groups which are abelian for $n \geq 2$. There is a space X, with base point x_0, such that $\pi_n(X,x_0) \approx \pi_n$ for $n \geq 1$.*

PROOF By corollary 5, for each $n \geq 1$ there is a space Y_n, with base point y_n, such that $\pi_q(Y_n,y_n) = 0$ for $q \neq n$ and $\pi_n(Y_n,y_n) \approx \pi_n$. Then the product space $\times\, Y_n$ with base point (y_n) has the desired properties. ∎

The last result can be strengthened so that if π_1 acts as a group of operators on π_n for every $n \geq 2$, then the sequence is realized as the sequence of homotopy groups of a space X in such a way that the action of π_1 on π_n corresponds to the action of $\pi_1(X,x_0)$ on $\pi_n(X,x_0)$ of theorem 7.3.8.

7 LEMMA *Let $F\colon H \to H'$ be a natural transformation between homotopy functors which induces an isomorphism of their qth coefficient groups for $q < n$ and a surjection of their nth coefficient groups (where $1 \leq n \leq \infty$). For any path-connected pointed CW complex W the map*

$$F(W)\colon H(W) \to H'(W)$$

is a bijection if $\dim W \leq n - 1$ *and a surjection if* $\dim W \leq n$.

PROOF Let $u \in H(Y)$ and $u' \in H'(Y')$ be universal elements for H and H', respectively, and let $f\colon Y \to Y'$ be a map such that $H'(f)(u') = F(Y)(u)$. For any CW complex W there is a commutative square

$$\begin{array}{ccc} [W;Y] & \xrightarrow{f_\#} & [W;Y'] \\ T_u\downarrow & & \downarrow T_{u'} \\ H(W) & \xrightarrow{F(W)} & H'(W) \end{array}$$

in which, by theorem 7.7.14, both vertical maps are bijections. Since $F(S^q)\colon H(S^q) \to H'(S^q)$ is an isomorphism for $q < n$ and a surjection for $q = n$, it follows that $f_\#\colon \pi_q(Y) \to \pi_q(Y')$ is an isomorphism for $q < n$ and a surjection for $q = n$. Since Y and Y' are path-connected pointed spaces, the map f

is an n-equivalence. The result follows from corollary 7.6.23 and the commutativity of the above square. ■

We use this last result to obtain the following classification theorem, which is a converse of lemma 4.

8 THEOREM *Let π be an abelian group, Y a space of type (π,n), and $\iota \in H^n(Y,y_0;\pi)$ an n-characteristic element for Y. Let $\psi\colon \pi^Y \to H^n(\cdot;\pi)$ be the natural transformation defined by $\psi[f] = f^*\iota$ for $[f] \in [X;Y]$. Then ψ is a natural equivalence on the category of path-connected pointed CW complexes.*

PROOF By lemma 7, it suffices to verify that ψ induces an isomorphism of all coefficient groups of the two homotopy functors π^Y and $H^n(\cdot\,;\pi)$. The only nonzero coefficient groups are $\pi_n(Y,y_0)$ and $H^n(S^n,p_0;\pi)$, and we need only verify that

$$\psi(S^n)\colon \pi_n(Y,y_0) \to H^n(S^n,p_0;\pi)$$

is an isomorphism. If $\nu\colon H^n(S^n,p_0;\pi) \approx \pi$ is defined by $\nu(v) = h(v)(\varphi[1_{S^n}])$ (as in the proof of lemma 4), then $\nu \circ \psi(S^n) = h(\iota) \circ \varphi$. Because ι is n-characteristic for Y, $\nu \circ \psi(S^n)$ is an isomorphism, and thus so is $\psi(S^n)$. ■

9 THEOREM *Let Y be a space of type $(\pi,1)$ and let H be the functor which assigns to a pointed space X the set of homomorphisms from $\pi_1(X,x_0)$ to $\pi_1(Y,y_0)$. Let $\bar\psi\colon \pi^Y \to H$ be the natural transformation defined by $\bar\psi[f] = f_\#$ for $[f] \in [X;Y]$. Then $\bar\psi$ is a natural equivalence on the category of path-connected pointed CW complexes.*

PROOF By lemma 7, it suffices to verify that

$$\bar\psi(S^1)\colon \pi_1(Y,y_0) \to H(S^1,p_0)$$

is an isomorphism. Let $\bar\nu\colon H(S^1,p_0) \approx \pi_1(Y,y_0)$ be the isomorphism defined by $\bar\nu(\gamma) = \gamma([1_{S^1}])$ for $\gamma\colon \pi_1(S^1,p_0) \to \pi_1(Y,y_0)$. Then $\bar\nu$ is an inverse of $\bar\psi(S^1)$, showing that $\bar\psi(S^1)$ is an isomorphism. ■

Note that if $\pi_1(Y,y_0)$ is abelian in theorem 9, the set of homomorphisms from $\pi_1(X,x_0)$ to $\pi_1(Y,y_0)$ is in one-to-one correspondence with the group

$$\text{Hom}\,(\pi_1'(X,x_0), \pi_1(Y,y_0)) \approx \text{Hom}\,(H_1(X,x_0), \pi_1(Y,y_0)) \approx H^1(X,x_0;\pi_1(Y,y_0))$$

and so theorems 8 and 9 agree in this case.

We now consider the free homotopy classes of maps from X to Y. Since any 0-cell x_0 of a CW complex X is a nondegenerate base point (because, by theorem 7.6.12, the inclusion map $x_0 \subset X$ is a cofibration), it follows from corollary 7.3.4 that there is an action of $\pi_1(Y,y_0)$ on the set $[X,x_0;\, Y,y_0]$. Furthermore, if Y and X are path connected and this action is trivial, then the map from base-point-preserving homotopy classes to free homotopy classes

$$[X,x_0;\, Y,y_0] \to [X;Y]$$

is a bijection. In case Y is a space of type (π,n), with $n > 1$, then $\pi_1(Y,y_0) = 0$, and so there is a bijection

$$[X,x_0;\ Y,y_0] \approx [X;Y]$$

In case Y is a space of type $(\pi,1)$, the action of $\pi_1(Y,y_0)$ on $[X,x_0;\ Y,y_0]$ corresponds under the bijection $\bar{\psi}$ of theorem 9 to the action of $\pi_1(Y,y_0)$ on $H(X,x_0)$ by conjugation. Thus, if π is abelian, there is a bijection

$$[X,x_0;\ Y,y_0] \approx [X;Y]$$

10 THEOREM *If π is an abelian group, Y is a space of type (π,n), and $\iota \in H^n(Y,y_0;\ \pi)$ is n-characteristic for Y, then for any relative CW complex (X,A) the map*

$$\psi\colon [X,A;\ Y,y_0] \to H^n(X,A;\ \pi)$$

is a bijection.

PROOF In case A is empty and X is path connected, it follows from theorem 8 and the observation above that there is a commutative square

$$\begin{array}{ccc} [X,x_0;\ Y,y_0] & \xrightarrow{\approx} & [X;Y] \\ \psi\downarrow\approx & & \downarrow\psi \\ H^n(X,x_0;\ \pi) & \xrightarrow{\approx} & H^n(X;\pi) \end{array}$$

and so $\psi\colon [X;Y] \approx H^n(X,\pi)$. In case A is empty and X is not path connected, let $\{X_\lambda\}$ be the set of path components of X. The result follows from the first case on observing that $[X;Y] \approx \times [X_\lambda;Y]$ and $H^n(X;\pi) \approx \times H^n(X_\lambda;\pi)$. In case A is not empty, let $k\colon (X,A) \to (X/A,x_0)$ be the collapsing map. Then the result follows from the already established bijection $\psi\colon [X/A;Y] \approx H^n(X/A;\pi)$ and the commutative diagram

$$\begin{array}{ccccc} [X,A;\ Y,y_0] & \xleftarrow[\approx]{k^\#} & [X/A,x_0;\ Y,y_0] & \xrightarrow{\approx} & [X/A;Y] \\ \psi\downarrow & & \psi\downarrow & & \approx\downarrow\psi \\ H^n(X,A;\ \pi) & \xleftarrow[\approx]{k^*} & H^n(X/A,x_0;\ \pi) & \xrightarrow{\approx} & H^n(X/A;\pi) \end{array}$$ ∎

11 THEOREM *Let Y be a space of type $(\pi,1)$. For any path-connected CW complex X the set of free homotopy classes of maps from X to Y is in one-to-one correspondence with the set of conjugacy classes of homomorphisms $\pi_1(X,x_0) \to \pi_1(Y,y_0)$ under the map $[f] \to f_\#$.*

PROOF This follows from theorem 9 and the remark above covering the action of $\pi_1(Y,y_0)$ on $[X,x_0;\ Y,y_0]$. ∎

12 THEOREM *Let Y be a space of type (π,n), with $n \geq 1$ and π abelian, and let $\iota \in H^n(Y,y_0;\ \pi)$ be n-characteristic for Y. If (X,A) is a relative CW complex, a map $f\colon A \to Y$ can be extended over X if and only if $\delta f^*(\iota) = 0$ in $H^{n+1}(X,A;\pi)$*

PROOF Assume $f = g \circ i$, where $i\colon A \subset X$ and $g\colon X \to Y$. Then $\delta f^*(\iota) = \delta i^* g^*(\iota) = 0$, because $\delta i^* = 0$. Hence, if f can be extended over X, then $\delta f^*(\iota) = 0$.

Conversely, assume $\delta f^*(\iota) = 0$. To extend f over X we need only extend f over each path component of X, and therefore there is no loss of generality in assuming X to be path connected (and A to be nonempty). Let Y' be the space obtained from the disjoint union $X \cup Y$ by identifying $a \in A$ with $f(a) \in Y$ for all $a \in A$. Then Y is imbedded in Y', the pair (Y',Y) is a relative CW complex, and there is a cellular map $j: (X,A) \to (Y',Y)$ which induces an isomorphism $j^*: H^*(Y',Y) \approx H^*(X,A)$ such that there is a commutative square

$$\begin{array}{ccc} H^n(Y,y_0) & \xrightarrow{\delta} & H^{n+1}(Y',Y) \\ f^*\downarrow & & \approx\downarrow j^* \\ H^n(A) & \xrightarrow{\delta} & H^{n+1}(X,A) \end{array}$$

Since $\delta f^*(\iota) = 0$, it follows that $\delta(\iota) = 0$, and there is $v \in H^n(Y',y_0;\pi)$ such that $v \mid (Y,y_0) = \iota$. Since X and Y are path connected and A is nonempty, Y' is path connected.

Let $\tilde{Y} = Y' \vee I$ (that is, $y_0 \in Y'$ is identified with $0 \in I$) and let $\tilde{y}_0 = 1 \in \tilde{Y}$. Then $\tilde{Y}$ is a path-connected space with nondegenerate base point $\tilde{y}_0$. Let $r: (\tilde{Y},I) \to (Y',y_0)$ be the retraction which collapses I to y_0 and let $\tilde{v} = r^*(v) \mid (\tilde{Y},\tilde{y}_0) \in H^n(\tilde{Y},\tilde{y}_0;\pi)$. By theorem 7.7.11, there is an imbedding of $\tilde{Y}$ in a space Y'' which is a classifying space for the nth cohomology functor with coefficients π and which has a universal element $\tilde{u} \in H^n(Y'',\tilde{y}_0;\pi)$ such that $\tilde{u} \mid (\tilde{Y},\tilde{y}_0) = \tilde{v}$. Then Y'' is a space of type (π,n), and there is a unique n-characteristic element $u \in H^n(Y'',y_0;\pi)$ such that $u \mid Y'' = \tilde{u} \mid Y''$. Then $u \mid (Y,y_0) = \iota$, and it follows from theorem 8 and the commutativity of the diagram

$$\begin{array}{ccc} [S^q,p_0;Y,y_0] & \to & [S^q,p_0;Y'',y_0] \\ & \psi_\iota\searrow\approx \quad \approx\swarrow\psi_u & \\ & H^n(S^q,p_0;\pi) & \end{array}$$

that $Y \subset Y''$ is a weak homotopy equivalence. Since the composite $X \xrightarrow{j \mid X} Y' \subset Y''$ is an extension of the composite $A \xrightarrow{f} Y \subset Y''$, it follows from theorem 7.6.22 that f can be extended to a map $X \to Y$. ■

We now show that cohomology operations are closely related to the cohomology of Eilenberg-MacLane spaces. Let $\Theta(n,q;\pi,G)$ be the group of all cohomology operations of type $(n,q;\pi,G)$. Thus π and G are abelian groups and an element $\theta \in \Theta(n,q;\pi,G)$ is a natural transformation from the singular cohomology functor $H^n(\cdot\,;\pi)$ to the singular cohomology functor $H^q(\cdot\,;G)$.

13 THEOREM *Let π be an abelian group and let Y be a space of type (π,n), with an n-characteristic element $\iota \in H^n(Y,y_0;\pi)$. There is an isomorphism*

$$\gamma: \Theta(n,q;\pi,G) \approx H^q(Y,y_0;G)$$

defined by $\gamma(\theta) = \theta(\iota)$ for $\theta \in \Theta(n,q;\pi,G)$.

PROOF Since, by theorem 7.8.1, every pair has a relative CW approximation, a cohomology operation corresponds bijectively to a cohomology operation on the category of relative CW complexes. To define an inverse to γ, given $u \in H^q(Y,y_0; G)$, let θ_u be the cohomology operation of type $(n,q; \pi,G)$ defined for a relative CW complex (X,A) by

$$\theta_u(v) = f_v^*(u) \qquad v \in H^n(X,A; \pi)$$

where $f_v: (X,A) \to (Y,y_0)$ is a map such that $f_v^*(\iota) = v$ (f_v exists and is unique up to homotopy, by theorem 10). Then

$$\gamma(\theta_u) = \theta_u(\iota) = 1_Y^*(u) = u$$

showing that the map $u \to \theta_u$ is a right inverse of γ. To show that it is also a left inverse of γ, let (X,A) be a relative CW complex and let $v \in H^n(X,A; \pi)$. We must show that $\theta_{\gamma(\theta)}(v) = \theta(v)$. Let $f_v: (X,A) \to (Y,y_0)$ be such that $f_v^*(\iota) = v$. Then we have

$$\theta(v) = \theta(f_v^*(\iota)) = f_v^*(\theta(\iota)) = f_v^*(\gamma(\theta)) = \theta_{\gamma(\theta)}(v) \quad ■$$

We present one application of this result.

14 COROLLARY *Let θ be a cohomology operation of type $(n,q; \pi,G)$. For any relative CW complex (X,A) the map*

$$\theta: H^n((X,A) \times (I,\dot I); \pi) \to H^q((X,A) \times (I,\dot I); G)$$

is a homomorphism.

PROOF The collapsing map

$$k: (X \times I, A \times I \cup X \times \dot I) \to X \times I/(A \times I \cup X \times \dot I)$$

induces isomorphisms in cohomology. Furthermore, $X \times I/(A \times I \cup X \times \dot I)$ is homeomorphic to $S(X/A)$ (where X/A is understood to be the disjoint union of X and a base point x_0 in case A is empty). Thus it suffices to show that if X' is any pointed CW complex, then the map

$$\theta: H^n(SX',x_0'; \pi) \to H^q(SX',x_0'; G)$$

is a homomorphism.

Let Y be a CW complex of type (π,n), with n-characteristic element ι, and let Y' be a space of type (G,q), with q-characteristic element ι'. Let $f: Y \to Y'$ be a map such that $f^*\iota' = \theta(\iota)$. There is then a commutative diagram

$$\begin{array}{ccc} [SX',x_0'; Y,y_0] & \xrightarrow{f_\#} & [SX',x_0'; Y',y_0'] \\ \psi\downarrow\approx & & \approx\downarrow\psi \\ H^n(SX',x_0'; \pi) & \xrightarrow{\theta} & H^q(SX',x_0'; G) \end{array}$$

It is trivial that $f_\#$ is a homomorphism when the top two sets are given group structures by the H cogroup structure of SX'. By lemma 7.7.6, it follows that both vertical maps are homomorphisms. Hence the bottom map θ is a homomorphism. ■

Let $\bar{I} \in H^1(I,\dot{I}; Z)$ be a generator and define an isomorphism

$$\tau: H^r(X,A; G') \approx H^{r+1}((X,A) \times (I,\dot{I}); G')$$

by $\tau(u) = u \times \bar{I}$. Given a cohomology operation θ of type $(n,q; \pi,G)$, its *suspension* $S\theta$ is the cohomology operation of type $(n-1, q-1; \pi,G)$ defined by $(S\theta)(u) = \tau^{-1}\theta\tau(u)$ for $u \in H^{n-1}(X,A; \pi)$. Then corollary 14 implies that the suspension of any cohomology operation is an additive cohomology operation.

We now extend theorems 10 and 12 to other spaces Y by restricting the dimension of the relative CW complex (X,A). Let Y be an n-simple $(n-1)$-connected pointed space for some $n \geq 1$ [if $n = 1$ then $\pi_1(Y,y_0)$ is abelian]. If $\iota \in H^n(Y,y_0; \pi)$ is an n-characteristic element for Y, an argument similar to that in theorem 12 shows that Y can be imbedded in a space Y' of type (π,n) having an n-characteristic element $u \in H^n(Y',y_0; \pi)$ such that $u \mid Y = \iota$. It follows that the inclusion map $Y \subset Y'$ is an $(n+1)$-equivalence. Then theorems 7.6.22 and 10 yield the following generalization of theorem 10.

15 THEOREM *Let $\iota \in H^n(Y,y_0; \pi)$ be n-characteristic for an n-simple $(n-1)$-connected pointed space Y and let (X,A) be a relative CW complex. The map*

$$\psi_\iota: [X,A; Y,y_0] \to H^n(X,A; \pi)$$

defined by $\psi_\iota[f] = f^*(\iota)$ is a bijection if $\dim (X - A) \leq n$ *and a surjection if* $\dim (X - A) \leq n + 1$. ■

For the special case $Y = S^n$ let $s^* \in H^n(S^n,p_0; \mathbf{Z})$ be a generator. Then s^* is an n-characteristic element of S^n, and we obtain the following *Hopf classification theorem.*[1]

16 COROLLARY *Let (X,A) be a relative CW complex, with* $\dim (X - A) \leq n$, *where $n \geq 1$. If $s^* \in H^n(S^n,p_0; \mathbf{Z})$ is a generator, there is a bijection*

$$\psi_{s^*}: [X,A; S^n,p_0] \approx H^n(X,A; \mathbf{Z})$$

defined by $\psi_{s^*}([f]) = f^*(s^*)$. ■

Similarly, we obtain the following generalization of theorem 12.

17 THEOREM *Let $\iota \in H^n(Y,y_0; \pi)$ be n-characteristic for an n-simple $(n-1)$-connected pointed space Y and let (X,A) be a relative CW complex, with* $\dim (X - A) \leq n + 1$. *A map $f: A \to Y$ can be extended over X if and only if $\delta f^*(\iota) = 0$ in $H^{n+1}(X,A; \pi)$.* ■

This specializes to the following *Hopf extension theorem.*

[1] See H. Hopf, Die Klassen der Abbildungen der n-dimensionalen Polyeder auf die n-dimensionale Sphäre, *Commentarii Mathematici Helvetici,* vol. 5, pp. 39–54, 1933, and H. Whitney, The maps of an n-complex into an n-sphere, *Duke Mathematical Journal,* vol. 3, pp. 51–55, 1937.

18 COROLLARY *Let (X,A) be a relative CW complex, with* $\dim (X - A) \leq n + 1$, *and let $s^* \in H^n(S^n,p_0; \mathbf{Z})$ be a generator. A map $f: A \to S^n$ can be extended over X if and only if $\delta f^*(s^*) = 0$ in $H^{n+1}(X,A; \mathbf{Z})$.* ■

2 PRINCIPAL FIBRATIONS

This section is concerned with fibrations whose fiber is an Eilenberg-MacLane space. We shall develop an obstruction theory for the lifting problem of maps of relative *CW* complexes to such fibrations. In the next section we shall show that many maps can be factored up to weak homotopy type as infinite composites of such fibrations. In this way the obstruction theory for these special fibrations leads to an obstruction theory for arbitrary maps.

For any pointed space B' there is the path fibration $PB' \xrightarrow{p'} B'$, where PB' is the space of paths in B' beginning at the base point b_0'. Under the exponential correspondence there is a one-to-one correspondence between homotopies $H: X \times I \to B'$ such that $H(x,0) = b_0'$ and maps $H': X \to PB'$, the correspondence defined by $H'(x)(t) = H(x,t)$. This easily implies the following result (which is dual to lemma 7.1.1).

1 LEMMA *A map $X \to B'$ is null homotopic if and only if it can be lifted to the path fibration $PB' \to B'$.* ■

If $\theta: B \to B'$ is a base-point-preserving map, there is a fibration $p_\theta: E_\theta \to B$ induced from the path fibration $PB' \to B'$. This induced fibration is called the *principal fibration induced by θ* and has fiber $p_\theta^{-1}(b_0) = b_0 \times \Omega B'$. A straightforward verification shows that there is a covariant functor from the category of base-point-preserving maps between pointed spaces to the subcategory of fibrations which assigns to θ the principal fibration induced by θ.

Let (X,A) be a pair and let $i: A \subset X$ be the inclusion map. Let $p_\theta: E_\theta \to B$ be the principal fibration induced by $\theta: B \to B'$. Recall that a map pair $f: i \to p_\theta$ (defined in Sec. 7.8) is a commutative square

$$\begin{array}{ccc} A & \xrightarrow{f''} & E_\theta \\ {\scriptstyle i}\downarrow & & \downarrow{\scriptstyle p_\theta} \\ X & \xrightarrow{f'} & B \end{array}$$

The set of homotopy classes $[i;p_\theta]$ of map pairs from i to p_θ is the object function of a functor of two variables contravariant in pairs (X,A) and covariant in base-point-preserving maps θ. We are interested in studying in more detail the relative-lifting problem (that is, the map $\rho: [X;E_\theta] \to [i;p_\theta]$) for this situation. Because p_θ is an induced fibration, the relative-lifting problem is equivalent to an extension problem, as shown below.

Let $p_\theta: E_\theta \to B$ be induced by $\theta: B \to B'$. For any space W a map $f: W \to E_\theta$ consists of a pair $f_1: W \to B$ and $f_2: W \to PB'$ such that $p' \circ f_2 = \theta \circ f_1$. By the exponential correspondence, f_2 corresponds to a homotopy $F: W \times I \to B'$ from the constant map to $\theta \circ f_1$. Thus, given a map $f_1: W \to B$, there is a one-to-one correspondence between liftings $f: W \to E_\theta$

of f_1 and homotopies $F\colon W \times I \to B'$ from the constant map to $\theta \circ f_1$.

Let (X,A) be a pair with inclusion map $i\colon A \subset X$ and let $f\colon i \to p_\theta$ be a map pair consisting of maps $f''\colon A \to E_\theta$ and $f'\colon X \to B$ such that $p_\theta \circ f'' = f' \circ i$. We define a map

$$\theta(f)\colon (A \times I \cup X \times \dot{I},\, X \times 0) \to (B',b_0')$$

by the conditions $\theta(f)(x,0) = b_0'$, $\theta(f)(x,1) = \theta f'(x)$, for $x \in X$, and $\theta(f) \mid A \times I$ is the homotopy from the constant map $A \to b_0'$ to the map $\theta \circ f' \circ i$ corresponding to the lifting f'' of $f' \circ i$. There is then a one-to-one correspondence between liftings of f and extensions of $\theta(f)$ over $X \times I$.

We now specialize to the case where B' is a space of type (π,n), with $n \geq 1$ and π abelian, and we let $\iota \in H^n(B',b_0';\pi)$ be n-characteristic for B'. In this case, if $\theta\colon B \to B'$ is a base-point-preserving map, the induced fibration $p_\theta\colon E_\theta \to B$ is called a *principal fibration of type* (π,n). If (X,A) is a relative CW complex, then $(X,A) \times (I,\dot{I})$ is also a relative CW complex, and given a map $g\colon A \times I \cup X \times \dot{I} \to B'$, it follows from theorem 8.1.12 that g can be extended over $X \times I$ if and only if $\delta g^*(\iota) = 0$ in $H^{n+1}((X,A) \times (I,\dot{I});\pi)$. In particular, given a map pair $f\colon i \to p_\theta$, there is a lifting of f if and only if $\delta\theta(f)^*(\iota) = 0$. The *obstruction to lifting* f, denoted by $c(f) \in H^n(X,A;\pi)$, is defined by

$$\delta\theta(f)^*(\iota) = (-1)^n \tau(c(f))$$

where $\tau\colon H^n(X,A;\pi) \approx H^{n+1}((X,A) \times (I,\dot{I});\pi)$ is the map $\tau(u) = u \times \bar{I}$, defined in Sec. 8.1 [$\bar{I} \in H^1(I,\dot{I};\mathbf{Z})$ is the generator such that if $\bar{0} \in H^0(\{0\};\mathbf{Z})$ and $\bar{1} \in H^0(\{1\};\mathbf{Z})$ are the respective unit integral cohomology classes, then, identifying $H^0(\dot{I};\mathbf{Z}) \approx H^0(\{0\};\mathbf{Z}) \oplus H^0(\{1\};\mathbf{Z})$, we have $\delta\bar{1} = \bar{I} = -\delta\bar{0}$].

2 EXAMPLE In case A is empty, a map pair $f\colon i \to p_\theta$ is just a map $f'\colon X \to B$. In this case $\theta(f)\colon X \times \dot{I} \to B'$ is such that $\theta(f)(x,0) = b_0'$ and $\theta(f)(x,1) = \theta f'(x)$. Then $\theta(f)^*(\iota) = f'^*\theta^*(\iota) \times \bar{1}$, and so, by statement 5.6.6,

$$\delta\theta(f)^*(\iota) = (-1)^n f'^*\theta^*(\iota) \times \bar{I} = (-1)^n \tau f'^*\theta^*(\iota)$$

Therefore, in this case $c(f) = f'^*\theta^*(\iota)$.

It is clear from the definition that the obstruction to lifting f is functorial in i and θ and that it vanishes if and only if there is a lifting of f. We obtain a similar cohomological criterion for the existence of a homotopy relative to f of two liftings of f.

Let $f\colon i \to p_\theta$ be a map pair, where (X,A) is a relative CW complex, with $i\colon A \subset X$, and p_θ is a principal fibration of type (π,n). Given two liftings $f_0, f_1\colon X \to E_\theta$ of f, let $g\colon i' \to p_\theta$ be the map pair consisting of the commutative square

$$\begin{array}{ccc} A \times I \cup X \times \dot{I} & \xrightarrow{g''} & E_\theta \\ {\scriptstyle i'}\downarrow & & \downarrow{\scriptstyle p_\theta} \\ X \times I & \xrightarrow{g'} & B \end{array}$$

where g' is the composite $X \times I \to X \xrightarrow{f'} B$ and g'' is the map such that $g''(x,0) = \bar{f}_0(x)$ and $g''(x,1) = \bar{f}_1(x)$ for $x \in X$ and $g''(a,t) = f''(a)$ for $a \in A$ and $t \in I$. Then $\bar{f}_0$ and $\bar{f}_1$ are homotopic relative to f if and only if g can be lifted. The obstruction to lifting g is an element $c(g) \in H^n((X,A) \times (I,\dot{I}); \pi)$, and we define the *difference between* $\bar{f}_0$ *and* $\bar{f}_1$, denoted by $d(\bar{f}_0,\bar{f}_1) \in H^{n-1}(X,A; \pi)$, by

$$c(g) = (-1)^n \tau(d(\bar{f}_0,\bar{f}_1))$$

[so $\delta\theta(g)^*(\iota) = \tau^2(d(\bar{f}_0,\bar{f}_1))$]. Then $\bar{f}_0$ and $\bar{f}_1$ are homotopic relative to f if and only if $d(\bar{f}_0,\bar{f}_1) = 0$. The difference $d(\bar{f}_0,\bar{f}_1)$ is functorial and has the following fundamental properties.

3 LEMMA *Given a map pair $f: i \to p_\theta$ and liftings $\bar{f}_0, \bar{f}_1, \bar{f}_2: X \to E_\theta$, then*

$$d(\bar{f}_0,\bar{f}_2) = d(\bar{f}_0,\bar{f}_1) + d(\bar{f}_1,\bar{f}_2)$$

PROOF Let $I_1 = [0,\frac{1}{2}]$, $\dot{I}_1 = \{0,\frac{1}{2}\}$, $I_2 = [\frac{1}{2},1]$, and $\dot{I}_2 = \{\frac{1}{2},1\}$ and define a map pair $G: \bar{i} \to p_\theta$ consisting of the commutative square

$$\begin{array}{ccc} A \times I \cup X \times (\dot{I}_1 \cup \dot{I}_2) & \xrightarrow{G''} & E_\theta \\ \downarrow i & & \downarrow p_\theta \\ X \times I & \xrightarrow{G'} & B \end{array}$$

where $G'(x,t) = f'(x)$, $G''(a,t) = f''(a)$, $G''(x,0) = \bar{f}_0(x)$, $G''(x,\frac{1}{2}) = \bar{f}_1(x)$, and $G''(x,1) = \bar{f}_2(x)$. Then $c(G) \in H^n((X,A) \times (I, \dot{I}_1 \cup \dot{I}_2); \pi)$, and by the naturality of $c(G)$ and the definition of d, we see that

$$c(G) \mid (X,A) \times (I,\dot{I}) = (-1)^n \tau(d(\bar{f}_0,\bar{f}_2))$$
$$c(G) \mid (X,A) \times (I_1,\dot{I}_1) = (-1)^n \tau_1(d(\bar{f}_0,\bar{f}_1))$$
$$c(G) \mid (X,A) \times (I_2,\dot{I}_2) = (-1)^n \tau_2(d(\bar{f}_1,\bar{f}_2))$$

where $\qquad \tau_1: H^{n-1}(X,A) \approx H^n((X,A) \times (I_1,\dot{I}_1))$

and $\qquad \tau_2: H^{n-1}(X,A) \approx H^n((X,A) \times (I_2,\dot{I}_2))$

are defined analogously to τ. From these properties, an argument similar to that used in proving that the Hurewicz homomorphism is a homomorphism (cf. theorem 7.4.3) shows that

$$\tau(d(\bar{f}_0,\bar{f}_2)) = \tau(d(\bar{f}_0,\bar{f}_1)) + \tau(d(\bar{f}_1,\bar{f}_2))$$

Since τ is an isomorphism, this is the result. ∎

4 THEOREM *Given a map pair $f: i \to p_\theta$, a lifting $\bar{f}_0: X \to E_\theta$ of f, and an element $v \in H^{n-1}(X,A; \pi)$, there is a lifting $\bar{f}_1: X \to E_\theta$ of f such that $d(\bar{f}_0,\bar{f}_1) = v$.*

PROOF The map $\theta(f): A \times I \cup X \times \dot{I} \to B'$ used in defining $c(f)$ admits an extension $h_0: X \times I \to B'$ which corresponds to the lifting $\bar{f}_0: X \to E_\theta$. We seek another extension of $\theta(f)$ which will correspond to the desired lifting $\bar{f}_1$ of f. Let $F: (A \times I \times I \cup X \times (0 \times I \cup I \times \dot{I}), X \times I \times 0) \to (B',b_0')$ be the map defined by $F(a,t,t') = \theta(f)(a,t')$ for $a \in A$ and t, $t' \in I$, and $F(x,0,t) = h_0(x,t)$, $F(x,t,0) = b_0'$, and $F(x,t,1) = h_0(x,1)$ for $x \in X$ and $t \in I$.

Because $X \times I \times 0$ is a strong deformation retract of the space $A \times I \times I \cup X \times (0 \times I \cup I \times \dot{I})$, there is a homotopy relative to $X \times I \times 0$ from F to the constant map F' from $A \times I \times I \cup X \times (0 \times I \cup I \times \dot{I})$ to b_0'.

Let $G: (X \times 1 \times I, A \times 1 \times I \cup X \times 1 \times \dot{I}) \to (B', b_0')$ be a map such that $G^*(\iota) = (-1)^{n-1}v \times \bar{1} \times \bar{I} \in H^n((X,A) \times \{1\} \times (I,\dot{I}); \pi)$ [such a map exists, by theorem 8.1.10, because $(X,A) \times \{1\} \times (I,\dot{I})$ is a relative CW complex]. There is a well-defined map

$$H': (A \times I^2 \cup X \times \dot{I}^2, A \times I \times I \cup X \times (0 \times I \cup I \times \dot{I})) \to (B', b_0')$$

such that $H' \mid X \times 1 \times I = G$. Then

$$H' \mid A \times I \times I \cup X \times (0 \times I \cup I \times \dot{I}) = F'$$

and because $(X,A) \times (I \times I, 0 \times I \cup I \times \dot{I})$ is a relative CW complex, the homotopy $F' \simeq F$ rel $X \times I \times 0$ extends to a homotopy $H' \simeq H$ rel $X \times I \times 0$, where

$$H: (A \times I \times I \cup X \times \dot{I} \times I \cup X \times I \times \dot{I}, X \times I \times 0) \to (B', b_0')$$

is an extension of F. Let $h_1: X \times I \to B'$ be defined by $h_1(x,t) = H(x,1,t)$. Since H is an extension of F, h_1 is an extension of $\theta(f)$, and hence h_1 corresponds to a lifting $\bar{f}_1$ of f.

We now show that $\bar{f}_1$ has the desired properties. The definition of the map pair $g: i' \to p_\theta$ used to define $d(\bar{f}_0,\bar{f}_1)$ is such that $\theta(g) = H$. Therefore

$$\tau^2(d(\bar{f}_0,\bar{f}_1)) = \delta H^*(\iota) = \delta H'^*(\iota)$$

H' is a map from $(A \times I^2 \cup X \times \dot{I}^2, A \times I^2 \cup X \times (0 \times I \cup I \times \dot{I}))$ to (B', b_0') whose restriction to $X \times 1 \times I$ is G. From the commutativity of the diagram [where the map μ is given by $\mu(w \times \bar{1} \times \bar{I}) = w \times \bar{I}$ for $w \in H^*(X,A)$]

$$\begin{array}{ccc} & H^n(A \times I^2 \cup X \times \dot{I}^2, A \times I^2 \cup X \times (0 \times I \cup I \times \dot{I})) & \\ & {\scriptstyle\approx}\swarrow \qquad \searrow{\scriptstyle\approx} & \\ H^n(A \times I^2 \cup X \times \dot{I}^2, X \times I \times 0) & & H^n(X \times 1 \times I, A \times 1 \times I \cup X \times 1 \times \dot{I}) \\ \delta\downarrow & & \mu\downarrow\approx \\ H^{n+1}((X,A) \times (I^2,\dot{I}^2)) & \xleftarrow{(-1)^{n-1}\tau} & H^n((X,A) \times (I,\dot{I})) \end{array}$$

it follows that

$$\delta H'^*(\iota) = (-1)^{n-1}\tau\mu G^*(\iota) = \tau(v \times \bar{I}) = \tau^2(v)$$

Since τ^2 is an isomorphism, $d(\bar{f}_0,\bar{f}_1) = v$. ∎

5 THEOREM *Let (X,A) be a relative CW complex and let (X',A) be a subcomplex, with inclusion maps $i: A \subset X$, $i': A \subset X'$, and $i'': X' \subset X$. Given a map pair $f: i \to p_\theta$ (consisting of $f'': A \to E_\theta$ and $f': X \to B$) and two liftings $\bar{g}_0, \bar{g}_1: X' \to E_\theta$ of $f \mid i': i' \to p_\theta$, let $g_0, g_1: i'' \to p_\theta$ be the map pairs consisting, respectively, of the commutative squares*

$$\begin{array}{ccc} X' & \xrightarrow{\bar g_0} & E_\theta \\ {\scriptstyle i''}\downarrow & & \downarrow{\scriptstyle p_\theta} \\ X & \xrightarrow{f'} & B \end{array} \qquad \begin{array}{ccc} X' & \xrightarrow{\bar g_1} & E_\theta \\ {\scriptstyle i''}\downarrow & & \downarrow{\scriptstyle p_\theta} \\ X & \xrightarrow{f'} & B \end{array}$$

Then

$$\delta d(\bar g_0,\bar g_1) = c(g_0) - c(g_1)$$

where δ: $H^{n-1}(X',A;\ \pi) \to H^n(X,X';\ \pi)$.

PROOF Let h: $\dot{i} \to p_\theta$ be the map pair defined by the commutative square

$$\begin{array}{ccc} A \times I \cup X' \times \dot I & \xrightarrow{h''} & E_\theta \\ {\scriptstyle \dot i}\downarrow & & \downarrow{\scriptstyle p_\theta} \\ X' \times I \cup X \times \dot I & \xrightarrow{h'} & B \end{array}$$

where $h''(a,t) = f''(a)$ for $a \in A$ and $t \in I$, $h''(x',0) = \bar g_0(x')$ and $h''(x',1) = \bar g_1(x')$ for $x' \in X'$, and $h'(x,t) = f'(x)$ for $(x,t) \in X' \times I \cup X \times \dot I$. Then $c(h) \in H^n(X' \times I \cup X \times \dot I, A \times I \cup X' \times \dot I;\ \pi)$. There is an isomorphism

$$H^n(X' \times I \cup X \times \dot I, A \times I \cup X' \times \dot I;\ \pi) \approx H^n((X',A) \times (I,\dot I);\ \pi) \oplus H^n((X,X') \times \dot I;\ \pi)$$

induced by restriction. By the naturality of the obstruction, $c(h)$ corresponds to $(-1)^n \tau d(\bar g_0,\bar g_1) = (-1)^n d(\bar g_0,\bar g_1) \times \bar I$ in the first summand and to $c(g_0) \times \bar 0 + c(g_1) \times \bar 1$ in the second summand.

Let $\bar h$: $\bar{\dot i} \to p_\theta$ be the map pair defined by the commutative square

$$\begin{array}{ccc} A \times I \cup X' \times \dot I & \xrightarrow{h''} & E_\theta \\ {\scriptstyle \bar{\dot i}}\downarrow & & \downarrow{\scriptstyle p_\theta} \\ X \times I & \xrightarrow{\bar h'} & B \end{array}$$

where $\bar h'(x,t) = f'(x)$ for $x \in X$ and $t \in I$. Then

$$c(\bar h) \in H^n(X \times I, A \times I \cup X' \times \dot I;\ \pi)$$

and by the naturality of the obstruction again,

$$c(\bar h) \mid (X' \times I \cup X \times \dot I, A \times I \cup X' \times \dot I) = c(h)$$

From the exactness of the sequence

$$H^n(X \times I, A \times I \cup X' \times \dot I) \to H^n(X' \times I \cup X \times \dot I, A \times I \cup X' \times \dot I) \xrightarrow{\delta} H^{n+1}(X \times I, X' \times I \cup X \times \dot I)$$

it follows that $\delta c(h) = 0$. Therefore, in $H^{n+1}((X,A) \times (I,\dot I);\ \pi)$ we have (using theorem 5.6.6)

$$\begin{aligned} 0 &= \delta[(-1)^n d(\bar g_0,\bar g_1) \times \bar I + c(g_0) \times \bar 0 + c(g_1) \times \bar 1] \\ &= (-1)^n \delta d(\bar g_0,\bar g_1) \times \bar I - (-1)^n c(g_0) \times \bar I + (-1)^n c(g_1) \times \bar I \end{aligned}$$

Therefore $\tau(\delta d(\bar g_0,\bar g_1) - c(g_0) + c(g_1)) = 0$, and since τ is an isomorphism, the result follows. ■

We compute the obstruction $c(f)$ explicitly for the case of a fibration $p'\colon \Omega B' \to b_0'$, where B' is a space of type (π,n), with $n > 1$. Then $\Omega B'$ is a space of type $(\pi, n-1)$, and if $\iota' \in H^{n-1}(\Omega B',\omega_0'; \pi)$ is $(n-1)$-characteristic for $\Omega B'$ and $\iota \in H^n(B',b_0'; \pi)$ is n-characteristic for B', then $\delta\iota'$ and $p^*\iota$ [where $\delta\colon H^{n-1}(\Omega B',\omega_0') \approx H^n(PB',\Omega B')$ and $p\colon (PB',\Omega B') \to (B',b_0')$] are both elements of $H^n(PB',\Omega B'; \pi)$. The characteristic elements ι and ι' are said to be *related* if $\delta\iota' = p^*\iota$. Given one of ι or ι', it is always possible to choose the other one (uniquely) so that the two are related.

6 THEOREM *Let $\iota \in H^n(B',b_0'; \pi)$ and $\iota' \in H^{n-1}(\Omega B',\omega_0'; \pi)$ be related characteristic elements. Let (X,A) be a relative CW complex, with inclusion map $i\colon A \subset X$. Given a map pair $f\colon i \to p'$, where $p'\colon \Omega B' \to b_0'$, then $c(f) = -\delta f''^*(\iota')$, where $f''\colon A \to \Omega B'$ is part of f.*

PROOF Let $\bar f\colon (A \times I, A \times \dot I) \to (PB',\Omega B')$ be the map defined by $\bar f(a,t)(t') = f''(a)(tt')$. Then

$$\theta(f)\colon (A \times I \cup X \times \dot I, X \times 0) \to (B',b_0')$$

is the map such that $\theta(f) \mid A \times I = p \circ \bar f$ and $\theta(f)(X \times \dot I) = b_0'$. Let $\bar{\bar f}\colon (A \times I \cup X \times \dot I, X \times \dot I) \to (B',b_0')$ be the map defined by $\theta(f)$ and let $\bar f'\colon (A \times \dot I, A \times 0) \to (\Omega B',\omega_0')$ be the map defined by $\bar f$. There is then a commutative diagram [in which j and j' are appropriate inclusion maps and $h_1\colon A \to (X \times \dot I, A \times 0)$ is defined by $h_1(a) = (a,1)$]

$$\begin{array}{ccccc}
 & & H^n(A \times I \cup X \times \dot I, X \times 0) & & \\
 & \overset{\theta(f)^*}{\nearrow} & j^*\uparrow & \searrow^{\delta} & \\
H^n(B',b_0') & \xrightarrow{\bar{\bar f}^*} & H^n(A \times I \cup X \times \dot I, X \times \dot I) & \xrightarrow{\delta} & H^{n+1}((X,A) \times (I,\dot I)) \\
p^*\downarrow & & j'^*\downarrow & & \approx\uparrow (-1)^{n-1}\tau \\
H^n(PB',\Omega B') & \xrightarrow{\bar f^*} & H^n(A \times I, A \times \dot I) & & H^n(X,A) \\
\delta\uparrow & & \delta\uparrow & \underset{\approx}{\nwarrow}^{(-1)^{n-1}\tau} & \uparrow\delta \\
H^{n-1}(\Omega B',\omega_0') & \xrightarrow{\bar f'^*} & H^{n-1}(A \times \dot I, A \times 0) & \underset{\approx}{\xrightarrow{h_1^*}} & H^{n-1}(A)
\end{array}$$

Furthermore, $\delta \circ \tau^{-1} \circ j'^* = \tau^{-1} \circ \delta\colon H^n(A \times I \cup X \times \dot I, X \times \dot I) \to H^n(X,A)$. Since $f'' = \bar f' \circ h_1$, then $f''^* = h_1^* \circ \bar f'^*$, and we have

$$(-1)^{n-1}\tau^{-1}\delta(\theta(f))^*(\iota) = \delta f''^*(\iota')$$

By definition, the left-hand side above equals $-c(f)$. ■

3 MOORE-POSTNIKOV FACTORIZATIONS

This section is devoted to a method of factorizing a large class of maps up to weak homotopy type as infinite composites of simpler maps, the simpler maps being of the same weak homotopy type as principal fibrations of type (π,n) for some π and n. The cohomological description of the lifting problem for these fibrations, given in the last section, will lead us ultimately to an iterative attack on general lifting problems.

Given a sequence of fibrations $E_0 \xleftarrow{p_1} E_1 \xleftarrow{p_2} \cdots$, we define

$$E_\infty = \lim_{\leftarrow} \{E_q, p_q\} = \{(e_q) \in \times E_q \mid p_q(e_q) = e_{q-1}\}$$

and we define $a_q: E_\infty \to E_q$ to be the projection of E_∞ to the qth coordinate. Then each map a_q is a fibration and $a_q = p_{q+1} \circ a_{q+1}$ for $q \geq 0$. For any space X a map $f: X \to E_\infty$ corresponds bijectively to a sequence of maps $\{f_q: X \to E_q\}_{q\geq 0}$ such that $f_q = p_{q+1} \circ f_{q+1}$ for $q \geq 0$ (given f, the sequence $\{f_q\}$ is defined by $f_q = a_q \circ f$). In particular, given a pair (X,A) with inclusion map $i: A \subset X$ and a map pair $f: i \to a_0$ consisting of the commutative square

$$\begin{array}{ccc} A & \xrightarrow{f''} & E_\infty \\ {\scriptstyle i}\downarrow & & \downarrow{\scriptstyle a_0} \\ X & \xrightarrow{f'} & E_0 \end{array}$$

a lifting $\bar{f}: X \to E_\infty$ corresponds bijectively to a sequence of maps $\{\bar{f}_q: X \to E_q\}_{q\geq 0}$ such that

(a) $\bar{f}_0 = f': X \to E_0$

(b) For $q \geq 1$ the map $\bar{f}_q: X \to E_q$ is a lifting of the map pair from i to p_q consisting of the commutative square

$$\begin{array}{ccc} A & \xrightarrow{a_q \circ f''} & E_q \\ {\scriptstyle i}\downarrow & & \downarrow{\scriptstyle p_q} \\ X & \xrightarrow{\bar{f}_{q-1}} & E_{q-1} \end{array}$$

In this way the relative-lifting problem for a map pair $f: i \to a_0$ corresponds to a sequence of relative-lifting problems for map pairs from i to p_q. In many cases the relative-lifting problems for the fibrations p_q may be simpler to deal with than the original relative-lifting problem for the fibration a_0.

A sequence of fibrations $E_0 \xleftarrow{p_1} E_1 \xleftarrow{p_2} \cdots$ is said to be *convergent* if for any $n < \infty$ there is N_n such that p_q is an n-equivalence for $q > N_n$.

Let $f: Y' \to Y$ be a map. A *convergent factorization* of f consists of a sequence $\{p_q, E_q, f_q\}_{q\geq 1}$ such that

(a) For $q > 1$, $p_q: E_q \to E_{q-1}$ is a fibration, and for $q = 1$, $p_1: E_1 \to Y$ is a fibration.

(b) For $q \geq 1$, $f_q: Y' \to E_q$ is a map, $f_q = p_{q+1} \circ f_{q+1}$ for $q \geq 1$, and $f = p_1 \circ f_1$.

(c) For any $n < \infty$ there is N_n such that f_q is an n-equivalence for $q > N_n$.

Conditions (a) and (b) imply that for $q \geq 1$, f equals the composite

$p_1 \circ \cdots \circ p_q \circ f_q$. The convergence condition (*c*) implies that, in a certain sense, the infinite composite $p_1 \circ p_2 \circ \cdots$ exists.

If $\{p_q,E_q,f_q\}_{q\geq 1}$ is a convergent factorization of a map $f\colon Y' \to Y$, then the sequence of fibrations $Y \xleftarrow{p_1} E_1 \xleftarrow{p_2} \cdots$ is convergent. The following theorem shows that any convergent sequence of fibrations is obtained in this way from a convergent factorization of some map.

1 THEOREM *If $E_0 \xleftarrow{p_1} E_1 \xleftarrow{p_2} \cdots$ is a convergent sequence of fibrations, then $\{p_q,E_q,a_q\}_{q\geq 1}$ is a convergent factorization of the map $a_0\colon E_\infty \to E_0$.*

PROOF Conditions (*a*) and (*b*) for a convergent factorization are clearly satisfied. To prove that the convergence condition (*c*) is also satisfied, given $1 \leq n < \infty$, choose N so that p_q is an $(n+1)$-equivalence if $q \geq N$. We prove that a_q is an n-equivalence for $q \geq N$. Because $a_q = p_{q+1} \circ a_{q+1}$, and p_{q+1} is an $(n+1)$-equivalence for $q \geq N$, it suffices to prove that a_N is an n-equivalence.

Let (P,Q) be a polyhedral pair such that $\dim P \leq n$ and let $\alpha\colon Q \to E_\infty$ and $\beta'_N\colon P \to E_N$ be maps such that $\beta'_N \mid Q = a_N \circ \alpha$. We now prove that there is an extension $\beta\colon P \to E_\infty$ of α such that $a_N \circ \beta = \beta'_N$. The map α corresponds to a sequence $\alpha_q = a_q \circ \alpha\colon Q \to E_q$ such that $\alpha_q = p_{q+1} \circ \alpha_{q+1}$, and to define a map $\beta\colon P \to E_\infty$ with the desired properties, we must obtain a sequence of maps $\beta_q\colon P \to E_q$ such that $\beta_q \mid Q = \alpha_q$, $\beta_q = p_{q+1} \circ \beta_{q+1}$, and $\beta_N = \beta'_N$. Such a sequence of maps $\{\beta_q\}$ is defined for $q \leq N$ by $\beta_q = p_{q+1} \circ \cdots \circ p_N \circ \beta'_N$, and for $q \geq N$ it is defined by induction on q as follows. Assuming β_q defined for some $q \geq N$, we use theorem 7.6.22 to find a map $\beta'_{q+1}\colon P \to E_{q+1}$ such that $\beta'_{q+1} \mid Q = \alpha_{q+1}$ and such that $\beta_q \simeq p_{q+1} \circ \beta'_{q+1}$ rel Q. We use the fact that p_{q+1} is a fibration (and theorem 7.2.6) to alter β'_{q+1} by a homotopy relative to Q to obtain a map $\beta_{q+1}\colon P \to E_{q+1}$ such that $\beta_{q+1} \mid Q = \alpha_{q+1}$ and such that $\beta_q = p_{q+1} \circ \beta_{q+1}$. Thus the sequence $\{\beta_q\}$ can be found, and hence a map $\beta\colon P \to E_\infty$ with the requisite properties exists.

Taking P to be a single point and Q to be empty, we see that a_N is surjective, and so a_N maps $\pi_0(E_\infty)$ surjectively to $\pi_0(E_N)$. Taking $(P,Q) = (I,\dot{I})$, we see that a_N maps $\pi_0(E_\infty)$ injectively to $\pi_0(E_N)$. Then a_N induces a one-to-one correspondence between the set of path components of E_∞ and the set of path components of E_N.

Let $e_* = (e_q) \in E_\infty$ be arbitrary and let $1 \leq k \leq n$. Taking $(P,Q) = (S^k,z_0)$ it follows that $a_{N\#}$ maps $\pi_k(E_\infty,e_*)$ epimorphically to $\pi_k(E_N,e_N)$. For $1 \leq k < n$, taking $(P,Q) = (E^{k+1},S^k)$, it follows that $\alpha_{N\#}$ maps $\pi_k(E_\infty,e_*)$ monomorphically to $\pi_k(E_N,e_N)$. Hence a_N is an n-equivalence. ∎

2 COROLLARY *Let $\{p_q,E_q,f_q\}_{q\geq 1}$ be a convergent factorization of a map $f\colon Y' \to Y$ and let $f'\colon Y' \to E_\infty$ be the map such that $a_q \circ f' = f_q$ for $q \geq 1$ and $a_0 \circ f' = f$. Then f' is a weak homotopy equivalence.*

PROOF For any $1 \leq n < \infty$ there is q such that a_q and f_q are both $(n+1)$-equivalences (by theorem 1). Then f' is also an n-equivalence (because $a_q \circ f' = f_q$). Since this is so for all n, f' is a weak homotopy equivalence. ∎

In particular, given a convergent factorization $\{p_q,E_q,f_q\}_{q\geq 1}$ of a weak fibration $p: E \to B$, there is a weak homotopy equivalence $g: p \to a_0$ consisting of the commutative square

$$\begin{array}{ccc} E & \xrightarrow{f'} & E_\infty \\ p\downarrow & & \downarrow a_0 \\ B & \xrightarrow{1} & B \end{array}$$

If (X,A) is a relative CW complex, with inclusion map $i: A \subset X$, it follows from theorem 7.8.12 that the relative-lifting problem for a map pair $h: i \to p$ is equivalent to the relative lifting problem for the map pair $g \circ h: i \to a_0$. We shall now add hypotheses which will ensure that the sequence of fibrations into which the fibration a_0 is factored (namely, the fibrations $\{p_q\}$) leads to relative-lifting problems which can be settled by the methods of the last section.

A *Moore-Postnikov sequence* of fibrations $E_0 \xleftarrow{p_1} E_1 \xleftarrow{p_2} \cdots$ is a convergent sequence of fibrations such that $p_q: E_q \to E_{q-1}$ is a principal fibration of type (G_q,n_q) for $q \geq 1$. A *Moore-Postnikov factorization* of a map $f: Y' \to Y$ is a convergent factorization $\{p_q,E_q,f_q\}_{q\geq 1}$ of f such that $E_0 \xleftarrow{p_1} E_1 \xleftarrow{p_2} \cdots$ is a Moore-Postnikov sequence of fibrations. A *Postnikov factorization* of a space Y' is a Moore-Postnikov factorization of the map $f: Y' \to Y$, where Y is the set of path components of Y' topologized by the quotient topology and f is the collapsing map. Thus, if Y' is path connected, a Postnikov factorization of Y' is a Moore-Postnikov factorization of the constant map $Y' \to y_0$.

A Moore-Postnikov factorization of a map is a factorization of the map (up to weak homotopy type) as an infinite composite of elementary maps. The relative-lifting problem associated to this sequence is thereby factored into an infinite sequence of elementary relative-lifting problems. We shall show that Moore-Postnikov factorizations exist for a large class of maps between path-connected spaces.

Let $f: Y' \to Y$ be a map between path-connected pointed spaces. For $n \geq 1$ an *n-factorization of f* is a factorization of f as a composite $Y' \xrightarrow{b'} E' \xrightarrow{p'} Y$ such that

(*a*) E' is a path-connected pointed space, p' is a fibration, and b' is a lifting of f (that is, $f = p' \circ b'$)
(*b*) $b'_\#: \pi_q(Y') \to \pi_q(E')$ is an isomorphism for $1 \leq q < n$ and an epimorphism for $q = n$ (that is, b' is an n-equivalence)
(*c*) $p'_\#: \pi_q(E') \to \pi_q(Y)$ is an isomorphism for $q > n$ and a monomorphism for $q = n$

A map $f: Y' \to Y$ between path-connected pointed spaces is said to be *simple* if $f_\#(\pi_1(Y'))$ is a normal subgroup of $\pi_1(Y)$ and the quotient group is abelian, and if (Z_f, Y') is n-simple for $n \geq 1$ (as defined in Sec. 7.3). We are heading toward a proof of the result that a simple map admits Moore-Postnikov factorizations. We need one more auxiliary concept.

Given a pointed pair (X,A) of path-connected spaces, a cohomology class $v \in H^n(X,A;\pi)$ is said to be *n-characteristic* for (X,A) if either of the following conditions hold:

(*a*) $n = 1$ and $i_\#(\pi_1(A))$ is a normal subgroup of $\pi_1(X)$ whose quotient group is mapped isomorphically onto π by the composite

$$\pi_1(X)/i_\#(\pi_1(A)) \xrightarrow{\varphi} H_1(X)/i_*(H_1(A)) \xrightarrow{j^*} H_1(X,A) \xrightarrow{h(v)} \pi$$

(*b*) $n > 1$ and the composite

$$\pi_n(X,A) \xrightarrow{\varphi} H_n(X,A) \xrightarrow{h(v)} \pi$$

is an isomorphism

In case $A = \{x_0\}$, the concept of n-characteristic element for the pair $(X,\{x_0\})$ agrees with the concept of n-characteristic element for the space X as defined in Sec. 8.1.

3 LEMMA *Let $i: A \subset X$ be a simple inclusion map between path-connected pointed spaces such that the pair (X,A) is $(n-1)$-connected, where $n \geq 1$. Then there exist cohomology classes $v \in H^n(X,A;\pi)$ which are n-characteristic for (X,A), where $\pi = \pi_1(X)/i_\#(\pi_1(A))$ for $n = 1$ and $\pi = \pi_n(X,A)$ for $n > 1$.*

PROOF If $n = 1$, it follows from the absolute Hurewicz isomorphism theorem applied to A and to X that there are isomorphisms

$$\pi_1(X)/i_\#(\pi_1(A)) \overset{\varphi}{\approx} H_1(X)/i_*(H_1(A)) \overset{j_*}{\approx} H_1(X,A)$$

By the universal-coefficient formula for cohomology, there is also an isomorphism

$$h: H^1(X,A;\pi) \approx \mathrm{Hom}\,(H_1(X,A),\pi)$$

Hence, if $\pi = \pi_1(X)/i_\#(\pi_1(A))$, there exist 1-characteristic elements $v \in H^1(X,A;\pi)$.

If $n > 1$, it follows from the relative Hurewicz isomorphism theorem and the universal-coefficient formula for cohomology that there are isomorphisms $\varphi: \pi_n(X,A) \approx H_n(X,A)$ and $h: H^n(X,A;\pi) \approx \mathrm{Hom}\,(H_n(X,A),\pi)$. Therefore, if $\pi = \pi_n(X,A)$, there are n-characteristic elements $v \in H^n(X,A;\pi)$. ∎

4 LEMMA *Let (X,A) be a pointed pair of path-connected spaces $(n-1)$-connected for some $n \geq 1$ and such that the inclusion map $i: A \subset X$ is simple. Then there is an n-factorization $A \xrightarrow{b'} E' \xrightarrow{p'} X$ of i such that p' is a principal fibration of type (π,n), where $\pi = \pi_1(X)/i_\#(\pi_1(A))$ if $n = 1$ and $\pi = \pi_n(X,A)$ if $n > 1$.*

PROOF By lemma 3, there is a class $v \in H^n(X,A;\pi)$ which is n-characteristic for (X,A). Let CA be the cone (nonreduced) over A and observe that $\{X,CA\}$ is an excisive couple in $X \cup CA$. Therefore there is an element $v' \in H^n(X \cup CA;\pi)$ corresponding to v under the isomorphisms

$$H^n(X \cup CA;\pi) \underset{\approx}{\leftarrow} H^n(X \cup CA, CA;\pi) \underset{\approx}{\rightarrow} H^n(X,A;\pi)$$

It is possible to imbed $X \cup CA$ in a space X' of type (π,n) having an n-characteristic element ι' such that $\iota' \mid X \cup CA = v'$. Let $p'\colon E' \to X$ be the principal fibration induced by the inclusion $X \subset X'$ and let $p_A'\colon E_A' \to A$ be the restriction of this fibration to A. There is a section $s\colon A \to E_A'$ such that $s(a) = (a,\omega_a)$ for $a \in A$, where ω_a is the path from x_0 to the vertex of CA followed by the path from the vertex of CA to a (that is, $\omega_a(t) = [x_0, 1 - 2t]$ for $0 \le t \le ½$ and $\omega_a(t) = [a, 2t - 1]$ for $½ \le t \le 1$). We define $b'\colon A \to E'$ to be the composite $A \xrightarrow{s} E_A' \overset{i_A}{\subset} E'$ and shall prove that $A \xrightarrow{b'} E' \xrightarrow{p'} X$ is an n-factorization of i.

The fiber of p' (and hence also of p_A') is $\Omega X'$, and we define $g\colon E_A' \to \Omega X'$ by $g(a,\omega) = \omega * (s(a))^{-1}$. Then $g \mid \Omega X'\colon \Omega X' \to \Omega X'$ is homotopic to the identity map. If $i''\colon \Omega X' \subset E_A'$ is the inclusion map, it follows from the exactness of the homotopy sequence of the fibration $p_A'\colon E_A' \to A$ that there is a direct-sum decomposition

$$\pi_q(E_A') \approx i''_\# \pi_q(\Omega X') \oplus s_\# \pi_q(A) \qquad q \ge 1$$

(This is a direct-product decomposition for $q = 1$, but we shall still write it additively.) We define a homomorphism $\lambda\colon \pi_q(X,A) \to \pi_{q-1}(\Omega X')$, where $q \ge 1$, to be the composite

$$\pi_q(X,A) \xrightarrow[\approx]{p_\#'^{-1}} \pi_q(E',E_A') \xrightarrow{\partial} \pi_{q-1}(E_A') \xrightarrow{g_\#} \pi_{q-1}(\Omega X')$$

We show that the following diagram commutes up to sign:

$$\begin{array}{ccccccc}
\pi_q(A) & \xrightarrow{i_\#} & \pi_q(X) & \xrightarrow{j_\#} & \pi_q(X,A) & \xrightarrow{\partial} & \pi_{q-1}(A) \\
\downarrow{\scriptstyle b_\#'} & & \downarrow{\scriptstyle =} & & \downarrow{\scriptstyle \lambda} & & \downarrow{\scriptstyle b_\#'} \\
\pi_q(E') & \xrightarrow{p_\#'} & \pi_q(X) & \xrightarrow{\bar\partial} & \pi_{q-1}(\Omega X') & \xrightarrow{i_\#'} & \pi_{q-1}(E')
\end{array}$$

In fact, the left-hand and middle squares are easily seen to be commutative. We shall show that $b_\#' \circ \partial = -i_\#' \circ \lambda$.

For $q = 1$ this is so because $\pi_0(A) = 0$ implies that $b_\#' \circ \partial$ is the trivial map and the fact that $j_\#$ is surjective and $i_\#' \circ \lambda \circ j_\# = i_\#' \circ \bar\partial = 0$ implies that $i_\#' \circ \lambda$ is also the trivial map. For $q > 1$ we have

$$\alpha = i''_\# g_\# \alpha + s_\# p_{A\#}' \alpha \qquad \alpha \in \pi_{q-1}(E_A')$$

Since the composite $\pi_q(E',E_A') \xrightarrow{\partial} \pi_{q-1}(E_A') \xrightarrow{i_{A\#}'} \pi_{q-1}(E')$ is trivial, it follows that for $\beta \in \pi_q(E',E_A')$

$$\begin{aligned} 0 = i_{A\#}' \partial\beta &= i_{A\#}' i''_\# g_\# \partial\beta + i_{A\#}' s_\# p_{A\#}' \partial\beta \\ &= i_\#' g_\# \partial\beta + b_\#' \partial p_\# \beta \end{aligned}$$

By definition of λ, we see that $\lambda p_\# \beta = g_\# \partial\beta$. Therefore

$$i_\#' \lambda p_\# \beta + b_\#' \partial p_\# \beta = 0$$

Since $p_\#\colon \pi_q(E',E_A') \approx \pi_q(X,A)$, this proves $b_\#' \circ \partial = -i_\#' \circ \lambda$.

A straightforward verification shows that λ is also the composite

$$\pi_n(X,A) \to \pi_n(X \cup CA, CA) \underset{\approx}{\leftarrow} \pi_n(X \cup CA) \to \pi_n(X') \underset{\approx}{\xrightarrow{\bar\partial}} \pi_{n-1}(\Omega X')$$

The construction of X' and $\iota' \in H^n(X',\pi)$ shows that there is a commutative diagram

$$\begin{array}{ccccccc}
\pi_n(X,A) & \to & \pi_n(X \cup CA, CA) & \underset{\approx}{\leftarrow} & \pi_n(X \cup CA) & \to & \pi_n(X') \\
\varphi\downarrow\approx & & \varphi\downarrow & & \varphi\downarrow & & \approx\downarrow\varphi \\
H_n(X,A) & \to & H_n(X \cup CA, CA) & \underset{\approx}{\leftarrow} & H_n(X \cup CA) & \to & H_n(X') \\
& {\scriptstyle h(v)}\searrow\approx & & \searrow & \swarrow{\scriptstyle h(v')} & & \approx\swarrow{\scriptstyle h(\iota')} \\
& & & \pi & & &
\end{array}$$

Therefore $\lambda: \pi_n(X,A) \approx \pi_{n-1}(\Omega X')$.

In case $n = 1$, $\bar\partial: \pi_1(X) \to \pi_0(\Omega X')$ is surjective [because $\pi_0(A) = 0$], and so E' is path connected. If $n > 1$, E' is path connected because $\pi_0(\Omega X') = 0$. Therefore E' is a path-connected pointed space. Since $\pi_q(\Omega X') = 0$ for $q \geq n$, it follows from the exactness of the homotopy sequence of the fibration $p': E' \to X$ that $p'_\#: \pi_q(E') \to \pi_q(X)$ is an isomorphism for $q > n$ and a monomorphism for $q = n$.

Because $\lambda: \pi_q(X,A) \to \pi_{q-1}(\Omega X')$ is a bijection for $q \leq n$ (the only nontrivial case in these dimensions being $q = n$), it follows from the five lemma and the commutativity up to sign of the diagram on page 442 that $b'_\#: \pi_q(A) \to \pi_q(E')$ is an isomorphism for $1 \leq q < n$ and an epimorphism for $q = n$. Therefore b' and p' have the properties required of an n-factorization of i. ■

5 COROLLARY *Let $g: X' \to X$ be a simple map between path-connected pointed spaces such that for some $n \geq 1$ the map $g_\#: \pi_q(X') \to \pi_q(X)$ is an isomorphism for $1 \leq q < n - 1$ and an epimorphism for $q = n - 1$. Then there is an n-factorization $X' \xrightarrow{b'} E' \xrightarrow{p'} X$ of g such that p' is a principal fibration of type (π,n) for some abelian group π.*

PROOF Let Z be the reduced mapping cylinder of g (that is, the mapping cylinder of $g \mid x'_0: x'_0 \to x_0$ has been collapsed to a point). Then (Z,X') is a pointed pair of path-connected spaces $(n-1)$-connected and with simple inclusion map $i: X' \subset Z$. By lemma 4, there is an n-factorization $X' \xrightarrow{b''} E'' \xrightarrow{p''} Z$ of i such that p'' is a principal fibration of type (π,n). Let $p': E' \to X$ be the restriction of p'' to X. Then $E' \subset E''$ is a homotopy equivalence, so there is a map $\bar b'': X' \to E'$ such that b'' is homotopic to the composite $X' \xrightarrow{b''} E' \subset E''$. Then $p' \circ \bar b''$ is easily seen to be homotopic to g. By the homotopy lifting property of p', there is a map $b': X' \to E'$ homotopic to $\bar b''$ such that $p' \circ b' = g$. Then $X' \xrightarrow{b'} E' \xrightarrow{p'} X$ is easily verified to have the requisite properties. ■

We are now ready to prove the existence of Moore-Postnikov factorizations of a simple map between path-connected pointed spaces.

6 THEOREM *Let $f\colon Y' \to Y$ be a simple map between path-connected pointed spaces. There is a Moore-Postnikov factorization $\{p_q,E_q,f_q\}_{q\geq 1}$ of f such that for $n \geq 1$ the sequence*

$$Y' \xrightarrow{f_n} E_n \xrightarrow{p_1 \circ \cdots \circ p_n} Y$$

is an n-factorization of f.

PROOF By induction on q, we prove the existence of a sequence $\{p_q,E_q,f_q\}_{q\geq 1}$ such that

(*a*) For $n = 1$ the sequence $Y' \xrightarrow{f_1} E_1 \xrightarrow{p_1} Y$ is a 1-factorization of f.

(*b*) For $n > 1$ the sequence $Y' \xrightarrow{f_n} E_n \xrightarrow{p_n} E_{n-1}$ is an n-factorization of f_{n-1}.

(*c*) For $n \geq 1$, p_n is a principal fibration of type (π_n,n) for some π_n.

Once such a sequence $\{p_q,E_q,f_q\}$ has been found, it is easy to verify that it is a Moore-Postnikov factorization of f with the desired property. Therefore we limit ourselves to proving the existence of such a sequence.

By corollary 5, with $n = 1$, there is a 1-factorization $Y' \xrightarrow{f_1} E_1 \xrightarrow{p_1} Y$ of f with p_1 a principal fibration of type $(\pi_1,1)$ for some π_1. This defines p_1, E_1, and f_1. Assume $\{p_q,E_q,f_q\}$ defined for $1 \leq q < n$, where $n > 1$, to satisfy (*a*), (*b*), and (*c*) above. By corollary 5, there is an n-factorization $Y' \xrightarrow{f_n} E_n \xrightarrow{p_n} E_{n-1}$ of f_{n-1} such that p_n is a principal fibration of type (π_n,n) for some π_n. Then p_n, E_n, and f_n have the desired properties. ∎

7 COROLLARY *Let Y' be a simple path-connected pointed space. Then Y' has a Postnikov factorization $\{p_q,E_q,f_q\}_{q\geq 1}$ in which $\pi_q(E_n) = 0$ for $q \geq n$ and $f_n\colon Y' \to E_n$ is an n-equivalence.*

PROOF If Y' is a simple space, the constant map $Y' \to y_0$ is a simple map. The result follows from theorem 6. ∎

In the above the spaces E_n approximate Y' in low dimensions. We now present an alternate method of approximating a space in high dimensions by killing low-dimensional homotopy groups.

8 COROLLARY *Let Y be a simple path-connected pointed space. There is a Moore-Postnikov sequence of fibrations $Y \xleftarrow{p_1} E_1 \xleftarrow{p_2} \cdots$ such that E_n is n-connected and $p_1 \circ \cdots \circ p_n\colon E_n \to Y$ induces isomorphisms $\pi_q(E_n) \approx \pi_q(Y)$ for $q > n$.*

PROOF If Y is a simple space, the inclusion map $y_0 \subset Y$ is a simple map. The result then follows from theorem 6. ∎

In the last result the fibration $p_1\colon E_1 \to Y$ has the homotopy properties of a universal covering space of Y. The fibration $p_1 \circ \cdots \circ p_n\colon E_n \to Y$ is a kind of "n-covering space."

4 OBSTRUCTION THEORY

In this section we show how to use Moore-Postnikov factorizations to study the relative-lifting problem. A sequence of obstructions to the existence of a lifting (or to the existence of a homotopy between two liftings) is defined iteratively, and we apply the general machinery to the special case where either the first one or the first two obstructions are the only ones that enter.

Let $p: E \to B$ be a fibration between path-connected pointed spaces and assume that p is a simple map. By theorem 8.3.6, there exist Moore-Postnikov factorizations $\{p_q, E_q, f_q\}_{q \geq 1}$ of p. By corollary 8.3.2, there is a map $p': E \to E_\infty$ which is a weak homotopy equivalence. Since $p = a_0 \circ p'$, where $a_0: E_\infty \to B$, if (X,A) is a relative CW complex, with $i: A \subset X$, it follows from theorem 7.8.12 that the relative-lifting problem for a map pair from i to p is equivalent to the relative-lifting problem for a corresponding map pair from i to a_0. Thus we are led to consider the relative-lifting problem for a map pair from i to a_0.

Let $E_0 \xleftarrow{p_1} E_1 \xleftarrow{p_2} \cdots$ be a sequence of fibrations with limit E_∞ and maps $a_q: E_\infty \to E_q$ and let (X,A) be a relative CW complex, with inclusion map $i: A \subset X$. A map pair $f: i \to a_0$ is a commutative square

$$\begin{array}{ccc} A & \xrightarrow{f''} & E_\infty \\ {\scriptstyle i}\downarrow & & \downarrow{\scriptstyle a_0} \\ X & \xrightarrow{f'} & E_0 \end{array}$$

where f'' corresponds to a collection $\{f_q'': A \to E_q\}_{q \geq 0}$ such that $p_{q+1} \circ f_{q+1}'' = f_q''$ for $q \geq 0$. For $q \geq 1$ let $f_q: i \to p_1 \circ \cdots \circ p_q$ be the map pair consisting of the commutative square

$$\begin{array}{ccc} A & \xrightarrow{f_q''} & E_q \\ {\scriptstyle i}\downarrow & & \downarrow{\scriptstyle p_1 \circ \cdots \circ p_q} \\ X & \xrightarrow{f'} & E_0 \end{array}$$

If $\bar{f}_q: X \to E_q$ is a lifting of f_q, then $p_q \circ \bar{f}_q$ is a lifting of f_{q-1} for $q > 1$ and a lifting $\bar{f}: X \to E_\infty$ of f corresponds to a sequence $\{\bar{f}_q: X \to E_q\}_{q \geq 1}$ such that

(*a*) $\bar{f}_q$ is a lifting of f_q for $q \geq 1$.
(*b*) $p_{q+1} \circ \bar{f}_{q+1} = \bar{f}_q$ for $q \geq 1$.

Given a lifting $\bar{f}_q: X \to E_q$ of f_q for $q \geq 1$, let $g(\bar{f}_q): i \to p_{q+1}$ be the map pair consisting of the commutative square

$$\begin{array}{ccc} A & \xrightarrow{f_{q+1}''} & E_{q+1} \\ {\scriptstyle i}\downarrow & & \downarrow{\scriptstyle p_{q+1}} \\ X & \xrightarrow{\bar{f}_q} & E_q \end{array}$$

A map $\bar{f}_{q+1}\colon X \to E_{q+1}$ is a lifting of $g(\bar{f}_q)$ if and only if it is a lifting of f_{q+1} such that $p_{q+1} \circ \bar{f}_{q+1} = \bar{f}_q$. Thus a sequence of maps $\{\bar{f}_q\colon X \to E_q\}_{q\geq 1}$ satisfies conditions (*a*) and (*b*) above if and only if it has the following properties:

(*c*) $\bar{f}_1$ is a lifting of f_1.
(*d*) For $q \geq 1$, $\bar{f}_{q+1}$ is a lifting of $g(\bar{f}_q)$.

We now add the hypothesis that $E_0 \xleftarrow{p_1} E_1 \xleftarrow{p_2} \cdots$ is a Moore-Postnikov sequence of fibrations. For each $q \geq 1$, p_q is then a principal fibration of type (π_q,n_q). It follows from Sec. 8.2 that f_1 can be lifted if and only if $c(f_1) \in H^{n_1}(X,A;\ \pi_1)$ is zero. The class $c(f_1)$ is called the *first obstruction to lifting f*.

Assume that for some $q > 1$ there exist liftings $\bar{f}_{q-1}\colon X \to E_{q-1}$ of the map pair $f_{q-1}\colon i \to p_1 \circ \cdots \circ p_{q-1}$. We then obtain map pairs $g(\bar{f}_{q-1})\colon i \to p_q$ and corresponding elements $c(g(\bar{f}_{q-1})) \in H^{n_q}(X,A;\ \pi_q)$. The collection $\{c(g(\bar{f}_{q-1}))\}$ corresponding to the set of all liftings $\bar{f}_{q-1}\colon X \to E_{q-1}$ of f_{q-1} is called the *qth obstruction to lifting f*. It is a subset of $H^{n_q}(X,A;\ \pi_q)$ and is defined if and only if f_{q-1} can be lifted. It is clear that there is a lifting of f_q if and only if the qth obstruction to lifting f is defined and contains the zero element of $H^{n_q}(X,A;\ \pi_q)$.

Corresponding to a Moore-Postnikov sequence of fibrations we have been led to a sequence of successive obstructions. The first obstruction is a single cohomology class, while the higher obstructions are subsets of cohomology groups. In some cases these obstructions can be effectively computed in terms of the given map pair $f\colon i \to a_0$, and this computation provides a solution of the lifting problem in these cases. In general, however, the determination of the successive obstructions involves an iterative procedure of increasing complexity and has not been effectively carried out in each case.

We illustrate this technique by applying it to the Postnikov factorization of a simple path-connected pointed space Y, given in corollary 8.3.7. There is a Postnikov factorization $\{p_q,E_q,f_q\}_{q\geq 1}$ of Y in which $\pi_q(E_m) = 0$ for $q \geq m$ and $f_m\colon Y \to E_m$ is an m-equivalence. We call this the *standard Postnikov factorization* of Y. By corollary 8.3.2, there is a weak homotopy equivalence $f'\colon Y \to E_\infty$, and so we consider the lifting problem for a map $i \to a_0$, where $i\colon A \subset X$ and $a_0\colon E_\infty \to y_0$. Since y_0 is a point, this is equivalent to the extension problem for a map $f''\colon A \to E_\infty$.

Thus we seek a sequence of maps $\bar{f}_q\colon X \to E_q$ such that $\bar{f}_1\colon X \to E_1$ is an extension of $a_1 \circ f''$ and $\bar{f}_{q+1}\colon X \to E_{q+1}$ for $q \geq 1$ is a lifting of the map pair $g(\bar{f}_q)\colon i \to p_{q+1}$ consisting of

$$\begin{array}{ccc} A & \xrightarrow{a_{q+1} \circ f''} & E_{q+1} \\ {\scriptstyle i}\downarrow & & \downarrow{\scriptstyle p_{q+1}} \\ X & \xrightarrow{\bar{f}_q} & E_q \end{array}$$

Since p_{q+1} is a principal fibration of type $(\pi_q(Y,y_0),\ q+1)$, the obstruction to lifting $g(\bar{f}_q)$ is an element of $H^{q+1}(X,A;\ \pi_q(Y,y_0))$. Hence there is defined a sequence of obstructions to extending $f''\colon A \to Y$, the $(q+1)$st obstruction being a subset of $H^{q+1}(X,A;\ \pi_q(Y,y_0))$. If Y is $(n-1)$-connected for some $n \geq 1$, the lowest-dimensional nontrivial obstruction is in $H^{n+1}(X,A;\ \pi_n(Y,y_0))$. If $\iota \in H^n(Y,y_0;\ \pi)$ is n-characteristic for such a space Y, it follows easily from theorem 8.2.6 that this lowest obstruction is $\pm\delta f''^{*}\iota$. This gives us the following generalization of theorem 8.1.17.[1]

1 THEOREM *Let $\iota \in H^n(Y,y_0;\ \pi)$ be n-characteristic for a simple $(n-1)$-connected pointed space Y, where $n \geq 1$, and let (X,A) be a relative CW complex such that $H^{q+1}(X,A;\ \pi_q(Y,y_0)) = 0$ for $q > n$. A map $f\colon A \to Y$ can be extended over X if and only if $\delta f^*(\iota) = 0$ in $H^{n+1}(X,A;\ \pi)$.*

PROOF We use the standard Postnikov factorization of Y. This leads to a sequence of obstructions to extending f which are subsets of $H^{q+1}(X,A;\ \pi_q(Y,y_0))$. Since these are all zero except $H^{n+1}(X,A;\ \pi_n(Y,y_0)) \approx H^{n+1}(X,A;\ \pi)$, the only obstruction to extending f is an element of $H^{n+1}(X,A;\ \pi)$. By the remarks above, this obstruction vanishes if and only if $\delta f^*(\iota) = 0$. ∎

Let $f_0, f_1\colon X \to Y$ be maps and define $g\colon X \times \dot{I} \to Y$ by $g(x,0) = f_0(x)$ and $g(x,1) = f_1(x)$. For any $u \in H^q(Y)$, $\delta g^*(u) = (-1)^q\tau(f_1^* u - f_0^* u)$ in $H^{q+1}(X \times I, X \times \dot{I})$. Therefore $\delta g^*(u) = 0$ if and only if $f_0^*(u) = f_1^*(u)$, and we obtain the following partial generalization of theorem 8.1.15 by applying theorem 1 to the pair $(X \times I, X \times \dot{I})$.

2 THEOREM *Let $\iota \in H^n(Y,y_0;\ \pi)$ be n-characteristic for a simple $(n-1)$-connected space Y, where $n \geq 1$, and let X be a CW complex such that $H^q(X;\ \pi_q(Y,y_0)) = 0$ for $q > n$. Then $f_0, f_1\colon X \to Y$ are homotopic if and only if $f_0^*(\iota) = f_1^*(\iota)$.* ∎

This last result gives a condition that the map $\psi_\iota\colon [X;Y] \to H^n(X,\pi)$ be injective. The condition that ψ_ι be surjective is that if $\{p_q,E_q,f_q\}_{q\geq 1}$ is the standard Postnikov factorization of Y, then any map $X \to E_{n+1}$ can be lifted. The obstructions to lifting such a map lie in $H^{q+1}(X;\ \pi_q(Y,y_0))$ for $q > n$. Therefore, by combining these, we have the following result.

3 THEOREM *Let $\iota \in H^n(Y,y_0;\ \pi)$ be n-characteristic for a simple $(n-1)$-connected space Y, where $n \geq 1$, and let X be a CW complex such that $H^q(X;\pi_q(Y)) = 0$ and $H^{q+1}(X;\pi_q(Y)) = 0$ for all $q > n$. Then there is a bijection*

$$\psi_\iota\colon [X;Y] \approx H^n(X;\pi) \qquad ∎$$

These last results have been derived by assuming hypotheses which ensure that the lowest-dimensional obstruction is the only nontrivial one. In this case we are essentially studying maps to a space of type (π,n). The case where the two lowest-dimensional obstructions are the only nontrivial obstructions is essentially the study of maps to a fibration $E \to B$ of type (G,q), where B is a

[1] See S. Eilenberg, Cohomology and continuous mappings, *Annals of Mathematics*, vol. 41, pp. 231–251, 1940.

space of type (π,n). Before we consider this, let us establish some cohomology properties of $X \times I$.

Define inclusion maps

$$A \times I \cup X \times 1 \overset{i_1}{\subset} A \times I \cup X \times \dot{I} \overset{j_1}{\subset} (A \times I \cup X \times \dot{I}, A \times I \cup X \times 1)$$

There is a weak retraction $r: A \times I \cup X \times \dot{I} \to A \times I \cup X \times 1$ defined by $r(x,t) = (x,1)$ for $(x,t) \in A \times I \cup X \times \dot{I}$ (that is, $r \circ i_1$ is homotopic to the identity map of $A \times I \cup X \times 1$). Using the exactness of the cohomology sequence of $(A \times I \cup X \times \dot{I}, A \times I \cup X \times 1)$, it follows that for an arbitrary element $u \in H^q(A \times I \cup X \times \dot{I})$ there is an associated unique element $u' \in H^q(A \times I \cup X \times \dot{I}, A \times I \cup X \times 1)$ such that

$$u = j_1^* u' + r^* i_1^* u$$

Let $h: (X,A) \to (A \times I \cup X \times \dot{I}, A \times I \cup X \times 1)$ be defined by $h(x) = (x,0)$ for $x \in X$. Then h induces an isomorphism

$$h^*: H^q(A \times I \cup X \times \dot{I}, A \times I \cup X \times 1) \approx H^q(X,A)$$

and we define an epimorphism

$$\Delta: H^q(A \times I \cup X \times \dot{I}) \to H^q(X,A)$$

by $\Delta(u) = h^* u'$, where $u' \in H^q(A \times I \cup X \times \dot{I}, A \times I \cup X \times 1)$ is the unique element associated to u. Then Δ is a natural transformation on the category of pairs (X,A).

4 LEMMA *Commutativity holds in the triangle*

$$\begin{array}{ccc} H^q(A \times I \cup X \times \dot{I}) & \xrightarrow{\delta} & H^{q+1}((X,A) \times (I,\dot{I})) \\ & \searrow^{\Delta} \quad \nearrow_{(-1)^{q+1}\tau} & \\ & H^q(X,A) & \end{array}$$

PROOF Let $\bar{r}: X \times I \to A \times I \cup X \times 1$ be defined by $\bar{r}(x,t) = (x,1)$. Then $\bar{r} \mid (A \times I \cup X \times \dot{I}) = r$, and so $r^* i_1^* u = (\bar{r}^* i_1^* u) \mid (A \times I \cup X \times \dot{I})$ for $u \in H^q(A \times I \cup X \times \dot{I})$. For any $v \in H^q(X \times I)$, $\delta(v \mid (A \times I \cup X \times \dot{I})) = 0$. Therefore, $\delta r^* i_1^* u = 0$, and to complete the proof it suffices to show that for $u' \in H^q(A \times I \cup X \times \dot{I}, A \times I \cup X \times 1)$, $\delta j_1^*(u') = (-1)^{q+1}\tau h^*(u')$. This follows from the commutativity of a diagram analogous to the one used in the proof of theorem 8.2.4. ■

5 COROLLARY *Let (X,A) be a relative CW complex, with inclusion map $i: A \subset X$, and let $p': \Omega B' \to b_0'$ be the constant map, where B' is a space of type $(\pi,n+1)$. Given a map pair $f: i \to p'$ and two liftings $f_0, f_1: X \to \Omega B'$ of f, let $g'': A \times I \cup X \times \dot{I} \to \Omega B'$ be defined by $g''(x,0) = f_0(x)$, $g''(x,1) = f_1(x)$, and $g''(a,t) = f_0(a)$. If $\iota' \in H^n(\Omega B',\omega_0'; \pi)$ and $\iota \in H^{n+1}(B',b_0'; \pi)$ are related characteristic elements, then $d(f_0,f_1) = -\Delta g''^*(\iota')$.*

PROOF Let $g: i' \to p'$ be the map pair consisting of the commutative square

$$\begin{array}{ccc} A \times I \cup X \times \dot{I} & \xrightarrow{g''} & \Omega B' \\ {\scriptstyle i'}\downarrow & & \downarrow{\scriptstyle p'} \\ X \times I & \xrightarrow{g} & b_0' \end{array}$$

From the definition of $d(f_0,f_1)$ we have $d(f_0,f_1) = (-1)^{n+1}\tau^{-1}(c(g))$. By theorem 8.2.6 $c(g) = -\delta g''^*(\iota')$, and therefore $d(f_0,f_1) = (-1)^n\tau^{-1}\delta g''^*(\iota')$. The result follows from this and lemma 4. ■

6 LEMMA *Let h_0, h_1: $(X,A) \to (A \times I \cup X \times \dot{I}, A \times I)$ be defined by $h_0(x) = (x,0)$ and $h_1(x) = (x,1)$. For any $u \in H^q(A \times I \cup X \times \dot{I}, A \times I)$*

$$\Delta(u \mid (A \times I \cup X \times \dot{I})) = h_0^*(u) - h_1^*(u)$$

PROOF There are inclusion maps

$$(A \times I \cup X \times 1, A \times I) \overset{i_1'}{\subset} (A \times I \cup X \times \dot{I}, A \times I) \overset{j_1'}{\subset}$$
$$(A \times I \cup X \times \dot{I}, A \times I \cup X \times 1)$$

and a weak retraction r': $(A \times I \cup X \times \dot{I}, A \times I) \to (A \times I \cup X \times 1, A \times I)$ defined by $r'(x,t) = (x,1)$. For $v \in H^q(A \times I \cup X \times \dot{I}, A \times I)$ there is an associated unique element $v' \in H^q(A \times I \cup X \times \dot{I}, A \times I \cup X \times 1)$ such that

$$v = j_1'^* v' + r'^* i_1'^* v$$

If k: $A \times I \cup X \times \dot{I} \subset (A \times I \cup X \times \dot{I}, A \times I)$, we then have

$$k^* v = k^* j_1'^* v' + k^* r'^* i_1'^* v = j_1^* v' + r^* i_1^* k^* v$$

Therefore $\Delta k^* v = h^* v'$. Since $h = j_1' \circ h_0$ and $h_1 = i_1' \circ r' \circ h_0$, we have

$$\Delta k^* v = h_0^* j_1'^* v' = h_0^* (v - r'^* i_1'^* v) = h_0^* v - h_1^* v \quad ■$$

7 COROLLARY *Given a map pair g: $i' \to p$, where (X,A) is a relative CW complex, i': $A \times I \subset A \times I \cup X \times \dot{I}$, and p: $E \to B$ is a principal fibration of type (G,q) induced by a map θ: $B \to B'$, let f_0, f_1: $i \to p$ be the map pairs from i: $A \subset X$ to p defined by restriction of g to $(X,A) \times 0$ and $(X,A) \times 1$, respectively. Then*

$$\Delta g'^* \theta^* (\iota) = c(f_0) - c(f_1)$$

where g': $A \times I \cup X \times \dot{I} \to B$ is part of the map pair g.

PROOF The obstruction $c(g) \in H^q(A \times I \cup X \times \dot{I}, A \times I; G)$ has the property that $c(g) \mid (A \times I \cup X \times \dot{I})$ is the obstruction to lifting g'. Therefore

$$c(g) \mid (A \times I \cup X \times \dot{I}) = g'^* \theta^* (\iota)$$

By the naturality of the obstruction, $h_0^* c(g) = c(f_0)$ and $h_1^* c(g) = c(f_1)$. The result now follows from lemma 6. ■

Let θ be a cohomology operation of type $(n,q;\ \pi,G)$. Given a cohomology class $u \in H^n(X;\pi)$, we define a map $\Delta(\theta,u)\colon H^n(X,A;\ \pi) \to H^q(X,A;\ G)$ by

$$\Delta(\theta,u)(v) = \Delta\theta(j_1^* h^{*-1}(v) + k^* u) \qquad v \in H^n(X,A;\ \pi)$$

where $k\colon A \times I \cup X \times \dot{I} \to X$ is defined by $k(x,t) = x$. In case θ is an additive cohomology operation, we have

$$\Delta(\theta,u)(v) = \Delta(j_1^* h^{*-1}\theta(v) + k^*\theta(u)) = \theta(v)$$

Therefore $\Delta(\theta,u) = \theta$ if θ is additive.

Given a cohomology operation θ of type $(n,q;\ \pi,G)$ and a cohomology class $u \in H^n(X;\pi)$, we define a map $S\Delta(\theta,u)\colon H^{n-1}(X,A;\ \pi) \to H^{q-1}(X,A;\ G)$ by the equation $S\Delta(\theta,u) = \tau^{-1} \circ \Delta(\theta,u') \circ \tau$, where $u' \in H^n(X \times I;\ \pi)$ is the image of u under the homomorphism induced by the projection $X \times I \to X$. If θ is an additive operation, then $S\Delta(\theta,u) = S\theta$. In any case, we have the following analogue of corollary 8.1.14.

8 LEMMA *If θ is a cohomology operation of type $(n,q;\ \pi,G)$ and $u \in H^n(X;\pi)$, the map*

$$S\Delta(\theta,u)\colon H^{n-1}(X,A;\ \pi) \to H^{q-1}(X,A;\ G)$$

is a homomorphism.

PROOF Let $I_1 = [0,\frac{1}{2}]$, $\dot{I}_1 = \{0,\frac{1}{2}\}$, $I_2 = [\frac{1}{2},1]$, and $\dot{I}_2 = \{\frac{1}{2},1\}$, and let $v_1,\ v_2 \in H^{n-1}(X,A;\ \pi)$. Let $v_1' = \tau_1(v_1) \in H^n((X,A) \times (I_1,\dot{I}_1))$ and let $v_2' = \tau_2(v_2) \in H^n((X,A) \times (I_2,\dot{I}_2))$, and let $v \in H^n((X,A) \times (I,\ \dot{I}_1 \cup \dot{I}_2))$ be the unique class such that $v \mid (X,A) \times (I_1,\dot{I}_1) = v_1'$ and $v \mid (X,A) \times (I_2,\dot{I}_2) = v_2'$. Then $v \mid (X,A) \times (I,\dot{I}) = \tau(v_1) + \tau(v_2)$. Since θ and Δ are both natural,

$$\Delta(\theta,u')(v) \mid (X,A) \times (I,\dot{I}) = \tau S\Delta(\theta,u)(v_1 + v_2)$$

and

$$\Delta(\theta,u')(v) \mid (X,A) \times (I_1,\dot{I}_1) = \tau_1 S\Delta(\theta,u)(v_1)$$
$$\Delta(\theta,u')(v) \mid (X,A) \times (I_2,\dot{I}_2) = \tau_2 S\Delta(\theta,u)(v_2)$$

Therefore, as in the proof of lemma 8.2.3,

$$\tau S\Delta(\theta,u)(v_1 + v_2) = \tau S\Delta(\theta,u)(v_1) + \tau S\Delta(\theta,u)(v_2)$$

Since τ is an isomorphism, this gives the result. ■

Let B be a space of type (π,n) and let $p\colon E \to B$ be a principal fibration of type (G,q) induced by a map $\bar{\theta}\colon B \to B'$. Let $\theta' = \bar{\theta}^*(\iota') \in H^q(B,b_0;\ G)$ correspond to a cohomology operation θ of type $(n,q;\ \pi,G)$ (that is, $\theta(\iota) = \theta'$). Given a CW complex X, a map $f\colon X \to B$ can be lifted to E if and only if $\theta(f^*(\iota)) = 0$. For any element $u \in H^n(X;\pi)$ such that $\theta(u) = 0$ it follows that there are liftings $f\colon X \to E$ such that $(p \circ f)^*(\iota) = u$. We shall determine how many homotopy classes of such liftings there are.

9 LEMMA *Let $f_0, f_1\colon X \to E$ be maps such that $p \circ f_0 = p \circ f_1$ (that is, f_0 and f_1 are liftings of the same map $X \to B$). Then $f_0 \simeq f_1$ if and only if there is $d \in H^{n-1}(X;\pi)$ such that $d(f_0,f_1) = S\Delta(\theta,u)(d)$, where $u = (p \circ f_0)^*(\iota)$.*

PROOF Let $F_0\colon \dot{\imath}' \to p$ be the map pair consisting of

$$\begin{array}{ccc} X \times \dot{I} & \xrightarrow{F_0''} & E \\ {\scriptstyle i'}\downarrow & & \downarrow{\scriptstyle p} \\ X \times I & \xrightarrow{F_0'} & B \end{array}$$

where $F_0''(x,0) = f_0(x)$, $F_0''(x,1) = f_1(x)$, and $F_0'(x,t) = pf_0(x)$. Then $d(f_0,f_1) = (-1)^q\tau^{-1}(c(F_0))$. It is clear that $f_0 \simeq f_1$ if and only if there is a homotopy $F_1'\colon X \times I \to B$ from $p \circ f_0$ to $p \circ f_1$ such that for the corresponding map pair $F_1\colon i' \to p$ we have $c(F_1) = 0$. Let $G'\colon (X \times \dot{I}) \times I \cup (X \times I) \times \dot{I} \to B$ be defined by $G'(x,0,t) = G'(x,1,t) = pf_0(x)$, $G'(x,t,0) = F_0'(x,t)$ and $G'(x,t,1) = F_1'(x,t)$. By corollary 7,

$$\Delta G'^*(\theta') = c(F_0) - c(F_1)$$

Thus $f_0 \simeq f_1$ if and only if there is a map $F_1'\colon X \times I \to B$ such that for the corresponding map G' we have

$$d(f_0,f_1) = (-1)^q\tau^{-1}(\Delta G'^*(\theta'))$$

It is easily verified that $G'^*(\iota) = j_1^* h^{*-1} \Delta G'^*(\iota) + k^* u'$, where $u' \in H^n(X \times I;\, \pi)$ is the image of $u = (p \circ f_0)^*(\iota)$ under the projection $X \times I \to X$. By definition,

$$\Delta G'^*(\theta') = \Delta G'^*\theta(\iota) = \Delta\theta G'^*(\iota) = \Delta(\theta,u')(\Delta G'^*(\iota))$$

Since $F_0', F_1'\colon X \times I \to B$ are two liftings of the map pair

$$\begin{array}{ccc} X \times \dot{I} & \to & B \\ \downarrow & & \downarrow \\ X \times I & \to & b_0 \end{array}$$

it follows from corollary 5 that $d(F_0',F_1') = -\Delta G'^*(\iota)$, and by theorem 8.2.4, given $d \in H^{n-1}(X;\pi)$, there is a homotopy $F_1'\colon X \times I \to B$ from $p \circ f_0$ to $p \circ f_1$ such that $\Delta G'^*(\iota) = (-1)^q\tau(d)$. Combining all of these, we see that $f_0 \simeq f_1$ if and only if there is $d \in H^{n-1}(X;\pi)$ such that

$$d(f_0,f_1) = \tau^{-1}\Delta(\theta,u')\tau(d) = S\Delta(\theta,u)(d) \quad ■$$

We summarize these results in the following classification theorem.

10 THEOREM *Let $p\colon E \to B$ be a principal fibration of type (G,q) over a space B of type (π,n) induced by a map $\bar\theta\colon B \to B'$ such that $\bar\theta^*(\iota') = \theta(\iota)$. Given a CW complex X, there is a map $\psi\colon [X;E] \to H^n(X;\pi)$ defined by $\psi[f] = (p \circ f)^*(\iota)$. Then $\operatorname{im} \psi = \{u \in H^n(X;\pi) \mid \theta(u) = 0\}$, and for every $u \in \operatorname{im} \psi$ the set $\psi^{-1}(u)$ is in one-to-one correspondence with*

$$H^{q-1}(X;G)/S\Delta(\theta,u)H^{n-1}(X;\pi)$$

PROOF We have already seen that $\operatorname{im} \psi$ is as described in the theorem. Given $u \in \operatorname{im} \psi$, let $f_0\colon X \to E$ be such that $\psi[f_0] = u$. Given any map

$f_1: X \to E$ such that $\psi[f_1] = u$, there is a map $f_1': X \to E$ homotopic to f_1 such that $p \circ f_1' = p \circ f_0$ (by the homotopy lifting property of p). To such a map f_1' we associate the element $d(f_0,f_1') \in H^{q-1}(X;G)$. In this way the set of maps $X \to E$ which are liftings of $p \circ f_0$ is mapped into $H^{q-1}(X;G)$, and by theorem 8.2.4, this map is surjective.

Two maps $f_1, f_2: X \to E$ such that $p \circ f_1 = p \circ f_0 = p \circ f_2$, are homotopic by lemma 9 if and only if $d(f_1,f_2) \in S\Delta(\theta,u)H^{n-1}(X;\pi)$. By lemma 8.2.3, $d(f_0,f_2) = d(f_0,f_1) + d(f_1,f_2)$, and so $f_1 \simeq f_2$ if and only if $d(f_0,f_1)$ and $d(f_0,f_2)$ belong to the same coset of $S\Delta(\theta,u)H^{n-1}(X;\pi)$ in $H^{q-1}(X;G)$. Hence the function which assigns the coset $d(f_0,f_1) + S\Delta(\theta,u)H^{n-1}(X;\pi)$ to a map $f_1: X \to E$ with $p \circ f_1 = p \circ f_0$ induces a bijection from $\psi^{-1}(u)$ to

$$H^{q-1}(X;G)/S\Delta(\theta,u)H^{n-1}(X;\pi) \quad \blacksquare$$

We now apply this to the complex projective space $P_m(\mathbf{C})$ for $m \geq 1$. There is a map $P_m(\mathbf{C}) \to P_\infty(\mathbf{C})$ and $P_\infty(\mathbf{C})$ is a space of type $(\mathbf{Z},2)$, by example 8.1.3. Furthermore, if ι is a characteristic element for $P_\infty(\mathbf{C})$ and B' is a space of type $(\mathbf{Z}, 2m + 2)$, there is a map $\bar\theta: P_\infty(\mathbf{C}) \to B'$ such that $\bar\theta^*(\iota') = (\iota)^{m+1}$. For the principal fibration $p: E \to P_\infty(\mathbf{C})$ induced by $\bar\theta$ there is a map $P_m(\mathbf{C}) \to E$ which is a $(2m + 2)$-equivalence. In this case the operation θ is the $(m + 1)$st-power operation, and therefore

$$\begin{aligned} S\Delta(\theta,u)(v) &= \tau^{-1}\Delta[j_1^* h^{*-1}(\tau(v)) + k^* u']^{m+1} \\ &= \tau^{-1}\Delta[(m + 1)k^*(u')^m \smile j_1^* h^{*-1}(\tau(v))] = (m + 1)u^m \smile v \end{aligned}$$

because $\tau(v) \smile \tau(v) = 0$. This gives the following application of theorem 10.

11 THEOREM *Let $\iota \in H^2(P_m(\mathbf{C});\mathbf{Z})$ be 2-characteristic for $P_m(\mathbf{C})$ and let X be a CW complex. Define ψ: $[X;P_m(\mathbf{C})] \to H^2(X;\mathbf{Z})$ by $\psi[f] = f^*(\iota)$. If* $\dim X \leq 2m + 2$, *then* $\operatorname{im} \psi = \{u \in H^2(X;\mathbf{Z}) \mid u^{m+1} = 0\}$. *If* $\dim X \leq 2m + 1$, *then ψ is surjective, and for a given $u \in H^2(X;\mathbf{Z})$, $\psi^{-1}(u)$ is in one-to-one correspondence with $H^{2m+1}(X;\mathbf{Z})/[(m + 1)u^m \smile H^1(X;\mathbf{Z})]$.* $\blacksquare$

5 THE SUSPENSION MAP

One of the most useful tools for the study of the homotopy groups of spaces is the suspension homomorphism from $\pi_q(X)$ to $\pi_{q+1}(SX)$. Iteration of this homomorphism yields a sequence of groups and homomorphisms

$$\pi_q(X) \to \pi_{q+1}(SX) \to \pi_{q+2}(S^2X) \to \cdots$$

This sequence has the stability property that from some point on, all the homomorphisms are isomorphisms. For a fixed X and q, therefore, there are only a finite number of different groups in the above sequence.

In this section we shall study the suspension map in some detail and establish the stability property. This will enable us to compute $\pi_{n+1}(S^n)$ for all n. Knowledge of these groups, combined with obstruction theory, will lead to the Steenrod classification theorem, which closes the section.[1]

We consider the category of pointed spaces and maps. There is a functorial suspension map $S: [X;Y] \to [SX;SY]$ such that $S[f] = [Sf]$. The exponential correspondence defines a natural isomorphism

$$[SX;SY] \approx [X;\Omega SY]$$

and we define $\bar{S}: [X;Y] \to [X;\Omega SY]$ to be the functorial map which is the composite of S with this isomorphism. The following result shows that $\bar{S}$ is induced by a map $Y \to \Omega SY$.

1 LEMMA *Let $\rho: Y \to \Omega SY$ be the map defined by $\rho(y)(t) = [y,t]$ for $y \in Y$ and $t \in I$. Then for any space X*

$$\bar{S} = \rho_{\#}: [X;Y] \to [X;\Omega SY]$$

PROOF The exponential correspondence takes the identity map $SY \subset SY$ to the map $\rho: Y \to \Omega SY$. Because of functorial properties of the exponential correspondence, it takes the composite

$$SX \xrightarrow{Sf} SY \subset SY$$

to the composite

$$X \xrightarrow{f} Y \xrightarrow{\rho} \Omega SY \quad \blacksquare$$

Thus, to study the suspension map S, we study the map ρ. To do this we use the fibration $PSY \to SY$, which has fiber ΩSY. With this in mind, let us investigate homology properties of fibrations over SY. We assume that $y_0 \in Y$ is a nondegenerate base point. We define $C_-Y = \{[y,t] \in SY \mid 0 \le t \le ½\}$ and $C_+Y = \{[y,t] \in SY \mid ½ \le t \le 1\}$. Then $SY = C_-Y \cup C_+Y$, and there is a homeomorphism $Y \approx C_-Y \cap C_+Y$ (sending y to $[y,½]$) by means of which we identify Y with $C_-Y \cap C_+Y$. Let $S'Y$ be the unreduced suspension defined to be the quotient space of $Y \times I$ in which $Y \times 0$ is collapsed to one point and $Y \times 1$ is collapsed to another point and let C'_-Y, C'_+Y be analogous subspaces of $S'Y$ (so $C'_-Y \cap C'_+Y = Y$). The map collapsing $S'y_0$ in $S'Y$ is a collapsing map $k: S'Y \to SY$ such that $k(C'_-Y) = C_-Y$ and $k(C'_+Y) = C_+Y$.

2 LEMMA *If y_0 is a nondegenerate base point, the collapsing map $k: S'Y \to SY$ defines a homotopy equivalence from any pair consisting of the spaces $S'Y$, C'_-Y, C'_+Y, and Y to the corresponding pair consisting of SY, C_-Y, C_+Y, and Y.*

PROOF Because y_0 is a nondegenerate base point of Y, it follows, as in the proof of lemma 7.3.2c, that $Y \times \dot{I} \cup y_0 \times I \subset Y \times I$ is a cofibration. Let $[y,t]' \in S'Y$ denote the point of $S'Y$ determined by $(y,t) \in Y \times I$ under the quotient map $k': Y \times I \to S'Y$. Let $H': (Y \times \dot{I} \cup y_0 \times I) \times I \to S'Y$ be the homotopy defined by $H'(y,0,t) = [y_0,t/2]'$, $H'(y,1,t) = [y_0, (2-t)/2]'$, and $H'(y_0,t',t) = [y_0, (1-t)t' + t/2]'$. Then H' can be extended to a homotopy

[1] The first detailed study of the suspension map appears in H. Freudenthal, Über die Klassen der Sphärenabbildungen I, *Compositio Mathematica*, vol. 5, pp. 299–314, 1937.

$H'': Y \times I \times I \to S'Y$ such that $H''(y,t,0) = k'(y,t)$. Since $H''(y,0,t) = H''(y',0,t)$ and $H''(y,1,t) = H''(y',1,t)$ for all $y, y' \in Y$, it follows that there is a homotopy $H: S'Y \times I \to S'Y$ such that $H([y,t]', t') = H''(y,t,t')$. Then H is a homotopy from the identity map of $S'Y$ to a map which collapses $S'y_0$ to a single point such that $H(S'y_0 \times I) \subset S'y_0$. Since $H(B \times I) \subset B$ if $B = C'_-Y$, C'_+Y, or Y, the result follows from lemma 7.1.5. ∎

3 COROLLARY *If Y is a path-connected space with nondegenerate base point, then SY is simply connected.*

PROOF By lemma 2, $S'Y$ and SY have the same homotopy type, so it suffices to prove that $S'Y$ is simply connected. It is clearly path connected, being the quotient of the path-connected space $Y \times I$.

Let $U_- = \{[y,t]' \in S'Y \mid t < 1\}$ and $U_+ = \{[y,t]' \in S'Y \mid 0 < t\}$. Then U_- and U_+ are each open and contractible subsets of $S'Y$. If ω is any closed path in $S'Y$ at $[y_0,\frac{1}{2}]'$, there is a partition of I, say, $0 = t_0 < t_1 < \cdots < t_n = 1$, such that for each $1 \le i \le n$ either $\omega([t_{i-1},t_i]) \subset U_-$ or $\omega([t_{i-1},t_i]) \subset U_+$. Furthermore, it can be assumed that $\omega(t_i) \in U_- \cap U_+$ for all $0 \le i \le n$ (if some $\omega(t_i)$ is not in $U_- \cap U_+$, t_i can be omitted from the partition to obtain another partition of I satisfying the original hypothesis, and iteration of this procedure will lead to a partition having the additional property demanded). Since $U_- \cap U_+$ is homeomorphic to $Y \times \mathbf{R}$, it is path connected. For each i let ω_i be a path in $U_- \cap U_+$ from $\omega(t_{i-1})$ to $\omega(t_i)$ and let ω' be the closed path at $[y_0,\frac{1}{2}]'$ defined by $\omega'(t) = \omega_i((t - t_{i-1})/(t_i - t_{i-1}))$ for $t_{i-1} \le t \le t_i$. Because U_- and U_+ are each simply connected, $\omega \mid [t_{i-1},t_i]$ is homotopic to $\omega' \mid [t_{i-1},t_i]$ relative to $\{t_{i-1},t_i\}$. Therefore $\omega \simeq \omega'$ rel $\dot I$. Since ω' is a closed path in U_+, it is null homotopic. Therefore ω is null homotopic and $S'Y$ is simply connected. ∎

4 COROLLARY *Let Y have a nondegenerate base point and let $p: E \to SY$ be a fibration. Then $\{p^{-1}(C_-Y),p^{-1}(C_+Y)\}$ is an excisive couple in E.*

PROOF Let $p': E' \to S'Y$ be the fibration induced from p by $k: S'Y \to SY$ and let $\bar k: E' \to E$ be the associated map. It follows from lemma 2 that $\bar k$ induces vertical isomorphisms in the commutative diagram

$$\begin{array}{ccc} H_*(p'^{-1}(C'_+Y),p'^{-1}(Y)) & \to & H_*(E',p'^{-1}(C'_-Y)) \\ \approx\downarrow & & \downarrow\approx \\ H_*(p^{-1}(C_+Y),p^{-1}(Y)) & \to & H_*(E,p^{-1}(C_-Y)) \end{array}$$

Since C'_+Y is a strong deformation retract of U_+ (with U_+ as defined in corollary 3) and Y is a strong deformation retract of $U_+ \cap C'_-Y$, it follows that $p'^{-1}(C'_+Y)$ and $p'^{-1}(Y)$ are strong deformation retracts of $p'^{-1}(U_+)$ and $p'^{-1}(U_+ \cap C'_-Y)$, respectively. This implies that $\{p'^{-1}(C'_-Y),p'^{-1}(C'_+Y)\}$ is an excisive couple. From the commutative diagram above, the result follows. ∎

Because C_+Y and C_-Y are contractible relative to y_0, it follows, as in Sec. 2.8, that for any fibration $p: E \to SY$ with fiber $F = p^{-1}(y_0)$ there are fiber homotopy equivalences $f_-: C_-Y \times F \to p^{-1}(C_-Y)$ and $g_+: p^{-1}(C_+Y) \to C_+Y \times F$ such that $f_- \mid y_0 \times F$ is homotopic to the map $(y_0,z) \to z$ and $g_+ \mid F$ is homotopic to the map $z \to (y_0,z)$. The corresponding *clutching function* $\mu: Y \times F \to F$ is defined by the equation

$$g_+f_-(y,z) = (y, \mu(y,z)) \qquad y \in Y, z \in F$$

Then $\mu \mid y_0 \times F$ is homotopic to the map $(y_0,z) \to z$.

5 THEOREM *Let $p: E \to SY$ be a fibration with $F = p^{-1}(y_0)$, where y_0 is a nondegenerate base point of Y. If $\mu: Y \times F \to F$ is a clutching function for p, there are exact sequences (any coefficient module)*

$$\cdots \to H_q(E) \to H_q(C_-Y \times F, Y \times F) \xrightarrow{\mu_*\partial} H_{q-1}(F) \xrightarrow{i_*} H_{q-1}(E) \to \cdots$$

$$\cdots \to H^q(E) \xrightarrow{i^*} H^q(F) \xrightarrow{\delta\mu^*} H^{q+1}(C_-Y \times F, Y \times F) \to H^{q+1}(E) \to \cdots$$

PROOF Consider the exact homology sequence of (E,F)

$$\cdots \to H_q(F) \xrightarrow{i_*} H_q(E) \to H_q(E,F) \xrightarrow{\partial} H_{q-1}(F) \to \cdots$$

Using homotopy properties and corollary 4, there are isomorphisms induced by inclusion maps

$$H_q(E,F) \underset{\approx}{\to} H_q(E,p^{-1}(C_+Y)) \underset{\approx}{\leftarrow} H_q(p^{-1}(C_-Y),p^{-1}(Y))$$

There is also a homotopy equivalence

$$f_-: (C_-Y \times F, Y \times F) \to (p^{-1}(C_-Y), p^{-1}(Y))$$

and a commutative diagram

$$\begin{array}{ccccccc}
H_q(E,F) & \underset{\approx}{\to} & H_q(E,p^{-1}(C_+Y)) & \underset{\approx}{\leftarrow} & H_q(p^{-1}(C_-Y),p^{-1}(Y)) & \underset{\approx}{\xleftarrow{f_{-*}}} & H_q((C_-Y,Y) \times F) \\
\downarrow\partial & & \downarrow\partial & & \downarrow\partial & & \downarrow\partial \\
H_{q-1}(F) & \underset{\approx}{\xrightarrow{j_*}} & H_{q-1}(p^{-1}(C_+Y)) & \leftarrow & H_{q-1}(p^{-1}(Y)) & \underset{\approx}{\xleftarrow{(f_- \mid Y \times F)_*}} & H_{q-1}(Y \times F)
\end{array}$$

There is also a homotopy equivalence $g_+: p^{-1}(C_+Y) \to C_+Y \times F$ and isomorphisms

$$H_{q-1}(p^{-1}(C_+Y)) \underset{\approx}{\xrightarrow{g_{+*}}} H_{q-1}(C_+Y \times F) \underset{\approx}{\to} H_{q-1}(F)$$

where the right-hand homomorphism is induced by projection to the second factor. Because $g_+ \mid F$ is homotopic to the map $z \to (y_0,z)$, the above composite equals j_*^{-1}. By definition, μ is the composite

$$Y \times F \xrightarrow{f_- \mid Y \times F} p^{-1}(Y) \subset p^{-1}(C_+Y) \xrightarrow{g_+} C_+Y \times F \to F$$

Therefore there is a commutative diagram

$$\begin{array}{ccc}
H_q(E,F) & \underset{\approx}{\to} & H_q((C_-Y,Y) \times F) \\
\downarrow\partial & & \downarrow\partial \\
H_{q-1}(F) & \xleftarrow{\mu_*} & H_{q-1}(Y \times F)
\end{array}$$

The desired exact sequence for homology follows on replacing $H_q(E,F)$ by $H_q((C_-Y,Y) \times F)$ and ∂ by $\mu_* \partial$ in the homology sequence of (E,F). A similar argument establishes the exactness of the cohomology sequence. ■

Specializing to the case where $Y = S^{n-1}$, by lemma 1.6.6, $S(S^{n-1})$ is homeomorphic to S^n, and we obtain the following exact *Wang sequence* of a fibration over S^n.

6 COROLLARY *Let $p: E \to S^n$ be a fibration with fiber F. There are exact sequences*

$$\cdots \to H_q(F) \xrightarrow{i_*} H_q(E) \to H_{q-n}(F) \to H_{q-1}(F) \to \cdots$$

$$\cdots \to H^q(E) \xrightarrow{i^*} H^q(F) \xrightarrow{\theta} H^{q-n+1}(F) \to H^{q+1}(E) \to \cdots$$

If the second sequence has coefficients in a commutative ring with a unit, then

$$\theta(u \smile v) = \theta(u) \smile v + (-1)^{(n-1)\deg u} u \smile \theta(v)$$

PROOF Letting $Y = S^{n-1}$ in theorem 5, we have (C_-Y,Y) homeomorphic to (E^n,S^{n-1}). Therefore

$$H_q((C_-Y,Y) \times F) \approx H_q((E^n,S^{n-1}) \times F) \approx H_{q-n}(F)$$

and the exact sequences result from the exact sequences of theorem 5 on replacing $H_q(C_-Y \times F,\ Y \times F)$ and $H^q(C_-Y \times F,\ Y \times F)$ by $H_{q-n}(F)$ and $H^{q-n}(F)$, respectively. The additional fact concerning θ results from the observation that for the map $\mu^*: H^q(F) \to H^q(S^{n-1} \times F)$ the definitions are such that

$$\mu^*(u) = 1 \times u + s^* \times \theta(u)$$

where $s^* \in H^{n-1}(S^{n-1})$ is a suitable generator. Then, since $s^* \smile s^* = 0$,

$$\begin{aligned} 1 \times (u \smile v) + s^* \times \theta(u \smile v) &= \mu^*(u \smile v) \\ &= [1 \times u + s^* \times \theta(u)] \smile [1 \times v + s^* \times \theta(v)] \\ &= 1 \times (u \smile v) + s^* \times [\theta(u) \smile v + (-1)^{(n-1)\deg u} u \smile \theta(v)] \end{aligned}$$

This implies the multiplicative property of θ. ■

We now specialize to the path fibration $p: PSY \to SY$ with fiber ΩSY. In this case there is the following simple expression for a clutching function.

7 LEMMA *Let $s_-: C_-Y \to p^{-1}(C_-Y)$ and $s_+: C_+Y \to p^{-1}(C_+Y)$ be sections of the fibration $p: PSY \to SY$ such that $s_-(y_0)$ and $s_+(y_0)$ are both null homotopic loops. Then the map $\mu: Y \times \Omega SY \to \Omega SY$ defined by*

$$\mu(y,\omega) = (\omega * s_-(y)) * s_+(y)^{-1}$$

is a clutching function for p.

PROOF Such sections exist because C_-Y and C_+Y are contractible relative to y_0. We define fiber-preserving maps

$$f_-: C_-Y \times \Omega SY \to p^{-1}(C_-Y) \qquad g_-: p^{-1}(C_-Y) \to C_-Y \times \Omega SY$$
$$f_+: C_+Y \times \Omega SY \to p^{-1}(C_+Y) \qquad g_+: p^{-1}(C_+Y) \to C_+Y \times \Omega SY$$

by $f_-(z,\omega) = \omega * s_-(z)$ and $g_-(\omega) = (p(\omega), \omega * (s_-p(\omega))^{-1})$ and $f_+(z,\omega) = \omega * s_+(z)$ and $g_+(\omega) = (p(\omega), \omega * (s_+p(\omega))^{-1})$, respectively. It is easy to verify that $g_- \circ f_-$ is fiber homotopic to the identity map of $C_-Y \times \Omega SY$ and $f_- \circ g_-$ is fiber homotopic to the identity map of $p^{-1}(C_-Y)$. Therefore f_- is a fiber homotopy equivalence. Similarly, g_+ is a fiber homotopy equivalence. Furthermore, $f_-(y_0,\omega) = \omega * s_-(y_0)$ is homotopic to the map $(y_0,\omega) \to \omega$ because $s_-(y_0)$ is null homotopic. Similarly, for $\omega \in \Omega SY$, $g_+(\omega) = (y_0, \omega * s_+(y_0)^{-1})$ is homotopic to the map $\omega \to (y_0,\omega)$. Therefore the composite

$$Y \times \Omega SY \xrightarrow{f_-} p^{-1}(Y) \xrightarrow{g_+} Y \times \Omega SY \to \Omega SY$$

is a clutching function for p. This composite is the map

$$(y,\omega) \to (\omega * s_-(y)) * s_+(y)^{-1} \quad \blacksquare$$

Let s_- and s_+ be sections as in lemma 7 and let $\mu': Y \to \Omega SY$ be defined by $\mu'(y) = s_-(y) * s_+(y)^{-1}$. μ' is called a *characteristic map* for the fibration $p: PSY \to SY$.

8 COROLLARY *Let $\mu': Y \to \Omega SY$ be a characteristic map for the fibration $p: PSY \to SY$. The map $Y \times \Omega SY \to \Omega SY$ sending (y,ω) to $\omega * \mu'(y)$ is homotopic to a clutching function for p.*

PROOF This follows from lemma 7, because the map

$$(y,\omega) \to (\omega * s_-(y)) * s_+(y)^{-1}$$

is clearly homotopic to the map $(y,\omega) \to \omega * (s_-(y) * s_+(y)^{-1}) = \omega * \mu'(y)$. $\blacksquare$

The following theorem is the main part of the proof of the suspension theorem.

9 THEOREM *Let Y be n-connected for some $n \geq 0$ and let y_0 be a nondegenerate base point of Y. If $\mu': Y \to \Omega SY$ is a characteristic map for the fibration $p: PSY \to SY$, then μ' induces an isomorphism*

$$\mu'_*: H_q(Y) \approx H_q(\Omega SY) \qquad q \leq 2n + 1$$

PROOF By corollary 3, SY is simply connected. By corollary 4, $\{C_-Y, C_+Y\}$ is an excisive couple, and from the exactness of the reduced Mayer-Vietoris sequence, $\tilde{H}_q(SY) \approx \tilde{H}_{q-1}(Y)$. Combining these with the absolute Hurewicz isomorphism theorem, SY is $(n+1)$-connected. Therefore ΩSY is n-connected. Because PSY is contractible, it follows from the version of theorem 5, using reduced modules, that there is an isomorphism

$$\mu_* \partial: H_q((C_-Y,Y) \times \Omega SY) \approx \tilde{H}_{q-1}(\Omega SY)$$

If ω_0 is the constant loop, then because ΩSY is n-connected and (C_-Y,Y) is $(n+1)$-connected, it follows from the Künneth theorem that the inclusion map $(C_-Y,Y) \times \omega_0 \subset (C_-Y,Y) \times \Omega SY$ induces an isomorphism

$$H_q((C_-Y,Y) \times \omega_0) \approx H_q((C_-Y,Y) \times \Omega SY) \qquad q \leq 2n + 2$$

Let $\mu: Y \times \Omega SY \to \Omega SY$ be a clutching function which is homotopic to the map $(y,\omega) \to \omega * \mu'(y)$ (such a μ exists, by corollary 8). Since $\mu(y,\omega_0)$ is homotopic to the map $y \to \mu'(y)$, there is a commutative diagram

$$\begin{array}{ccccc}
H_q(C_-Y,Y) & \xrightarrow{\approx} & H_q((C_-Y,Y) \times \omega_0) & \to & H_q((C_-Y,Y) \times \Omega SY) \\
\downarrow \partial & & \downarrow \partial & & \downarrow \partial \\
\tilde{H}_{q-1}(Y) & \xrightarrow{\approx} & \tilde{H}_{q-1}(Y \times \omega_0) & \to & \tilde{H}_{q-1}(Y \times \Omega SY) \\
& \searrow{\scriptstyle \mu'_*} & & \swarrow{\scriptstyle \mu_*} & \\
& & \tilde{H}_{q-1}(\Omega SY) & &
\end{array}$$

The result follows from the commutativity of this diagram. ∎

10 COROLLARY *Let Y have a nondegenerate base point. If Y is n-connected for $n \geq 0$, the map $\rho: Y \to \Omega SY$ induces an isomorphism*

$$\rho_*: H_q(Y) \approx H_q(\Omega SY) \qquad q \leq 2n + 1$$

PROOF Let $s_-: C_-Y \to p^{-1}(C_-Y)$ and $s_+: C_+Y \to p^{-1}(C_+Y)$ be the sections defined by $s_-[y,t](t') = [y,tt']$ and $s_+[y,t](t') = [y, 1 - t' + tt']$. The corresponding characteristic map is equal to the map $\rho: Y \to \Omega SY$. The result follows from theorem 9. ∎

We are now ready for the following *suspension theorem.*[1]

11 THEOREM *Let Y be n-connected for $n \geq 1$ with a nondegenerate base point and let X be a pointed CW complex. Then the suspension map*

$$S: [X;Y] \to [SX;SY]$$

is surjective if $\dim X \leq 2n + 1$ *and bijective if* $\dim X \leq 2n$.

PROOF Because Y and ΩSY are simply connected, it follows from corollary 10 and the Whitehead theorem that ρ is a $(2n + 1)$-equivalence. The result follows from corollary 7.6.23 and lemma 1. ∎

Let Y be a space with a nondegenerate base point. Then SY also has a nondegenerate base point and is path connected, S^2Y is simply connected, and S^mY is $(m - 1)$-connected. If X is a CW complex, so is S^mX, and $\dim (S^mX) = m + \dim X$. Hence, if X is finite dimensional and $m \geq 2 + \dim X$, it follows from theorem 11 that $S: [S^mX; S^mY] \approx [S^{m+1}X; S^{m+1}Y]$. Therefore, for any finite-dimensional CW complex X the sequence

$$[X;Y] \xrightarrow{S} [SX;SY] \xrightarrow{S} \cdots \xrightarrow{S} [S^mX;S^mY] \xrightarrow{S} \cdots$$

[1] For a general relative form of this theorem see E. Spanier and J. H. C. Whitehead, The theory of carriers and S-theory, in "Algebraic Geometry and Topology" (a symposium in honor of S. Lefschetz), Princeton University Press, Princeton, N.J., 1957, pp. 330–360.

consists of isomorphisms from some point on. Taking $X = S^{n+k}$ and $Y = S^n$ and recalling that the suspension of a sphere is a sphere, we see that there is a sequence

$$\pi_{n+k}(S^n) \xrightarrow{S} \pi_{n+k+1}(S^{n+1}) \xrightarrow{S} \cdots$$

consisting of isomorphisms from some point on. The direct limit of this sequence is called the *k-stem*. It follows from theorem 11 that the k-stem is isomorphic to $\pi_{2k+2}(S^{k+2})$. In particular, the 0-stem is infinite cyclic. The following result determines the 1-stem.

12 THEOREM $\pi_4(S^3) \approx \mathbf{Z}_2$.

PROOF Let $u_0 \in H^0(\Omega S^3)$ be the unit integral class and define generators $u_i \in H^{2i}(\Omega S^3)$, by induction on i from the exactness of the Wang sequence in corollary 6 for the fibration $PS^3 \to S^3$, by the equation

$$\theta(u_{i+1}) = u_i \qquad i \geq 0$$

Because θ is a derivation, $\theta(u_1 \smile u_1) = 2u_1$, whence $u_1 \smile u_1 = 2u_2$. We know $\pi_2(\Omega S^3) \approx \pi_3(S^3)$ is infinite cyclic. It follows that ΩS^3 can be imbedded in a space X of type $(\mathbf{Z},2)$ such that the inclusion map $\Omega S^3 \subset X$ induces an isomorphism $\pi_2(\Omega S^3) \approx \pi_2(X)$. Since $P_\infty(\mathbf{C})$ is also a space of type $(\mathbf{Z},2)$, it follows that $H^*(X) \approx H^*(P_\infty(\mathbf{C})) \approx \lim_{\leftarrow} \{H^*(P_j(\mathbf{C}))\}$ is a polynomial algebra with a single generator $v \in H^2(X)$, and v can be chosen so that $v \mid \Omega S^3 = u_1$.

An easy computation using the exact cohomology sequence of $(X,\Omega S^3)$ establishes that $H^q(X,\Omega S^3) = 0$ for $q < 5$ and $H^5(X,\Omega S^3) \approx \mathbf{Z}_2$. By the universal-coefficient formula, $H_q(X,\Omega S^3) = 0$ for $q < 4$ and $H_4(X,\Omega S^3) \approx \mathbf{Z}_2$. By the relative Hurewicz isomorphism theorem, $\pi_4(X,\Omega S^3) \approx \mathbf{Z}_2$. Because $\pi_3(X) = 0 = \pi_4(X)$, we have $\pi_4(X,\Omega S^3) \overset{\partial}{\approx} \pi_3(\Omega S^3) \approx \pi_4(S^3)$. ∎

The $(n-2)$-fold suspension of a generator of $\pi_3(S^2)$ is a generator of $\pi_{n+1}(S^n)$ (because $S: \pi_3(S^2) \to \pi_4(S^3)$ is an epimorphism, by theorem 11). Attaching a cell to S^n by this map must, therefore, kill $\pi_{n+1}(S^n)$. The resulting CW complex has the same homotopy type as the $(n-2)$-fold suspension of the complex projective plane $P_2(\mathbf{C})$. Therefore we have proved the following result.

13 COROLLARY $\pi_{n+1}(S^{n-2}(P_2(\mathbf{C}))) = 0 \qquad n \geq 2$ ∎

We want to classify maps of an $(n+1)$-complex into S^n. For $n = 2$ this is given by the case $m = 1$ of theorem 8.4.11. By using the standard Postnikov factorization of S^n, we are reduced to classifying maps of an $(n+1)$-complex into E, where $p: E \to B$ is a principal fibration of type $(\mathbf{Z}_2, n+2)$, with base space B a space of type $(\mathbf{Z},n)$. This fibration determines a cohomology operation θ_n of type $(n, n+2; \mathbf{Z},\mathbf{Z}_2)$.

14 LEMMA *For $n > 2$ the cohomology operation θ_n is $Sq^2 \circ \mu_*$, where $\mu_*: H^n(X;\mathbf{Z}) \to H^n(X;\mathbf{Z}_2)$ is induced by the coefficient homomorphism $\mu: \mathbf{Z} \to \mathbf{Z}_2$.*

PROOF $S^n \subset S^{n-2}(P_2(\mathbf{C}))$ is not a retract, by theorem 12 and corollary 13. Therefore θ_n: $H^n(S^{n-2}(P_2(\mathbf{C}));\mathbf{Z}) \to H^{n+2}(S^{n-2}(P_2(\mathbf{C}));\mathbf{Z}_2)$ is nontrivial (if θ_n were trivial, there would be a map f: $S^{n-2}(P_2(\mathbf{C})) \to S^n$ such that

$$f^*: H^n(S^n;\mathbf{Z}) \approx H^n(S^{n-2}(P_2(\mathbf{C}));\mathbf{Z})$$

is inverse to the restriction map $H^n(S^{n-2}(P_2(\mathbf{C}));\mathbf{Z}) \approx H^n(S^n;\mathbf{Z})$, and such a map f would be homotopic to a weak retraction). Since $Sq^2 \circ \mu_*$ is also nontrivial, it follows that $\theta_n = Sq^2 \circ \mu_*$ in the space $S^{n-2}(P_2(\mathbf{C}))$.

The rest of the argument follows by showing that $S^{n-2}(P_2(\mathbf{C}))$ is universal for θ_n and $Sq^2 \circ \mu_*$. Let X be any CW complex of dimension $\leq n+2$ and let $u \in H^n(X;\mathbf{Z})$. Because $\pi_{n+1}(S^{n-2}(P_2(\mathbf{C}))) = 0$, there is a map f: $X \to S^{n-2}(P_2(\mathbf{C}))$ such that $f^*v = u$, where v is a generator of $H^n(S^{n-2}(P_2(\mathbf{C})))$. By the naturality of θ_n and $Sq^2 \circ \mu_*$, it follows that

$$\theta_n(u) = \theta_n f^* v = f^* \theta_n v = f^* Sq^2 \mu_* v = Sq^2 \mu_*(u)$$

Since this is true for every CW complex of dimension $\leq n+2$ and θ_n and $Sq^2 \circ \mu_*$ are operations of type $(n, n+2; \mathbf{Z},\mathbf{Z}_2)$, it is true for every CW complex. ■

Combining lemma 14 with theorem 8.4.10 yields the following *Steenrod classification theorem.*[1]

15 THEOREM *Let $s^* \in H^n(S^n;\mathbf{Z})$ be a generator, where $n > 2$, and let X be a CW complex. Then the map ψ: $[X;S^n] \to H^n(X;\mathbf{Z})$ has image equal to $\{u \in H^n(X;\mathbf{Z}) \mid Sq^2\mu_*(u) = 0\}$ if $\dim X \leq n+2$, and if $\dim X \leq n+1$, $\psi^{-1}(u)$ is in one-to-one correspondence with $H^{n+1}(X;\mathbf{Z}_2)/Sq^2\mu_* H^{n-1}(X;\mathbf{Z})$.* ■

EXERCISES

A SPACES OF TYPE (π,n)

1 For p an integer let $L_n(p)$ be the generalized lens space $L_n(p) = L(p, \overbrace{1, \ldots, 1}^{n})$. Show that $L_n(p) \subset L_{n+1}(p)$ and that $L_\infty(p) = \bigcup_n L_n(p)$ topologized with the topology coherent with $\{L_n(p)\}$ is a space of type $(\mathbf{Z}_p,1)$.

2 If X is a CW complex of type (π,n) for $n > 1$ and Y is a CW complex, prove that

$$\pi_n(X \vee Y) \approx \pi_n(Y) \oplus \bigoplus_{\lambda \in \pi_1(Y)} \pi_\lambda$$

where $\pi_\lambda = \pi$ for each $\lambda \in \pi_1(Y)$.

3 Given a sequence of groups $\{\pi_q\}_{q \geq 1}$, with π_q abelian for $q > 1$, and given an action of π_1 as a group of operators on π_q for $q > 1$, prove that there is a space Y which realizes this sequence (that is, $\pi_q(Y) \approx \pi_q$ and $\pi_1(Y)$ acting on $\pi_q(Y)$ corresponds to the action of π_1 on π_q).

[1] See N. E. Steenrod, Products of cocycles and extensions of mappings, *Annals of Mathematics*, vol. 48, pp. 290–320, 1947.

B EXACT SEQUENCES CONTAINING $g_{\#}$

Let $g\colon (Y,B) \to (Y',B')$ be a base-point-preserving map and let $g' = g \mid Y\colon Y \to Y'$ and $g'' = g \mid B\colon B \to B'$.

1 Prove that $E_{g''}$ is a subspace of $E_{g'}$ and $p_{g''} = p_{g'} \mid E_{g''}$.

2 Define $p\colon (E_{g'},E_{g''}) \to (Y,B)$ so that $p \mid E_{g'} = p_{g'}$ and $j\colon (\Omega Y',\Omega B') \to (E_{g'},E_{g''})$ so that $j(\omega) = (y_0,\omega)$. Prove that there is an exact sequence

$$(\Omega Y,\Omega B) \xrightarrow{\Omega g} (\Omega Y',\Omega B') \xrightarrow{j} (E_{g'},E_{g''}) \xrightarrow{p} (Y,B) \xrightarrow{g} (Y',B')$$

3 Prove that there is an exact sequence

$$\cdots \xrightarrow{\Omega^n j} \Omega^n(E_{g'},E_{g''}) \xrightarrow{\Omega^n p} \Omega^n(Y,B) \xrightarrow{\Omega^n g} \Omega^n(Y',B') \to \cdots \xrightarrow{g} (Y',B')$$

4 Define a map $(\Omega Y' \times E_{g'}, \Omega B' \times E_{g''}) \to (E_{g'},E_{g''})$ sending $\omega \times (y_0,\omega')$ to $(y_0, \omega * \omega')$ and use this to define an action $a \top b$ of $[X,A; \Omega Y',\Omega B']$ on the left on $[X,A; E_{g'},E_{g''}]$. Prove that $p_{\#}(b_1) = p_{\#}(b_2)$ for $b_1, b_2 \in [X,A; E_{g'},E_{g''}]$ if and only if there is $a \in [X,A; \Omega Y',\Omega B']$ such that $b_1 = a \top b_2$.

5 Prove that $j_{\#}(a_1) = j_{\#}(a_2)$ for $a_1, a_2 \in [X,A; \Omega Y',\Omega B']$ if and only if there is $c \in [X,A; \Omega Y,\Omega B]$ such that $a_1 = a_2(\Omega g)_{\#}(c)$.

C EXAMPLES

1 Find an example of an n-dimensional polyhedron X, with $n > 1$, and a map $f\colon X \to S^n$ such that $f_*\colon \tilde{H}_*(X) \to \tilde{H}_*(S^n)$ is trivial but f is not homotopic to a constant map.

2 Let X be an n-dimensional polyhedron. Prove that $f, g\colon X \to S^n$ are homotopic if and only if $f_* = g_*\colon H_n(X;G) \to H_n(S^n;G)$ for $G = \mathbf{Z}_p$ with p a prime, and for $G = \mathbf{R}$.

3 Compute the cohomotopy group $\pi^{2m-1}(P_m(\mathbf{C}))$ for $m \geq 2$.

4 Let (Y,B) be a pair which is $(n-1)$-connected for $n \geq 2$, with a simple inclusion map $B \subset Y$, and let $\iota \in H^n(Y,B;\pi)$ be n-characteristic for (Y,B). If (X,A) is a relative CW complex and $f\colon (X,A) \to (Y,B)$, prove that $f^*(\iota) \in H^n(X,A;\pi)$ is the first obstruction to deforming f relative to A to a map from X to B.

D SUSPENSION

1 Let X be an $(n-1)$-connected CW complex of dimension $\leq 2n-1$. Prove that there is a CW complex Y such that SY has the same homotopy type as X. [*Hint:* Show that X has the same homotopy type as a CW complex X', with $(X')^{n-1}$ a single point. Construct Y inductively by desuspending the attaching maps of the cells of X'.]

2 Let A and B be closed subsets of a space X such that $X = A \cup B$. Assume that $f, g\colon X \to Y$ are such that $f(A) = y_0 = g(B)$ and define $h\colon X \to Y$ so that $h \mid A = g \mid A$ and $h \mid B = f \mid B$. Prove that, in $[SX;SY]$,

$$[Sf][Sg] = [Sh]$$

3 Let X and Y be path-connected pointed CW complexes. Prove that a map $f\colon X \to Y$ has the property that $S^k f\colon S^k X \to S^k Y$ is a homotopy equivalence for some $k \geq 0$ if and only if $Sf\colon SX \to SY$ is a homotopy equivalence. [*Hint:* Show that either condition is equivalent to the condition $f_*\colon H_*(X) \approx H_*(Y)$.]

4 Let X and Y be path-connected pointed CW complexes and let $p_1\colon X \times Y \to X$ and $p_2\colon X \times Y \to Y$ be the projections and $k\colon X \times Y \to X \# Y = X \times Y / X \vee Y$ the collapsing map. Regard all three as maps into $X \vee Y \vee (X \# Y)$ and prove that

$$((Sp_1) * (Sp_2)) * (Sk): S(X \times Y) \to S(X \vee Y \vee X \# Y)$$

is a homotopy equivalence.

5 Show that there exist CW complexes with different homotopy type whose suspensions have the same homotopy type.

E THE SUSPENSION CATEGORY

Let $\{X,A;\ Y,B\} = \lim_{\to} [S^kX,S^kA;\ S^kY,S^kB]$, and for q an integer (positive or negative) let $\{X,A;\ Y,B\}_q = \lim_{\to} [S^{k+q}X,S^{k+q}A;\ S^kY,S^kB]$. If α: $S^{k+q}(X,A) \to S^k(Y,B)$, then $\{\alpha\}$ will denote the corresponding element of $\{X,A;\ Y,B\}_q$.

1 Prove that there is a pairing

$$\{Y,B;\ Z,C\}_p \otimes \{X,A;\ Y,B\}_q \to \{X,A;\ Z,C\}_{p+q}$$

sending $\{\alpha\} \otimes \{\beta\}$ to $\{\alpha \circ \beta\}$, where

$$S^{p+q+k}(X,A) \xrightarrow{\beta} S^{p+k}(Y,B) \xrightarrow{\alpha} S^k(Z,C)$$

2 If A is closed in X and (X,A) has a nondegenerate base point, prove that $\{(C_-X,C_-A), (C_+X,C_+A)\}$ is an excisive couple of subsets. Let S: $H_q(X,A) \approx H_{q+1}(SX,SA)$ and S: $H^q(X,A) \approx H^{q+1}(SX,SA)$ be the isomorphisms of the corresponding relative Mayer-Vietoris sequences.

3 Prove that there are pairings

$$\{X,A;\ Y,B\}_p \otimes H_q(X,A) \to H_{p+q}(Y,B)$$
$$\{X,A;\ Y,B\}_p \otimes H^r(Y,B) \to H^{r-p}(X,A)$$

sending $\{\alpha\} \otimes z$ to $S^{-k}(\alpha_*(S^{k+p}z))$ and $\{\alpha\} \otimes u$ to $S^{-k-p}(\alpha^*(S^ku))$ for $z \in H_q(X,A)$, $u \in H^r(Y,B)$, and α: $S^{k+p}(X,A) \to S^k(Y,B)$.

4 If (X,A) is a pointed pair, with $A \subset X$ a cofibration, and Y is a pointed space, prove that there is an exact sequence

$$\cdots \to \{X;Y\}_q \to \{A;Y\}_q \to \{X/A;Y\}_{q-1} \to \{X;Y\}_{q-1} \to \cdots$$

5 Let X be a pointed space and (Y,B) a pointed pair, with $B \subset Y$ a cofibration. If f: $X \to Y$ is such that the composite $X \xrightarrow{f} Y \xrightarrow{k} Y/B$ is null homotopic, prove that Sf is homotopic to the composite $SX \xrightarrow{f'} SB \subset SY$ for some f'. Deduce the existence of an exact sequence

$$\cdots \to \{X;B\}_q \to \{X;Y\}_q \to \{X;Y/B\}_q \to \{X;B\}_{q-1} \to \cdots$$

F DUALITY IN THE SUSPENSION CATEGORY[1]

In this group of exercises all spaces are assumed to be finite CW complexes with base points. An *n-duality* is an element $u \in \{X^* \# X;\ S^0\}_{-n}$ such that the map sending $\{\alpha\} \in \{S^0;X^*\}_q \approx \{S^q;X^*\}$ to $u \circ (\{\alpha\} \# \{1_X\}) \in \{S^q \# X;\ S^0\}_{-n} \approx \{X;S^0\}_{q-n}$ is an isomorphism

$$D_u: \{S^0;X^*\}_q \approx \{X;S^0\}_{q-n}$$

and the map sending $\{\beta\} \in \{S^0;X\}_q \approx \{S^q;X\}$ to $u \circ (\{1_{X^*}\} \# \{\beta\}) \in \{X^* \# S^q;\ S^0\}_{-n} \approx \{X^*;S^0\}_{q-n}$ is an isomorphism

[1] See E. Spanier, Function spaces and duality, *Annals of Mathematics*, vol. 70, pp. 338–378, 1959, for a different development of this topic. The one given in the text is based on a suggestion of P. Freyd and has also been considered by D. Husemoller.

$$D^u\colon \{S^0;X\}_q \approx \{X^*;S^0\}_{q-n}$$

1 If $f\colon S^p \# S^q \to S^{p+q}$ is a homeomorphism, prove that $\{f\} \in \{S^p \# S^q;\ S^0\}_{-p-q}$ is a $(p+q)$-duality.

2 If $u \in \{X^* \# X;\ S^0\}_{-n}$ is an n-duality, prove that the element $u' \in \{X \# X^*;\ S^0\}_{-n}$ corresponding to u under the homeomorphism $X \# X^* \to X^* \# X$ is also an n-duality.

3 If $u \in \{X^* \# X;\ S^0\}_{-n}$ is an n-duality, prove that for any Y and Z there are isomorphisms

$$D_u\colon \{Y;\ Z \# X^*\}_q \approx \{Y \# X;\ Z\}_{q-n}$$
$$D^u\colon \{Y;\ X \# Z\}_q \approx \{X^* \# Y;\ Z\}_{q-n}$$

such that $D_u\{\alpha\} = (\{1_Z\} \# u) \circ (\{\alpha\} \# \{1_X\})$ for $\{\alpha\} \in \{Y;\ Z \# X^*\}_q$ and $D^u\{\beta\} = (u \# \{1_Z\}) \circ (\{1_{X^*}\} \# \{\beta\})$ for $\{\beta\} \in \{Y;\ X \# Z\}_q$. (*Hint:* If Y and Z are spheres, this is true by definition of n-duality. For arbitrary Y and Z use induction on the number of cells and the five lemma.)

Given n-dualities $u \in \{X^* \# X;\ S^0\}_{-n}$ and $v \in \{Y^* \# Y;\ S^0)_{-n}$, define an isomorphism

$$D(u,v)\colon \{X;Y\}_q \approx \{Y^*;X^*\}_q$$

so that the following diagram is commutative:

$$\begin{array}{ccc} \{X;Y\}_q & \xrightarrow{D(u,v)} & \{Y^*;X^*\}_q \\ & {\scriptstyle D^v}\searrow\approx \quad \approx\swarrow{\scriptstyle D_u} & \\ & \{Y^* \# X;\ S^0\}_{q-n} & \end{array}$$

4 Prove that $D(v',u') = (D(u,v))^{-1}\colon \{Y^*;X^*\}_q \approx \{X;Y\}_q$.

5 If $u \in \{X^* \# X;\ S^0\}_{-n}$, $v \in \{Y^* \# Y;\ S^0\}_{-n}$, and $w \in \{Z^* \# Z;\ S^0\}_{-n}$ are n-dualities and $\{\alpha\} \in \{X;Y\}_p$ and $\{\beta\} \in \{Y;Z\}_q$, prove that, in $\{Z^*;X^*\}_{p+q}$,

$$D(u,w)(\{\beta\} \circ \{\alpha\}) = (D(u,v)\{\alpha\}) \circ (D(v,w)\{\beta\})$$

Assume that $f\colon X^* \# X \to S^n$ and $g\colon Y^* \# Y \to S^n$ are such that $\{f\}$ and $\{g\}$ are n-dualities and let $\alpha\colon X \to Y$ and $\beta\colon Y^* \to X^*$ be maps such that

$$f \circ (\beta \# 1_X) \simeq g \circ (1_{Y^*} \# \alpha)\colon Y^* \# X \to S^n$$

[which implies $D(\{f\},\{g\})\{\alpha\} = \{\beta\}$]. Let C_α and C_β be the mapping cones of α and β, respectively, and consider the coexact sequences

$$X \xrightarrow{\alpha} Y \xrightarrow{i} C_\alpha \xrightarrow{k} SX \xrightarrow{S\alpha} SY$$
$$Y^* \xrightarrow{\beta} X^* \xrightarrow{i'} C_\beta \xrightarrow{k'} SY^* \xrightarrow{S\beta} SX^*$$

6 Prove that there is a map $h\colon C_\beta \# C_\alpha \to S^{n+1}$ such that the following squares are homotopy commutative:

$$\begin{array}{ccc} X^* \# C_\alpha & \xrightarrow{1 \# k} X^* \# SX \leftrightarrow & S(X^* \# X) \\ {\scriptstyle i' \# 1}\downarrow & & \downarrow{\scriptstyle Sf} \\ C_\beta \# C_\alpha & \xrightarrow{h} & S^{n+1} \end{array} \qquad \begin{array}{ccc} C_\beta \# Y & \xrightarrow{1 \# i} & C_\beta \# C_\alpha \\ {\scriptstyle k' \# 1}\downarrow & & \downarrow{\scriptstyle h} \\ SY^* \# Y \leftrightarrow S(Y^* \# Y) & \xrightarrow{Sg} & S^{n+1} \end{array}$$

Deduce that $\{h\} \in \{C_\beta \# C_\alpha;\ S^0\}_{-n-1}$ is an $(n+1)$-duality.

7 For any X there is an integer n for which there exists a space X^* and an n-duality $u \in \{X^* \# X;\ S^0\}_{-n}$. (*Hint:* Prove this by induction on the number of cells of X, using exercises 1 and 6 above.)

CHAPTER NINE

SPECTRAL SEQUENCES AND HOMOTOPY GROUPS OF SPHERES

THE TECHNIQUE OF OBSTRUCTION THEORY DEVELOPED IN THE LAST CHAPTER focuses attention on the computation of homotopy groups. In this chapter we obtain some results about the homotopy groups of spheres. The method we follow is due to Serre[1] and uses the technical tool known as a spectral sequence. This algebraic concept is introduced for the study of the homology and cohomology properties of arbitrary fibrations, but it has other important applications in algebraic topology, and the number of these is constantly increasing. Some indication of the power of spectral sequences will be apparent from the results obtained by its use here.

Section 9.1 contains the definition of a spectral sequence, and in Sec. 9.2 the homology spectral sequence of a fibration is established. This is used in Sec. 9.3 to prove generalizations of the Gysin and Wang exact homology sequences. There is also a proof of the homotopy excision theorem, which is used in connection with the Hopf invariant to study in more detail the suspension map for homotopy groups of spheres.

[1] See two basic papers, J.-P. Serre, Homologie singulière des espaces fibrés, *Annals of Mathematics*, vol. 54, pp. 425–505, 1951, and Groupes d'homotopie et classes de groupes abéliens, *Annals of Mathematics*, vol. 58, pp. 258–294, 1953.

Cohomology spectral sequences are considered in Sec. 9.4, and the cohomology spectral sequence of a fibration is established. The multiplicative property of cohomology spectral sequences is applied in Sec. 9.5 to obtain stronger results than were obtained with the homology spectral sequence. Serre classes of abelian groups are introduced in Sec. 9.6, and some technical results, based on spectral sequences, are derived for isomorphisms of groups modulo a Serre class.

In Sec. 9.7 the machinery based on Serre classes is used to prove that all the homotopy groups of spheres are finitely generated and are finite except for stated exceptions. There are also some results concerning p-primary components of homotopy groups of spheres. Further information about homotopy groups of spheres appears in the exercises at the end of the chapter.

1 SPECTRAL SEQUENCES

Corresponding to a subcomplex of a chain complex, there is an associated exact sequence of homology modules. Hence, corresponding to an increasing sequence of subcomplexes of a chain complex, there is an associated sequence of exact sequences of homology modules. This sequence of exact sequences constitutes a new algebraic object, known as a spectral sequence, which provides information about the homology of the chain complex in terms of the homology of the quotient complexes of the sequence of subcomplexes. A spectral sequence consists of a sequence of chain complexes each of which is the homology module of the preceding one. There is an associated limit module, and the spectral sequence itself is viewed as a sequence of approximations to this limit module. In this section we shall define the algebraic concepts involved. In the next section we shall apply these concepts to study the homology of a fibration.

Let us consider modules over a fixed principal ideal domain R. A *bigraded module* E *(over* R*)* is an indexed collection of R modules $E_{s,t}$ for every pair of integers s and t. A *differential* $d\colon E \to E$ of *bidegree* $(-r, r-1)$ is a collection of homomorphisms $d\colon E_{s,t} \to E_{s-r,t+r-1}$, for all s and t, such that $d^2 = 0$. The *homology module* $H(E)$ is the bigraded module defined by

$$H_{s,t}(E) = [\ker (d\colon E_{s,t} \to E_{s-r,t+r-1})]/d(E_{s+r,t-r+1})$$

Note that if E_q is defined to equal $\bigoplus_{s+t=q} E_{s,t}$, the differential d defines a homomorphism $\partial\colon E_q \to E_{q-1}$ such that $\{E_q,\partial\}$ is a chain complex. Furthermore, the qth homology module of this chain complex equals $\bigoplus_{s+t=q} H_{s,t}(E)$.

An E^k *spectral sequence* E is a sequence $\{E^r,d^r\}$ for $r \geq k$ such that

(*a*) E^r is a bigraded module and d^r is a differential of bidegree $(-r, r-1)$ on E^r.

(*b*) For $r \geq k$ there is given an isomorphism $H(E^r) \approx E^{r+1}$.

Note that the spectral sequence begins with E^k, and the only role of the integer k is to specify where the spectral sequence starts. In our applications it will usually turn out that $k = 1$ or 2. Clearly, any E^k spectral sequence defines an $E^{k'}$ spectral sequence for every $k' \geq k$.

A *homomorphism* $\varphi: E \to E'$ from one E^k spectral sequence to another is a collection of homomorphisms $\varphi^r: E^r_{s,t} \to E'^r_{s,t}$ for $r \geq k$ and all s and t commuting with the differentials and such that $\varphi^r_*: H(E^r) \to H(E'^r)$ corresponds to $\varphi^{r+1}: E^{r+1} \to E'^{r+1}$ under the isomorphisms of the spectral sequence. The composite of homomorphisms is a homomorphism, and so there is a category of E^k spectral sequences (for fixed k) and homomorphisms.

To define the limit term of a spectral sequence, for $r \geq k$ we regard E^{r+1} as identified with $H(E^r)$ by the isomorphisms of the spectral sequence. Let Z^k be the bigraded module $Z^k_{s,t} = \ker (d^k: E^k_{s,t} \to E^k_{s-k,t+k-1})$ and let B^k be the bigraded module $B^k_{s,t} = d^k(E^k_{s+k,t-k+1})$. Then $B^k \subset Z^k$ and $E^{k+1} = Z^k/B^k$. Let $Z(E^{k+1})$ be the bigraded module $Z(E^{k+1})_{s,t} = \ker (d^{k+1}: E^{k+1}_{s,t} \to E^{k+1}_{s-k-1,t+k})$ and let $B(E^{k+1})$ be the bigraded module $B(E^{k+1})_{s,t} = d^{k+1}(E^{k+1}_{s+k+1,t-k})$. By the Noether isomorphism theorem, there exist bigraded submodules Z^{k+1} and B^{k+1} of Z^k containing B^k such that $Z(E^{k+1})_{s,t} = Z^{k+1}_{s,t}/B^k_{s,t}$ and $B(E^{k+1})_{s,t} = B^{k+1}_{s,t}/B^k_{s,t}$ for all s and t. It follows that $B^{k+1} \subset Z^{k+1}$, and we have

$$B^k \subset B^{k+1} \subset Z^{k+1} \subset Z^k$$

Continuing by induction, we obtain submodules for $r \geq k$

$$B^k \subset B^{k+1} \subset \cdots \subset B^r \subset \cdots \subset Z^r \subset \cdots \subset Z^{k+1} \subset Z^k$$

such that $E^{r+1} = Z^r/B^r$. We define bigraded modules $Z^\infty = \bigcap_r Z^r$, $B^\infty = \bigcup_r B^r$, and $E^\infty = Z^\infty/B^\infty$. The bigraded module E^∞ is called the *limit* of the spectral sequence E, and the terms E^r of the spectral sequence are successive approximations to E^∞.

The spectral sequence E is said to *converge* if for every s and t there exists an integer $r(s,t) \geq k$ such that for $r \geq r(s,t)$, $d^r: E^r_{s,t} \to E^r_{s-r,t+r-1}$ is trivial. Then $E^{r+1}_{s,t}$ is isomorphic to a quotient of $E^r_{s,t}$ and $E^\infty_{s,t}$ is isomorphic to the direct limit of the sequence

$$E^{r(s,t)}_{s,t} \to E^{r(s,t)+1}_{s,t} \to \cdots$$

It is frequently the case that the spectral sequence converges and for given s and t there exists $r(s,t)$ such that all homomorphisms in the sequence displayed above are isomorphisms (so $E^r_{s,t} \approx E^\infty_{s,t}$ for $r \geq r(s,t)$). For example, if E has the property that for some r there exist integers N and N' such that $E^r_{s,t} = 0$ for $s < N$ or $t < N'$, the same is true for $E^{r'}_{s,t}$ for $r' \geq r$. Then, given s and t, if $r' \geq r$ is chosen so that $r' > \sup (s - N, t - N' + 1)$, we have

$$E^{r'}_{s+r',t-r'+1} \xrightarrow{d^{r'}} E^{r'}_{s,t} \xrightarrow{d^{r'}} E^{r'}_{s-r',t+r'-1}$$

The first module equals 0 because $t - r' + 1 < N'$ and the last module equals 0 because $s - r' < N$. Therefore, if r' is large enough,

$$E^{r'}_{s,t} \approx E^{r'+1}_{s,t} \approx \cdots \approx E^\infty_{s,t},$$

and E is convergent in this stronger sense.

A particular example of such spectral sequences is a *first-quadrant* spectral sequence, which is defined to be a spectral sequence E having the property that for some r, $E^r_{s,t} = 0$ if $s < 0$ or $t < 0$. Such a spectral sequence is convergent in the strong sense, and for any q there are only a finite number of nontrivial modules $E^\infty_{s,t}$ with $s + t = q$. A first-quadrant spectral sequence is conveniently represented by attaching $E^r_{s,t}$ to the lattice point (s,t) in the first quadrant of the plane and representing the differential d^r by oblique arrows:

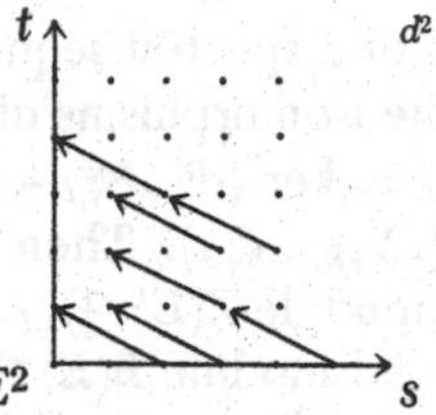

Then E^{r+1} is the quotient of the kernel of the arrow which originates at (s,t) by the image of the arrow which terminates at (s,t).

A homomorphism $\varphi: E \to E'$ between E^k spectral sequences induces a homomorphism $\varphi^\infty: E^\infty \to E'^\infty$ between their limit terms. Therefore there is a covariant functor from the category of E^k spectral sequences to the category of bigraded modules which assigns to every spectral sequence its limit. The following useful result is an easy consequence of the fact that a chain transformation which is an isomorphism induces an isomorphism of the corresponding homology modules.

1 THEOREM *Let $\varphi: E \to E'$ be a homomorphism of E^k spectral sequences which is an isomorphism for some $r \geq k$. Then φ is an isomorphism for all $r' \geq r$. Furthermore, if E and E' converge, φ^∞ is an isomorphism of their limits.* ■

An (*increasing*) *filtration* F on an R module A is a sequence of submodules F_sA for all integers s such that $F_sA \subset F_{s+1}A$. If A is a graded module (that is, $A = \{A_t\}$), the filtration F is required to be compatible with the gradation (that is, F_sA is graded by $\{F_sA_t\}$). Given a filtration F on A, the *associated graded module* $G(A)$ is defined by $G(A)_s = F_sA/F_{s-1}A$. If A is a graded module, the associated module $G(A)$ is bigraded by the modules $G(A)_{s,t} = F_sA_{s+t}/F_{s-1}A_{s+t}$. In this case, s is called the *filtered degree*, t the *complementary degree*, and $s + t$ the *total degree* of an element of $G(A)_{s,t}$. The sequence

$$\cdots \subset F_{s-1}A \subset F_sA \subset F_{s+1}A \subset \cdots$$

is an infinite normal series for A, and the associated module consists of the quotients of this normal series.

A filtration F on A is said to be *convergent* if $\bigcap_s F_sA = 0$ and $\bigcup_s F_sA = A$. For convergent filtrations the associated module $G(A)$ is more closely tied to A than in the case of an arbitrary filtration. However, even if the filtration is finite in the sense that $F_sA = 0$ for some s and $F_{s'}A = A$ for some s', it is not true that $G(A)$ determines A. In the latter case $G(A)$ determines A up to a finite number of module extensions.

A filtration F on a chain complex C is a filtration compatible with the gradation of C and with the differential of C (that is, F_sC is a chain subcomplex of C consisting of $\{F_sC_t\}$). The filtration F on C induces a filtration F on $H_*(C)$ defined by

$$F_sH_*(C) = \text{im}\,[H_*(F_sC) \to H_*(C)]$$

Because the homology functor commutes with direct limits, if F is a convergent filtration of C, it follows that $\bigcup_s F_sH_*(C) = H_*(C)$; however, it is not generally true that $\bigcap_s F_sH_*(C) = 0$. Thus, to ensure that F be a convergent filtration on $H_*(C)$ we need a stronger assumption about the original filtration on C.

A filtration F on a graded module A is said to be *bounded below* if for any t there is $s(t)$ such that $F_{s(t)}A_t = 0$. It is clear that if F is a filtration bounded below on a chain complex C, then the induced filtration on $H_*(C)$ is also bounded below. Thus, if F is convergent and bounded below on C, the same is true of the induced filtration on $H_*(C)$.

The following theorem associates a spectral sequence to a filtration on a chain complex.

2 THEOREM *Let F be a convergent filtration bounded below on a chain complex C. There is a convergent E^1 spectral sequence with*

$$E^1_{s,t} \approx H_{s+t}(F_sC/F_{s-1}C)$$

[and d^1 corresponding to the connecting homomorphism of the triple $(F_sC, F_{s-1}C, F_{s-2}C)$] and E^∞ isomorphic to the bigraded module $GH_*(C)$ (associated to the filtration $F_sH_*(C) = \text{im}\,[H_*(F_sC) \to H_*(C)]$).

PROOF For arbitrary r we define

$$Z^r_s = \{c \in F_sC \mid \partial c \in F_{s-r}C\}$$
$$Z^\infty_s = \{c \in F_sC \mid \partial c = 0\}$$

These are graded modules with $Z^r_{s,t} = \{c \in F_sC_{s+t} \mid \partial c \in F_{s-r}C\}$ and $Z^\infty_{s,t} = \{c \in F_sC_{s+t} \mid \partial c = 0\}$. We then have a sequence of graded modules

$$\cdots \subset \partial Z^{-1}_{s-1} \subset \partial Z^0_s \subset \partial Z^1_{s+1} \subset \cdots \subset \partial C \cap F_sC \subset Z^\infty_s \subset \cdots$$
$$\cdots \subset Z^1_s \subset Z^0_s = F_sC$$

We define

$$E^r_s = Z^r_s/(Z^{r-1}_{s-1} + \partial Z^{r-1}_{s+r-1})$$
$$E^\infty_s = Z^\infty_s/(Z^\infty_{s-1} + \partial C \cap F_sC)$$

The map ∂ sends Z^r_s to Z^r_{s-r} and $Z^{r-1}_{s-1} + \partial Z^{r-1}_{s+r-1}$ to ∂Z^{r-1}_{s-1}. Therefore it induces a homomorphism

$$d^r\colon E^r_s \to E^r_{s-r}$$

Then E^r is a bigraded module and d^r is a differential of bidegree $(-r, r-1)$ on it. For $r < 0$, $d^r = 0$ and $E_s^r = F_sC/F_{s-1}C$. Therefore

$$E_{s,t}^0 = F_sC_{s+t}/F_{s-1}C_{s+t} = G(C)_{s,t}$$

and $d^0\colon F_sC_{s+t}/F_{s-1}C_{s+t} \to F_sC_{s+t-1}/F_{s-1}C_{s+t-1}$ is just the boundary operator of the quotient complex $F_sC/F_{s-1}C$. Furthermore,

$$E_{s,t}^1 = Z_{s,t}^1/(Z_{s-1,t+1}^0 + \partial Z_{s,t+1}^0),$$

where $Z_{s,t}^1 = \{c \in F_sC_{s+t} \mid \partial c \in F_{s-1}C_{s+t-1}\}$. Therefore $Z_{s,t}^1/Z_{s-1,t+1}^0$ is the module of $(s+t)$-cycles of $F_sC/F_{s-1}C$ and $(Z_{s-1,t+1}^0 + \partial Z_{s,t+1}^0)/Z_{s-1,t+1}^0$ is the module of $(s+t)$-boundaries of $F_sC/F_{s-1}C$. By the Noether isomorphism theorem, $E_{s,t}^1 \approx H_{s+t}(F_sC/F_{s-1}C)$. The fact that under this isomorphism d^1 corresponds to the boundary operator of the triple $(F_sC, F_{s-1}C, F_{s-2}C)$ is proved by direct verification, using the definitions.

We prove that $E = \{E^r\}_{r\geq 1}$ is a spectral sequence by computing the homology of E^r with respect to d^r. We have

$$\begin{aligned}\{c \in Z_s^r \mid \partial c \in Z_{s-r-1}^{r-1} + \partial Z_{s-1}^{r-1}\}&\\ &= \{c \in Z_s^r \mid \partial c \in F_{s-r-1}C\} + \{c \in Z_s^r \mid \partial c \in \partial Z_{s-1}^{r-1}\}\\ &= Z_s^{r+1} + (Z_{s-1}^{r-1} + Z_s^\infty) = Z_s^{r+1} + Z_{s-1}^{r-1}\end{aligned}$$

Therefore $\ker(d^r\colon E_s^r \to E_{s-r}^r) = (Z_s^{r+1} + Z_{s-1}^{r-1})/(Z_{s-1}^{r-1} + \partial Z_{s+r-1}^{r-1})$. By definition, $\operatorname{im}(d^r\colon E_{s+r}^r \to E_s^r) = (\partial Z_{s+r}^r + Z_{s-1}^{r-1})/(Z_{s-1}^{r-1} + \partial Z_{s+r-1}^{r-1})$. Hence, by the Noether isomorphism theorem, in E_s^r we have

$$\begin{aligned}\ker d^r/\operatorname{im} d^r &\approx (Z_s^{r+1} + Z_{s-1}^{r-1})/(\partial Z_{s+r}^r + Z_{s-1}^{r-1}) \approx Z_s^{r+1}/[Z_s^{r+1} \cap (\partial Z_{s+r}^r + Z_{s-1}^{r-1})]\\ &= Z_s^{r+1}/(\partial Z_{s+r}^r + Z_{s-1}^r) = E_s^{r+1}\end{aligned}$$

Therefore there is an isomorphism $H_*(E^r) \approx E^{r+1}$, and E is a spectral sequence.

We now compute the limit of this spectral sequence. By definition and the Noether isomorphism theorem,

$$E_s^r = Z_s^r/(Z_{s-1}^{r-1} + \partial Z_{s+r-1}^{r-1}) \approx (Z_s^r + F_{s-1}C)/(F_{s-1}C + \partial Z_{s+r-1}^{r-1})$$

In the last expression the numerators decrease as r increases and the denominators increase as r increases. By definition, the limit equals

$$\begin{aligned}\textstyle\bigcap_r (Z_s^r + F_{s-1}C)/\bigcup_r (F_{s-1}C + \partial Z_{s+r-1}^{r-1}) &=\\ &(\textstyle\bigcap_r Z_s^r + F_{s-1}C)/(F_{s-1}C + \bigcup_r \partial Z_{s+r-1}^{r-1})\end{aligned}$$

Since $\bigcup_s F_sC = C$, then $\bigcup_r \partial Z_{s+r-1}^{r-1} = \partial C \cap F_sC$. For given t, $\bigcap_r Z_{s,t}^r = Z_{s,t}^\infty$, because $F_sC_t = 0$ for s small enough. Therefore the limit term equals

$$(Z_s^\infty + F_{s-1}C)/(F_{s-1}C + \partial C \cap F_sC) = Z_s^\infty/(Z_{s-1}^\infty + \partial C \cap F_sC) = E_s^\infty$$

To show that the spectral sequence converges, note that because the filtration is bounded below, for fixed $s+t$, $E_{s,t}^r = 0$ for s small enough. Therefore, for fixed s and t there exists r such that for $r' \geq r$, $E_{s,t}^{r'+1}$ is a quotient of $E_{s,t}^{r'}$, and the spectral sequence converges.

To complete the proof we interpret the limit E^∞ as $GH_*(C)$. By definition, $GH_*(C)_{s,t} = F_sH_{s+t}(C)/F_{s-1}H_{s+t}(C)$, where

$$F_sH_{s+t}(C) = \operatorname{im}[H_{s+t}(F_sC) \to H_{s+t}(C)]$$

Therefore the graded module $F_sH_*(C) = Z_s^\infty/\partial C \cap F_sC$, and

$$\begin{aligned} F_sH_*(C)/F_{s-1}H_*(C) &= (Z_s^\infty/\partial C \cap F_sC)/(Z_{s-1}^\infty/\partial C \cap F_{s-1}C) \\ &\approx Z_s^\infty/(Z_{s-1}^\infty + \partial C \cap F_sC) \\ &= E_s^\infty \quad \blacksquare \end{aligned}$$

In theorem 2 note that even in the most favorable circumstances E^∞ does not determine $H_*(C)$ completely, but only up to module extensions. Note that we have, in fact, defined an E^0 spectral sequence. The theorem was stated in terms of the corresponding E^1 spectral sequence because the E^1 term contains more information than the E^0 term.

It should be observed that the spectral sequence of theorem 2 is functorial on the category of chain complexes with a convergent filtration bounded below. Combining this with theorem 1, we obtain the following result.

3 THEOREM *Let τ: $C \to C'$ be a chain map preserving filtration between chain complexes having convergent filtrations bounded below. If for some $r \geq 1$ the induced map τ^r: $E^r \to E'^r$ is an isomorphism, then τ induces an isomorphism*

$$\tau_*: H_*(C) \approx H_*(C')$$

PROOF By theorem 1, τ^∞ is an isomorphism. We have a commutative diagram with exact rows

$$\begin{array}{ccccccccc} 0 & \to & F_{s-1}H_n(C) & \to & F_sH_n(C) & \to & E_{s,n-s}^\infty & \to & 0 \\ & & \downarrow{\scriptstyle\tau_*} & & \downarrow{\scriptstyle\tau_*} & & \downarrow{\scriptstyle\tau^\infty} & & \\ 0 & \to & F_{s-1}H_n(C') & \to & F_sH_n(C') & \to & E'^\infty_{s,n-s} & \to & 0 \end{array}$$

For fixed n, $F_{s-1}H_n(C)$ and $F_{s-1}H_n(C')$ are both 0 for s small enough (because the filtrations are bounded below). It follows by induction on s, using the five lemma and the fact that τ^∞ is an isomorphism, that $\tau_*: F_sH_n(C) \approx F_sH_n(C')$ for all s. Because the filtrations are convergent, $H_n(C) = \bigcup_s F_sH_n(C)$ and $H_n(C') = \bigcup_s F_sH_n(C')$, and so $\tau_*: H_n(C) \approx H_n(C')$. $\blacksquare$

We present some examples of spectral sequences.

4 Let C' and C'' be free nonnegative chain complexes with boundary operators ∂' and ∂'', respectively, and let $C = C' \otimes C''$ be their tensor product, with boundary operator ∂. There is a convergent filtration bounded below on C defined by $F_sC = \bigoplus_{q\leq s} C'_q \otimes C''$. For the corresponding spectral sequence,

$$E_{s,t}^1 \approx C'_s \otimes H_t(C'')$$

and for $r \geq 2$ $E^r_{s,t} \approx H_s(C' \otimes H_t(C''))$. A similar result is obtained by filtering the tensor product by the gradation of the second factor.

5 An (*increasing*) *filtration* of a topological pair (X,A) is a sequence of subspaces X_s containing A such that $X_s \subset X_{s+1}$. Such a filtration on (X,A) induces a filtration F on the chain complex $\Delta(X,A)$ by $F_s(\Delta(X,A)) = \Delta(X_s,A)$. If $X_s = A$ for some s, the induced filtration is bounded below. If $X = \bigcup X_s$ and every compact subset of X is contained in some X_s, then $\bigcup_s F_s(\Delta(X,A)) = \Delta(X,A)$. Therefore, if the filtration $\{X_s\}$ has both the above properties, the induced filtration on $\Delta(X,A)$ is convergent and bounded below so there is a convergent E^1 spectral sequence with $E^1_{s,t} \approx H_{s+t}(X_s,X_{s-1})$ and in which d^1 corresponds to the boundary operation of the triple (X_s,X_{s-1},X_{s-2}). The limit term of the spectral sequence is the bigraded module associated to the corresponding filtration of $H(X,A)$.

In particular, if (X,A) is a relative CW complex, $X_s = (X,A)^s$ is the s-dimensional skeleton for $s \geq 0$, and $X_s = A$ for $s < 0$, then $E^1_{s,t} \neq 0$ only if $t = 0$, and $E^1_{s,0} \approx H_s(X_s,X_{s-1})$. Therefore, for $r \geq 2$, $E^r_{s,0}$ is the homology of the chain complex $C = \{C_s,\partial\}$, where $C_s = H_s(X_s,X_{s-1})$, and $\partial\colon C_s \to C_{s-1}$ is the boundary operator of the triple (X_s,X_{s-1},X_{s-2}).

Our next example is an alternate description of the spectral sequence of example 5 whose construction does not involve the chain modules. It can also be applied to obtain a spectral sequence corresponding to any sequence of functors having the exactness properties of the homology functors.

6 Let $\{X_s\}$ be an increasing filtration of a pair (X,A). For each s there is an exact homology sequence of (X_s,X_{s-1}) (with some coefficient module)

$$\cdots \to H_q(X_{s-1}) \xrightarrow{i_*} H_q(X_s) \xrightarrow{j_*} H_q(X_s,X_{s-1}) \xrightarrow{\partial} H_{q-1}(X_{s-1}) \to \cdots$$

Therefore we have a sequence of exact sequences. These combine to form a commutative diagram

$$\begin{array}{ccccccccc}
 & & \vdots & & & & \vdots & & \\
 & & \downarrow & & & & \downarrow & & \\
\cdots & \xrightarrow{\partial} & H_q(X_{s-1}) & \xrightarrow{j_*} & H_q(X_{s-1},X_{s-2}) & \xrightarrow{\partial} & H_{q-1}(X_{s-2}) & \xrightarrow{j_*} & \cdots \\
 & & \downarrow{\scriptstyle i_*} & & & & \downarrow{\scriptstyle i_*} & & \\
\cdots & \xrightarrow{\partial} & H_q(X_s) & \xrightarrow{j_*} & H_q(X_s,X_{s-1}) & \xrightarrow{\partial} & H_{q-1}(X_{s-1}) & \xrightarrow{j_*} & \cdots \\
 & & \downarrow{\scriptstyle i_*} & & & & \downarrow{\scriptstyle i_*} & & \\
\cdots & \xrightarrow{\partial} & H_q(X_{s+1}) & \xrightarrow{j_*} & H_q(X_{s+1},X_s) & \xrightarrow{\partial} & H_{q-1}(X_s) & \xrightarrow{j_*} & \cdots \\
 & & \downarrow & & & & \downarrow & & \\
 & & \vdots & & & & \vdots & &
\end{array}$$

in which any sequence consisting of a vertical map i_* followed by two horizontal maps j_* and ∂ and then a vertical map i_* followed again by j_* and ∂ and iteration of this (one possible such sequence is indicated by heavy arrows in the diagram) is exact. From this diagram there is obtained an E^1 spectral sequence in which $E^1_{s,t} = H_{s+t}(X_s,X_{s-1})$ and for $r \geq 2$, $E^r_{s,t}$ is defined to be the quotient $Z^r_{s,t}/B^r_{s,t}$, where

$$Z^r_{s,t} = \partial^{-1}(i_*{}^{r-1}H_{s+t-1}(X_{s-r}))$$
$$B^r_{s,t} = j_*(\ker [i_*{}^{r-1}: H_{s+t}(X_s) \to H_{s+t}(X_{s+r-1})])$$

This spectral sequence converges if for fixed q, $H_q(X) \approx \lim_\to \{H_q(X_s)\}$, and for s small enough $H_q(X_s,A) = 0$. In this case the limit is the bigraded module associated to the filtration of $H(X,A)$ defined by $F_sH(X,A) = \text{im}\,[H(X_s,A) \to H(X,A)]$.

It is not hard to verify that the E^1 spectral sequence of example 6 is the same as the E^1 spectral sequence of example 5. The same process can be applied to obtain a spectral sequence from any diagram having the exactness properties of the diagram in example 6. These properties have been formalized in the concept of *exact couple*,[1] but we omit the precise definition.

2 THE SPECTRAL SEQUENCE OF A FIBRATION

One of the most fruitful applications of spectral sequences is to the homology of fibrations. With a suitable orientability assumption on the fibration, there is a spectral sequence converging to the bigraded module associated to a filtration on the homology of the total space and whose E^2 term is isomorphic to the homology of the base space with coefficients in the homology of the fiber. This section is devoted to a construction of the spectral sequence. It depends on a study of the homology of the total space of a fibration whose base space is a relative CW complex utilizing the filtration of the total space consisting of the inverse images of the skeleta of the base space. By using CW approximations, an E^2 spectral sequence is defined for a fibration over any path-connected base space. Some applications of this spectral sequence will be given in the next section.

Let $p: E \to B$ be a fibration. For a subspace $A \subset B$ let $E_A = p^{-1}(A) \subset E$. Then p maps (E,E_A) to (B,A). Assume that (B,A) is a relative CW complex and let $E_s = p^{-1}((B,A)^s)$ be the part of E lying over the s-dimensional skeleton of B for $s \geq 0$ and $E_s = E_A$ if $s < 0$. Then $E_s \subset E_{s+1}$, so $\{E_s\}$ is an increasing filtration of (E,E_A). Furthermore, $E_{-1} = E_A$, $\cup_s E_s = E$, and every compact subset of E is contained in E_s for some s. By the method of example 9.1.5, we have the following result.

1 THEOREM *Let $p: E \to B$ be a fibration over a relative CW complex (B,A). For singular homology with any coefficient module G there is a convergent E^1 spectral sequence with $E^1_{s,t} \approx H_{s+t}(E_s,E_{s-1};\, G)$, d^1 the boundary*

[1] See W. S. Massey, Exact couples in algebraic topology, parts I and II, *Annals of Mathematics*, vol. 56, pp. 364–396. 1952, and parts III, IV, and V, *Annals of Mathematics*, vol. 57, pp. 248–286, 1953.

operator of the triple (E_s,E_{s-1},E_{s-2}), and E^∞ the bigraded module associated to the filtration of $H_*(E,E_A;\, G)$ defined by

$$F_sH_*(E,E_A;\, G) = \operatorname{im}\, [H_*(E_s,E_A;\, G) \to H_*(E,E_A;\, G)] \quad \blacksquare$$

To apply this result we need to compute the module $H_n(E_s,E_{s-1};\, G)$. We let $\{e_\lambda\}$ be the collection of s-cells of $B - A$.

2 LEMMA *The inclusion maps $i_\lambda\colon (p^{-1}(e_\lambda),p^{-1}(\dot e_\lambda)) \subset (E_s,E_{s-1})$ induce a direct-sum representation*

$$\{i_{\lambda *}\}\colon \bigoplus_\lambda H_n(p^{-1}(e_\lambda),\, p^{-1}(\dot e_\lambda)) \approx H_n(E_s,E_{s-1})$$

and a direct-product representation

$$\{i_\lambda^*\}\colon H^n(E_s,E_{s-1}) \approx \times\, H^n(p^{-1}(e_\lambda),p^{-1}(\dot e_\lambda))$$

PROOF For each λ let e_λ' be a simplex of dimension s contained in $e_\lambda - \dot e_\lambda$. Then the inclusion maps $((B,A)^s,\, (B,A)^{s-1}) \subset ((B,A)^s,\, (B,A)^s - \bigcup_\lambda (e_\lambda' - \dot e_\lambda'))$ and $(e_\lambda,\dot e_\lambda) \subset (e_\lambda,\, e_\lambda - (e_\lambda' - \dot e_\lambda'))$ are homotopy equivalences. Therefore the corresponding inclusion maps $(E_s,E_{s-1}) \subset (E_s,\, E_s - \bigcup_\lambda p^{-1}(e_\lambda' - \dot e_\lambda'))$ and $(p^{-1}e_\lambda,p^{-1}\dot e_\lambda) \subset (p^{-1}e_\lambda,\, p^{-1}(e_\lambda - (e_\lambda' - \dot e_\lambda')))$ are homotopy equivalences. There is a commutative diagram induced by inclusion maps

$$\begin{array}{ccc}
\bigoplus_\lambda H_n(p^{-1}e_\lambda,p^{-1}\dot e_\lambda) & \xrightarrow{\{i_{\lambda *}\}} & H_n(E_s,E_{s-1}) \\
\downarrow & & \downarrow \\
\bigoplus_\lambda H_n(p^{-1}e_\lambda,p^{-1}(e_\lambda - (e_\lambda' - \dot e_\lambda'))) & \longrightarrow & H_n(E_s,E_s - \bigcup_\lambda p^{-1}(e_\lambda' - \dot e_\lambda')) \\
\uparrow & & \uparrow \\
\bigoplus_\lambda H_n(p^{-1}e_\lambda',p^{-1}\dot e_\lambda') & \longrightarrow & H_n(\bigcup_\lambda p^{-1}e_\lambda',\, \bigcup_\lambda p^{-1}\dot e_\lambda')
\end{array}$$

in which all the vertical maps are isomorphisms, the top two because they are induced by homotopy equivalences and the bottom two because they are induced by suitable excision maps. Since e_λ' is disjoint from e_μ' if $\lambda \neq \mu$, the bottom map is an isomorphism because it is induced by a chain isomorphism. This proves the first part of the lemma. A similar argument proves the result for cohomology. $\blacksquare$

Before proceeding further with the computation of $H_n(E_s,E_{s-1})$ and the boundary operator of the triple (E_s,E_{s-1},E_{s-2}), we introduce a subcomplex of the singular chain complex of a relative CW complex which is chain equivalent to the singular complex itself.

3 THEOREM *Let $\{X_s\}$ be an increasing sequence of subspaces of a space X and let $\bar\Delta(X)$ be the subcomplex of $\Delta(X)$ generated by singular simplexes $\sigma\colon \Delta^q \to X$ such that $\sigma(\Delta^q)^k \subset X_k$ for all k. If (X,X_{s-1}) is $(s-1)$-connected for all s, then the inclusion map $\bar\Delta(X) \subset \Delta(X)$ is a chain equivalence.*

PROOF We shall use lemma 7.4.7. We associate to every singular simplex $\sigma: \Delta^q \to X$ a map $P(\sigma): \Delta^q \times I \to X$ such that the hypotheses of lemma 7.4.7 are satisfied [with $C = \bar{\Delta}(X)$]. This is done by induction on q. If $q = 0$ and $\sigma(\Delta^0) \subset X_0$, define $P(\sigma)$ to be the composite $\Delta^0 \times I \to \Delta^0 \xrightarrow{\sigma} X$. If $\sigma(\Delta^0) \not\subset X_0$, there is a path $\omega: I \to X$ from $\sigma(\Delta^0)$ to some point of X_0 [because (X,X_0) is 0-connected]. Then $P(\sigma): \Delta^0 \times I \to X$ is defined by $P(\sigma)(v_0,t) = \omega(t)$.

Let $q > 0$ and assume inductively that $P(\sigma)$ has been defined to satisfy lemma 7.4.7 for all singular simplexes σ of dimension $< q$. Let $\sigma: \Delta^q \to X$ be a singular q-simplex of X. If $\sigma(\Delta^q)^k \subset X_k$ for all k, define $P(\sigma)$ to be the composite $\Delta^q \times I \to \Delta^q \xrightarrow{\sigma} X$. If $\sigma(\Delta^q)^k \not\subset X_k$ for some k, then a and c of lemma 7.4.7 define $P(\sigma)$ on $\Delta^q \times 0 \cup \dot{\Delta}^q \times I$ and b of lemma 7.4.7 ensures that on $\dot{\Delta}^q \times 1$ the resulting map sends $(\dot{\Delta}^q)^k \times 1$ into X_k for all k.

It is clearly possible to find a homeomorphism $\Delta^q \times I \approx E^q \times I$ which takes $(\Delta^q \times 0 \cup \dot{\Delta}^q \times I,\ \dot{\Delta}^q \times 1)$ onto $(E^q,S^{q-1}) \times 0$ and $\Delta^q \times 1$ onto $S^{q-1} \times I \cup E^q \times 1$. Because (X,X_q) is q-connected, it follows that the given map from $(\Delta^q \times 0 \cup \dot{\Delta}^q \times I,\ \dot{\Delta}^q \times 1)$ to (X,X_q) extends to a map $P(\sigma): \Delta^q \times I \to X$ such that $P(\sigma)(\Delta^q \times 1) \subset X_q$. Then $P(\sigma) \mid \Delta^q \times 1: \Delta^q \times 1 \to X$ is a map such that $(\Delta^q)^k$ is sent into X_k for all k, and so $P(\sigma)$ can be defined for all σ to satisfy the hypotheses of lemma 7.4.7. ■

Note that theorem 3 applies to the filtration defined by the skeleta of a relative CW complex (X,A). Hence, if $\bar{\Delta}(X) \subset \Delta(X)$ is the subcomplex of cellular singular simplexes, then $\bar{\Delta}(X) \subset \Delta(X)$ is a chain equivalence. Furthermore, if (X',A) is a subcomplex of (X,A), then $\bar{\Delta}(X') = \Delta(X') \cap \bar{\Delta}(X)$. Using theorem 4.6.10 and the five lemma, we see that the inclusion map $\bar{\Delta}(X,X') \subset \Delta(X,X')$ is a chain equivalence. In particular,

$$\bar{\Delta}((X,A)^s, (X,A)^{s-1}) \subset \Delta((X,A)^s, (X,A)^{s-1})$$

is a chain equivalence for any s.

4 COROLLARY *Given a relative CW complex (X,A), let $C(X,A) = \{C_s,\partial\}$ be the chain complex, with*

$$C_s = H_s(\bar{\Delta}((X,A)^s, (X,A)^{s-1}))$$

and with $\partial: C_s \to C_{s-1}$ the boundary operator of the triple $((X,A)^s, (X,A)^{s-1}, (X,A)^{s-2})$. Then $H_*(C(X,A)) \approx H_*(X,A)$.

PROOF Let F be the filtration on $\bar{\Delta}(X,A)$ defined by $F_s\bar{\Delta}(X,A) = \bar{\Delta}((X,A)^s, A)$. Then the corresponding spectral sequence has the property that $E^1_{s,t} \approx H_{s+t}((X,A)^s, (X,A)^{s-1})$ and d^1 corresponds to the boundary operator of the triple $((X,A)^s, (X,A)^{s-1}, (X,A)^{s-2})$. By application of lemma 2 to the trivial fibration $X \to X$, it follows that there is an isomorphism

$$\bigoplus_\lambda H_q(e_\lambda,\dot{e}_\lambda) \approx H_q((X,A)^s, (X,A)^{s-1})$$

where $\{e_\lambda\}$ is the collection of s-cells of (X,A). Then $H_q((X,A)^s, (X,A)^{s-1}) = 0$ if $q \neq s$, and so $E^1_{s,t} = 0$ if $t \neq 0$ and $E^1_{s,0} \approx C_s$. This implies that $E^2_{s,t} = 0$

if $t \neq 0$ and $E^2_{s,0} \approx H_s(C(X,A))$. Therefore, by induction on r, we see that $E^r_{s,t} = 0$ for $t \neq 0$ and $E^r_{s,0} = E^2_{s,0}$ for $r \geq 2$. Hence $E^\infty_{s,t} = 0$ for $t \neq 0$ and $E^\infty_{s,0} \approx H_s(C(X,A))$. Since E^∞ is the bigraded module associated to a filtration on $H(X,A)$, we have $H_s(C(X,A)) \approx H_s(X,A)$. ∎

Recall that, by theorem 2.8.12, there is a contravariant functor from the fundamental groupoid of B to the homotopy category which assigns to $b \in B$ the fiber F_b over b and to a path class $[\omega]$ in B the homotopy class $h[\omega] \in [F_{\omega(0)};F_{\omega(1)}]$. Therefore, for fixed R there is a contravariant functor from the fundamental groupoid of B to the category of graded R modules which assigns to $b \in B$ the module $H_*(F_b;R)$ and to a path class $[\omega]$ the homomorphism $h[\omega]_*: H_*(F_{\omega(0)};R) \to H_*(F_{\omega(1)};R)$. The fibration is said to be *orientable over R* if for any closed path ω in B, $h[\omega]_* = 1$. This is a generalization of the concept of orientability of a sphere bundle. (In fact, a sphere bundle ξ is orientable as a sphere bundle if and only if $\dot{p}_\xi: \dot{E}_\xi \to B$ is orientable as a fibration.)

5 THEOREM *(a) A fibration over a simply connected base space is orientable over any R.*

(b) A fibration induced from a fibration orientable over R is itself orientable over R.

PROOF The first statement is immediate from the definition. For the second, let $p': E' \to B'$ be induced from $p: E \to B$ by a map $f: B' \to B$ and let $g': E' \to E$ be the associated map. For any path class $[\omega']$ in B' let $g'_0: F'_{\omega'(0)} \to F_{g\omega'(0)}$ and $g'_1: F'_{\omega'(1)} \to F_{g\omega'(1)}$ be the homeomorphisms defined by g'. It is then easy to verify that $[g'_1] \circ h[\omega'] = h[g \circ \omega'] \circ [g'_0]$, and this implies (b). ∎

A map $f: p^{-1}(b_0) \to p^{-1}(b_1)$ between fibers of a fibration $p: E \to B$ is said to be *admissible* if there is some path class $[\omega]$ from b_0 to b_1 such that $[f] = h[\omega]$. The following facts are immediate from the definition.

6 *An admissible map is a homotopy equivalence.* ∎

7 *The composite of admissible maps is an admissible map.* ∎

8 *A homotopy inverse of an admissible map is an admissible map.* ∎

9 *If B is path connected, there is an admissible map between any two fibers over B.* ∎

10 *If $p: E \to B$ is orientable over R, any two admissible maps from $p^{-1}(b_0)$ to $p^{-1}(b_1)$ induce the same homomorphism from $H_*(p^{-1}(b_0);R)$ to $H_*(p^{-1}(b_1);R)$.* ∎

Let $b_0 \in B$ be a base point and let $F = p^{-1}(b_0)$. Given a map $\alpha: X \to B$, an *admissible lifting of* α is a map $\bar{\alpha}: X \times F \to E$ such that $p\bar{\alpha}(x,z) = \alpha(x)$ for $x \in X$ and $z \in F$ and such that for any $x \in X$ the map $f_x: F \to p^{-1}(\alpha(x))$ defined by $f_x(z) = \bar{\alpha}(x,z)$ for $z \in Z$ is an admissible map. The following result is a useful criterion for the admissibility of a lifting.

11 LEMMA *Let $p: E \to B$ be a fibration and let X be a path-connected space. Given maps $\alpha: X \to B$ and $\bar{\alpha}: X \times F \to E$ such that $p\bar{\alpha}(x,z) = \alpha(x)$, then $\bar{\alpha}$ is an admissible lifting of α if and only if there is some $x_0 \in X$ such that $f_{x_0}: F \to p^{-1}(\alpha(x_0))$ is admissible.*

PROOF The necessity of the condition is obvious. To prove the sufficiency, let $x_1 \in X$ and let $\omega: I \to X$ be a path from x_0 to x_1. Since f_{x_0} is admissible, there is a path ω' in B from b_0 to $\alpha(x_0)$ such that $[f_{x_0}] = h[\omega']$. It is then easy to verify that $[f_{x_1}] = h[\omega' * (\alpha \circ \omega)]$, and so f_{x_1} is admissible. ∎

We want to prove the existence of admissible liftings in certain cases. The following is an alternate version of corollary 7.2.7 valid for a nonpolyhedral pair (X,A).

12 LEMMA *Let $p: E \to B$ be a fibration and let X be a space and A a strong deformation retract of X. Given maps $f: A \to E$ and $g: X \to B$ such that $p \circ f = g \mid A$, there is a map $\bar{g}: X \to E$ such that $p \circ \bar{g} = g$ and $\bar{g} \mid A$ is fiber homotopic to f.*

PROOF Let $D: X \times I \to X$ be a homotopy relative to A from some retraction $r: X \to A$ to 1_X (D exists, because A is a strong deformation retract of X). Then $g \circ D: X \times I \to B$ and $f \circ r: X \to E$ are maps such that $gD(x,0) = gr(x) = pfr(x)$. By the homotopy lifting property of p, there exists a map $F: X \times I \to E$ such that $p \circ F = g \circ D$ and $F(x,0) = fr(x)$. Let $\bar{g}: X \to E$ be defined by $\bar{g}(x) = F(x,1)$. Then

$$p\bar{g}(x) = pF(x,1) = gD(x,1) = g(x)$$

and $F \mid A \times I$ is a fiber homotopy from f to $\bar{g} \mid A$. Therefore $\bar{g}$ has the requisite properties. ∎

Let $p: E \to B$ be a fibration over a path-connected base space and let (B,A) be a relative CW complex. Let $b_0 \in B$ be a base point and $F = p^{-1}(b_0)$. For any s, v_0 is a strong deformation retract of Δ^s, and so $v_0 \times F$ is a strong deformation retract of $\Delta^s \times F$. It follows from lemma 12 that, given a singular simplex

$$\sigma: (\Delta^s,\dot{\Delta}^s) \to ((B,A)^s, (B,A)^{s-1})$$

in $\bar{\Delta}((B,A)^s)$, there exist maps $\bar{\sigma}: (\Delta^s,\dot{\Delta}^s) \times F \to (E_s,E_{s-1})$ such that $p\bar{\sigma}(x,z) = \sigma(x)$ for $x \in \Delta^s$ and $z \in F$ and such that $\bar{\sigma} \mid v_0 \times F: F \to p^{-1}(\sigma(v_0))$ is admissible. By lemma 11, $\bar{\sigma}$ is an admissible lifting of σ.

If $\bar{\sigma}_0, \bar{\sigma}_1: (\Delta^s,\dot{\Delta}^s) \times F \to (E_s,E_{s-1})$ are two admissible liftings of σ, there is an admissible map $f: F \to F$ such that $\bar{\sigma}_0 \mid v_0 \times F \underset{p}{\simeq} (\bar{\sigma}_1 \mid v_0 \times F) \circ f$. Let $g: \Delta^s \times F \times 0 \cup v_0 \times F \times I \cup \Delta^s \times F \times 1 \to E_s$ be the map defined by $g(x,z,0) = \bar{\sigma}_0(x,z)$, $g(x,z,1) = \bar{\sigma}_1(x,f(z))$, and $g \mid v_0 \times F \times I: \bar{\sigma}_0 \mid v_0 \times F \underset{p}{\simeq} (\bar{\sigma}_1 \mid v_0 \times F) \circ f$. Since $p \circ g$ can be extended to a map $\Delta^s \times F \times I \to B$ [by sending (x,z,t) to $\sigma(x)$] and $\Delta^s \times F \times 0 \cup v_0 \times F \times I \cup \Delta^s \times F \times 1$ is a strong deformation retract of $\Delta^s \times F \times I$, it follows from lemma 12 that

$g \mid \Delta^s \times F \times 0 \underset{p}{\simeq} g \mid \Delta^s \times F \times 1$. Therefore $\bar{\sigma}_0$ is homotopic to the composite

$$(\Delta^s,\dot{\Delta}^s) \times F \xrightarrow{1 \times f} (\Delta^s,\dot{\Delta}^s) \times F \xrightarrow{\bar{\sigma}_1} (E_s,E_{s-1})$$

In case $p: E \to B$ is orientable over R, then $f_*: H_*(F) \to H_*(F)$ is the identity map, and so the composite

$$H_*((\Delta^s,\dot{\Delta}^s) \times F) \xrightarrow{(1 \times f)_*} H_*((\Delta^s,\dot{\Delta}^s) \times F) \xrightarrow{\bar{\sigma}_{1*}} H_*(E_s,E_{s-1})$$

equals $\bar{\sigma}_{1*}$. Therefore $\bar{\sigma}_{0*} = \bar{\sigma}_{1*}$, and we have proved the following result.

13 THEOREM *Let $p: E \to B$ be an orientable fibration over a path-connected relative CW complex (B,A), with $F = p^{-1}(b_0)$. For $\sigma: (\Delta^s,\dot{\Delta}^s) \to ((B,A)^s, (B,A)^{s-1})$ there is a well-defined homomorphism*

$$\bar{\sigma}_*: H_*((\Delta^s,\dot{\Delta}^s) \times F) \to H_*(E_s,E_{s-1})$$

defined to be the homomorphism induced by any admissible lifting $\bar{\sigma}: (\Delta^s,\dot{\Delta}^s) \times F \to (E_s,E_{s-1})$ of σ. ■

The identity map $\xi_s: \Delta^s \subset \Delta^s$ is a cycle modulo $\dot{\Delta}^s$, and its homology class $\{\xi_s\}$ generates $H_s(\Delta^s,\dot{\Delta}^s; R)$. Given $w \in H_n(F;G)$, then $\{\xi_s\} \times w \in H_{n+s}((\Delta^s,\dot{\Delta}^s) \times F;\, G)$ and $\bar{\sigma}_*(\{\xi_s\} \times w) \in H_{n+s}(E_s,E_{s-1};\, G)$. It is clear that for fixed σ the map $w \to \bar{\sigma}_*(\{\xi_s\} \times w)$ is a homomorphism from $H_n(F;G)$ to $H_{n+s}(E_s,E_{s-1};\, G)$. Because the cellular σ's form a basis of the free module $\bar{\Delta}_s(B) = \bar{\Delta}_s((B,A)^s)$, there is a homomorphism

$$\psi: \bar{\Delta}_s((B,A)^s) \otimes H_n(F;G) \to H_{n+s}(E_s,E_{s-1};\, G)$$

defined by $\psi(\sigma \otimes w) = \bar{\sigma}_*(\{\xi_s\} \times w)$. If $\sigma(\Delta^s) \subset (B,A)^{s-1}$, then $\bar{\sigma}(\Delta^s \times F) \subset E_{s-1}$ for any admissible lifting of σ. Therefore $\bar{\sigma}_*(\{\xi_s\} \times w) = 0$, and so ψ defines a homomorphism

$$\psi: \bar{\Delta}_s((B,A)^s, (B,A)^{s-1}) \otimes H_n(F;G) \to H_{n+s}(E_s,E_{s-1};\, G)$$

Next we show that ψ induces a homomorphism from the module $H_s((B,A)^s, (B,A)^{s-1}; H_n(F;G))$.

14 LEMMA *The composite*

$$\begin{array}{c} \bar{\Delta}_{s+1}((B,A)^s, (B,A)^{s-1}) \otimes H_n(F;G) \\ \downarrow{\scriptstyle \partial \otimes 1} \\ \bar{\Delta}_s((B,A)^s, (B,A)^{s-1}) \otimes H_n(F;G) \\ \downarrow{\scriptstyle \psi} \\ H_{n+s}(E_s,E_{s-1};G) \end{array}$$

is trivial.

PROOF Let $\sigma: (\Delta^{s+1},(\Delta^{s+1})^{s-1}) \to ((B,A)^s,(B,A)^{s-1})$ be a cellular $(s+1)$-simplex of $(B,A)^s$ and let

$$\bar{\sigma}: (\Delta^{s+1},(\Delta^{s+1})^{s-1}) \times F \to (E_s,E_{s-1})$$

be an admissible lifting of σ. For $0 \le i \le s+1$ let $e^i_{s+1}: \Delta^s \to \Delta^{s+1}$ omit the ith vertex. Then $\sigma^{(i)} = \sigma \circ e^i_{s+1}$, and the composite

$$\Delta^s \times F \xrightarrow{e^i_{s+1} \times 1} \Delta^{s+1} \times F \xrightarrow{\sigma'} E_s$$

where $\sigma' = \bar{\sigma} \mid \Delta^{s+1} \times F$, is an admissible lifting of $\sigma^{(i)}$. Therefore

$$\psi(\sigma^{(i)} \otimes w) = \sigma'_*(e^i_{s+1} \times 1)_*(\{\xi_s\} \times w) = \sigma'_*(\{e^i_{s+1}\} \times w)$$

where $\{e^i_{s+1}\} \in H_s(\dot{\Delta}^{s+1},(\Delta^{s+1})^{s-1})$. It follows that

$$\psi(\partial\sigma \times w) = \sigma'_*(\{\sum (-1)^i e^i_{s+1}\} \times w)$$

However, in $\Delta(\Delta^{s+1})$ we have the relation $\partial\xi_{s+1} = \sum (-1)^i e^i_{s+1}$. Hence, if $j: (\dot{\Delta}^{s+1},(\Delta^{s+1})^{s-1}) \subset (\Delta^{s+1},(\Delta^{s+1})^{s-1})$, then $j_* \{\sum (-1)^i e^i_{s+1}\} = 0$. Because σ' equals the composite

$$\dot{\Delta}^{s+1} \times F \overset{j \times 1}{\subset} \Delta^{s+1} \times F \xrightarrow{\bar{\sigma}} E_s$$

it follows that

$$\psi(\partial\sigma \otimes w) = \bar{\sigma}_*(j \times 1)_*(\{\sum (-1)^i e^i_{s+1}\} \times w) = \bar{\sigma}_*(0) = 0 \quad \blacksquare$$

Every element of $\bar{\Delta}_s((B,A)^s, (B,A)^{s-1}) \otimes H_n(F;G)$ is an s-dimensional cycle of the chain complex $\bar{\Delta}((B,A)^s, (B,A)^{s-1}) \otimes H_n(F;G)$, and the s-dimensional boundaries are the elements in the image of

$$\partial \otimes 1: \bar{\Delta}_{s+1}((B,A)^s, (B,A)^{s-1}) \otimes H_n(F;G) \to \bar{\Delta}_s((B,A)^s, (B,A)^{s-1}) \otimes H_n(F;G)$$

It follows from lemma 14 that ψ induces a homomorphism

$$\psi_*: H_s((B,A)^s, (B,A)^{s-1}; H_n(F;G)) \to H_{n+s}(E_s,E_{s-1}; G)$$

The computation of the E^1 term of the spectral sequence is completed by the following result.

15 THEOREM *(a) For all $s \ge 0$ there is an isomorphism*

$$\psi_*: H_s((B,A)^s, (B,A)^{s-1}; H_n(F;G)) \approx H_{n+s}(E_s,E_{s-1}; G)$$

(b) For $s \ge 1$ there is a commutative square

$$\begin{array}{ccc} H_s((B,A)^s, (B,A)^{s-1}; H_n(F;G)) & \xrightarrow{\psi_*} & H_{n+s}(E_s,E_{s-1}; G) \\ \partial\downarrow & & \downarrow\partial \\ H_{s-1}((B,A)^{s-1}, (B,A)^{s-2}; H_n(F;G)) & \xrightarrow{\psi_*} & H_{n+s-1}(E_{s-1},E_{s-2}; G) \end{array}$$

PROOF (a) Because of direct-sum properties of both modules (the right-hand one by lemma 2) and an obvious naturality property of ψ_*, it suffices to prove that for an s-cell e of $B - A$ the map

$$\psi_*: H_s(e,\dot{e}; H_n(F;G)) \to H_{n+s}(p^{-1}(e),p^{-1}(\dot{e}); G)$$

is an isomorphism. Let $f: (E^s,S^{s-1}) \to (e,\dot{e})$ be a characteristic map for e and let $p': E' \to E^s$ be the induced fibration over E^s with corresponding map f': $(E', p'^{-1}(S^{s-1})) \to (p^{-1}(e), p^{-1}(\dot{e}))$. Then there is a commutative square

$$\begin{array}{ccc} H_s(E^s,S^{s-1};\, H_n(F;G)) & \xrightarrow{\psi_*} & H_{n+s}(E',p'^{-1}(S^{s-1});\, G) \\ {\scriptstyle f_*}\downarrow & & \downarrow{\scriptstyle f'_*} \\ H_s(e,\dot{e};\, H_n(F;G)) & \xrightarrow{\psi_*} & H_{n+s}(p^{-1}(e),p^{-1}(\dot{e});\, G) \end{array}$$

in which the vertical maps are isomorphisms (by excision and homotopy properties). Therefore it suffices to prove the result for a trivial fibration over E^s, and for such a fibration ψ_* is an isomorphism by the Künneth theorem.

(*b*) Given $\sigma\colon (\Delta^s,\dot{\Delta}^s) \to ((B,A)^s, (B,A)^{s-1})$, let $\{\sigma \otimes w\}$ be the element of $H_s((B,A)^s, (B,A)^{s-1}; H_n(F;G))$ determined by the cycle $\sigma \otimes w$. Then in $H_{s-1}((B,A)^{s-1}, (B,A)^{s-2}; H_n(F;G))$, we have $\partial\{\sigma \otimes w\} = \{\Sigma\,(-1)^i\sigma^{(i)} \otimes w\}$. Let $\bar{\sigma}\colon (\Delta^s,\dot{\Delta}^s) \times F \to (E_s,E_{s-1})$ be an admissible lifting of σ. For $0 \le i \le s$ the composite

$$(\Delta^{s-1},\dot{\Delta}^{s-1}) \times F \xrightarrow{e_s^i \times 1} (\dot{\Delta}^s,(\Delta^s)^{s-2}) \times F \xrightarrow{\sigma'} (E_{s-1},E_{s-2})$$

where $\sigma' = \bar{\sigma} \mid (\dot{\Delta}^s,(\Delta^s)^{s-2})$, is an admissible lifting of $\sigma^{(i)}$. Therefore

$$\begin{aligned} \psi_*\partial\{\sigma \otimes w\} &= \Sigma\,(-1)^i\sigma'_*(e_s^i \times 1)_*(\{\xi_{s-1}\} \times w) = \sigma'_*(\{\Sigma\,(-1)^ie_s{}^i\} \times w) \\ &= \sigma'_*(\partial\{\xi_s\} \times w) = \sigma'_*\partial(\{\xi_s\} \times w) = \partial\bar{\sigma}_*(\{\xi_s\} \times w) \\ &= \partial\psi_*\{\sigma \otimes w\} \quad \blacksquare \end{aligned}$$

Because $H_s((B,A)^s, (B,A)^{s-1})$ is a free module, it follows from the universal-coefficient theorem that

$$\begin{aligned} H_s((B,A)^s, (B,A)^{s-1}; H_n(F;G)) &\approx H_s((B,A)^s, (B,A)^{s-1}) \otimes H_n(F;G) \\ &= C_s(B,A) \otimes H_n(F;G) \end{aligned}$$

Under this isomorphism, it is easy to see that the boundary operator of the triple $((B,A)^s, (B,A)^{s-1}, (B,A)^{s-2})$ corresponds to the map

$$\partial \otimes 1\colon C_s(B,A) \otimes H_n(F;G) \to C_{s-1}(B,A) \otimes H_n(F;G)$$

Therefore theorem 15 can be interpreted as asserting that ψ induces an isomorphism of the bigraded chain complex $C_*(B,A) \otimes H_*(F;G)$, with the E^1 term of the spectral sequence of theorem 1. This, together with corollary 4, gives the following result about the E^2 term.

16 THEOREM *Let $p\colon E \to B$ be an orientable fibration over a path-connected relative CW complex (B,A) and let $F = p^{-1}(b_0)$. There is a convergent E^2 spectral sequence with $E^2_{s,t} \approx H_s(B,A; H_t(F;G))$ and E^∞ the bigraded module associated to the filtration of $H_*(E,E_A; G)$ defined by*

$$F_sH_*(E,E_A; G) = \operatorname{im}\,[H_*(E_s,E_A; G) \to H_*(E,E_A; G)] \quad \blacksquare$$

Note that the spectral sequence of theorem 16 is a first-quadrant spectral sequence and is functorial on the category of orientable fibrations $p\colon E \to B$ over a path-connected relative CW complex (B,A) and fiber-preserving maps $f'\colon E' \to E$ such that the base space pair is mapped by a cellular map $f\colon (B',A') \to (B,A)$.

To extend the spectral sequence to fibrations with more general base spaces, let $p: E \to B$ be an orientable fibration over a path-connected base space B and let $A \subset B$. Let $f: (B',A) \to (B,A)$ be a relative CW approximation to (B,A) (which exists by theorem 7.8.1). Let $p': E' \to B'$ be the induced fibration and $f': E' \to E$ the fiber-preserving map induced by f. It follows from the exactness of the homotopy sequence of a fibration and the five lemma that f' is a weak homotopy equivalence. Therefore f' induces an isomorphism of the homology sequence of (E',E'_A) with the homology sequence of (E,E_A). Because B' is path connected and $p': E' \to B'$ is orientable, there is a convergent E^2 spectral sequence with

$$E^2_{s,t} \approx H_s(B',A;\ H_t(F;G)) \approx H_s(B,A;\ H_t(F;G))$$

and E^∞ associated to a filtration of $H_*(E',E'_A;\ G) \approx H_*(E,E_A;\ G)$. If $g: (B'',A) \to (B,A)$ is another relative CW approximation to (B,A), there is a cellular map $h: (B'',A) \to (B',A)$ such that $f \circ h \simeq g$ rel A. The map h induces an isomorphism of the E^2 spectral sequences of $p'': E'' \to B''$ and $p': E' \to B'$ (but not an isomorphism of the E^1 terms). It follows that the filtration of $H_*(E,E_A;\ G)$ induced by the isomorphisms $H_*(E',E'_A;\ G) \approx H_*(E,E_A;\ G)$ and $H_*(E'',E''_A;\ G) \approx H_*(E,E_A;\ G)$ correspond, and we have the following main result.

17 THEOREM *Let $p: E \to B$ be an orientable fibration with B path connected and fiber F over $b_0 \in B$. Given $A \subset B$, there is a convergent E^2 spectral sequence, with $E^2_{s,t} \approx H_s(B,A;\ H_t(F;G))$ and E^∞ the bigraded module associated to some filtration of $H_*(E,E_A;\ G)$. This spectral sequence is a first-quadrant spectral sequence functorial on the category of orientable fibrations and fiber-preserving maps.* ∎

3 APPLICATIONS OF THE HOMOLOGY SPECTRAL SEQUENCE

In this section we shall consider applications of the spectral sequence of a fibration and show that it leads to generalized Gysin and Wang homology sequences in case the fiber or base is a homology sphere. We shall also use the spectral sequence in the proof of the homotopy excision theorem. The section concludes with a definition of the Hopf invariant homomorphism and an exact sequence connecting it and the suspension homomorphism of homotopy groups of spheres.

1 THEOREM *Let $p: E \to B$ be a fibration which is orientable over a field with path-connected base and with fiber F. Assume that the Euler characteristics $\chi(F)$ and $\chi(B)$ are defined (over the field). Then $\chi(E)$ is defined, and $\chi(E) = \chi(B)\chi(F)$.*

PROOF We use the spectral sequence of theorem 9.2.17. For a finitely generated bigraded module E^r we define the Euler characteristic $\chi(E^r) = \Sigma_{s,t}(-1)^{s+t} \dim E^r_{s,t}$. Because we are considering a field as coefficients, it follows from the Künneth formula that

$$E^2_{s,t} \approx H_s(B;H_t(F)) \approx H_s(B) \otimes H_t(F)$$

Therefore $\chi(E^2) = \chi(B)\chi(F)$. Because $E^{r+1} \approx H(E^r)$, it follows (as in theorem 4.3.14) that

$$\chi(E^2) = \chi(E^3) = \cdots = \chi(E^r)$$

Because $E^2_{s,t} = 0$ if s and t are large enough, the same is true of $E^r_{s,t}$ for any r. Therefore $E^\infty = E^r$ for large enough r, and so $\chi(E^\infty) = \chi(B)\chi(F)$. By a standard property of dimension,

$$\dim [H_n(E)] = \sum_{s+t=n} \dim E^\infty_{s,t}$$

and so $\chi(E) = \chi(E^\infty) = \chi(B)\chi(F)$. ■

We now compute the homomorphism induced by $i: F \subset E$ in terms of the spectral sequence. For $r \geq 2$, $E^{r+1}_{0,t}$ is a quotient of $E^r_{0,t}$ (because $E^r_{-r,t+r-1} = 0$ in a first-quadrant spectral sequence). Therefore there is an epimorphism $E^2_{0,t} \to E^\infty_{0,t}$. Because B is path connected, there is an isomorphism $H_t(F;G) \approx H_0(B;\ H_t(F;G))$. By using the spectral sequence of the fibration $F \to b_0$ and the functorial property of the spectral sequence, it follows that $i_*: H_t(F;G) \to H_t(E;G)$ is the composite

$$H_t(F;G) \approx H_0(B;\ H_t(F;G)) \approx E^2_{0,t} \to E^\infty_{0,t} = F_0H_t(E;G) \subset H_t(E;G)$$

This leads to the following *generalized Wang homology sequence.*

2 THEOREM *Let $p: E \to B$ be a fibration, with fiber F and simply connected base B which is a homology n-sphere (over R) for some $n \geq 2$ [that is, $H_q(B) = 0$ if $q \neq 0$ or n and $H_0(B) \approx R \approx H_n(B)$]. Then there is an exact sequence*

$$\cdots \to H_t(F;G) \xrightarrow{i_*} H_t(E;G) \to H_{t-n}(F;G) \to H_{t-1}(F;G) \xrightarrow{i_*} \cdots$$

PROOF Because $H_*(B)$ has no torsion, $E^2_{s,t} \approx H_s(B) \otimes H_t(F;G)$ in the spectral sequence of p. Therefore $E^2_{s,t} = 0$ unless $s = 0$ or n, and the only nonzero differential is $d^n: E^2_{n,t} \to E^2_{0,t+n-1}$. Hence there are exact sequences

$$0 \to E^\infty_{n,t} \to E^2_{n,t} \xrightarrow{d^n} E^2_{0,t+n-1} \to E^\infty_{0,t+n-1} \to 0$$

and

$$0 \to E^\infty_{0,t} \to H_t(E;G) \to E^\infty_{n,t-n} \to 0$$

These fit together into an exact sequence

$$\cdots \to H_t(E;G) \to E^2_{n,t-n} \xrightarrow{d^n} E^2_{0,t-1} \to H_{t-1}(E;G) \to \cdots$$

The result follows on observing that

$$E^2_{n,t-n} \approx H_n(B) \otimes H_{t-n}(F;G) \approx H_{t-n}(F;G)$$
$$E^2_{0,t-1} \approx H_0(B) \otimes H_{t-1}(F;G) \approx H_{t-1}(F;G)$$

and that on replacing $E^2_{0,t-1}$ by $H_{t-1}(F;G)$ in the exact sequence, the resulting map $H_{t-1}(F;G) \to H_{t-1}(E;G)$ is i_*. ∎

Let $p: E \to B$ be an orientable fibration with path-connected base and let $B' \subset B$ and $E' = p^{-1}(B')$. We now show how the homomorphism induced by $p: (E,E') \to (B,B')$ is determined from the spectral sequence. For $r \geq 2$, $E^{r+1}_{s,0}$ is a submodule of $E^r_{s,0}$ (because $E^r_{s+r,-r+1} = 0$). Therefore there is a monomorphism $E^\infty_{s,0} \to E^2_{s,0}$. The augmentation homomorphism $H_0(F;G) \to G$ induces a homomorphism $H_s(B,B';\ H_0(F;G)) \to H_s(B,B';\ G)$. By using the spectral sequence of the fibration $B \subset B$ and the functorial property of the spectral sequence, it follows that $p_*: H_s(E,E';\ G) \to H_s(B,B';\ G)$ is the composite

$$H_s(E,E';\ G) = F_sH_s(E,E';\ G) \to E^\infty_{s,0} \to E^2_{s,0} \approx H_s(B,B';\ H_0(F;G)) \to H_s(B,B';\ G)$$

This leads to the following *generalized Gysin homology sequence.*

3 THEOREM *Let $p: E \to B$ be an orientable fibration with path-connected base space and with fiber F a homology n-sphere (over R), where $n \geq 1$. If $B' \subset B$ and $E' = p^{-1}(B')$, there is an exact sequence*

$$\cdots \to H_s(E,E';\ G) \xrightarrow{p_*} H_s(B,B';\ G) \to H_{s-n-1}(B,B';\ G) \to H_{s-1}(E,E';\ G) \xrightarrow{p_*} \cdots$$

PROOF Because, in the spectral sequence of p,

$$E^2_{s,t} \approx H_s(B,B';\ H_t(F;G)) = 0 \qquad t \neq 0 \text{ or } n$$

the only nonzero differential is $d^{n+1}: E^2_{s,0} \to E^2_{s-n-1,n}$. Hence there are exact sequences

$$0 \to E^\infty_{s,0} \to E^2_{s,0} \xrightarrow{d^{n+1}} E^2_{s-n-1,n} \to E^\infty_{s-n-1,n} \to 0$$

and

$$0 \to E^\infty_{s-n,n} \to H_s(E,E';\ G) \to E^\infty_{s,0} \to 0$$

These fit together into an exact sequence

$$\cdots \to H_s(E,E';\ G) \to E^2_{s,0} \xrightarrow{d^{n+1}} E^2_{s-n-1,n} \to H_{s-1}(E,E';\ G) \to \cdots$$

The result follows on observing that

$$E^2_{s,0} \approx H_s(B,B';\ H_0(F;G)) \approx H_s(B,B';\ G)$$
$$E^2_{s-n-1,n} \approx H_{s-n-1}(B,B';\ H_n(F;G)) \approx H_{s-n-1}(B,B';\ G)$$

and that on replacing $E^2_{s,0}$ by $H_s(B,B';\ G)$ in the exact sequence, the resulting map $H_s(E,E';\ G) \to H_s(B,B';\ G)$ is p_*. ∎

4 LEMMA *Let $p: E \to B$ be an orientable fibration with path-connected base space and with path-connected fiber F. Assume that $H_q(B,B') = 0$ for $q < n$ and $H_q(F) = 0$ for $0 < q < m$ (all coefficients R). Then the homomor-*

phism $p_*: H_q(E,E') \to H_q(B,B')$ *is an isomorphism for* $q \le n + m - 1$ *and an epimorphism for* $q = n + m$.

PROOF For the spectral sequence we have

$$E^2_{s,t} \approx H_s(B,B'; H_t(F)) \approx H_s(B,B') \otimes H_t(F) \oplus H_{s-1}(B,B') * H_t(F)$$

By the hypotheses, $E^2_{s,t} = 0$ if $s < n$ or $0 < t < m$. Therefore, if $q \le n + m - 1$, then $E^2_{s,q-s} = 0$, except possibly for the term $E^2_{q,0}$. It follows that $E^r_{s,q-s} = 0$, except for the term $E^r_{q,0}$, and $E^r_{q,0} \approx E^2_{q,0}$. Therefore $E^\infty_{q,0} \approx E^2_{q,0}$ and $E^\infty_{s,q-s} = 0$ if $s \ne q$. Hence

$$H_q(E,E') \approx H_q(B,B'; H_0(F)) \approx H_q(B,B')$$

and the isomorphism is induced by p_*.

If $q = n + m$, then $E^2_{s,n+m-s} = 0$ except for the terms $E^2_{n+m,0}$ and $E^2_{n,m}$. Since $E^2_{n+m-r,r-1} = 0$ for $r \ge 2$, it follows that

$$E^\infty_{n+m,0} \approx E^2_{n+m,0} \approx H_{n+m}(B,B'; H_0(F)) \approx H_{n+m}(B,B')$$

Therefore $p_*(H_{n+m}(E,E')) = H_{n+m}(B,B')$. ∎

We use this to prove the following *homotopy excision theorem.*[1]

5 THEOREM *Let A, B, and A ∩ B be path-connected subspaces of a simple space X such that*

(*a*) *Either* $X = \text{int } A \cup \text{int } B$, *or* $X = A \cup B$ *where A and B are closed subsets of X such that A ∩ B is a strong deformation retract of some neighborhood in A (or in B).*

(*b*) *A ∩ B, A, B, and X have isomorphic fundamental groups.*

(*c*) *(A, A ∩ B) is n-connected and (B, A ∩ B) is m-connected, where* $n, m \ge 1$.

Then the homomorphism

$$j_\#: \pi_q(A, A \cap B) \to \pi_q(X,B)$$

induced by the excision map $j: (A, A \cap B) \subset (X,B)$ *is an isomorphism for* $q \le n + m - 1$ *and an epimorphism for* $q = n + m$.

PROOF First we reduce consideration to the case $X = \text{int } A \cup \text{int } B$. If A and B are closed in X and $A \cap B$ is a strong deformation retract of some neighborhood U in B, let $A' = A \cup U$ and observe that A is a strong deformation retract of A'. Furthermore, $A' \cap B = U$, and the inclusion map $(A, A \cap B) \subset (A', A' \cap B)$ is a homotopy equivalence, so that $(A', A' \cap B)$ is n-connected. By the exactness of the homotopy sequence of the triple $(B, A' \cap B, A \cap B)$ and the fact that $(A' \cap B, A \cap B)$ is k-connected for all k, we see that $(B, A' \cap B)$ is m-connected. Note that

$$X = A \cup (B - A) \subset \text{int } A' \cup \text{int } B,$$

[1] A more general form of this theorem can be found in A. L. Blakers and W. S. Massey, The homotopy groups of a triad, II, *Annals of Mathematics,* vol. 55, pp. 192–201, 1952.

and so A' and B satisfy conditions (*a*), (*b*), and (*c*). Since there is a commutative triangle

$$\begin{array}{ccc} \pi_q(A, A \cap B) & \xrightarrow{\approx} & \pi_q(A', A' \cap B) \\ & {}_{j_\#}\searrow \quad \swarrow_{j'_\#} & \\ & \pi_q(X,B) & \end{array}$$

we are reduced to proving that $j'_\#$ has the desired properties.

Similarly, if $A \cap B$ is a strong deformation retract of some neighborhood V in A, let $B' = V \cup B$ and observe that B is a strong deformation retract of B'. Then $A \cap B' = V$, and it follows, as in the case above, that $(A, A \cap B')$ is n-connected and $(B', A \cap B')$ is m-connected. Since $X = (A - B) \cup B$ is contained in int A ∪ int B', we see that A and B' satisfy conditions (*a*), (*b*), and (*c*). From the commutativity of the square

$$\begin{array}{ccc} \pi_q(A, A \cap B) & \xrightarrow{\approx} & \pi_q(A, A \cap B') \\ {\scriptstyle j_\#}\downarrow & & \downarrow{\scriptstyle j''_\#} \\ \pi_q(X,B) & \xrightarrow{\approx} & \pi_q(X,B') \end{array}$$

we are reduced to proving that $j''_\#$ has the desired properties.

In either case we have shown that it suffices to prove the theorem under the hypothesis that $X = \text{int } A \cup \text{int } B$, and we make this assumption now. By corollary 8.3.8, there is a fibration $p: E \to X$ such that E is simply connected and $p_\#: \pi_q(E) \approx \pi_q(X)$ for $q > 1$. Let E_A and E_B be the parts of E over A and B, respectively, and note that $E_A \cap E_B$ is the part of E over $A \cap B$. From theorem 7.2.8 it follows that $(E_A, E_A \cap E_B)$ is n-connected and $(E_B, E_A \cap E_B)$ is m-connected. Using (*b*) and the exactness of the homotopy sequence of a fibration, it is easy to see that $E_A \cap E_B$, E_A, and E_B are all simply connected. Since it is obvious that $E \subset p^{-1}(\text{int } A) \cup p^{-1}(\text{int } B) \subset \text{int } E_A \cup \text{int } E_B$, we have reduced the theorem to the case where all the spaces in question are simply connected by virtue of the commutativity of the square

$$\begin{array}{ccc} \pi_q(E_A, E_A \cap E_B) & \xrightarrow{\approx} & \pi_q(A, A \cap B) \\ {\scriptstyle \bar{j}_\#}\downarrow & & \downarrow{\scriptstyle j_\#} \\ \pi_q(E,E_B) & \xrightarrow{\approx} & \pi_q(X,B) \end{array}$$

Thus, assume $X = \text{int } A \cup \text{int } B$ and that $A \cap B$, A, B, and X are all simply connected. We replace the inclusion map $A \subset X$ by the homotopically equivalent mapping path fibration $p: P \to X$ as in theorem 2.8.9. Then P is the space of paths $\omega: (I,0) \to (X,A)$ in the compact-open topology, and $p(\omega) = \omega(1)$. The fiber F of p over a point $a_0 \in A \cap B$ is the space of paths in X which start in A and end at a_0. If $p': PX \to X$ is the path fibration of all paths in X which end at a_0 and $p'(\omega) = \omega(0)$, then $F = p'^{-1}(A)$. Since PX is contractible, there are isomorphisms

$$\pi_q(X,A) \underset{\approx}{\xleftarrow{p'_\#}} \pi_q(PX,F) \underset{\approx}{\xrightarrow{\partial}} \pi_{q-1}(F)$$

Because $X = \operatorname{int} A \cup \operatorname{int} B$, the excision map $j'\colon (B, A \cap B) \subset (X,A)$ induces isomorphisms in homology. It follows from the relative Hurewicz isomorphism theorem and the m-connectedness of $(B, A \cap B)$ that (X,A) is also m-connected. Therefore F is $(m-1)$-connected, and so $H_q(F) = 0$ for $0 < q < m$.

Let $E' = p^{-1}(B)$ and observe that since X is simply connected, the fibration $p\colon P \to X$ is orientable. Since $j_*\colon H_q(A, A \cap B) \approx H_q(X,B)$, it follows that $H_q(X,B) = 0$ for $q < n + 1$. By lemma 4, the homomorphism

$$p_*\colon H_q(P,E') \to H_q(X,B)$$

is an isomorphism for $q \le n + m$ and an epimorphism for $q = n + m + 1$. The map $j\colon (A, A \cap B) \subset (X,B)$ has a lifting $\bar{j}\colon (A, A \cap B) \to (P,E')$, where $\bar{j}(a)$ is the constant path at a for all $a \in A$. There is a commutative triangle

$$\begin{array}{ccc} H_q(A, A \cap B) & \xrightarrow{\bar{j}_*} & H_q(P,E') \\ & {\scriptstyle j_*}\searrow\approx \quad \swarrow{\scriptstyle p_*} & \\ & H_q(X,B) & \end{array}$$

Therefore $\bar{j}_*$ is an isomorphism for $q \le n + m$. Since $\bar{j} \mid A\colon A \to P$ is a homotopy equivalence, it follows from the five lemma that the homomorphism

$$(\bar{j} \mid A \cap B)_*\colon H_q(A \cap B) \to H_q(E')$$

is an isomorphism for $q \le n + m - 1$.

Because $\pi_1(E') \approx \pi_1(F) \approx \pi_2(X,A)$, and the latter group is a quotient group of $\pi_2(X)$ since $\pi_1(A) \approx \pi_1(X)$, we see that E' has an abelian fundamental group. Since $A \cap B$ is simply connected, it follows from the absolute Hurewicz isomorphism theorem that E' is also simply connected. By the Whitehead theorem, the homomorphism

$$(\bar{j} \mid A \cap B)_\#\colon \pi_q(A \cap B) \to \pi_q(E')$$

is an isomorphism for $q \le n + m - 2$ and an epimorphism for $q = n + m - 1$. Since $\bar{j} \mid A\colon A \to E$ is a homotopy equivalence, it follows from the five lemma that the homomorphism

$$\bar{j}_\#\colon \pi_q(A, A \cap B) \to \pi_q(P,E')$$

is an isomorphism for $q \le n + m - 1$ and an epimorphism for $q = n + m$. The result follows from this and the commutativity of the triangle

$$\begin{array}{ccc} \pi_q(A, A \cap B) & \xrightarrow{\bar{j}_\#} & \pi_q(P,E') \\ & {\scriptstyle j_\#}\searrow \quad \approx\swarrow{\scriptstyle p_\#} & \\ & \pi_q(X,B) & \end{array} \quad \blacksquare$$

It should be noted that the main argument above involved the case where A and B satisfy (*c*), satisfy (*b*) in the stronger form that all the spaces in question are simply connected, and satisfy the condition that $\{A, B\}$ is an excisive couple of subsets of X, which is a weak form of (*a*). It should also be observed that if A and B satisfy condition (a) of theorem 5, then if $(A, A \cap B)$ is n-connected [or $(B, A \cap B)$ is m-connected], it is easy to show that (X,B) is also n-connected [or (X,A) is m-connected]. Furthermore, if A and B satisfy a and c and $A \cap B$ is simply connected, then it follows that A and B are each simply connected and also that X is simply connected. Hence condition b is also satisfied, and theorem 5 is valid in this case.

6 COROLLARY *Let (X,A) be an n-connected relative CW complex, where $n \geq 2$, such that A is m-connected, where $m \geq 1$. Then the collapsing map $k: (X,A) \to (X/A,x_0)$ induces a homomorphism*

$$k_\#: \pi_q(X,A) \to \pi_q(X/A)$$

which is an isomorphism for $q \leq m + n$ and an epimorphism for $q = m + n + 1$.

PROOF Let CA be the unreduced cone over A and regard it as a space whose intersection with X is A. Since A is m-connected and CA is contractible, it follows that (CA,A) is $(m + 1)$-connected. We shall apply theorem 5, with A and B replaced by X and CA, respectively. Since $X \cap CA = A$ is a strong deformation retract of some neighborhood in CA, a of theorem 5 is satisfied. Since A is simply connected and c is also satisfied, it follows, as in the remarks above, that b is satisfied too. Hence the hypotheses of theorem 5 are satisfied, and it follows that $j: (X,A) \subset (X \cup CA, CA)$ induces a homomorphism

$$j_\#: \pi_q(X,A) \to \pi_q(X \cup CA, CA)$$

which is an isomorphism for $q \leq n + m$ and an epimorphism for $q = n + m + 1$. It follows from lemma 7.1.5 that the collapsing map $k': (X \cup CA, CA) \to (X \cup CA, CA)/CA$ is a homotopy equivalence. The result follows from the commutativity of the triangle

$$\begin{array}{ccc} \pi_q(X,A) & \xrightarrow{j_\#} & \pi_q(X \cup CA, CA) \\ & {\scriptstyle k_\#}\searrow \quad \swarrow{\scriptstyle \approx\, k'_\#} & \\ & \pi_q(X/A) & \end{array}$$

■

7 COROLLARY *Let $f: (X',A') \to (X,A)$ be a relative homeomorphism between relative CW complexes both of which are n-connected, with $n \geq 2$, and such that A' and A are m-connected, with $m \geq 1$. Then f induces an isomorphism*

$$f_\#: \pi_q(X',A') \approx \pi_q(X,A) \qquad q \leq n + m$$

PROOF Let $k': (X',A') \to (X'/A',x_0')$ and $k: (X,A) \to (X/A,x_0)$ be the collapsing maps. Then f induces a homeomorphism $f': X'/A' \to X/A$ such that $f' \circ k' = k \circ f$. Since f' induces isomorphisms of the homotopy groups in all dimensions, the result follows from corollary 6. ■

We use this last result to study the suspension map

$$S: \pi_q(S^n) \to \pi_{q+1}(S^{n+1})$$

in more detail. Since $S^{n+1} = S(S^n)$, there is a characteristic map $\mu'\colon S^n \to \Omega S^{n+1}$ for the path fibration $PS^{n+1} \to S^{n+1}$. From the commutativity of the triangle

$$\begin{array}{ccc} \pi_q(S^n) & \xrightarrow{\mu'_\#} & \pi_q(\Omega S^{n+1}) \\ & S\searrow \quad \nearrow\approx \partial & \\ & \pi_{q+1}(S^{n+1}) & \end{array}$$

it suffices to study the map $\mu'_\#$.

Let X^{2n} be the space obtained from $S^n \times S^n$ by identifying (z,z_0) with (z_0,z) for all $z \in S^n$ (where z_0 is a base point of S^n). We regard S^n as imbedded in X^{2n} as the set of points corresponding to $S^n \times z_0$ in $S^n \times S^n$. Then X^{2n} is a CW complex consisting of S^n and a single $2n$-cell attached by a map $\alpha_n\colon S^{2n-1} \to S^n$.

8 LEMMA *There is a map $g\colon X^{2n} \to \Omega S^{n+1}$, where $n \geq 2$, which is a $(3n-1)$-equivalence such that $g \mid S^n = \mu'$.*

PROOF Let $\mu\colon S^n \times \Omega S^{n+1} \to \Omega S^{n+1}$ be the map defined by $\mu(z,\omega) = \omega * \mu'(z)$. By corollary 8.5.8, μ is homotopic to a clutching function for the fibration $PS^{n+1} \to S^{n+1}$. Let $f\colon S^n \times S^n \to \Omega S^{n+1}$ be defined by $f(z,z') = \mu'(z') * \mu'(z)$. There is a commutative diagram

$$\begin{array}{ccccc} H_{n+1}(C_-S^n,S^n) \otimes H_n(\Omega S^{n+1}) & \xrightarrow{\approx} & H_{2n+1}((C_-S^n,S^n) \times \Omega S^{n+1}) & & \\ \partial\otimes 1 \downarrow\approx & & & \approx\searrow \mu_*\partial & \\ H_n(S^n) \otimes H_n(\Omega S^{n+1}) & \xrightarrow{\approx} & H_{2n}(S^n \times \Omega S^{n+1}) & \xrightarrow{\mu_*} & H_{2n}(\Omega S^{n+1}) \\ 1\otimes\mu'_* \uparrow\approx & & (1\times\mu')_* \nwarrow & & \nearrow f_* \\ H_n(S^n) \otimes H_n(S^n) & \xrightarrow{\approx} & & H_{2n}(S^n \times S^n) & \end{array}$$

Therefore $f_*\colon H_{2n}(S^n \times S^n) \approx H_{2n}(\Omega S^{n+1})$. Since $f \mid S^n \vee S^n$ is homotopic to the map sending (z,z_0) to $\mu'(z)$ and (z_0,z) to $\mu'(z)$, f is homotopic to a map f' such that $f'(z,z_0) = \mu'(z) = f'(z_0,z)$. Then f' defines a map $g\colon X^{2n} \to \Omega S^{n+1}$ such that $g \circ k = f'$, where $k\colon S^n \times S^n \to X^{2n}$ is the quotient map. Then $g \mid S^n = \mu'$, and since $H_n(S^n) \approx H_n(X^{2n})$, $g_*\colon H_n(X^{2n}) \approx H_n(\Omega S^{n+1})$. Since $k_*\colon H_{2n}(S^n \times S^n) \approx H_{2n}(X^{2n})$, it follows that $g_*\colon H_{2n}(X^{2n}) \approx H_{2n}(\Omega S^{n+1})$. The only nontrivial homology groups of X^{2n} are in degrees, 0, n, and $2n$, and in degrees $< 3n$ the only nontrivial homology groups of ΩS^{n+1} are in degrees 0, n, and $2n$. Therefore $g_*\colon H_q(X^{2n}) \approx H_q(\Omega S^{n+1})$ for $q < 3n$. Since $n \geq 2$, X^{2n} and ΩS^{n+1} are both simply connected. By the Whitehead theorem, the homomorphism

$$g_\#\colon \pi_q(X^{2n}) \to \pi_q(\Omega S^{n+1})$$

is an isomorphism for $q < 3n - 1$ and an epimorphism for $q = 3n - 1$. ∎

Let $\bar{\alpha}_n\colon (E^{2n},S^{2n-1}) \to (X^{2n},S^n)$ be the characteristic map for the $2n$-cell of X^{2n} corresponding to the attaching map $\alpha_n\colon S^{2n-1} \to S^n$. Then $\bar{\alpha}_n$ is a relative homeomorphism between $(2n-1)$-connected pairs such that S^{2n-1} and S^n are both $(n-1)$-connected. It follows from corollary 7 that $\bar{\alpha}_{n\#}\colon \pi_q(E^{2n},S^{2n-1}) \to \pi_q(X^{2n},S^n)$ is an isomorphism for $q \le 3n-2$.

The *Hopf invariant*[1] is the homomorphism

$$H\colon \pi_{q+1}(S^{n+1}) \to \pi_{q-1}(S^{2n-1}) \qquad q \le 3n-2$$

defined so that the following diagram is commutative [where $j\colon X^{2n} \subset (X^{2n},S^n)$]:

$$\begin{array}{ccccc}
\pi_{q+1}(S^{n+1}) & \xrightarrow[\approx]{\tilde{\partial}} & \pi_q(\Omega S^{n+1}) & \xrightarrow[\approx]{g_\#^{-1}} & \pi_q(X^{2n}) \\
{\scriptstyle H}\downarrow & & & & \downarrow{\scriptstyle j_\#} \\
\pi_{q-1}(S^{2n-1}) & \xleftarrow[\approx]{\partial} & \pi_q(E^{2n},S^{2n-1}) & \xleftarrow[\approx]{\bar{\alpha}_{n\#}^{-1}} & \pi_q(X^{2n},S^n)
\end{array}$$

The Hopf invariant plays an important role in the study of the suspension homomorphism by virtue of the following exactness property.[2]

9 THEOREM *For $n \ge 2$ there is an exact sequence*

$$\pi_{3n-2}(S^n) \xrightarrow{S} \cdots \to \pi_q(S^n) \xrightarrow{S} \pi_{q+1}(S^{n+1}) \xrightarrow{H} \pi_{q-1}(S^{2n-1}) \xrightarrow{\alpha_{n\#}} \pi_{q-1}(S^n) \to \cdots$$

PROOF The result follows from the exactness of the homotopy sequence of (X^{2n},S^n) and the commutativity of the following diagram:

$$\begin{array}{ccccc}
 & & \pi_q(S^n) & & \\
 & {\scriptstyle S}\swarrow & \downarrow{\scriptstyle \mu'_\#} & \searrow{\scriptstyle i_\#} & \\
\pi_{q+1}(S^{n+1}) & \xrightarrow[\approx]{\tilde{\partial}} & \pi_q(\Omega S^{n+1}) & \xrightarrow[\approx]{g_\#^{-1}} & \pi_q(X^{2n}) \\
{\scriptstyle H}\downarrow & & & & \downarrow{\scriptstyle j_\#} \\
\pi_{q-1}(S^{2n-1}) & \xleftarrow[\approx]{\partial} & \pi_q(E^{2n},S^{2n-1}) & \xleftarrow[\approx]{\bar{\alpha}_{n\#}^{-1}} & \pi_q(X^{2n},S^n) \\
 & {\scriptstyle \alpha_{n\#}}\searrow & & \swarrow{\scriptstyle \partial} & \\
 & & \pi_{q-1}(S^n) & &
\end{array}$$

■

If G is an infinite cyclic group, we define a function $|\cdot|$ from G to the set of nonnegative integers by the condition $|g| = m$ if and only if there is a generator $g' \in G$ such that $g = mg'$. Since $\pi_{2n-1}(S^{2n-1}) \approx \mathbf{Z}$, we can define $|H[\alpha]|$ for $[\alpha] \in \pi_{2n+1}(S^{n+1})$. The following is an interpretation of $|H[\alpha]|$

10 THEOREM *Let $\alpha\colon S^{2n+1} \to S^{n+1}$ be a base-point-preserving map and let $E_\alpha \to S^{2n+1}$ be the principal fibration induced by α. Then $|H[\alpha]| = m$ if and only if the integral homology group $H_{2n}(E_\alpha)$ is isomorphic to $\mathbf{Z}_m$ (where $\mathbf{Z}_0 = \mathbf{Z}$).*

[1] See H. Hopf, Über die Abbildungen von Sphären auf Sphären niedrigerer Dimension, *Fundamenta Mathematica*, vol. 25, pp. 427–440, 1935, and G. W. Whitehead, A generalization of the Hopf invariant, *Annals of Mathematics*, vol. 51, pp. 192–237, 1950.

[2] See G. W. Whitehead, On the Freundenthal theorems, *Annals of Mathematics*, vol. 57, pp. 209–228, 1953.

PROOF From the definition of H and the naturality of the Hurewicz homomorphism φ, it is easily seen that $|H[\alpha]| = |\varphi\bar{\partial}[\alpha]|$, where

$$\varphi\bar{\partial}[\alpha] \in H_{2n}(\Omega S^{n+1}) \approx \mathbf{Z}$$

Since α induces a map $\bar{\alpha}\colon (E_\alpha,\Omega S^{n+1}) \to (PS^{n+1},\Omega S^{n+1})$, there is a commutative diagram

$$\begin{array}{ccc} \pi_{2n+1}(S^{2n+1}) & & \\ \alpha_\# \downarrow & \overset{\bar{\partial}}{\searrow} & \pi_{2n}(\Omega S^{n+1}) \xrightarrow{\varphi} H_{2n}(\Omega S^{n+1}) \\ & \underset{\partial}{\nearrow} & \\ \pi_{2n+1}(S^{n+1}) & & \end{array}$$

Therefore $|\varphi\bar{\partial}[\alpha]| = |\varphi\bar{\partial}\alpha_\#[1_{S^{2n+1}}]| = |\varphi\bar{\partial}[1_{S^{2n+1}}]|$. There is also a commutative diagram

$$\begin{array}{ccccc} \pi_{2n+1}(S^{2n+1}) & \overset{\approx}{\leftarrow} & \pi_{2n+1}(E_\alpha,\Omega S^{n+1}) & \xrightarrow{\partial} & \pi_{2n}(\Omega S^{n+1}) \\ \varphi\downarrow\approx & & \varphi\downarrow & & \downarrow\varphi \\ H_{2n+1}(S^{2n+1}) & \overset{\approx}{\leftarrow} & H_{2n+1}(E_\alpha,\Omega S^{n+1}) & \xrightarrow{\partial} & H_{2n}(\Omega S^{n+1}) \end{array}$$

from which it follows that $|\varphi\bar{\partial}[1_{S^{2n+1}}]| = |\partial(z)|$, where z is a generator of $H_{2n+1}(E_\alpha,\Omega S^{n+1})$. By lemma 4, $H_{2n}(E_\alpha,\Omega S^{n+1}) \approx H_{2n}(S^{2n+1}) = 0$, and so

$$H_{2n}(E_\alpha) \approx H_{2n}(\Omega S^{n+1})/\partial H_{2n+1}(E_\alpha,\Omega S^{n+1})$$

and this gives the result. ∎

4 MULTIPLICATIVE PROPERTIES OF SPECTRAL SEQUENCES

This section is devoted to pairings from two spectral sequences to a third. This will be applied, by means of the cross product, to pair the homology spectral sequences of two fibrations to the spectral sequence of their product. We shall also consider cohomology spectral sequences. There is a cohomology spectral sequence for a fibration and a cross-product pairing of the cohomology spectral sequences of two fibrations to the cohomology spectral sequence of their product. The diagonal map then endows the cohomology spectral sequence with a multiplicative structure, which will be applied in the next section.

Let $p\colon (E,E_A) \to (B,A)$ and $p'\colon (E',E'_{A'}) \to (B',A')$ be fibrations over relative CW complexes and let $p''\colon E \times E' \to B \times B'$ be the product fibration (that is, $p'' = p \times p'$). There is a filtration of the pair $(E \times E', E_A \times E' \cup E \times E'_{A'})$ defined by $(E \times E')_k = E_A \times E' \cup E \times E'_{A'} \cup \bigcup_{i+j=k} E_i \times E'_j$, where $\{E_i\}$ and $\{E'_j\}$ are the filtrations of (E,E_A) and $(E',E'_{A'})$ corresponding to the skeleta of (B,A) and (B',A'), respectively. Then $E \times E' = \bigcup_k (E \times E')_k$, and every compact subset of $E \times E'$ is contained in $(E \times E')_k$ for some k. By the method of example 9.1.5, there is a convergent E^1 spectral sequence with

$E^1_{s,t} \approx H_{s+t}((E \times E')_s, (E \times E')_{s-1}; G)$ and E^∞ the bigraded module associated to the filtration of $H_* = H_*((E,E_A) \times (E',E'_{A'}); G)$ defined by

$$F_sH_* = \text{im } [H_*((E \times E')_s, E \times E'_{A'} \cup E_A \times E'; G) \to H_*]$$

We relate this spectral sequence to the cross product of the spectral sequences of p and of p'. If E, E', and E'' are E^k spectral sequences, a *pairing* from E and E' to E'' is a sequence of homomorphisms

$$h^r: E^r_{s,t} \otimes E'^r_{s',t'} \to E''^r_{s+s',t+t'}$$

for all $r \geq k$ such that for $x \in E^r_{s,t}$

$$d^rh^r(x \otimes y) = h^r(d^rx \otimes y) + (-1)^{s+t}h^r(x \otimes d'^ry)$$

and such that h^{r+1} is the composite

$$E^{r+1} \otimes E'^{r+1} \approx H(E^r) \otimes H(E'^r) \to H(E^r \otimes E'^r) \xrightarrow{h^r_*} H(E''^r) \approx E''^{r+1}$$

For the sequence of submodules used to define E^∞

$$B^k \subset B^{k+1} \subset \cdots \subset B^r \subset \cdots \subset Z^r \subset \cdots \subset Z^{k+1} \subset Z^k$$

it is clear that h^k pairs Z^k and Z'^k to Z''^k in such a way that $Z^r \otimes Z'^r$ is mapped to Z''^r and $B^r \otimes Z'^r + Z^r \otimes B'^r$ is mapped to B''^r for all $r \geq k$. It follows that h^k maps $Z^\infty \otimes Z'^\infty = (\bigcap Z^r) \otimes (\bigcap Z'^r)$ to $\bigcap Z''^r = Z''^\infty$ and maps $B^\infty \otimes Z'^\infty + Z^\infty \otimes B'^\infty = \bigcup_r (B^r \otimes \bigcap_j Z'^j) + \bigcup_r (\bigcap_j Z^j \otimes B'^r)$ to $\bigcup B''^r = B''^\infty$. There is induced a pairing

$$h^\infty: E^\infty \otimes E'^\infty \to E''^\infty$$

which is compatible with the pairings $\{h^r\}$.

1 THEOREM *Let $p: E \to B$ and $p': E' \to B'$ be orientable fibrations over path-connected relative CW complexes (B,A) and (B',A') with fibers $F = p^{-1}(b_0)$ and $F' = p'^{-1}(b'_0)$, respectively. There is a pairing $\{h^r\}$ from the E^1 spectral sequences of p and p' to the E^1 spectral sequence of $p \times p'$ such that h^∞ is induced by the cross-product pairing*

$$H_*(E,E_A; G) \otimes H_*(E',E'_{A'}; G') \to H_*((E,E_A) \times (E',E'_{A'}); G \otimes G')$$

PROOF An Eilenberg-Zilber chain map

$$\Delta(E,E_A) \otimes \Delta(E',E'_{A'}) \to \Delta((E,E_A) \times (E',E'_{A'}))$$

induces a map from $F_s\Delta(E,E_A) \otimes F_{s'}\Delta(E',E'_{A'})$ to $F_{s+s'}\Delta((E,E_A) \times (E',E'_{A'}))$ for all s and s'. Therefore it induces in a natural way a pairing of the corresponding spectral sequences. Since an Eilenberg-Zilber chain map induces the homology cross product, the result follows. ∎

To interpret this result on the E^2 level, let $C_*(B,A)$ and $C_*(B',A')$ be the chain complexes of the relative CW complexes (B,A) and (B',A'), respectively, defined as in corollary 9.2.4. If $\sigma \in \bar{\Delta}_s((B,A)^s, (B,A)^{s-1})$, then

$$\{\sigma\} \in H_s(\bar{\Delta}((B,A)^s, (B,A)^{s-1})) = C_s(B,A)$$

and these elements $\{\sigma\} \in C_s(B,A)$ generate $C_s(B,A)$. We define a homomorphism

$$\psi'': C_s(B,A) \otimes C_{s'}(B',A') \otimes H_n(F \times F'; G'') \to H_{s+s'+n}((E \times E')_{s+s'}, (E \times E')_{s+s'-1}; G'')$$

by $$\psi''(\{\sigma\} \otimes \{\sigma'\} \otimes w) = (\bar\sigma \times \bar\sigma')_* T_*(\{\xi_s\} \times \{\xi_{s'}\} \times w)$$

where $\bar\sigma: (\Delta^s,\dot\Delta^s) \times F \to (E_s,E_{s-1})$ and $\bar\sigma': (\Delta^{s'},\dot\Delta^{s'}) \times F' \to (E'_{s'},E'_{s'-1})$ are admissible liftings of σ and σ', respectively, and

$$T: (\Delta^s,\dot\Delta^s) \times (\Delta^{s'},\dot\Delta^{s'}) \times F \times F' \to (\Delta^s,\dot\Delta^s) \times F \times (\Delta^{s'},\dot\Delta^{s'}) \times F'$$

is the map which interchanges the second and third coordinates. The fact that ψ'' is well-defined follows by an argument similar to that of lemma 9.2.14.

2 LEMMA *The map ψ'' induces an isomorphism*

$$\psi'': [C_*(B,A) \otimes C_*(B',A')]_s \otimes H_n(F \times F'; G'') \approx E''^1_{s,n}$$

such that $\psi'' \circ (\partial \otimes 1) = d^1 \circ \psi''$ and such that there is a commutative square

$$\begin{array}{ccc} C_s(B,A) \otimes H_t(F;G) \otimes C_{s'}(B,A) \otimes H_{t'}(F';G') & \xrightarrow{\psi \otimes \psi'} & E^1_{s,t} \otimes E'^1_{s',t'} \\ \downarrow \varphi & & \downarrow h^1 \\ C_s(B,A) \otimes C_{s'}(B',A') \otimes H_{t+t'}(F \times F'; G'') & \xrightarrow{\psi''} & E''^1_{s+s',t+t'} \end{array}$$

where $\varphi(c \otimes w \otimes c' \otimes w) = (-1)^{ts'} c \otimes c' \otimes (w \times w')$, with G and G' paired to G''.

PROOF The first part follows by an argument similar to that of theorem 9.2.15. For the second part we have

$$\begin{aligned} \psi''\varphi(\{\sigma\} \otimes w \otimes \{\sigma'\} \otimes w') &= (-1)^{ts'}\psi''(\{\sigma\} \otimes \{\sigma'\} \otimes (w \times w')) \\ &= (-1)^{ts'}(\bar\sigma \times \bar\sigma')_* T_*(\{\xi_s\} \times \{\xi_{s'}\} \times (w \times w')) \\ &= (\bar\sigma \times \bar\sigma')_*((\{\xi_s\} \times w) \times (\{\xi_{s'}\} \times w')) \\ &= h^1(\psi(\{\sigma\} \otimes w) \otimes \psi'(\{\sigma'\} \otimes w')) \quad ■ \end{aligned}$$

It follows from theorem 1 and lemma 2 that

$$E''^2_{s,t} \approx H_s((B,A) \times (B',A'); H_t(F \times F'; G''))$$

and that the pairing h^2 from $E^2_{s,t}$, $E'^2_{s',t'}$ to $E''^2_{s+s',t+t'}$ corresponds to $(-1)^{ts'}$ times the pairing given by cross product

$$H_s(B,A; H_t(F;G)) \otimes H_{s'}(B',A'; H_{t'}(F';G')) \to H_{s+s'}((B,A) \times (B',A'); H_{t+t'}(F \times F'; G''))$$

where the coefficients are themselves paired by the cross product. That is, the left-hand side is isomorphic to

$$H_s((B,A)^s,(B,A)^{s-1}) \otimes H_t(F;G) \otimes H_{s'}((B',A')^{s'}, (B',A')^{s'-1}) \otimes H_{t'}(F';G')$$

the right-hand side is isomorphic to

$$H_{s+s'}(((B,A) \times (B',A'))^s, ((B,A) \times (B',A'))^{s-1}) \otimes H_{t+t'}(F \times F'; G'')$$

and the map sends $x \otimes y \otimes x' \otimes y'$ to $(-1)^{ts'}(x \times x') \otimes (y \times y')$.

3 THEOREM *Let $p: E \to B$ and $p': E' \to B'$ be orientable fibrations with path-connected base spaces and with fibers F and F', respectively. Let $A \subset B$ and $A' \subset B'$ and assume that $\{B \times A', A \times B'\}$ is an excisive couple in $B \times B'$ and $\{E_A \times E', E \times E'_{A'}\}$ is an excisive couple in $E \times E'$. Given a pairing $G \otimes G' \to G''$, there is a pairing of the E^2 spectral sequences of p and p' to the E^2 spectral sequence of $p \times p'$, which on E^2 corresponds to $(-1)^{ts'}$ times the cross-product pairing*

$$H_s(B,A; H_t(F;G)) \otimes H_{s'}(B',A'; H_{t'}(F';G')) \to H_{s+s'}((B,A) \times (B',A'); H_{t+t'}(F \times F'; G''))$$

and on E^∞ is compatible with the cross-product pairing

$$H_n(E,E_A; G) \otimes H_{n'}(E',E'_{A'}; G') \to H_{n+n'}((E,E_A) \times (E',E'_{A'}); G'')$$

PROOF Let $f:(X,A) \to (B,A)$ and $f': (X',A') \to (B',A')$ be relative CW approximations to (B,A) and (B',A'), respectively. Let E_X and $E'_{X'}$ be the induced fibrations over X and X', respectively, with corresponding maps $\bar f: (E_X,E_A) \to (E,E_A)$ and $\bar f': (E'_{X'},E'_{A'}) \to (E',E'_{A'})$. The excisiveness hypotheses ensure that the Künneth formula can be applied to deduce isomorphisms

$$(f \times f')_*: H_*((X,A) \times (X',A')) \approx H_*((B,A) \times (B',A'))$$
$$(\bar f \times \bar f')_*: H_*((E_X,E_A) \times (E'_{X'},E'_{A'})) \approx H_*((E,E_A) \times (E',E'_{A'}))$$

The result now follows from application of theorem 1 and lemma 2 to the fibrations $E_X \to X$ and $E'_{X'} \to X'$ and from the remarks made above about the pairing induced on the E^2 terms (the resulting E^2 spectral sequence being independent of the choices of X and X'). ∎

The pairing of theorem 3 has properties analogous to those of the cross-product pairing. In particular, it is functorial on fiber-preserving maps and commutes up to sign with the homomorphism induced by interchanging the factors of $p \times p'$.

We next consider cohomology spectral sequences. Let $C^* = \{C^q,\delta\}$ be a cochain complex. A (*decreasing*) *filtration* F on C^* is a sequence of subcomplexes F^sC^* such that $F^sC^* \supset F^{s+1}C^*$ for all s. The filtration is *convergent* if $\bigcup F^sC^* = C^*$ and $\bigcap F^sC^* = 0$. It is *bounded above* if for each n there is $s(n)$ such that $F^{s(n)}C^n = 0$. Given a convergent filtration bounded above on a cochain complex C^*, there is an analogue of theorem 9.1.2 which asserts the existence of a convergent E_1 spectral sequence $\{E_r,d_r\}$, where E_r is bigraded by $E_r^{s,t}$ and d_r is a differential on E_r of bidegree $(r, 1 - r)$. Furthermore, we have $E_1^{s,t} \approx H^{s+t}(F^sC^*/F^{s+1}C^*)$ and d_1 corresponds to the coboundary operator of the triple $(F^sC^*,F^{s+1}C^*,F^{s+2}C^*)$. The limit term E_∞ is the bigraded module associated to the filtration on $H^*(C^*)$ defined by

$$F^sH^*(C^*) = \ker [H^*(C^*) \to H^*(F^{s-1}C^*)]$$

(that is, $E_\infty^{s,t} \approx \ker [H^{s+t}(C^*) \to H^{s+t}(F^{s-1}C^*)]/\ker [H^{s+t}(C^*) \to H^{s+t}(F^sC^*)]$).

4 EXAMPLE Let $\{X_s\}$ be an increasing filtration of a pair (X,A) and let $\bar\Delta(X,A)$ be the subcomplex of $\Delta(X,A)$ generated by singular simplexes $\sigma: \Delta^q \to X$ such that $\sigma((\Delta^q)^k) \subset X_k$ for all k. Let $\bar C^* = \operatorname{Hom}(\bar\Delta(X,A), G)$. A decreasing filtration on $\bar C^*$ is defined by

$$F^s\bar C^* = \{c \in \bar C^* \mid c \mid \bar\Delta(X_{s-1},A) = 0\}$$

where $\bar\Delta(X_{s-1},A) = \bar\Delta(X,A) \cap \Delta(X_{s-1},A)$. Since $\bar\Delta_s(X,A) = \bar\Delta_s(X_s,A)$, it follows that $F^{s+1}\bar C^s = 0$, and so the filtration is bounded above. In case the original filtration on (X,A) is bounded below (that is, $X_s = A$ for some s), then $\bigcup F^s\bar C^* = \{c \in \bar C^* \mid c \mid \bar\Delta(A,A) = 0\} = \bar C^*$. Hence, in the latter case there is an associated convergent E_1 spectral sequence. In case the inclusion maps $\bar\Delta(X,A) \subset \Delta(X,A)$ and $\bar\Delta(X_s,A) \subset \Delta(X_s,A)$ are chain equivalences, this spectral sequence has the property that $E_1^{s,t} \approx H^{s+t}(X_s,X_{s-1};\ G)$ and E_∞ is the bigraded module associated to the filtration on $H^*(X,A;\ G)$ defined by

$$F^sH^*(X,A;\ G) = \ker [H^*(X,A;\ G) \to H^*(X_{s-1},A;\ G)]$$

In particular, if (X,A) is a relative CW complex, $X_s = (X,A)^s$ if $s \geq 0$, and $X_s = A$ if $s < 0$, it follows from theorem 9.2.3 that the hypotheses are satisfied and that

$$E_1^{s,t} \approx H^{s+t}((X,A)^s, (X,A)^{s-1};\ G) = 0 \qquad t \neq 0$$

Therefore the spectral sequence collapses and $H^s(X,A;\ G)$ is isomorphic to $E_2^{s,0} \approx H^*(C^*)$, where $C^* = \{C^q,\delta\}$ is the cochain complex

$$C^q = H^q((X,A)^q, (X,A)^{q-1};\ G)$$

and δ is the coboundary operator of the triple $((X,A)^q, (X,A)^{q-1}, (X,A)^{q-2})$. By the universal-coefficient theorem for cohomology, $C^* = \operatorname{Hom}(C_*(X,A), G)$. Hence we have proved that $H^*(X,A;\ G) \approx H^*(C_*(X,A);\ G)$.

5 THEOREM *Let $p: E \to B$ be a fibration over a relative CW complex (B,A). There is a convergent E_1 cohomology spectral sequence, with $E_1^{s,t} \approx H^{s+t}(E_s,E_{s-1};\ G)$ and E_∞ the bigraded module associated to the filtration of $H^*(E,E_A;\ G)$ defined by*

$$F^sH^*(E,E_A;\ G) = \ker [H^*(E,E_A;\ G) \to H^*(E_{s-1},E_A;\ G)]$$

PROOF Since $(B, (B,A)^s)$ is s-connected for all s, it follows easily from theorem 7.2.8 that (E,E_s) is s-connected for all s. By theorem 9.2.3, the chain complex $\bar\Delta(E,E_A)$ is chain equivalent to $\Delta(E,E_A)$ and $\bar\Delta(E_s,E_A)$ is chain equivalent to $\Delta(E_s,E_A)$. The result now follows by the method of example 4. ∎

To compute $E_1^{s,t}$ we assume that B is path connected and that $p: E \to B$ is an orientable fibration. Let $F = p^{-1}(b_0)$ and let $\sigma: (\Delta^s,\dot\Delta^s) \to ((B,A)^s, (B,A)^{s-1})$

be a singular simplex in $\bar{\Delta}(B,A)$. If $\bar{\sigma}$: $(\Delta^s,\dot{\Delta}^s) \times F \to (E_s,E_{s-1})$ is an admissible lifting of σ, the homomorphism

$$\bar{\sigma}^*: H^n(E_s,E_{s-1}; G) \to H^n((\Delta^s,\dot{\Delta}^s) \times F; G)$$

depends only on σ and not the particular choice of the lifting $\bar{\sigma}$ (because the fibration is orientable). Let $\{\xi_s\}^* \in H^s(\Delta^s,\dot{\Delta}^s)$ be the generator characterized by the condition $\langle\{\xi_s\}^*,\{\xi_s\}\rangle = 1$. It follows from theorem 5.6.1 that the map $v \to \{\xi_s\}^* \times v$ is an isomorphism

$$H^q(F;G) \approx H^{s+q}((\Delta^s,\dot{\Delta}^s) \times F; G)$$

As in theorem 9.2.13 and lemma 9.2.14, it can be shown that there is a well-defined homomorphism

$$\psi^*: H^n(E_s,E_{s-1}; G) \to H^s((B,A)^s, (B,A)^{s-1}; H^{n-s}(F;G))$$

characterized by the equation

$$\{\xi_s\}^* \times \langle\psi^*(u),\{\sigma\}\rangle = \bar{\sigma}^*(u)$$

where σ: $(\Delta^s,\dot{\Delta}^s) \to ((B,A)^s, (B,A)^{s-1})$ and $\bar{\sigma}$: $(\Delta^s,\dot{\Delta}^s) \times F \to (E_s,E_{s-1})$ is an admissible lifting of σ, and $\langle\psi^*(u),\{\sigma\}\rangle \in H^{n-s}(F;G)$. Analogous to theorem 9.2.15 is the result that ψ^* is an isomorphism (this uses the second part of lemma 9.2.2 instead of the first part) and that it commutes with the differentials d_1 and the coboundary operator of the triple $((B,A)^s, (B,A)^{s-1}, (B,A)^{s-2})$. Using the technique of relative CW approximation, we have the following analogue of theorem 9.2.17.

6 THEOREM *Let p: $E \to B$ be an orientable fibration over a path-connected base and let $F = p^{-1}(b_0)$. Given $A \subset B$, there is a convergent E_2 cohomology spectral sequence, with $E_2^{s,t} \approx H^s(B,A; H^t(F;G))$ and E_∞ the bigraded module associated to some filtration of $H^*(E,E_A; G)$. This spectral sequence is a first-quadrant spectral sequence functorial on the category of orientable fibrations and fiber-preserving maps.* ∎

For the multiplicative properties of cohomology spectral sequences we shall use the following result about pairings of cohomology spectral sequences.

7 THEOREM *Let p: $E \to B$ and p': $E' \to B'$ be orientable fibrations over path-connected relative CW complexes (B,A) and (B',A'), with fibers F and F', respectively. There is a pairing $\{h_r\}$ from the E_1 cohomology spectral sequences of p and p' to the E_1 cohomology spectral sequence of $p \times p'$ such that h_2 is induced by $(-1)^{ts'}$ times the cross-product pairing*

$$H^s(B,A; H^t(F;G)) \otimes H^{s'}(B',A'; H^{t'}(F';G')) \to H^{s+s'}((B,A) \times (B',A'); H^{t+t'}(F \times F'; G''))$$

and h_∞ is induced by the cross-product pairing

$$H^*(E,E_A; G) \otimes H^*(E',E'_{A'}; G') \to H^*((E,E_A) \times (E',E'_{A'}); G'')$$

where G and G' are paired to G''.

PROOF There are chain equivalences

$$\bar{\Delta}(E,E_A) \otimes \bar{\Delta}(E',E'_{A'}) \subset \Delta(E,E_A) \otimes \Delta(E',E'_{A'}) \to \Delta((E,E_A) \times (E',E'_{A'}))$$

and therefore an isomorphism

$$H^*((E,E_A) \times (E',E'_{A'}); G'') \approx H^*(\bar{\Delta}(E,E_A) \otimes \bar{\Delta}(E',E'_{A'}); G'')$$

We define a filtration on $C^* = \text{Hom}\,(\bar{\Delta}(E,E_A) \otimes \bar{\Delta}(E',E'_{A'}), G'')$ by

$$F^sC^* = \{c \in C^* \mid c \mid \bar{\Delta}(E_i,E_A) \otimes \bar{\Delta}(E'_j, E'_{A'}) = 0, i + j = s\}$$

Then the cross product

$$\text{Hom}\,(\bar{\Delta}(E,E_A), G) \otimes \text{Hom}\,(\bar{\Delta}(E',E'_{A'}), G') \to C^*$$

maps $F^s \otimes F^{s'}$ to $F^{s+s'}C^*$. It follows easily that there is an induced pairing of the corresponding cohomology spectral sequences and that h_∞ has the stated property.

To prove the statement about the pairing h_2, let C_* and C'_* be the chain complexes of (B,A) and (B',A'), respectively, and let $C''_* = C_* \otimes C'_*$. We define a homomorphism

$$\psi''^*: E_1''^{s,t} \to \text{Hom}\,(C''_s, H^t(F \times F'; G''))$$

by the condition

$$\{\xi_i\}^* \times \{\xi_j\}^* \times \langle \psi''^*(u), \{\sigma\} \times \{\sigma'\}\rangle = (\bar{\sigma} \times \bar{\sigma}')^*(u)$$

where $\sigma: (\Delta^i,\dot{\Delta}^i) \to ((B,A)^i, (B,A)^{i-1})$, $\sigma': (\Delta^j,\dot{\Delta}^j) \to ((B',A')^j, (B',A')^{j-1})$, with $i + j = s$, $\bar{\sigma}: (\Delta^i,\dot{\Delta}^i) \times F \to (E_i,E_{i-1})$, and $\bar{\sigma}': (\Delta^j,\dot{\Delta}^j) \times F' \to (E'_j,E'_{j-1})$ are admissible liftings, and where $u \in H^{s+t}((E \times E')_s, (E \times E')_{s-1}; G \otimes G')$. Then ψ'' is an isomorphism taking d_1 into the coboundary operator of the cochain complex $\text{Hom}\,(C''_*, H^*(F \times F'; G''))$.

Furthermore, if $v \in E_1^{s,t}$ and $v' \in E_1'^{s',t'}$, then $v \times v' \in E_1^{s+s',t+t'}$, and from the definitions we have

$$\begin{aligned}
&\{\xi_s\}^* \times \{\xi_{s'}\}^* \times \langle \psi''^*(v \times v'), \{\sigma\} \times \{\sigma'\}\rangle \\
&\quad = (\bar{\sigma} \times \bar{\sigma}')^*(v \times v') = \bar{\sigma}^*(v) \times \bar{\sigma}'^*(v') \\
&\quad = (\{\xi_s\}^* \times \langle \psi^*(v),\{\sigma\}\rangle) \times (\{\xi_{s'}\}^* \times \langle \psi'^*(v'),\{\sigma'\}\rangle) \\
&\quad = (-1)^{ts'}\{\xi_s\}^* \times \{\xi_{s'}\}^* \times \langle \psi^*(v) \times \psi'^*(v'), \{\sigma\} \times \{\sigma'\}\rangle
\end{aligned}$$

Therefore $\psi''^*(v \times v') = (-1)^{ts'}\psi^*(v) \times \psi'^*(v')$, and this implies the result about the pairing h_2. ∎

This gives the following important multiplicative property for the cohomology spectral sequence of a fibration.

8 THEOREM *Let $p: E \to B$ be an orientable fibration over a path-connected base, with fiber F. Let $\{A_1,A_2\}$ be an excisive couple of path-connected subspaces of B such that $\{E_{A_1},E_{A_2}\}$ is an excisive couple in E. Then there is a functorial pairing of the E_2 cohomology spectral sequences of (E,E_{A_1}) and (E,E_{A_2}) to the E_2 cohomology spectral sequence of $(E, E_{A_1} \cup E_{A_2})$, which on E_2 is isomorphic to $(-1)^{ts'}$ times the cup-product pairing (G and G' paired to G'')*

$$H^s(B,A_1; H^t(F;G)) \otimes H^{s'}(B,A_2; H^{t'}(F;G')) \to H^{s+s'}(B, A_1 \cup A_2; H^{t+t'}(F;G''))$$

and on E_∞ *is induced by the cup-product pairing*

$$H^*(E,E_{A_1}; G) \otimes H^*(E,E_{A_2}; G') \to H^*(E,E_{A_1} \cup E_{A_2}; G'')$$

PROOF We begin by showing that there exists a CW complex X, with subcomplexes X_1 and X_2, and a weak homotopy equivalence $f: X \to B$ such that $f | X_1: X_1 \to A_1$ and $f | X_2: X_2 \to A_2$ are also weak homotopy equivalences. In fact, let $g: Y \to B$, $g_1: Y_1 \to A_1$, and $g_2: Y_2 \to A_2$ be CW approximations. Then there exist maps $g_1': Y_1 \to Y$ and $g_2': Y_2 \to Y$ (which can be taken to be cellular) such that $g \circ g_1': Y_1 \to B$ and $g \circ g_2': Y_2 \to B$ are homotopic, respectively, to the composites $Y_1 \xrightarrow{g_1} A_1 \overset{i}{\subset} B$ and $Y_2 \xrightarrow{g_2} A_2 \overset{i'}{\subset} B$. Let X be the CW complex obtained from the disjoint union $Y_1 \times I \cup Y \cup Y_2 \times I$ by identifying $(y_1,0)$ with $g_1'(y_1) \in Y$ for all $y_1 \in Y_1$ and $(y_2,0)$ with $g_2'(y_2) \in Y$ for all $y_2 \in Y_2$. Let $k: Y_1 \times I \cup Y \cup Y_2 \times I \to X$ be the collapsing map and define a map $f: X \to B$ such that $(f \circ k) | Y = g$, $(f \circ k) | Y_1 \times I$ is a homotopy from $g \circ g_1'$ to $i \circ g_1$, and $(f \circ k) | Y_2 \times I$ is a homotopy from $g \circ g_2'$ to $i' \circ g_2$. Let $X_1 = k(Y_1 \times 1)$ and $X_2 = k(Y_2 \times 1)$ and observe that X_1 and X_2 are subcomplexes of X such that $f | X_1: X_1 \to A_1$ and $f | X_2: X_2 \to A_2$ are weak homotopy equivalences. Furthermore, $k(Y)$ is a strong deformation retract of X, and since $f | k(Y): k(Y) \to B$ is a weak homotopy equivalence, so is $f: X \to B$. Therefore the map $f: X \to B$ has the desired properties.

The excisiveness assumption about $\{A_1,A_2\}$ implies that f induces an isomorphism

$$f^*: H^*(B, A_1 \cup A_2) \approx H^*(X, X_1 \cup X_2)$$

Let $p': E_X \to X$ be the induced fibration over X and let $\bar{f}: E_X \to E$ be the corresponding map. Then $\bar{f}$ induces isomorphisms

$$H^*(E,E_{A_1}) \approx H^*(E_X,E_{X_1}) \qquad \text{and} \qquad H^*(E,E_{A_2}) \approx H^*(E_X,E_{X_2})$$

The excisiveness assumption about $\{E_{A_1},E_{A_2}\}$ ensures that $\bar{f}$ also induces an isomorphism

$$H^*(E, E_{A_1} \cup E_{A_2}) \approx H^*(E_X, E_{X_1} \cup E_{X_2})$$

By theorem 7, there is a pairing of the E_2 cohomology spectral sequences of (E_X,E_{X_1}) and (E_X,E_{X_2}) to the E_2 cohomology spectral sequence of $(E_X,E_{X_1}) \times (E_X,E_{X_2})$, which corresponds to cross product on the E_2 and E_∞ terms. There is a commutative square (whose horizontal maps are diagonal maps)

$$\begin{array}{ccc} E_X & \to & E_X \times E_X \\ {\scriptstyle p'}\downarrow & & \downarrow{\scriptstyle p' \times p'} \\ X & \longrightarrow & X \times X \end{array}$$

Let $d: X \to X \times X$ be a cellular approximation to the diagonal map having the property that $d(X_1) \subset X_1 \times X_1$ and $d(X_2) \subset X_2 \times X_2$ (such maps exist). It follows that there is a lifting $\bar{d}: E_X \to E_X \times E_X$ of $d \circ p': E_X \to X \times X$

which is homotopic to the diagonal map $E_X \to E_X \times E_X$. Then $\bar{d}$ maps the filtration of $(E_X, E_{X_1} \cup E_{X_2})$ into the filtration of $(E_X, E_{X_1}) \times (E_X, E_{X_2})$ and so induces a homomorphism from the E_2 cohomology spectral sequence of $(E_X, E_{X_1}) \times (E_X, E_{X_2})$ into the E_2 cohomology spectral sequence of $(E_X, E_{X_1} \cup E_{X_2})$. Since $\bar{d}$ takes cross products in $E_X \times E_X$ into cup products in E_X, the composite of this homomorphism with the pairing above is a pairing from the spectral sequences of (E_X, E_{X_1}) and (E_X, E_{X_2}) to the spectral sequence of $(E_X, E_{X_1} \cup E_{X_2})$, which is induced by $\pm$ cup product on the E_2 and E_∞ terms.

By means of the isomorphisms induced by f and $\bar{f}$, this gives a pairing from the E_2 cohomology spectral sequences of (E, E_{A_1}) and (E, E_{A_2}) to the E_2 cohomology spectral sequence of $(E, E_{A_1} \cup E_{A_2})$, which is induced by $\pm$ cup product on the E_2 and E_∞ terms. The resultant pairing is independent of the choice of X. ∎

9 COROLLARY *Let $p: E \to B$ be an orientable fibration with path-connected base B, with fiber F. For any $A \subset B$ there is a convergent E_2 cohomology spectral sequence of bigraded algebras with $E_2^{s,t} \approx H^s(B,A; H^t(F;R))$ and E_∞ the bigraded algebra associated to some filtration of $H^*(E,E_A; R)$. This spectral sequence is functorial on the category of such fibrations and fiber-preserving maps.* ∎

5 APPLICATIONS OF THE COHOMOLOGY SPECTRAL SEQUENCE

Because the cohomology spectral sequence of a fibration has a multiplicative structure, it is a more powerful tool than the homology spectral sequence. We shall use it in deriving the generalized Wang and Gysin cohomology sequences and then apply the cohomology spectral sequence to obtain another description of the Hopf invariant in a particular dimension. The section closes with some results about the homology and cohomology of spaces of type $(\pi,1)$.

Let $p: E \to B$ be an orientable fibration over a path-connected base and with fiber F. First we shall determine $i^*: H^*(E;G) \to H^*(F;G)$, where $i: F \subset E$, in terms of the cohomology spectral sequence of E. Because this is a first-quadrant spectral sequence, there is a monomorphism $E_\infty^{0,t} \to E_2^{0,t}$. Since B is path connected, there is an isomorphism $H^0(B; H^t(F;G)) \approx H^t(F;G)$. Using the fact that the cohomology spectral sequence is functorial, it follows that i^* maps the spectral sequence of $E \to B$ to the spectral sequence of $F \to b_0$. Therefore $i^*: H^*(E;G) \to H^*(F;G)$ is the composite

$$H^t(E;G) = F^0H^t(E;G) \to E_\infty^{0,t} \to E_2^{0,t} \approx H^0(B; H^t(F;G)) \approx H^t(F;G)$$

This leads to the following *generalized Wang cohomology sequence.*

1 THEOREM *Let $p: E \to B$ be a fibration, with fiber F and simply connected base B, which is a cohomology n-sphere (over R) for some $n \geq 2$. There is an exact sequence*

$$\cdots \to H^t(E;G) \xrightarrow{i^*} H^t(F;G) \xrightarrow{\theta} H^{t-n+1}(F;G) \to H^{t+1}(E;G) \xrightarrow{i^*} \cdots$$

in which $\theta(u \smile v) = \theta(u) \smile v + (-1)^{(n+1)\deg u}\, u \smile \theta(v)$, *the coefficients being suitably paired.*

PROOF Since B has no torsion, for the cohomology spectral sequence of $E \to B$ we have

$$E_2^{s,t} \approx H^s(B) \otimes H^t(F;G) = 0 \qquad s \neq 0, n$$

As in the proof of theorem 9.3.2, this leads to an exact sequence

$$\cdots \to H^t(E;G) \to E_2^{0,t} \xrightarrow{d_n} E_2^{n,t-n+1} \to H^{t+1}(E;G) \to \cdots$$

Let $1 \in H^0(B)$ be the unit class and let $w \in H^n(B)$ be a generator of $H^n(B)$. The map $u \to 1 \otimes u$ is an isomorphism of $H^t(F;G)$, with $E_2^{0,t}$, and the map $v \to w \otimes v$ is an isomorphism of $H^{t-n+1}(F;G)$, with $E_2^{n,t-n+1}$. Define $\theta\colon H^t(F;G) \to H^{t-n+1}(F;G)$ by the condition

$$d_n(1 \otimes u) = w \otimes \theta(u)$$

Then the desired exact sequence is obtained from the exact sequence above on replacing $E_2^{0,t}$ by $H^t(F;G)$ and $E_2^{n,t-n+1}$ by $H^{t-n+1}(F;G)$ and interpreting the resulting homomorphisms. To verify that θ has the stated multiplicative property, we use the fact that d_n is a derivation. Then we have

$$\begin{aligned} w \otimes \theta(u \smile v) &= d_n(1 \otimes (u \smile v)) = d_n(1 \otimes u \smile 1 \otimes v) \\ &= d_n(1 \otimes u) \smile 1 \otimes v + (-1)^{\deg u}\, 1 \otimes u \smile d_n(1 \otimes v) \\ &= w \otimes [\theta(u) \smile v + (-1)^{(n+1)\deg u}\, u \smile \theta(v)] \quad \blacksquare \end{aligned}$$

Let $p\colon E \to B$ be an orientable fibration with path-connected base and let $B' \subset B$ and $E' = p^{-1}(B')$. We show how the homomorphism

$$p^*\colon H^*(B,B';\, G) \to H^*(E,E';\, G)$$

can be interpreted in terms of the cohomology spectral sequence of (E,E'). Because the spectral sequence is a first-quadrant spectral sequence, there is an epimorphism $E_2^{s,0} \to E_\infty^{s,0}$. The augmentation $G \to H^0(F;G)$ induces a homomorphism $H^s(B,B';\, G) \to H^s(B,B';\, H^0(F;G))$. Using the spectral sequence of the fibration $B \subset B$ and the functorial property of the cohomology spectral sequence, it follows that $p^*\colon H^*(B,B';\, G) \to H^*(E,E';\, G)$ is the composite

$$H^s(B,B';\, G) \to H^s(B,B';\, H^0(F;G)) \approx E_2^{s,0} \to E_\infty^{s,0} \approx F^sH^s(E,E';\, G) \subset H^s(E,E';\, G)$$

This leads to the following *generalized Gysin cohomology sequence.*

2 THEOREM *Let* $p\colon E \to B$ *be an orientable fibration with path-connected base space and with fiber* F *a cohomology* n*-sphere (over* R*), with* $n \geq 1$. *If* $B' \subset B$ *and* $E' = p^{-1}(B')$, *there is an exact sequence*

$$\cdots \xrightarrow{p^*} H^s(E,E';\, G) \to H^{s-n}(B,B';\, G) \xrightarrow{\Psi} H^{s+1}(B,B';\, G) \xrightarrow{p^*} H^{s+1}(E,E';\, G) \to \cdots$$

in which $\Psi(u) = u \smile \Omega$ *for some* $\Omega \in H^{n+1}(B;R)$. *If* n *is even,* $2\Omega = 0$.

PROOF For the cohomology spectral sequence of (E,E') we have

$$E_2^{s,t} \approx H^s(B,B';\, H^t(F;G)) = 0 \qquad t \neq 0, n$$

As in the proof of theorem 9.3.3, this leads to an exact sequence

$$\cdots \to H^s(E,E';\, G) \to E_2^{s-n,n} \xrightarrow{d_{n+1}} E_2^{s+1,0} \to H^{s+1}(E,E';\, G) \to \cdots$$

Let $1 \in H^0(F;R)$ be the unit class and let $w \in H^n(F;R)$ be a generator of $H^n(F;R)$. Corresponding to these generators are isomorphisms $G \approx H^0(F;G)$ and $G \approx H^n(F;G)$. Thus we have isomorphisms

$$H^s(B,B';\, G) \approx H^s(B,B';\, H^0(F;G)) \approx E_2^{s,0}$$

whose composite will be denoted by α: $H^s(B,B';\, G) \approx E_2^{s,0}$, and

$$H^s(B,B';\, G) \approx H^s(B,B';\, H^n(F;G)) \approx E_2^{s,n}$$

whose composite will be denoted by β: $H^s(B,B';\, G) \approx E_2^{s,n}$. Define the homomorphism Ψ: $H^{s-n}(B,B';\, G) \to H^{s+1}(B,B';\, G)$ by the equation

$$\alpha\Psi(u) = (-1)^{\deg u}\, d_{n+1}\beta(u)$$

The desired exact sequence is obtained from the exact sequence above on replacing $E_2^{s-n,n}$ by $H^{s-n}(B,B';\, G)$, $E_2^{s+1,0}$ by $H^{s+1}(B,B';\, G)$ and interpreting the resulting homomorphisms.

In the spectral sequence of E with coefficients R there are similar isomorphisms α: $H^s(B,R) \approx E_2^{s,0}$ and β: $H^s(B;R) \approx E_2^{s,n}$. Let 1 also denote the unit class of $H^0(B;R)$ and define $\Omega \in H^{n+1}(B;R)$ by the equation

$$\alpha(\Omega) = d_{n+1}\beta(1)$$

To verify that $\Psi(u) = u \smile \Omega$, we use the cup-product pairing from the spectral sequence of (E,E') with coefficients G, and the spectral sequence of E with coefficients R, to the spectral sequence of (E,E') with coefficients G. Then

$$\begin{aligned}\alpha\Psi(u) &= (-1)^{\deg u} d_{n+1}\beta(u) = (-1)^{\deg u} d_{n+1}(\alpha(u) \smile \beta(1)) \\ &= \alpha(u) \smile d_{n+1}\beta(1) = \alpha(u) \smile \alpha(\Omega) = \alpha(u \smile \Omega)\end{aligned}$$

Therefore $\Psi(u) = u \smile \Omega$. Since $w \smile w = 0$, $\beta(1) \smile \beta(1) = 0$ in the spectral sequence of E. Therefore, if n is even,

$$0 = d_{n+1}(\beta(1) \smile \beta(1)) = \alpha(\Omega) \smile \beta(1) + \beta(1) \smile \alpha(\Omega) = \beta(2\Omega)$$

showing that $2\Omega = 0$. ∎

We use the cohomology spectral sequence to give another interpretation of the integer $|H[\alpha]|$, where $[\alpha] \in \pi_{2n+1}(S^{n+1})$ and H: $\pi_{2n+1}(S^{n+1}) \to \pi_{2n-1}(S^{2n-1})$ is the Hopf invariant defined in Sec. 9.3.

3 THEOREM *Let α: $S^{2n+1} \to S^{n+1}$ be a base-point-preserving map and let Y_α be the CW complex obtained by attaching a $(2n+2)$-cell to S^{n+1} by the map α. Then $H^{n+1}(Y_\alpha)$ and $H^{2n+2}(Y_\alpha)$ are both infinite cyclic, and if u and v are generators, respectively, then $u \smile u = \pm|H[\alpha]|v$.*

PROOF If Z is the mapping cylinder of α, then $Y_\alpha = Z/S^{2n+1}$, and so $H^*(Y_\alpha) \approx H^*(Z,S^{2n+1})$. Let $u \in H^{n+1}(Z,S^{2n+1})$ and $v \in H^{2n+2}(Z,S^{2n+1})$ be respective generators. It suffices to prove that $u \smile u = \pm|H[\alpha]|v$.

Let $r\colon Z \to S^{n+1}$ be the retraction and let $E \to Z$ be the principal fibration induced by r. Since r is a homotopy equivalence, the induced map $E \to PS^{n+1}$ induces isomorphisms of homology. Therefore $\tilde{H}_*(E) = 0$ and $\tilde{H}^*(E) = 0$. The restriction of E to S^{2n+1} is the principal fibration $E_\alpha \to S^{2n+1}$ induced by α. By theorem 9.3.10, $|H[\alpha]| = m$ if and only if $H_{2n}(E_\alpha) \approx \mathbf{Z}_m$. From the following portion of the Wang homology sequence of E_α

$$0 \to H_{2n+1}(E_\alpha) \to H_0(\Omega S^{n+1}) \to H_{2n}(\Omega S^{n+1}) \to H_{2n}(E_\alpha) \to 0$$

it follows that if $m \neq 0$, $H_{2n+1}(E_\alpha) = 0$, and if $m = 0$, then $H_{2n+1}(E_\alpha) \approx \mathbf{Z}$. By the universal-coefficient formula for cohomology, $H^{2n+1}(E_\alpha) \approx \mathbf{Z}_m$ no matter whether $m = 0$ or not (recall that we have adopted the convention that $\mathbf{Z}_0 = \mathbf{Z}$).

Since $\tilde{H}^*(E) = 0$, there is an isomorphism

$$\delta\colon H^{2n+1}(E_\alpha) \approx H^{2n+2}(E,E_\alpha)$$

and so $H^{2n+2}(E,E_\alpha) \approx \mathbf{Z}_m$, where $m = |H[\alpha]|$. We compute the order of $H^{2n+2}(E,E_\alpha)$ by using the cohomology spectral sequence.

For $s + t = 2n + 2$ the only nonzero term $E_2^{s,t}$ is the term

$$E_2^{2n+2,0} \approx H^{2n+2}(Z,S^{2n+1}) \otimes H^0(\Omega S^{n+1}),$$

and for $s + t = 2n + 1$ the only nonzero term $E_2^{s,t}$ is the term

$$E_2^{n+1,n} \approx H^{n+1}(Z,S^{2n+1}) \otimes H^n(\Omega S^{n+1}).$$

It follows that

$$H^{2n+2}(E,E_\alpha) \approx E_\infty^{2n+2,0} \approx E_2^{2n+2,0}/d_{n+1}(E_2^{n+1,n})$$

Let $u' \in H^{n+1}(Z)$ be the generator defined by $u' = u \mid Z$. Then, since $\tilde{H}^*(E) = 0$, there is a generator $w \in H^n(\Omega S^{n+1})$ such that in the spectral sequence of E we have $d_{n+1}(1 \otimes w) = u' \otimes 1$. Using the pairing of the cohomology spectral sequences of (E,E_α) and E_α to that of (E,E_α), we see that

$$\begin{aligned} d_{n+1}(u \otimes w) &= d_{n+1}(u \otimes 1 \smile 1 \otimes w) = \pm u \otimes 1 \smile d_{n+1}(1 \otimes w) \\ &= \pm u \otimes 1 \smile u' \otimes 1 = \pm(u \smile u') \otimes 1 \\ &= \pm(u \smile u) \otimes 1 \end{aligned}$$

Therefore $H^{2n+2}(E,E_\alpha)$ is infinite cyclic if and only if $u \smile u = 0$, and $H^{2n+2}(E,E_\alpha)$ has order m if and only if $u \smile u = \pm\, mv$. Comparing this with the earlier calculation of $H^{2n+2}(E,E_\alpha)$ gives the result. ■

4 COROLLARY *For any integer $m \geq 1$ the Hopf invariant*

$$H\colon \pi_{4m+1}(S^{2m+1}) \to \pi_{4m-1}(S^{4m-1})$$

is the trivial homomorphism.

PROOF For any α: $S^{4m+1} \to S^{2m+1}$, if Y_α is the CW complex obtained by attaching a $(4m + 2)$-cell to S^{2m+1} by the map α and if $u \in H^{2m+1}(Y_\alpha)$ is arbitrary, then $u \smile u = -u \smile u$, and so $u \smile u = 0$. By theorem 3, $|H[\alpha]| = 0$, and so $H[\alpha] = 0$ for all $[\alpha] \in \pi_{4m+1}(S^{2m+1})$. ∎

5 COROLLARY *For any $m \geq 1$, if α_{2m}: $S^{4m-1} \to S^{2m}$ is the map used in forming the CW complex X^{4m}, then $|H[\alpha_{2m}]| = 2$.*

PROOF Recall the definition of $X^{4m} = Y_{\alpha_{2m}}$ in Sec. 9.3. There is a collapsing map k: $S^{2m} \times S^{2m} \to X^{4m}$ with the property that if $u' \in H^{2m}(S^{2m})$ is a generator, there are generators $u \in H^{2m}(X^{4m})$ and $v \in H^{4m}(X^{4m})$ such that $k^* u = u' \times 1 + 1 \times u'$ and $k^* v = u' \times u'$. Then

$$k^*(u \smile u) = (u' \times 1 + 1 \times u') \smile (u' \times 1 + 1 \times u') = 2u' \times u'$$

Since $k^*: H^{4m}(X^{4m}) \approx H^{4m}(S^{2m} \times S^{2m})$, it follows that $u \smile u = 2v$, and the result follows from theorem 3. ∎

If π is a group, we define $H_*(\pi)$ [and $H^*(\pi)$] to be the integral homology [and cohomology] groups of a space of type $(\pi,1)$. Since any two spaces of type $(\pi,1)$ are easily seen to have the same weak homotopy type, these groups are independent (up to canonical isomorphism) of the space of type $(\pi,1)$ chosen. Furthermore, any homomorphism $\pi \to \pi'$ induces homomorphisms $H_*(\pi) \to H_*(\pi')$ and $H^*(\pi') \to H^*(\pi)$. We use the cohomology spectral sequence to obtain information about these groups.

6 THEOREM *For $n > 1$ there are isomorphisms*

$$H^q(\mathbf{Z}_n) \approx \begin{cases} 0 & q \text{ odd} \\ \mathbf{Z} & q = 0 \\ \mathbf{Z}_n & q \text{ even}, q > 0 \end{cases}$$

PROOF Let X be a CW complex of type $(\mathbf{Z},2)$ and let $PX \to X$ be the path fibration. Then the fiber ΩX of this fibration is a space of type $(\mathbf{Z},1)$. Therefore ΩX is a cohomology 1-sphere, and since PX is contractible, it follows from theorem 2 that $H^*(X)$ is a polynomial algebra on a generator Ω of degree 2, characterized by the equation $\Omega \otimes 1 = d_2(1 \otimes w)$ [where w is a generator of $H^1(\Omega X)$ and d_2 is the differential operator in E_2 of the cohomology spectral sequence of the fibration $PX \to X$]. Let f: $X \to X$ be a map such that $f^*\iota = n\iota$ for some 2-characteristic element $\iota \in H^2(X)$ (such a map exists, by theorem 8.1.10). It follows that $f^*(u) = nu$ for any $u \in H^2(X)$ and $f_\#$: $\pi_2(X) \to \pi_2(X)$ is the homomorphism $f_\#[\alpha] = n[\alpha]$. Let p: $E \to X$ be the principal fibration induced by f. Then p has fiber ΩX, and from the functorial property of the cohomology spectral sequence, we have $d_2(1 \otimes w) = f^*\Omega \otimes 1 = n\Omega \otimes 1$ in the spectral sequence of p: $E \to X$. Therefore, in the Gysin sequence of p the homomorphism

$$\Psi: H^{s-1}(X) \to H^{s+1}(X)$$

equals the cup product by $n\Omega$, and so

$$\ker \Psi = 0 \qquad \operatorname{coker} \Psi \approx \mathbf{Z}_n, \text{ for odd } s$$

Therefore $H^q(E) = 0$ unless q is even, and $H^0(E) \approx \mathbf{Z}$ and $H^q(E) \approx \mathbf{Z}_n$ if q is even and $q > 0$.

It merely remains to verify that E is a space of type $(\mathbf{Z}_n,1)$. This follows from the following commutative diagram with exact rows:

$$\begin{array}{ccccccccc}
\cdots \to & \pi_q(X) & \xrightarrow{\partial} & \pi_{q-1}(\Omega X) & \to & \pi_{q-1}(E) & \xrightarrow{p_\#} & \pi_{q-1}(X) & \to \cdots \\
& \downarrow f_\# & & \downarrow = & & \downarrow & & \downarrow & \\
\cdots \to & \pi_q(X) & \xrightarrow{\partial} & \pi_{q-1}(\Omega X) & \to & \pi_{q-1}(PX) & \to & \pi_{q-1}(X) & \to \cdots
\end{array}$$ ∎

7 COROLLARY *For $n > 1$ there are isomorphisms*

$$H_q(\mathbf{Z}_n) \approx \begin{cases} 0 & q \text{ even, } q \neq 0 \\ \mathbf{Z} & q = 0 \\ \mathbf{Z}_n & q \text{ odd, } q > 0 \end{cases}$$

PROOF This will follow from theorem 6 and the universal-coefficient formula 5.5.12 once we have verified that $H_*(\mathbf{Z}_n)$ is of finite type. We use the particular space E of type $(\mathbf{Z}_n,1)$ constructed in the proof of theorem 6. Since the fiber ΩX of the fibration $p\colon E \to X$ is a homology 1-sphere, there is, by theorem 9.3.3, an exact Gysin homology sequence

$$\cdots \to H_s(E) \xrightarrow{p_*} H_s(X) \to H_{s-2}(X) \to H_{s-1}(E) \to \cdots$$

Since $H_*(X)$ is of finite type [in fact, $H_q(X) = 0$ if q is odd and $H_q(X) \approx \mathbf{Z}$ if q is even and $q > 0$, as can be seen by using the Gysin sequence of the fibration $PX \to X$], it follows that $H_*(E)$ is of finite type. ∎

Because S^1 is a space of type $(\mathbf{Z},1)$, we now know $H_*(\pi)$ if π is a cyclic group. The groups $H_*(\pi)$ for π a finite direct sum of cyclic groups can be computed by induction, using the Künneth formula and the following result.

8 LEMMA *If Y and Y' are spaces of type $(\pi,1)$ and $(\pi',1)$, respectively, then $Y \times Y'$ is a space of type $(\pi \times \pi', 1)$.*

PROOF It is easily verified from the definitions of the homotopy groups that $\pi_q(Y \times Y') \approx \pi_q(Y) \times \pi_q(Y')$. ∎

In this way we can determine $H_*(\pi)$ if π is a finitely generated abelian group. The following result gives information about $H_*(\pi)$ for an arbitrary abelian group π.

9 THEOREM *Let $\{\pi_\alpha\}$ be the family of finitely generated subgroups directed by inclusion of an abelian group π. Then*

$$H_*(\pi) \approx \varinjlim \{H_*(\pi_\alpha)\}$$

PROOF For each element $\lambda \in \pi$ let S_λ^1 be a 1-sphere. Let $X^1 = \bigvee S_\lambda^1$ and define a homomorphism $\beta\colon \pi_1(X^1) \to \pi$ by the condition $\beta[\omega_\lambda] = \lambda$, where

$[\omega_\lambda] \in \pi_1(X')$ is determined by the inclusion map $S_\lambda^1 \subset X^1$ ($\pi_1(X')$ is the free group generated by the collection $\{[\omega_\lambda]\}_\lambda$). For every base-point-preserving map ω: $S^1 \to X^1$ such that $\beta[\omega] = 0$, attach a 2-cell to X^1 and let X^2 be the space obtained by adjoining all these 2-cells to X^1. Continue inductively, defining X^m for $m \geq 3$ to be the space obtained from X^{m-1} by attaching m-cells for every map $S^{m-1} \to X^{m-1}$. Let X be the CW complex whose m-skeleton is X^m for all $m \geq 1$ and whose 0-skeleton is the base point of X^1. Then X is a space of type $(\pi,1)$.

For any finite subset a of π let X_a be the largest subcomplex of X such that $X_a{}^1 = \bigvee_{\lambda \in a} S_\lambda{}^1$. Then it is clear from the construction of X that X_a is a space of type $(\pi_a,1)$, where π_a is the subgroup of π generated by the set a. Since every compact subset of X is contained in X_a for some finite subset a of π, it follows that

$$H_*(\pi) \approx H_*(X) \approx \lim_{\rightarrow} \{H_*(X_a)\} \approx \lim_{\rightarrow} \{H_*(\pi_a)\}$$

Since π_a is a finitely generated subgroup of π and every finitely generated subgroup of π is of this form, the right-hand side above is isomorphic to $\lim_{\rightarrow} \{H_*(\pi_\alpha)\}$. ■

These results on $H_*(\pi)$ will be used in the next section in the proof of the generalized Hurewicz isomorphism theorem.

6 SERRE CLASSES OF ABELIAN GROUPS

The spectral sequence of a fibration is well suited for inductive arguments based on the lowest (or highest) dimension in which a particular phenomenon occurs. Such arguments can be simplified further by systematically neglecting certain abelian groups in order to carry along just that portion of a given group which is relevant to the phenomenon in question. For example, in studying the p-primary components of finitely generated abelian groups, it is convenient to neglect finite summands whose order is not divisible by p. The process of neglecting certain groups will be formalized in this section by means of a study of groups "modulo a Serre class of abelian groups." The machinery will be applied, by means of the spectral sequence of a fibration, to the study of the homotopy groups of a space. In particular, the section closes with interesting generalizations of the Hurewicz and Whitehead theorems.

A *Serre class* of abelian groups is a nonempty class $\mathcal{C}$ of abelian groups having the property that for any exact three-term sequence of abelian groups $A \to B \to C$, if $A, C \in \mathcal{C}$, then $B \in \mathcal{C}$.

1 THEOREM *A class $\mathcal{C}$ of abelian groups is a Serre class if and only if it has the following properties:*

(a) *$\mathcal{C}$ contains a trivial group.*
(b) *If $A \in \mathcal{C}$ and $A \approx A'$, then $A' \in \mathcal{C}$.*
(c) *If $A \subset B$ and $B \in \mathcal{C}$, then $A \in \mathcal{C}$.*

(*d*) *If* $A \subset B$ *and* $B \in \mathcal{C}$, *then* $B/A \in \mathcal{C}$.

(*e*) *If* $0 \to A \to B \to C \to 0$ *is a short exact sequence, with* $A, C \in \mathcal{C}$, *then* $B \in \mathcal{C}$.

PROOF If $\mathcal{C}$ is a Serre class, it is nonempty, and if $A \in \mathcal{C}$, then (*a*) follows from the exactness of $A \to 0 \to A$. Properties (*b*), (*c*), and (*d*) follow from (*a*) and the exactness of the sequences $0 \to A' \to A$, $0 \to A \to B$, and $B \to B/A \to 0$, respectively, while (*e*) follows from the defining property of a Serre class.

Conversely, if $\mathcal{C}$ satisfies properties (*a*) to (*e*), then $\mathcal{C}$ is nonempty, by (*a*). If $A \xrightarrow{\alpha} B \xrightarrow{\beta} C$ is an exact sequence, there is a short exact sequence

$$0 \to \operatorname{im} \alpha \to B \to \operatorname{coker} \alpha \to 0$$

and isomorphisms $A/\ker \alpha \approx \operatorname{im} \alpha$ and $\operatorname{coker} \alpha \approx \operatorname{im} \beta \subset C$. If $A \in \mathcal{C}$, it follows from properties (*d*) and (*b*) that $\operatorname{im} \alpha \in \mathcal{C}$. If $C \in \mathcal{C}$, it follows from properties (*c*) and (*b*) that $\operatorname{coker} \alpha \in \mathcal{C}$. If $A, C \in \mathcal{C}$, it follows on using (*e*) that $B \in \mathcal{C}$. Hence, $\mathcal{C}$ is a Serre class. ■

Note that it follows from *a* and *b* of theorem 1 that a Serre class does not form a set. We list some examples of Serre classes.

2 The class of all abelian groups

3 The class of trivial groups

4 The class of finitely generated abelian groups

5 The class of finite abelian groups

6 The class of torsion abelian groups

7 The class of *p*-groups for a given prime *p*

8 The class of torsion groups having no element with order a positive power of a given prime *p*

Given a class $\mathcal{C}$, we are interested in computing modulo groups in $\mathcal{C}$. Thus a homomorphism $\varphi: A_1 \to A_2$ is defined to be a $\mathcal{C}$-*monomorphism* (or $\mathcal{C}$-*epimorphism*) if $\ker \varphi \in \mathcal{C}$ (or $\operatorname{coker} \varphi \in \mathcal{C}$) and is a $\mathcal{C}$-*isomorphism* if both conditions are satisfied. It is easily verified that the composite of $\mathcal{C}$-isomorphisms is a $\mathcal{C}$-isomorphism. Two abelian groups A_1 and A_2 are said to be $\mathcal{C}$-*isomorphic*, denoted by $A_1 \underset{\mathcal{C}}{\approx} A_2$, if there exists an abelian group A and $\mathcal{C}$-isomorphisms $A \to A_1$ and $A \to A_2$. Note the similarity between the definition of $\mathcal{C}$-isomorphic abelian groups and the definition of spaces of the same weak homotopy type.

9 **LEMMA** *The relation of being* $\mathcal{C}$-*isomorphic is an equivalence relation.*

PROOF The relation is clearly reflexive and symmetric. To show that it is transitive, assume $A_1 \underset{\mathcal{C}}{\approx} A_2$ and $A_2 \underset{\mathcal{C}}{\approx} A_3$. There exist abelian groups B and B' and $\mathcal{C}$-isomorphisms $\varphi_1: B \to A_1$, $\varphi_2: B \to A_2$, $\varphi_2': B' \to A_2$, and $\varphi_3': B' \to A_3$.

Let $C = \{(b,b') \in B \oplus B' \mid \varphi_2(b) = \varphi_2'(b')\}$ and let $p: C \to B$ and $p': C \to B'$ be the projections (C is the fibered product of φ_2 and φ_2' in the category of abelian groups). Because there is an exact sequence

$$\ker \varphi_2' \to C \xrightarrow{p} B \to \operatorname{coker} \varphi_2'$$

it follows that p is a $\mathcal{C}$-isomorphism. Similarly, p' is a $\mathcal{C}$-isomorphism. Therefore the composites $C \xrightarrow{p} B \xrightarrow{\varphi_1} A_1$ and $C \xrightarrow{p'} B' \xrightarrow{\varphi_3'} A_3$ are $\mathcal{C}$-isomorphisms, showing that $A_1 \underset{\mathcal{C}}{\approx} A_3$. ∎

A topological space X is said to be *$\mathcal{C}$-acyclic* if its integral homology groups $H_q(X) \in \mathcal{C}$ for $q > 0$. In order to ensure that the product of two $\mathcal{C}$-acyclic spaces be $\mathcal{C}$-acyclic, we need $\mathcal{C}$ to have the additional property that $A, B \in \mathcal{C}$ imply $A \otimes B, A * B \in \mathcal{C}$. A Serre class with this additional property is called a *ring of abelian groups.*

A pair (X,X') with X' nonempty is said to be *$\mathcal{C}$-acyclic* if the integral groups $H_q(X,X') \in \mathcal{C}$ for all q. In order to ensure that the product of a $\mathcal{C}$-acyclic pair and an arbitrary space is a $\mathcal{C}$-acyclic pair, we need $\mathcal{C}$ to have the property that $A \in \mathcal{C}$ implies $A \otimes B$, $A * B \in \mathcal{C}$ for arbitrary B. A Serre class $\mathcal{C}$ with this additional property is called an *ideal of abelian groups.* Obviously, an ideal of abelian groups is a ring of abelian groups. Examples 2, 3, 6, 7, and 8 are ideals of abelian groups, while examples 4 and 5 are rings of abelian groups which are not ideals of abelian groups.

In the sequel some of the results will be valid for a ring of abelian groups and somewhat stronger results will be valid for an ideal of abelian groups. The results will usually be stated in pairs, one for a ring of abelian groups, and the other for an ideal of abelian groups. The proofs of the two pairs will usually differ only in minor details. The following generalization of lemma 9.3.4 is the main result obtained from a spectral-sequence argument.

10 THEOREM *Let $p: E \to B$ be an orientable fibration with path-connected fiber F and path-connected base B and let B' be a nonempty subspace of B. Define $E' = p^{-1}(B')$ and let $\mathcal{C}$ be a Serre class. We assume that $H_i(B,B'; R) \in \mathcal{C}$ for $0 \le i < n$ and $H_j(F;G) \in \mathcal{C}$ for $0 < j < m$. We define an integer $r \ge 0$ as follows:*

(a) If $\mathcal{C}$ is a ring of abelian groups and $H_1(B,B'; R) = 0$, let $r = \inf(n, m+1)$.
(b) If $\mathcal{C}$ is an ideal of abelian groups, let $r = n + m - 1$.

Then the homomorphism $p_*: H_q(E,E'; G) \to H_q(B,B ; G)$ is a $\mathcal{C}$-monomorphism for $q \le r$ and a $\mathcal{C}$-epimorphism for $q \le r + 1$.

PROOF We use the spectral sequence of (E,E') and show first that $E^2_{s,t} \in \mathcal{C}$ if $s + t \le r$ and $t \ge 1$. We know that

$$E^2_{s,t} \approx H_s(B,B'; R) \otimes H_t(F;G) \oplus H_{s-1}(B,B'; R) * H_t(F;G)$$

In case (*a*), because $H_0(B,B'; R) = 0 = H_1(B,B'; R)$, it follows that $E^2_{s,t} \in \mathcal{C}$ if

$s = 0$ or 1. If $s > 1$, then $t < m$ (because $s + t \leq m + 1$). Therefore $H_t(F;G) \in \mathcal{C}$, and because it is also true that $s < n$, $H_s(B,B'; R)$ and $H_{s-1}(B,B'; R)$ are both in $\mathcal{C}$. Because $\mathcal{C}$ is a ring of abelian groups, $E^2_{s,t} \in \mathcal{C}$.

In case (*b*),

$$s + t \leq r = n + m - 1$$

implies $s \leq n - 1$ or $t \leq m - 1$. Because $\mathcal{C}$ is an ideal of abelian groups, it again follows that $E^2_{s,t} \in \mathcal{C}$.

To complete the proof, note that the spectral sequence gives a normal series

$$0 \subset D_0 \subset D_1 \subset \cdots \subset D_q = H_q(E,E'; G)$$

where $D_0, D_1/D_0, \ldots, D_q/D_{q-1}$ are the limit terms of the spectral sequence (that is, $D_i/D_{i-1} \approx E^\infty_{i,q-i}$). Therefore

$$H_q(E,E'; G)/D_{q-1} \approx E^\infty_{q,0} \subset E^2_{q,0} = H_q(B,B'; G)$$

and the kernel of the homomorphism $p_*: H_q(E,E'; G) \to H_q(B,B'; G)$ equals the kernel of the map $D_q \to D_q/D_{q-1}$. To show that p_* is a $\mathcal{C}$-monomorphism, therefore, we must show that $D_{q-1} \in \mathcal{C}$. By a simple induction (on D_k for $0 \leq k \leq q - 1$), it suffices to show that $E^\infty_{s,t} \in \mathcal{C}$ for $s + t \leq r$ and $t \geq 1$. This follows from the corresponding property of $E^2_{s,t}$ already established.

To prove that p_* is a $\mathcal{C}$-epimorphism, we must show that $E^2_{q,0}/E^\infty_{q,0} \in \mathcal{C}$ if $q \leq r + 1$. However, there is a sequence

$$E^2_{q,0} \supset E^3_{q,0} \supset \cdots \supset E^{q+1}_{q,0} = E^\infty_{q,0}$$

and again by a simple induction, it suffices to show that $E^k_{q,0}/E^{k+1}_{q,0} \in \mathcal{C}$ for $q \leq r + 1$ and $k \geq 2$. By definition,

$$E^{k+1}_{q,0} \approx \ker\,(d^k: E^k_{q,0} \to E^k_{q-k,k-1})$$

Therefore $E^k_{q,0}/E^{k+1}_{q,0}$ is isomorphic to a submodule of $E^k_{q-k,k-1}$, and it suffices to show that $E^k_{q-k,k-1} \in \mathcal{C}$ for $q \leq r + 1$ and $k \geq 2$. This follows from the fact (already established) that $E^2_{q-k,k-1} \in \mathcal{C}$ for $q \leq r + 1$ and $k \geq 2$. ∎

By specializing to the case where B' is a point, we get the following interesting applications of this last result.

11 COROLLARY *Let $p: E \to B$ be a fibration with path-connected fiber F and simply connected base B. Assume that E is $\mathcal{C}$-acyclic and $H_i(B) \in \mathcal{C}$ for $0 < i < n$. Then $H_i(F) \in \mathcal{C}$ for $0 < i < n - 1$, and*

(*a*) *If $\mathcal{C}$ is a ring of abelian groups, $H_{n-1}(F) \underset{\mathcal{C}}{\approx} H_n(B)$.*

(*b*) *If $\mathcal{C}$ is an ideal of abelian groups, $H_i(F) \underset{\mathcal{C}}{\approx} H_{i+1}(B)$ for $i < 2n - 2$.*

PROOF Let $B' = \{b_0\}$ and $E' = p^{-1}(b_0) = F$ and use induction on n. Inductively we can assume $H_i(F) \in \mathcal{C}$ for $0 < i < n - 1$. We apply theorem 10, with $m = n - 1$. In case (*a*) $r = n$ and in case (*b*) $r = 2n - 2$, and $H_i(E,F) \underset{\mathcal{C}}{\approx} H_i(B,b_0)$ for $i \leq r$. Because E is $\mathcal{C}$-acyclic, $H_i(E,F) \underset{\mathcal{C}}{\approx} H_{i-1}(F)$ for

$i \geq 2$, and because b_0 is a point, $H_i(B) \approx H_i(B,b_0)$ for $i > 0$. The result follows by combining these $\mathcal{C}$-isomorphisms. ∎

12 COROLLARY *Let $p: E \to B$ be a fibration with path-connected fiber F and simply connected base B. If $\mathcal{C}$ is a ring of abelian groups and two of the three spaces E, B, and F are $\mathcal{C}$-acyclic, so is the third.*

PROOF If B and F are $\mathcal{C}$-acyclic, let $B' = \{b_0\}$ and $E' = p^{-1}(b_0) = F$ and apply theorem 10*a*, with $n = m = \infty$. We find that (E,F) is $\mathcal{C}$-acyclic, and, since F is $\mathcal{C}$-acyclic, E is $\mathcal{C}$-acyclic. If E and B are $\mathcal{C}$-acyclic, it follows from corollary 11*a* that F is $\mathcal{C}$-acyclic. If E and F are $\mathcal{C}$-acyclic, let $n \geq 2$ be the smallest integer such that $H_n(B) \notin \mathcal{C}$ (if such integers exist). By corollary 11*a*, $H_{n-1}(F) \notin \mathcal{C}$, which is a contradiction. Therefore B is $\mathcal{C}$-acyclic. ∎

The following special case of this last result is worth explicit mention.

13 COROLLARY *Let X be a simply connected space and let $\mathcal{C}$ be a ring of abelian groups. Then X is $\mathcal{C}$-acyclic if and only if its loop space ΩX is $\mathcal{C}$-acyclic.* ∎

For our next application of the spectral sequence of a fibration (namely, to prove the generalized Hurewicz isomorphism theorems) we need another property of Serre classes of abelian groups. A Serre class $\mathcal{C}$ of abelian groups is said to be an *acyclic class* if any space of type $(\pi,1)$ with $\pi \in \mathcal{C}$ is $\mathcal{C}$-acyclic. Thus $\mathcal{C}$ is an acyclic class if and only if $\pi \in \mathcal{C}$ implies $H_q(\pi) \in \mathcal{C}$ for $q > 0$. From the remarks and results at the end of Sec. 9.5, it follows that each of examples 2 to 8 is an acyclic class.

The loop space of a space of type (π,n), with $n \geq 2$, is a space of type $(\pi, n-1)$. Hence we have the following result by induction on n from corollary 13.

14 LEMMA *If $\mathcal{C}$ is an acyclic ring of abelian groups, any space of type (π,n), with $n \geq 1$ and $\pi \in \mathcal{C}$, is $\mathcal{C}$-acyclic.* ∎

Using this and a Postnikov system for X, it can be shown that if X is a simply connected space whose homotopy groups belong to an acyclic ring $\mathcal{C}$ of abelian groups, then X is $\mathcal{C}$-acyclic. This is also a consequence of the following *generalized absolute Hurewicz isomorphism theorem.*

15 THEOREM *Let $\mathcal{C}$ be an acyclic ring of abelian groups and let X be a simply connected space. The following are equivalent:*

(*a*) $\pi_i(X) \in \mathcal{C}$ *for* $2 \leq i < n$.
(*b*) $H_i(X) \in \mathcal{C}$ *for* $2 \leq i < n$.

Furthermore, either implies that the Hurewicz homomorphism

$$\varphi: \pi_i(X) \to H_i(X)$$

is a $\mathcal{C}$-isomorphism for $i \leq n$.

PROOF It clearly suffices to prove that (*b*) implies $\varphi: \pi_n(X) \underset{\mathcal{C}}{\approx} H_n(X)$. For

$n = 2$ this is a consequence of the absolute Hurewicz isomorphism theorem. We assume $n \geq 3$ and prove the result by induction on n. It follows that $\pi_i(X) \in \mathcal{C}$ for $i < n$.

By corollary 8.3.8, there exists a sequence of fibrations

$$E_{n-1} \xrightarrow{p_{n-1}} \cdots \xrightarrow{p_2} E_1 = X$$

such that E_j is j-connected and $p_j: E_j \to E_{j-1}$ has a fiber F_j which is a space of type $(\pi_j(X), j-1)$.

It follows from the acyclicity of $\mathcal{C}$ that F_j is $\mathcal{C}$-acyclic for $2 \leq j \leq n-1$ [because $\pi_j(X) \in \mathcal{C}$ for $j < n$]. By induction on j, for $2 \leq j \leq n-1$, we prove that $p_{j*}: H_i(E_j) \underset{\mathcal{C}}{\approx} H_i(E_{j-1})$ for $i \leq n$. Assuming it for $j-1$, where $j > 1$, we see that $H_i(E_{j-1},e_0) \in \mathcal{C}$ for $i < n$ and $H_i(F) \in \mathcal{C}$ for $0 < i$. We deduce from theorem 10a that $p_{j*}: H_i(E_j,F_j) \underset{\mathcal{C}}{\approx} H_i(E_{j-1})$ for $0 < i \leq n$. Since F_j is $\mathcal{C}$-acyclic, this implies that $p_{j*}: H_i(E_j) \underset{\mathcal{C}}{\approx} H_i(E_{j-1})$ for $i \leq n$. This completes the induction.

Therefore the composite $f = p_2 \circ \cdots \circ p_{n-1}: E_{n-1} \to X$ has the property that

$$f_*: H_i(E_{n-1}) \underset{\mathcal{C}}{\approx} H_i(X) \qquad i \leq n$$

We have a commutative diagram

$$\begin{array}{ccc} \pi_n(E_{n-1}) & \xrightarrow{\varphi} & H_n(E_{n-1}) \\ {\scriptstyle f_\#}\downarrow & & \downarrow{\scriptstyle f_*} \\ \pi_n(X) & \xrightarrow{\varphi} & H_n(X) \end{array}$$

and, by the absolute Hurewicz isomorphism theorem, the top homomorphism is an isomorphism. Since both vertical maps are $\mathcal{C}$-isomorphisms, the result follows. ∎

This theorem clearly implies that a simply connected space is $\mathcal{C}$-acyclic (for an acyclic ring $\mathcal{C}$ of abelian groups) if and only if all its homotopy groups are in $\mathcal{C}$. Taking $\mathcal{C}$ to be the acyclic ring of all finitely generated abelian groups, we obtain the following result.

16 COROLLARY *A simply connected space has finitely generated homology groups in every dimension if and only if it has finitely generated homotopy groups in every dimension.* ∎

In particular, it follows from corollary 16 that any sphere S^n has finitely generated homotopy groups. Corollary 16 is not true if $\pi_1(X)$ is assumed to be finitely generated (instead of 0), as shown by the following example.

17 EXAMPLE Let $X = S^2 \vee S^1$. Then $\pi_1(X) \approx \mathbf{Z}$ is finitely generated and $H_i(X)$ is finitely generated for all i, but $\pi_2(X)$ is a free abelian group on a countable set of generators and so is not finitely generated.

Nevertheless, there is a generalization of corollary 16 valid for spaces which are not assumed to be simply connected. If X is a path-connected space, let $j: X \subset B$ be an imbedding of X in a space B of type $(\pi_1(X),1)$ such that $j_\#: \pi_1(X) \approx \pi_1(B)$. Let $p: P_j \to B$ be the mapping path fibration corresponding to j, as defined in Sec. 2.8 (so X and P_j have the same homotopy type). The space X is said to be *strongly simple* if $\pi_1(X)$ is abelian and if $p: P_j \to B$ is orientable over $\mathbf{Z}$.

18 EXAMPLE Assume that X is a space such that for every element $a \in \pi_1(X)$ there is a map $\bar{\omega}: S^1 \times X \to X$, with $\bar{\omega} \mid S^1 \times x_0$ representing a and $\bar{\omega} \mid p_0 \times X$ homotopic to 1_X. Then X is strongly simple (because P_j also has the same property as X). In particular, any H-space is strongly simple.

19 LEMMA *Let $\mathcal{C}$ be an acyclic ring of abelian groups and assume that X is a strongly simple space such that $\pi_1(X) \in \mathcal{C}$ and $H_i(X) \in \mathcal{C}$ for $0 < i < n$, where $n \geq 2$. If F is the fiber of the fibration $p: P_j \to B$, then $H_q(F) \to H_q(P_j)$ is a $\mathcal{C}$-isomorphism for $q \leq n$.*

PROOF Since X and P_j have the same homotopy type, $H_i(P_j) \in \mathcal{C}$ for $0 < i < n$. Let $m < n$ and assume inductively that $H_q(F) \underset{\mathcal{C}}{\approx} H_q(P_j)$ for $q \leq m$. Then $H_q(F) \in \mathcal{C}$ for $0 < q \leq m$.

We now prove that $H_{m+1}(F) \underset{\mathcal{C}}{\approx} H_{m+1}(P_j)$. From the spectral sequence of the fibration (the fibration being orientable, since X is strongly simple), there is a composition series

$$0 \subset D_0 \subset D_1 \subset \cdots \subset D_{m+1} = H_{m+1}(P_j)$$

where $D_s/D_{s-1} \approx E^\infty_{s,m+1-s}$ and $D_0 = \text{im}\,[H_{m+1}(F) \to H_{m+1}(P_j)]$. To show that $H_{m+1}(F) \to H_{m+1}(P_j)$ is a $\mathcal{C}$-epimorphism, it suffices to show that $E^\infty_{s,m+1-s} \in \mathcal{C}$ for $s > 0$. This will be so if $E^2_{s,m+1-s} \in \mathcal{C}$ for $s > 0$. However,

$$E^2_{s,m+1-s} \approx H_s(B) \otimes H_{m+1-s}(F) \oplus H_{s-1}(B) * H_{m+1-s}(F)$$

Since $\mathcal{C}$ is an acyclic Serre class, $H_s(B) \in \mathcal{C}$ for $s > 0$ [and, of course, $H_0(B) \approx \mathbf{Z}$]. Since, by the inductive hypothesis, $H_{m+1-s}(F) \in \mathcal{C}$ for $s > 0$, we see that $E^2_{s,m+1-s} \in \mathcal{C}$ for $s \geq 0$ because $\mathcal{C}$ is a ring of abelian groups.

To show that $H_{m+1}(F) \to H_{m+1}(P_j)$ is a $\mathcal{C}$-monomorphism, we have a sequence of homomorphisms

$$H_{m+1}(F) \approx E^2_{0,m+1} \to E^3_{0,m+1} \to \cdots \to E^\infty_{0,m+1} \approx D_0$$

and it suffices to prove $\ker\,(E^r_{0,m+1} \to E^{r+1}_{0,m+1}) \in \mathcal{C}$ for $r \geq 2$. This is equivalent to showing that $d^r(E^r_{r,m+2-r}) \subset E^r_{0,m+1}$ is in $\mathcal{C}$ for $r \geq 2$. This will be true if $E^2_{r,m+2-r} \in \mathcal{C}$ for $r \geq 2$. However,

$$E^2_{r,m+2-r} \approx H_r(B) \otimes H_{m+2-r}(F) \oplus H_{r-1}(B) * H_{m+2-r}(F)$$

and because $m + 2 - r \leq m$, $H_{m+2-r}(F) \in \mathcal{C}$, by the inductive assumption. The result follows because $\mathcal{C}$ is an acyclic ring of abelian groups. ∎

We now have the following strengthened version of theorem 15.

20 THEOREM *Let X be a strongly simple space and let $\mathcal{C}$ be an acyclic ring of abelian groups. If $\pi_1(X) \in \mathcal{C}$, the following are equivalent:*

(a) $\pi_i(X) \in \mathcal{C}$ *for* $2 \le i < n$.
(b) $H_i(X) \in \mathcal{C}$ *for* $2 \le i < n$.

Either implies that $\varphi: \pi_i(X) \to H_i(X)$ *is a* $\mathcal{C}$*-isomorphism for* $i \le n$.

PROOF It suffices to prove that (b) implies $\varphi: \pi_n(X) \underset{\mathcal{C}}{\approx} H_n(X)$. Let F be the fiber of the fibration $p: P_j \to B$. Since X and P_j have the same homotopy type, there is a map $f: F \to X$, equivalent to $F \subset P_j$. Since $\pi_i(F) \approx \pi_i(P_j)$ for $i \ge 2$, it follows that $f_\#: \pi_i(F) \approx \pi_i(X)$ for $i \ge 2$. By lemma 19, $f_*: H_i(F) \to H_i(X)$ is a $\mathcal{C}$-isomorphism for $i \le n$. Since F is simply connected, it follows from theorem 15 that $\varphi: \pi_n(F) \underset{\mathcal{C}}{\approx} H_n(F)$. Since $\varphi \circ f_\# = f_* \circ \varphi$, this gives the result. ■

We use this result to establish the following *generalized relative Hurewicz isomorphism theorem.*

21 THEOREM *Let* $\mathcal{C}$ *be an acyclic ideal of abelian groups, let* $A \subset X$, *and assume that* A *and* X *are simply connected. The following are equivalent:*

(a) $\pi_i(X,A) \in \mathcal{C}$ *for* $2 \le i < n$.
(b) $H_i(X,A) \in \mathcal{C}$ *for* $2 \le i < n$.

Either property implies

(c) $\varphi: \pi_n(X,A) \underset{\mathcal{C}}{\approx} H_n(X,A)$.

PROOF It suffices to prove that (b) implies (c) by induction on n. For $n = 2$ this follows from the relative Hurewicz isomorphism theorem. Therefore we assume $n \ge 3$ and $\pi_i(X,A) \in \mathcal{C}$ for $i < n$. Let $x_0 \in A$, let PX be the space of paths in X with origin x_0, and let $p: PX \to X$ be the fibration sending a path to its terminal point. The fiber $p^{-1}(x_0)$ is the loop space ΩX. By theorem 7.2.8, $p_\#: \pi_k(PX,p^{-1}(A)) \approx \pi_k(X,A)$ for $k \ge 1$. Because PX is contractible, $\pi_k(PX,p^{-1}(A)) \approx \pi_{k-1}(p^{-1}(A))$ for $k \ge 2$ and $H_k(PX,p^{-1}(A)) \approx H_{k-1}(p^{-1}(A))$ for $k \ge 2$. For $i \ge 2$ there is a commutative diagram

$$\begin{array}{ccccc}
\pi_{i-1}(p^{-1}(A)) & \overset{\partial}{\underset{\approx}{\leftarrow}} & \pi_i(PX,p^{-1}(A)) & \overset{p_\#}{\underset{\approx}{\rightarrow}} & \pi_i(X,A) \\
\downarrow \varphi & & \downarrow \varphi & & \downarrow \varphi \\
H_{i-1}(p^{-1}(A)) & \overset{\partial}{\underset{\approx}{\leftarrow}} & H_i(PX,p^{-1}(A)) & \overset{p_*}{\rightarrow} & H_i(X,A)
\end{array}$$

Applying theorem 10*b*, where $H_i(X,A) \in \mathcal{C}$ for $i < n$, and taking $m = 1$ (ΩX is path connected because X was assumed to be simply connected), we see that $p_*: H_i(PX,p^{-1}(A)) \underset{\mathcal{C}}{\approx} H_i(X,A)$ for $i \le n$. Therefore, for $i = n$ all the horizontal maps in the above diagram are $\mathcal{C}$-isomorphisms, and to complete the proof it suffices to prove $\varphi: \pi_{n-1}(p^{-1}(A)) \underset{\mathcal{C}}{\approx} H_{n-1}(p^{-1}(A))$. This will follow from theorem 20 once we have verified that $p^{-1}(A)$ is strongly simple.

Because $p^{-1}(A)$ is a principal fibration with fiber ΩX, there is a continuous map $\Omega X \times p^{-1}(A) \to p^{-1}(A)$. Since $\pi_2(X) \to \pi_2(X,A)$ is an epimorphism, so is $\pi_1(\Omega X) \to \pi_1(p^{-1}(A))$. Therefore the existence of the map

$$\Omega X \times p^{-1}(A) \to p^{-1}(A)$$

implies that $p^{-1}(A)$ is strongly simple, as in example 18. ∎

By using the mapping cylinder (as in the proof of theorem 7.5.9), the following *generalized Whitehead theorem* can be deduced from theorem 21.

22 THEOREM *Let $\mathcal{C}$ be an acyclic ideal of abelian groups and let $f: X \to Y$ be a map between simply connected spaces. For $n \geq 1$ the following are equivalent:*

(a) $f_{\#}: \pi_i(X) \to \pi_i(Y)$ is a $\mathcal{C}$-isomorphism for $i \leq n$ and a $\mathcal{C}$-epimorphism for $i = n + 1$.
(b) $f_*: H_i(X) \to H_i(Y)$ is a $\mathcal{C}$-isomorphism for $i \leq n$ and a $\mathcal{C}$-epimorphism for $i = n + 1$. ∎

7 HOMOTOPY GROUPS OF SPHERES

The results of the last section were obtained by using the homology spectral sequence of a fibration. In this section we shall use the multiplicative properties of the cohomology spectral sequence to obtain some specific results about the homotopy groups of spheres. These homotopy groups are finitely generated, and we shall obtain information about their p-primary components. The first main result is that the only homotopy groups of S^n which are infinite are $\pi_n(S^n)$, and if n is even, $\pi_{2n-1}(S^n)$. The next main result concerns the double suspension. It will be shown that for odd n the double suspension

$$S^2: \pi_m(S^n) \to \pi_{m+2}(S^{n+2})$$

induces an isomorphism of the p-primary components of these groups for a wider range of values of m and n (depending on p) than the range for which it is an isomorphism between the groups. Combining this with specific computations of p-primary components of $\pi_m(S^3)$, we determine the lowest dimension $m > n$ for which $\pi_m(S^n)$ for n odd has a nontrivial p-primary component.

We begin with the following useful technical result about the cohomology spectral sequence of a fibration.

1 LEMMA *Let X be a simply connected space and assume that there is an element $u \in H^n(X;R)$, with $n \geq 2$, such that $u^{m-1} \neq 0$ for some $m \geq 2$ and $\{1,u,u^2, \ldots ,u^{m-1}\}$ form a basis for $H^*(X;R)$ in degrees $< mn$. Then there is an element $v \in H^{n-1}(\Omega X;R)$ such that $\{1,v\}$ form a basis for $H^*(\Omega X;R)$ in degrees $< mn - 2$.*

PROOF We use the spectral sequence of theorem 9.4.7, with A empty, for the fibration $PX \to X$. Because PX is contractible, $E_\infty^{s,t} = 0$ if $(s,t) \neq (0,0)$, and because X has no torsion in degrees $< mn$, $E_2^{s,t} \approx H^s(X) \otimes H^t(\Omega X)$ for $s < mn$ (all coefficients R). Then we have $E_2^{s,t} = 0$ if $s < mn$ and

$s \neq 0, n, 2n, \ldots, (m-1)n$. Because d_r has bidegree $(r, 1-r)$, it follows that for $s < mn$, $d_r\colon E_r^{s,t} \to E_r^{s+r,t-r+1}$ is zero unless $r = n, 2n, \ldots, (m-1)n$. Therefore $E_n^{s,t} \approx E_2^{s,t}$ for $s < mn$. If $t < n-1$, $E_n^{0,t} \approx E_\infty^{0,t}$, and if $0 < t < n-1$, we see that

$$H^t(\Omega X) \approx E_2^{0,t} \approx E_\infty^{0,t} = 0$$

and so $H^t(\Omega X) = 0$ for $0 < t < n-1$. Furthermore, there is an exact sequence

$$0 \to E_\infty^{0,n-1} \to H^0(X) \otimes H^{n-1}(\Omega X) \xrightarrow{d_n} H^n(X) \otimes H^0(\Omega X) \to E_\infty^{n,0} \to 0$$

Because $E_\infty^{0,n-1} = 0 = E_\infty^{n,0}$, it follows that there is an element $v \in H^{n-1}(\Omega X)$ such that $d_n(1 \otimes v) = u \otimes 1$. Because d_n is a derivation, $d_n(u^k \otimes v) = (-1)^{kn} u^{k+1} \otimes 1$. The assumption about the cohomology of X ensures that for $s < mn$ the map $d_n\colon E_n^{s-n,n-1} \to E_n^{s,0}$ is an isomorphism. Because d_n is a differential, the composite

$$E_n^{s-2n,2n-2} \xrightarrow{d_n} E_n^{s-n,n-1} \xrightarrow{d_n} E_n^{s,0}$$

is trivial. Therefore $d_n\colon E_n^{s-2t,2n-2} \to E_n^{s-n,n-1}$ is trivial for $s < mn$ and $E_{n+1}^{s-n,n-1} = 0 = E_{n+1}^{s,0}$ for $s < mn$. Hence

$$E_r^{s,t} = 0 \qquad s < mn,\ t \le n-1,\ r \ge n+1$$
$$E_{n+1}^{0,2n-2} = E_2^{0,2n-2}$$

Assume the lemma false and let q be the smallest integer such that $n-1 < q < mn-2$ and $H^q(\Omega X) \neq 0$. We shall show that

$$E_\infty^{0,q} = E_2^{0,q} \approx H^q(\Omega X),$$

which is a contradiction. We know that $E_n^{0,q} \approx E_2^{0,q}$. Furthermore, $d_n\colon E_n^{0,q} \to E_n^{n,q-n+1}$ is trivial, because if $q-n+1 \neq n-1$, then $H^{q-n+1}(\Omega X) = 0$ and $E_r^{s,q-n+1} = 0$ for all r and s, and if $q-n+1 = n-1$, then $E_{n+1}^{0,2n-2} = E_2^{0,2n-2}$. Therefore $E_{n+1}^{0,q} \approx E_n^{0,q}$. From the assumption that q is the smallest degree larger than $n-1$ for which $H^q(\Omega X) \neq 0$, it follows that $E_r^{s,t} = 0$ if $s < mn$, $t < q$, and $r \ge n+1$ (in case $t \le n-1$ this was noted above). Therefore $d_r\colon E_r^{0,q} \to E_r^{r,q-r+1}$ is trivial for all $r \ge n+1$ and $E_\infty^{0,q} \approx E_{n+1}^{0,q}$. Hence we have the isomorphisms

$$E_\infty^{0,q} \approx E_{n+1}^{0,q} \approx E_n^{0,q} \approx E_2^{0,q} \quad \blacksquare$$

By using the generalized Gysin sequence of theorem 9.5.2, it is easy to show that if ΩX is a cohomology n-sphere for some odd $n \ge 1$, then $H^*(X)$ is a polynomial algebra on a generator of degree $n+1$. The following converse is an immediate consequence of lemma 1 for the case $m = \infty$.

2 COROLLARY *Let X be a simply-connected space such that $H^*(X;R)$ is a polynomial algebra on a generator of degree n (n is then necessarily even). Then the loop space ΩX is a cohomology $(n-1)$-sphere.* ∎

We shall also need the following consequence of the generalized Wang sequence of theorem 9.5.1.

3 LEMMA *Let X be a simply connected space which is a cohomology n-sphere for some odd $n > 1$. Then the cohomology ring $H^*(\Omega X)$ of its loop space ΩX has a basis consisting of elements $\{1, u_1, u_2, \ldots\}$ with degree $u_k = k(n-1)$ and $u_p \smile u_q = [(p+q)!/p!q!]u_{p+q}$.*

PROOF We use the Wang exact sequence of the fibration $PX \to X$. Because PX is contractible, the map

$$\theta\colon H^t(\Omega X) \to H^{t-n+1}(\Omega X)$$

is an isomorphism for $t \neq 0$. Define $u_k \in H^{k(n-1)}(\Omega X)$ for $k \geq 0$ by induction by the equations

$$u_0 = 1$$
$$\theta(u_k) = u_{k-1} \qquad k > 0$$

Then the set $\{1, u_1, u_2, \ldots\}$ is a basis for $H^*(\Omega X)$, and we verify that it has the stated multiplicative property by double induction on p and q. If $i = 0$ or $j = 0$, then $u_i \smile u_j = u_{i+j}$. Let $p > 0$ and $q > 0$ and assume that $u_i \smile u_j = [(i+j)!/i!j!]u_{i+j}$ if $i + j < p + q$, $i \geq 0$, and $j \geq 0$. Because n is odd,

$$\begin{aligned}\theta(u_p \smile u_q) &= \theta(u_p) \smile u_q + u_p \smile \theta(u_q) = u_{p-1} \smile u_q + u_p \smile u_{q-1} \\ &= \left[\frac{(p+q-1)!}{(p-1)!q!} + \frac{(p+q-1)!}{p!(q-1)!}\right] u_{p+q-1} = \frac{(p+q)!}{p!q!} u_{p+q-1}\end{aligned}$$

Because θ is a monomorphism,

$$u_p \smile u_q = \frac{(p+q)!}{p!q!} u_{p+q} \quad \blacksquare$$

It follows from lemma 3 that $(u_1)^p = p! u_p$. Over a field of characteristic 0, the elements $\{1, u_1, u_1^2, \ldots\}$ also form a basis of $H^*(\Omega X)$, and so we obtain the next result.

4 COROLLARY *Let X be a simply connected space which is a rational cohomology n-sphere for some odd $n > 1$. The rational cohomology algebra of the loop space ΩX is a polynomial algebra with one generator of degree $n - 1$.* ∎

Let X be a space of type $(\mathbf{Z},3)$ and let $f\colon S^3 \to X$ be a map such that $f_\#\colon \pi_3(S^3) \approx \pi_3(X)$. Let $p\colon E \to S^3$ be the principal fibration induced by f. Then the fiber F of $p\colon E \to S^3$ is a space of type $(\mathbf{Z},2)$. We shall need the following computation of the homology groups of E.

5 LEMMA *Let $p\colon E \to S^3$ be a fibration with fiber F a space of type $(\mathbf{Z},2)$ such that $\partial\colon \pi_3(S^3) \approx \pi_2(F)$. Then the integral homology of E is given by*

$$H_q(E) \approx \begin{cases} 0 & q \text{ odd} \\ \mathbf{Z} & q = 0 \\ \mathbf{Z}_n & q = 2n > 0 \end{cases}$$

PROOF We know that $H^*(F)$ is a polynomial algebra with one generator u of degree 2. Because $\bar{\partial}\colon \pi_3(S^3) \approx \pi_2(F)$, it follows that E is 3-connected, and so $H^2(E) = 0$. By the exact Wang sequence of the fibration, $\theta\colon H^2(F) \approx H^0(F)$. Without loss of generality, we can assume that u has been chosen so that $\theta(u) = 1$. Then $\theta(u^n) = nu^{n-1}$. By the exact Wang sequence again, $H^q(E) = 0$ if q is even and $q < 0$, and $H^{2n+1}(E) \approx \mathbf{Z}_n$ if $n \geq 1$. The result then follows from the universal-coefficient theorem. ∎

If $\mathcal{C}$ is the Serre class of groups having no element with order a positive power of a given prime p, then $H_i(E) \in \mathcal{C}$ for $0 < i < 2p$. By theorem 9.6.15, $\pi_i(E) \in \mathcal{C}$ for $i < 2p$ and $\pi_{2p}(E) \underset{\mathcal{C}}{\approx} \mathbf{Z}_p$. Because $\pi_i(E) \approx \pi_i(S^3)$ for $i > 3$, we obtain the following result.

6 COROLLARY *The p-primary component of $\pi_i(S^3)$ is zero if $3 < i < 2p$ and is $\mathbf{Z}_p$ if $i = 2p$.* ∎

We are now ready to prove the finiteness of the higher homotopy groups of odd-dimensional spheres.

7 THEOREM *If n is odd, $\pi_m(S^n)$ is finite for $m \neq n$.*

PROOF We use induction on n. If $n = 1$, we know that $\pi_m(S^1) = 0$ if $m \neq 1$, and the result is valid in this case. For $n = 3$, if E is the space of lemma 5 and $\mathcal{C}$ is the Serre class of finite groups, then E is $\mathcal{C}$-acyclic. By theorem 9.6.15, $\pi_i(E) \in \mathcal{C}$ for all i. Because $\pi_i(E) \approx \pi_i(S^3)$ for $i > 3$, $\pi_i(S^3)$ is finite for $i > 3$.

Assume $n > 3$ and $\pi_m(S^{n-2})$ finite for $m \neq n - 2$. We compute the rational cohomology algebras of ΩS^n and $\Omega^2 S^n$. By corollary 4, that of ΩS^n is a polynomial algebra with one generator of degree $n - 1$. By corollary 2, $\Omega^2 S^n$ is a rational cohomology $(n - 2)$-sphere. By the universal-coefficient theorem, the integral group $H_i(\Omega^2 S^n)$ is a torsion group if $0 < i \neq n - 2$, and $H_{n-2}(\Omega^2 S^n)$ is isomorphic to a direct sum of $\mathbf{Z}$ and a torsion group. Furthermore, $\pi_k(\Omega^2 S^n) \approx \pi_{k+2}(S^n)$ for all k. Therefore $\Omega^2 S^n$ is $(n - 3)$-connected and $\varphi\colon \pi_{n-2}(\Omega^2 S^n) \approx H_{n-2}(\Omega^2 S^n)$. If $\alpha\colon S^{n-2} \to \Omega^2 S^n$ is a generator and $\mathcal{C}$ is the Serre class of torsion groups, it follows that $\alpha_*\colon H_i(S^{n-2}) \underset{\mathcal{C}}{\approx} H_i(\Omega^2 S^n)$ for all i. Because $\mathcal{C}$ is an acyclic ideal of abelian groups and S^{n+2} and $\Omega^2 S^n$ are both simply connected, we can apply the generalized Whitehead theorem 9.6.22 to deduce that $\alpha_\#\colon \pi_i(S^{n-2}) \underset{\mathcal{C}}{\approx} \pi_i(\Omega^2 S^n)$. By the inductive assumption $\pi_i(S^{n-2})$ is finite for $i > n - 2$. Therefore $\pi_i(\Omega^2 S^n) \underset{\mathcal{C}}{\approx} \pi_{i+2}(S^n)$ is a torsion group for $i > n - 2$. Because $\pi_m(S^n)$ is known to be finitely generated, $\pi_m(S^n)$ is finite for $m \neq n$. ∎

We want to establish a result similar to theorem 7 for even-dimensional spheres. This will be done by considering a suitable $(n - 1)$-sphere bundle over S^n. Let W^{2n-1} be the subspace of $R^n \times R^n$ consisting of pairs of unit vectors (z_1,z_2) which are orthogonal and let $p\colon W^{2n-1} \to S^n$ map (z_1,z_2) to z_1. Then $p\colon W^{2n-1} \to S^n$ is a fiber bundle with fiber S^{n-1} (it is the unit tangent bundle of S^n), as can be verified by constructing an explicit homeomorphism $p^{-1}(U) \approx U \times S^{n-1}$ for any proper open subset $U \subset S^n$.

8 LEMMA *If n is even, the integral homology groups of W^{2n-1} are all finite except for $H_0(W^{2n-1})$ and $H_{2n-1}(W^{2n-1})$, which are infinite cyclic.*

PROOF Because n is even, there is no map $f\colon S^n \to S^n$ which sends each point of S^n to an orthogonal point of S^n (by corollary 4.7.11). It follows that $p\colon W^{2n-1} \to S^n$ has no section. If $[\alpha] \in \pi_n(S^n)$ is a generator, $[\alpha]$ is not in the image of

$$p_\#\colon \pi_n(W^{2n-1}) \to \pi_n(S^n)$$

(because there is no section). Therefore, $\bar{\partial}[\alpha] \neq 0$ in $\pi_{n-1}(S^{n-1})$. Because $\pi_{n-1}(S^{n-1})$ is infinite cyclic, $\bar{\partial}\colon \pi_n(S^n) \to \pi_{n-1}(S^{n-1})$ is a monomorphism and $\pi_{n-1}(W^{2n-1}) \approx \pi_{n-1}(S^{n-1})/\bar{\partial}(\pi_n(S^n))$ is a finite group. Because W^{2n-1} is $(n-2)$-connected, $\pi_{n-1}(W^{2n-1}) \approx H_{n-1}(W^{2n-1})$. Therefore $H_{n-1}(W^{2n-1})$ is a finite group, and by the exact Wang sequence

$$0 \to H_n(W^{2n-1}) \to H_0(S^{n-1}) \xrightarrow{\theta} H_{n-1}(S^{n-1}) \to H_{n-1}(W^{2n-1}) \to 0$$

we see that $H_n(W^{2n-1}) = 0$. From this exact sequence it also follows that $H_i(W^{2n-1}) = 0$ for $n < i < 2n-1$ and $H_{2n-1}(W^{2n-1})$ is infinite cyclic. ■

9 THEOREM *If n is even, $\pi_m(S^n)$ is finite for $m \neq n$ and $m \neq 2n-1$, and $\pi_{2n-1}(S^n)$ is the direct sum of an infinite cyclic group and a finite group.*

PROOF Let $\mathcal{C}$ be the acyclic ideal of abelian groups consisting of the torsion groups. By lemma 8, $H_i(W^{2n-1}) \in \mathcal{C}$ for $0 < i < 2n-1$. By theorem 9.6.15, $\pi_{2n-1}(W^{2n-1}) \underset{\mathcal{C}}{\approx} H_{2n-1}(W^{2n-1})$. Because $\pi_{2n-1}(W^{2n-1})$ is finitely generated (by corollary 9.6.16) and $H_{2n-1}(W^{2n-1})$ is infinite cyclic, $\pi_{2n-1}(W^{2n-1})$ is a direct sum of an infinite cyclic group and a finite group. If $\alpha\colon S^{2n-1} \to W^{2n-1}$ represents a generator of the infinite cyclic summand of $\pi_{2n-1}(W^{2n-1})$, then $\alpha_*\colon H_i(S^{2n-1}) \underset{\mathcal{C}}{\approx} H_i(W^{2n-1})$ for all i. By the generalized Whitehead theorem, $\alpha_\#\colon \pi_i(S^{2n-1}) \underset{\mathcal{C}}{\approx} \pi_i(W^{2n-1})$ for all i. Using this and theorem 7, $\pi_i(W^{2n-1})$ is finite for $i \neq 2n-1$. The theorem now follows from the exact homotopy sequence of the fibration $W^{2n-1} \to S^n$ and the fact that, by theorem 7, $\pi_i(S^{n-1})$ is finite for $i \neq n-1$. ■

We now consider the double suspension

$$\pi_i(S^n) \xrightarrow{S} \pi_{i+1}(S^{n+1}) \xrightarrow{S} \pi_{i+2}(S^{n+2})$$

where n is odd. This involves a study of the composite

$$S^n \xrightarrow{\rho} \Omega S^{n+1} \xrightarrow{\Omega\rho} \Omega(\Omega S^{n+2})$$

We begin with the following partial computation of the $\mathbf{Z}_p$ homology of $\Omega^2 S^{n+2}$.

10 LEMMA *Let n be odd and p be prime. Then $H_q(\Omega^2 S^{n+2};\mathbf{Z}_p) = 0$ for $n < q < p(n+1) - 2$.*

PROOF By lemma 3, the set of elements $\{1, u_1, u_1^2, \ldots, u_1^{p-1}\}$ forms a basis for $H^*(\Omega S^{n+2};\mathbf{Z}_p)$ in degrees $< p(n+1)$. By lemma 1, there is an element $v \in H^n(\Omega^2 S^{n+2};\mathbf{Z}_p)$ such that $\{1,v\}$ forms a basis for $H^*(\Omega^2 S^{n+2};\mathbf{Z}_p)$ in degrees $< p(n+1) - 2$. The lemma then follows by the universal-coefficient theorem. ∎

This implies the following result about the double suspension.

11 THEOREM *Let n be an odd integer, p a prime, and $\mathcal{C}$ the acyclic ideal of torsion groups with trivial p-primary component. Then*

$$S^2\colon \pi_i(S_n) \to \pi_{i+2}(S^{n+2})$$

is a $\mathcal{C}$-isomorphism for $i < p(n+1) - 3$ and a $\mathcal{C}$-epimorphism for $i = p(n+1) - 3$.

PROOF The composite

$$S^n \xrightarrow{\rho} \Omega S^{n+1} \xrightarrow{\Omega\rho} \Omega^2 S^{n+2}$$

induces an isomorphism of $\pi_n(S^n)$ with $\pi_n(\Omega^2 S^{n+2})$ and, by the Whitehead theorem, an isomorphism of $H_n(S^n)$ with $H_n(\Omega^2 S^{n+2})$. From this and lemma 10 it follows that the above composite induces an isomorphism of $H_q(S^n;\mathbf{Z}_p)$, with $H_q(\Omega^2 S^{n+2};\mathbf{Z}_p)$ for $q \le p(n+1) - 3$. By the universal-coefficient theorem, it induces a $\mathcal{C}$-isomorphism of $H_q(S^n)$, with $H_q(\Omega^2 S^{n+2})$ for $q \le p(n+1) - 3$. From the generalized Whitehead theorem, it induces a $\mathcal{C}$-isomorphism of $\pi_q(S^n)$ with $\pi_q(\Omega^2 S^{n+2})$ for $q < p(n+1) - 3$, and a $\mathcal{C}$-epimorphism of $\pi_{p(n+1)-3}(S^n)$ to $\pi_{p(n+1)-3}(\Omega^2 S^{n+2})$. The theorem follows from the fact that S^2 corresponds to the above induced homomorphism under the isomorphism

$$\pi_q(\Omega^2 S^{n+2}) \approx \pi_{q+2}(S^{n+2}) \quad ∎$$

12 COROLLARY *Let $n \ge 3$ be odd and p prime. Then $\pi_i(S^n)$ and $\pi_{i-n+3}(S^3)$ have isomorphic p-primary components if $i < 4p + n - 6$.*

PROOF We use induction on n. If $n = 3$, there is nothing to be proved. If $n \ge 5$, we need only prove that $S^2\colon \pi_{i-2}(S^{n-2}) \to \pi_i(S^n)$ induces an isomorphism of p-primary components. By theorem 11, this will be true if $i - 2 < p(n-1) - 3$. Hence we need only verify that

$$4p + n - 6 \le p(n-1) - 1$$

But this is equivalent to $(p-1)(n-5) \ge 0$. ∎

Combining corollary 12 with corollary 6, we have the following result.

13 COROLLARY *Let $n \ge 3$ be odd and p prime. For $0 < m < 2p - 3$, $\pi_{n+m}(S^n)$ has trivial p-primary component and $\pi_{n+2p-3}(S^n)$ has $\mathbf{Z}_p$ as p-primary component.* ∎

EXERCISES

A SPECTRAL SEQUENCES AND SUSPENSION

In this group of exercises all spaces will be assumed to be finite pointed CW complexes and all pairs will be finite pointed CW pairs.

1 Prove that $\{X;Y\}$ is finitely generated.

2 For spaces X and Y prove that there is a convergent E^2 spectral sequence $\{E^r\}$ with

$$E^2_{s,t} \approx H_s(Y; \{X;S^0\}_t)$$

and with E^∞ isomorphic to the graded group associated to the increasing filtration on $\{X;Y\}_*$ defined by

$$F_s\{X;Y\}_* = \operatorname{im}(\{X;Y^s\}_* \to \{X;Y\}_*)$$

3 For spaces X and Y prove that there is a convergent E_2 spectral sequence $\{E_r\}$ with

$$E_2^{s,t} \approx H^s(X; \{S^0;Y\}_{-t})$$

and with E_∞ isomorphic to the graded group associated to the decreasing filtration of $\{X;Y\}_*$ defined by

$$F^s\{X;Y\}_* = \ker(\{X;Y\}_* \to \{X^{s-1};Y\}_*)$$

B THE TRANSGRESSION HOMOMORPHISM

Let $p: E \to B$ be a fibration with path-connected base and path-connected fiber $F = p^{-1}(b_0)$. Consider the homomorphisms

$$H^q(F;G) \xrightarrow{\delta} H^{q+1}(E,F;G) \xleftarrow{p^*} H^{q+1}(B,b_0;G) \xrightarrow{j^*} H^{q+1}(B;G)$$

The *transgression* τ is the homomorphism [from a subgroup of $H^q(F;G)$ to a quotient group of $H^{q+1}(B;G)$]

$$\tau: \delta^{-1}(\operatorname{im} p^*) \to H^{q+1}(B;G)/j^*(\ker p^*)$$

defined by $\tau(u) = j^*p^{*-1}\delta(u)$, where $u \in H^q(F;G)$ is such that $\delta(u) \in p^*(H^{q+1}(B,b_0;G))$.

1 Prove that τ commutes with the Steenrod squaring operations Sq^i and with induced homomorphisms for induced fibrations.

2 Assume that B is $(n-1)$-connected for $n \geq 2$ and consider the path fibration $p: PB \to B$ with fiber ΩB. Prove that $\tau: H^{n-1}(\Omega B;G) \approx H^n(B;G)$ and that $\iota \in H^{n-1}(\Omega B;G)$ is $(n-1)$-characteristic for ΩB if and only if $\tau(\iota)$ is n-characteristic for B.

For the remainder of this group of exercises we assume that the fibration is orientable over R and the coefficient module is R.

3 For the spectral sequence of the fibration prove the following:

(a) $\delta^{-1}(\operatorname{im} p^*) \approx E^{0,q}_{q+1} \subset E^{0,q}_2 \approx H^q(F)$.
(b) $H^{q+1}(B)/j^*(\ker p^*) \approx E^{q+1,0}_{q+1}$ and is a quotient of $E^{q+1,0}_2 \approx H^{q+1}(B;G)$.
(c) Under these isomorphisms τ corresponds to $d_{q+1}: E^{0,q}_{q+1} \to E^{q+1,0}_{q+1}$.

4 If $\tilde H^*(E) = 0$, prove that $H^i(F) = 0$ for $0 < i < q$ if and only if $H^i(B) = 0$ for $0 < i < q+1$ and, in this case, $\tau: H^q(F) \approx H^{q+1}(B)$.

5 Assume that $H^i(B) = 0$ for $0 < i < s$ and $H^j(F) = 0$ for $0 < j < t$. Prove the exactness of the following *Serre cohomology sequence:*

$$\cdots \to H^q(F) \xrightarrow{\tau} H^{q+1}(B) \xrightarrow{p^*} H^{q+1}(E) \xrightarrow{i^*} H^{q+1}(F) \to \cdots \to H^{s+t-1}(F)$$

Consider the homomorphisms

$$H_q(B) \xrightarrow{j_*} H_q(B,b_0) \xleftarrow{p_*} H_q(E,F) \xrightarrow{\partial} H_{q-1}(F)$$

and define the *homology transgression*

$$\tau_*: j_*^{-1}(\text{im } p_*) \to H_{q-1}(F)/\partial(\ker p_*)$$

by $\tau_*(z) = \partial p_*^{-1} j_*(z)$, where $z \in H_q(B)$ is such that $j_*(z) \in p_*(H_q(E,F))$.

6 If $H_i(B) = 0$ for $0 < i < s$ and $H_j(F) = 0$ for $0 < j < t$, prove the exactness of the following *Serre homology sequence:*

$$H_{s+t-1}(F) \to \cdots \to H_q(F) \xrightarrow{i_*} H_q(E) \xrightarrow{p_*} H_q(B) \xrightarrow{\tau_*} H_{q-1}(F) \to \cdots$$

C SERRE CLASSES OF ABELIAN GROUPS

A *chain complex modulo* $\mathcal{C}$ is a graded group $C = \{C_q\}$ and a sequence of homomorphisms $\{\partial_q: C_q \to C_{q-1}\}$ such that $(\partial_{q-1} \circ \partial_q)(C_q) \in \mathcal{C}$ for all q. The *homology group of* C is the graded group $H(C) = \{H_q(C)\}$, where

$$H_q(C) = \ker \partial_q/(\ker \partial_q \cap \text{im } \partial_{q+1}) \approx (\ker \partial_q \cup \text{im } \partial_{q+1})/\text{im } \partial_{q+1}$$

A three-term sequence of groups and homomorphisms

$$G' \xrightarrow{\alpha} G \xrightarrow{\beta} G''$$

is said to be $\mathcal{C}$-*exact* if $(\text{im } \alpha \cup \ker \beta)/\text{im } \alpha \in \mathcal{C}$ and if $(\text{im } \alpha \cup \ker \beta)/\ker \beta \in \mathcal{C}$. Longer sequences are $\mathcal{C}$-*exact* if every three-term sequence is $\mathcal{C}$-exact.

1 Let C be a chain complex modulo $\mathcal{C}$, let C' be a subcomplex of C (that is, $C'_q \subset C_q$ and $\partial'_q = \partial_q \mid C'_q$ for all q), and define the quotient complex $C/C' = \{C_q/C'_q, \partial''_q\}$, where ∂''_q is induced from ∂_q. Prove that there is a $\mathcal{C}$-exact sequence

$$\cdots \to H_q(C') \to H_q(C) \to H_q(C/C') \to H_{q-1}(C') \to \cdots$$

2 Let $0 \to C' \xrightarrow{\alpha} C \xrightarrow{\beta} C'' \to 0$ be a short $\mathcal{C}$-exact sequence of chain complexes modulo $\mathcal{C}$ and chain maps (α and β commute with the boundary homomorphisms of the chain complexes). Prove that there is a $\mathcal{C}$-exact sequence

$$\cdots \to H_q(C') \xrightarrow{\alpha_*} H_q(C) \xrightarrow{\beta_*} H_q(C'') \xrightarrow{\partial_*} H_{q-1}(C') \to \cdots$$

3 Prove the *five lemma modulo* $\mathcal{C}$. That is, given a commutative diagram

$$\begin{array}{ccccccccc} G_5 & \xrightarrow{\alpha_5} & G_4 & \xrightarrow{\alpha_4} & G_3 & \xrightarrow{\alpha_3} & G_2 & \xrightarrow{\alpha_2} & G_1 \\ \downarrow{\scriptstyle \gamma_5} & & \downarrow{\scriptstyle \gamma_4} & & \downarrow{\scriptstyle \gamma_3} & & \downarrow{\scriptstyle \gamma_2} & & \downarrow{\scriptstyle \gamma_1} \\ H_5 & \xrightarrow{\beta_5} & H_4 & \xrightarrow{\beta_4} & H_3 & \xrightarrow{\beta_3} & H_2 & \xrightarrow{\beta_2} & H_1 \end{array}$$

with $\mathcal{C}$-exact rows such that γ_1, γ_2, γ_4, and γ_5 are $\mathcal{C}$-isomorphisms, prove that γ_3 is also a $\mathcal{C}$-isomorphism.

For the rest of this group of exercises assume that $p: E \to B$ is a fibration with path-connected fiber and simply connected base space and that $\mathcal{C}$ is an ideal of abelian groups.

4 If $H_i(B) \in \mathcal{C}$ for $0 < i$, prove that $H_i(F) \underset{\mathcal{C}}{\approx} H_i(E)$ for all i.

5 *Vietoris-Begle mapping theorem modulo* $\mathcal{C}$. If F is $\mathcal{C}$-acyclic, prove that $H_i(E) \underset{\mathcal{C}}{\approx} H_i(B)$ for all i.

D HOMOTOPY GROUPS OF SPHERES

1 If S^n is an H-space, prove that there is a short exact sequence

$$0 \to \pi_q(S^n) \xrightarrow{S} \pi_{q+1}(S^{n+1}) \xrightarrow{H} \pi_{q-1}(S^{2n-1}) \to 0 \qquad q \leq 3m - 2$$

2 Prove that $\pi_7(S^4) \approx \pi_6(S^3) \oplus \pi_7(S^7)$ and that the order of $\pi_8(S^5)$ is twice the order of $\pi_6(S^3)$.

3 Let X_p^n be a CW complex consisting of an n-sphere with an $(n + 1)$-cell attached by a map of degree p. If $n \geq 2$ and p is a prime, prove that $\pi_q(X_p^n)$ is a finite p-group for all q, and if $q \leq 2n - 2$, prove there is an exact sequence

$$0 \to \pi_q(S^n) \otimes \mathbf{Z}_p \to \pi_q(X_p^n) \to \pi_{q-1}(S^n) * \mathbf{Z}_p \to 0$$

4 Prove that $S(X_p^n)$ has the same homotopy type as X_p^{n+1} and $\{X_p^n;X_p^n\} \approx \mathbf{Z}_p$ if $p \neq 2$ and $\{X_2^n;X_2^n\}$ is a group of order 4.

5 Let $p\colon E \to S^3$ be a fibration, with fiber F a space of type $(\mathbf{Z},2)$, such that $\bar{\partial}\colon \pi_3(S^3) \approx \pi_2(F)$, as in lemma 9.7.5. Let $f\colon X_2^4 \to E$ be a map such that $f_\#\colon \pi_4(X_2^4) \approx \pi_4(E)$ [such a map exists, because $\pi_4(E) \approx \mathbf{Z}_2$]. Prove that $f_\#\colon \pi_5(X_2^4) \approx \pi_5(E)$ and $f_\#$ is a monomorphism of $\pi_6(X_2^4)$ onto the 2-primary component of $\pi_6(E)$. [*Hint:* Show that $f_*\colon H_q(X_2^4) \to H_q(E)$ is an isomorphism of 2-primary components for $q < 8$ and use the generalized Whitehead theorem.]

6 Prove that $\pi_{n+2}(S^n) \approx \mathbf{Z}_2$ for $n \geq 2$.

7 Prove the following:

(*a*) $\pi_5(S^2) \approx \mathbf{Z}_2$.
(*b*) $\pi_6(S^3)$ is a group of order 12.
(*c*) $\pi_7(S^4) \approx \pi_6(S^3) \oplus \mathbf{Z}$.
(*d*) $\pi_{n+3}(S^n)$ is a group of order 24 for $n \geq 5$.

INDEX

9 780387 944265